INTRODUCTION TO THE MECHANICS OF SPACE ROBOTS

SPACE TECHNOLOGY LIBRARY

Published jointly by Microcosm Press and Springer

For further volumes:
www.springer.com/series/6575

Giancarlo Genta

Introduction to the Mechanics of Space Robots

Springer

Prof. Dr. Giancarlo Genta
Department of Mechanical and
Aerospace Engineering
Politecnico di Torino
Corso Duca degli Abruzzi 24
Torino 10129
Italy
giancarlo.genta@polito.it

ISBN 978-94-007-1795-4 e-ISBN 978-94-007-1796-1
DOI 10.1007/978-94-007-1796-1
Springer Dordrecht Heidelberg London New York

Library of Congress Control Number: 2011935862

Cover design: VTeX UAB, Lithuania

Printed on acid-free paper

Springer is part of Springer Science+Business Media (www.springer.com)

To Franca and Alessandro

Preface

This text started as a collection of notes of the lectures on Space Robotics given by the author to the students of the International Master on Space Exploration and Development Systems (SEEDS). The aim of the course was the study of the automatic machines aimed to operate both autonomously and as a support to astronauts in space exploration and exploitation missions, with particular attention to the devices designed for planetary environment, including small planets, comets and asteroids.

This material was then completed and made more systematic so that it can hopefully be useful not only to the students of that course but also to those who have an interest in the wide and much interdisciplinary field of space robotics, and in particular in its mechanical aspects.

The focus is drawn mainly on the mechanics of space robots: the author is well aware that, even in this specific field, it is far from being complete and that robots, like all mechatronic systems, are so integrated that no single aspect can be dealt separately. Many important aspects are either dealt with only marginally or altogether left out. The very important topics of the control and the behavior of robots, for instance, are only marginally touched, even if their influence on the mechanical aspect to which this book is dedicated is not at all marginal.

The structure of the book is so organized:

- Chapter 1: a very short introductory overview of human and robotic space exploration, stressing the need for man-machine cooperation in exploration. The various types of robotic missions in LEO, deep space and on planets and their basic requirements are shortly summarized.
- Chapter 2 deals in a synthetic way with the main characteristics of the environments space robots are facing and will face in the future. Since space environment is a specialized subject, dealt with in many books, this subject is only briefly summarized.
- The configurations of robot arms and the basic kinematic and dynamic relationships needed for their design are described in Chap. 3.
- Chapter 4 is devoted to the study of mobility on planetary surfaces, using different kind of supporting devices, like wheels, legs and aerodynamic or aerostatic devices.

- The basic characteristics of wheeled robots and vehicles are summarized in Chap. 5. The behavior of wheeled devices is studied in its various aspects, like longitudinal, lateral and suspension dynamics. The consequences of operating wheeled machines in the various environments are analyzed in some detail. The chapter is concluded by a description of the only vehicle that successfully carried humans on the surface of the Moon, the *Apollo* Lunar Roving Vehicle.
- Vehicles and robots that use legs, tracks or other devices to move on a solid surface are described in Chap. 6. Since a great number of different architectures were proposed and sometimes even used in the past, not all the possible configurations are illustrated: the choice was based on the actual existing applications and on the perspectives of future use.
- Chapter 7 is devoted to a short overview of the transducers used for actuation and sensing in space robots.
- A short overview of the energy sources and storage devices that can be used for space robots is reported in Chap. 8.

The book includes two appendices summarizing the theoretical formulations allowing to write mathematical models of space robots including a variety of mechanical components, such as arms, legs, etc. The author found it necessary to include them, since the participants to the course in Space Exploration and Development Systems have a much varied background and what may seem obvious to some students, could be difficult for other ones. In a similar way, some of the readers of this book may not be familiar with the concepts of analytical mechanics or dynamics of deformable bodies used in the text, mainly in Chaps. 3 and 5.

The author is grateful to colleagues and students of the Mechanics Department and the Mechatronics Laboratory of the Politecnico di Torino for their suggestions, criticism and general exchange of ideas. Students, in particular postgraduate students, cooperated to this book with their thesis work and their questions, but mainly with their very presence that compelled me to clarify my own ideas and to work out all details. To all of them goes my gratitude.

Last, but far from least, this book could not have been written without the support, encouragement and patience by my wife Franca—advisor, critic, editor, companion and best friend since 44 years.

A Note on the Illustrations I have made every effort to seek permission from the original copyright holders of the figures, and I apologize if there are cases where I have not been able to achieve my objective. This applies in particular to figures taken from the web, like Figs. 4.10, 4.35, 4.36, 4.40, 5.2, 6.1, 6.15, 6.17, 6.21, 6.23b, 6.24, 6.25a, 6.28, 6.31a, b, c, 7.18, 7.19 and 7.21.

Torino, Italy Giancarlo Genta

Contents

Symbols and Acronyms

Symbols

a	length of the contact area; acceleration distance between center of mass and front axle
$\mathbf{a}$	acceleration vector
b	width of contact area; distance between center of mass and rear axle; wingspan
c	cohesive bearing strength; viscous damping coefficient; wing chord
c_{cr}	critical damping
c_{opt}	optimal damping
d	soil deformation; diameter
$\mathbf{d}$	direct piezoelectric matrix
d_i	second DH parameter: offsett
e	energy
$\mathbf{e}$	error
$\mathbf{e}_i$	unit vector of the ith axis
f	friction coefficient; rolling coefficient
f_0	rolling coefficient at zero speed
f_r	rollover factor
f_s	sliding factor
g	gravitational acceleration
$\mathbf{g}$	gravitational acceleration vector
h	sinking in the ground
h_c	convection coefficient
i	grade of the road; imaginary unit ($i = \sqrt{-1}$); current
i_t	transversal grade of the road
k	stiffness; modulus of soil deformation
k_c	cohesive modulus
k_ϕ	frictional modulus
l	length of the arm; wheelbase
l_i	third DH parameter: length
m	mass

m_e	equivalent mass
m_s	sprung mass
m_u	unsprung mass
p	pressure
$\mathbf{p}$	generalized momenta
p, q, r	angular velocities in the xyz frame
p_s	bearing capacity of the soil with no sinking
p_0	bearing capacity of the soil
q	eigenfunction
$\mathbf{q}$	vector of the generalized coordinates; eigenvector
r	radius
$\mathbf{r}$	vector
s	laplace variable
t	time; track; pneumatic trail; thickness
u	displacement
$\mathbf{u}$	displacement vector
u, v, w	velocities in the xyz frame
v	volume
v_g	velocity of the ground due to slip
z	sinking; number of teeth
xyz	body-fixed reference frame
$\mathbf{x}$	coordinate vector
$\mathbf{z}$	state vector
A	area
$\mathbf{A}$	dynamic matrix in the state space
B_r	magnetic remanence
$\mathbf{B}$	input gain matrix
C	cornering stiffness; capacitance
C_D	drag coefficient
C_f	force coefficient
C_L	lift coefficient
C_S	side force coefficient
C_γ	camber stiffness
C_σ	longitudinal force coefficient
$\mathbf{C}$	damping matrix; output gain matrix
D	aerodynamic drag; displacement
$\mathbf{D}$	direct link matrix; dynamic matrix in the configuration space
E	Young's modulus; modulus of deformation (soil); aerodynamic efficiency
$\mathbf{E}$	stiffness matrix of the material
F	force
$\mathbf{F}$	force vector
F_n	normal force
$\mathcal{F}_r$	Froude number
F_t	tangential force

G	shear modulus; gravitational constant
$\mathbf{G}$	gyroscopic matrix
H_c	coercitive magnetic field
$\mathcal{H}$	Hamiltonian function
$\mathbf{H}$	circulatory matrix
$\mathbf{I}$	identity matrix; inertia matrix
J	moment of inertia
$\mathbf{J}$	Jacobian matrix
K	stiffness
K_B	back EMF constant
K_T	torque constant
$\mathbf{K}$	stiffness matrix; matrix of the control gains
$\mathbf{K}_d$	derivative gains matrix
$\mathbf{K}_i$	integrative gains matrix
$\mathbf{K}_p$	proportional gains matrix
L	reference length; aerodynamic lift
$\mathcal{L}$	Lagrangian function
M	mass
$\mathbf{M}$	mass matrix; moment
$\mathcal{M}$	molecular mass
$\mathcal{M}_a$	Mach number
N	number of turns
N_u	Nusselt number
$\mathbf{N}$	matrix of the shape functions
P	power
Q	flow
R	radius of the wheel (unloaded); radius of the trajectory; universal gas constant; resistance to motion; electric resistance
$\mathcal{R}$	reluctance
$\mathbf{R}$	rotation matrix
R_c	radius of the trajectory (low speed conditions)
$\mathcal{R}_e$	Reynolds number
R_e	effective rolling radius
R_l	radius under load
S	first order mass moment; aerodynamic side force; reference surface
T	temperature; torque
$\mathbf{T}$	torque vector; homogeneous transformation matrix
$\mathcal{T}$	kinetic energy
$\mathcal{U}$	potential energy
V	vehicle speed; volume; voltage
$\mathbf{V}$	velocity vector
V_f	velocity of the foot relative to the body
V_r	velocity relative to the atmosphere
V_s	velocity of sound
V_B	back electromotive force

W	work
XYZ	inertial frame
α	sideslip angle; grade angle of the road; angle of attack
α_i	fourth DH parameter: twist
α_t	transversal grade angle of the road
β	sideslip angle of the vehicle; duty factor
γ	camber angle; inclination angle
δ	steering angle; aerodynamic sideslip angle; resistivity
δ_c	steering angle (low speed steering)
$\delta\mathcal{L}$	virtual work
$\delta\boldsymbol{\theta}$	virtual displacement
ϵ	strain; deformation of the soil
$\boldsymbol{\epsilon}$	strain vector
ϵ_f	efficiency of the brake
η	efficiency; modal coordinate
η_i	modal coordinates vector
θ	pitch angle; thermal resistance
θ_i	first DH parameter: rotation angle
$\boldsymbol{\theta}$	vector of the generalized coordinates at the joints
λ	thermal conductivity
μ	dynamic viscosity; traction coefficient
μ_0	magnetic permeability of vacuum
μ^*	friction coefficient
μ_r	relative magnetic permeability
μ_x	longitudinal force coefficient
μ_{x_p}	longitudinal traction coefficient
μ_{x_s}	sliding longitudinal traction coefficient
μ_y	cornering force coefficient
μ_{y_p}	lateral traction coefficient
μ_{y_s}	sliding lateral traction coefficient
ν	Poisson's ratio; kinematic viscosity
ρ	density
σ	normal pressure; stress; longitudinal slip
$\boldsymbol{\sigma}$	stress vector
τ	shear stress; transmission ratio; time delay; nondimensional time
ϕ	roll angle; friction angle ($\phi = \text{atan}(\mu)$)
χ	torsional stiffness
ψ	yaw angle
ω	frequency; circular frequency
ω_n	natural frequency
Γ	rotational damping coefficient
Δh	increase of sinking
Π	torsional stiffness of the tires of an axle
$\boldsymbol{\Phi}$	matrix of the eigenvectors
Ω	angular velocity

$\mathbf{\Omega}$	angular velocity vector
$\Im$	imaginary part
$\Re$	real part
$_{,x}$	differentiation with respect to variable x
∇	Laplace operator

Subscripts

d	derivative
i	inner; integrative
o	outer
p	proportional
t	tangential

Values of Some Physical Constants

G	Gravitation constant $6.67259 \times 10^{-11}\ \mathrm{m^3\,kg^{-1}\,s^{-2}}$
R	Universal gas constant $8.314510\ \mathrm{J\,mol^{-1}\,K^{-1}}$
μ	Magnetic permeability of vacuum $1.257 \times 10^{-6}\ \mathrm{H\,m^{-1}}$

Acronyms

ACE	Advanced Composition Explorer
ACR	Anomalous Cosmic Rays
AFC	Alkaline Fuel Cell
AGV	Automatically Guided Vehicles
AI	Artificial Intelligence
AU	Astronomical Unit
BAP	Body Armour Powered
BEMF	Back ElecroMotive Force
CRP	Carbon Reinforced Plastics
CVT	Continuously Variable Transmission
DC	Direct Current
DH	Denavit–Hartenberg
DMFC	Direct Methanol Fuel Cell
ECU	Electronic Control Unit
EMF	ElectroMotive Force
EVA	Extra Vehicular Activity
FEM	Finite Element Method
GCR	Galactic Cosmic Rays
GEO	Geostationary Earth Orbit
GPS	Global Positioning System
GRP	Glass Reinforced Plastics
GTO	Geostationary transfer orbit
HMI	Human–Machine Interface
ICE	Internal Combustion Engine
ICME	Interplanetary Coronal Mass Ejections

IMF	Interplanetary Magnetic Field
ISO	International Standards Organization
ISRU	In Situ Resource Utilization
ISS	International Space Station
KBO	Kuiper Belt Object
LEO	Low Earth Orbit
LRV	Lunar Roving Vehicle
LVDT	Linear Variable Differential Transformer
MCFC	Molten Carbonate Fuel Cell
MEMD	Motional Electromagnetic Damper
MEMS	Micro Electromechanical System
MER	Mars Exploration Rover
MK	natural undamped (system)
NEA	Near Earth Asteroids
NEC	Near Earth Comets
NEO	Near Earth Objects
NEM	Near Earth Meteoroid
ODE	Ordinary Differential Equation
PAFC	Phosphoric Acid Fuel Cell
PD	Proportional, Derivative
PDE	Partial derivatives Differential Equation
PEMFC	Proton Exchange Membrane Fuel Cell
PHA	Potentially Hazardous Asteroids
PID	Proportional, Integral, Derivative
PWM	Pulse Width Modulation
RB	Rocker Bogie
R/C	Radio Controlled
RHU	Radioisotope Heat Unit
rms	Root Mean Square
RTG	Radioisotope Thermoelectric Generator
RVDT	Rotary Variable Differential Transformer
S/C	Spacecraft
SAA	South Atlantic Anomaly
SAE	Society of the Automotive Engineers
SAR	Synthetic Aperture Radar
SCARA	Selective Compliance Articulated (or Assembly) Robot Arm
SMA	Shape Memory Alloy
SNAP	System for Nuclear Auxiliary Power
SOFC	Solid Oxide Fuel Cell
SRG	Stirling Radioisotope Generator
SRMS	Shuttle Remote Manipulator System
SSRMS	Space Station Remote Manipulator System
TEMD	Transformer Electromagnetic Damper
UAV	Unmanned Aerial Vehicle
UGV	Unmanned Ground Vehicle

UV	UltraViolet
WEB	Warm Electronic Box
VDC	Vehicle Dynamic Control
4WD	Four Wheel Drive
4WDS	Four Wheel Drive and Steering
4WS	Four Wheel Steering

Chapter 1
Introduction

1.1 Robots in Space

Books dealing with robots often start with the definition of what robots are.

The origin of the word robot dates back from 1920, when the Czech writer Karel Kapec published his science fiction play R.U.R. (Rossum's Universal Robots) dealing with artificial men built for performing work in place of human beings. He invented the word *Robota*, from the base robot-, as in robota, compulsory labor, or robotník, peasant owing such labor. The robots described in the play are humanoid.

The ISO (International Standards Organization) 8373 standard defines a robot as *an automatically controlled, reprogrammable, multipurpose, manipulator programmable in three or more axes, which may be either fixed in place or mobile for use in industrial automation applications*. The International Federation of Robotics, the European Robotics Research Network (EURON), and many national standards committees use this definition.

A broader definition is given by the Robotics Institute of America (RIA): a *reprogrammable multi-functional manipulator designed to move materials, parts, tools, or specialized devices through variable programmed motions for the performance of a variety of tasks*. The RIA subdivides robots into four classes

1. devices that manipulate objects with manual control,
2. automated devices that manipulate objects with predetermined cycles,
3. programmable and servo-controlled robots with continuous point-to-point trajectories, and
4. robots of the last type which also acquire information from the environment and move intelligently in response.

The New Oxford American Dictionary defines a robot as *a machine able of carrying out a complex series of actions automatically, esp. one programmable by a computer*. Here the ability of manipulating objects is not required.

According to Encyclopaedia Britannica, a robot is *any automatically operated machine that replaces human effort, though it may not resemble human beings in appearance or perform functions in a human-like manner*. Here the anthropomorphic appearance emerges, at least for being explicitly negated.

G. Genta, *Introduction to the Mechanics of Space Robots*, Space Technology Library 26,
DOI 10.1007/978-94-007-1796-1_1, © Springer Science+Business Media B.V. 2012

The definition from Random House Webster's dictionary is *a machine that resembles a human and does mechanical, routine tasks on command* (here the anthropomorphic shape is needed, autonomy not) or *a machine that resembles a human and does mechanical, routine tasks on command.*

Another definition, from Encyclopaedia Britannica, is a *machine that looks like a human being and performs various complex acts (as walking or talking) of a human being.*

In some definitions also virtual software agents performing the same tasks are defined as robots (in some cases the latter are simply called *bots*), but here we will strictly require that a robot is a material object. A mathematical model of a robot implemented on a computer is thus a mathematical model of a robot, but not a robot in itself.

While the word robot comes from science fiction, the idea of a more or less anthropomorphic machine or, in general, an anthropomorphic being able to substitute humans is much older. They are present in ancient literature, as early as the *Iliad* (the self-operating tools used by the god Hephaestus), and many attempts to build automata were performed since Greek–Roman times.

In many cases these fictional robots were biological (from the Golem to the 'thing' of the *Frankenstein* novel, and more recently the replicants of the novel *Do Androids Dream of Electric Sheep* by Philip K. Dick from which the movie Blade Runner was taken) and in others they were mechanical, from the various automata described by Heron of Alexandria, Ctesibius of Alexandria and Philo of Byzantium to the modern droids of the *Star Wars* saga.

In all cases the attempt was that of describing, and in some case building, artificial human beings, and anthropomorphism has always played a role in them. Although in most cases the goal was an artificial being looking like, moving as and behaving as a human, there are instances in which the similarity with humans was restricted to the behavior or even to "the way it thinks" (mental anthropomorphism).

The real-world situation is much different. Usually robots are electromechanical (or electrohydraulic, or electropneumatic) machines, controlled by a programmable computer, so that they are able to do some task on their own. It is still open the discussion whether more or less anthropomorphic devices controlled directly by humans are robots. They can be simple mechanical devices, like the arms used to manipulate radioactive substances, but also very sophisticated systems able to perform complex tasks under the complete telecontrol from human operators. The robotic arm of the *Space Shuttle* is a good example. The term *telemanipulator* or *tele-agent* seem more adequate than robot in this case.

Since the very beginning, the devices launched into space had to perform, either automatically or under control from Earth, a number of tasks. With the increasing complexity of space missions these requirements became more demanding and the complexity of automatic spacecraft increased.

The habit of calling these complex space devices robots became widespread. So an automatic probe started to be called a robotic probe, even if it has no manipulator arms, a small autonomy in taking decisions and obviously it has no human-like appearance. Rovers operating on the Moon or on a planet are invariably defined robots or robotic rovers, even when it would be better to call them automatic rovers.

The very world robot may be in this case misleading owing to the overtones in our cultural background and in itself is saying very little. When most of us hear the word robot his/her mind goes to the pages of so many science fiction books or to scenes of movies. Even if the general attitude toward robots is very different in the various cultures, with Orientals being much more positive and Westerners more hostile, the image of a robot, particularly if it is small and looks friendly, and is used in some obviously innocuous activity, generates sympathetic reactions. The same object that would generate little interest if referred to as an automatic machine, is immediately humanized and regarded with sympathy (or fear and hostility, if big and looking menacing) if called a robot. Sometimes it seems that these feelings are purposely evoked to create interest or support toward a given mission.

While for specialists this problem about names creates no problem, since they know well the performance and the limitations of these machines, misusing the term 'robot' can be misleading for the general public and can create expectations beyond the possibilities of present technology. This has in turn a potentially dangerous aspect: if the public has disproportionate expectation, it can become difficult to have people to appreciate the importance of the small steps we painstakingly take in space, and to support the expenditures required for technological and scientific advancement. After all, why spending money to build and design better spacecraft with more advanced technology, when we have, or at least we can build, robots that can perform almost all important tasks in the solar system?

1.2 Humans and Robots

From the very beginning of space exploration (and even earlier), the advisability of human beings participating directly to space missions has stirred many debates and continues to do so. There is no doubt that the presence of people on board a space vehicle makes its design much more complex and challenging, causing a large increase in costs. It has been thus ever since the beginning of the space age, and is even more so today.

First, the requirements for safety of space vehicles are greatly increased. In the case of automatic missions, it is possible to increase the number of space probes which are launched, decreasing their reliability. The cost reduction and the increase of the results/costs ratio which can be achieved in this way may be important. The large lead time between the starting of the design of a mission and its completion may also be reduced in this way. It is inevitable that the very high reliability needed when humans are on board increases the costs greatly.

The performance of the life-support systems required by the people aboard a space vehicle must be guaranteed. They are usually heavy, bulky and costly; and their complexity increases for long duration missions. The miniaturization of instruments and all electronic devices makes things worse from this viewpoint. As technology advances, robotic probes become lighter, smaller and more convenient than manned vehicles. This leads to a reduction of the size, and cost, of the launch

Fig. 1.1 An astronaut performing ExtraVehicular Activity (EVA) to repair the Hubble Space Telescope (NASA photo)

vehicles. Furthermore, advances in electronics and computer science allow increasingly complex tasks to be entrusted to robots.

However, experience has shown that the presence of humans in space is not only popular with the public, but above all is useful; there are many cases when direct intervention by an astronaut was essential to correct the malfunction of an automatic device. The ability to react to unexpected situations and the ability to perform a wide variety of tasks are two human characteristics which are precious in space missions. Astronauts and cosmonauts have proved that they can adapt to conditions of weightlessness and work in space without encountering too many problems. The operations to repair and later to upgrade the *Hubble Space Telescope* are perhaps the best example, but they are just two of many (Fig. 1.1).

This is even more true in the case of deep space missions. If the human exploration of Mars is a very difficult enterprise, robotic exploration is not much simpler. Any automatic probe moving on the surface of that planet must work autonomously. While in the case of the Moon it is possible for someone on the ground to teleoperate a probe, since the two-way link time is only about 3 seconds, the same cannot be done on Mars. Several minutes elapse between the instant the camera of a rover detects an obstacle in its path and that when the course correction commands arrive from Earth. The automatic vehicles crawling on the surface of the planet, such as the *Sojourner* (Fig. 1.2) rover of the *Mars Pathfinder* mission of 1997 or the more recent

Fig. 1.2 The robotic rover Sojourner (NASA image)

Spirit and *Opportunity*, are very slow and must have at least some operational autonomy. However, these rovers were manually teleoperated for many of their functions in spite of the long time needed to receive the control inputs.

Many of the promises of artificial intelligence are still far from being fulfilled. Although the power of computers goes on increasing at a quick rate, the construction of machines simulating human logical reasoning is being delayed to an increasingly distant future. The more the performance of computers improves, the more is it realized how difficult it is to build machines displaying actual logical abilities. Although nobody has yet succeeded in defining exactly what intelligence is, or perhaps mainly for that reason, today the term "intelligent" is applied to a variety of situations—intelligent machines, intelligent structures, intelligent weapons, and even intelligent suspensions in motor vehicle technology. And perhaps that is also with good reason, as these devices are capable of performance which was unthinkable a few years ago. But the term 'intelligent' must be properly understood: many machines can perform a wide variety of tasks in an autonomous way, but the intelligence expected from a human being is completely different.

Similar considerations apply in the industrial world. Many discussions have taken place on fully automated factories, in which all operations are performed without any human intervention, and forecasts of the complete substitution of workers by robots in many technological processes have been made. Today these perspectives are being revised. All machines, even the smartest ones, must cooperate with humans and classical robots are often unsuitable for such a task. The word 'cobot', from 'collaborative robot', has been invented to designate an intelligent (in the above sense) machine capable of helping a human operator without replacing him.

The relative new field of domestic robotics or, more in general, service robotics, is gaining momentum. The personal robot may in the future duplicate the success of the personal computer and the goal of a robot in every house seems now to be not less meaningful than the goal of a computer in every house was twenty years ago. To work in homes, offices, hospitals and man-carrying vehicles, robots must become more reliable, user friendly and autonomous than they are at present and must become able to mix with people, a thing industrial robots are not able—and allowed—to do.

Similar, and contradictory, trends are also apparent in the space field. Tasks which were in the past entrusted only to machines are now sometimes performed by human beings, perhaps with the aim of using simpler and less costly devices, or to obtain better performance. Attempts in this direction, even clumsy and dangerous ones, were performed to avoid costly automatic devices. A much publicized accident occurred to the *Mir* space station when a Progress automatic cargo vehicle failed a docking attempt under manual guidance, hitting a solar panel and causing much damage. In that case the docking manoeuvre was performed manually to cut costs, without providing the pilots with the required instrumentation to perform it safely.

In other cases to *put the man in the control loop* is a welcome simplification, which lowers the cost of a mission without compromising its safety. The lunar probes planned by ESA at the end of the 1990s, for instance, were meant to be piloted by a human operator on the ground to a greater extent than previous lunar probes. In this perspective, the added costs due to the presence of humans on board a spacecraft can at least partially be compensated for by a reduction of the cost of the control systems. Many operations, which were meant to be performed under completely automatic control, can be performed more efficiently by astronauts, perhaps helped by their 'cobots'. The man–machine relationship, which sometimes tends to become conflicting, must evolve toward a closer cooperation.

An example of this man–machine cooperation is the *Mars Outposts* approach to Mars exploration launched by the Planetary Society. Here a number of robotic research stations, equipped with permanent communications and navigational systems, would be sent to the red planet. They would perform research, and establish infrastructure needed to prepare future landing sites and return vehicles for the exploration of Mars by humans.

A reduction of the cost of launching payloads into Earth orbit—of the cost of getting out of the Earth gravitational well—is essential. This will only result from marked progress in new launch technologies, and will make it easier for human beings to participate directly in space exploration. And there is a cascade effect: the more humans will live in space and settle on other celestial bodies, the less will be the number of people and the amount of materials to be launched into space from the Earth. What is required in our generation is a "bootstrap" effort which will slowly gain momentum.

If space is more than a place to build automatic laboratories and to start some industrial enterprises in the immediate vicinity of our planet, the presence of humans is essential. Humans must learn not only how to work but also how to live in space for many years. They must learn how to travel through space toward destinations which will be not only scientific bases but also places to live. In other words what humankind can do—and in the future could decide to do—is to colonize space, and not only to send robotic devices to explore it.

The role of robots will nevertheless be essential and it is easy to predict that, wherever humans will go in space, they will be preceded, accompanied and tended by automatic machinery, which will likely be referred to as robots.

1.3 Artificial Intelligence

The term Artificial Intelligence (AI) was first used in a conference held in the summer of 1956 at the Dartmouth College, New Hampshire, and has been widely used since then. Although no definition of Artificial Intelligence (or, for that matter, even of human intelligence) exist, there is a wide agreement in accepting that it consists of imitating human intelligent behavior by a machine. The well known Turing test is what is closest to a definition: a machine is intelligent if it is impossible to tell it from a human being during an interaction based on exchanging messages on any subject.

At the beginning, two ways were followed to reach this goal. The first consists in writing dedicated software running on conventional, although very powerful, computers, able to manipulate symbols following well established rules. At the base of these attempts there is the assumption that intelligence is based on algorithms to perform logical operations by manipulating symbols. The human brain is then considered as a biological computer, and the human mind is the result of a sort of software running on it. Similar results could then be obtained by using a non-biological computer, provided that it is powerful enough to allow a suitable software to run.

The second approach is the neural one, based on the construction of a network of artificial neurons, simulating the structure of animal, and then also human, brains. The Artificial Neural Networks (ANN) so devised are not running programs, but operate by learning.

These two approaches are not as much different as they look, since very often neural networks are simulated on computers, i.e. they are reduced to a software running on conventional machines. In this way the second approach seems to reduce to a particular case of the first.

At any rate the neural approach seemed to be haunted by unsolvable problems and at the end of the 1960s it seemed to be a dead end. On the contrary the algorithmic approach seemed to obtain encouraging results.

By the mid 1980s the neural approach regained momentum due to the solution of the mentioned problems, but the goal of building intelligent machines proved to be much harder than predicted. While in the 1960s it was a common opinion that by the year 2000 they would have been an established technological result, at present (2010) their realization seems to be still far and many cast doubts on their possibility, at least with present day technology.

A tentative scheme of the path leading from inert matter to intelligent and conscious systems is shown in Fig. 1.3.

Following this scheme, a material system, able to manipulate energy, is a dynamic system. If it can also receive signals and manipulate information, it can be considered as an automatic device, and so on.

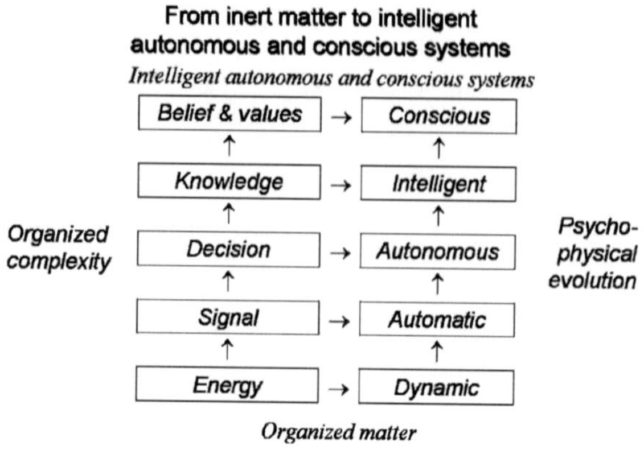

Fig. 1.3 Tentative scheme of the path leading from inert matter to intelligent autonomous and conscious systems

It is debatable whether the *decision* box should be over or under the *knowledge* box. Here it is assumed that a being (living or robotic) can take decisions reacting to the inputs from the outer world even without building an internal model of it: this point has been controversial, but here it is assumed that a positive answer to this problem is realistic. Moreover, sometimes it seems that there is no complete agreement on the meaning of the terms in the boxes, so that the answer depends on the exact interpretations of what the words knowledge and decision mean.

The boxes on the right should not be considered as separate steps in a ladder, but rather as levels in a continuous evolutionary process, and there is an infinity of shades between each of them.

Telemanipulators, i.e. remote controlled agents able to perform well determined tasks, are without any doubt at the second level, automatic systems, like are many other automatic machines of various kind. In many cases, telemanipulators display some form of limited autonomy, being able of taking low-level decisions, while being controlled by humans for higher level tasks.

The earliest manipulators, used for the preparation of radioactive materials, were purely mechanical devices, made by an arm and a gripper, able to duplicate exactly the motion of the arm and the hand of a human operator. Later an increasing number of movements were performed under autonomous control, with the human controller simply taking decisions of higher level. An analogy can be that of the gearbox of a motor car: in basic transmissions the driver has to control the clutch and the gearbox lever, supplying the power needed to perform the action. In semi-automatic transmissions the driver takes the decision about which gear to engage and the device operates the clutch and the gearbox, using a source of power. In fully automatic gearboxes it is still the driver who controls the speed through the accelerator pedal and doing so causes the gear switching. Here all low level decisions are taken by the device, but the high level decision is still taken by the driver in real time.

To qualify as a true robot, the device must possess a good degree of autonomy, being able to perform its tasks without direct, real time, human intervention. A robot must also interact with the environment and perform its tasks in a flexible, easily reprogrammable, way. It belongs then at least to the third level in Fig. 1.3.

Ideally a robot should be even more autonomous from human intervention and should perform also in an intelligent way or even to be conscious. Both these characteristics are still far from being present in actual robots.

An important point is: how much artificial intelligence is needed in space exploration?

Generally speaking, it can be stated that the autonomy of robots must increase with the distance they operate from the human controller. While it is possible to conceive telemanipulators for all tasks to be performed on the Moon, the autonomy required for Mars exploration (at least until humans will be present on that planet or on its satellites) must be more developed. The distance of satellites and planets of the outer solar system is such that unmanned exploration requires true robots. The distances unmanned devices will operate when interstellar exploration will be undertaken may make it necessary to resort to intelligent machines, in the sense defined above.

Strong AI is based on the assumption that all human characteristics can be duplicated by machines and consequently that they not only will be intelligent but also they will possess a true mind, with the related consciousness. This is, however, an unproved statement and, particularly the last part, quite a controversial one.

In particular, a controversial point is the use of *Von Neumann* machines, i.e. intelligent machines able to self-replicate. A *Von Neumann* probe is a space probe endowed with intelligence that can replicate itself.

Once a probe of this kind reaches its target, it could land on a particular celestial body, an asteroid, a planet with a solid surface or a planetary satellite, and start building a copy of itself. A strategy for space exploration based on probes of this kind has been proposed by Tipler.[1] A Von Neumann probe could be launched toward a nearby star with a comparatively simple propulsion system so that, after several hundreds, or even thousands, years it reaches its destination. The probe would land and start to produce other probes, which would then leave that extrasolar system, heading off for other nearby stars. Once its primary task of continuing the expansion to other solar systems had been fulfilled, the probes would begin their scientific tasks, sending reports back to Earth. Eventually, most of our galaxy would be settled by these probes. A single intelligent species could even begin to explore the whole Universe using Von Neumann probes. Such intelligent machines might not just explore, but could also reproduce organic life.

But even if a Von Neumann machine can be built, we will never be sure that, after many replications of itself, errors would not creep in. After all, this is one of the mechanisms by which evolution creates new living beings. Moreover, it is impossible to be sure that a probe programmed on Earth will always perform correctly in the new environments it will find in other planetary systems. Checking, or even modifying, the programming of the probe by radio from Earth is possible only for the first few replications. Then the distances in both space and time become so large that everything must be done by the on-board artificial intelligence systems.

[1] F.J. Tipler, *The Physics of Immortality*, Macmillan, Basingstoke, 1994.

It is impossible to say what might be the outcome of such machines, once they stop behaving exactly as their builders envisaged, owing to random modifications of their genetic code.

Another, more important point has to be addressed. Assuming that such intelligent machines can be built, is it morally acceptable to do so? Should self-replicating machines fill the Universe? That question has caused fierce arguments. Carl Sagan believed the answer to be no: the advisable line for a technological civilization is that of banning the construction of interstellar Von Neumann machines and strictly limiting their use on its home planet. If the argument of Carl Sagan is accepted, such an invention would jeopardize the whole Universe and the control and the destruction of interstellar Von Neumann machines would then become a task with which all civilized countries—the more technologically advanced, in particular—would in some way have to be involved.

Frank Tipler's is of the opposite idea: if humankind abdicates that role, it will miss all chances of colonizing, first, nearby solar systems and then the Universe. Humankind will betray its cosmic duty, and condemn himself to extinction. By Frank Tipler's reasoning, the dissemination throughout the Universe of Von Neumann machines may be considered as another aspect of that evolutionary process which produced humankind and which may in the future produce other intelligent species to take its place. The ultimate task of humankind would then be to create intelligent machines, i.e. to move the evolutionary line from beings based on the biology of carbon to beings based on the chemistry of silicon.

But these are problems for a distant future.

At present the point is not whether intelligent robots are advisable, but whether existing, or short term, robots are adequate to space exploration tasks.

It is questionable, for example, whether one of the most important scientific goals of space exploration, namely the search for extraterrestrial life, can be pursued using robots. There is no doubt that it is possible to program robots for searching for living matter of terrestrial type, but it is difficult that they can recognize life of more exotic type. It is true that it is questionable whether even humans can perform this task, but the probability that robots succeed is even less.

Some examples are self explanatory. The astrobiological experiments performed by the *Viking* probes on Mars were inconclusive, and are still stirring strong debates. While it is positive that they did not find life, it is questionable whether they proved that there was no life in the landing zones. An even more striking example is that of the alleged fossil life-forms found in the ALH 84001 meteorite from Mars. Almost 15 years after the discovery of those microscopic formations, although the meteorite was studied in the better equipped labs and by the most qualified scientists available on this planet, the outcome is still controversial. How can we expect that a robot, working in difficult conditions, can succeed where human scientists working at ease in well equipped labs failed?

Instead of building increasingly complex robots, it may be possible in the future to build a number of simpler machines able to interact with each other so that the swarm can exhibit a sort of collective intelligence. This is the approach that evolution followed with social insects and with other animals (birds, fish) that display quite a simple behavior at the individual level, but can perform complex tasks when operating in swarms, flocks or schools. The behavior of these animals is a popular subject of study in the field of the so-called science of complexity, an there are scholars that believe that from these studies a new approach to space and planetary robotics will evolve.

1.4 Missions for Robots and Manipulators

The types of missions requiring the use of robots and manipulators in space or on celestial bodies can be tentatively subdivided in the following classes

1. Robotic exploration missions;
2. Robotic commercial and exploitation missions;
3. Robotic missions to prepare the way to human missions;
4. Human exploration missions with robotic devices to help humans in exploration duties;
5. Human exploration missions with vehicles to enhance human mobility;
6. Human exploration–exploitation missions requiring construction–excavation devices.

As already stated, in the case of unmanned missions, a very important factor is how far from Earth the device must operate: only in the case of Lunar missions (with perhaps an extension to missions to near-Earth asteroids and comets, NEO in general) true teleoperation is possible. In all other cases a sufficient autonomy must be considered. In the case of the Moon there may be little difference between a machine for missions of type (1) and (4) since there is not a very large advantage in having a human controller close by instead of having him on Earth, while for Mars the difference is huge, to the point that a manned mission in which the astronauts land on a Mars satellite to teleoperate from there machines exploring the planet has been suggested. It must be noted that it is not just a point of communication delay due to distance (2–3 seconds for the Moon, up to half an hour for Mars and much more for main belt asteroids and outer planet satellites) but mainly to the availability of radio relays in orbit around the planet and/or Earth. This can be increased by putting communication satellites in orbit around Mars, a thing that will be at any rate needed if serious exploration of the Red Planet is undertaken.

Human presence on site greatly simplifies this issue, even if some autonomy will at any rate be needed. For instance, low level control should not be entrusted to humans, to allow them to perform more important tasks, but this may be relatively easy, since high level path planning functions and occasional supervision and recovery tasks are the most difficult to be performed automatically.

The devices required for missions of type (1) and (4) may in general be quite small, since the instruments that constitute their payload can be light and not much bulky and will be increasingly miniaturized in the future. There are two factors limiting miniaturization: the need to collect or dig out and carry back samples and the fact that in general the mobility and the range of a small vehicle are more limited than those of a large one. This is, however, strictly linked to the type of environment, since there are places that can be managed by a small vehicle and not by a large one. The size and the mass of the robot is also limited by the space and mass allowance for the spacecraft.

Not considering missions of type (2) that involve small automatic spacecraft like telecommunication, meteorological or Earth resources satellites, commercial and exploitation missions may require a wide variety of different spacecraft and rovers. They include, for instance, 'virtual tourism' on the Moon, which requires a medium size rover with many panoramic cameras that is telecontrolled from Earth and machines able to land on the Moon or an asteroid to dig out resources from the soil and to process them. The size of a machine to do this may vary from medium to very large size; the long time goal being to carry huge asteroid mining robots on NEA or main belt asteroids, able to extract, process and send to Earth or possibly to a Lunar or Mars colony the resources that are abundantly present in those places. Again, the autonomy required depends on the distance: while exploitation of the Moon and some NEOs can be performed using teleoperators, more distance places need increasing autonomy.

The automatic machines that will be used to prepare the sites for human missions include again a wide variety of different devices. On the Moon the point is mainly positioning and anchoring the habitats on the ground, digging the regolith needed to shield them from radiation, and doing all the required preparatory work. On Mars the tasks include also producing the propellant for the return journey and their energy requirement may be much larger. The rover may thus be required to unload a nuclear reactor from the lander, locate it in a suitable site, set it in operation and then to assist the ISRU plant. Other tasks may be preparing the terrain for a landing pad and possibly to prepare tracks on which the vehicles can move in the vicinity of the outpost.

Missions of type (5) and (6) require large vehicles, able to carry humans on board or earth-moving machines. All these machines may be directly operated by astronauts, teleoperated or may display some grade of autonomy. In any case the more autonomous they are, the less they require the astronaut to spend time in more or less menial, or at least less repetitive and sometime dangerous, tasks. Since the astronaut's time will be one of the most precious commodities in outposts and space colonies, there will be an increasing interest in autonomous operation also in missions of this type.

Space robots can be used in three distinct types of environments:

- Low Earth orbit (LEO)
- Deep space
- Planetary surfaces

1.4.1 Low Earth Orbit (LEO)

Robotic Spacecraft

Many unmanned satellites operating in low orbit work without (or with very little) human intervention. However, most scientific and commercial satellites, although operating autonomously, are more automatic machines than robots, since they operate along fixed lines, performing their tasks in a rigid way. When decisions must be made the close proximity of Earth makes it possible to teleoperate them, again in a simple way. Only very complex satellites may be considered as robots, like for example the *Hubble Space Telescope*.

They will not be dealt with here.

Robotic Arms (Telemanipulators)

Telemanipulators find a large use in manned missions. The best known of such devices is the *Canadarm* of the *Space Shuttle*, used for a wide variety of tasks, from satellite deployment, retrieval and maintenance to assembly of space structures and assistance to Extra Vehicular Activities (EVA). The *International Space Station* has a multipurpose robotic arm, whose main task is maintenance of the station itself, but can be used for many other tasks. The degree of autonomy of these devices is different, but usually they are controlled by the astronauts.

EVA Assistants

Up to now the task of helping astronauts in their extra vehicular activity has been performed by robotic arms, controlled directly by other astronauts (Fig. 1.1). However, free flying robots, more or less controlled by astronauts, have been proposed and will be used in the future. Free flyers of the same kind can be used also for other tasks, like space station maintenance, servicing and towing satellites, etc.

Robotic Space Suits

Space suits are heavy, a thing that in microgravity may not constitute a great disadvantage, but are also stiff and not easily operated, a thing that hampers the mobility of astronauts in EVA. This is particularly true for the parts of the space suit covering the arms and the hands. A motorized space suit that operates like a telemanipulator under the direct control of the astronaut wearing it could improve greatly his/her performance in EVA, reducing fatigue and improving safety. Several man-amplifier exoskeletons were designed and built in the past, starting from the Hardyman, built by GE in 1965, all aimed to be used on Earth and often with military applications in mind. The concept can be adapted both to EVA in space and to individual mobility on planetary surfaces, where the presence of a gravitational field makes them even more useful.

1.4.2 Deep Space

If with deep space we mean space beyond the Van Allen belts, i.e. where the space environment is dominated by the Sun and not by the presence of Earth, also satellites in GEO (Geostationary Earth Orbit) must be considered in deep space. They are mostly telecommunication satellites and for them the same things said for commercial satellites in LEO could be repeated.

Robotic Probes

Deep space probes are usually considered as robots, since they must operate autonomously, performing a wide variety of tasks. The farther from Earth they must operate, the more autonomous they must be. The maximum distance space probes reached up to now is well beyond the orbit of Pluto, and they reached the boundary of the solar system, i.e. the heliopause, where is located the interface between the region of space dominated by the Sun and the interstellar medium.

Many probes must follow complex trajectories, with deep space manoeuvres to perform planetary flybys to get energy from planetary assists, sometimes also entering planetary orbits or navigating for years in the system of the satellites of a giant planet. All this must be performed keeping the antenna well oriented, with strict tolerances, toward the Earth and the sensors and cameras toward the scientific targets.

1.4.3 Planetary Surfaces

Landers

Robots performing planetary exploration (with this term here we mean exploration of any celestial body, be it a true planet, a satellite or even an asteroid or a comet) must at first do a controlled descent. Unmanned landers must perform autonomously on all bodies except the Moon, where teleoperation from Earth is possible. Depending on the type of descent planned (by parachute, ballute, airbag, or rockets) the control, particularly during the final phase, is more or less complex. During atmospheric entry a communication blackout phase may be present.

Scientific operations may start before landing, taking images and measurements. After landing scientific operations on the surface will start. Up to now, landers performed many different tasks, like deploying rovers, taking samples of the ground, performing scientific experiments and sending images back to Earth. Future landers may perform many other tasks, like deploying exploration aircraft or balloons, return vehicles for sending back samples, exploiting the local resources (ISRU), etc.

Rovers

A wide variety of robotic rovers have been designed, tested and some actually used in planetary missions. All rovers actually used where wheeled machines, with the exception of early Mars rovers that moved on skis and jumping rovers aimed to the satellite of Mars Phobos that never reached their destination. Rovers with tracks, legs or jumping devices were tested on Earth, like also flying exploration vehicles based on aerodynamic forces (aircraft) or aerostatic forces (balloons).

As already stated, except for the case of the Moon, where full teleoperation is possible, in all other cases a good degree of autonomy is needed. However, since at the present state of technology this is still a critical requirement, all rovers used on Mars were very slow, so that high level control could be supplied in any case from Earth.

Peculiar rovers will be the construction machines that will prepare the ground infrastructure for the arrival of astronauts or will start the exploitation of asteroids and other celestial bodies. Apart from the mobility devices, they will be supplied by surveying instruments, to find what they are looking for, digging and earth-moving devices and arms to move various equipment.

The size of rovers can go from much less than one meter to very large rovers able to move and to set up habitats, nuclear reactors and other infrastructures, to grade landing pads and roads, etc. Considering their size and mass, rovers can be classified as

- nanorovers (mass less than 5 kg),
- micorovers (mass between 5 and 30 kg),
- minirovers (mass between 30 and 150 kg), and
- macrorovers (mass above 150 kg).

Vehicles

The only vehicle used by astronauts on the Moon, the Lunar Roving Vehicle (LRV) of the last *Apollo* missions was not a robot, since it was completely controlled by astronauts. Future man-carrying vehicles will have a certain degree of autonomy, being able to perform also as teleoperators (i.e. to move under control of an astronaut not on board) or as more or less autonomous vehicles. The autonomy may go from low level autonomy, something just little more than ordinary vehicles with automatic gear, to full autonomy, being able to carry humans under a loose control from the latter.

Generally speaking, human-carrying vehicles can be subdivided into two categories:

- simple mobility devices, without any life-support facility, whose task is to carry humans protected by their own space suit, and
- vehicles providing a true shirt-sleeve environment.

The LRV belonged to the first type. These vehicles can be as small as a city car and can be very simple and light. Vehicles of the second type are true mobile habitats and are very complex machines, in a way similar to military personnel carrying vehicles able to maintain inside an atmosphere not contaminated by possible chemical or bacteriological weapons. A modular approach may be followed and a small habitat can be carried by a vehicle of the first type.

Astronaut Assistants

Robots designed to assist astronauts on the surface of planets may span from microrovers similar to those seen for autonomous exploration, but operating under the high level guidance of human explorers to go into the most difficult or dangerous places, to large construction machines to build habitats and bases, to mining and transportation machines for planetary exploitation.

Their autonomy and degree of teleoperation can be very different from case to case. In general they will benefit from the studies aimed to build similar machines for use in difficult and dangerous workplaces on Earth, like mines or heavy industries, and for dangerous military tasks, like demining, defusing unexploded weapons and reconnaissance.

Assistants that will help astronauts in their exploration tasks may be similar, as already stated, to the present exploration rovers, but with an important difference: not to impair the movements of the astronaut they need to be much faster, at least as fast as a walking human, and at least autonomous enough to follow their 'master' without needing direct guidance. These features are posing challenges still to be met.

To work together with humans in a structured environment (an outpost, a base or even a large spacecraft) a robot of this kind may have a humanoid shape. Like in service robotics a humanoid body allows to perform well in environments studied for humans and to use tools designed for them without any modification. A shape that seems convenient in EVA is that of the so-called *centaurs*, legged or, more often, wheeled rovers with four or more wheels (or legs) having in the front part a human-like torso with arms and a head.

Another form of astronaut assistants are the already mentioned robotic space suits, adapted to the various planetary environments. Their aim can be that of improving the mobility usually hampered by the stiffness of the suit, compensating for the weight of the latter, to allow the astronaut to perform tasks that he could not perform even in Earth conditions and to offer a better protection from the environment than that offered by the usual space suits.

1.5 Open Problems

The deficiencies of artificial intelligence is not the only open problem in space robotics. Practically all aspects involved are still needing much research. The open problems can be classified in the following themes:

- Control
- Mechanics
- Transducers
- Power
- Communications

Many of the open problems are not much dissimilar to those still open in standard robotics, but often they are carried to higher levels. If robotic industry will actually develop as some specialists predict (some studies assess that the industry of personal robots will have in the 21th century a role similar to that played by automotive industry in the 20th), the large scale of production will make it possible to perform very deep studies at a reasonable cost. Space robotics will then benefit of a technology transfer from hi-tech non-space applications, while the latter will find in space application a driver for the most demanding solutions.

1.5.1 Control

Both the control hardware and software need much development. Hardware is rapidly developing, and the most recent and powerful devices are not yet qualified for space use. Space environment, particularly for what radiations are concerned, is very hard for electronic devices, mainly when they must operate beyond Low Earth Orbit (LEO). Generally speaking, space robots must rely on less powerful and older hardware than their counterparts on Earth.

Software is also critical, particularly for unconventional applications, like highly autonomous robots, cooperative devices and, in telemanipulators, for man–machine interfaces that can be used in extreme environment.

1.5.2 Mechanics

The mechanical and structural components of space robots are even more difficult than those of standard robots. Low weight and high reliability are prerequisites for anything to be launched in space, as well as the possibility of working in space environment. A space robot must withstand strong vibration at launch and possibly re-entry, hard vacuum (with the ensuing lubrication problems), large temperature variations, radiation, extreme dust on some planets, etc. for long periods of time without maintenance.

Much work in the field of deployable and inflatable structures is still needed.

1.5.3 Transducers

Sensors and actuators are critical components in all robotic applications. While sensors may be similar to those of standard robots, actuators are much limited by the environment and by the low power usually available in space.

Some types of actuators are not usable at all, like brush DC motors that do not operate correctly in vacuum or shape memory alloys (SMA) that need cooling and have very low efficiency. Also pneumatic and hydraulic devices are seldom considered, the first ones for their high air consumption and low efficiency, the second for the need of heavy ancillary equipment. The actuators considered for most applications are restricted to electric motors (usually brushless) and electromagnetic and piezoelectric actuators.

1.5.4 Power

Power is a very critical issue in all robotic applications, but in space things are much worse. The transducers and the control electronics are usually energized by batteries, but the latter must be kept in charged conditions, since only for very short missions it is possible to use primary (non rechargeable) batteries.

The most common solution is the use of solar panels, but they can supply only low power and their output decreases fast getting far from the Sun. On Mars their output is half than on Earth, and they cannot practically be used in the outer solar system (beyond the orbit of Jupiter). A good solution can be fuel cells, but they can be used only either for short missions or in connection with the production of fuel and oxidizer from locally available materials (in situ resource utilization, ISRU). RTGs (Radioisotope Thermoelectric Generators) can supply low power for very long time and are usually the best choice for robotic probes operating in deep space, but have some limits in planetary applications. Nuclear reactors are likely to be the best choice for producing large amounts of power both for deep space and planetary installations, and can be used to keep the batteries of robots charged.

RHUs (Radioisotope Heat Units) can be used in thermal control to keep warm robots, particularly in the outer solar system or on planets, like during the long lunar nights. Supercapacitors are very good candidates to supply high peaks of power for short times, but must be recharged.

1.5.5 Communications

The need for communications depends on the type of machine considered. Telemanipulators need to receive all commands, possibly in real (or almost-real) time. The more autonomous is the robot, the lower the need of receiving communications is. All devices need to communicate back to Earth the results of their work with the possible exception of highly autonomous robots able to operate by themselves, storing information and downloading them after their return. Present technology is adequate to send back information from the frontiers of the solar system with the low power available on board of probes, but this requires to maintain large and costly receiving equipment on Earth. To communicate with robots on the surface of planets or on the far side of the Moon requires the availability of relay stations, for instance in the Lagrange points of the Earth–Moon system or in orbit around the planet.

Some problems related with keeping the antennas of the spacecraft pointed precisely enough toward Earth and to fold and deploy them have been successfully solved several times but are not trivial at all.

Chapter 2
Space and Planetary Environment

The environments in which exploration and exploitation robots operate can be roughly subdivided into four types

- Low Earth orbit
- Deep space
- Interstellar medium
- Planets and small celestial bodies

The latter can be further subdivided in

- Moon
- Rocky planets
- Giant planets
- Satellites of giant planets
- Small bodies

2.1 Low Earth Orbit Environment

Usually Low Earth Orbits (LEO) are orbits with a height lower than 1,000 km. However, the environment encountered in LEO is roughly the same encountered in orbits up to the Van Allen belts (Fig. 2.1) surrounding our planet.

The Earth magnetosphere protects our planet and all space below the Van Allen belts from most radiations coming from the Sun and deep space. A diagram of the Earth's magnetosphere is shown in Fig. 2.2.

Owing to the protection by the magnetosphere, radiation is moderate in LEO, even during strong solar activity (solar flares). However, there is an anomaly in Earth magnetic field off the cost of Brazil and in that zone much stronger radiation reach the upper atmosphere. The Van Allen radiation belts are symmetric about the Earth's magnetic axis, which is tilted with respect to the Earth's rotational axis by an angle of $\approx$11 degrees. This tilt, together with an offset of $\approx$450 kilometers causes the inner Van Allen belt to be closer to the Earth's surface over the south Atlantic

G. Genta, *Introduction to the Mechanics of Space Robots*, Space Technology Library 26, DOI 10.1007/978-94-007-1796-1_2, © Springer Science+Business Media B.V. 2012

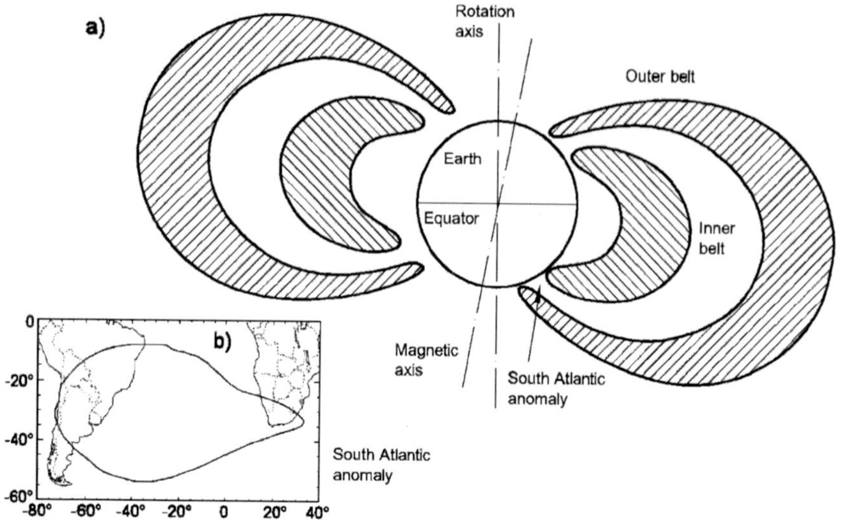

Fig. 2.1 (**a**) Schematic cross section of the Van Allen radiation belts. (**b**) Earth zone interested by the South Atlantic Anomaly

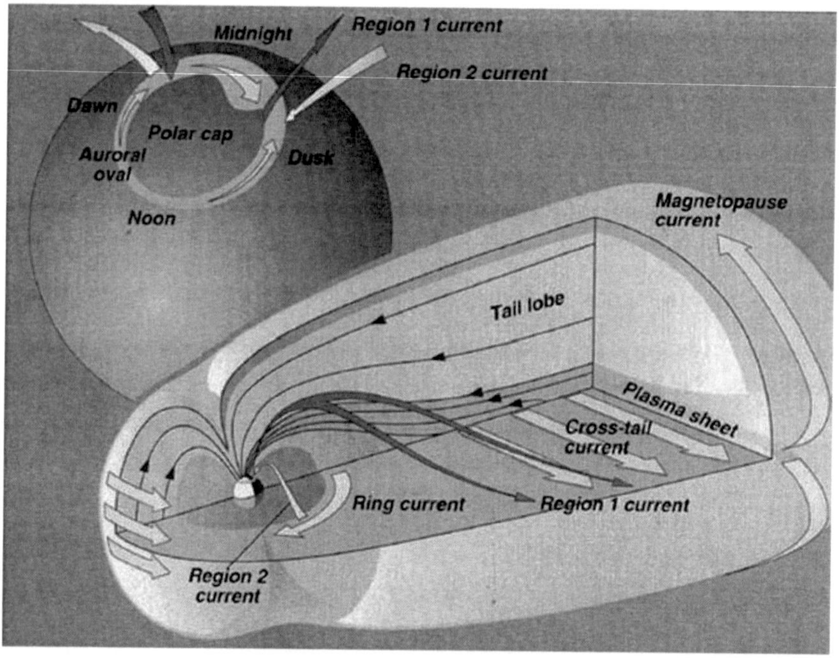

Fig. 2.2 Diagram of the Earth's magnetosphere—the volume occupied by the geomagnetic field in space—and, *in the upper left*, the northern polar region. Large electric currents (millions of Amperes) flow through space (from G. Genta, M. Rycroft, *Space, The Final Frontier*? Cambridge University Press, Cambridge, 2003)

Table 2.1 Some values of atmospheric temperature, pressure and density in LEO from the US Standard atmosphere

Altitude (km)	T (K)	p (Pa)	ρ (kg/m^3)
0	288.15	1.01×10^{5}	1.23
100	195.08	3.20×10^{-2}	5.60×10^{-7}
200	854.56	8.47×10^{-5}	2.08×10^{-9}
300	976.01	8.77×10^{-6}	1.92×10^{-11}
400	995.83	1.45×10^{-6}	2.80×10^{-12}

ocean, causing the South Atlantic Anomaly (SAA), and farthest from the Earth's surface over the north Pacific ocean.

Remark 2.1 The boundaries of the SAA vary with altitude and its shape changes over time. At an altitude of 500 km, the SAA ranges between −90° and +40° in longitude and −50° to 0° in latitude. Its extent increases with increasing altitude.

The characteristics of the upper atmosphere and of space above it are quite variable, both with the altitude and time. In the zone between 90 and 1,000 km the average values of pressure, density and temperature stated by the US Standard atmosphere can be used (see Table 2.1).

The atoms in the upper atmosphere are strongly ionized and the atmospheric layer between 50 and 600 km altitude is referred to as the ionosphere. They are mostly oxygen ions, but above 300 km the composition changes gradually to mostly hydrogen ions.

The conditions in that zone of space are well known, but much variable depending on the *space weather*, a term commonly used to define the phenomena involving ambient plasma, magnetic fields, radiation and other matter in a region of space. The space weather close to the Earth is a consequence of the behavior of the Sun, the Earth's magnetic field, and our location in the solar system and affects deeply not only the space activities of humankind, but also our planet and our technological activities on Earth.

In the whole solar system space weather is greatly influenced by the speed and density of the solar wind (see Sect. 2.2). Data on the current solar wind speed and density and on solar flares are continuously available, for instance on www.spaceweather.com.

The density reaches maxima at the peak of solar activity cycles, every 11 years. During such periods the drag on satellites gets much stronger and the danger of losing altitude and deorbiting increases. To remain in LEO between 200 and 400 km satellites require periodical reboost, particularly if their surface/mass ratio is large. The International Space Station (ISS) is particularly subject to high atmosphere drag, owing to its large solar panels. Reboost becomes even more important in high solar activity periods.

At greater altitudes the pressure and density decrease fast; at about 1000 km a satellite may remain in orbit indefinitely (at least with reference to normal service time of man-made machines) without reboost.

Apart from plasma, space is full of debris of various type, both natural and artificial. Natural ones consist mainly in very small meteorites, micrometeorites and dust grains that are captured by Earth's gravitational field and enter the atmosphere to be destroyed by air drag. Occasionally, larger meteorites able to reach the surface occur.

Most items of space debris in LEO are, however, artificial. The large pieces of space debris are accurately tracked using radars and telescopes; their number and positions are well known. In 1996 there were about 4,000 detectable objects, and now there are some 8,500 objects with sizes about 10 cm or greater. Although new debris is always being produced, the older debris decays owing to upper atmospheric drag and eventually re-enter the atmosphere, being completely destroyed before reaching the ground. Only meter-sized objects have some chance of getting to the ground, and so constituting a danger for people, a danger which is much lower than that due to natural objects.

Remark 2.2 The 11 year solar cycle has a strong effect on space debris since, as already stated, the density of the high atmosphere is much greater near solar maximum conditions than at solar minimum. A periodic clean-up of debris in the lowest orbits thus occurs.

Smaller pieces of debris are produced by the explosion, whether accidental or intentional, of upper stages or satellites. It has been computed that about half of the centimeter-sized debris has been produced in this way. Military satellites are most responsible for this type of pollution. Dangerous debris has been produced when the core of the nuclear reactor of military satellites is jettisoned in order to be placed in a safe (higher) orbit after the satellite is decommissioned. While doing this, the cooling system of the reactor lets swarms of droplets of the coolant, a liquid sodium-potassium alloy, to escape. These liquid droplets are dangerous sub-centimetric projectiles, which can penetrate the skin of satellites.

International treaties forbidding the intentional explosion of satellites are being prepared; they state that precautions must be taken against accidental events which may produce space debris. They also state that decommissioned satellites must be de-orbited and destroyed in the atmosphere or, if this is impossible as in the case of geosynchronous satellites, moved into a less used orbit.

Remark 2.3 The most critical orbits are those lying between 1,000 and 1,400 km altitude, where the air drag is insufficient to cause debris to decay and re-enter.

The ultimate danger is a situation in which there are so many objects that collisions are frequent enough to produce new fragments continuously. This sort of chain reaction would end up creating a debris belt in which no object could survive. But this nightmare scenario will not occur for several centuries.

Apart from the much publicized accident when a *Progress* cargo craft hit a solar panel of the *Mir* space station (but in this case the cause was a wrong rendezvous manoeuvre), only two space collision between two unrelated objects involving a working satellite occurred to date. The first was when a suitcase-sized fragment from the explosion of the upper stage of an *Ariane* rocket hit, after ten years in space, the small French satellite *Cerise*. The damage in that case was not too large—a boom protruding from the satellite for stabilization purposes was cut off—but the satellite might well have been wiped out.

The second, occurring on February 10, 2009, involved a US commercial Iridium satellite and an inactive Russian satellite Cosmos 2251. Both were completely destroyed, generating a cloud of thousands of pieces of debris. Two other collisions between unrelated non-working objects were also recorded. One occurred in late December 1991, and involved a Russian decommissioned satellite (Cosmos 1934) and a piece of debris from Cosmos 929; the other is more recent (January 17, 2005) and involved a 31 years old US Thor rocket body and a fragment from a Chinese CZ-4 launch vehicle.

The probability that a large piece of debris will come closer than 100 m from a satellite in LEO is about one in 100 years. In the case of a space station, a large piece of debris identified in advance can be avoided by suitable manoeuvres. Very small objects do not cause damage and debris smaller than a few mm should be stopped by the shields of the International Space Station. The most dangerous particles are those which cannot be detected from the ground but are larger than a few mm. It has been computed that the probability that a 1 cm sized particle will pierce through the hull of the ISS is about 1% in its 20 years life.

There are well consolidated design practices for commercial and scientific satellites in LEO, so this can be considered an environment that does not pose great problems.

2.2 Interplanetary Medium

Space beyond the Van Allen Belts is pervaded by the *solar wind*, which fills the whole solar system The solar wind is mainly made of hydrogen ions (protons) flowing at high speed out of the Sun's corona. The temperature of the corona is so high that the coronal gases are accelerated to a velocity of about 400 km/s. This component of the solar wind, the so called *slow solar wind*, has a temperature of $(1.4–1.6) \times 10^6$ K and a composition similar to that of the corona. Over coronal holes the speed of solar wind can reach 750–800 km/s and a temperature of about 8×10^5 K. The composition of this *fast solar* wind is closer to that of Sun's photosphere. Over those parts of the outer layers of the Sun that are colder, the solar wind can be as slow as 300 km/s.

While the slow solar wind is mostly ejected from the regions around the equator of the Sun, up to latitudes of 30°–35°, the fast solar wind originates from the coronal holes, which are located mostly in the regions about the Sun's magnetic poles.

The interaction of the particles with different velocities and the rotation of the Sun causes the solar wind to be quite unsteady and the space weather in the whole solar system to be much variable. In 1997 the Advanced Composition Explorer (ACE) satellite was launched and placed into an orbit about the L1 point between the Earth and the Sun. This point, located at about 1.5 million km from the Earth on the line connecting the Earth with the Sun, is characterized by the equilibrium between solar and Earth's attraction, so that a body located there (or better in a halo orbit about L1, since in this point the equilibrium is unstable), can remain between the Earth and the Sun indefinitely. The ACE monitors continuously the solar wind, providing real-time information on space weather.

From time to time, fast-moving bursts of plasma called Interplanetary Coronal Mass Ejections (ICME) may disrupt the standard pattern of the solar wind, launching in the surrounding space electromagnetic waves and fast particles (mostly protons and electrons) to form showers of ionizing radiation. When these ejections impact the magnetosphere of a planet they temporarily deform the planet's magnetic field. On Earth they induce large electrical ground currents and send protons and electrons toward the atmosphere, where they form the aurora.

Remark 2.4 Solar flares are one of the causes of interplanetary coronal mass ejections. They constitute a danger to spacecraft, manned and unmanned.

Owing to the motion of these charged particles, an interplanetary magnetic field (IMF) pervades the whole solar system.

The interplanetary medium is filled by radiation, *cosmic radiation*, not only from the Sun but also from extrasolar objects. The cosmic radiation that enter Earth's atmosphere are made up mostly by protons (90%), plus about 9% helium nuclei (alpha particles) and about 1% of electrons (β particles), plus photons and neutrinos. Their energy is in the range of over 1000 eV.

Apart from the radiation from the Sun, *Galactic cosmic rays* (GCRs) come from outside the solar system but generally from within our Milky Way. They are atomic nuclei trapped by the galactic magnetic field with all of the surrounding electrons taken away during their travel through the galaxy at a speed close to the speed of light. As they travel through the very thin gas of interstellar space they emit gamma rays. Their composition is similar to the composition of the Earth and solar system.

Another component of the cosmic radiation are the Anomalous Cosmic Rays (ACRs) They are due to the neutral atoms of the interstellar matter that flow through the solar system (the charged particle are kept outside the heliosphere (see Sect. 2.3) by the interplanetary magnetic field), at a speed of about 25 km/sec. When closer to the Sun, these atoms undergo the loss of one electron in photo-ionization or by charge exchange, and are then accelerated by the Sun's magnetic field and the solar wind. ACRs include large quantities of helium, oxygen, neon, and other elements with high ionization potentials.

Apart from these heavy particles, there is also the cosmic microwave background radiation, consisting of very low energy photons (energy of about 2.78 Kelvin) which are remnants from the time when the universe was only about 200,000 years old. Neutrinos, photons of different energies (produced by the Sun, other stars, quasi-stellar objects, black-hole accretion disks, gamma-ray bursts and so on), electrons, muons, and other particles are also present.

Remark 2.5 All these particles are not dangerous on Earth since they are deflected by the Earth's magnetic field or are stopped by the atmosphere. On other celestial bodies, which have no magnetosphere and whose atmosphere is thin (the Moon, Mars, etc.) they reach down to the surface and constitute a danger.

Giant planets, like Jupiter, have a strong magnetosphere, shielding the planet but causing zones of strong radiation, like the Van Allen belts of Earth. Spacecraft entering these zones must be designed taking this factor into account.

Apart from plasma, there is also a tiny amount of neutral hydrogen: at the distance of Earth's orbit from the Sun, the concentration of neutral hydrogen is about 10^4 atoms per m^3. As said above, some of these atoms come from interstellar space.

A relatively small amount of dust particles—micrometeroids—exist in the solar system. Much of this dust is thought to have been produced in collisions between asteroids and in the shedding of material from comets while passing close to the Sun. About 30,000 tons of interplanetary dust particles are estimated to enter Earth's upper atmosphere annually.

The vacuum is much higher than in LEO, and hydrogen ions from the Sun substitute oxygen ions from Earth's atmosphere. So, while the environment in LEO is oxidizing, that in deep space is reducing.

2.3 Interstellar Medium

The Sun moves, with the planets and all the bodies of the solar system, through the very rarefied medium that fills the interstellar space, the *interstellar medium*. Most of this medium is gas (about 99%), and the remaining is dust. Although very rarefied, it constitutes about the 15% of the matter of our Milky Way. The density and composition of the interstellar medium is quite variable from place to place. Its density ranges from a few thousand to a few hundred million particles per cubic meter with an average value of a million particles per cubic meter.

The gas is roughly 89% hydrogen (either molecular of atomic), 9% helium and 2% heavier elements. It forms cold clouds of neutral atomic or molecular hydrogen. Hot newly born stars ionize the gas with their ultraviolet light, giving way to hot ionized regions. In the outer regions of the galaxy, the interstellar medium blends smoothly into the surrounding intergalactic medium.

Interstellar dust is made of small particles, with a size of a fraction of a micron, of silicates, carbon, ice, and iron compounds.

In its motion through the interstellar gas, the interplanetary medium creates a shock wave ahead of it, in a way which is similar to what happens when a supersonic aircraft flies through the air. This shock wave is called the *bow shock*. A sketch of the bow shock is shown in Fig. 2.3a.

The solar system is contained in a magnetic bubble, the *heliosphere*, that extends well beyond the orbit of Neptune. Although electrically neutral atoms from interstellar space can penetrate this bubble, virtually all of the material in the heliosphere emanates from the Sun itself.

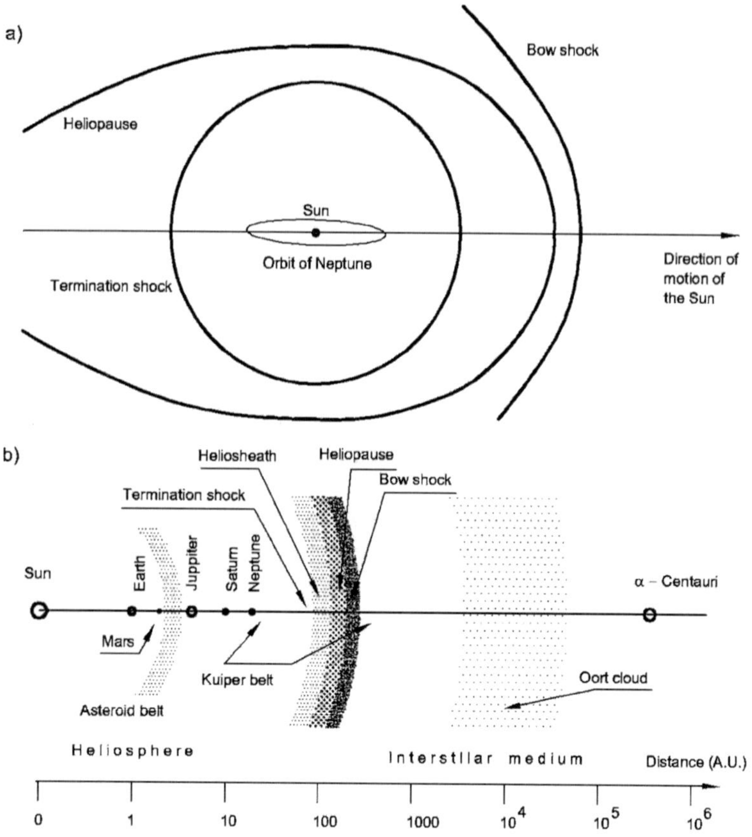

Fig. 2.3 (**a**) Diagram of the heliosphere, showing the termination shock, the heliopause and the bow shock. (**b**) Distances (in Astronomical Units) from the Sun of the various objects up to the nearest star. Pseudo-logarithmic scale (in a true logarithmic scale the origin cannot be represented)

The solar wind travels at supersonic speed; as it approaches the heliopause, it slows suddenly, forming a shock wave where it passes from supersonic to subsonic. This standing shock wave is called the *termination shock*.

The termination shock is believed to be 75 to 90 AU from the Sun. In 2007, *Voyager 2* passed through the Sun's termination shock at about 84 AU from the Sun. Since the position of the termination shock is not fixed (its distance from the Sun changes due to changes in the solar activity), *Voyager 2* actually passed through the termination shock several times. Similar symptoms of being through the termination shock were shown by *Voyager 1* in 2004 at 94 AU. from the Sun. The termination shock may be also irregularly shaped, apart from being variable in time.

Once subsonic, the solar wind may be affected by the ambient flow of the interstellar medium. This subsonic region is called the *heliosheath*. The limit of the heliosheath is at about 80 to 100 AU. at its closed point. Since it has a comet-like shape, its tail trails behind the Sun for a distance of several hundreds AU.

The outer surface of the heliosheath, where the heliosphere meets the interstellar medium, is called the *heliopause*. The outer limit of the heliopause is the bow shock. The heliopause and the termination shock are shown in the sketch of Fig. 2.3a.

A region very rich in hot hydrogen, named the *hydrogen wall*, is assumed to be located between the bow shock and the heliopause. The wall is composed of interstellar material interacting with the edge of the heliosphere.

The solar system, in logarithmic scale, showing the outer extent of the heliosphere, the Oort cloud and out to Alpha Centauri is shown in Fig. 2.3b. Note that the scale is not a true logarithmic scale, since it should not show the zero, and that α Centauri is actually not in the direction of motion of the Sun.

Remark 2.6 Most of the galactic cosmic radiation is deflected by the heliosphere, which protects the whole solar system from this harsh galactic environment. The Earth has thus two protective layers against cosmic radiation: its own magnetosphere and the heliosphere. Outside the heliosphere the radiation environment is thus much worse than in the solar system in general.

2.4 Lunar Environment

The lunar environment is dealt with separately, since it is much better known than other planetary environments and the recent decision to return to the Moon makes it more important.

Since the Moon is gravitationally locked in its orbit around the Earth, it shows always the same side (the near side) to the Earth. The other side (the far side) cannot be seen from our planet. However, small variations (libration) in the angle from which the Moon is seen allow about 59% of its surface to be seen from the Earth. The rotation and orbital periods are equal to 27.322 days, while the average length of lunar day is 29.531 days.

The gravitational acceleration at the surface of the Moon is $1.62\ \text{m/s}^2$ (at the equator), however, the lunar gravity field is not completely uniform due to mass concentrations (mascons) beneath the surface, associated at least in part with the presence of dense basaltic lava flows in some of the giant impact basins. Although the variations in acceleration due to these irregularities are generally less than one-thousandth of the surface gravity, they greatly influence the orbit of spacecraft about the Moon.

The escape velocity is 2.38 km/s at the equator.

The atmosphere is very thin and can practically be considered as a vacuum: the total mass of gases surrounding the Moon has been evaluated as only 104 kg. The mean atmospheric pressure is about 3×10^{-15} bar, but it has large variations mainly with the night and day cycle. The gases come mostly from surface outgassing and are lost in space owing to solar light pressure and, for ionized particles, solar wind magnetic field. The elements detected, both using Earth based spectrometers and instruments on space probes, are hydrogen, helium-4, sodium, potassium, radon-222, polonium-210, argon-40 plus molecular gases like oxygen, methane, nitrogen, carbon monoxide and carbon dioxide. Likely, hydrogen and helium come from the solar wind, while argon originates from the lunar interior. Vacuum must be accounted for when choosing the materials and many common plastics and rubbers are unsuitable, as their strength and flexibility is reduced by outgassing of their volatile components. Even materials approved for use in LEO may not be suitable for use on the Moon.

The thin atmosphere does not refract light, but the moon is not a black and white world: very deep black and blinding white are accompanied by a vast range of browns, tans, and grays, with a few shades of purple and rusty reds. There are no blues and greens, except from objects human will bring with them.

Due to the lack of a substantial atmosphere and the long duration of the day, the surface temperature varies greatly from day to night and also from place to place. The average temperature is −23°C, but during the day the average surface temperature is 107°C, with maximum values as high as 123°C and up to 280°C at noon on the equator. At night, the average temperature is −153°C but can fall as low as −233°C in the areas of the south basin that are permanently shaded. A typical non-polar minimum temperature is −181°C (at the Apollo 15 site). During the month long day, quite strong temperature variation rates can occur. Average temperature also changes about 6°C between aphelion and perihelion. Temperatures just below the surface remain relatively constant; at a depth of 1 m, it is almost constant at −35°C.

Objects on the lunar surface are heated during the day more by the hot ground than by direct sunlight: the sky is seen as a heat sink with a tiny hot spot, while the ground is seen as a large hot surface, emitting strong infrared radiation.

There have been reports of lunar geothermal activity at a few locations, like the Aristarchus Crater region, where "glowing clouds" have been reported. Photographic evidence of recent small-scale volcanic activity has also been recorded. If these findings are confirmed, warm magma close to the surface might be exploited as an energy source.

The Moon has a very weak magnetic field, from 3×10^{-3} to 0.33 μT (as a comparison, the Earth's magnetic field is 30–60 μT—more than hundred times larger) While on Earth the magnetic field is dipolar, generated by a geodynamo in its core, this effect is completely absent, at present, on the Moon, and the lunar magnetic field originates in the crust and is local in nature. Possibly, it is a remnant of a global magnetic field, which was present in the past when a geodynamo effect was active, but this theory is debatable. The fact that the largest crustal magnetizations appear to be located near the antipodes of the giant impact basins seems to suggest, on the contrary, an impact origin.

Additionally, on the lunar surface there is an external magnetic field, of about 5×10^{-9}–10×10^{-9} T, due to the solar wind.

The main features of the lunar surface are the dark and relatively featureless plains called maria (singular mare, Latin for sea), the lighter-colored, hilly regions called terrae (singular terra, Latin for land), or more commonly highlands, and the craters, almost absent on maria, but abundant on terrae. The far side almost completely lacks maria, which represent just about 2.6% of the far side, against 31.2%

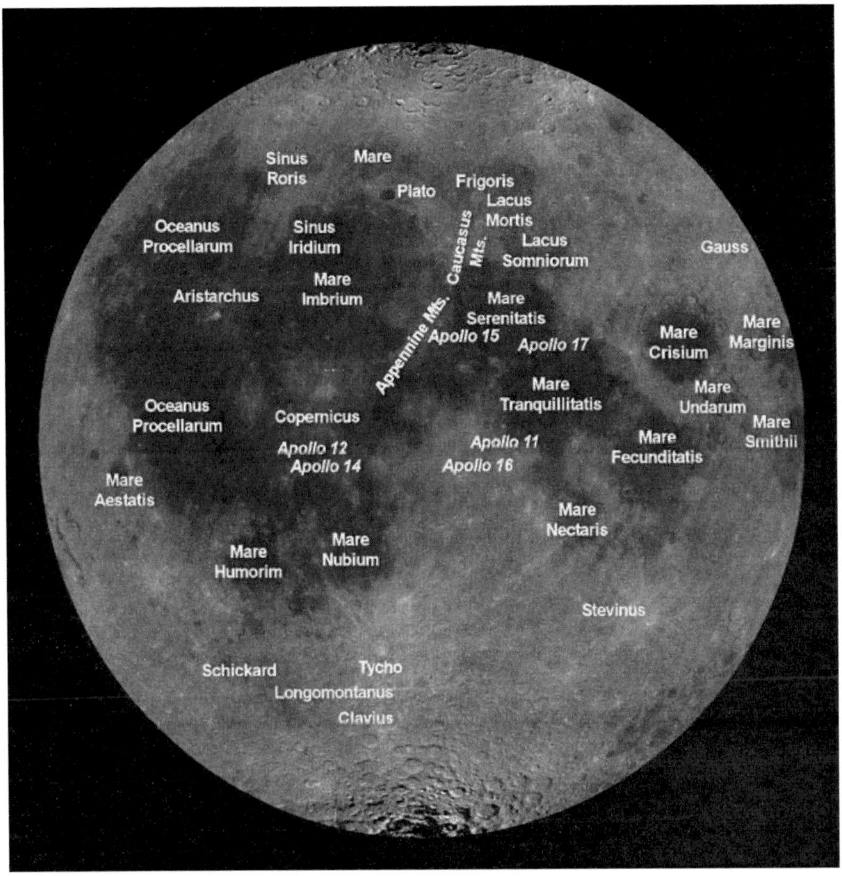

Fig. 2.4 Simplified map of the near side, with the landing point of some American spacecraft

of the near side. A map of the near side, with the landing point of some American spacecraft, is shown in Fig. 2.4. A map of the far side is shown in Fig. 2.5.

Maria are vast solidified pools of ancient basaltic lava, containing iron, titanium, and magnesium. In most cases these lava outflows are attributed to impacts due to large meteorites, whose impact basins can be identified. At the periphery of these giant impact basins many of which have been filled by basalt maria, many large mountain ranges can be found. These are believed to be the surviving remnants of the impact basin's outer rims. No major lunar mountains are believed to have formed as a result of tectonic events like on Earth. Slopes are mild, as an average, with grade angles of 15°–20° (compared to 30°–35° on Earth) in spite of the low lunar gravity which would allow steeper grades to be stable. Faults and depressions on the Moon are called rilles and clefts. Snaking rilles are lava channels or collapsed lava tubes, which were probably active during the maria formation. Lava tubes are

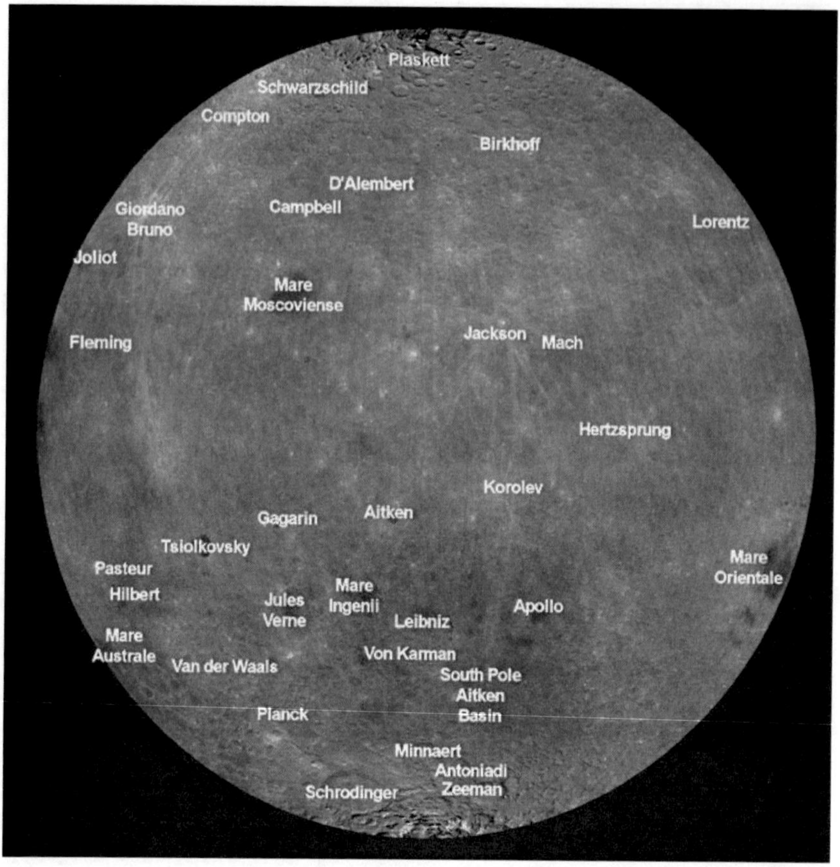

Fig. 2.5 Simplified map of the far side of the Moon

supposed to be frequent and much larger than similar formations on Earth. They may be hundreds of meters across.

Some peaks at the north pole are always in sunlight, owing to the small tilt of the lunar axis on the ecliptic plane. Similarly, at the south pole, the rim of Shackleton crater is illuminated for about 80% of the lunar day. The bottom of some craters in the polar regions are always in the shade.

Highlands are thickly cratered, owing to collisions of asteroids and comets: about half a million craters have diameters greater than 1 km. Since impact craters accumulate at a nearly constant rate, the number of craters per unit area can be used to estimate the age of the surface. The lack of an atmosphere, weather and recent geological processes ensures that many of these craters have remained relatively well preserved in comparison to those on Earth. The largest crater on the Moon, which is also the largest known crater in the Solar System, is the Aitken basin on the far side, between the South Pole and equator; its diameter is about 2,240 km and its depth is 13 km. It has the lowest elevation of the whole Moon.

Table 2.2 Grain-size distribution of lunar regolith taken from the landing site of Apollo 11

Grain size (mm)	% by weight
10–4	1.67
4–2	2.39
2–1	3.20
1–0.5	4.01
0.5–0.25	7.72
0.25–0.15	8.23
0.15–0.090	11.51
0.090–0.075	4.01
0.075–0.045	12.40
0.045–0.020	18.02
Less than 0.020	26.85

Both maria and terrae are covered by regolith, i.e. pulverized rock with grain size between less than 20 and 270 μm, most grains being of the smallest size. The grain-size distribution from specimens taken by the Apollo 11 astronauts in Mare Tranquillitatis is reported in Table 2.2 (from Calina C. Seybold, http://www.tsgc.utexas.edu/tadp/1995/spects/environment.html).

The regolith of older surfaces is generally thicker than for younger surfaces. The thickness of the regolith layer varies from about 3–5 m in the maria, and about 10–20 m in the highlands. Beneath the finely comminuted regolith layer is what is generally referred to as the megaregolith. This layer is much thicker (on the order of tens of kilometers) and comprises highly fractured bedrock.

The lunar regolith is rich in sulfur, iron, magnesium, manganese, calcium, and nickel. Many of these elements are found in oxides such as FeO, MnO, MgO, etc. Ilmenite ($FeTiO_3$), most common in the maria regions, is the best source of in situ oxygen. The carbon, hydrogen, helium and nitrogen found in the soil are almost entirely due to implantation by the solar winds.

This dust is electrically charged, so it sticks to everything and constitutes a problem for both equipment and humans. The *Apollo* astronauts noted that after excursions dust was carried on board (they stated that is smells like spent gunpowder) and this may constitute a health risk. The dust is mostly made of silicon dioxide glass (SiO_2), most likely created from the meteors that have crashed into the Moon's surface. It also contains calcium and magnesium. Electrostatic devices for collecting the dust and preventing it from entering mechanisms and lungs of people are being studied.

Particles in lunar regolith are very jagged and abrasive, which causes them to interlock and increase their harmful effects on health when breathed. Unlike Earth sand grains, they do not flow over each other. Subjected to pressure, they jam against each other and resist like solid rock. All mechanisms must be properly sealed, particularly because, in absence of atmosphere and due to electric charge, once raised from the ground, the particles tend to fly on parabolic trajectories. Because of the hard vacuum they do not remain in the 'atmosphere', but quickly return to the surface. They usually reach no more than 1.5 m from the surface, and once they get on any surface, they tend to stick to it and are difficult to remove.

Dust can be lifted off the lunar surface by the thrusters of a landing shuttle, the wheels of a rover, impacts by meteorites and also by electric charging from ultraviolet radiation from the Sun.

The porosity at the surface is about 40–43%, but decreases quickly with depth. Owing to porosity, the density at the surface is quite low, about 1000 kg/m^3 but increases quickly with depth. At 200 mm below the surface it is of 2000 kg/m^3.

The cohesive bearing strength is about 300 Pa at the surface and 3 kPa at a depth of 200 mm. Some tests on material brought by the Luna 16 and Luna 20 probes showed a compression resistance up to 1.2 kPa. Then the regolith flows as if the edges on the grains start to break, to lock again for pressures above 50 kPa. Lunar regolith has thus a good carrying capacity for vehicles and structures, as seen during the excursions in the *Apollo* missions, but can make digging and grading quite difficult. Techniques based on vibrating devices may be worth experimenting.

The Moon received large quantities of water mostly by cometary bombardment, but liquid and solid (ice) water are not stable in the conditions present on most of the lunar surface and would quickly change to water vapor. Sunlight splits much of this water into hydrogen and oxygen, which then escape into space, together with water vapor, over time, because of the Moon's weak gravity. However, since there are regions that are permanently in the shadow, water ice can be stable for long periods of time in some selected places.

Both the Clementine and the Lunar Prospector missions showed evidence of small pockets of water ice just below the surface. Estimates for the total quantity of water ice are close to one cubic kilometer, however, the question of how much water there is on the Moon is still open.

Because the atmosphere is too weak to provide protection, sunlight, solar wind and cosmic radiation strikes the lunar surface with its full strength (the latter is, however, already screened by the heliosphere). The lack of a magnetic field leaves the surface without protection to radiation, so that the radiation environment is similar to that of deep space. Since the atmosphere is so thin, it provides no protection against micrometeorite impacts. Lunar regolith is a good radiation shield: solar wind penetrates much less than a micron, and only radiation from solar flares is able to penetrate the surface by one centimeter. Hard cosmic radiation may penetrate a few meters.

Due to the lack of plate tectonics, seismic activity is quite limited. The 500 detectable moonquakes occurring each year fall within the 1–2 magnitude range on the Richter scale and then are not detectable without instruments (by comparison on earth there are 10,000 earthquakes/year). Only a few moonquakes up to 4 on the Richter scale have been observed.

Seismic activity is usually caused by tidal forces and secondary effects from impacts. Other, non-seismic activity includes astronaut activity and impacts (both meteorite and artificial). Owing to the low damping, seismic activity is registered over long distances. The good propagation of seismic waves suggests a possible communication way between places on the Moon. The very good wave propagation may be a danger too, since seismic activity can create widespread secondary effects, such as crater wall collapses and landslides.

The Moon is considered as a lifeless world; that not withstanding NASA implemented strict quarantine procedures for the *Apollo* landings, and the same is likely to be done in the future. The only fossil life-forms that may be found on the Moon are ancient life-forms from Earth, which may have been carried there on board of meteorites, detached from Earth by impacts and then impacting the lunar surface. Finding them may be the only way of obtaining information on very ancient terrestrial life, particularly if it started earlier than expected, when the initial meteoritic bombardment subsided and then was wiped out by a huge impact.

The two-way communication delay with the lunar surface is 2.5–3 s, but to communicate with the far side relay satellites, either in lunar orbit or in one of the Earth–Moon Lagrange points, must be used. In the latter case a longer delay is present.

2.5 Rocky Planets

Apart from Earth, the planets of the solar system with a hard surface are just three: Mercury, Venus and Mars. A miscellany of data regarding these three planets is reported in Table 2.3. The values related to Earth are reported for comparison.

2.5.1 *Mars*

Mars will be dealt with first, since it is the most interesting planet for both robotic and manned missions. Mars has the largest volcano in the solar system, Olympus Mons, 25 km tall (but there is no active volcano on Mars now), and a canyon, Marineris Vallis, which is likely the deepest and widest in the Solar system. It bears the traces of impressive and dramatic events in the past, which have remodeled its surface—the northern lowlands, Vastitas Borealis, probably due to the impact of a large meteorite, the Tharsis Bulge, probably of volcanic origin, with three huge volcanoes (Pavonis, Arsia and Ascraeus Mons), and a huge number of impact craters, chasms, and mountains. If Vastitas Borealis is considered an impact basing, it is the largest found in the solar system, four times the size of the lunar Aitken basin.

The poles are covered by ice caps which shrink in the summer and grow in the winter. The northern ice cap is made mainly of water ice, while the southern one has a frozen carbon dioxide upper layer and an underlying layer of water ice. About 25 to 30% of the atmosphere condenses during a polar winter, forming thick slabs of CO_2 ice, to sublime again when the pole is again exposed to sunlight. This creates huge wind storms from the poles with wind velocities up to 400 km/h. Both polar caps contain frozen carbon dioxide as well as water ice, and a figure of two million cubic kilometers of water ice has been forwarded for the norther ice cap.

Table 2.3 Main characteristics of the rocky planets of the solar system (the values regarding Earth are reported for comparison). The surface acceleration at the equator, taking into account the planet's rotation, is also included

	Mercury	Venus	Mars	Earth
Mass (10^{24} kg)	0.3302	4.8685	0.64185	5.9736
Volume (10^{10} km^3)	6.083	92.843	16.318	108.321
Equatorial radius (km)	2,439.7	6,051.8	3,396.2	6,378.1
Polar radius (km)	2,439.7	6,051.8	3,376.2	6,356.8
Ellipticity (flattening)	0.00	0.000	0.00648	0.00335
Topographic range (km)	–	15	30	20
Mean density (kg/m^3)	5,427	5,243	3,933	5,515
Surface gravity (m/s^2)	3.70	8.87	3.71	9.81
Surface acceleration (m/s^2)	3.70	8.87	3.69	9.78
Escape velocity (km/s)	4.3	10.36	5.03	11.19
Solar irradiance (W/m^2)	9,126.6	2,613.9	589.2	1,367.6
Orbit semimajor axis (10^6 km)	57.91	108.21	227.92	149.60
Sidereal orbit period (days)	87.969	224.701	686.980	365.256
Perihelion (10^6 km)	46.00	107.48	206.62	147.09
Aphelion (10^6 km)	69.82	108.94	249.23	152.10
Synodic period (days)	115.88	583.92	779.94	
Mean orbital velocity (km/s)	47.87	35.02	24.13	29.78
Max. orbital velocity (km/s)	58.98	35.26	26.50	30.29
Min. orbital velocity (km/s)	38.86	34.79	21.97	29.29
Orbit inclination (deg)	7.00	3.39	1.850	0.000
Orbit eccentricity	0.2056	0.0067	0.0935	0.0167
Sidereal rotation period (hrs)	1,407.6	−5,832.5	24.6229	23.9345
Length of day (hrs)	4,222.6	2,802.0	24.6597	24.0000
Inclination of rot. axis (deg)	≈0	177.36	25.19	23.45
Min. dist. from Earth (10^6 km)	77.3	38.2	55.7	
Max. dist. from Earth (10^6 km)	221.9	261.0	401.3	

The difference between the two ice caps is due to the fact that the orbit of Mars around the Sun is much more elliptical than is the Earth's and the inclination of the planet's axis of rotation makes the seasons far more extreme in the southern hemisphere than in the northern: Mars is near perihelion when in the southern hemisphere it is summer and near aphelion when it is winter. This is also believed to be the explanation for the occurrence of violent dust storms which can last for months at a time.

The martian day, usually referred to as a *sol*, is slightly longer than Earth's day (24 h, 39 min, 35 s). The tilt of the rotation axis is similar to that of the Earth's axis

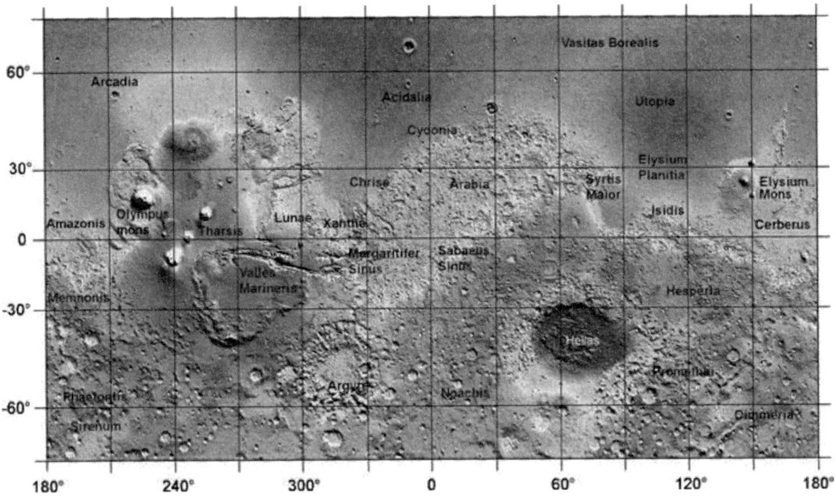

Fig. 2.6 Simplified map of the Martian surface (from a NASA image)

(25° while the latter is inclined of 23°), producing seasons similar to ours, although they last almost twice as long, owing to the much longer duration of the year.

Mars is smaller than Earth, its surface area being about as large as the sum of all the continents of our planet. A simplified map of the planet is given in Fig. 2.6.

Martian atmosphere is very thin: the pressure at the ground is less than one hundredth of the atmospheric pressure on Earth (it roughly equals atmospheric pressure on Earth at 35 km altitude) and has much variability with altitude and latitude. It varies from a minimum of around 0.3 millibar on Olympus Mons to over 11.6 millibar in the depths of Hellas Planitia, with a mean surface level pressure of 6.36 millibar, variable from 4.0 to 8.7 millibar depending on season The average surface density is about 0.020 kg/m^3.

The composition by volume of the atmosphere is 95.32% carbon dioxide, 2.7% nitrogen, 1.6% argon, oxygen 0.13%, carbon monoxide 0.08%, water 210 ppm; nitrogen oxide 100 ppm; neon 2.5 ppm, heavy water (in form of deuterium protium oxide HDO) 0.85 ppm; krypton 0.3 ppm and traces of methane, the latter concentrated in a few places during the northern summer. Since methane is broken down by ultraviolet radiation, for it to be present a mechanism producing that gas on the planet, like volcanic activity, cometary impacts or the presence of methanogenic microbial life forms, is needed. The mean molecular weight of the atmosphere is 43.34 g/mole.

Clouds of water ice were photographed by the *Opportunity* rover in 2004.

The atmosphere is quite dusty, containing particulates about 1.5 μm in diameter which give the Martian sky a tawny color when seen from the surface.

At the *Viking* landing sites, wind speeds of 2–7 m/s were recorded in the summer, 5–10 m/s in the fall, with occasional 17–30 m/s dust storms.

Remark 2.7 Although the wind speeds are high, the aerodynamic forces exerted by the wind are modest owing to the low atmospheric density: vehicles and structures on the Mars surface are thus not expected to be much loaded by the strong winds.

Winds carry large quantities of dusts that is even finer than the dust on the moon. The danger to machinery and human beings due to dust rich in iron oxide must be accounted for when planning missions to the Red Planet. Mars plains are frequently crossed by dust devils: the solar panels of the Spirit and Opportunity rovers were cleaned more than once by them, contributing to maintain these device operational for a long time.

The average temperature is $-63°C$, with marked variation with the time of the day and of the year. In the site of the *Viking 1* landing diurnal variations between -89 and $-31°C$ were recorded. Larger variations, from -120 to $-14°C$ were recorded during the years-long *Viking* missions. In the summer, in the southern hemisphere, temperatures as high as 20 to 30°C have been recorded. Liquid water cannot exist on the surface at these temperature and pressure combinations and most of the frost depicted in the images from the *Viking* landers is frozen carbon dioxide.

Martian atmosphere offers very little protection from the Sun's ultraviolet radiation and there is only very limited protection from cosmic rays due to the almost complete absence of a planetary magnetic field after Mars lost its magnetosphere about 4 billion years ago. There is evidence that at the beginning Mars had plate tectonics and a planetary dynamic effect, producing a global magnetic field. Some remnants are still found in the form of local magnetization.

Remark 2.8 From the point of view of radiation, Mars is only a slightly better place than the Moon, even if the thin atmosphere scatters light and the look from the surface is more like that on a planet than that in space.

The geography of planet Mars is complex. The main features are shield volcanoes, lava plains (mostly in the northern hemisphere) and highlands with a large number of impact craters and deep canyons. The four largest volcanoes, all extinct, have already been mentioned. A total number of 43,000 craters with a diameter of 5 km or greater have been found, together with a large number of smaller ones.

The largest canyon, Vallis Marineris, has a length of 4,000 km and a depth of up to 7 km. It was formed due to the swelling of the Tharsis area which caused the crust in the area of Vallis Marineris to collapse. Also Ma'adim Vallis is a canyon much bigger than the Grand Canyon of Earth.

Pictures of entrances of large caves, 100 to 250 m wide, were transmitted by the probes; moreover also on Mars it is possible that lava tubes exist, and are larger than those on Earth owing to the lower gravity. The interiors of caverns and lava tubes may be protected from micrometeoroids, UV radiation, solar flares and high energy particles that bombard the planet's surface and thus are good candidates for the search of liquid water and signs of life, apart from being also possible locations for human settlements.

Martian rocks seem mostly basalt, although a portion of the Martian surface seems more rich in silica than typical basalt. The plains are similar to rocky deserts on Earth, covered by red sand, with rocks and boulders scattered all around. The most interesting places are, however, the steep slopes of the mountains and canyons, which are very difficult to manage for wheeled and even tracked machines.

The soil is essentially regolith, rich in finely-grained iron oxide dust, which is thinner than on the Moon. Its granulometry and composition is more variable from place to place, due to water erosion in ancient times when water was flowing on the surface, and to wind erosion.

The soil of Mars is basic, and values of pH of 8.3 were obtained by the *Phoenix* lander. This, together with the results of the old *Viking* probes seem to exclude the possibility of finding life-forms on the surface of the planet. Even if some life-form might be found in the future, in particular in places shaded from radiation and direct sunlight, like the bottom of canyons or caves, it is easy to state that products of biological origin are not a widespread constituent of the surface of the planet.

The geological history of Mars is subdivided into three main periods, namely

- Noachian epoch (named after Noachis Terra), from 3.8 billion to 3.5 billion years ago.
- Hesperian epoch (named after Hesperia Planum): 3.5 billion years ago to 1.8 billion years ago.
- Amazonian epoch (named after Amazonis Planitia): 1.8 billion years ago to present.

The Tharsis bulge formation and extensive flooding by liquid water are ascribed to the Noachian epoch. Extensive lava plains are supposed to have been formed in the Hesperian epoch, while the Olympus Mons formed during the Amazonian epoch, along with lava flows elsewhere on Mars.

Owing to the in situ observations performed by robotic spacecraft, it is now certain that in ancient geological periods Mars had extensive water coverage, with liquid water running on the surface and geyser-like water flows. At that time the atmosphere was much thicker too.

Large quantities of water are thought to be trapped underground. In the northern hemisphere an ice permafrost mantle stretches down from the pole to latitudes of about 60° and large quantities of water ice have been observed both at the poles and at mid-latitudes. A large release of liquid water is thought to have occurred when the Vallis Marineris formed early in Mars's history, forming massive outflow channels. A much more recent (5 million years ago) outflow of water is supposed to have occurred when the Cerberus Fossae chasm formed. There are also clues of more recent flows of water on the surface, at least for short periods of time, but these findings are still debated.

Once the planet was much more suitable for living organisms than today, which does not imply that life actually existed. If so, fossil remnants may still exist. The controversial finding of fossils in the ALH84001 meteorite, which supposedly was blasted into space by a meteorite strike and wandered through space for 15 million years to land finally on Earth, still stirs much debate.

Table 2.4 Data of the satellites of Mars Phobos and Deimos

	Phobos	Deimos
Orbit's semimajor axis (km)	9,378	23,459
Sidereal orbit period (days)	0.31891	1.26244
Orbital inclination (deg)	1.08	1.79
Orbital eccentricity	0.0151	0.0005
Major axis radius (km)	13.4	7.5
Median axis radius (km)	11.2	6.1
Minor axis radius (km)	9.2	5.2
Mass (10^{15} kg)	10.6	2.4
Mean density (kg/m^3)	1,900	1,750

The two-way communication delay with Mars is much variable depending on the positions of the two planets. It is up to 30 minutes or more. However, things are usually much worse: taking into account the relative motions of the two planets and the unavailability of telecommunication satellites in Mars orbit, the *Pathfinder* probe had a communication window of just 5 minutes per day.

Mars has two small, irregularly shaped, moons, Phobos and Deimos, which orbit close to the planet. Their characteristics are reported in Table 2.4. They may be captured asteroids, similar to 5261 Eureka, a Martian Trojan asteroid, but their capture by an almost airless world is difficult to explain. Phobos' orbit is lower than synchronous orbit: it rises in the west, sets in the east, and rises again in just 11 hours. Deimos' orbit is just above synchronous orbit and then its apparent speed in the sky is low. Even if the period of its orbit is 30 hours, it takes 2.7 days from rising in the east to setting in the west.

Remark 2.9 Phobos' orbit is unstable: it decays owing to the tidal forces and it will either crash on the planet in about 50 million years or fragment producing a ring around it.

2.5.2 Mercury

Mercury is an extremely hostile environment. It is a small planet, the smallest in the solar system, very close to the Sun. Each square meter of its surface receives seven times the energy received by a square meter of the Earth surface. Its rate of rotation is very slow, with a sidereal rotation period of 58.7 Earth days, yielding a duration of the day of 175.94 Earth days. It performs three rotations about its axis every two orbits (is locked in a 3:2 resonance).

It has no true atmosphere and its general aspect is similar to that of the Moon with many craters and some smooth plains (Fig. 2.7). The faint gases that surround it are made by hydrogen, helium, oxygen, sodium, calcium, magnesium, silicon and

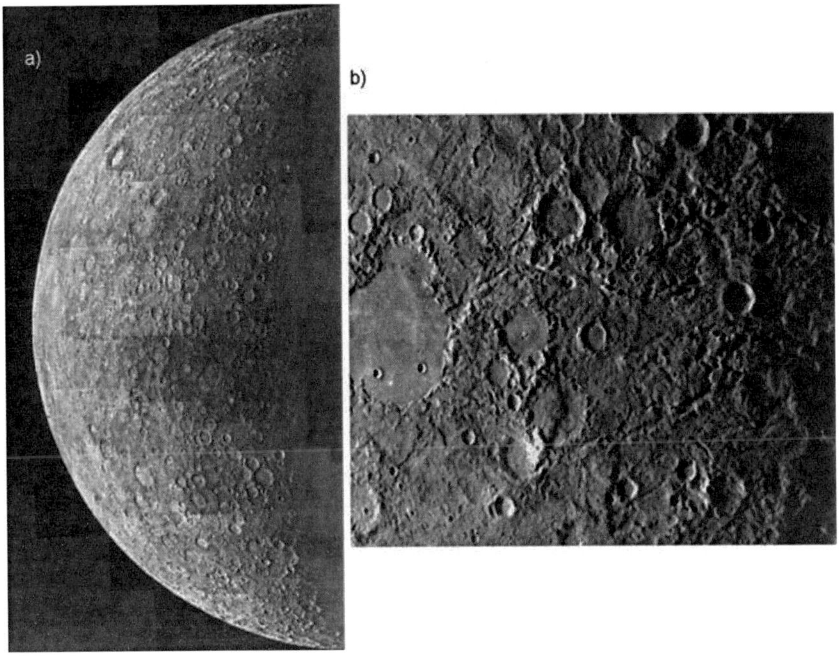

Fig. 2.7 Mercury as observed by Mariner 10 in 1974 (NASA photo)

potassium, with some water vapor and traces of other elements. This faint atmosphere is unstable, in the sense that the matter it contains is continuously lost to space and substituted by new matter from the solar wind, diffusing into Mercury's magnetosphere, and from the planet's surface. Atmospheric pressure on the surface is about 10^{-15} bar and the total mass of the atmosphere is evaluated in less than 1000 kg.

Its surface, likely covered with regolith like the Moon, has a mean temperature of 169.35°C. The slow rotation and the lack of an atmosphere cause extreme variations of temperature, which range from 427°C on the side facing the Sun (the hottest point being that directly facing the Sun) to -183°C during the night (the coldest points being in the bottom of the polar craters).

Remark 2.10 Space probes have found water ice in the bottom of polar craters, where the light of the Sun never reaches and the temperature remains below -171°C. A figure of 10^{14} to 10^{15} kg of ice has been forwarded.

The recent studies (2009) performed by the Messenger probe showed traces of recent volcanic activity. Mercury is thus not a geologically dead planet, as was believed. As can be deduced from a comparison of the chemical composition of the surfaces of Mercury and of the Moon, the two bodies seem to have evolved in different ways. Also, some findings suggest that the folds, widespread on the whole

Fig. 2.8 Venus, as seen by the Magellan probe using its synthetic aperture radar (NASA image)

surface, which were attributed to a planet's contraction, might have been caused by tectonic activity, although this is still a hypothesis.

Unlike the moon, Mercury has a large iron, likely molten, core, which generates a weak global magnetic field. Its strength is about 1% of that of Earth. The magnetic field strength at the equator is about 300 nT. Mercury's magnetic field is dipolar, but unlike Earth's magnetic field, its magnetic axis is nearly aligned with the planet's spin axis. Mercury's magnetic field is strong enough to create a magnetosphere which deflects the solar wind about the planet and to supply some protection to the surface.

Owing to the high eccentricity of the orbit, an observer would be able to see the Sun rise about halfway, then reverse and set before rising again, all within the same Mercurian day.

2.5.3 Venus

Venus is always covered by thick layers of clouds and its surface cannot be seen from space by optical observation. However, the Magellan probe mapped its surface accurately using a Synthetic Aperture Radar (SAR) (Fig. 2.8). The clouds of Venus extend from about 50 to 70 km and may be divided into three distinct layers. Below the clouds is a layer of haze down to about 30 km and below that it is clear.

The rotation of Venus about its axis is extremely slow, the day being about 243 Earth days long.

The planet has a very dense atmosphere: the atmospheric pressure at the surface is 92 bar and its density is 65 kg/m^3. Its main constituent is carbon dioxide (96.5% by volume), with about 3.5% of nitrogen. Minor constituents are sulfur dioxide (150 ppm), argon (70 ppm), water (20 ppm) carbon monoxide (17 ppm), helium (12 ppm) and neon (7 ppm). The mean molecular weight is 43.45 g/mole.

The presence of sulfur dioxide (sulfur is likely to have a volcanic origin, dating back from the beginning of the planet's evolution) gives way to sulfuric acid clouds in the high atmosphere but sulfur dioxide is also present in the haze layer.

Venus has no global magnetic field, and solar and cosmic radiation penetrates the high atmosphere. The surface should, however, be protected by the thick and dense gas layers.

The surface of Venus is hot, the average temperature being 464°C. Owing to the high density and opacity of the atmosphere, temperature variations with both time and place are limited, with maximum values of about 480°C, hot enough to melt lead.

The velocity of the wind at the surface is low, between 0.3 and 2.0 m/s, but in the upper layers of the atmosphere, above the clouds, jet streams, faster at the equator and slower at the poles, as fast as 200–400 km/h, are present.

The present environment on Venus is believed to be the outcome of a runaway greenhouse effect. An increase of the Venusian temperature caused the evaporation of the seas, if they ever existed, and the production of carbon dioxide from the carbonates in the soil. The increasing amounts of water vapor and carbon dioxide in the atmosphere caused a further increase of the greenhouse effect, and the very high temperatures which we now observe. The water vapor was decomposed by the Sun's light into oxygen and hydrogen, the latter light gas disappearing into space. Venus does not have more carbon dioxide than Earth; the point is that it is all in the atmosphere instead of being fixed in the soil and absorbed in the oceans.

The environment on the surface of Venus is harsh and the only two robotic probes that landed on it worked for just minutes, as they were designed to do. However, Venus is an interesting place to study and it would be interesting to perform robotic missions on its surface.

Remark 2.11 There are little chances that life could evolve in Venus' difficult environment, but there is some possibility that bacteria-like life-forms survive in the relatively cool high atmosphere or at least that life evolved when conditions were milder.

2.6 Giant Planets

The giant planets are essentially huge gas balls, without a true solid surface. The density of the gas constituting the outer part of the planet (whether there

Table 2.5 Main characteristics of the giant planets of the solar system. Radii, gravity and escape velocity are referred at the level at which the atmospheric pressure is equal to 1 bar. The surface acceleration at the equator, taking into account the planet's rotation, is also included

	Jupiter	Saturn	Uranus	Neptune
Mass (10^{24} kg)	1,898.6	568.46	86.832	102.43
Volume (10^{10} km^3)	82,713	6,833	16.318	6,254
Equatorial radius (1 bar level) (km)	71,492	60,268	25,559	24,764
Polar radius (km)	66,854	54,3648	24,973	24,341
Volumetric mean radius (km)	69,911	58,232	25,362	0.01708
Ellipticity (flattening)	0.06487	0.09796	0.02293	0.00335
Mean density (kg/m^3)	1,326	687	1,270	1,638
Surface gravity (m/s^2)	24.79	10.44	8.87	11.15
Surface acceleration (m/s^2)	23.12	8.96	8.69	11.00
Escape velocity (km/s)	59.5	35.5	21.3	23.5
Solar irradiance (W/m^2)	50.50	14.90	3.71	1.51
Black-body temperature (K)	110.0	81.1	58.2	46.6
Orbit semimajor axis (10^6 km)	778.57	1,433.53	2,872.46	4,495.06
Sidereal orbit period (days)	4,332.589	10,759.22	30,685.4	60,189.
Perihelion (10^6 km)	740.52	1,352.55	2,741.30	4,444.45
Aphelion (10^6 km)	816.62	1,514.50	3,003.62	4,545.67
Synodic period (days)	398.88	583.92	369.66	367.49
Mean orbital velocity (km/s)	13.07	9.69	6.81	5.43
Max. orbital velocity (km/s)	13.72	10.18	7.11	5.50
Min. orbital velocity (km/s)	12.44	9.09	6.49	5.37
Orbit inclination (deg)	1.304	2.485	0.772	1.769
Orbit eccentricity	0.0489	0.0565	0.0457	0.0113
Sidereal rotation period (hrs)	9.9250	10.656	−17.24	16.11
Length of day (hrs)	9.9259	10.656	17.24	16.11
Inclination of rot. axys (deg)	3.12	26.73	97.86	29.56
Min. dist. from Earth (10^6 km)	588.5	1,195.5	2,581.9	4,305.9
Max. dist. from Earth (10^6 km)	968.1	1,658.5	3,157.3	4,687.3

is a solid core is still debatable) increases going inside, as well as the temperature. Although the environment of the giant planets is extreme, there is no doubt that there is a region where temperature and pressure are manageable by robotic probes.

The characteristics of the four gas giants of the Solar system are summarized in Table 2.5. Since there is no true surface, the radii, gravity and escape velocity are referred at the level at which the atmospheric pressure is equal to 1 bar.

2.6.1 *Jupiter*

Jupiter is the only giant planet whose atmosphere was directly probed by a spacecraft, the *Galileo* probe, which sent back many data before being crushed by the pressure. It is the largest planet in the Solar System and its mass is equal to 70% of all the planets combined.

It rotates so fast about its axis that its shape is an oblate spheroid. Its atmosphere does not rotate as a rigid body, and the period of rotation changes with the latitude, with the equator having a period about 5 min shorter than the poles. The conventional period of rotation is that of the magnetosphere.

The major constituents of the atmosphere are molecular hydrogen (89.8% in volume) and helium (10.2%). Minor constituents are methane 3000 ppm, ammonia 260 ppm, hydrogen deuteride (HD) 28 ppm, ethane 5.8 ppm, water 4 ppm. The mean molecular weight is 2.22 g/mole. To these gas fractions some aerosols must be added, particularly in the outer layers, where they form belts of clouds: ammonia ice, water ice, ammonia hydrosulfite. Darker orange and brown clouds at lower levels may contain sulfur, as well as simple organic compounds.

The clouds are located in the tropopause and are arranged into bands of different latitudes, known as tropical regions. These are subdivided into lighter-hued zones and darker belts. The interactions of these circulation patterns cause storms and turbulence with wind speeds of 100 m/s being common. The zones have been observed to vary in width, color and intensity from year to year, but they have remained sufficiently stable for astronomers to give them identifying designations.

The cloud layer is only about 50 km deep, and consists of at least two decks of clouds: a thick lower deck and a thin clearer region. There may also be a thin layer of water clouds underlying the ammonia layer. Powerful electrical discharges may be present where water clouds form thunderstorms driven by the heat rising from thc interior.

The largest storm, which has lasted since centuries ago and may be a permanent feature of the planet, is the Great Red Spot, a persistent anticyclonic storm located 22° south of the equator that is larger than 3 times the Earth. The Great Red Spot's dimensions are (24,000–40,000) km × (12,000–14,000) km and it rotates counterclockwise, with a period of about six days (Fig. 2.9). Its maximum altitude is about 8 km above the surrounding cloud tops.

The temperatures at 1 and 0.1 bar levels are, respectively, −108 and −161°C. The density at 1 bar level is 0.16 kg/m^3. The wind speed is up to 150 m/s in the equatorial region (latitude less than 30°), and up to 40 m/s at higher latitudes.

Temperature and pressure increase going inwards in the atmosphere, and are so high in the interior of the planet that matter behaves not as a gas but as a supercritical fluid. Pressures well above 1,000 bar are predicted. At a radius of about 78% of the planet's radius hydrogen becomes a metallic liquid. Likely Jupiter has a molten rock core, of unknown composition, located inside this thick metallic hydrogen layer. The temperature and the pressure at the core are estimated at about 36,000 K and 3,000–4,500 GPa.

Fig. 2.9 A picture of Jupiter in which the great red spot is well visible (NASA image, based on a 1979 picture from the *Voyager 1* spacecraft)

Jupiter has a strong global magnetic field, 14 times as strong as the Earth's. Its strength ranges between 0.42 mT at the equator to 1.0–1.4 mT at the poles. It creates a large magnetosphere. The structure of the magnetosphere follows a pattern not much different from what is seen for the solar magnetic field with a bow shock at about 75 Jupiter radii from the planet. Around Jupiter's magnetosphere is a magnetopause, at the inner edge of a magnetosheath, where the planet's magnetic field becomes weak and the solar wind interacts with the matter from the planet. All four largest moons of Jupiter orbit within the magnetosphere.

Even if the presence of organic compounds has been verified, it is highly unlikely that there is any Earth-like life on Jupiter owing to the scarcity of water in the atmosphere. Moreover, any possible solid surface deep within Jupiter would be under extremely high pressures. However, there are hypotheses that ammonia- or water-based life could evolve in Jupiter's upper atmosphere.

Jupiter has four large moons, Io, Europa, Ganymede, and Callisto, discovered by Galileo in 1610, and a host of minor satellites, out of which 59 are known. It has also a faint planetary ring system composed of three main segments, likely made of dust rather than ice. The rings are related with particular satellites like Adrastea and Metis (main ring) or Thebe and Amalthea (outer ring).

2.6.2 Saturn

Saturn is the second largest planet in the Solar System. The compositions of the two inner gas giants are similar and that of Saturn is not unlike what was seen for

Jupiter: mostly hydrogen, with small proportions of helium and trace elements. Also its interior is thought to consist of a small core of rock and ice, surrounded by a thick layer of metallic hydrogen and a gaseous outer layer. The temperatures at the core are very high, although lower than those of Jupiter.

Also Saturn rotates fast and its shape is oblate; moreover, the rotation period depends on the latitude. Its average density is 0.69 g/cm^3: it is the only planet of the Solar System that is less dense than water.

The atmosphere consists of about 96.3% molecular hydrogen and 3.25% helium (the latter remarkably less abundant than on Jupiter), with trace amounts of methane (4,500 ppm), ammonia (125 ppm), hydrogen deuteride (HD, 110 ppm), ethane (7 ppm), acetylene (C_2H_2) and phosphine (PH_3). Also aerosols, ammonia ice, water ice, ammonia hydrosulfite are present and form clouds, like on Jupiter, which give a band pattern to the atmosphere of Saturn too, even if the bands are much fainter.

The mean molecular weight of the atmosphere is 2.07 g/mole.

The upper clouds on Saturn are composed of ammonia crystals, while the lower level clouds, about 50 km thick and with a temperature of −93°C, appear to be composed of either ammonium hydrosulfite or water. Over that, extending for 10 km and with a temperature of −23°C, is a layer made up of water ice. The temperature and density at the 1 bar altitude are −139°C and 0.19 kg/m^3. The temperature at the 0.1 bar altitude is −189°C. The pressure at the surface is greater than 1000 bar.

Saturn's winds are among the Solar System's fastest. Voyager data indicate peak easterly winds of 500 m/s (1,800 km/h). Average speed is 400 m/s in the equatorial zone (latitude less than 30°) and 150 m/s at higher latitudes. A Great White Spot, a unique but short-lived phenomenon which occurs once every Saturnian year, or roughly every 30 Earth years, around the time of the northern hemisphere's summer solstice, has been observed.

Saturn has a planetary magnetic field intermediate in strength between that of Earth and that of Jupiter. Its strength at the equator is 20 μT, slightly weaker than Earth's magnetic field. Its magnetosphere is much smaller than the Jovian and extends slightly beyond the orbit of Titan.

The most impressive characteristic of Saturn is its system of rings, consisting mostly of ice particles (containing 93% percent of water ice with tholin impurities, and 7% amorphous carbon) with a smaller amount of rocky debris and dust. They extend from 6,630 km to 120,700 km above Saturn's equator, with an average thickness of approximately 20 meters in thickness (Fig. 2.10a).

Sixty-one known moons orbit the planet, not counting hundreds of moonlets within the rings.

The rings have an intricate structure of thousands of thin gaps and ringlets. This structure is controlled by the gravitational pull of Saturn's many moons. Some gaps are cleared out by the passage of tiny moonlets, and some ringlets seem to be maintained by the gravitational effects of small shepherd satellites. Other gaps arise from resonances between the orbital period of particles in the gap and that of a more massive moon further out. Still more structure in the rings consists of spiral waves raised by the moons' periodic gravitational perturbations.

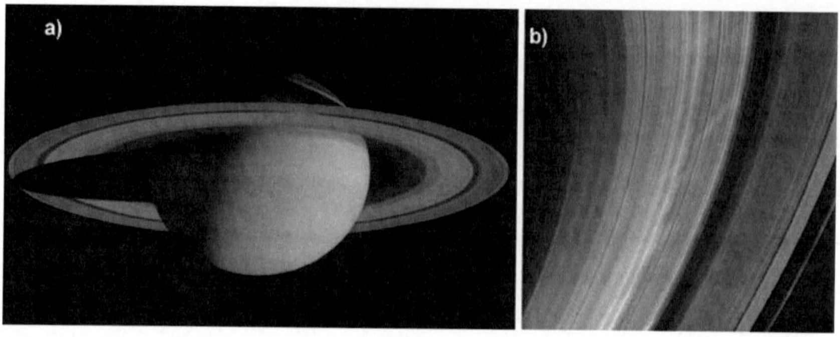

Fig. 2.10 (**a**) Mosaic image obtained from 126 images acquired in two hours on October 6, 2004 by the *Cassini* probe from a distance of about 6.3 million kilometers from Saturn. (**b**) Image of bright spokes in Saturn's B ring taken by *Cassini* in August 2009 (NASA images)

The rings of Saturn possess their own atmosphere, independent of that of the planet itself, composed of molecular oxygen gas produced when ultraviolet light from the Sun interacts with water ice in the rings, with some molecular hydrogen too. This atmosphere is extremely thin, so thin that if it were condensed onto the rings, its thickness would be about one atom thick.

Radial features, dubbed spokes, were discovered in some rings and explained exclusively as the action of gravitational forces (Fig. 2.10b). However, it seems that some cannot be explained in this way and the precise mechanism generating the spokes is still unknown.

The origin of the rings of Saturn (and the other Jovian planets) is unknown. Though they may have had rings since their formation, the ring systems are not stable and must be regenerated by some ongoing process, perhaps the breakup of larger satellites. The current set of rings may be only a few hundred million years old.

2.6.3 Uranus

Uranus and the other giant planet Neptune, although generally similar to the gas giants Jupiter and Saturn, have some significant differences, to the point that for them the term *ice giants* was formulated. Uranus consists of three layers: a small rocky core in the center, an icy mantle in the middle and an outer gaseous hydrogen and helium atmosphere.

Following the most common (but not universally accepted) model of the planet, the core has a density of about 9,000 kg/m^3, a pressure of 800 GPa and a temperature of 5,000 K. The bulk of the planet is represented by the mantle, composed of a hot and dense fluid consisting of water, ammonia and other volatiles. This fluid layer is sometimes called a water–ammonia ocean. Uranus has thus no solid surface: the gaseous atmosphere gradually transitions into the internal liquid layers.

The Uranian atmosphere, the coldest planetary atmosphere in the Solar System (colder than Neptune's, with a minimum temperature of −224°C), can be divided into three layers: the troposphere, between altitudes of −300 km (the zero-altitude level is that where the pressure equals 1 bar) and 50 km and pressures from 100 to 0.1 bar; the stratosphere, spanning altitudes between 50 and 4,000 km and pressures of between 0.1 and 10^{-10} bar and the thermosphere extending from 4,000 km to as high as 50,000 km from the surface.

The temperature and density at 1 bar level are −197°C and 0.42 kg/m^3; at the 0.1 bar level the temperature is −220°C.

The atmosphere is made primarily of molecular hydrogen (82.5%), helium (15.2%) and methane (2.3%), with traces of hydrogen deuteride (148 ppm). The abundance of less volatile compounds such as ammonia, water and hydrogen sulfide in the atmosphere is poorly known. Trace amounts of various hydrocarbons are found in the stratosphere, including ethane, acetylene, methylacetylene and diacetylene together with traces of water vapor, carbon monoxide and carbon dioxide.

Like in the atmospheres of the other giant planets, also in the atmosphere of Uranus there are aerosols such as water ice, ammonia ice, ammonia hydrosulfite and methane. Uranus has a complex, layered cloud structure, with the lowest clouds likely made of water and the uppermost layer of clouds, possibly made of methane.

Uranus has the shape of an oblate spheroid, with different rotation periods at different latitudes. What is unique is the tilt of the rotational axis of rotation, which is 97.86°. Seasonal changes are completely unlike those of the other major planets. Near the time of Uranian solstices, one pole faces the Sun continuously while the other pole faces away. Only a narrow strip around the equator experiences a rapid day–night cycle, but with the Sun very low over the horizon as in the Earth's polar regions. Each pole gets around 42 years of continuous sunlight, followed by 42 years of darkness Near the time of the equinoxes, the Sun faces the equator of Uranus giving day–night cycles similar to those seen on most of the other planets. The extreme axial tilt results in extreme seasonal weather variations, with changes in brightness and strong thunderstorms.

Uranus has a global magnetic field, but it is quite peculiar: it is tilted at 59° from the axis of rotation and is not centered on the planet's geometric center. The magnetosphere is highly asymmetric and the magnetic field strength on the surface in the southern hemisphere can be as low as 10 μT, while in the northern hemisphere it can be as high as 110 μT, with an average field at the surface of 23 μT. In other respects the Uranian magnetosphere is like those of other planets: it has a bow shock located at about 23 Uranian radii ahead of it, a magnetopause at 18 Uranian radii, a fully developed magnetotail and radiation belts.

Like the other giant planets, Uranus has a ring system, made of at least 13 distinct rings. Owing to the orientation of the axis of the planet, the rings are almost perpendicular to the orbit, and from time to time are seen from Earth as circling the planet. All rings of Uranus except two are extremely narrow, being a few kilometers wide. Some of them appear to be gray, some red and some are blue, but mostly made of quite dark matter.

Uranus has a number of moons, 27 of which were given a name.

2.6.4 Neptune

Neptune is similar in composition and structure to Uranus. Its interior, like that of Uranus, the other ice giant, is primarily composed of ices and rock. The pressure and the temperature at its center are about 700 GPa and 5,400 K, respectively, following the model at present considered as the most realistic.

Neptune's atmosphere is made of molecular hydrogen (80.0%), helium (19.0%) and methane (1.5%) with traces of hydrogen deuteride (192 ppm) and ethane (1.5 ppm). It contains aerosols like ammonia ice, water ice, ammonia hydrosulfite, methane ice in a higher proportion than that of gas giants. The absorption of red light by the atmospheric methane is thought to be a cause of the blue color of the planet. The difference of hue with Uranus is accredited to the fact that, being denser and heavier, more of the methane from the mantle leaks to the surface, giving it a richer color.

The temperature and density at the 1 bar level are $-201°$C and $0.45\ \text{kg/m}^3$. At 0.1 bar the temperature is $-218°$C. Neptune's outer atmosphere is one of the coldest places in the Solar System, although warmer than Uranus's. The mean molecular weight of the atmosphere is 2.53–2.69 g/mole.

Neptune's atmosphere is divided into three main regions; the lower troposphere, where temperature decreases with altitude, and the stratosphere, where temperature increases with altitude. Their boundary, the tropopause, occurs at a pressure of 10 kPa. The stratosphere then gives way to the thermosphere at a pressure lower than 1 to 10 Pa.

Like the other giant planets, also Neptune is an oblate spheroid and undergoes differential rotation. Since the differential rotation is the most pronounced of any planet in the Solar System, stronger latitudinal wind shear are present. Its axial tilt is 29.56°, not dissimilar to that of Earth and Mars. Climate variations are thus not so extreme as those on Uranus.

The atmosphere of Neptune carries visible weather patterns. The Great Dark Spot is an anticyclonic storm system spanning $13{,}000 \times 6{,}600$ km and is comparable to the Great Red Spot on Jupiter. Other climate patterns were dubbed Scooter and Small Dark Spot, a southern cyclonic storm. These weather patterns, which seem not being permanent (they are thought to last for several months), are driven by the strongest sustained winds of any planet in the Solar System, with average speeds from 400 m/s along the equator to 250 m/s at the poles. Wind speeds as high as 600 to 2,100 km/h were, however, recorded. Most of the winds on Neptune move in a direction opposite the planet's rotation at low latitudes, while blowing in the rotation direction at high latitudes. These patterns are thought to be driven by internal heat generation. Both Uranus and Neptune radiate more heat than that received by the Sun, but Neptune much more than Uranus. The process in which this heat is generated is not yet known.

Neptune's troposphere is banded by clouds of varying compositions depending on altitude. The upper-level clouds occur at pressures below one bar (100 kPa), where the temperature is suitable for methane to condense. For pressures between 100 and 500 kPa clouds of ammonia and hydrogen sulfide are believed to form.

Above a pressure of 500 kPa the clouds may consist of ammonia, ammonium sulfide, hydrogen sulfide and water. Deeper clouds of water ice should be found at pressures of about 5.0 MPa, where the temperature reaches 0°C. Underneath, clouds of ammonia and hydrogen sulfide may be found. High-altitude clouds on Neptune have been observed casting shadows on the opaque cloud deck below. There are also high-altitude cloud bands that wrap around the planet at constant latitude. These circumferential bands, with widths of 50–150 km, lie about 50–110 km above the cloud deck.

The magnetosphere resembles that of Uranus, with a magnetic field strongly tilted relative to its rotational axis at 47° and offset at least 0.55 radii from the planet's geometric center. The dipole component of the magnetic field at the magnetic equator of Neptune is about 14 μT, but the magnetic field has a complex geometry that includes relatively large contributions from non-dipolar components, including a strong quadrupole moment that may exceed the dipole moment in strength.

Neptune's bow shock occurs at a distance of 34.9 times the radius of the planet. The magnetopause lies at a distance of 23–26.5 times the radius of Neptune. The tail of the magnetosphere extends out to at least 72 times the radius of Neptune, and very likely farther.

Neptune too has a planetary ring system. The rings may consist of ice particles coated with silicates or carbon-based material, and have a reddish hue. The rings were given names like Adams (at 63,000 km from the center) Le Verrier Ring, (at 53,000 km), Galle (42,000 km), Lassell (a faint outward extension to the Le Verrier Ring) and Arago (at 57,000 km). Other larger incomplete rings were named Courage, Liberté, Egalité 1, Egalité 2 and Fraternité. Their shape can be explained by the gravitational effects of Galatea, a moon just inward.

2.7 Satellites of Giant Planets

As already stated, the giant planets have a large number of satellites, some of which are larger than Earth's Moon and even than Planet Mercury. The majority of them, however, are similar to asteroids and some are very small and have an irregular shape. Actually, the number of the small satellites of giant planets is unknown, and new ones are continuously discovered.

Jupiter has 63 satellites (data referred to the beginning of 2009), but only 4 of them, the Galilean satellites, have a diameter larger than 500 km (Fig. 2.11). The fifth one in size, Amalthea, is quite irregular, having dimensions 250 × 146 × 128 km.

In order of distance from the planet they are Metis, Adrastea, Amalthea, Thebe, Io, Europa, Ganymede, Callisto, Themisto, Leda, Himalia, Lysithea, Elara, S/2000 J 11, Carpo, S/2003 J 12, Euporie, S/2003 J 3, S/2003 J 18, Thelxinoe, Euanthe, Helike, Orthosie, Iocaste, S/2003 J 16, Praxidike, Harpalyke, Mneme, Hermippe, Thyone, Ananke, S/2003 J 17, Aitne, Kale, Taygete, S/2003 J 19, Chaldene, S/2003 J 15, S/2003 J 10, S/2003 J 23, Erinome, Aoede, Kallichore, Kalyke, Carme, Callirrhoe, Eurydome, Pasithee, Kore, Cyllene, Eukelade, S/2003 J 4, Pasiphaë,

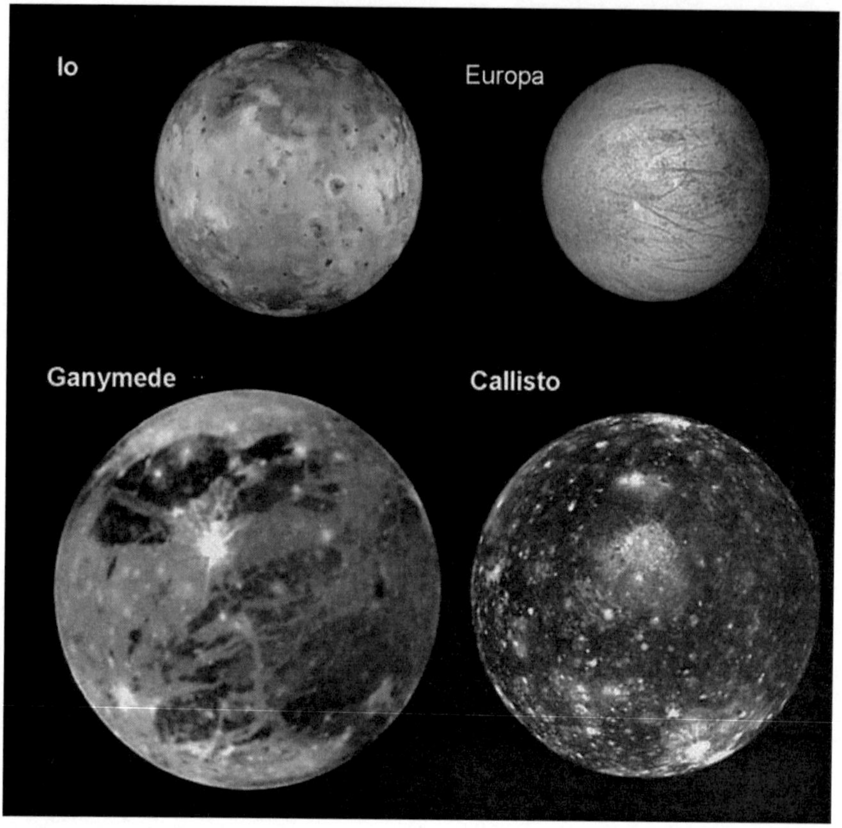

Fig. 2.11 Images of the four Galilean Satellites of Jupiter shown in the same scale. Pictures taken by the Galileo probe (NASA images)

Hegemone, Arche, Isonoe, S/2003 J 9, S/2003 J 5, Sinope, Sponde, Autonoe, Megaclite, S/2003 J 2.

Saturn has 61 confirmed satellites: Pan, Daphnis, Atlas, Prometheus, Pandora, Epimetheus, Janus, Aegaeon, Mimas, Methone, Anthe, Pallene, Enceladus, Tethys, Telesto, Calypso, Dione, Helene, Polydeuces, Rhea, Titan, Hyperion, Iapetus, Kiviuq, Ijiraq, Phoebe, Paaliaq, Skathi, Albiorix, S/2007 S 2, Bebhionn, Erriapus, Skoll, Siarnaq, Tarqeq, S/2004 S 13, Greip, Hyrrokkin, Jarnsaxa, Tarvos, Mundilfari, S/2006 S 1, S/2004 S 17, Bergelmir, Narvi, Suttungr, Hati, S/2004 S 12, Farbauti, Thrymr, Aegir, S/2007 S 3, Bestla, S/2004 S 7, S/2006 S 3, Fenrir, Surtur, Kari, Ymir, Loge, Fornjot. To these a number of moonlets must be added.

Many of the Saturn's moons are very small: 34 are less than 10 km in diameter, and another 14 less than 50 km. Only seven are massive enough to have collapsed into hydrostatic equilibrium under their own gravitation. Out of them, 6 (Enceladus, Tethys, Dione, Rhea, Titan and Iapetus) have a diameter larger than 500 km and Mimas is only slightly smaller.

Table 2.6 Diameter d, mass m, gravitational acceleration g, escape velocity V_e and orbital data (semimajor axis a, period T, inclination i and eccentricity e) of the satellites of the giant planets having a diameter larger than 500 km. A negative orbital period indicates retrograde motion

Name	d (km)	m (10^{21} kg)	g (m/s^2)	V_e (km/s)	a (km)	T (days)	I (deg)	e
Jupiter								
Io	3,643	89	1.80	2.56	421,700	1.769	0.050	0.0041
Europa	3,122	48	1.32	2.02	671,034	3.551	0.471	0.0094
Ganymede	5,362	150	1.43	2.74	1,070,412	7.154	0.204	0.0011
Callisto	4,821	110	1.24	2.45	1,882,709	16.689	0.205	0.0074
Saturn								
Enceladus	504	0.108			237,950	1.370	0.010	0.0047
Tethys	1,066	0.617			294,619	1.887	0.168	0.0001
Dione	1,123	1.095			377,396	2.737	0.002	0.0022
Rhea	1,529	2.307			527,108	4.518	0.327	0.0013
Titan	5,151	134.520	1.35	1.86	1,221,930	15.945	0.3485	0.0280
Iapetus	1,472	1.806			3,560,820	79.321	7.570	0.0286
Uranus								
Ariel	1,157.8	1.35			191,020	2.520	0.260	0.0012
Umbriel	1,169.4	1.17			266,300	4.144	0.205	0.0000
Titania	1,577.8	3.53			435,910	8.706	0.340	0.0011
Oberon	1,522.8	3.01			583,520	13.463	0.058	0.0014
Neptune								
Triton	2,707	21.4	0.78	1.45	354,800	−5.877	156.8	

Uranus has 27 known satellites, whose names are chosen from characters from the works of Shakespeare and Alexander Pope. They are Cordelia, Ophelia, Bianca, Cressida, Desdemona, Juliet, Portia, Rosalind, Cupid, Belinda, Perdita, Puck, Mab, Miranda, Ariel, Umbriel, Titania, Oberon, Francisco, Caliban, Stephano, Trinculo, Sycorax, Margaret, Prospero, Setebos and Ferdinand.

Only 4 of them (Ariel, Umbriel, Titania and Oberon) are larger than 500 km diameter, while Miranda is slightly smaller.

Neptune has 13 satellites: Naiad, Thalassa, Despina, Galatea, Larissa, Proteus, Triton, Nereid, Halimede, Sao, Laomedeia, Psamathe and Neso. Only Triton exceeds 500 km diameter, while Proteus is slightly smaller.

The main characteristics of the satellites of the four giant planets larger than 500 km diameter are listed in Table 2.6.

The largest satellites are characterized by a gravitational acceleration similar to that of the Moon.

The satellites of the giant planets are quite different from each other and many of them are interesting targets for exploration missions. Some are here dealt with in some detail.

2.7.1 Io

The orbits of Io, Europa, and Ganymede form a pattern known as a Laplace resonance; for every four orbits that Io makes around Jupiter, Europa makes exactly two orbits and Ganymede makes exactly one. This resonance causes the gravitational effects of the three large moons to distort their orbits into elliptical shapes. The tidal force from Jupiter, on the other hand, works to circularize their orbits. The eccentricity of their orbits causes strong tidal effects that heat the moons' interiors via friction. This is seen most dramatically in the extraordinary volcanic activity of innermost Io.

Io orbits close to the cloud tops of Jupiter, within an intense radiation belt. The magnetosphere of Jupiter strips away about 1 ton per second of volcanic gases and other materials from the satellite, which remain in orbit about the planet, inflating its magnetosphere. Io is supposed to have a metallic (iron, nickel) core surrounded by a partially melted silicate rich mantle and a thin rocky crust.

Owing to the heat generated by tidal deformations, Io is the most volcanic body in the solar system. Volcanic plumes extending for more than 100 km have been recorded. The plumes are rich in sulfur dioxide, which forms clouds that rapidly freeze and snow back to the surface. The dark areas in the floors of the calderas that were discovered by the *Voyager* and *Galileo* probes may be pools of molten sulfur, a very dark form of sulfur.

Apparently about 15% of Io is made of water ice, and the presence of hot parts suggests that zones where liquid water can form may exist on the surface or underground. Owing to the absence of an atmosphere, water should evaporate quickly on the surface, but underground lakes cannot be excluded. Io is thus a candidate for the search of life-forms in the solar system, provided that these underground pockets supply some protection from the strong radiations due to the Jupiter magnetosphere.

2.7.2 Europa

Europa is very different from Io. It is covered by a layer of water ice that reminds us of our polar ice-pack and contains large quantities of water, probably 15% or more of the total mass of the satellite. Its surface is thus among the brightest in the solar system, owing to the fact that this ice layer is relatively young and smooth.

The high resolution images taken by the *Galileo* probe show that the ice is fractured in a way suggesting that some plates can move, floating on an underlying layer of water or more plastic ice, as happens with glaciers on Earth. Studies performed using radar altimeters and measuring the magnetic and gravitational fields also suggest that there is an ocean of water under the layer of ice, although neither the thickness of the ice nor the depth of the ocean are known.

The satellite is supposed to have a metallic core made, possibly, of iron and nickel, surrounded by a shell of rock and then by a layer of water in liquid form or ice. This ocean of liquid water, perhaps hundreds of kilometers deep, is then surrounded by a layer of ice whose thickness may be some kilometers, even ten or more.

A very tenuous oxygen atmosphere has been recently detected.

Europa is thus another candidate for the search of extraterrestrial life, in spite of its distance from the Sun, being heated by tidal effects caused by the proximity of Jupiter.

2.7.3 Ganymede

The two last Galilean satellites, Ganymede and Callisto, have a heavily cratered appearance.

Ganymede, the largest moon of Jupiter and in our solar system, is most likely composed of a rocky core with a very thick (about 50% of the outer radius) water/ice mantle and a crust of rock and ice. It has no known atmosphere, but recently a thin ozone layer was detected at its surface. This suggests that also Ganymede probably has a thin tenuous oxygen atmosphere.

The geological history of Ganymede is complex, with mountains, valleys, craters and lava flows and it shows light and dark regions. Contrary to the Moon, where the darker regions are the most recent, here the dark regions are more cratered, implying ancient origin. The bright regions show a grooved terrain with ridges and troughs.

At the center a dense metallic core seems to exist, which is the source of a global magnetic field discovered by the *Galileo* probe. Surrounding the metallic core, there should be a rock layer, overlaid by a deep layer of warm soft ice. Above it the surface is made by a water ice crust; images show features showing geological and tectonic disruption of the surface in the past.

2.7.4 Callisto

Callisto is about the same size as Mercury. It orbits just beyond Jupiter's main radiation belt. Its icy crust is very ancient and dates back 4 billion years, just shortly after the solar system was formed and is the most heavily cratered surface in the solar system. The largest craters have been erased by the flow of the icy crust over geologic time. Two enormous concentric ring shaped impact basins are Valhalla, with a bright central region 600 kilometers in diameter and rings extending to 3,000 kilometers in diameter, and Asgard, about 1,600 kilometers in diameter.

Callisto has the lowest density (1,860 kg/m^3) of the Galilean satellites.

The surface layer of ice is thought to be about 200 kilometers thick. Beneath this crust it is possible that a salty ocean more than 10 kilometers deep exists. The clues of this ocean are the Callisto's magnetic field variations in response to the background magnetic field generated by Jupiter.

Beneath the ocean, Callisto seems to have a not entirely uniform interior, composed of compressed rock and ice with the percentage of rock increasing toward the center.

2.7.5 Enceladus, Tethys, Dione, Rhea and Iapetus

These satellites of Saturn are heavily cratered icy worlds. Tethys has a large impact crater and many valleys and troughs stretching three quarters of the way around the satellite.

Dione and Rhea have bright, heavily cratered leading hemispheres and darker trailing hemispheres with thin streaks that are thought to be produced by deposits of ice inside surface troughs or cracks. The latter may have a tenuous ring system of its own.

Iapetus, the outermost of the large icy satellites, has a dark leading hemisphere and a bright trailing hemisphere.

Encedalus is the innermost of the large Saturn's satellites and is more heated than the others by the planet's tidal effects. The *Cassini* probe found evidence of liquid water reservoirs that erupt in geysers. Images had also shown particles of water in its liquid state being emitted by icy jets and towering plumes.

Many of the icy satellites of the giant planets, which are all more or less heated by tidal effects, may have liquid water oceans under a kilometers deep icy crust. In the case of Encedalus the icy crust above the ocean, or at least local liquid water pockets, may be no more than tens of meters deep below the surface.

Some broad regions of Enceladus are uncratered, showing that geological activity has resurfaced areas of the satellite within the last 100 million years.

At the beginning of the space Age, in 1958, when planning the Project Orion, Freeman Dyson chose Encedalus as the destination for a manned mission with a 10,000 tons interplanetary ship propelled by 15 kiloton nuclear explosions.

2.7.6 Titan

Titan, by far the largest of the satellites of Saturn, is larger than the planet Mercury and is the only moon in the Solar System to possess a significant atmosphere.

While apparently being similar to Earth's environment, the surface of Titan is completely different. The role played on the former by rock, on the latter is played by water ice. The hard cobblestones in the floodplain in the pictures taken by the *Huygens* probe that landed on its surface in 2005 (Fig. 2.12a) are actually chunks of

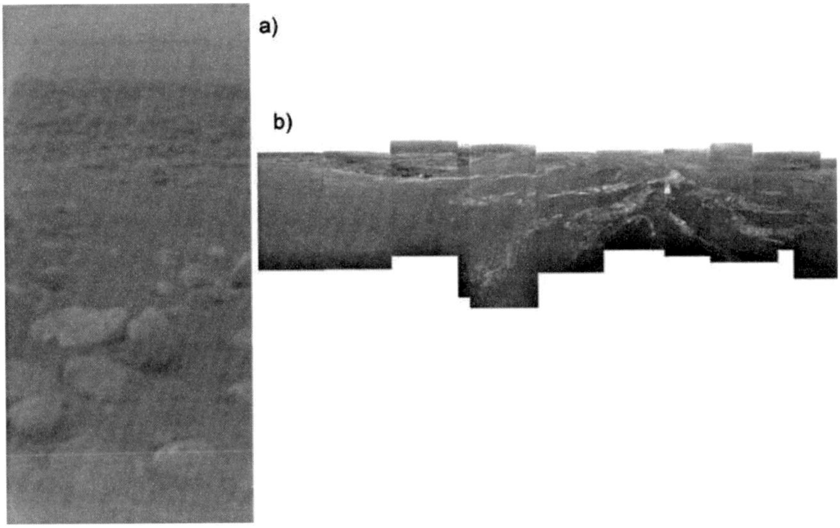

Fig. 2.12 Images of Titan taken by the *Huygens* probe. (**a**) Image of the ground taken after landing. (**b**) Image taken during the descent (images ESA/NASA/Univ. of Arizona)

ice, and the volcanoes (cold volcanoes) that remodel the surface erupt water mixed to ice (slush) which is the local equivalent of lava.

The liquids on the surface are hydrocarbons (methane and ethane), which show an evaporation-condensation cycle similar to the water cycle on Earth. The ground in the landing site of Huygens is soaked in liquid methane, but free from the dark hydrocarbon particles that fall from the sky, like snow on Earth. These hydrocarbon grains collect in some areas and are driven by the wind forming tall dunes.

Likely there are no tall mountains, although a range 150 km long, 30 km wide and 1.5 km high was discovered in the southern hemisphere by *Cassini* in 2006. Some impact craters have been discovered, but their small number shows that the surface is relatively young.

The temperature is in the range of −180°C. At this temperature water ice does not sublimate or evaporate, so the atmosphere is nearly free of water vapor. The haze in Titan's atmosphere cause an effect that is opposite to the greenhouse effect, reflecting sunlight back into space and making its surface significantly colder than its upper atmosphere. The clouds on Titan, probably composed of methane, ethane or other simple organic materials, are scattered and variable, punctuating the overall haze. This atmospheric methane conversely creates a greenhouse effect on Titan's surface, without which Titan would be far colder.

Hydrocarbon lakes, seas and rivers, mostly of liquid methane but with ethane and dissolved nitrogen, were discovered near Titan's north pole; the largest of the seas is almost the size of the Caspian Sea.

Titan's atmosphere is denser than Earth's, with a surface pressure more than one and a half times that of our planet. It supports opaque haze layers that block most visible light from the Sun and other sources. Titan's lower gravity means that its atmosphere is far more extended than Earth's; even at a distance of 975 km.

The atmosphere is 98.4% nitrogen with the remaining 1.6% composed of methane and trace amounts of other gases such as hydrocarbons (including ethane, diacetylene, methylacetylene, acetylene, propane), cyanoacetylene, hydrogen cyanide, carbon dioxide, carbon monoxide, cyanogen, argon and helium.

Titan has no magnetic field, although it seems to retain remnants of Saturn's magnetic field in occasions when it passes outside the Saturn's magnetosphere and is directly exposed to the solar wind.

2.7.7 Miranda, Ariel, Umbriel, Titania and Oberon

These moons of Uranus are ice-rock conglomerates composed of roughly fifty percent ice and fifty percent rock. The ice may include ammonia and carbon dioxide. Ariel appears to have the youngest surface with the fewest impact craters, while Umbriel's appears oldest. Miranda possesses fault canyons 20 kilometers deep, terraced layers, and a chaotic variation in surface ages and features. Miranda's past geologic activity is believed to have been driven by tidal heating at a time when its orbit was more eccentric than currently, probably as a result of a former 3:1 orbital resonance with Umbriel. Extensional processes associated with upwellings of ice are the likely origin of the moon's racetrack-like coronae. Similarly, Ariel is believed to once have been held in a 4:1 resonance with Titania.

2.7.8 Triton

Triton, the only large satellite of Neptune, has a retrograde orbit, indicating that it was captured rather than formed in place; it probably was once a small body of the Kuiper belt. It is close enough to Neptune to be locked into a synchronous rotation, and it is slowly spiraling inward because of tidal effects. It will eventually fall on the planet, or more likely be torn apart, in about 3.6 billion years, when it reaches the Roche limit. No other large satellite in the solar system has a retrograde motion.

A value of the surface temperature of −235°C was estimated for Triton, the lowest temperature recorded in the solar system, although it is likely that the Kuiper belt objects are colder.

It has an extremely thin atmosphere, with a pressure at the surface of about 15 microbars. Nitrogen ice particles might form thin clouds a few kilometers above the surface.

The higher density suggest that Triton contains more rock in its interior than the icy satellites of Saturn and Uranus.

Triton is scarred by enormous cracks. Active geyser-like eruptions spewing nitrogen gas and dark dust particles several kilometers into the atmosphere were imaged by *Voyager 2*.

2.8 Small Bodies

Apart from the satellites of the planets, there are many other small objects in the solar system. They can be roughly subdivided into some categories:

- Main belt asteroids
- Kuiper belt objects (KBO)
- Trojan asteroids
- Other asteroids
- Comets

The asteroids and Kuiper belt objects come in different sizes. The largest are spherical bodies, with characteristics not dissimilar to small planets or large satellites. The largest main belt asteroids are Ceres, Vesta, Pallas, and Hygiea, whose total mass amounts roughly to half the mass of all main belt objects. Ceres is now classified as a dwarf planet. Also a number of objects of the Kuiper belt, like Pluto. Quaoar, Eris, Sedna, Haumea, and Makemake are large enough to be spherical and to be given the status of dwarf planets. Pluto was considered the ninth planet of the solar system until the large objects of the Kuiper belt were discovered. Owing to their distance from the Sun, they are poorly seen, and it is likely that many other, yet undiscovered, large KBO exist.

The characteristics of some large minor bodies orbiting the Sun (dwarf planets and large asteroids) are reported in Table 2.7.

Smaller objects have an irregular shape.

2.8.1 Main Belt Asteroids

With a diameter of about 950 km, Ceres is by far the largest and most massive body in the asteroid belt and is assumed to be a surviving protoplanet, which formed 4.57 billion years ago in the asteroid belt. Like the planets, its structure should consist of a crust, a mantle and a core. The amount of water on Ceres is larger than that on Earth; it is possible that some of this water is (or at least was) in liquid form, an ocean located between the rocky core and ice mantle like the one that may exist on Europa. Ammonia may be dissolved in the water. Like for Europa, the possible existence of this liquid water layer suggests the possibility of finding extraterrestrial life.

The surface composition contains hydrated materials, iron-rich clays and carbonates which are common minerals in carbonaceous chondrite meteorites.

The surface of Ceres is relatively warm, with temperatures ranging from -106 to -38°C.

Ceres may have a tenuous atmosphere and water frost on the surface, which is expected to sublimate when exposed directly to solar radiation. Its rotational period is 9 hours and 4 minutes and the tilt of its rotation axis is about 3°.

Vesta, too, has a differentiated interior, though it is devoid of water and its composition is mainly basaltic rock such as olivine.

Table 2.7 Diameter d, mass m, gravitational acceleration g, escape velocity V_e and orbital data (semimajor axis a, period T, inclination to the ecliptic i and eccentricity e) of some large minor bodies orbiting the Sun (dwarf planets and large asteroids)

Name	d (km)	m (10^{21} kg)	g (m/s^2)	V_e (km/s)	a (10^6 km)	T (days)	i (deg)	e
Main belt								
Ceres	974.6	0.943	0.27	0.51	413.833	1680.5	10.585	0.079
Vesta	560	0.267	0.22	0.35	353.268	1325.2	7.135	0.089
Pallas	556	0.211	0.18	0.32	414.737	1686.0	34.838	0.231
Hygiea	407	0.0885	0.091	0.21	469.580	2,031.01	3.842	0.117
Kuiper belt								
Pluto	2,303	13.14	0.6		5,906.438	90,589	17.140	0.249
Haumea	1,518	4.01	0.44	0.84	6,452.000	103,468	28.22	0.195
Quaoar	1,260				6,524.262	105,196	7.985	0.037
Makemake	1,276	4.18			6,850.086	113,191	28.963	0.159
Eris	2,600	16.7			10,120.000	203,600	44.187	0.441
Sedna	≈1,500				78,668.000	4,404,480	11.934	0.855

Vesta's shape is relatively close to a gravitationally relaxed oblate spheroid, but its irregularity precluded it from being considered a dwarf planet. It has a very large impact basin of 460 kilometers diameter at its southern pole. Its width is 80% of the diameter of Vesta, its floor is about 13 kilometers deep, and its rim rises 4–12 km above the surrounding terrain. A central peak rises 18 kilometers above the crater floor.

Vesta is thought to consist of a metallic iron–nickel core, an overlying rocky olivine mantle, with a surface crust. The impact crater at the south pole is so deep that it exposed part of the inner core. The eastern and western hemispheres are markedly different; the former seems to be similar to lunar highlands while large regions of the latter are covered with more recent basalts, perhaps analogous to the lunar maria.

The rotation period is 5.342 h, with a tilt of the rotation axis of 29°. Temperatures on the surface have been estimated between −190°C and a maximum of −20°C.

Pallas is unusual in that it rotates on its side (the axis is tilted by 78°, with some uncertainty), like Uranus and its rotation period is 0.326 days. Its orbit has an unusually high inclination to the plane of the main asteroid belt, and eccentricity, the latter being nearly as large as that of Pluto. As a consequence, Pallas is a difficult target for space probes. Its composition is similar to that of Ceres: high in carbon and silicon.

Hygiea is largest of C-type asteroids with a carbonaceous surface and, unlike the other largest asteroids, lies relatively close to the plane of the ecliptic. Its rotation period is 27.623 h, and its surface temperature is between −26 and −109°C. Water ice might have been present on its surface in the past.

Hundred thousands asteroids are currently known, over 200 of them being larger than 100 km, and the total number ranges in the millions or more.

Remark 2.12 The asteroids are spread over such a large volume that the main belt is mostly empty: the popular idea of the main asteroid belt as a portion of space full of objects is thus a misconception. To reach an asteroid requires very careful navigation, and hitting one by chance is unlikely.

Most main belt asteroids belong to three wide categories: C-type or carbonaceous asteroids, S-type or silicate asteroids, and M-type or metallic asteroids.

Over 75% of the visible asteroids are C-type asteroids; they are rich in carbon and dominate the belt's outer regions. They are more red in hue than the other asteroids and have a very low albedo. Their surface composition is similar to carbonaceous chondrite meteorites.

About 17% of the asteroids are S-type. They are more common toward the inner region of the belt. Their spectra show the presence of silicates and some metal, but no significant carbonaceous compounds. They have a relatively high albedo.

Less than 10% of the asteroids are M-type; their spectra being similar to those of iron–nickel meteorites. However, there are also some silicate compounds that can produce a similar appearance. For example, the large M-type asteroid 22 Kalliope does not appear to be primarily composed of metal. M-type asteroids at any rate seem to contain a huge quantity of useful, and even precious, metals.

Remark 2.13 A standard 1 km diameter M-type asteroid is assumed to contain 200 million tonnes iron, 30 million tonnes nickel, 1.5 million tonnes cobalt and 7,500 tonnes of metals of the platinum group. Only the latter are worth 150 billion $ at current prices.

Clearly, if such asteroids can be mined, the price of precious metals would decrease sharply, but this would allow to use them for many technical applications at present impossible owing to their high cost. Mining asteroids is a possible future use for space robots.

Within the main belt, the number distribution of M-type asteroids peaks at a semimajor axis of about 2.7 AU. It is not yet clear whether all M-types are compositionally similar, or whether they are simply asteroids which do not fit neatly into the main C and S classes.

There are many other types, including G-type asteroids, similar to the C-type objects, but with a slightly different spectrum, D-type, A-type and V-type, or basaltic, asteroids. The latter were assumed to be the products of the collision that produced the large crater on Vesta and thus to be younger than other asteroids (about 1 billion years or less). This origin is, however, now less certain.

Some objects in the outer part of the asteroid belt show cometary activity. Since their orbits are different from the orbits of classical comets, it is likely that many asteroids in that region are icy, with the ice occasionally exposed to sublimation through small impacts.

The structure of irregular asteroids is not known in general. In the past, asteroids were often assumed to be mostly rubble piles, but now it is certain that many of them are more or less solid bodies.

The asteroid belt is quite wide, so that the temperature of the asteroids varies with the distance from the Sun. The temperature of dust particles located at the distance from the Sun of the asteroid belt ranges from −73°C at 2.2 AU to −108°C at 3.2 AU. The actual temperature of the surface of asteroids, however, changes while the asteroid rotates.

The orbits are not evenly distributed: orbital resonances with Jupiter destabilize some orbits, giving way to gaps, named Kirkwood gaps. They occur where the period of revolution about the Sun is an integer fraction of Jupiter's orbital period.

Collisions between main belt bodies with a mean radius of 10 km are expected to occur about once every 10 million years, which can be defined frequently, on astronomical time scales. As a result, an asteroid may fragment into numerous smaller pieces, leading to the formation of a new asteroid family. If the relative speed is low, the two asteroids may also join together.

Some asteroids, even small ones, have one or more satellites, usually just tiny rocks orbiting about them.

2.8.2 Kuiper Belt Objects

In the outer solar system there is a region where orbits have a 2:3 resonance with Neptune. These asteroids are said to be *Plutinos*, since are in the same resonance as Pluto. Other asteroids are in orbits with a 1:2 resonance with Neptune's orbit (they are said *twotinos*). All these asteroids, and others with orbits outside that of Neptune out to about 55 AU from the Sun are said to be Kuiper Belt Objects (KBOs). The region where they orbit is said the *Kuiper Belt*.

Many Plutinos, including Pluto, have orbits crossing that of Neptune, though their resonance means they can never collide. The eccentricity of their orbit is large, suggesting that they are not native to their current positions but migrated there under the gravitational pull of Neptune.

Apart from twotinos, whose orbits have a semimajor axes of about 47.7 AU, other resonances also exist at 3:4, 3:5, 4:7 and 2:5.

Kuiper belt objects are icy worlds, usually larger than the asteroids in the main belt. Like Jupiter's gravity dominates the asteroid belt, Neptune's gravity dominates the Kuiper belt. Also here there are zones where the orbits are unstable, and the Kuiper belt's structure has gaps, like the region between 40 and 42 AU.

Three of the KBO—Pluto, Haumea and Makemake—are classified as dwarf planets. They, and many smaller and irregular objects orbiting so far from the Sun, are different from the main belt asteroids: the former are mostly rock and metal, while the Kuiper belt objects are composed largely of frozen volatiles such as light hydrocarbons (like methane), ammonia and water, a composition not much different from that of comets. These substances are in ice form since the temperature of the belt is only about −220°C. Also the signature of amorphous carbon was discovered in the spectra of some Kuiper belt objects.

To distinguish these objects from asteroids and from comets, they are often referred to as cometoids.

At present over a thousand KBOs have been discovered and more than 70,000 other objects over 100 km in diameter are believed to exist between the orbit of Neptune and 100 AU from the Sun. Triton, the largest satellite of Neptune, is believed to be a captured KBO.

Pluto has a rotation period of 6.387 days. It has three satellites, the largest of which, Charon, has a diameter of 1,207 km, a mass of 1.9×10^{21} kg and could be considered as a dwarf planet if it were not orbiting Pluto. The Pluto–Charon system can be considered as a double planet, like other KBO. Nix follows a circular orbit in the same plane as Charon and, with an orbital period of 24.9 days is close to a 1:4 orbital resonance with Charon. Hydra orbits in the same plane as Charon and Nix, but on a less circular orbit. Its orbital period of 38.2 days is close to a 1:6 orbital resonance with Charon.

Also Eris and Haumea have satellites. Haumea has a very elongated orbit, and is the farthest object in the solar system, except for comets.

2.8.3 Trojan Asteroids

A *trojan* minor planet, asteroid or satellite is a celestial body that shares an orbit with a larger planet or satellite, but does not collide with it because it is located in one of the two stable Lagrangian points L4 or L5, which lead or lag by 60° the larger body. Actually, a trojan is not located exactly in the Lagrangian point, but orbits about it with a complex dynamics, also for the perturbations due to other bodies. Three-dimensional halo orbits are periodic, while other types of orbits are quasi periodic. This allows a large number of bodies to coexist in different orbits about the same Lagrange point without colliding. Trojans located at a certain Lagrange point are, at any moment, actually spread out in a wide arc on the orbit of the main body.

The term originally referred to two groups of asteroids orbiting the Sun in the same orbit as Jupiter. The first of them, lagging Jupiter by 60°, was discovered in 1904 by the German astronomer Max Wolf and named Achilles. The others were named after Greek (those lagging Jupiter) and Trojan heroes (those leading Jupiter) from the Iliad, with the exceptions of Hector and Patroclus, which are in the 'enemy camp'. Those later discovered on the orbits of other planets or satellites could be designated in a more generic way as *Lagrangian asteroids*, but the term trojans at the end prevailed. Currently over 1,800 trojan asteroids associated with Jupiter are known. About 60% of them are in L4, leading Jupiter in its orbit, while the other 40% are about L5 and trail the planet, like Achilles.

Eureka, discovered in 1990, is the first Mars trojan to be found (5 of them, all located at L5, are known at present). Six Neptune trojans were then discovered and now is believed they are much more numerous than those of Jupiter. Apparently

Saturn and the Earth have no trojans, but clouds of dust were fond at the Lagrange points L4 and L5 on the Earth's orbit and even on the Moon's orbit.

Four examples of trojan satellites are known: Telesto and Calypso, satellites of Saturn, are trojans to Tethys, and Helene and Polydeuces, are trojans to another Saturn's satellite, Dione.

The composition of Trojan asteroids is similar to that of the nearby asteroids, so those on Jupiter's orbit are likely to be similar to those of the outer main belt and those on Neptune's orbit are likely to be similar to KBOs. Some trojans, like Patroclus, seem to have a comet-like composition. Achilles, Hector and other Jupiter's trojans are D-type asteroids (a rare, low albedo, carbon rich, type). Eureka is an A-type asteroid.

2.8.4 *Other Asteroids*

Among the other asteroid, not belonging to the above categories, Near-Earth Asteroids (NEA), and more in general, Near-Earth Objects (NEO), i.e. objects whose orbit passes close to that of Earth. are the most important. NEOs having a diameter less than 50 meters are said to be Near-Earth Meteoroids (NEMs).

All NEOs have a perihelion distance less than 1.3 AU, and include (at the end of 2008), 5,857 NEAs and 82 Near-Earth Comets (NECs), for a total of 5,939 NEOs. Their number is, however, quickly growing with new discoveries. Out of them, up to 1,000 have a diameter equal to or larger than one km and are potentially able to cause a global catastrophe on Earth. 943 asteroids have been classified as potentially dangerous (PHA, Potentially Hazardous Asteroids), because they could get dangerously close to Earth.

To assess the potential danger posed by PHAs, a scale, named the Torino Scale, was devised. This scale, of the type of the Richter scale for earthquakes, aims to classify the danger represented by the asteroids and comets that are discovered. It has 11 degrees (from 0, no risk, to 10, certainty of generalized destruction), and five colors, (from white, certainty of no impact, to red, certainty of impact). The usefulness of this type of scale derives from the fact that when an asteroid is discovered, its orbit can only be calculated in an approximate way and the potential danger it represents may only be assessed in statistical terms. Even when the orbit is better known, it can be changed by gravitational perturbations of the planets or even of the Earth in such a way that the danger it represents changes. These perturbations cannot be computed with the required precision: a variation of a few thousand km, a trifle on an astronomic scale, may transform a harmless asteroid into a serious danger.

None of the known objects has a degree higher than one on the Torino scale. An example of an asteroid with a Torino Scale value equal to 1 is 2004 MN4 Apophis, which will make several close passes, coming very close in 2029, and has some probability of hitting the Earth on April 13, 2036. Its estimated diameter is 390 m.

Some NEOs are easier to reach than the Moon, since they require a lower ΔV. Not only they may present interesting scientific targets for direct geochemical and astronomical investigation, but are also potentially economical sources of extraterrestrial materials for human exploitation. Two near-Earth objects have been visited by spacecraft: 433 Eros, by NASA's Near Earth Asteroid Rendezvous probe, and 25143 Itokawa, by the JAXA Hayabusa mission.

Depending on their orbits, NEA are subdivided into three families

- Atens, orbiting inside Earth orbit (average orbital radii closer to the Sun than one AU and aphelia outside Earth's perihelion, at 0.983 A.U.). 453 Atens were known in May 2008.
- Apollos, whose orbit has an average orbital radius greater than that of the Earth and perihelion less than Earth's aphelion at 1.017 A.U. 2,053 Apollos were known in May 2008.
- Amors, orbiting outside Earth orbit (average orbital radii in between the orbits of Earth and Mars and perihelia slightly outside Earth's orbit (1.017–1.3 AU). They often cross the orbit of Mars. 2,894 Amors were known in May 2008.

Remark 2.14 Many Atens and all Apollos have orbits that cross that of the Earth, so they are a threat to impact the Earth on their current orbits. Amors do not cross the Earth's orbit and are not immediate impact threats.

However, their orbits may evolve into Earth-crossing orbits in the future. NEOs' orbits are actually the result of the gravitational perturbations of the outer planets, mainly Jupiter, and these objects come from the asteroid belt or even the outer solar system. The Kirkwood gaps, where orbital resonances with Jupiter occur, are the places from where most of them come. Their orbits are bound to change again in the future: new asteroids are constantly moved into near-Earth orbits and the present ones are moved away, even ejected from the solar system or sent into the Sun.

2.8.5 Comets

Comets are small bodies which, when get close enough to the Sun, exhibit a visible coma, and sometimes a tail, both because of the effects of solar radiation upon the comet's nucleus. Comet nuclei are themselves loose collections of ice, dust and small rocky particles, ranging from 100 m to more than 40 km across.

Remark 2.15 Comet nuclei are often described as 'dirty snowballs', but observation by spacecraft showed that their surface is made of dry dust or rock, suggesting that the ices are hidden beneath the crust.

The results of the *Deep Impact* probe suggest that the majority of a comet's water ice is below the surface, and that these reservoirs feed the jets of vaporized water that form the coma.

The nucleus is made of rock, dust, water ice, and frozen gases such as carbon monoxide, carbon dioxide, methane and ammonia. It contains also a variety of other organic compounds including methanol, hydrogen cyanide, formaldehyde, ethanol and ethane, and perhaps more complex molecules such as long-chain hydrocarbons and amino acids. Cometary nuclei are among the darkest objects known to exist in the solar system: Halley's Comet nucleus, for instance, reflects approximately 4% of the incoming light, and the surface of Borrelly's Comet is even darker, reflecting 2.4 to 3.0% of the incoming light. This darkness is ascribed to complex organic compounds.

Like small asteroids, comets have irregular shapes.

A wide variety of orbits of comets exists. Some of them dwell in the inner solar system and in the Kuiper belt, with orbital periods of a few years, while others get quite far from the Sun, with periods of hundreds of thousands years. Some are believed to pass only once through the inner Solar System before being thrown out into interstellar space.

Short-period comets are thought to originate in the Kuiper Belt, beyond the orbit of Neptune. Long-period comets are believed to originate in the Oort cloud, consisting of debris left over from the condensation of the solar nebula, located well beyond the Kuiper Belt. Comets are thrown toward the Sun and the inner solar system by gravitational perturbations from the outer planets (in the case of Kuiper Belt objects) or nearby stars (in the case of Oort Cloud objects), or as a result of collisions between objects within these regions.

The difference between asteroids and comets lies in their orbit and composition, but the latter distinction is more and more blurred, since outer solar system asteroid are now understood to contain much frozen volatile substances. In particular, centaurs are comet-like bodies that orbit among the gas giants, between Jupiter and Neptune. Their orbits are unstable and have dynamic lifetimes of a few million years. Until they remain so far from the Sun, they have no coma and seem to be asteroids, but when such bodies enter the inner solar system their cometary nature unfolds. Chiron, for instance, and other centaurs are listed both as asteroids and as comets.

In May 2009, 3,648 known comets had been reported, about 400 being short-period comets. This number is believed to represent only a tiny fraction of the total potential comet population: the outer solar system may contain something like one trillion cometary bodies.

As a comet approaches the inner solar system, solar radiation causes the volatile materials within the comet to vaporize and stream out of the nucleus, carrying dust away with them. Since the gases are not ejected from the comet in an isotropic way, they exert a thrust, so that the orbit of comets is affected by this sort of rocket propulsion.

The streams of dust and gas form a huge, extremely tenuous atmosphere around the comet and, under the pressure of the Sun's light, an enormous tail pointing away from the Sun. The streams of dust and gas form two distinct tails, pointing in slightly different directions.

Even if the coma and the tail are very thin by the standards we are used to, the region of space close to an active comet is believed to be dangerous for any spacecraft.

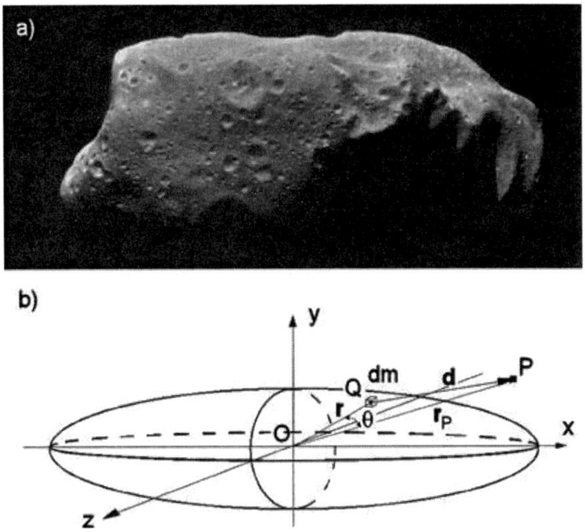

Fig. 2.13 (**a**) Asteroid 243 Ida imaged by Galileo spacecraft on August 28, 1993. The asteroid is 52 km long, and the image is a mosaic of five pictures acquired from distances from 3,057 to 3,821 kilometers (NASA-JPL image). (**b**) Ellipsoidal body attracting gravitationally point P

While the solid nucleus of comets is generally less than 50 km across, the coma may be larger than the Sun, and ion tails have been observed to extend 1 astronomical unit (150 million km) or more. Ionized gas particles attain a positive electrical charge which in turn gives rise to an 'induced magnetosphere' around the comet. A bow shock is thus formed upstream of the comet, in the flow direction of the solar wind.

Comets are also known to break up into fragments, as happened with Comet 73P/Schwassmann-Wachmann 3 starting in 1995. This breakup may be triggered by tidal gravitational forces from the Sun or a large planet, by an explosion of volatile material, or for other reasons not fully explained.

2.8.6 Gravitational Acceleration on the Surface of Non-regular Asteroids

The irregular shape of all asteroids, but the largest ones, causes the gravitational acceleration to be variable from place to place and to be not perpendicular to the ground. For instance, the gravitational acceleration on Eros varies between 0.0023 and 0.0055 m/s^2. A picture of asteroid Ida as imaged by *Galileo* probe is shown in Fig. 2.13a.

It is well known that in case of spherical bodies (with mass axial symmetry) the gravitational acceleration at the surface and above is exactly the same as that of a point mass located at its center. All major bodies and the dwarf planets are close enough to be spherical, to the point that it is possible to consider their mass concentrated at their center, at least for first approximation evaluations. When the precision required is such that the deviation from a spherical shape (and constant density) must be taken into account, it is possible to proceed using a perturbation approach, considering the gravitational field of a sphere and applying subsequent small correction terms, obtained from the potential of the gravitational field.

Consider the generic body in Fig. 2.13b (in the figure the body is plotted as an ellipsoid, but the computation holds in general) and assume that a generic point Q inside the body has coordinates x, y, and z. The external point P where the field has to be computed has coordinates x_P, y_P, and z_P. The reference frame $Oxyz$ is centered in the center of mass of the body and vectors $\overline{(P-O)}$, $\overline{(Q-O)}$ and $\overline{(P-Q)}$ are indicated as $\mathbf{r}_P$, $\mathbf{r}$, and $\mathbf{d}$.

The gravitational potential per unit mass in point P due to mass dm located in Q is

$$d\mathcal{U} = \frac{G}{|\mathbf{d}|} dm, \tag{2.1}$$

where G is the gravitation constant.

The potential per unit mass in point P due to the whole body is

$$\mathcal{U} = \int_M \frac{G}{|\mathbf{d}|} dm. \tag{2.2}$$

Since

$$|\mathbf{d}| = \sqrt{(x-x_P)^2 + (y-y_P)^2 + (z-z_P)^2} = \sqrt{|\mathbf{r}_P|^2 + |\mathbf{r}|^2 - 2|\mathbf{r}||\mathbf{r}_P|\cos(\theta)}, \tag{2.3}$$

where θ is the angle defined in Fig. 2.13b, it follows that

$$\mathcal{U} = \frac{G}{\mathbf{r}_P} \int_V \frac{G\rho}{\sqrt{1+\alpha^2 - 2\alpha q}} dV, \tag{2.4}$$

where ρ is the density of the body, in general a function of space coordinates, and

$$\alpha = \frac{|\mathbf{r}|}{|\mathbf{r}_P|}, \qquad q = \cos(\theta) = \frac{x x_P + y y_P + z z_P}{|\mathbf{r}_P||\mathbf{r}|}.$$

Since $q \le 1$ and, in case of points outside the body, $\alpha < 1$ the square root at the denominator can be written as a Taylor series in α. The potential can thus be written as[1]

$$\mathcal{U} = \frac{G}{|\mathbf{r}_P|} \sum_{i=0}^{\infty} \int_V \rho P_i \alpha^i \, dV = \sum_{i=0}^{\infty} \mathcal{U}_i. \tag{2.5}$$

The first term is easily computed, obtaining

$$P_0 = 1, \qquad \mathcal{U}_0 = \frac{GM}{|\mathbf{r}_P|}. \tag{2.6}$$

[1] A.E. Roy, *Orbital Motion*, Adam Hilger, Bristol, 1991.

The first term is thus the potential of a point mass with the same mass of the body, located in its center.

Since O is the center of mass of the body, the second term ($i = 1$) can be shown to vanish:

$$P_1 = q, \qquad \mathcal{U}_1 = 0. \tag{2.7}$$

The third term can still be obtained with straightforward computations, obtaining

$$P_2 = \frac{1}{2}(3q^2 - 1), \qquad \mathcal{U}_2 = \frac{G}{2|\mathbf{r}_P|^3}(J_x + J_y + J_z - 3J_{OP}), \tag{2.8}$$

where the various J_k are the moments of inertia about the axes and about line OP.

Note that also this term vanishes for spherical bodies.

The fourth term can be shown to vanish if the body is symmetrical about the coordinate planes, like in the case of a homogeneous ellipsoid. The same can be shown to happen for the sixth, eighth, etc. terms, i.e. all terms $\mathcal{U}_i$ with odd i. If the body is a revolution solid, but is pear-shaped, like in the case of Earth, these terms do not vanish.

This approach, based on the series expansion of the potential, is increasingly difficult when the higher order terms become important, i.e. when

- point P is close to the surface
- the shape of the body departs from a sphere by a non small quantity or the density is not constant

When computing the gravitational acceleration on the surface of very small bodies, like asteroids and comets, a large number of terms should be accounted for. Even if their shape is approximated as a spheroid (something that can be done only as a theoretical example), this approach becomes unpractical.

A different approach will be followed here to show that the non-spherical shape causes the gravitational acceleration at the surface not only to change from point to point, but also to be not perpendicular to the surface even in the case of a homogeneous spheroid.

Using the same notation as above (Fig. 2.13b), the gravitational force exerted on a mass m_1 located in point P outside the body can be computed as

$$\mathbf{F} = m_1 \nabla \mathcal{U} = -G m_1 \int_M \frac{1}{|\mathbf{d}|^3} \mathbf{d}\, dm. \tag{2.9}$$

The local gravitational acceleration at point P is

$$\mathbf{g} = -G \int_M \frac{1}{|\mathbf{d}|^3} \mathbf{d}\, dm. \tag{2.10}$$

If the coordinates of point P are x_P, y_P and z_P, it follows that

$$\mathbf{g} = -G \int_V \frac{\rho}{[\sqrt{(x - x_P)^2 + (y - y_P)^2 + (z - z_P)^2}]^3} \begin{Bmatrix} x - x_P \\ y - y_P \\ z - z_P \end{Bmatrix} dx\, dy\, dz, \tag{2.11}$$

were V is the volume occupied by the body.

If the body is a homogeneous sphere with radius R and ρ is a constant, the integration can be performed in closed form, although not in rectangular coordinates. Using cylindrical coordinates, and setting x axis along the line connecting point P with the center of the sphere, it follows that

$$\mathbf{g} = \frac{4\pi G\rho R^3}{3x_{\mathrm{P}}^2}\begin{Bmatrix}1\\0\\0\end{Bmatrix} = \frac{MG}{x_{\mathrm{P}}^2}\begin{Bmatrix}1\\0\\0\end{Bmatrix}. \tag{2.12}$$

This amounts to state that the body behaves as a point mass located at its center, as stated above.

Consider a body whose shape is an ellipsoid with semi-axes a, b and c, respectively, in the direction of x, y and z axes, and assume that its density is constant. The gravitational acceleration is thus

$$\mathbf{g} = -G\rho\int_{-a}^{a}\int_{-b\sqrt{1-(\frac{x}{a})^2}}^{b\sqrt{1-(\frac{x}{a})^2}}\int_{-c\sqrt{1-(\frac{x}{a})^2-(\frac{y}{b})^2}}^{c\sqrt{1-(\frac{x}{a})^2-(\frac{y}{b})^2}}\frac{1}{|\mathbf{d}|^3}\begin{Bmatrix}x-x_{\mathrm{P}}\\y-y_{\mathrm{P}}\\z-z_{\mathrm{P}}\end{Bmatrix}dx\,dy\,dz. \tag{2.13}$$

The innermost integral can be performed in closed form, yielding

$$\mathbf{g} = -G\rho\int_{-a}^{a}\int_{-b\sqrt{1-(\frac{x}{a})^2}}^{b\sqrt{1-(\frac{x}{a})^2}}\begin{Bmatrix}\frac{(x-x_{\mathrm{P}})}{d_1^2}\left(\frac{A}{\sqrt{d_1^2+A^2}}+\frac{B}{\sqrt{d_1^2+B^2}}\right)\\\frac{(y-y_{\mathrm{P}})}{d_1^2}\left(\frac{A}{\sqrt{d_1^2+A^2}}+\frac{B}{\sqrt{d_1^2+B^2}}\right)\\\frac{1}{\sqrt{d_1^2+B^2}}-\frac{1}{\sqrt{d_1^2+A^2}}\end{Bmatrix}dx\,dy, \tag{2.14}$$

where

$$A = c\sqrt{1-\frac{x^2}{a^2}-\frac{y^2}{b^2}}-z_{\mathrm{P}},$$

$$B = c\sqrt{1-\frac{x^2}{a^2}-\frac{y^2}{b^2}}+z_{\mathrm{P}}, \tag{2.15}$$

$$d_1 = \sqrt{(x-x_{\mathrm{P}})^2+(y-y_{\mathrm{P}})^2}.$$

The other two integrals cannot be performed in closed form, but there is no difficulty in obtaining a numerical result, once the position of P and the geometrical characteristics of the ellipsoid are known.

For example, the gravitational acceleration on the surface of a prolate spheroid,[2] with the larger axis in x direction (i.e. $b = c < a$) is reported as a function of the

[2] A spheroid is an ellipsoid with two equal axes. If the third axis is smaller than the other two, the spheroid is oblate; if it is longer it is prolate.

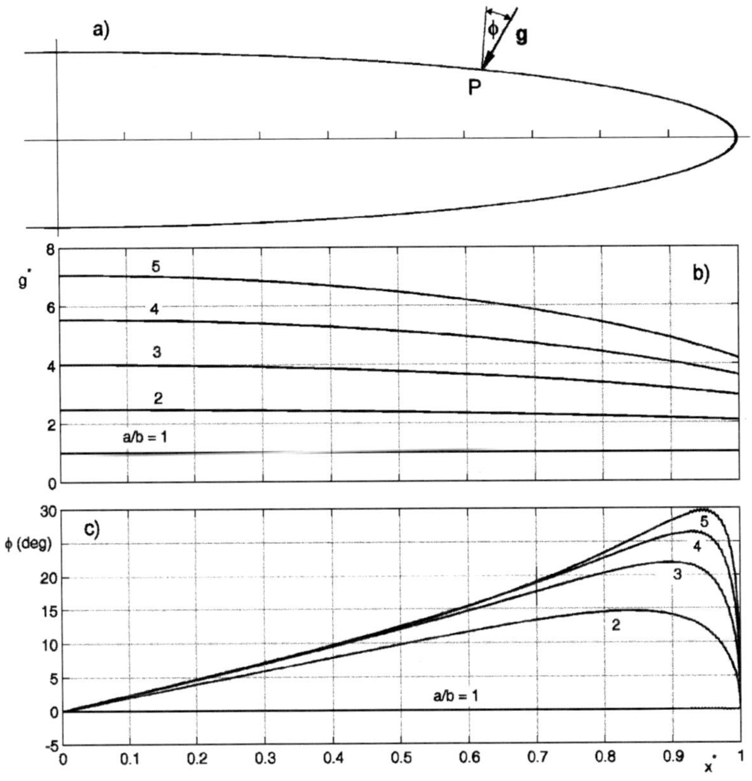

Fig. 2.14 Gravitational acceleration on the surface of a prolate spheroid. (**a**) Sketch; nondimensional value of the gravitational acceleration (**b**) and angle between the perpendicular to the ground and the vertical direction (i.e. the direction of the gravitational acceleration) (**c**) as functions of the nondimensional coordinate x^*

nondimensional $x^* = x/a$ coordinate of point P in Fig. 2.14b. The nondimensional value of the gravitational acceleration is

$$g^* = g\frac{a^2}{GM} \tag{2.16}$$

i.e. is made nondimensional by dividing it by the gravitational acceleration at the surface of a spherical body with the same mass and a radius equal to the semimajor axis of the spheroid.

The angle between the perpendicular to the ground and the vertical direction (i.e. the direction of the gravitational acceleration) is reported as a function of the nondimensional coordinate x^* in Fig. 2.14c.

It is clear that on a spheroidal body what matters is not so much the variability of the gravitational acceleration as the angle between the vertical direction and the perpendicular to the ground. Angles as large as 20° or even 30° are present.

Remark 2.16 An angle of 20° corresponds to a 18% grade. In places where the perpendicular to the flat ground has such an inclination with respect to the local vertical, any modest terrain irregularity may be very difficult or even impossible to overcome.

Remark 2.17 Actual asteroids are much more irregular than this and above all their shape may not be convex everywhere. There may be large zones where locomotion on the ground is very difficult and places that may be accessed, but from which it is quite difficult to get out.

Chapter 3
Manipulatory Devices

Space robots are often provided with manipulatory devices, which are essential to perform tasks like grasping spacecraft or specimens, operating tools or cameras for inspection and many other duties. In most cases these manipulatory devices are open kinematic chains, which may bear some similarity with human arms or at least animal limbs. The generic term *arm* is used for manipulators that follow the scheme of an open kinematic chain, even if their structure is not anthropomorphic.

The arms of space robots are similar, at least conceptually, to those of industrial robots. The main task of arms is carrying an end effector of some sort, able to perform the required task, moving it to a prescribed point in space, with a given orientation and often following a well determined trajectory.

The same anthropomorphic nomenclature used in industrial robots applies also to space robots: an arm starts at the shoulder, the middle joint is an elbow and the joint at the end effector is a wrist. If the latter is a manipulator, it is defined as a hand, and usually has fingers.

In some specialized applications, the wrist carries a specific tool to perform a determined job instead of a generic hand, and there are cases where different effectors may be mounted in an automatic way.

3.1 Degrees of Freedom and Workspace

An arm is usually assumed to be an open kinematic chain, made of rigid bodies (*links*) connected to each other by hinges (*joints*). The first link is hinged to the *base*, the last carries an *end effector* of some sort.

The position of the end effector is defined by a point P, characterized by coordinates

$$\mathbf{X} = [X \quad Y \quad Z]^T \tag{3.1}$$

in a reference frame fixed to the base.

If the arm must reach a generic point in the three-dimensional space, it must have a minimum of three degrees of freedom. The corresponding generalized coordinates,

G. Genta, *Introduction to the Mechanics of Space Robots*, Space Technology Library 26,
DOI 10.1007/978-94-007-1796-1_3,

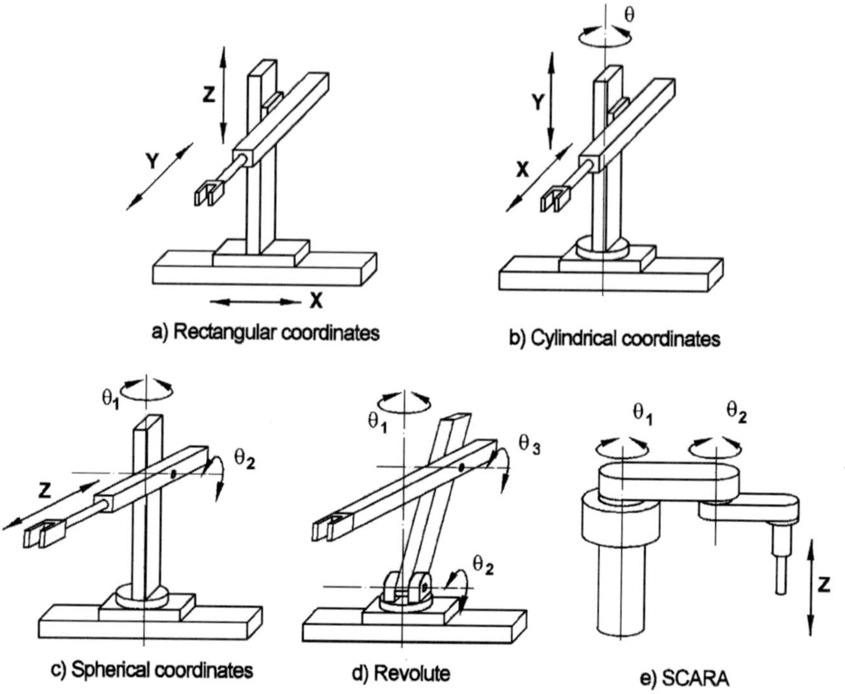

Fig. 3.1 Some possible arrangements of robotic arms to let them to reach a point in three-dimensional space. The three generalized coordinates are also shown

either rotational or translational coordinates, defining the positions of the joints, can be written in a vector

$$\boldsymbol{\theta} = [\theta_1 \quad \theta_2 \quad \theta_3]^T. \tag{3.2}$$

They are referred to as the *joint coordinates* of the arm.

A human arm, from the shoulder to the wrist (the latter not included), has three degrees of freedom: two rotational degrees of freedom at the shoulder plus another rotational degree of freedom at the elbow.

The joints can be materialized with cylindrical hinges (rotoidal joints) or linear motion sliders, and the corresponding generalized coordinates can consequently be angles or linear displacements. If two cylindrical hinges with orthogonal axes are located in the same point, the resulting articulation is a spherical hinge, an example being the human shoulder. Usually, a spherical hinge is modeled as two coincident cylindrical hinges. This implies that the link between the two hinges has zero length.

The possible general configurations for three degrees of freedom arms are many; some of them are shown in Fig. 3.1:

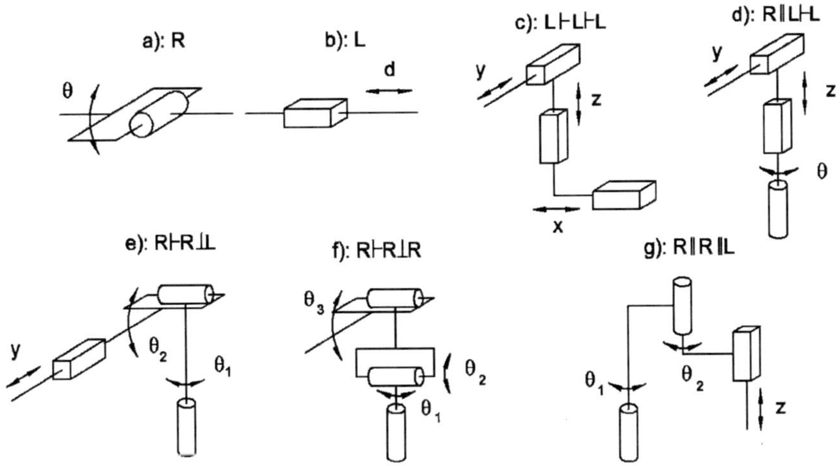

Fig. 3.2 Symbols for a revolute (**a**) and a prismatic (**b**) joint. From (**c**) to (**g**): symbolic representation of the arms of Fig. 3.1

- Rectangular coordinates arm: all joint coordinates are displacements, and the position of the end effector is directly expressed by a set of three Cartesian coordinates. It is also called a Cartesian arm.
- Cylindrical coordinates arm: the whole arm is pivoted on its support, and the first coordinate is thus a rotation angle. The other two coordinates are linear displacements and can be thought as Cartesian coordinates in a plane identified by the two arm segments. The position of the end effector is expressed in cylindrical coordinates.
- Spherical coordinates arm: the first two coordinates are two angles, while the third one is a displacement that can be materialized by a telescopic arm. The first link can be very short or even, as a limiting case, have vanishing length and the first two cylindrical joints become a spherical hinge. In this case the position of the end effector is expressed in spherical coordinates.
- Revolute arm: the first two coordinates are angles at the shoulder, while the third one is an angle too (at the elbow). It is often called an anthropomorphic arm, since the arrangement is that of a human limb. Again, the axes of the two cylindrical hinges at the shoulder may intersect, and the shoulder can be thought as a spherical hinge.
- Selective Compliance Articulated Robot Arm (SCARA): the first two coordinates are angles, while the third one is the linear displacement of the end effector. The axes of the two cylindrical hinges are parallel.

Usually the layout of an arm is represented in a symbolic way by a diagram in which cylindrical hinges and sliders (prismatic joints) are represented by the symbols of Figs. 3.2a and b. Another way is by stating the sequence of rotational (R) hinges and sliders (L), indicating whether the various axes are parallel (∥), orthogonal (⊢, i.e. intersecting at a right angle) or perpendicular (⊥, i.e. at right angle to

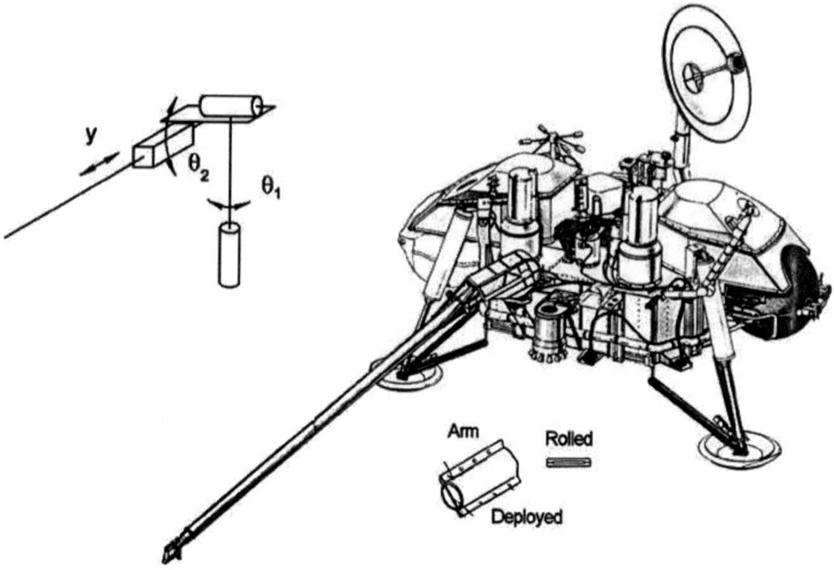

Fig. 3.3 The lander of the *Viking* probe with a spherical coordinates arm. The detail of the extensible arm is also shown (NASA image)

their common normal). A revolute arm is thus R⊢R⊥R,[1] while a Cartesian arm is L⊢L⊢L. The layouts of Fig. 3.1 are shown in symbolic form in Fig. 3.2 from c to g.

An example of spherical coordinates arm is the arm of the *Viking* lander (Fig. 3.3). It is pivoted at the shoulder so that it can be rotated about a vertical and a horizontal axis. The third motion is obtained by extending and contracting the arm, like in telescoping devices. However, it does not have the limitations of telescopic devices for what the ratio between the folded and the extended length is concerned: the boom is made by two seam-welded steel strips that can be rolled on a reel and when unrolled, the cross section springs back becoming almost circular and regaining its stiffness. Each one of the two halves of the arm is similar to a metal measuring tape.

The end effector cannot obviously reach any point in space: the part of the tridimensional space that can be actually reached is called *workspace*. The workspace of an arm based on a number of revolute joints is shown in Fig. 3.4. The workspace is a torus, whose cross section is the curved quadrilateral shown with dashed lines.

[1] Actually two parallel axes are, following this definition, also perpendicular, since they are, by definition, at right angle to their common normal. A revolute arm can thus be R⊢R⊥R, but also R⊢R ∥ R. Parallel axes are thus a particular case of perpendicular axes, as defined above. To avoid this potentially confusing definition, often only two cases are defined: parallel (∥), and perpendicular (⊥) axes, the latter with the meaning we here give to normal axes. However, in this way general skew axes cannot be included.

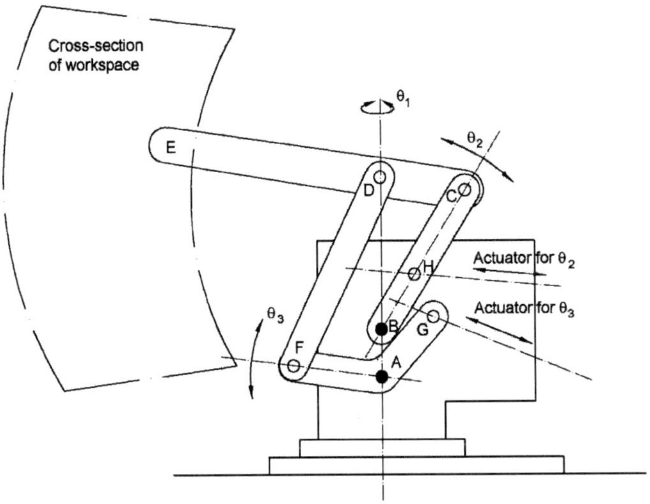

Fig. 3.4 Workspace of an arm based on three revolute joints

At first sight, the arm in the figure is not a revolute arm: the four bar linkage transforms the rotations about the two hinges A and B in a different way than in Fig. 3.1 and is not possible to identify an elbow. Moreover, the linkage is not an open kinematic chain, and the approach seen below cannot be used.

Remark 3.1 As stated above, a robotic arm is, when possible, modeled as an open kinematic chain, constituted by a number of rigid links. Even cases that apparently are not such, in many instances can be reduced to this model.

The arm of Fig. 3.4 can be reduced to a revolute arm, made by the two rigid links BC and CE. The hinge in B is the second hinge of the shoulder, while that in C is the elbow. The links GAF and FD are just a transmission system connecting the motor M_2 to the second link, while the first motor M_1 actuates directly the first link in H.

3.2 End Effectors

The aim of a robotic arm is usually either to perform some operation on a workpiece or to move objects. In the first case, very common in industrial robotics, the end effector is a tool. There are specialized robots, like welding or painting robots, that carry just one type of tool; in this case the tool can be directly a part of the arm. In other cases, the robot may carry different tools to perform different tasks. The arm may be able of performing the change of tool by itself, in a more or less automatic way.

Fig. 3.5 The NASA Robonaut has two anthroporphic arms provided with human-like hands (NASA image)

Space robots often perform the general task of picking up and moving objects, and in this case the end effector is a gripper. Also in this case the gripper may be specialized, to get only a class of objects, perhaps provided with a suitable fixture that matches the gripper, or be designed to get objects of various type or size.

There are cases where the gripper is modeled after a human hand, which is considered a very good universal gripper. Anthropomorphic robots, like the NASA Robonaut, are provided with hands that are as close as possible in shape, but above all in function, to human hands (Fig. 3.5).

The main feature of the human hand is to have a thumb that opposes to the other fingers, and this is replicated by most anthropomorphic hands. The number of fingers may by less than four, to simplify the design and to reduce the number of degrees of freedom. Solutions with two fingers plus the thumb, are quite common.

The control of the gripper is straightforward in case of a telemanipulator, while the difficulties are still large if it must be performed autonomously.

For the manipulation of large objects or for performing difficult tasks the use of two or more cooperating arms may be needed. Telemanipulators with two cooperating gripper arms controlled by a human operator with his two hands were common in the nuclear industry since the 1960s; however, the difficulties linked with the construction of autonomous cooperating arms are still large.

One of the reasons is that when two or more arms grip an object the kinematic chain contains loops and it is difficult to avoid that large forces circulate in these loops. In other words, it is difficult to make the arms really to cooperate and not to work one against the other.

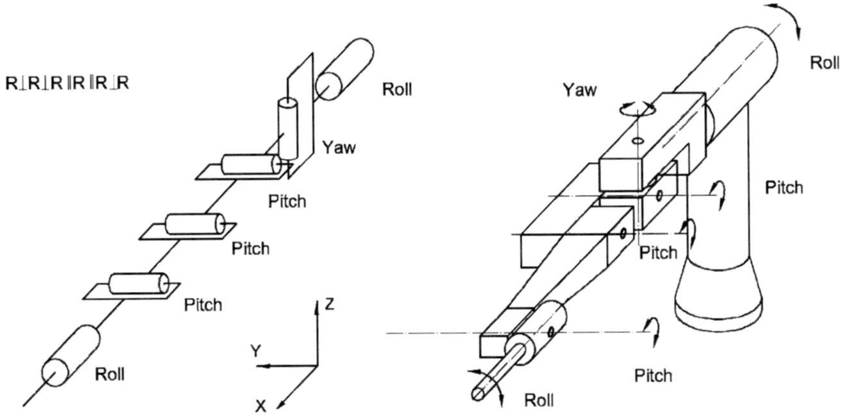

Fig. 3.6 Sketch of a six-degrees of freedom arm, based on revolute joints

3.3 Orientation of the End Effector

Up to now the arm was considered as a device whose goal is to position the end effector in a certain position in space. If the end effector is considered as a rigid body, its orientation in space must be considered too. Since a rigid body has six degrees of freedom in the tridimensional space, vector $\mathbf{X}$ in (3.1), in which the generalized coordinates of the end effector are listed has now six components.

The association of an orientation to a position is usually defined a *pose*.

To achieve a general pose of the end effector in any point of the workspace a robotic arm must have six degrees of freedom (joint degrees of freedom, Fig. 3.6). In a human arm, for instance, the wrist has three additional degrees of freedom to obtain the required orientation of the hand, although the amplitude of the rotation about the axis perpendicular to the plane of the hand is not large.

Unlike the human arm, it is not said that three degrees of freedom are associated with the arm, and three with the wrist: as shown in the figure, many different arrangements are possible.

In the figure, the rotations are named

- roll, when about the x axis,
- pitch, when about the y axis,
- yaw, when about the z axis.

They are referred to a fixed xyz frame and to the arm in a reference position.

An example of a six-degrees of freedom arm is the Shuttle Remote Manipulator System (SRMS), or Canadarm (Fig. 3.7). It is a telemanipulator attached to the cargo bay of the *Space Shuttle*, used to move a payload from the cargo bay to its deployment position and then to release it, or to catch a free-flying payload and store it in the orbiter. As clearly visible from the figure, the degrees of freedom are

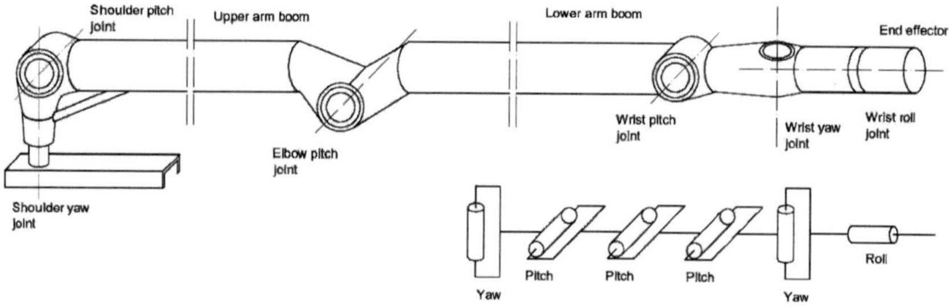

Fig. 3.7 Sketch of the telemanipulator of the Space Shuttle. It is shown in the rest position, where it is supported at 3 location (plus the shoulder) on the orbiter longeron

yaw and pitch rotation of the shoulder, pitch of the elbow and pitch, yaw and roll of the wrist. It is a fully anthropomorphic arm.

The Canadarm is 15.2 m long and 380 mm in diameter, has a mass of 410 kg and is capable of deploying or retrieving payloads up to 29,000 kg in space. However, its motors are unable to lift the arm's own weight when on the ground.

Often, however, the rotation of the end effector about an axis located in the direction of the last part of the arm is not important. In this case, the arm has only five degrees of freedom.

3.4 Redundant Degrees of Freedom

With six joint degrees of freedom it is possible to put the end effector in any place within the workspace and to orient it in the required way, i.e. to reach the required pose. To each set of joint degrees of freedom corresponds a pose of the end effector.

Remark 3.2 In many cases, owing to the nonlinearity of the kinematic equations, there may be more than one set of the joint coordinates yielding a given pose of the end effector.

If the space where the arm operates is not free, the arm must act to avoid obstacles or forbidden areas. In this case a larger flexibility than that allowed by the basic six joint degrees of freedom must be present and a larger number of degrees of freedom must be provided.

If the number of degrees of freedom of the joints θ_i is larger than the number of degrees of freedom of the end effector, the kinematic relationships cannot be inverted (see below). This means that, although once the coordinates of the joints θ_i are stated it is possible to find the position and the orientation of the end effector (direct kinematics), if the latter is stated an infinity of different sets of joint coordinates θ_i can be found. This is exactly the flexibility that was sought to go around obstacles: there are many possible positions of the arm yielding the same position of the end effector.

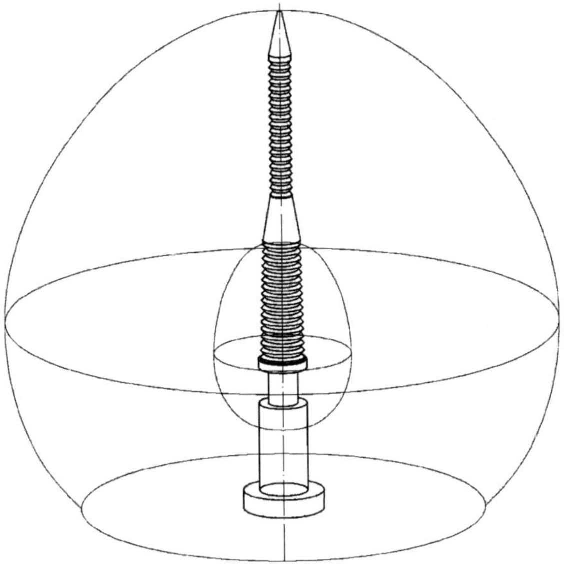

Fig. 3.8 Serpentine robot arm, shown in vertical position. Also the workspace is shown

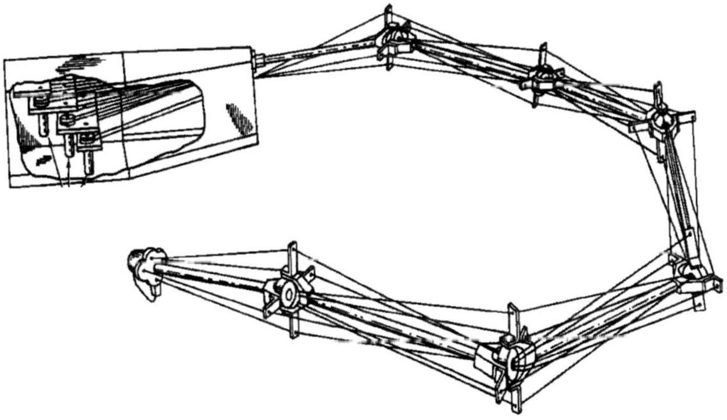

Fig. 3.9 The NASA Space Crane: a serpentine robot arm for space applications (NASA image)

To choose one of them, other conditions must be stated.

There is no limit to the number of degrees of freedom that can be used, other than the ensuing mechanical and control complexity. An example from nature of a device with redundant degrees of freedom is the trunk of elephants. A robot arm based on a serpentine configuration is shown in Fig. 3.8, together with its workspace.

A simple example of an arm with many degrees of freedom is the Space Crane designed by NASA, although never actually built (Fig. 3.9). Each degree of freedom is controlled by an electric motor at the shoulder through tendons. If the links in the arm are seven, like in the figure, and no part has a roll degree of freedom, the total number of degrees of freedom is 16.

In space robotics arms are usually installed on movable carriers (e.g. space vehicles or rovers), so that the motion of the carrier can help in obtaining the required position and orientation of the end effector. Even if there are obstacles, it is possible to avoid them by changing the position of the platform. For this reason, the arms of moving robots can in many cases avoid the complexities of redundant coordinates. This simplifies not only the mechanical structure and the actuators, with a related mass saving, but also the control hardware and software. After all, this is the case of human limbs, which do not have redundant degrees of freedom: to reach difficult positions with our hand, we often displace our whole body or change our posture.

3.5 Arm Layout

A robot arm is essentially a device that must carry a payload to a determined position, perhaps following a given trajectory, in a certain time under the effect of given forces acting on it.

In industrial robotics the payload is usually a tool, but may also be a an object that the end effector has picked up. In space robotics very often the payload is an object the arm is manipulating. If the space around the arm is empty often only the final position must be stated and the trajectory followed to reach it has little importance. If on the contrary there are obstacles, the trajectory needs to be stated, at least in terms of a number of waypoints that must be reached in succession.

An important factor is the time required to perform the task, which determines the speed at which the various joints must move. In space robotics the speed is generally less important than in industrial robotics, where fast movements are generally required for productivity reasons. With few exceptions, in space the speed is not an important factor and space robots are usually slow.

Apart from inertia forces, the forces acting on the arm are usually of two types: the weight of the payload and of the arm itself and the forces due to the end effector. The latter usually are present only when the arm has reached the final position and the tool has started its work. There is a large difference between arms that must work in space, where the prevalent conditions are microgravity (with the exception of arms that must operate during propelled phases of spaceflight), and those intended for planetary surfaces. However, in the most interesting locations, gravitational acceleration is much lower than on Earth.

When the arm moves, it is loaded by inertia forces that increase fast with increasing speed of the arm. In space, particularly for manipulators, inertia forces may be the only forces acting on the arm and low operating speed may be mandatory to keep the stressing and deformations of the arms within reasonable limits without increasing the mass of the arm and not to exert large inertia forces on the spacecraft or the rover. Moreover, increasing the strength and stiffness of the arm results in increasing its mass and hence the inertia forces.

The stiffness of the arm is an important requirement, both to allow to position precisely the end effector under the loads due to gravity and forces applied to the end effector and to avoid vibration, or at least to increase their frequency. Low stiffness can be compensated, up to a certain point, by suitable control actions aimed at compensating for static deflection under load and actively damping vibration. The dynamic behavior of the arm strongly influences its speed, since little damped vibrations may force to slow down the motion or to wait that vibration dies out before starting to perform the tasks after the arm has reached a certain position.

The joints may be powered by rotational or linear actuators. In a way this is a consequence of the type of motion the joint must perform, but only up to a certain point. A rotary joint may be powered by a linear actuator, like in the example in Fig. 3.4, or a linear motion can be obtained by a rotary motor, like in the extending arm of the *Viking* lander (Fig. 3.3) that unreels from a spool.

As already stated, space robots are mostly powered by electric motors, which can drive directly rotary joints, possibly through a reduction gear or through a screw, in general a ball or planetary roller screw. The motor, reduction gear and screw assembly is usually referred to as an electric cylinder, since it performs the same task as pneumatic or hydraulic cylinders.

The torque required at the revolute joints, and often also on the screw of electric cylinders, is usually quite high and is accompanied by a low rotational speed. In this case a reduction gear is often used to avoid massive electric motors. High ratio gears may be of the planetary or harmonic drives types, the latter in particular where cost is not a major problem. The development of low speed, high torque motors (often referred to as torque motors) is an important issue in robotics (see Chap. 7).

The actuator is in general a massive components, and it is better placed where it must not be moved much. Instead of locating actuators directly at the joints (distributed actuators) it may be advisable to put the actuators on the fixed part of the arm, and to drive the joints through a linkage, as in Fig. 3.4, or through tendons (Fig. 3.9).

3.6 Position of a Rigid Body in Tridimensional Space

As already stated, the end effector must be put in a certain position in space, with a certain orientation, i.e. in a certain pose.

Consider a rigid body free in tridimensional space. Define a fixed[2] reference frame $OXYZ$ and a frame $Gxyz$ fixed to the body and centered in a point G that may be its center of mass, but may be any other point. The position of the rigid body is defined once the pose of frame $Gxyz$ is defined with respect to $OXYZ$,

---

[2]Here the generic term 'fixed' is used. In dynamics, the equations of motion are usually written with reference to an inertial frame, however, here frame $OXYZ$ is not required to be such. It is simply a frame that does not follow the rigid body in its motion, and in which the motion of the body is described.

which is once the transformation leading $OXYZ$ to coincide with Gxyz is defined. It is well known that the motion of the second frame can be considered as the sum of a displacement plus a rotation and then the parameters to be defined are six: three components of the displacement, two of the components of the unit vector defining the rotation axis (the third component needs not to be defined and may be computed from the condition that the unit vector has unit length) and the rotation angle. A rigid body has thus six degrees of freedom in tridimensional space.

There is no problem in defining the generalized coordinates for the translational degrees of freedom, since the coordinates of G (that may be the center of mass) in any fixed reference frame (in particular, in frame $OXYZ$) are usually the simplest, and the most obvious, choice. For the other generalized coordinates the choice is much more complicated. It is possible to resort, for instance, to two coordinates of a second point and to one of the coordinates of a third point, not on a straight line through the other two, but this choice is far from being the most expedient.

An obvious way to define the rotation of frame Gxyz with respect to $OXYZ$ is to express directly the rotation matrix linking the two reference frames. It is a square matrix of size 3×3 (in tridimensional space) and thus has nine elements. Three of them are independent, while the other six may be obtained from the first three using suitable equations.

Alternatively, the orientation of the body-fixed frame can be defined with a sequence of three rotations about the axes. Since rotations are not vectors, the order in which they are performed must be specified.

Start rotating, for instance, the fixed frame about X-axis. The second rotation may be performed about axes Y or Z (obviously in the position they take after the first rotation), but not about X-axis, since in the latter case the two rotation would simply add to each other and would amount to a single rotation. Assume for instance to rotate the frame about Y-axis. The third rotation may occur about either X-axis or Z-axis (in the new position, taken after the second rotation), but not about Y-axis.

The possible rotation sequences are 12, but may be subdivided into two types: those like $X \rightarrow Y \rightarrow X$ or $X \rightarrow Z \rightarrow X$, where the third rotation occurs about the same axis as the first one, and those like $X \rightarrow Y \rightarrow Z$ or $X \rightarrow Z \rightarrow Y$, where the third rotation is performed about a different axis.

In the first cases the angles are said to be *Euler angles*, since they are of the same type of the angles Euler proposed to study the motion of gyroscopes (precession ϕ about Z-axis, nutation θ about X-axis and rotation ψ, again about Z-axis). In the second case they are said to be *Tait–Bryan angles*.[3]

The possible rotation sequences are reported in the following table

First	X				Y				Z			
Second	Y		Z		X		Z		X		Y	
Third	X	Z	X	Y	Y	Z	Y	X	Z	Y	Z	X
Type	E	TB	E	TB	E	TB	E	TB	E	TB	E	TB

[3] Sometimes all sets of three ordered angles are said to be Euler angles. With this wider definition also Tait–Brian angles are considered as Euler angles.

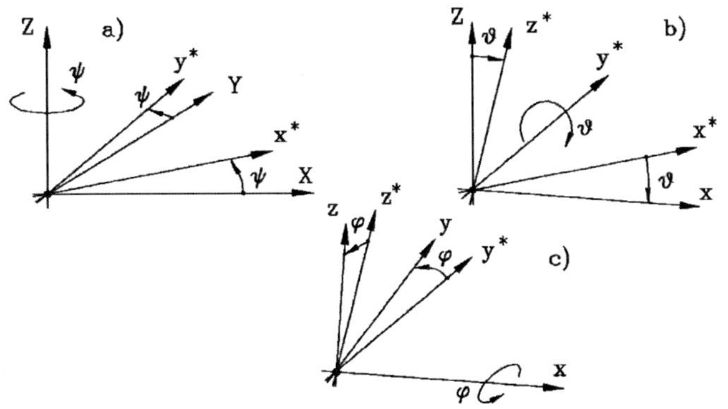

Fig. 3.10 Definition of the yaw ψ (**a**), pitch θ (**b**) and roll ϕ (**c**) angles

Remark 3.3 Euler angles have the drawback of being indeterminate when plane $x_i x_j$ of the rigid body is parallel to $X_i X_j$-plane of the inertial frame (assuming that the first rotation occurs about X_k axis.

Often Euler angles yield indications that are less intuitively clear than Tait–Bryan angles.

In the study of robots the most common approach is that of using Tait–Bryan angles of the type $Z \rightarrow Y \rightarrow X$ so defined (Fig. 3.10):

- Rotate frame XYZ about Z-axis until axis X coincides with the projection of x-axis on plane XY (Fig. 3.10a). Such a position of X-axis can be indicated as x^*; The rotation angle between axes X and x^* is the *yaw* angle ψ. The rotation matrix allowing to pass from x^*y^*Z frame, which will be defined as *intermediate frame*, to the inertial frame XYZ is

$$\mathbf{R}_1 = \begin{bmatrix} \cos(\psi) & -\sin(\psi) & 0 \\ \sin(\psi) & \cos(\psi) & 0 \\ 0 & 0 & 1 \end{bmatrix}. \tag{3.3}$$

- The second rotation is the *pitch* rotation θ about y^*-axis, leading axis x^* in the position of x-axis (Fig. 3.10b). The rotation matrix is

$$\mathbf{R}_2 = \begin{bmatrix} \cos(\theta) & 0 & \sin(\theta) \\ 0 & 1 & 0 \\ -\sin(\theta) & 0 & \cos(\theta) \end{bmatrix}. \tag{3.4}$$

- The third rotation is the *roll* rotation ϕ about x-axis, leading axes y^* and z^* to coincide with axes y and z (Fig. 3.10c). The rotation matrix is

$$\mathbf{R}_3 = \begin{bmatrix} 1 & 0 & 0 \\ 0 & \cos(\phi) & -\sin(\phi) \\ 0 & \sin(\phi) & \cos(\phi) \end{bmatrix}. \tag{3.5}$$

The rotation matrix allowing one to rotate any vector from the body-fixed frame xyz to the inertial frame XYZ is clearly the product of the three matrices

$$\mathbf{R} = \mathbf{R}_1 \mathbf{R}_2 \mathbf{R}_3. \tag{3.6}$$

Performing the product of the rotation matrices, it follows that

$$\mathbf{R} = \begin{bmatrix} c(\psi)c(\theta) & c(\psi)s(\theta)s(\phi) - s(\psi)c(\phi) & c(\psi)s(\theta)c(\phi) + s(\psi)s(\phi) \\ s(\psi)c(\theta) & s(\psi)s(\theta)s(\phi) + c(\psi)c(\phi) & s(\psi)s(\theta)c(\phi) - c(\psi)s(\phi) \\ -s(\theta) & c(\theta)s(\phi) & c(\theta)c(\phi) \end{bmatrix}, \tag{3.7}$$

where symbols cos and sin have been substituted by c and s.

The columns of the rotation matrix are nothing else than the unit vectors of the three body-fixed axes

$$e_x = \begin{Bmatrix} 1 \\ 0 \\ 0 \end{Bmatrix}, \qquad e_y = \begin{Bmatrix} 0 \\ 1 \\ 0 \end{Bmatrix} \quad \text{and} \quad e_z = \begin{Bmatrix} 0 \\ 0 \\ 1 \end{Bmatrix}$$

expressed in the fixed reference frame $OXYZ$.

Conversely, the rows of the rotation matrix are the unit vectors of the three fixed axes e_X, e_Y and e_Z expressed in the body-fixed frame.

An inverse relationship allowing to obtain the Tait–Bryan angles from the rotation matrix can be obtained. From the elements R_{11} and R_{21} of the matrix it is possible to obtain value of the yaw angle

$$\psi = \operatorname{atan}\left[\frac{R_{21}}{R_{11}}\right]. \tag{3.8}$$

Similarly, from the elements R_{32} and R_{33} it follows that

$$\phi = \operatorname{atan}\left[\frac{R_{32}}{R_{33}}\right] \tag{3.9}$$

and from the elements R_{31} and R_{21} it follows that

$$\theta = \operatorname{atan}\left[-\frac{R_{31}\sin(\psi)}{R_{21}}\right] = \operatorname{atan}\left[\frac{-R_{31}}{\sqrt{R_{11}^2 + R_{21}^2}}\right]. \tag{3.10}$$

Remark 3.4 These relationships may yield indeterminate relationships for certain values of the angles. In this case they must be substituted by other relationships obtained from the nonzero elements of the rotation matrix.

3.7 Homogeneous Coordinates

Rotation matrix **R** allows to transform any vector **x** written in the body-fixed frame $Gxyz$ into a vector **X** written in the fixed frame $OYYZ$. The coordinates of any point in the fixed frame can thus be expressed starting from the position in the body-fixed frame by the relationship

$$\mathbf{X} = \mathbf{R}\mathbf{x} + \mathbf{d}, \tag{3.11}$$

where

$$\mathbf{X} = [X \quad Y \quad Z]^T, \qquad \mathbf{x} = [x \quad y \quad z]^T, \qquad \mathbf{d} = [X_G \quad Y_G \quad Z_G]^T$$

are, respectively, the position of the point in the fixed frame, the position of the same point in the body-fixed frame and the displacement of the second frame with respect to the first. Note that this relationship is linear in the translational coordinates X_G, Y_G and Z_G but strongly nonlinear with respect to the rotational coordinates ψ, θ and ϕ.

To perform the whole coordinate transformation, which includes a displacement and a rotation, homogeneous coordinates have been introduced. Equation (3.11) can be written as

$$\begin{Bmatrix} \mathbf{X} \\ 1 \end{Bmatrix} = \begin{bmatrix} \mathbf{R} & \mathbf{d} \\ \mathbf{0} & 1 \end{bmatrix} \begin{Bmatrix} \mathbf{x} \\ 1 \end{Bmatrix}. \tag{3.12}$$

The four-elements vectors (or 4-vectors)

$$\mathbf{X} = [X \quad Y \quad Z \quad 1]^T, \qquad \mathbf{x} = [x \quad y \quad z \quad 1]^T$$

are the homogeneous coordinates of the point in the fixed and in the body-fixed frame, respectively, and the 4×4 matrix

$$\mathbf{T} = \begin{bmatrix} \mathbf{R} & \mathbf{d} \\ \mathbf{0} & 1 \end{bmatrix}$$

is the homogeneous transformation matrix.

Remembering that the inverse of a rotation matrix coincides with its transpose, i.e. $\mathbf{R}^{-1} = \mathbf{R}^T$, the inverse transformation matrix allowing to express a 4-vector expressed in the fixed frame into the body-fixed frame, is

$$\mathbf{T}^{-1} = \begin{bmatrix} \mathbf{R}^T & -\mathbf{R}^T\mathbf{d} \\ \mathbf{0} & 1 \end{bmatrix}. \tag{3.13}$$

3.8 Denavit–Hartenberg Parameters

As already stated, each link of the chain is considered as a rigid body, and a reference frame is attached to it. If the arm is made by n links there are thus $n + 1$ frames:

- frame $x_0y_0z_0$ (base frame), fixed to the base of the arm, which is assumed as link 0;
- frames $x_iy_iz_i$ (with $1 \le i \le n$), fixed to each one of the n links (Fig. 3.11).

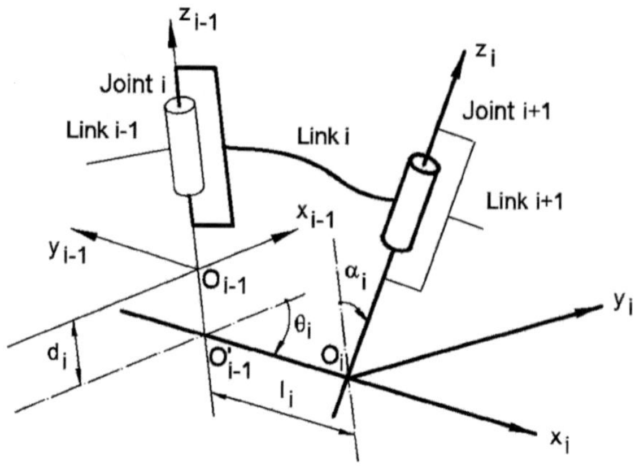

Fig. 3.11 Sketch of the ith link with its DH parameters

Frame $x_n y_n z_n$ is often defined as the *tool* or *end effector* frame, assuming that the end effector is rigidly attached to the last link.

Assume that the z_i axis has the direction in space of the rotation axis of the joint between the ith and the $(i + 1)$th link. The ith link is thus hinged at one end on the z_{i-1}-axis, and at the other end at the z_i-axis. In general these two axes are skew in space. Since it is possible to define a common perpendicular between two skew straight lines, this common perpendicular can be assumed as x_i axis. A third axis y_i can thus be stated and the reference frame $O_i x_i y_i z_i$ is obtained.

Remark 3.5 The origin of this frame lies on the $(i + 1)$th hinge axis (z_i axis); x_i axis is not said to lie within the link or to be its longitudinal axis.

The directions of x_0 and y_0 axes are arbitrary, but can be chosen so that x_0 coincides with x_1 in a particular position of the arm ($\theta_1 = 0$ in that position, see below).

Each link is characterized by its four Denavit–Hartenberg (DH) parameters:

- Angle θ_i, defined as the rotation about z_{i-1} axis leading axis x_{i-1} to be parallel to axis x_i. It is called the rotation angle of the ith link.
- Distance d_i, between points O_{i-1} and O'_{i-1} or, better, a translation along. z_{i-1} axis to bring point O_{i-1} on point O'_{i-1}. It is called the offset of the ith link.
- Distance l_i, between points O'_{i-1} and O_i or, better, a translation along. x_i axis to bring point O'_{i-1} on point O_i. It is called the length of the ith link.
- Angle α_i, defined as the rotation about x_i axis leading axis z_{i-1} to be parallel to axis z_i. It is called the twist of the ith link.

To pass from the frame $O_{i-1}x_{i-1}y_{i-1}z_{i-1}$ to frame $O_i x_i y_i z_i$ four operations are needed:

- a rotation of angle θ_i about z_{i-1} axis;
- a translation of distance d_i along z_{i-1} axis;
- a translation of distance l_i along x_i axis;
- a rotation of angle α_i about x_i axis.

In terms of homogeneous coordinates, the position of a point expressed in frame $O_{i-1}x_{i-1}y_{i-1}z_{i-1}$ can be obtained from the position of the same point expressed in frame $O_i x_i y_i z_i$ by the relationship

$$\begin{Bmatrix} \mathbf{x}_{i-1} \\ 1 \end{Bmatrix} = \begin{bmatrix} \cos(\theta_i) & -\sin(\theta_i) & 0 & 0 \\ \sin(\theta_i) & \cos(\theta_i) & 0 & 0 \\ 0 & 0 & 1 & 0 \\ 0 & 0 & 0 & 1 \end{bmatrix} \begin{bmatrix} 1 & 0 & 0 & 0 \\ 0 & 1 & 0 & 0 \\ 0 & 0 & 1 & d_i \\ 0 & 0 & 0 & 1 \end{bmatrix}$$
$$\times \begin{bmatrix} 1 & 0 & 0 & l_i \\ 0 & 1 & 0 & 0 \\ 0 & 0 & 1 & 0 \\ 0 & 0 & 0 & 1 \end{bmatrix} \begin{bmatrix} 1 & 0 & 0 & 0 \\ 0 & \cos(\alpha_i) & -\sin(\alpha_i) & 0 \\ 0 & \sin(\alpha_i) & \cos(\alpha_i) & 0 \\ 0 & 0 & 0 & 1 \end{bmatrix} \begin{Bmatrix} \mathbf{x}_i \\ 1 \end{Bmatrix}, \tag{3.14}$$

i.e.

$$\begin{Bmatrix} \mathbf{x}_{i-1} \\ 1 \end{Bmatrix} = \mathbf{T}_i \begin{Bmatrix} \mathbf{x}_i \\ 1 \end{Bmatrix}$$
$$= \begin{bmatrix} \cos(\theta_i) & -\sin(\theta_i)\cos(\alpha_i) & \sin(\theta_i)\sin(\alpha_i) & l_i\cos(\theta_i) \\ \sin(\theta_i) & \cos(\theta_i)\cos(\alpha_i) & -\cos(\theta_i)\sin(\alpha_i) & l_i\sin(\theta_i) \\ 0 & \sin(\alpha_i) & \cos(\alpha_i) & d_i \\ 0 & 0 & 0 & 1 \end{bmatrix} \begin{Bmatrix} \mathbf{x}_i \\ 1 \end{Bmatrix}, \tag{3.15}$$

where $\mathbf{T}_i$ is the homogeneous transformation matrix of the ith link.

The two rotations and two translations are thus equivalent to a single rotation with matrix

$$\begin{bmatrix} \cos(\theta_i) & -\sin(\theta_i)\cos(\alpha_i) & \sin(\theta_i)\sin(\alpha_i) \\ \sin(\theta_i) & \cos(\theta_i)\cos(\alpha_i) & -\cos(\theta_i)\sin(\alpha_i) \\ 0 & \sin(\alpha_i) & \cos(\alpha_i) \end{bmatrix}$$

followed by a translation

$$\begin{Bmatrix} a_i\cos(\theta_i) \\ a_i\sin(\theta_i) \\ d_i \end{Bmatrix}. \tag{3.16}$$

The inverse transformation can be easily computed using (3.13)

$$\begin{Bmatrix} \mathbf{x}_i \\ 1 \end{Bmatrix} = \mathbf{T}_i^{-1} \begin{Bmatrix} \mathbf{x}_{i-1} \\ 1 \end{Bmatrix}$$
$$= \begin{bmatrix} \cos(\theta_i) & \sin(\theta_i) & 0 & -l_i \\ -\sin(\theta_i)\cos(\alpha_i) & \cos(\theta_i)\cos(\alpha_i) & \sin(\alpha_i) & -d_i\sin(\alpha_i) \\ \sin(\theta_i)\sin(\alpha_i) & -\cos(\theta_i)\sin(\alpha_i) & \cos(\alpha_i) & -d_i\cos(\alpha_i) \\ 0 & 0 & 0 & 1 \end{bmatrix} \begin{Bmatrix} \mathbf{x}_{i-1} \\ 1 \end{Bmatrix}. \tag{3.17}$$

The link of Fig. 3.11 has two rotational joints at its end. In this case 3 of the DH parameters, namely d_i, l_i and α_i are constants defined once the geometry of the link is stated. The remaining DH parameter, θ_i, is one of the generalized coordinates of the system, that corresponding to the ith degree of freedom.

The same approach can be used for a link connected to the preceding links through a prismatic joint. In this case θ_i is a constant, defined by the geometry of the system, while l_i is the ith generalized coordinate.

3.9 Kinematics of the Arm

The position of the end effector is defined by a point P, characterized by the coordinates shown in (3.1) expressed in a fixed frame.

As already stated, the positions point P can actually take during the motion of the arm, define the workspace of the arm.

If the pose of the end effector is considered, the position and orientation of the end effector are defined by six coordinates. If the yaw, pitch and roll angles are chosen as coordinates for the rotational degrees of freedom, the vector defining the pose of the end effector is

$$\mathbf{X} = [X \quad Y \quad Z \quad \phi \quad \theta \quad \psi]. \tag{3.18}$$

The six-dimensional vector **X** defines the *task space* or *operational space*.

The volume of space the end effector can reach with at least one orientation is usually referred to as the *reachable workspace*, while the volume of space it can reach with any one orientation is the *dexterous workspace*. The latter is a subspace of the former.

The order of the rotations are in a way arbitrary, and their name too. In some texts the name roll is given to the rotation about z axis, since this is the axis about which the link rotates, when using the DH convention. Here the pose of the end effector is referred to a fixed frame of the kind shown in Fig. 3.6, with the same name for the rotations. In this case it does not follow the DH conventions for the first frame $x_0y_0z_0$, whose z_0 axis is horizontal.

The six generalized coordinates at the joints are

$$\boldsymbol{\theta} = [\theta_1 \quad \theta_2 \quad \theta_3 \quad \theta_4 \quad \theta_5 \quad \theta_6]. \tag{3.19}$$

The components of $\boldsymbol{\theta}$ may be angles or distances, depending on the type of joints. The coordinates at the joints can be chosen in different ways; for instance in Fig. 3.4 the angles about points A and B, together with the angle of rotation of the while arm about the vertical axis, seem to be a natural choice, but also the lengths of the linear actuators driving the joints or even the rotation of the electric motors driving the actuators can be used.

The n-dimensional space defined by the n joint coordinates is called the *joint space*. Also in the joint space there is a zone that can be reached and a zone that cannot, owing to the limited rotations or displacements of the joints.

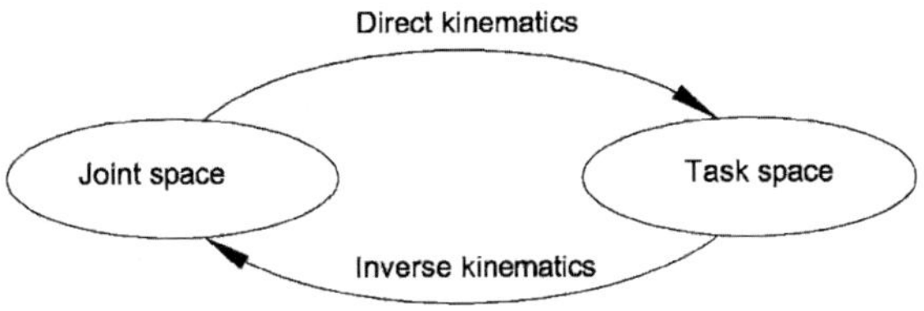

Fig. 3.12 Kinematics and inverse kinematics

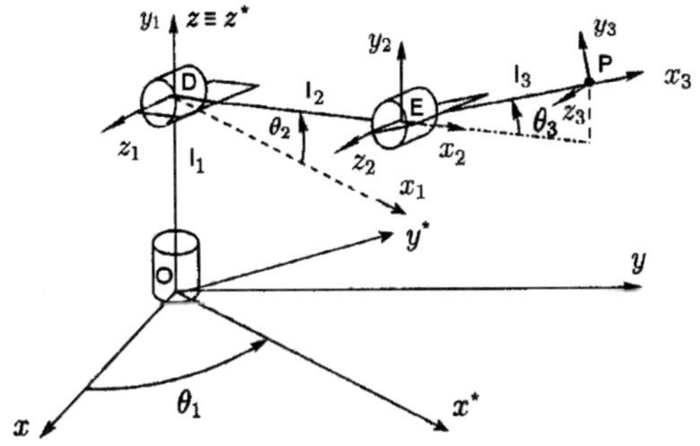

Fig. 3.13 Revolute arm: geometrical definitions and degrees of freedom

The relationship linking $\mathbf{X}$ to $\boldsymbol{\theta}$

$$\mathbf{X} = f(\boldsymbol{\theta}) \tag{3.20}$$

defines the *direct kinematics* (or simply kinematics) of the arm. The kinematics of the arm allows thus to compute the pose of the end effector in the physical space (i.e. its position in the task space), once the joint coordinates (i.e. the position in the joint space) are known (Fig. 3.12).

The kinematic relationship allowing to compute the position of the end effector once that the coordinates at the joints are stated, is usually nonlinear.

If (3.20) is solved in $\boldsymbol{\theta}$

$$\boldsymbol{\theta} = f^{-1}(\mathbf{X}) \tag{3.21}$$

the *inverse kinematic*, i.e. the relationship yielding the values of $\boldsymbol{\theta}$ needed to reach a given point, is obtained. Since the relationship is nonlinear, it is not said that such a solution exists for all values of $\boldsymbol{\theta}$, and, in general, this solution exists only if point P belongs to the dexterous workspace.

If a solution exists, it may be non unique and in general it cannot be obtained in closed form.

Example 3.1 Compute the kinematics and inverse kinematics of the revolute arm shown in Fig. 3.13.

Let the position of point P to be expressed in the fixed reference frame Oxyz (coinciding with frame O$x_0y_0z_0$, i.e. with the base frame), and the joint coordinates be the rotations θ_i. The reference frame Oxyz may be an inertial frame, but, particularly in the case the arm is installed on a rover or a space vehicle, may be a moving frame too.

Define an auxiliary frame $Ox^*y^*z^*$ by rotating frame $Oxyz$ about z-axis by an angle θ_1. Points O and D define the shoulder and point E defines the elbow.

The coordinates of the elbow in x^*z^* plane are

$$\begin{Bmatrix} x^* \\ z^* \end{Bmatrix}_E = \begin{Bmatrix} l_2\cos(\theta_2) \\ l_1 + l_2\sin(\theta_2) \end{Bmatrix}.$$

In a similar way, the coordinates of point P in the same reference frame are

$$\begin{Bmatrix} x^* \\ z^* \end{Bmatrix}_P = \begin{Bmatrix} l_2\cos(\theta_2) + l_3\cos(\theta_2+\theta_3) \\ l_1 + l_2\sin(\theta_2) + l_3\sin(\theta_2+\theta_3) \end{Bmatrix}.$$

The coordinates of P in the reference frame $Oxyz$ are

$$\begin{Bmatrix} x \\ y \\ z \end{Bmatrix}_P = \begin{Bmatrix} x^*\cos(\theta_1) \\ x^*\sin(\theta_1) \\ z^* \end{Bmatrix},$$

i.e.

$$\begin{Bmatrix} x \\ y \\ z \end{Bmatrix}_P = \begin{Bmatrix} [l_2\cos(\theta_2) + l_3\cos(\theta_2+\theta_3)]\cos(\theta_1) \\ [l_2\cos(\theta_2) + l_3\cos(\theta_2+\theta_3)]\sin(\theta_1) \\ l_1 + l_2\sin(\theta_2) + l_3\sin(\theta_2+\theta_3) \end{Bmatrix}.$$

The inverse kinematics can be solved in closed form. By dividing the second equation (3.9) by the first, angle θ_1 is obtained

$$\theta_1 = \text{artg}\left(\frac{y}{x}\right).$$

x^* can thus be computed and, remembering that $z^* = z$, (3.9) can be used to compute θ_2 and θ_3.

Equation (3.9) can be rewritten as

$$\begin{cases} x^* - l_2\cos(\theta_2) = l_3\cos(\theta_2+\theta_3), \\ z^* - l_1 - l_2\sin(\theta_2) = l_3\sin(\theta_2+\theta_3), \end{cases}$$

i.e.

$$\begin{cases} x^{*2} + l_2^2\cos^2(\theta_2) - 2x^*l_2\cos(\theta_2) = l_3^2\cos^2(\theta_2+\theta_3), \\ (z^* - l_1)^2 + l_2^2\sin^2(\theta_2) - 2(z^* - l_1)l_2\sin(\theta_2) = l_3^2\sin^2(\theta_2+\theta_3). \end{cases}$$

By adding the two equations and rearranging it follows that

$$2l_2\left[x^*\cos(\theta_2) + \left(z^* - l_1\right)\sin(\theta_2)\right] = x^{*2} + \left(z^* - l_1\right)^2 + l_2^2 - l_3^2.$$

By remembering that

$$\cos(\theta_2) = \frac{1-t^2}{1+t^2}, \qquad \sin(\theta_2) = \frac{2t}{1+t^2},$$

where

$$t=\tan\left(\frac{\theta_2}{2}\right)$$

and stating

$$\alpha=\frac{x^{*2}+(z^*-l_1)^2+l_2^2-l_3^2}{2l_2},$$

(3.9) becomes

$$x^*\left(1-t^2\right)+2\left(z^*-l_1\right)t=\alpha\left(1-t^2\right),$$

i.e.

$$\left(x^*+\alpha\right)t^2+2\left(z^*-l_1\right)t-x^*+\alpha=0.$$

The solution in t is thus

$$t=\frac{(z^*-l_1)\pm\sqrt{(z^*-l_1)^2+x^{*2}-\alpha^2}}{x^*+\alpha}.$$

The value of θ_2 is thus

$$\theta_2=2\mathrm{artg}\sqrt{\frac{(z^*-l_1)\pm\sqrt{(z^*-l_1)^2+x^{*2}-\alpha^2}}{x^*+\alpha}}.$$

From the first equation (3.9) the value of θ_3 can thus be obtained

$$\theta_3=\arccos\frac{[x^*-l_1\cos(\theta_2)]}{l_2}-\theta_2.$$

The $\pm$ sign shows that there are two sets of angles θ_i that yield the same position $\mathbf{x}$ of the end effector.

Note that the inverse kinematics is fairly complex, but it is at any rate possible, in this case, to solve it in closed form.

If an arm consists of an open kinematic chain made of n links, the transformation between the reference frame attached to the last link (the end effector frame) and that fixed to the base (base frame) is

$$\begin{Bmatrix}\mathbf{x}_0\\1\end{Bmatrix}=\mathbf{T}_1\mathbf{T}_2\mathbf{T}_3\cdots\mathbf{T}_n\begin{Bmatrix}\mathbf{x}_n\\1\end{Bmatrix}=\prod_{i=1}^{n}\mathbf{T}_i\begin{Bmatrix}\mathbf{x}_n\\1\end{Bmatrix}. \tag{3.22}$$

The global transformation matrix is thus the product of the transformation matrices of all the links. It is a function of the geometrical parameters of the system and of the generalized coordinates of the various links θ_i for links with rotational joints and l_i for those having prismatic joints. The coordinates $\mathbf{x}_n$ are constant values, defining the position of a point in the reference frame of the last link.

Usually the origin of the last reference frame is located in the end effector, so that the coordinates of the end effector in the end effector frame and in the base frame are

$$\begin{Bmatrix} \mathbf{x}_n \\ 1 \end{Bmatrix} = [0 \quad 0 \quad 0 \quad 1]^T, \tag{3.23}$$

$$\begin{Bmatrix} \mathbf{x}_0 \\ 1 \end{Bmatrix} = \prod_{i=1}^{n} \mathbf{T}_i [0 \quad 0 \quad 0 \quad 1]^T = \begin{Bmatrix} \mathbf{f}(\theta_i) \\ 1 \end{Bmatrix}, \tag{3.24}$$

where all joint coordinates (rotational as well as translational) are indicated with symbol θ_i.

This relationship defines the first three relationships of the direct kinematics of the arm.

Example 3.2 Repeat the computation of the kinematics of the revolute arm shown in Fig. 3.13 seen in Example 3.1, using the DH parameters.

To be fully consistent with the DH conventions, the axes of the first hinge should be written as $x_0y_0z_0$ and not xyz. The DH parameters are listed in the following table

Link	Variable	α_i	l_i	d_i
1	θ_1	90°	0	l_1
2	θ_2	0	l_2	0
3	θ_3	0	l_3	0

The end effector is located at the end of the third link, i.e. in the origin of the reference frame $x_3y_3z_3$. The coordinates of the end effector in the reference frame $x_0y_0z_0$ are easily computed through the homogeneous transformation matrices of the three links:

$$\begin{Bmatrix} \mathbf{x}_0 \\ 1 \end{Bmatrix} = \begin{bmatrix} \cos(\theta_1) & 0 & \sin(\theta_1) & 0 \\ \sin(\theta_1) & 0 & -\cos(\theta_1) & 0 \\ 0 & 1 & 0 & d_1 \\ 0 & 0 & 0 & 1 \end{bmatrix}$$

$$\times \begin{bmatrix} \cos(\theta_2) & -\sin(\theta_2) & 0 & l_2\cos(\theta_2) \\ \sin(\theta_2) & \cos(\theta_2) & 0 & l_2\sin(\theta_2) \\ 0 & 0 & 1 & 0 \\ 0 & 0 & 0 & 1 \end{bmatrix}$$

$$\times \begin{bmatrix} \cos(\theta_3) & -\sin(\theta_3) & 0 & l_3\cos(\theta_3) \\ \sin(\theta_3) & \cos(\theta_3) & 0 & l_3\sin(\theta_3) \\ 0 & 0 & 1 & 0 \\ 0 & 0 & 0 & 1 \end{bmatrix} \begin{Bmatrix} 0 \\ 0 \\ 0 \\ 1 \end{Bmatrix}.$$

The total transformation matrix is thus

$$\begin{bmatrix} c(\theta_1)c(\theta_t) & -c(\theta_1)s(\theta_t) & s(\theta_1) & c(\theta_1)[l_2c(\theta_2)+l_3c(\theta_t)] \\ s(\theta_1)c(\theta_t) & -s(\theta_1)s(\theta_t) & -c(\theta_1) & s(\theta_1)[l_2c(\theta_2)+l_3c(\theta_t)] \\ s(\theta_t) & c(\theta_2+\theta_3) & 0 & l_1+l_2s(\theta_2)+l_3s(\theta_t) \\ 0 & 0 & 0 & 1 \end{bmatrix},$$

where c and s stand for cos and sin, respectively, and $\theta_t = \theta_2 + \theta_3$.

By multiplying it by the vector $[0\ 0\ 0\ 1]^T$, the direct kinematics is immediately obtained:

$$\begin{Bmatrix} x_0 \\ y_0 \\ z_0 \end{Bmatrix} = \begin{Bmatrix} \cos(\theta_1)[l_2\cos(\theta_2)+l_3\cos(\theta_2+\theta_3)] \\ \sin(\theta_1)[l_2\cos(\theta_2)+l_3\cos(\theta_2+\theta_3)] \\ l_1+l_2\sin(\theta_2)+l_3\sin(\theta_2+\theta_3) \end{Bmatrix},$$

that coincides with the direct kinematics of the arm computed in Example 3.1.

The rotation matrix is made by the first three rows and columns of the total transformation matrix. The orientation of the end effector is thus immediately obtained:

$$\psi = \operatorname{atan}\left[\frac{R_{21}}{R_{11}}\right] = \theta_1,$$

$$\theta = \operatorname{atan}\left[-\frac{R_{31}\sin(\psi)}{R_{21}}\right] = -(\theta_2+\theta_3),$$

$$\phi = \operatorname{atan}\left[\frac{R_{32}}{R_{33}}\right] = 90°.$$

Remark 3.6 As clearly shown in Example 3.2, the component **d** of the total transformation matrix (elements 14, 24 and 34) yields the first three elements of vector **X**, while the component **R** (upper left 3×3 submatrix) yields the orientation, i.e. the last three elements of vector **X**.

In the case of six degrees of freedom arms, vector **X**, expressing the pose of the end effector, must be obtained by multiplying six homogeneous transformation matrices. Although the computations are more complex, they can be performed in closed form, obtaining the equations yielding the pose as a function of the six joint coordinates, i.e. the direct kinematics of the arm.

The inverse kinematics equation (3.21) may have multiple solutions, since the equations are nonlinear, meaning that different sets of joint coordinates may yield the same pose of the end effector. For instance, in the case of an arm with six degrees of freedom, it has been shown that if the joints are all revolute there is a maximum of 16 solutions. The inverse kinematic equation may also have no solution, as when the required position of the end effector lies outside the workspace, or if the orientation requested is inconsistent with the possibilities of the arm (in general, when the position required lies outside the dexterous workspace).

If the arm has redundant degrees of freedom, the inverse kinematics is undetermined, and an infinity of solutions exist. This, as already said, can be exploited to avoid obstacles or, generally, to increase the flexibility of the system.

The inverse kinematic of a revolute arm was obtained in closed form in Example 3.1. This is, however, not a general case and closed form inverse kinematics relationships can be obtained only in special cases.

There are cases where the inverse kinematic problem can be split into two distinct subproblems, one involving the orientation that depends only on the generalized coordinates of the wrist, and one regarding the position, depending on the generalized coordinates of the arm.

A case of this type is that of a revolute arm with a spherical wrist, i.e. a wrist in which the rotation axes converge in a single point: the orientation is solved first and then the equations already seen for the position can be used.

When no analytical solution exists, a numerical approach must be used. In general numerical methods are iterative, and start from an initial guess, converging to one of the solutions, if multiple solutions exist. The simplest numerical method is the Newton–Raphson algorithm for solving sets of nonlinear equations. The relationship linking the solution $\boldsymbol{\theta}^{(k+1)}$ after the kth iteration with the solution $\boldsymbol{\theta}^{(k)}$ before the same iteration can be obtained by developing (3.20) in Tailor series truncated after the first term

$$\mathbf{X} = \mathbf{f}\big(\boldsymbol{\theta}^{(k)}\big) + \mathbf{J}^{(k)}\big(\boldsymbol{\theta}^{(k+1)} - \boldsymbol{\theta}^{(k)}\big), \tag{3.25}$$

where the Jacobian matrix at the ith iteration

$$\mathbf{J}^{(k)} = \left(\frac{\partial \mathbf{f}(\boldsymbol{\theta})}{\partial \boldsymbol{\theta}}\right)_{\boldsymbol{\theta}=\boldsymbol{\theta}^{(k)}} \tag{3.26}$$

can be computed from the forward kinematic equations.

Equation (3.25) can be solved in $\boldsymbol{\theta}^{(k+1)}$ obtaining

$$\boldsymbol{\theta}^{(k+1)} = \boldsymbol{\theta}^{(k)} - \big(\mathbf{J}^{(k)}\big)^{-1}\big[\mathbf{f}\big(\boldsymbol{\theta}^{(k)}\big) - \mathbf{X}\big]. \tag{3.27}$$

The Newton–Raphson method is reliable, although the basins of attraction of the solutions are usually complex and have a fractal geometry. There are also cases where the iterative procedure locks in a loop, without reaching any solution, but in these cases usually it is enough to change the initial guess, repeating the computation. Convergence can be slow when the equations are highly nonlinear and, close to a singularity, the inverse of the Jacobian is ill-conditioned and may cause the algorithm to fail.

This approach can, however, be applied only when the number of links (and thus of unknowns) is equal to the number of components of the pose vector $\mathbf{X}$ (and thus of equations). In this case the Jacobian matrix is square and can be inverted, provided it is not singular.

In case of redundant arms, the Jacobian matrix is not square (it has as many row as the number of elements in $\mathbf{X}$ and as many columns as the number of generalized

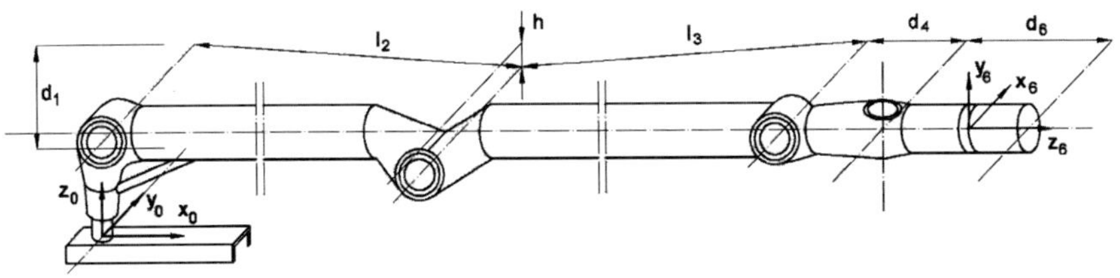

Fig. 3.14 Sketch and dimensions of an arm similar to the Canadarm

degrees of freedom θ_i) and cannot clearly be inverted. A possibility of overcoming this difficulty is by resorting to the pseudo-inverse

$$\mathbf{J}^{\dagger} = \mathbf{J}^T\left(\mathbf{J}\mathbf{J}^T\right)^{-1}$$

instead of the inverse. This yields one of the infinite solutions, restricting the flexibility granted by the redundant configuration. An additional problem of this approach is the computational complexity linked with the computation of the pseudo-inverse. An approximated approach aimed at solving this problem is substituting the inverse with the transpose of the Jacobian matrix.

Alternatively, the inverse kinematics problem can be converted into a differential equation in terms of $\boldsymbol{\theta}$ and $\dot{\boldsymbol{\theta}}$ or into a nonlinear optimization problem.[4]

Example 3.3 Compute the direct kinematics of an arm with the same configuration as the Canadarm. Verify the results computing the direct kinematics in the rest position, and then compute the kinematics and the inverse kinematics in the positions defined below.

The origin of the reference frames of the joints 4 and 5 coincide, and as a consequence both l_5 and d_5 vanish.

With reference to Fig. 3.14, the DH parameters, with their numerical values (they are not the actual values of the Canadarm), and the data for the direct kinematics computations are

Link	Variable	α_i	l_i	d_i	Values of θ_i at rest
1	θ_1	90°	0	$d_1 = 0.8$ m	0
2	θ_2	0	$l_2 = 7$ m	0	$-\mathrm{asin}(h/l_2) = -1.637$
3	θ_3	0	$l_3 = 6.5$ m	0	$\mathrm{asin}(h/l_2) + \mathrm{asin}(h/l_3) = 3.4005$
4	θ_4	−90°	$l_4 = 0.4$ m	0	$-\mathrm{asin}(h/l_3) = -1.763$
5	θ_5	90°	0	0	90°
6	θ_6	0	0	$d_6 = 1$ m	0

The offset h of the third joint is $h = 0.2$ m.

[4]See, for instance, D. Tolani, A. Goswami, N.I. Badler, *Real-Time Inverse Kinematics Techniques for Anthropomorphic Limbs*, Graphical Models, Vol. 62, pp. 353–388, 2000.

The values of the joint coordinates for the direct kinematics computations are $\theta_1 = 30°$, $\theta_2 = 45°$, $\theta_3 = -10°$, $\theta_4 = 100°$, $\theta_5 = 20°$ and $\theta_6 = -35°$. The pose of the end effector for the computation of the inverse kinematics is $x_0 = 7$ m, $y_0 = 2$ m, $z_0 = 5$ m, $\phi = 90°$, $\theta = 0$, $\psi = 90°$.

Performing the relevant computations, the elements R_{ij} of the rotation matrix (upper 3×3 submatrix of the total transformation matrix) are (c and s stand for cos and sin):

$$
\begin{aligned}
R_{11} &= c(\theta_1)\big[c(\theta_6)c(\theta_5)c(\theta_t) - s(\theta_6)s(\theta_t)\big] - c(\theta_6)s(\theta_1)s(\theta_5),\\
R_{21} &= s(\theta_1)\big[c(\theta_6)c(\theta_5)c(\theta_t) - s(\theta_6)s(\theta_t)\big] + c(\theta_6)c(\theta_1)s(\theta_5),\\
R_{31} &= c(\theta_5)c(\theta_6)s(\theta_t) + s(\theta_6)c(\theta_t),\\
R_{12} &= c(\theta_1)\big[-s(\theta_6)c(\theta_5)c(\theta_t) - c(\theta_6)s(\theta_t)\big] + s(\theta_6)s(\theta_1)s(\theta_5),\\
R_{22} &= s(\theta_1)\big[-s(\theta_6)c(\theta_5)c(\theta_t) - c(\theta_6)s(\theta_t)\big] - s(\theta_6)c(\theta_1)s(\theta_5),\\
R_{32} &= -c(\theta_5)s(\theta_6)s(\theta_t) + c(\theta_6)c(\theta_t),\\
R_{13} &= s(\theta_5)c(\theta_1)c(\theta_t) + s(\theta_1)c(\theta_5),\\
R_{23} &= s(\theta_5)s(\theta_1)c(\theta_t) - c(\theta_1)c(\theta_5),\\
R_{33} &= s(\theta_t)s(\theta_5),
\end{aligned}
$$

where

$$\theta_t = \theta_2 + \theta_3 + \theta_4.$$

From the fourth column of the total transformation matrix the first three functions $f_i(\theta_i)$ are readily obtained

$$
\begin{aligned}
f_1(\theta_i) &= d_6\big[s(\theta_5)c(\theta_1)c(\theta_t) + s(\theta_1)c(\theta_5)\big]\\
&\quad + l_4c(\theta_1)c(\theta_t) + l_3c(\theta_1)c(\theta_2+\theta_3) + l_2c(\theta_1)c(\theta_2),\\
f_2(\theta_i) &= d_6\big[s(\theta_5)s(\theta_1)c(\theta_t) - c(\theta_1)c(\theta_5)\big]\\
&\quad + l_4s(\theta_1)c(\theta_t) + l_3s(\theta_1)c(\theta_2+\theta_3) + l_2s(\theta_1)c(\theta_2),\\
f_3(\theta_i) &= d_6s(\theta_5)s(\theta_t) + l_4s(\theta_t) + l_3s(\theta_2+\theta_3) + l_2s(\theta_2) + d_1.
\end{aligned}
$$

The last three functions $f_i(\theta_i)$ defining the orientation of the end effector in the fixed reference frame are

$$
\begin{aligned}
f_4(\theta_i) &= \phi = \operatorname{atan}\left[\frac{R_{32}}{R_{33}}\right] = \operatorname{atan}\left[\frac{-c(\theta_5)s(\theta_6)s(\theta_t) + c(\theta_6)c(\theta_t)}{s(\theta_t)s(\theta_5)}\right],\\
f_5(\theta_i) &= \theta = \operatorname{atan}\left(-\frac{R_{31}}{\sqrt{R_{11}^2 + R_{21}^2}}\right),\\
f_6(\theta_i) &= \psi = \operatorname{atan}\left[\frac{R_{21}}{R_{11}}\right].
\end{aligned}
$$

The closed form computation of the kinematic is quite complex, and the numerical computation of the inverse kinematics requiring the computation of the Jacobian matrix of functions $f_i(\theta_i)$ is even more complex, even in the case of non-redundant arms.

The position of the end effector at rest is

$$\begin{Bmatrix} x_0 \\ y_0 \\ z_0 \end{Bmatrix} = \begin{Bmatrix} d_6 + l_4 + \sqrt{l_3^2 - h^2} + \sqrt{l_2^2 - h^2} \\ 0 \\ d_1 \end{Bmatrix} = \begin{Bmatrix} 14.894 \\ 0 \\ 0.8 \end{Bmatrix} \text{ m.}$$

The rotation matrix of the end effector is

$$\begin{bmatrix} 0 & 0 & 1 \\ 1 & 0 & 0 \\ 0 & 1 & 0 \end{bmatrix}.$$

The first column is the unit vector of the x axis of the end effector (x_6): it coincides with axis y_0.. In a similar way axes y_6 and z_6 coincide with axes z_0 and x_0. The angles giving the orientation of the end effector are thus

$$\phi = \text{atan}\left[\frac{R_{32}}{R_{33}}\right] = 90^\circ,$$

$$\theta = \text{atan}\left(\frac{-R_{31}}{\sqrt{R_{11}^2 + R_{21}^2}}\right) = 0,$$

$$\psi = \text{atan}\left[\frac{R_{21}}{R_{11}}\right] = 90^\circ.$$

The direct kinematic computations for the case with the values of θ_i reported above yield the following rotation matrix:

$$\begin{bmatrix} -0.2602 & -0.9298 & 0.2604 \\ 0.1733 & -0.3103 & -0.9347 \\ 0.9499 & -0.1981 & 0.2418 \end{bmatrix},$$

$$[x_0 \quad y_0 \quad z_0]^T = [8.913 \quad 4.061 \quad 10.003]^T \text{ m.}$$

The angles defining the orientation of the end effector are

$$\phi = -39.32^\circ, \qquad \theta = -71.78^\circ, \qquad \psi = 146.34^\circ.$$

The inverse kinematics computations were performed using the Newton–Raphson technique. The Jacobian matrix is computed numerically, by first computing the values of the function $f_i(\theta_i)$ with the relevant values of the unknowns. Then each one of one the unknowns is incremented by 0.001 rad and the computation is repeated and the relevant row of the Jacobian is computed as the ratio between the increment of the functions and that of the independent variable.

To avoid an initial guess too close to a singular configuration, the initial unknown vector assumed was

$$\boldsymbol{\theta}^{(0)} = [10 \quad 10 \quad 10 \quad 10 \quad 100 \quad 10]^T \text{ deg.}$$

After one iteration we found:

$$\boldsymbol{\theta}^{(1)} = [13.49 \quad -305.22 \quad 687.63 \quad -384.04 \quad 83.46 \quad -12.08]^T \text{ deg.}$$

After 5 iterations, the error, computed as the sum of the squares of the differences between the results at the current iteration and at the previous one, is below 0.001 rad. The result is

$$\boldsymbol{\theta}^{(5)} = [18.43 \quad -270.48 \quad 604.96 \quad -334.48 \quad 71.57 \quad 0.00]^T \text{ deg,}$$

i.e., unwrapping the angles,

$$\boldsymbol{\theta}^{(5)} = [18.43 \quad 89.52 \quad -115.04 \quad 25.51 \quad 71.57 \quad 0.00]^T \text{ deg.}$$

As a confirmation, the direct kinematic computation with these values of the angles yields $x_0 = 7.0001$ m, $y_0 = 1.9995$ m, $z_0 = 4.9991$ m, $\phi = 90.01°$, $\theta = 0.003$, $\psi = 90.000°$, a result quite close to the correct one.

This solution is not, however, unique. If the starting configuration were

$$\boldsymbol{\theta}^{(0)} = [30 \quad 30 \quad 30 \quad 30 \quad 90 \quad 30]^T \text{ deg}$$

the result would have been

$$\boldsymbol{\theta}^{(6)} = [18.43 \quad -18.85 \quad 115.04 \quad -96.18 \quad 71.57 \quad 0.00]^T \text{ deg}$$

which satisfies the requirements of the problem. No wonder that such a nonlinear problem has multiple solutions, and that the Newton–Raphson algorithm converges on different results.

3.10 Velocity Kinematics

The velocity of a point located on the kth link is easily obtained by differentiating its position:

$$\begin{Bmatrix} \dot{\mathbf{X}} \\ 0 \end{Bmatrix} = \frac{d}{dt}\left[\left(\prod_{i=1}^{k} \mathbf{T}_i\right)\begin{Bmatrix} \mathbf{x}_k \\ 1 \end{Bmatrix}\right]. \tag{3.28}$$

By introducing in this equation the coordinates of the end effector and extending the product to the whole arm, from (3.24) it follows that

$$\dot{\mathbf{X}} = \dot{\mathbf{f}}(\boldsymbol{\theta}) = \mathbf{J}(\boldsymbol{\theta})\dot{\boldsymbol{\theta}}, \tag{3.29}$$

where $\mathbf{J}(\boldsymbol{\theta})$ is the Jacobian matrix defined in the previous section computed in the relevant position. This relationship holds both for the case in which the number of joint coordinates is equal to that of the components of $\mathbf{X}$ and for redundant arms. Both $\dot{\mathbf{X}}$ and $\dot{\boldsymbol{\theta}}$ are generalized velocities: if the first three components of $\mathbf{X}$ define the position and the last three define the orientation, the first three velocities are linear velocities and the last three are angular velocities. In the same way, the elements of $\dot{\boldsymbol{\theta}}$ corresponding to rotational joints are angular velocities, those corresponding to sliders correspond to linear velocities.

The Jacobian matrix defines the *velocity kinematics* of the arm, i.e. allows to compute the velocities in the task space once the joint velocities are known. This is clearly an instant velocity, and depends on the position of the arm.

In general (3.29) can be written as

$$[\dot{X} \quad \dot{Y} \quad \dot{Z} \quad \dot{\phi} \quad \dot{\theta} \quad \dot{\psi}]^T = \begin{bmatrix} \mathbf{J}_D(\boldsymbol{\theta}) \\ \mathbf{J}_R(\boldsymbol{\theta}) \end{bmatrix} \dot{\boldsymbol{\theta}}, \tag{3.30}$$

where $\mathbf{J}_D(\boldsymbol{\theta})$ and $\mathbf{J}_R(\boldsymbol{\theta})$ are, respectively, the displacement and the rotational Jacobian matrices.

To compute the angular velocity of the end effector instead of the derivatives of the Tait–Brian angles, the rotational Jacobian matrix can be computed by matrix $\mathbf{A}^T$ defined in (A.146)

$$[\dot{X} \quad \dot{Y} \quad \dot{Z} \quad \Omega_x \quad \Omega_y \quad \Omega_z]^T = \begin{bmatrix} \mathbf{J}_D(\boldsymbol{\theta}) \\ \mathbf{A}^T \mathbf{J}_R(\boldsymbol{\theta}) \end{bmatrix} \dot{\boldsymbol{\theta}}. \tag{3.31}$$

Matrix $\mathbf{A}^T$ depends explicitly on ϕ and θ, which must be computed through the direct kinematic of the arm.

Consider an arm with a spherical wrist. In this case the position of the end effector is determined by the arm and its orientation by the wrist. The displacement Jacobian matrix depends only on the generalized coordinates of the arm, and the displacement velocity is

$$[\dot{X} \quad \dot{Y} \quad \dot{Z}]^T = \mathbf{J}_D^*(\theta_1, \theta_2, \theta_3)[\dot{\theta}_1 \quad \dot{\theta}_2 \quad \dot{\theta}_3]^T, \tag{3.32}$$

where $\mathbf{J}_D^*$ is a 3×3 matrix containing the nonzero elements of $\mathbf{J}_D$.

The angular velocity, on the contrary, depends on all joint generalized coordinates

$$[\Omega_x \quad \Omega_y \quad \Omega_z]^T = \mathbf{A}^T \mathbf{J}_R(\boldsymbol{\theta})\dot{\boldsymbol{\theta}}. \tag{3.33}$$

If the Jacobian matrix is not singular, it is possible to write the inverse velocity kinematics of the arm

$$\dot{\boldsymbol{\theta}} = \mathbf{J}^{-1}(\boldsymbol{\theta})\dot{\mathbf{X}}. \tag{3.34}$$

The points where the Jacobian matrix is singular are the singular points of the arm, and can be of two kinds:

- boundary singular points, located at the boundaries of the workspace
- internal singular points

Often the arms are lined up in the singular points.

Example 3.4 Compute the velocity kinematics of the revolute arm studied in Example 3.1 and find its singular points.

The position of point P in the fixed frame was computed in the mentioned example. By differentiating the expressions there obtained with respect to the joint coordinates θ_1, θ_2 and θ_3, the Jacobian matrix is obtained:

$$J = \begin{bmatrix} -a\sin(\theta_1) & b\cos(\theta_1) & -l_3\sin(\theta_2+\theta_3)\cos(\theta_1) \\ a\cos(\theta_1) & b\sin(\theta_1) & -l_3\sin(\theta_2+\theta_3)\sin(\theta_1) \\ 0 & c & l_3\cos(\theta_2+\theta_3) \end{bmatrix},$$

where

$$\left\{ \begin{array}{l} a = [l_2\cos(\theta_2) + l_3\cos(\theta_2+\theta_3)] \\ b = -[l_2\sin(\theta_2) + l_3\sin(\theta_2+\theta_3)] \\ c = l_2\cos(\theta_2) + l_3\cos(\theta_2+\theta_3) \end{array} \right\}.$$

The determinant of the Jacobian matrix is

$$\det(J) = -l_3 l_2 \sin(\theta_3)\big[l_2\cos(\theta_2) + l_3\cos(\theta_2+\theta_3)\big].$$

The determinant vanishes when either

$$\sin(\theta_3) = 0$$

or

$$l_2\cos(\theta_2) + l_3\cos(\theta_2+\theta_3) = 0.$$

The first equation leads to the condition

$$\theta_3 = 0, \qquad \theta_3 = 180°.$$

These are boundary singular points located at the outer and inner surfaces of the workspace, which is an hollow sphere with outer radius $l_2 + l_3$ and inner radius $|l_2 - l_3|$.

The other condition is satisfied by all points laying on the Z axis, and hence are internal singular points.

3.11 Forces and Moments

Consider a robot arm at standstill in any given position and neglect the weight of the arm, the end effector and the payload. A force **F** and a moment **M** act on the end effector and a number of generalized forces $\mathbf{M}_\theta$ (in general moments) act on the joints. The former can be said to be the contact forces on the end effector, while the second are the joint torques.

If a virtual displacement $\delta\boldsymbol{\theta}$ is given to the joints, the virtual displacement or the end effector is

$$\delta\mathbf{X} = \mathbf{J}(\theta_i)\delta\boldsymbol{\theta}. \tag{3.35}$$

By listing the components of forces and moments in the generalized force vector $\mathbf{F}$, the virtual work done by forces at the end effector

$$\delta\mathcal{L} = \delta\mathbf{X}^T\mathbf{F} = \delta\boldsymbol{\theta}^T\mathbf{J}^T\mathbf{F} \tag{3.36}$$

must be equal to the virtual work done by the moments at the joints

$$\delta\mathcal{L} = \delta\boldsymbol{\theta}^T\mathbf{M}_\theta. \tag{3.37}$$

In case of a prismatic joint, the relevant $\mathbf{M}_\theta$ is a force instead of being a moment, but the equation is unchanged.

By equating the two expressions of the virtual work a relationship yielding the joint torques as functions of the contact forces is obtained

$$\mathbf{M}_\theta = \mathbf{J}^T\mathbf{F}. \tag{3.38}$$

3.12 Dynamics of Rigid Arms

In the previous sections only the kinematics of the robotic arm was considered, and when forces and torques were introduced, they were referred to a situation in which the arm was not moving. When studying the motion of the arm, on the contrary, the effects of the motion must be accounted for and inertia forces play an important role.

The dynamic study can be formulated in two different ways. In the direct problem the time histories of the joint torques $\mathbf{M}_\theta(t)$ are stated, and the aim is to compute the trajectory of the arm either in the joint space (and thus functions $\boldsymbol{\theta}(t)$, $\dot{\boldsymbol{\theta}}(t)$ and $\ddot{\boldsymbol{\theta}}(t)$ are the unknowns) or in the task space, obtaining functions $\mathbf{X}(t)$, $\dot{\mathbf{X}}(t)$ and $\ddot{\mathbf{X}}(t)$.

In the inverse problem the trajectory is stated (i.e. $\boldsymbol{\theta}(t)$, $\dot{\boldsymbol{\theta}}(t)$ and $\ddot{\boldsymbol{\theta}}(t)$ or $\mathbf{X}(t)$, $\dot{\mathbf{X}}(t)$ and $\ddot{\mathbf{X}}(t)$ are known) and the aim of the study is to obtain the torques at the joints $\mathbf{M}_\theta(t)$ that cause the arm to follow the required trajectory.

In the present section the links will be assumed to be rigid bodies, and no allowance is taken for their compliance.

An arm made of rigid parts can be modeled in two basic ways. Each part can be considered as a rigid body with its six degrees of freedom in the three-dimensional space. After stating a set of six generalized coordinates to define its position, there is no difficulty in writing the six relevant equations of motion. The equations of motion so obtained contain more coordinates than the degrees of freedom of the arm and must be complemented with a number of constraint equations.

As an example, in the case of an anthropomorphic arm (the revolute arm in Fig. 3.1) there are two rigid elements, so a set of 12 differential equations in 12 generalized coordinates must be written. Clearly, since the arm has three degrees of freedom, nine of such generalized coordinates are redundant and must be eliminated by writing the constraint equations. The shoulder joint constrains four degrees of freedom, while the elbow constrains five, so the constraints equations are nine and allow one to eliminate all the redundant degrees of freedom.

This approach is usually referred to as multibody approach and there are several commercial codes that operate along these lines.

However, a robot arms can usually be modeled as a kinematic chain in which the bodies are connected in series and the constraints between them are holonomic. In this case it is possible to consider the bodies one after the other, assuming that they have only the degrees of freedom allowed by the constraints. In this way a minimum set of equations is obtained and no constraint equations and redundant coordinates are used.

To write the n equations of motion for an arm with n links, in terms of joint variables, the kinetic and potential energies of the system can be written in terms of the joint variables and velocities $\boldsymbol{\theta}(t)$, $\dot{\boldsymbol{\theta}}(t)$ and then introduced into the Lagrange Equations.

Each link is a rigid body, and thus the kinetic energy of the system is simply

$$\mathcal{T} = \sum_{i=1}^{n} \frac{1}{2} m_i \mathbf{V}_{G_i}^T \mathbf{V}_{G_i} + \sum_{i=1}^{n} \frac{1}{2} \boldsymbol{\Omega}_i^T \mathbf{I}_i \boldsymbol{\Omega}_i , \tag{3.39}$$

where

- m_i is the mass of the ith link,
- $\mathbf{V}_{G_i}$ is the velocity of point G_i, center of mass of the link, referred to an inertial reference frame,
- $\mathbf{I}_i$ is the inertia tensor of the link, referred to its own reference frame,
- $\boldsymbol{\Omega}_i$ is the absolute angular velocity of the link.

The position of G_i in the base frame is simply

$$\begin{Bmatrix} \mathbf{x}_0 \\ 1 \end{Bmatrix}_{G_i} = \prod_{k=1}^{i} \mathbf{T}_i \begin{Bmatrix} \mathbf{x}_i \\ 1 \end{Bmatrix}_{G_i} , \tag{3.40}$$

and, assuming that the base frame is an inertial frame, its velocity is

$$\begin{Bmatrix} \mathbf{V}_{G_i} \\ 0 \end{Bmatrix}_{G_i} = \frac{d}{dt} \left(\prod_{k=1}^{i} \mathbf{T}_i \right) \begin{Bmatrix} \mathbf{x}_i \\ 1 \end{Bmatrix}_{G_i} . \tag{3.41}$$

Proceeding in the same way as seen for the end effector, it follows that

$$\mathbf{V}_{G_i} = \mathbf{J}_{Di}(\boldsymbol{\theta})\dot{\boldsymbol{\theta}}, \tag{3.42}$$

where $\mathbf{J}_{Di}(\theta_j)$ is the displacement Jacobian matrix referred to the center of mass of the ith link defined as in (3.30).

Since each joint rotates about its own z axis, the angular velocity of the ith link, referred to its own reference frame, is

$$\boldsymbol{\Omega}_i = R_{i-1\to i}\begin{Bmatrix} 0 \\ 0 \\ \dot{\theta}_i \end{Bmatrix} + R_{i-2\to i}\begin{Bmatrix} 0 \\ 0 \\ \dot{\theta}_{i-1} \end{Bmatrix} + R_{i-3\to i}\begin{Bmatrix} 0 \\ 0 \\ \dot{\theta}_{i-2} \end{Bmatrix} + \cdots, \tag{3.43}$$

where the rotation matrix $R_{i-j\to i}$ is the rotation matrix allowing to express a vector written in the frame fixed to the $(i-j)$th body, in the frame of the ith body. It is the transpose of the rotation matrix obtained by multiplying the transformation matrices from the $(i-j)$th to the ith body and taking the upper left 3×3 submatrix.

The angular velocity is thus

$$\boldsymbol{\Omega}_i = \mathbf{P}_i(\boldsymbol{\theta})\dot{\boldsymbol{\theta}}, \tag{3.44}$$

where matrix $\mathbf{P}_i$ is a matrix that corresponds to matrix $\mathbf{A}^T\mathbf{J}_R$ defined in (3.31). Matrix $\mathbf{A}^T$ is, however, now defined with reference to the joint space, while in (3.31) it was defined with reference to the Tait–Brian angles.

The kinetic energy of the arm is thus

$$\mathcal{T} = \frac{1}{2}\sum_{i=1}^{n} m_i \dot{\boldsymbol{\theta}}^T \mathbf{J}_{Di}^T(\boldsymbol{\theta})\mathbf{J}_{Di}(\boldsymbol{\theta})\dot{\boldsymbol{\theta}} + \frac{1}{2}\sum_{i=1}^{n} \dot{\boldsymbol{\theta}}^T \mathbf{P}_i^T(\boldsymbol{\theta})\mathbf{I}_i\mathbf{P}_i(\boldsymbol{\theta})\dot{\boldsymbol{\theta}} = \frac{1}{2}\dot{\boldsymbol{\theta}}^T\mathbf{M}(\boldsymbol{\theta})\dot{\boldsymbol{\theta}}. \tag{3.45}$$

Matrix $\mathbf{M}(\boldsymbol{\theta})$ is the mass matrix of the arm.

Remark 3.7 The kinetic energy is a quadratic form in the joint velocities and the mass matrix is a positive defined matrix, function of the joint variables.

The gravitational potential energy is easier to compute:

$$\mathcal{U} = -\sum_{i=1}^{n} m_i \mathbf{g}^T \mathbf{x}_{0G_i} = f_g(\boldsymbol{\theta}), \tag{3.46}$$

where $\mathbf{g}$ is a vector containing the gravitational acceleration in the base frame.

The equations of motion can be written directly using Lagrange equations

$$\frac{d}{dt}\left(\frac{\partial \mathcal{T}}{\partial \dot{\theta}_j}\right) - \frac{\partial \mathcal{T}}{\partial \theta_j} + \frac{\partial \mathcal{U}}{\partial \theta_j} = \mathbf{Q}_j. \tag{3.47}$$

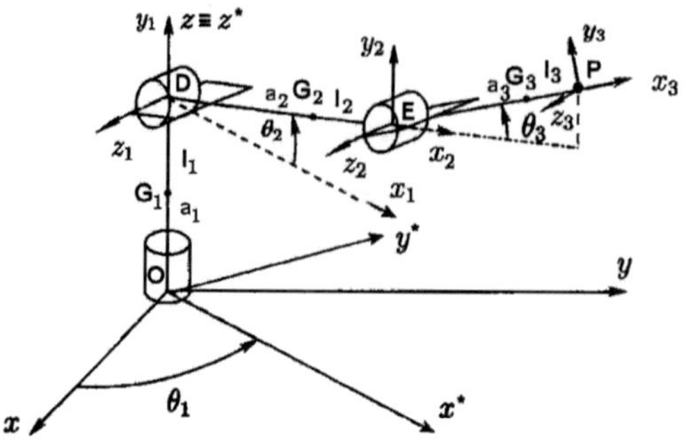

Fig. 3.15 Revolute arm with three degrees of freedom

Since the generalized variables are the rotations at the joints, the generalized forces $\mathbf{Q}_j$ are the torques applied at the joints $\mathbf{M}_{\theta j}$.

By introducing the expressions of the potential and kinetic energies, it follows that

$$\frac{d}{dt}[\mathbf{M}(\boldsymbol{\theta})\dot{\boldsymbol{\theta}}] - \frac{1}{2}\frac{\partial}{\partial \theta_j}[\dot{\boldsymbol{\theta}}^T \mathbf{M}(\boldsymbol{\theta})\dot{\boldsymbol{\theta}}] + \frac{\partial \mathcal{U}}{\partial \theta_j}[f_g(\boldsymbol{\theta})] = \mathbf{M}. \tag{3.48}$$

The structure of this equation of motion is

$$\mathbf{M}(\boldsymbol{\theta})\ddot{\boldsymbol{\theta}} + B(\boldsymbol{\theta})\{\dot{\boldsymbol{\theta}}_i\dot{\boldsymbol{\theta}}_j\} + \mathbf{C}(\boldsymbol{\theta})\{\dot{\boldsymbol{\theta}}^2\} + \mathbf{G}(\boldsymbol{\theta}) = \mathbf{M}_\theta, \tag{3.49}$$

where

- $\mathbf{B}(\boldsymbol{\theta})$ is a matrix with n rows and $n(n-1)/2$ columns. Its terms are usually referred to as *Coriolis coefficients*;
- $\{\dot{\boldsymbol{\theta}}_i\dot{\boldsymbol{\theta}}_j\}$ is a vector with $n(n-1)/2$ rows containing the products (but not the squares) of the joint velocities:

$$\{\dot{\boldsymbol{\theta}}_i\dot{\boldsymbol{\theta}}_j\} = [\dot{\boldsymbol{\theta}}_1\dot{\boldsymbol{\theta}}_2\ \dot{\boldsymbol{\theta}}_1\dot{\boldsymbol{\theta}}_3 \ldots \dot{\boldsymbol{\theta}}_2\dot{\boldsymbol{\theta}}_3 \ldots \dot{\boldsymbol{\theta}}_{n-1}\dot{\boldsymbol{\theta}}_n]^T; \tag{3.50}$$

- $\mathbf{C}(\boldsymbol{\theta})$ is an $n \times n$ square matrix. Its terms are usually referred to as *centrifugal coefficients*;
- $\{\dot{\boldsymbol{\theta}}^2\}$ is a vector with n rows containing the squares of the joint velocities;
- $\mathbf{G}(\theta_j)$ is a vector with n rows containing the gravity terms.

Remark 3.8 The equation of motion is thus nonlinear, and its dependence on the generalized coordinates may be complex.

Example 3.5 Write the dynamic equation of the revolute arm studied in Example 3.1 (Fig. 3.13) The position of the centers of mass of the various links are shown in Fig. 3.15 and the directions of the body fixed axes of the links are parallel to those of the baricentric principal axes of inertia.

Link. 1

The reference frame attached to the first link is frame $x_1y_1z_1$. Since its origin is at the end of the first link, the position of the center of mass G_1 is

$$\mathbf{x}_{1G_1} = [0 \quad a_1 - l_1 \quad 0]^T.$$

The transformation matrix for the first link is the first transformation matrix in Example 3.15. The position of G_1 (in terms of 4-vectors) in the base frame is thus

$$\mathbf{x}_{0G_1} = \begin{bmatrix} \cos(\theta_1) & 0 & \sin(\theta_1) & 0 \\ \sin(\theta_1) & 0 & -\cos(\theta_1) & 0 \\ 0 & 1 & 0 & l_1 \\ 0 & 0 & 0 & 1 \end{bmatrix} \begin{Bmatrix} 0 \\ a_1 - l_1 \\ 0 \\ 1 \end{Bmatrix} = \begin{Bmatrix} 0 \\ 0 \\ a_1 \\ 1 \end{Bmatrix}.$$

The corresponding velocity is equal to zero

$$\mathbf{V}_{0G_1} = \mathbf{0}.$$

The angular velocity of the first link is

$$\boldsymbol{\Omega}_1 = \begin{bmatrix} \cos(\theta_1) & 0 & \sin(\theta_1) \\ \sin(\theta_1) & 0 & -\cos(\theta_1) \\ 0 & 1 & 0 \end{bmatrix}^T \begin{Bmatrix} 0 \\ 0 \\ \dot{\theta}_1 \end{Bmatrix}$$

$$= \begin{bmatrix} 0 & 0 & 0 \\ 1 & 0 & 0 \\ 0 & 0 & 0 \end{bmatrix} \begin{Bmatrix} \dot{\theta}_1 \\ \dot{\theta}_2 \\ \dot{\theta}_3 \end{Bmatrix}.$$

Assuming that the axes of the frame fixed to the first link are principal axes of inertia, the inertia tensor is diagonal

$$\mathbf{I}_1 = \begin{bmatrix} J_{x1} & 0 & 0 \\ 0 & J_{y1} & 0 \\ 0 & 0 & J_{z1} \end{bmatrix},$$

the kinetic energy of the first link is

$$\mathcal{T}_1 = \frac{1}{2} \begin{Bmatrix} \dot{\theta}_1 \\ \dot{\theta}_2 \\ \dot{\theta}_3 \end{Bmatrix}^T \begin{bmatrix} 0 & 0 & 0 \\ 1 & 0 & 0 \\ 0 & 0 & 0 \end{bmatrix}^T \begin{bmatrix} J_{x1} & 0 & 0 \\ 0 & J_{y1} & 0 \\ 0 & 0 & J_{z1} \end{bmatrix}$$

$$= \begin{bmatrix} 0 & 0 & 0 \\ 1 & 0 & 0 \\ 0 & 0 & 0 \end{bmatrix} \begin{Bmatrix} \dot{\theta}_1 \\ \dot{\theta}_2 \\ \dot{\theta}_3 \end{Bmatrix} = \frac{1}{2} \begin{Bmatrix} \dot{\theta}_1 \\ \dot{\theta}_2 \\ \dot{\theta}_3 \end{Bmatrix}^T \begin{bmatrix} J_{y1} & 0 & 0 \\ 0 & 0 & 0 \\ 0 & 0 & 0 \end{bmatrix} \begin{Bmatrix} \dot{\theta}_1 \\ \dot{\theta}_2 \\ \dot{\theta}_3 \end{Bmatrix}.$$

The gravitational acceleration vector is

$$\mathbf{g} = [0 \quad 0 \quad -g]^T.$$

The potential energy of the first link is thus

$$\mathcal{U}_1 = -m_1[0 \quad 0 \quad -g][0 \quad 0 \quad a_1]^T = m_1 a_1 g.$$

Link. 2

The origin of frame $x_2y_2z_2$ is again at the end of the link and the position of the center of mass G_2 is

$$\mathbf{x}_{2G_2} = [a_2 - l_2 \quad 0 \quad 0]^T.$$

The transformation matrix for the second link is the product of the first and second transformation matrices in Example 3.15. The position of G_2 in the base frame is thus

$$\mathbf{x}_{0G_2} = \begin{bmatrix} \cos(\theta_1) & 0 & \sin(\theta_1) & 0 \\ \sin(\theta_1) & 0 & -\cos(\theta_1) & 0 \\ 0 & 1 & 0 & l_1 \\ 0 & 0 & 0 & 1 \end{bmatrix}$$

$$\times \begin{bmatrix} \cos(\theta_2) & -\sin(\theta_2) & 0 & l_2\cos(\theta_2) \\ \sin(\theta_2) & \cos(\theta_2) & 0 & l_2\sin(\theta_2) \\ 0 & 0 & 1 & 0 \\ 0 & 0 & 0 & 1 \end{bmatrix}$$

$$\times \begin{Bmatrix} a_2 - l_2 \\ 0 \\ 0 \\ 1 \end{Bmatrix} = \begin{Bmatrix} a_2\cos(\theta_1)\cos(\theta_2) \\ a_2\sin(\theta_1)\cos(\theta_2) \\ l_1 + a_2\sin(\theta_2) \\ 1 \end{Bmatrix}.$$

The corresponding velocity is

$$\mathbf{V}_{0G_2} = \begin{Bmatrix} -a_2\dot{\theta}_1\sin(\theta_1)\cos(\theta_2) - a_2\dot{\theta}_2\cos(\theta_1)\sin(\theta_2) \\ a_2\dot{\theta}_1\cos(\theta_1)\cos(\theta_2) - a_2\dot{\theta}_2\sin(\theta_1)\sin(\theta_2) \\ a_2\dot{\theta}_2\cos(\theta_2) \end{Bmatrix}$$

$$= \begin{bmatrix} -a_2\sin(\theta_1)\cos(\theta_2) & -a_2\cos(\theta_1)\sin(\theta_2) & 0 \\ a_2\cos(\theta_1)\cos(\theta_2) & -a_2\sin(\theta_1)\sin(\theta_2) & 0 \\ 0 & a_2\cos(\theta_2) & 0 \end{bmatrix} \begin{Bmatrix} \dot{\theta}_1 \\ \dot{\theta}_2 \\ \dot{\theta}_3 \end{Bmatrix}.$$

The translational kinetic energy of the second link is

$$\mathcal{T}_{2R} = \frac{1}{2} m_2 a_2^2 \begin{Bmatrix} \dot{\theta}_1 \\ \dot{\theta}_2 \\ \dot{\theta}_3 \end{Bmatrix}^T \begin{bmatrix} \cos^2(\theta_2) & 0 & 0 \\ 0 & 1 & 0 \\ 0 & 0 & 0 \end{bmatrix} \begin{Bmatrix} \dot{\theta}_1 \\ \dot{\theta}_2 \\ \dot{\theta}_3 \end{Bmatrix}.$$

The angular velocity of the second link is

$$\boldsymbol{\Omega}_2 = \begin{bmatrix} \cos(\theta_2) & -\sin(\theta_2) & 0 \\ \sin(\theta_2) & \cos(\theta_2) & 0 \\ 0 & 0 & 1 \end{bmatrix}^T \begin{Bmatrix} 0 \\ 0 \\ \dot{\theta}_2 \end{Bmatrix}$$

$$+ \begin{bmatrix} \cos(\theta_1)\cos(\theta_2) & -\cos(\theta_1)\sin(\theta_2) & \sin(\theta_1) \\ \sin(\theta_1)\cos(\theta_2) & \sin(\theta_1)\sin(\theta_2) & -\cos(\theta_1) \\ \sin(\theta_2) & \cos(\theta_2) & 0 \end{bmatrix}^T \begin{Bmatrix} 0 \\ 0 \\ \dot{\theta}_1 \end{Bmatrix}$$

$$= \begin{Bmatrix} \dot{\theta}_1 \sin(\theta_2) \\ \dot{\theta}_1 \cos(\theta_1) \\ \dot{\theta}_2 \end{Bmatrix} = \begin{bmatrix} \sin(\theta_2) & 0 & 0 \\ \cos(\theta_2) & 0 & 0 \\ 0 & 1 & 0 \end{bmatrix} \begin{Bmatrix} \dot{\theta}_1 \\ \dot{\theta}_2 \\ \dot{\theta}_3 \end{Bmatrix}.$$

The inertia tensor of the link is again a diagonal matrix, and thus the rotational kinetic energy of the second link is

$$\mathcal{T}_{2R} = \frac{1}{2} \begin{Bmatrix} \dot{\theta}_1 \\ \dot{\theta}_2 \\ \dot{\theta}_3 \end{Bmatrix}^T \begin{bmatrix} \sin(\theta_2) & 0 & 0 \\ \cos(\theta_2) & 0 & 0 \\ 0 & 1 & 0 \end{bmatrix}^T \begin{bmatrix} J_{x2} & 0 & 0 \\ 0 & J_{y2} & 0 \\ 0 & 0 & J_{z2} \end{bmatrix}$$

$$\times \begin{bmatrix} \sin(\theta_2) & 0 & 0 \\ \cos(\theta_2) & 0 & 0 \\ 0 & 1 & 0 \end{bmatrix} \begin{Bmatrix} \dot{\theta}_1 \\ \dot{\theta}_2 \\ \dot{\theta}_3 \end{Bmatrix}$$

$$= \frac{1}{2} \begin{Bmatrix} \dot{\theta}_1 \\ \dot{\theta}_2 \\ \dot{\theta}_3 \end{Bmatrix}^T \begin{bmatrix} J_{x2}\sin^2(\theta_2) + J_{y2}\cos^2(\theta_2) & 0 & 0 \\ 0 & J_{z2} & 0 \\ 0 & 0 & 0 \end{bmatrix} \begin{Bmatrix} \dot{\theta}_1 \\ \dot{\theta}_2 \\ \dot{\theta}_3 \end{Bmatrix}.$$

The potential energy of the second link is

$$\mathcal{U}_2 = -m_2 \begin{Bmatrix} 0 \\ 0 \\ -g \end{Bmatrix}^T \begin{Bmatrix} a_2 \cos(\theta_1)\cos(\theta_2) \\ a_2 \sin(\theta_1)\cos(\theta_2) \\ l_1 + a_2 \sin(\theta_2) 1 \end{Bmatrix} = m_2 g \left[l_1 + a_2 \sin(\theta_2) \right].$$

Link. 3

In the same way seen for the other links, the position of the center of mass G_3 is

$$\mathbf{x}_{3G_3} = [a_3 - l_3 \quad 0 \quad 0]^T.$$

The transformation matrix for the third link is the total transformation matrices in Example 3.15. By performing the product, the position of G_3 in the base frame is

$$\mathbf{x}_{0G_3} = \begin{Bmatrix} \cos(\theta_1)[l_2\cos(\theta_2) + a_3\cos(\theta_2 + \theta_3)] \\ \sin(\theta_1)[l_2\cos(\theta_2) + a_3\cos(\theta_2 + \theta_3)] \\ l_1 + l_2\sin(\theta_2) + a_3\sin(\theta_2 + \theta_3) \\ 1 \end{Bmatrix}.$$

The corresponding velocity is

$$\mathbf{V}_{0G_3} = \begin{Bmatrix} -\dot{\theta}_1 a \sin(\theta_1) - \dot{\theta}_2 b \cos(\theta_1) - \dot{\theta}_3 \cos(\theta_1)[a_3 \sin(\theta_2 + \theta_3)] \\ \dot{\theta}_1 a \cos(\theta_1) - \dot{\theta}_2 b \sin(\theta_1) - \dot{\theta}_3 \sin(\theta_1)[a_3 \sin(\theta_2 + \theta_3)] \\ \dot{\theta}_2 a + \dot{\theta}_3 [a_3 \cos(\theta_2 + \theta_3)] \end{Bmatrix}$$

$$= \begin{bmatrix} -a \sin(\theta_1) & -b \cos(\theta_1) & -a_3 \cos(\theta_1) \sin(\theta_2 + \theta_3) \\ a \cos(\theta_1) & -b \sin(\theta_1) & -a_3 \sin(\theta_1) a \sin(\theta_2 + \theta_3) \\ 0 & a & a_3 \cos(\theta_2 + \theta_3) \end{bmatrix} \begin{Bmatrix} \dot{\theta}_1 \\ \dot{\theta}_2 \\ \dot{\theta}_3 \end{Bmatrix},$$

where

$$a = l_2 \cos(\theta_2) + a_3 \cos(\theta_2 + \theta_3),$$

$$b = l_2 \sin(\theta_2) + a_3 \sin(\theta_2 + \theta_3).$$

The translational kinetic energy of the third link is

$$\mathcal{T}_{3R} = \frac{1}{2} m_3 \begin{Bmatrix} \dot{\theta}_1 \\ \dot{\theta}_2 \\ \dot{\theta}_3 \end{Bmatrix}^T$$

$$\times \begin{bmatrix} a^2 & 0 & 0 \\ 0 & l_2^2 + a_3^2 + 2a_3 l_2 \cos(\theta_3) & a_3^2 + a_3 l_2 \cos(\theta_3) \\ 0 & a_3^2 + a_3 l_2 \cos(\theta_3) & a_3^2 \end{bmatrix} \begin{Bmatrix} \dot{\theta}_1 \\ \dot{\theta}_2 \\ \dot{\theta}_3 \end{Bmatrix}.$$

The angular velocity of the third link is

$$\mathbf{\Omega}_3 = \begin{bmatrix} \cos(\theta_3) & -\sin(\theta_3) & 0 \\ \sin(\theta_3) & \cos(\theta_3) & 0 \\ 0 & 0 & 1 \end{bmatrix}^T \begin{Bmatrix} 0 \\ 0 \\ \dot{\theta}_3 \end{Bmatrix}$$

$$+ \begin{bmatrix} \cos(\theta_2 + \theta_3) & -\sin(\theta_2 + \theta_3) & 0 \\ \sin(\theta_2 + \theta_3) & \cos(\theta_2 + \theta_3) & 0 \\ 0 & 0 & 1 \end{bmatrix}^T \begin{Bmatrix} 0 \\ 0 \\ \dot{\theta}_2 \end{Bmatrix}$$

$$+ \begin{bmatrix} \cos(\theta_1)\cos(\theta_2 + \theta_3) & -\cos(\theta_1)\sin(\theta_2 + \theta_3) & \sin(\theta_1) \\ \sin(\theta_1)\cos(\theta_2 + \theta_3) & \sin(\theta_1)\sin(\theta_2 + \theta_3) & -\cos(\theta_1) \\ \sin(\theta_2 + \theta_3) & \cos(\theta_2 + \theta_3) & 0 \end{bmatrix}^T \begin{Bmatrix} 0 \\ 0 \\ \dot{\theta}_1 \end{Bmatrix}$$

$$= \begin{Bmatrix} \dot{\theta}_1 \sin(\theta_2 + \theta_3) \\ \dot{\theta}_1 \cos(\theta_1 + \theta_3) \\ \dot{\theta}_2 + \dot{\theta}_3 \end{Bmatrix} = \begin{bmatrix} \sin(\theta_2 + \theta_3) & 0 & 0 \\ \cos(\theta_2 + \theta_3) & 0 & 0 \\ 0 & 1 & 1 \end{bmatrix} \begin{Bmatrix} \dot{\theta}_1 \\ \dot{\theta}_2 \\ \dot{\theta}_3 \end{Bmatrix}.$$

The inertia tensor of the link is again a diagonal matrix, and thus the rotational kinetic energy of the second link is

$$\mathcal{T}_{2R} = \frac{1}{2}\begin{Bmatrix}\dot\theta_1\\ \dot\theta_2\\ \dot\theta_3\end{Bmatrix}^T \begin{bmatrix}\sin(\theta_2+\theta_3) & 0 & 0\\ \cos(\theta_2+\theta_3) & 0 & 0\\ 0 & 1 & 1\end{bmatrix}^T \begin{bmatrix}J_{x2} & 0 & 0\\ 0 & J_{y2} & 0\\ 0 & 0 & J_{z2}\end{bmatrix}$$

$$\times \begin{bmatrix}\sin(\theta_2+\theta_3) & 0 & 0\\ \cos(\theta_2+\theta_3) & 0 & 0\\ 0 & 1 & 1\end{bmatrix}\begin{Bmatrix}\dot\theta_1\\ \dot\theta_2\\ \dot\theta_3\end{Bmatrix}$$

$$= \frac{1}{2}\begin{Bmatrix}\dot\theta_1\\ \dot\theta_2\\ \dot\theta_3\end{Bmatrix}^T \begin{bmatrix}J_{x3}\sin^2(\theta_2+\theta_3)+J_{y3}\cos^2(\theta_2+\theta_3) & 0 & 0\\ 0 & J_{z3} & J_{z3}\\ 0 & J_{z3} & J_{z3}\end{bmatrix}\begin{Bmatrix}\dot\theta_1\\ \dot\theta_2\\ \dot\theta_3\end{Bmatrix}.$$

The potential energy of the third link is

$$\mathcal{U}_3 = -m_3\begin{Bmatrix}0\\ 0\\ -g\end{Bmatrix}^T \begin{Bmatrix}\cos(\theta_1)[l_2\cos(\theta_2)+a_3\cos(\theta_2+\theta_3)]\\ \sin(\theta_1)[l_2\cos(\theta_2)+a_3\cos(\theta_2+\theta_3)]\\ l_1+l_2\sin(\theta_2)+a_3\sin(\theta_2+\theta_3)\\ 1\end{Bmatrix}$$

$$= m_3 g\big[l_1+l_2\sin(\theta_2)+a_3\sin(\theta_2+\theta_3)\big].$$

Dynamics of the arm

The mass matrix is then easily obtained by adding together all the expressions of the kinetic energy

$$M = \begin{bmatrix}J_1^*(\theta_i) & 0 & 0\\ 0 & J_2^*+2J_{23}^*\cos(\theta_3) & J_3^*+J_{23}^*\cos(\theta_3)\\ 0 & J_3^*+J_{23}^*\cos(\theta_3) & J_3^*\end{bmatrix},$$

where

$$J_1^*(\theta_i) = J_{11}^*+J_{12}^*\cos^2(\theta_2)+J_{13}^*\cos^2(\theta_2+\theta_3)+2J_{23}^*\cos(\theta_2)\cos(\theta_2+\theta_3),$$

$$J_{11}^* = J_{y1}+J_{x2}+J_{x3}, \qquad J_2^* = J_{z2}+J_{z3}+m_2a_2^2+m_3\big(a_3^2+l_2^2\big),$$

$$J_{12}^* = J_{y2}-J_{x2}+m_2a_2^2+m_3l_2^2, \qquad J_3^* = J_{z3}+m_3a_3^2,$$

$$J_{13}^* = J_{y3}-J_{x3}+m_3a_3^2, \qquad J_{23}^* = m_3a_3l_2.$$

The total potential energy, written neglecting the constant terms, which have no influence on the equations of motion, is

$$\mathcal{U} = g[m_2a_2+m_3l_2]\sin(\theta_2)+m_3ga_3\sin(\theta_2+\theta_3).$$

The derivatives entering the Lagrange equations are

$$\frac{d}{dt}\left(\frac{\partial \mathcal{T}}{\partial \dot{\theta}_1}\right) = \frac{d}{dt}\left[J_1^*(\theta_i)\dot{\theta}_1\right] = J_1^*(\theta_i)\ddot{\theta}_1 + B_{11}(\theta_i)\dot{\theta}_1\dot{\theta}_2 + B_{12}(\theta_i)\dot{\theta}_1\dot{\theta}_3,$$

where

$$\begin{aligned} B_{11}(\theta_i) &= -2\big[J_{12}^*\sin(\theta_2)\cos(\theta_2) + J_{13}^*\sin(\theta_2+\theta_3)\cos(\theta_2+\theta_3) \\ &\quad + J_{23}^*\sin(2\theta_2+\theta_3)\big], \\ B_{12}(\theta_i) &= -2\big[J_{13}^*\sin(\theta_2+\theta_3)\cos(\theta_2+\theta_3) + J_{23}^*\cos(\theta_2)\sin(\theta_2+\theta_3)\big], \\ \frac{\partial \mathcal{T}}{\partial \theta_1} &= \frac{\partial \mathcal{U}}{\partial \theta_1} = 0, \\ \frac{d}{dt}\left(\frac{\partial \mathcal{T}}{\partial \dot{\theta}_2}\right) &= \left[J_2^* + 2J_{23}^*\cos(\theta_3)\right]\ddot{\theta}_2 + \left[J_3^* + J_{23}^*\cos(\theta_3)\right]\ddot{\theta}_3 \\ &\quad + B_{23}(\theta_i)\dot{\theta}_2\dot{\theta}_3 + C_{23}(\theta_i)\dot{\theta}_3^2, \end{aligned}$$

where

$$\begin{aligned} B_{23}(\theta_i) &= -2J_{23}^*\sin(\theta_3), \\ C_{23}(\theta_i) &= -J_{23}^*\sin(\theta_3), \\ \frac{\partial \mathcal{T}}{\partial \theta_2} &= -C_{21}(\theta_i)\dot{\theta}_1^2, \end{aligned}$$

where

$$\begin{aligned} C_{21}(\theta_i) &= J_{12}^*\sin(\theta_2)\cos(\theta_2) + J_{13}^*\sin(\theta_2+\theta_3)\cos(\theta_2+\theta_3) \\ &\quad + J_{23}^*\cos(\theta_2)\sin(2\theta_2+\theta_3), \\ \frac{\partial \mathcal{U}}{\partial \theta_2} &= G_2(\theta_i) = g[m_2a_2 + m_3l_2]\cos(\theta_2) + m_3ga_3\cos(\theta_2+\theta_3), \\ \frac{d}{dt}\left(\frac{\partial \mathcal{T}}{\partial \dot{\theta}_3}\right) &= \left[J_2^* + J_{23}^*\cos(\theta_3)\right]\ddot{\theta}_2 + J_3^*\ddot{\theta}_3 + B_{32}^{(1)}(\theta_i)\dot{\theta}_2\dot{\theta}_3, \end{aligned}$$

where

$$\begin{aligned} B_{32}^{(1)}(\theta_i) &= -J_{23}^*\sin(\theta_3), \\ \frac{\partial \mathcal{T}}{\partial \theta_3} &= -C_{31}(\theta_i)\dot{\theta}_1^2 - C_{32}(\theta_i)\dot{\theta}_2^2 - B_{32}^{(2)}(\theta_i)\dot{\theta}_2\dot{\theta}_3, \end{aligned}$$

where

$$\begin{aligned} B_{32}^{(2)}(\theta_i) &= J_{23}^*\sin(\theta_3), \\ C_{31}(\theta_i) &= J_{13}^*\sin(\theta_2+\theta_3)\cos(\theta_2+\theta_3) + J_{23}^*\cos(\theta_2)\sin(\theta_2+\theta_3), \end{aligned}$$

$$C_{32}(\theta_i) = J_{23}^* \sin(\theta_3),$$

$$\frac{\partial \mathcal{U}}{\partial \theta_3} = G_3(\theta_i) = m_3 g a_3 \cos(\theta_2 + \theta_3).$$

The equations of motion are thus

$$\begin{cases} J_1^*(\theta_i)\ddot{\theta}_1 + B_{11}(\theta_i)\dot{\theta}_1\dot{\theta}_2 + B_{12}(\theta_i)\dot{\theta}_1\dot{\theta}_3 = M_1, \\ \left[J_2^* + 2J_{23}^*\cos(\theta_3)\right]\ddot{\theta}_2 + \left[J_3^* + J_{23}^*\cos(\theta_3)\right]\ddot{\theta}_3 + B_{23}(\theta_i)\dot{\theta}_2\dot{\theta}_3 \\ \quad + C_{23}(\theta_i)\dot{\theta}_3^2 + C_{21}(\theta_i)\dot{\theta}_1^2 + G_2(\theta_i) = M_2, \\ \left[J_2^* + J_{23}^*\cos(\theta_3)\right]\ddot{\theta}_2 + J_3^*\ddot{\theta}_3 + C_{31}(\theta_i)\dot{\theta}_1^2 + C_{32}(\theta_i)\dot{\theta}_2^2 + G_3(\theta_i) = M_3. \end{cases}$$

The equation of motion (3.49) is a set of nonlinear differential equations of order $2n$. In some instances it is written in a simpler form:

$$\mathbf{M}(\boldsymbol{\theta})\ddot{\boldsymbol{\theta}} + \mathbf{V}(\dot{\boldsymbol{\theta}}, \boldsymbol{\theta}) + \mathbf{G}(\boldsymbol{\theta}) = \mathbf{M}_\theta, \tag{3.51}$$

where all terms that above were included in the Coriolis and centrifugal terms are now all included in the vector function $\mathbf{V}(\dot{\boldsymbol{\theta}}, \boldsymbol{\theta})$.

By inverting the mass matrix and introducing the joint velocities as auxiliary coordinates, it is possible to transform it in the state space:

$$\dot{\mathbf{z}} = \mathbf{f}(z) + \mathbf{B}^*(\boldsymbol{\theta})u, \tag{3.52}$$

where

- $\mathbf{z} = [\dot{\boldsymbol{\theta}}^T \ \boldsymbol{\theta}^T]^T$ is the *state vector*, of order $2n$,
- vector

$$\mathbf{f}(z) = \begin{Bmatrix} -\mathbf{M}^{-1}(\boldsymbol{\theta})[\mathbf{B}(\boldsymbol{\theta})\{\dot{\boldsymbol{\theta}}_i\dot{\boldsymbol{\theta}}_j\} + \mathbf{C}(\boldsymbol{\theta})\{\dot{\boldsymbol{\theta}}^2\} + \mathbf{G}(\boldsymbol{\theta})] \\ \dot{\boldsymbol{\theta}} \end{Bmatrix} \tag{3.53}$$

  has $2n$ rows,
- matrix $\mathbf{B}^*(\boldsymbol{\theta})$ (not to be confused with matrix $\mathbf{B}(\boldsymbol{\theta})$) is the *input gain matrix* defined as

$$\mathbf{B}^*(\boldsymbol{\theta}) = \begin{bmatrix} \mathbf{M}^{-1}(\boldsymbol{\theta}) \\ \mathbf{0} \end{bmatrix} \tag{3.54}$$

  has $2n$ rows and n columns,
- $\mathbf{u}(t)$ is a vector with n rows containing inputs, i.e., in the present case, the joint torques.

Instead of writing the equations of motion in the joint space, it is possible to write them in the task space

$$\mathbf{M}_x(\boldsymbol{\theta})\ddot{\mathbf{X}} + \mathbf{V}_x(\dot{\boldsymbol{\theta}}, \boldsymbol{\theta}) + \mathbf{G}_x(\boldsymbol{\theta}) = \mathbf{F}, \tag{3.55}$$

where $\mathbf{X}$ is the position vector of the end effector (or the pose, if also the orientation is considered), $\mathbf{F}$ is the vector containing the force (and possibly the moments) applied to the end effector and matrices and vectors with subscript x are referred to the task space.

Since

$$\mathbf{M}_\theta = \mathbf{J}^T \mathbf{F} \tag{3.56}$$

if the Jacobian matrix can be inverted, (3.51) can be premultiplied by $\mathbf{J}^{-T}$, (the symbol $^{-T}$ indicates the inverse of the transpose), obtaining

$$\mathbf{J}^{-T}\mathbf{M}(\boldsymbol{\theta})\ddot{\boldsymbol{\theta}} + \mathbf{J}^{-T}\mathbf{V}(\dot{\boldsymbol{\theta}}, \boldsymbol{\theta}) + \mathbf{J}^{-T}\mathbf{G}(\boldsymbol{\theta}) = \mathbf{J}^{-T}\mathbf{M}_\theta = \mathbf{F}. \tag{3.57}$$

The acceleration in the joint and task space can be related to each other by writing (3.29)

$$\dot{\mathbf{X}} = \mathbf{J}(\boldsymbol{\theta})\dot{\boldsymbol{\theta}}$$

and differentiating with respect to time

$$\ddot{\mathbf{X}} = \dot{\mathbf{J}}(\boldsymbol{\theta})\dot{\boldsymbol{\theta}} + \mathbf{J}(\boldsymbol{\theta})\ddot{\boldsymbol{\theta}}. \tag{3.58}$$

Solving for the joint space accelerations

$$\ddot{\boldsymbol{\theta}} = \mathbf{J}^{-1}(\boldsymbol{\theta})\ddot{\mathbf{X}} - \mathbf{J}^{-1}(\boldsymbol{\theta})\dot{\mathbf{J}}(\boldsymbol{\theta})\dot{\boldsymbol{\theta}} \tag{3.59}$$

and substituting into (3.57)

$$\begin{aligned} &\mathbf{J}^{-T}\mathbf{M}(\boldsymbol{\theta})\mathbf{J}^{-1}(\boldsymbol{\theta})\ddot{\mathbf{X}} - \mathbf{J}^{-T}\mathbf{M}(\boldsymbol{\theta})\mathbf{J}^{-1}(\boldsymbol{\theta})\dot{\mathbf{J}}(\boldsymbol{\theta})\dot{\boldsymbol{\theta}} \\ &\quad + \mathbf{J}^{-T}\mathbf{V}(\dot{\boldsymbol{\theta}}, \boldsymbol{\theta}) + \mathbf{J}^{-T}\mathbf{G}(\boldsymbol{\theta}) = \mathbf{F}. \end{aligned} \tag{3.60}$$

By comparing (3.55) with (3.60) it follows

$$\begin{aligned} \mathbf{M}_x(\boldsymbol{\theta}) &= \mathbf{J}^{-T}\mathbf{M}(\boldsymbol{\theta})\mathbf{J}^{-1}(\boldsymbol{\theta}), \\ \mathbf{V}_x(\dot{\boldsymbol{\theta}}, \boldsymbol{\theta}) &= \mathbf{J}^{-T}\left[\mathbf{V}(\dot{\boldsymbol{\theta}}, \boldsymbol{\theta}) - \mathbf{M}(\boldsymbol{\theta})\mathbf{J}^{-1}(\boldsymbol{\theta})\dot{\mathbf{J}}(\boldsymbol{\theta})\dot{\boldsymbol{\theta}}\right], \\ \mathbf{G}_x(\boldsymbol{\theta}) &= \mathbf{J}^{-T}\mathbf{G}(\boldsymbol{\theta}). \end{aligned} \tag{3.61}$$

3.13 Low Level Control

The motion of the arm is determined by the actuators, which apply given forces or moments to the links. The result is a trajectory that should follow a pattern that can be pre-programmed or generated following some rules in the case of autonomous robots. In the case of a telemanipulator the trajectory is determined in real time by a human controller.

Modern robots are not controlled continuously, but at discrete time intervals: the trajectory generator supplies the coordinates of the end effector or the coordinates of the joints (depending whether it works in the task or in the joint, space) and the low level controller of the arm tries to comply with these instructions by applying the required forces to the various links. This is what is usually called *position control.*

There are instances, in particular when the end effector must follow the outer surface of a workpiece, where it is impossible to state a trajectory, and the control must be based on the force the end effector exert against the workspace. This way of operating is called *force control.*

It is possible to control some degrees of freedom following the force control strategy, while others are controlled in position: this is called *hybrid control.*

In any case, it is possible to use a linear control or a nonlinear control: owing to the nonlinearity of the system, theoretically a nonlinear control is needed. However, to simplify the control task, it is possible to use a linear control strategy, provided that it is robust enough to tolerate the changes in the linearized dynamics of the system due to the changes of the position of the links.

In this section a position control is assumed. The trajectory generator supplies the controller of the arm a number of subsequent sets of coordinates, which act as *reference inputs.* They may be in the task space or in the joint space, but in the former case the controller must perform the inverse kinematic computation to obtain the reference inputs for the joints.

3.13.1 Open Loop Control

Assume that the required trajectory in the joint space is defined by function $\boldsymbol{\theta}_r(t)$. It is possible to compute the joint torques $\mathbf{M}_\theta(t)$ using either (3.49) or (3.51). In this latter case the torques are

$$\mathbf{M}_\theta = \mathbf{M}(\boldsymbol{\theta}_r)\ddot{\boldsymbol{\theta}}_r + \mathbf{V}(\dot{\boldsymbol{\theta}}_r, \boldsymbol{\theta}_r) + \mathbf{G}(\boldsymbol{\theta}_r). \tag{3.62}$$

This type of control is said to be *open-loop* or *feedforward* control, since the actual trajectory is not measured and the torques exerted by the actuators on the joints are computed using a model of the robot. This type of control is bound to introduce errors in the actual motion of the robot, which are large (perhaps unacceptably so) if the model of the robot is inaccurate. In practice a fully feedforward cannot accommodate for the unavoidable unmodeled dynamics present in all actual machines and for the uncertainties of many parameters.

3.13.2 Closed-Loop Control

Robot controllers work mostly in *closed-loop* or *feedback*: the actual position and possibly also the velocity of the joints are measured, and the joint actuators are controlled so that the position, and possibly the velocity are corrected toward the values required to follow the stated trajectory.

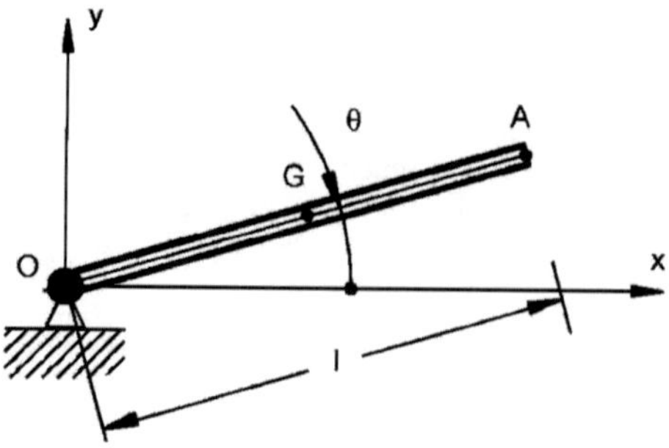

Fig. 3.16 Prismatic, homogeneous beam hinged at one end

Assume that each joint is provided with a sensor able to measure the coordinate θ_i and that the controller has the goal of obtaining the reference value θ_{i0} of such a coordinate.

An error can be defined as

$$\mathbf{e} = \boldsymbol{\theta} - \boldsymbol{\theta}_0. \tag{3.63}$$

If also the velocity of the joint is measured, and the reference input includes also the velocities, also the velocity error

$$\dot{\mathbf{e}} = \dot{\boldsymbol{\theta}} - \dot{\boldsymbol{\theta}}_0 \tag{3.64}$$

is known.

The simplest linear position controller is an ideal PID (proportional, integral, derivative) controller, i.e. a controller that causes the motors to supply a torque vector

$$\mathbf{M}_\theta = -\mathbf{K}_p\mathbf{e} - \mathbf{K}_d\dot{\mathbf{e}} - \mathbf{K}_i \int \mathbf{e}\, dt, \tag{3.65}$$

where $\mathbf{K}_p$, $\mathbf{K}_d$ and $\mathbf{K}_i$ are the matrices where the proportional, derivative and integrative control gains are listed. If they are diagonal, each degree of freedom is controlled independently from the others and the control is said to be *decentralized*.

Example 3.6 To understand the effect of the various control gains, consider the prismatic, homogeneous beam hinged at one end shown in Fig. 3.16. Study the cases in which the hinge is controlled by a PD and a PID controller. The data are $l = 1$ m, $m = 5$ kg, $g = 9.81$ m/s^2.

Since the beam is prismatic, the center of mass is at mid-length and the moment of inertia about the hinge is

$$J = \frac{ml^2}{3}.$$

The equation of motion is

$$J\ddot{\theta} + \frac{mgl}{2}\cos(\theta) = M.$$

The controller is required to bring the beam at the reference angle θ_0 and to keep it there.

PD controller

Using the expression of the error given by (3.63) and remembering that the reference θ_0 is constant, the control torque is

$$T = -K_p(\theta - \theta_0) - K_d\dot{\theta}.$$

The equation of motion of the controlled system is thus

$$J\ddot{\theta} + K_d\dot{\theta} + K_p\theta + \frac{mgl}{2}\cos(\theta) = K_p\theta_0.$$

The derivative gain plays the same role as that of a damping coefficient and the proportional gain as that of a stiffness. The larger is the second one, the quicker is the tendency toward the reference position, but also the stronger is the oscillatory behavior of the system. Derivative damping is needed to avoid strong oscillation. Neglecting the nonlinear part of the system, the natural frequency is

$$\omega_n = \sqrt{\frac{K_p}{J}}.$$

The damping ratio of the linearized system is

$$\zeta = \frac{K_d}{2\sqrt{K_p J}}.$$

These two relationships can be used to design the controller, i.e. to chose the values of the gains.
The position at rest can be computed by assuming that $\ddot{\theta}$ and $\dot{\theta}$ vanish, obtaining

$$\theta + \frac{mgl}{2K_p}\cos(\theta) - \theta_0.$$

A PD controller is then unable of reaching the reference position, if the system is subjected to external forces. The final position can be written as

$$\theta = \theta_0 + \Delta\theta,$$

where $\Delta\theta$ is the error in the final position. The equation yielding the equilibrium position can be written as

$$\Delta\theta + \frac{mgl}{2K_p}\cos(\theta_0 + \Delta\theta) = 0.$$

If $\Delta\theta$ is small,

$$\cos(\theta_0 + \Delta\theta) \approx \cos(\theta_0) - \Delta\theta\sin(\theta_0)$$

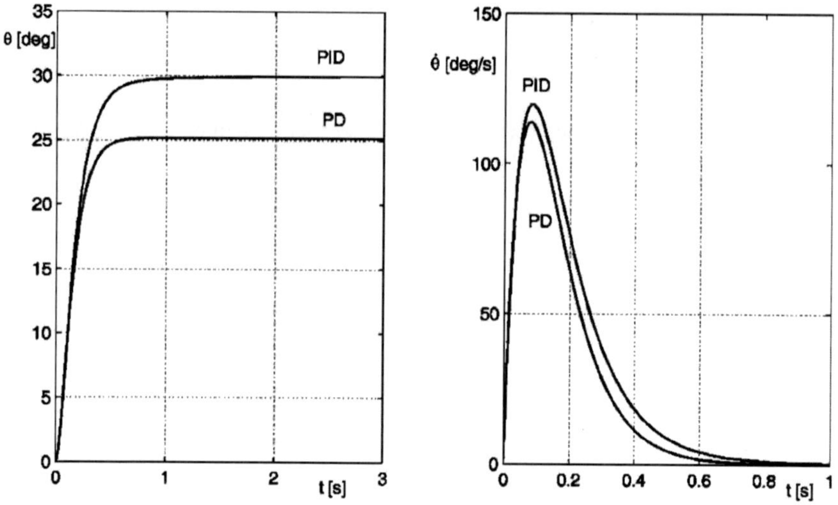

Fig. 3.17 Arm controlled using a PD and a PID control: time histories of the position and the velocity (the latter is reported for a shorter time). Reference value $\theta_0 = 30°$, initial value $\theta = 0$

and thus the error can be computed easily

$$\Delta\theta = -\frac{mgl\cos(\theta_0)}{2K_p - mgl\sin(\theta_0)}.$$

To compute the proportional gain assume a natural frequency of $\omega_n = 2$ Hz $= 12.57$ rad/s. The value of the proportional gain is thus $K_p = 263$ Nm/rad. Assuming that the system is critically damped ($\zeta = 1$), the value of the derivative gain is $K_d = 41.9$ Nm s/rad.

Assume a reference value is $\theta_0 = 30°$. The value of angle θ computed using the equation above is $\theta = 25.17°$, and then the steady-state error is $\Delta\theta = 4.83°$. The approximated value of $\Delta\theta$ obtained using the approximated relationship is 4.85°, a very good approximation.

Consider the arm at rest with $\theta = 0$ and apply the reference $\theta_0 = 30°$. The results of the numerical integration are reported in Fig. 3.17. The steady-state value of 25.17° is quickly reached.

The poles of the linearized system are $s_1 = -12.12$ 1/s and $s_2 = -13.02$ 1/s. They are almost equal, since the system is only very slightly overdamped ($\zeta = 1.0006$).

PID controller

Using again the expression of the error given by (3.63) and remembering that the reference θ_0 is constant, the control torque is

$$M_\theta = -K_p(\theta - \theta_0) - K_d\dot{\theta} - K_i\int_0^t (\theta - \theta_0)\,du.$$

The equation of motion of the controlled system is thus

$$\mathbf{J}\ddot{\theta} + K_d\dot{\theta} + K_p\theta + K_i\int_0^t \theta\,du + \frac{mgl}{2}\cos(\theta) = K_p\theta_0 + K_i\theta_0 t.$$

The equation of motion is then an integro-differential equation, and must be written in the state space. Introducing two auxiliary variables

$$v = \dot{\theta}, \qquad r = \int_0^t \theta\,du,$$

the equation becomes

$$J\dot{v} + K_d v + K_p\theta + K_i r + \frac{mgl}{2}\cos(\theta) = K_p\theta_0 + K_i\theta_0 t$$

or, in matrix form,

$$\begin{Bmatrix} \dot{v} \\ \dot{\theta} \\ \dot{r} \end{Bmatrix} = \begin{bmatrix} -\frac{K_d}{J} & -\frac{K_p}{J} & -\frac{K_i}{J} \\ 1 & 0 & 0 \\ 0 & 1 & 0 \end{bmatrix} \begin{Bmatrix} v \\ \theta \\ r \end{Bmatrix} + \begin{Bmatrix} \frac{mgl}{2J}\cos(\theta) + \frac{K_p}{J}\theta_0 + \frac{K_i}{J}\theta_0 t \\ 0 \\ 0 \end{Bmatrix}.$$

Add an integrative gain $K_i = 200$ Nm/rad s to the gains considered above, assume a reference value $\theta_0 = 30°$ and compute the time history of the angle with the arm starting with $\theta = 0$. The results of the numerical integration are reported in Fig. 3.17.

The steady-state value now coincides with the reference value and is quickly reached.

The equation of motion of the controlled system in the configuration space is

$$\begin{aligned} &\mathbf{M}(\boldsymbol{\theta})\ddot{\boldsymbol{\theta}} + \mathbf{V}(\dot{\boldsymbol{\theta}}, \boldsymbol{\theta}) + \mathbf{G}(\boldsymbol{\theta}) \\ &\quad = -\mathbf{K}_p(\boldsymbol{\theta} - \boldsymbol{\theta}_0) - \mathbf{K}_d(\dot{\boldsymbol{\theta}} - \dot{\boldsymbol{\theta}}_0) - \mathbf{K}_i\int(\boldsymbol{\theta} - \boldsymbol{\theta}_0)\,dt. \end{aligned} \tag{3.66}$$

The equation of motion can be written in the state space by introducing the auxiliary variables

$$\mathbf{v} = \dot{\boldsymbol{\theta}}, \qquad \mathbf{r} = \int_0^t \boldsymbol{\theta}\,du. \tag{3.67}$$

The state variables are thus $3n$.

Assuming a constant reference input $\boldsymbol{\theta}_0$, the state space equation of motion is

$$\begin{Bmatrix} \dot{\mathbf{v}} \\ \dot{\boldsymbol{\theta}} \\ \dot{\mathbf{r}} \end{Bmatrix} = \begin{bmatrix} -\mathbf{M}^{-1}\mathbf{K}_d & -\mathbf{M}^{-1}\mathbf{K}_p & -\mathbf{M}^{-1}\mathbf{K}_i \\ \mathbf{I} & \mathbf{0} & \mathbf{0} \\ \mathbf{0} & \mathbf{I} & \mathbf{0} \end{bmatrix} \begin{Bmatrix} \mathbf{v} \\ \boldsymbol{\theta} \\ \mathbf{r} \end{Bmatrix} + \begin{Bmatrix} \mathbf{M}^{-1}[\mathbf{V}+\mathbf{G}] \\ \mathbf{0} \\ \mathbf{0} \end{Bmatrix} + \begin{Bmatrix} \mathbf{M}^{-1}\mathbf{K}_p + \mathbf{M}^{-1}\mathbf{K}_i t \\ \mathbf{0} \\ \mathbf{0} \end{Bmatrix} \boldsymbol{\theta}_0. \tag{3.68}$$

Example 3.7 Consider the revolute arm studied in Example 3.5 (Fig. 3.15).

Study its motion in the vertical plane with angle θ_1 locked in any position when joints 2 and 3 are driven by a PID controller. Assume the following data: $l_1 = 1$ m, $l_2 = 0.6$ m, $m_1 = 5$ kg, $m_2 = 3$ kg, $g = 9.81$ m/s^2. Assume that the arms are prismatic and the hinges are at their ends. Assume that the booms can be modeled as one-dimensional objects (i.e., that their transversal dimensions are negligible) and, as a consequence, neglect the moment of inertia of the links about their longitudinal axis.

The moments of inertia of the links about their centers of mass are

$$J_{xi} = 0, \qquad J_{yi} = J_{zi} = \frac{m_i l_i^2}{12}.$$

Apply a decentralized controller, with proportional gains yielding natural frequencies of about 2 Hz and derivative gains yielding a damping close to critical.

Compute the time history of the generalized coordinates of the arm and the torques of the motors to perform a given manoeuver.

When the first link is locked, the mass matrix reduces to

$$M = \begin{bmatrix} J_2^* + 2J_{23}^* \cos(\theta_3) & J_3^* + J_{23}^* \cos(\theta_3) \\ J_3^* + J_{23}^* \cos(\theta_3) & J_3^* \end{bmatrix},$$

where

$$J_2^* = J_{z2} + J_{z3} + m_2 a_2^2 + m_3 \left(a_3^2 + l_2^2\right),$$

$$J_3^* = J_{z3} + m_3 a_3^2,$$

$$J_{23}^* = m_3 a_3 l_2.$$

Vector $\mathbf{V}(\dot{\boldsymbol{\theta}}_i \boldsymbol{\theta}_j) + \mathbf{G}(\boldsymbol{\theta}_j)$ reduces to

$$\mathbf{V}+\mathbf{G} = \begin{Bmatrix} -2J_{23}^* \sin(\theta_3)\dot{\theta}_2\dot{\theta}_3 - J_{23}^* \sin(\theta_3)(\theta_i)\dot{\theta}_3^2 + G_2 \\ J_{23}^* \sin(\theta_3)\dot{\theta}_2^2 + m_3 g a_3 \cos(\theta_2+\theta_3) \end{Bmatrix},$$

where

$$G_2 = g[m_2 a_2 + m_3 l_2]\cos(\theta_2) + m_3 g a_3 \cos(\theta_2+\theta_3).$$

The proportional gains are computed by assuming a natural frequency of 2 Hz for both uncoupled linearized systems. The two values of the gains so obtained are $K_{p1} = 794$ Nm/rad, $K_{p2} = 57$ Nm/rad.

Assuming a unit damping ratio for both uncoupled linearized systems it follows $K_{d1} = 126$ Nm s/rad, $K_{d2} = 9$ Nm s/rad.

The integrative gains are arbitrarily assumed: $K_{i1} = 1000$ Nm/rad s, $K_{i2} = 120$ Nm/rad s.

The matrices and vectors to be introduced into (3.65) are thus

$$\mathbf{K}_p = \begin{bmatrix} K_{p1} & 0 \\ 0 & K_{p2} \end{bmatrix}, \qquad \mathbf{K}_d = \begin{bmatrix} K_{d1} & 0 \\ 0 & K_{d2} \end{bmatrix},$$

$$\mathbf{K}_i = \begin{bmatrix} K_{i1} & 0 \\ 0 & K_{i2} \end{bmatrix}, \qquad \mathbf{e} = \begin{Bmatrix} \theta_1 - \theta_{10} \\ \theta_2 - \theta_{20} \end{Bmatrix}.$$

The equation of motion can be written in the state space by introducing the auxiliary variables of (3.67). The state variables to be introduced into the state equation (3.68) are thus six.

Since the equation is nonlinear, numerical integration in time must be performed. As an example, a maneuver aimed to bring the end effector to point with coordinates $x = 300$ mm, $y = 800$ mm starting from an horizontal position will be performed.

Since the destination point is within the workspace, the manoeuver is possible. The end values of the generalized coordinates can be computed from the inverse kinematic relationships and are, in degrees,

$$\boldsymbol{\theta}_0 = \begin{Bmatrix} 106.1 \\ -121.7 \end{Bmatrix}.$$

The steady-state torques at the motors are those needed to counterbalance the weight of the arms. They can be computed from the equation of motion by stating that all derivatives of the generalized coordinates vanish:

$$\begin{cases} M_{\theta 1} = g\big[(m_1 d_1 + m_2 l_1)\cos(\theta_1) + m_2 d_2 \cos(\theta_1 + \theta_2)\big], \\ M_{\theta 2} = g m_2 d_2 \cos(\theta_1 + \theta_2). \end{cases}$$

Using the values of $\boldsymbol{\theta}_0$ mentioned above, the values of the motor torques are $M_{\theta 1} = -6.50$ Nm and $M_{\theta 2} = 8.51$ Nm.

The time history of the torques can be obtained simply by extracting vector $\mathbf{M}_\theta$ during the numerical integration.

The results of the numerical integration are reported in Fig. 3.18.

The steady-state values now coincide with the reference values and the final position is quickly reached without oscillation, although with an overshot.

The trajectory of the end point and of the elbow are plotted in Fig. 3.19.

The control of the arm was performed giving the final position as a reference input, without any attempt to define the trajectory of the end effector.

This way of controlling an arm is very rough and can be used only as a simplified example: in any actual application the low level controller receives its reference from an higher level controller, the trajectory generator.

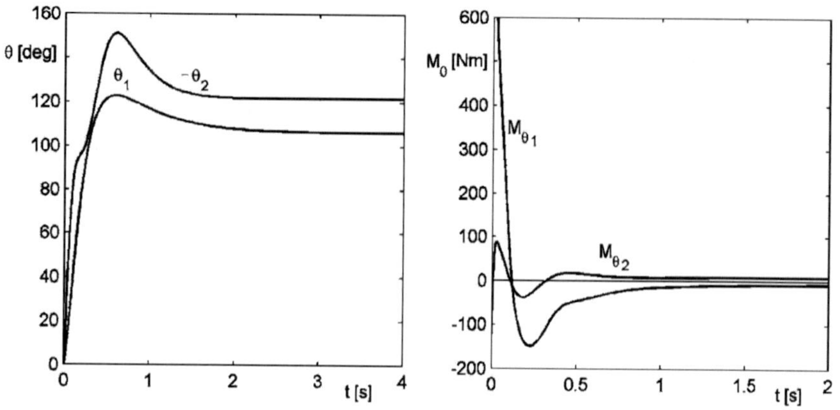

Fig. 3.18 Planar arm with two degrees of freedom controlled using a PID control. Final position $x = 300$ mm, $y = 80$ mm. Time histories of the angles and of the motor torques

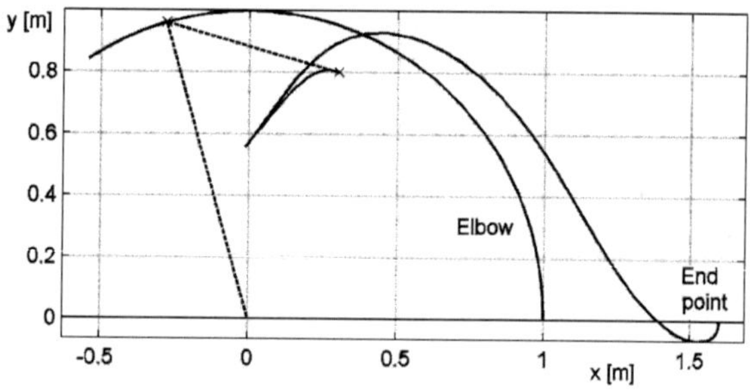

Fig. 3.19 Trajectory of the end point of the arm and of the elbow in the maneuver described in the previous figure

The motors are required to supply large torques during the first phase of the manoeuver, when the errors are large, and later they settle at about the steady-state values that are reached within a few seconds.

Example 3.8 Repeat the computations of the previous example, assuming that the motors cannot supply the large torques needed to follow the PID strategy.

Assume the following values for the saturation torques: $M_{\theta 1_{\max}} = 110$ Nm, $M_{\theta 2_{\max}} = 20$ Nm.

The problem becomes more nonlinear and, as a consequence, the gains cannot be introduced in the dynamic matrix. The motor torques $\mathbf{M}_\theta$ must be computed explicitly at each integration step, taking into account also saturation, and then introduced into the state space equation

$$\begin{Bmatrix} \dot{\mathbf{v}} \\ \dot{\boldsymbol{\theta}} \\ \dot{\mathbf{r}} \end{Bmatrix} = \begin{bmatrix} \mathbf{0} & \mathbf{0} & \mathbf{0} \\ \mathbf{I} & \mathbf{0} & \mathbf{0} \\ \mathbf{0} & \mathbf{I} & \mathbf{0} \end{bmatrix} \begin{Bmatrix} \mathbf{v} \\ \boldsymbol{\theta} \\ \mathbf{r} \end{Bmatrix} + \begin{Bmatrix} \mathbf{M}^{-1}\mathbf{f} \\ \mathbf{0} \\ \mathbf{0} \end{Bmatrix} + \begin{Bmatrix} \mathbf{M}^{-1}\mathbf{M}_\theta \\ \mathbf{0} \\ \mathbf{0} \end{Bmatrix}.$$

The results are shown in Fig. 3.20.

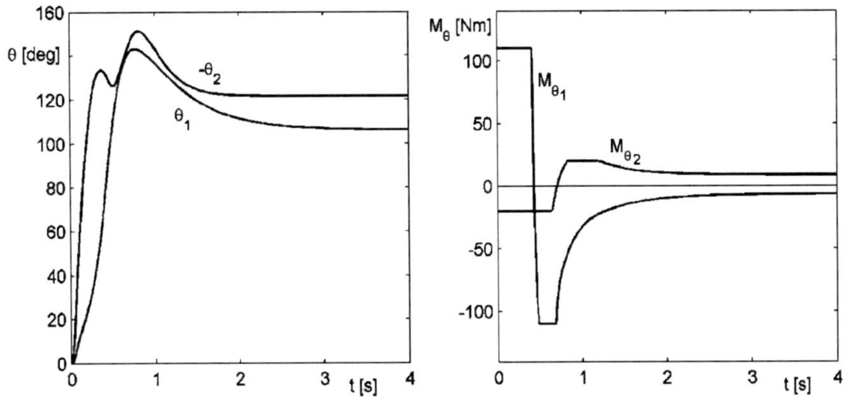

Fig. 3.20 Planar arm with two degrees of freedom controlled using a PID control with saturation. Final position $x = 300$ mm, $y = 80$ mm. Time history of the angles and of the motor torques

Saturation slows down the reaching of the final position, as expected, but its effect is mainly restricted to the first moments. After the arm has started, the effect of saturation on the motion reduces, while the effect on the torques remains large.

3.13.3 Model-Based Feedback Control

The equation of motion (3.51) can be used to improve the performance of the controller. Since at each instant the values of the joint coordinates are known, the actuators can supply a set of control moments equal to $\mathbf{G}(\boldsymbol{\theta})$: this amounts to have the actuators exerting torques compensating for the gravitational forces. In a similar way, if also the joint velocities are known, the actuators can supply a set of torques equal to $\mathbf{V}(\dot{\boldsymbol{\theta}}_i, \boldsymbol{\theta})$, compensating for Coriolis and centrifugal forces too. The joint torques are thus

$$\mathbf{M}_\theta = \mathbf{V}(\dot{\boldsymbol{\theta}}, \boldsymbol{\theta}) + \mathbf{G}(\boldsymbol{\theta}) + \mathbf{M}_c, \tag{3.69}$$

where $\mathbf{M}_c$ are the control torques, i.e. the torques exerted by the actuators that exceed those needed to compensate for gravitational, Coriolis and centrifugal forces. In the simplest case, the control torques can be obtained through a PID algorithm (Fig. 3.21) and thus the joint torques are

$$\mathbf{M}_\theta = \mathbf{V}(\dot{\boldsymbol{\theta}}, \boldsymbol{\theta}) + \mathbf{G}(\boldsymbol{\theta}) - \mathbf{K}_p\mathbf{e} - \mathbf{K}_d\dot{\mathbf{e}} - \mathbf{K}_i \int \mathbf{e}\,dt, \tag{3.70}$$

The equation of motion thus reduces to

$$\mathbf{M}(\boldsymbol{\theta})\ddot{\boldsymbol{\theta}} = M_c. \tag{3.71}$$

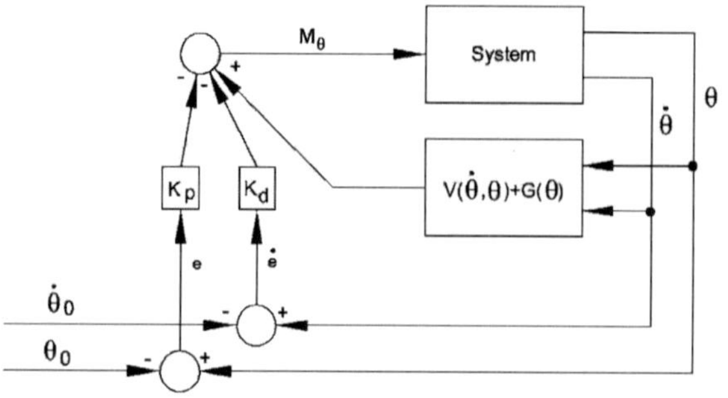

Fig. 3.21 Block diagram of a model-based controller. To simplify the scheme, the control is based on a PD strategy

The dynamics of the controlled system is thus described by the linear equation

$$\ddot{\boldsymbol{\theta}} = \mathbf{M}^{-1}(\boldsymbol{\theta})\mathbf{M}_c. \tag{3.72}$$

To perform this model-based control two conditions must be verified: the controller must be able to perform all the required computations in real time and the characteristics of the system must be known accurately. The first condition is now not difficult to be implemented, owing to the power of the microprocessors used to control robots, even if it must be remembered that the computers used in space applications are much less advanced than those used in industrial plants. A long time usually elapses before computers are space qualified and in applications beyond Earth orbit radiation hardened electronics must be used.

The second point is much more critical. As a first point, there are details of the dynamic of the arm that are difficult to model accurately. For instance, the resistance to motion of the joint has not been introduced explicitly in the models above: it was assumed implicitly that all friction and resistance torques in the joints were included in the joint torques $\mathbf{M}_\theta$. When building the mathematical model for the control of the arm, these effects must be included, and this can introduce large errors.

Another source of model uncertainties is the variability of the mass of the manipulator when the gripper picks up an object. Since the mass of the object may, in general, be unknown, this introduces an unmodeled dynamics.

However, even if the model-based controller does not completely compensate for nonlinearities, it makes anyway the system more linear than if no compensation at all were used, and the feedback control is usually able to manage the situation if its design is robust enough.

3.13.4 Mixed Feedforward and Feedback Control

The two basic strategies of feedforward and feedback control can be mixed to different extents. Assume for instance that the required trajectory in the joint space is defined by function $\boldsymbol{\theta}_r(t)$. In an open loop strategy the joint torques $\mathbf{M}_\theta(t)$ were computed using (3.62) from the desired trajectory. It is possible to add a feedback component to the open loop component, obtaining

$$\mathbf{M}_\theta = \mathbf{M}(\boldsymbol{\theta}_r)\ddot{\boldsymbol{\theta}}_r + \mathbf{V}(\dot{\boldsymbol{\theta}}_r, \boldsymbol{\theta}_r) + \mathbf{G}(\boldsymbol{\theta}_r) - \mathbf{K}_p\mathbf{e} - \mathbf{K}_d\dot{\mathbf{e}} - \mathbf{K}_i \int \mathbf{e}\,dt, \tag{3.73}$$

where the error is the same defined above and the feedback strategy is based on a PID controller.

While in the strategy described in the previous section the torque $\mathbf{V}(\dot{\boldsymbol{\theta}}, \boldsymbol{\theta}) + \mathbf{G}(\boldsymbol{\theta})$ was computed using the actual (measured) values of the joint coordinates and velocities, now the values used are the reference ones.

The equation of motion of the controlled system is

$$\begin{aligned}
&\mathbf{M}(\boldsymbol{\theta})\ddot{\boldsymbol{\theta}} + \mathbf{V}(\dot{\boldsymbol{\theta}}, \boldsymbol{\theta}) + \mathbf{G}(\boldsymbol{\theta}) \\
&\quad = \mathbf{M}(\boldsymbol{\theta}_r)\ddot{\boldsymbol{\theta}}_r + \mathbf{V}(\dot{\boldsymbol{\theta}}_r, \boldsymbol{\theta}_r) + \mathbf{G}(\boldsymbol{\theta}_r) \\
&\qquad - \mathbf{K}_p(\boldsymbol{\theta} - \boldsymbol{\theta}_r) - \mathbf{K}_d(\dot{\boldsymbol{\theta}} - \dot{\boldsymbol{\theta}}_r) - \mathbf{K}_i \int (\boldsymbol{\theta} - \boldsymbol{\theta}_r)\,dt.
\end{aligned} \tag{3.74}$$

A strategy of this kind can work only if the error is, in each instant of the motion, small enough to avoid that the feedforward control introduces forces that may cause further errors. For instance, it is impossible to use a strategy of this kind when giving directly the end position as a reference input at the beginning of the motion.

3.14 Trajectory Generation

In the previous examples no attempt to follow a given trajectory is done: only the point the end effector has to reach is defined, and the arm is left free to reach it (asymptotically) in any way. Usually a better definition of the motion is needed, and a trajectory must be computed, at least by defining a number of waypoints.

The simplest trajectory is a straight line in the physical space, with the end effector accelerating at constant acceleration for the first half of the motion and then decelerating at a constant deceleration, equal in absolute value to the previous one. This is possible only if the straight line connecting the starting and ending points (A and B) is all within the workspace and does not go through any singular points.

Assume that at time $t = 0$ the end effector is in A (position $\mathbf{X}_A$) and at time t_f it stops in B (position $\mathbf{X}_B$).

Owing to the assumption of constant acceleration $\mathbf{a}$, the velocity is easily computed

$$\begin{aligned}
&\dot{\mathbf{X}} = \mathbf{a}t \quad \text{for } 0 \le t \le \frac{t_f}{2}, \\
&\dot{\mathbf{X}} = \mathbf{a}\left(\frac{t_f}{2} - t + \frac{t_f}{2}\right) = \mathbf{a}(t_f - t) \quad \text{for } \frac{t_f}{2} \le t \le t_f.
\end{aligned} \tag{3.75}$$

By integrating, it follows that

$$\mathbf{X} = \mathbf{X}_\mathrm{A} + \frac{1}{2}\mathbf{a}t^2 \quad \text{for } 0 \le t \le \frac{t_f}{2},$$
$$\mathbf{X} = \mathbf{X}_\mathrm{A} + \mathbf{a}\left(tt_f - \frac{1}{4}t_f^2 - \frac{1}{2}t^2\right) \quad \text{for } \frac{t_f}{2} \le t \le t_f. \tag{3.76}$$

From the last equation it follows that

$$\mathbf{X}_\mathrm{B} = \mathbf{X}_\mathrm{A} + \frac{1}{4}\mathbf{a}t_f^2, \tag{3.77}$$

i.e.

$$\mathbf{a} = \frac{4(\mathbf{X}_\mathrm{A} - \mathbf{X}_\mathrm{B})}{t_f^2}. \tag{3.78}$$

The trajectory of the end effector can thus be computed, and easily transformed into the joint space.

The controller is supplied a variable reference in terms of $\boldsymbol{\theta}_0(t)$ and of $\dot{\boldsymbol{\theta}}_0(t)$. When using a PID controller, the variability of the reference input must be accounted for in the computation of the integral error, and thus the state space equation of motion (3.68) becomes

$$\begin{Bmatrix} \dot{\mathbf{v}} \\ \dot{\boldsymbol{\theta}} \\ \dot{\mathbf{r}} \end{Bmatrix} = \begin{bmatrix} -\mathbf{M}^{-1}\mathbf{K}_d & -\mathbf{M}^{-1}\mathbf{K}_p & -\mathbf{M}^{-1}\mathbf{K}_i \\ \mathbf{I} & \mathbf{0} & \mathbf{0} \\ \mathbf{0} & \mathbf{I} & \mathbf{0} \end{bmatrix} \begin{Bmatrix} \mathbf{v} \\ \boldsymbol{\theta} \\ \mathbf{r} \end{Bmatrix} + \begin{Bmatrix} \mathbf{M}^{-1}[\mathbf{V} + \mathbf{G}] \\ \mathbf{0} \\ \mathbf{0} \end{Bmatrix} + \begin{Bmatrix} \mathbf{M}^{-1}[\mathbf{K}_p\boldsymbol{\theta}_0(t) + \mathbf{K}_p\dot{\boldsymbol{\theta}}_0(t) + \mathbf{K}_i \int \boldsymbol{\theta}_0(t)\,dt] \\ \mathbf{0} \\ \mathbf{0} \end{Bmatrix}. \tag{3.79}$$

The time history of the trajectory assumed above is quadratic in time, so that the resulting acceleration is constant. Often this strategy is referred to as *bang–bang* control, and can be shown to yield the fastest motion for a given maximum value of the acceleration. However, it causes abrupt changes of acceleration (i.e. a theoretically infinite value of the *jerk*) at the beginning and at the end of the motion and when the shift from acceleration to braking occurs. To obtain a more gradual start and stop of the arm a time history that contains higher powers of time can be used, for instance a time history that is cubic in time.

In a similar way, also a trajectory that is not straight can be assumed. A common way to define smooth curved trajectories is by stating a number of waypoints and then by resorting to cubic splines passing through them.

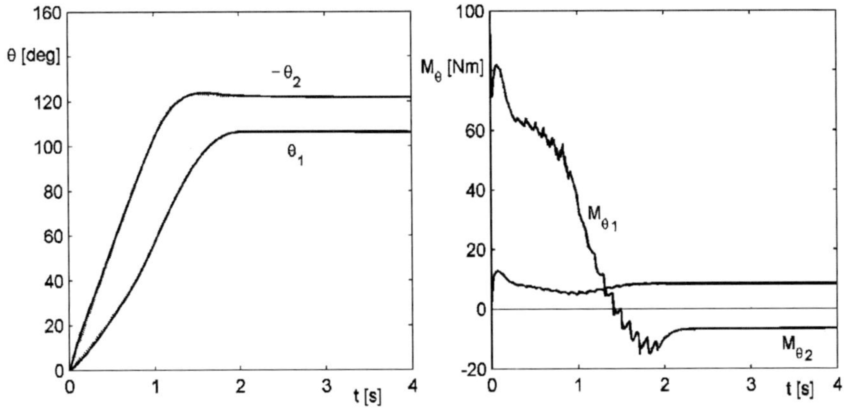

Fig. 3.22 Same arm with two degrees of freedom of the previous examples controlled using a PID control along a rectilinear trajectory of the end effector. Time histories of the angles and of the motor torques

Example 3.9 Repeat the computations of Example 3.7, assuming that the end effector must move between the initial and the end point along a straight line, in a time of 2 s.

The acceleration is easily computed

$$\mathbf{a} = \begin{Bmatrix} -1.3 \\ 0.8 \end{Bmatrix}.$$

Assume $[1{,}600\ 0]^T$ mm as coordinates of point A and $[300\ 800]^T$ mm as coordinates of point B and assume a duration of 2 s for the manoeuver. Compute the input reference in both displacement and velocity every 0.1 s, and supply these values to the controller of the arm.

Larger values of the gains are assumed, since now the arm must follow more promptly the reference input: $k_{p1} = 3{,}200$ Nm/rad, $k_{p2} = 227$ Nm/rad, $k_{d1} = 2{,}500$ Nm s/rad, $k_{d2} = 18$ Nm/rad, $k_{i1} = 2{,}000$ Nm/rad s, $k_{i2} = 240$ Nm/rad s.

The results are plotted in Fig. 3.22. In the figure the laws $\boldsymbol{\theta}_{01}(t)$ and $\boldsymbol{\theta}_{02}(t)$ are also reported as dotted lines, but the actual trajectories in the joint space are almost completely superimposed to the reference trajectories. The joint torques, and above all $M_{\theta 1}$, show a strong ripple: this is due to the fact that the reference input is given every 0.1 s and not in a continuous way. As expected, the joint torques are much smaller than those obtained by supplying the endpoint coordinates directly to the controller, leaving the arm free to move for the whole time.

The trajectory is shown in Fig. 3.23. The reference trajectory (a straight line) is plotted as a dotted line: the end effector follows quite accurately the reference trajectory.

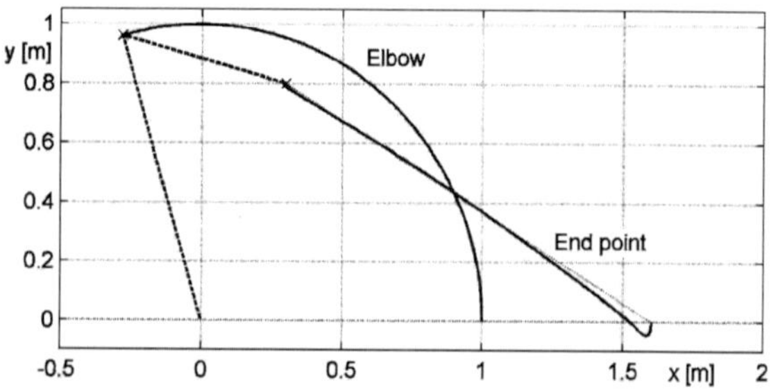

Fig. 3.23 Trajectory of the end point of the arm and of the elbow in the maneuver described in the previous figure

Remark 3.9 The trajectory can be much more complex than a straight line: several waypoints can be stated, and the trajectory can be obtained using splines or other geometrical interpolation curves.

3.15 Dynamics of Flexible Arms

No truly rigid body exists in the actual world and robot arms are no exception. In robotic arms two kinds of flexibility may be present: the flexibility of the structural parts like the beams constituting the various parts of the arm and that of the joints and the actuators. While the structural elements can often be modeled as linear systems, the joints contain many elements whose behavior is more or less nonlinear: bearings, gear wheels, chains, etc. In particular, clearances and nonlinear elasticity, possibly due to contact phenomena, are the main causes of nonlinearities.

In most cases the flexibility of the robot structure is neglected. This can be done if the structure is stiff enough so that the natural frequencies linked to deformation modes are much higher than those linked with the rigid-body and control dynamics. As a general rule, the lowest flexible body natural frequency must be at least twice the highest natural frequency linked with the controlled rigid manipulator.

While this is easily implemented for industrial robots, it is less easy for moving robots and above all for space robots. Very rigid arms are quite heavy and this requirement conflicts with the needs of weight reduction in all movable machines and, even more, in space hardware.

Some space manipulators have quite a long arm, which must operate in low gravity (or microgravity) conditions. The static loads are low and this allows one to design lightweight structures, that in turn are much more flexible. If also the control is designed to keep its natural frequencies correspondingly low, a very slow manipulator is obtained.

The needs of improving the performance of the manipulator while decreasing its mass result in the impossibility of neglecting the flexibility of the structural elements: the compliance of the arms and the joints must be accounted for both in the design of the mechanical subsystems and in the design of the control.

The dynamics of a flexible body can be studied by modeling it as a continuous system or by resorting to some discretization technique, like the Finite Element Method (FEM).

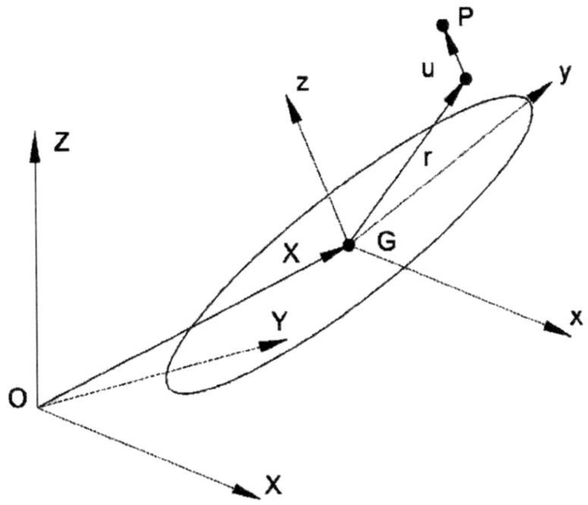

Fig. 3.24 Compliant link. Position of point P belonging to the ith link in the body fixed and base frames

Consider the generic link of a kinematic chain (the ith link, Fig. 3.24). At time t the position and the velocity of its center of mass, the rotation matrix and the angular velocity of the body-fixed reference frame are given by **X**, **V**, **R**, and **Ω**. The body-fixed frame may be a principal frame of inertia of the link, but this is actually not needed.

The position of a generic point P of the link in its deformed configuration is

$$\overline{(\mathrm{P}-\mathrm{O})} = \mathbf{X} + \mathbf{R}(\mathbf{r} + \mathbf{u}), \tag{3.80}$$

where

- $\mathbf{X} = [X\ Y\ Z]_{\mathrm{G}}^T$ is the vector defining the position of the center of mass G of the link in the base frame. Generally speaking, it is a function of the joint variables of the first i links, plus the deformation coordinates expressing the deformation of the endpoints of the previous $n-1$ links. In the three-dimensional space, the latter are generally speaking $6(n-1)$, if rotations are accounted for, and are usually considered as small quantities. Storing these variables in vector $\mathbf{x}_d$, it is possible to define an augmented vector of generalized coordinates

$$\boldsymbol{\theta}^* = \left[\boldsymbol{\theta}^T \quad \mathbf{x}_d^T\right]^T,$$

containing all coordinates defining the position of the center of mass of the ith link

$$\mathbf{X} = \mathbf{X}(\boldsymbol{\theta}^*). \tag{3.81}$$

- **R** is the rotation matrix of the link. It is a function of the same variables as **X**.
- $\mathbf{r} = [x_i\ y_i\ z_i]_{\mathrm{P}}^T$ is the vector defining the position of point P in its reference position, usually corresponding to the undeformed configuration if the link is not a rigid body. It is expressed in the body-fixed frame, which in the present case has its origin in the center of mass of the link.

- $\mathbf{u} = [u_x \ u_y \ u_z]_P^T$ is the displacement vector of the same point expressed in the same frame. In many cases, the displacement $\mathbf{u}$ may be considered as a small displacement.

The generalized coordinates of each point of the link are thus

$$\mathbf{q} = \begin{bmatrix} \boldsymbol{\theta}^{*T} & \mathbf{u}^T \end{bmatrix}^T. \tag{3.82}$$

The discrete part of the link has the same number of degrees of freedom as the components of vector $\boldsymbol{\theta}^*$, the joint coordinates of the links from the first to the ith, plus another $6(n-1)$ coordinates $\mathbf{x}_d$. Vector $\mathbf{u}(x, y, z, t)$ contains the generalized coordinates of the continuous part and, in three-dimensional space, has three components.

Remark 3.10 This formulation of the problem is partially written in terms of a discrete system and partially in terms of a continuous system. This formulation is usually referred to as hybrid.

The velocity of point P expressed in the body-fixed frame is

$$\mathbf{V}_P = \mathbf{R}^T\dot{\mathbf{X}} + \boldsymbol{\Omega}\Lambda(\mathbf{r} + \mathbf{u}) + \dot{\mathbf{u}}. \tag{3.83}$$

By expressing the vector product in matrix notation, it follows that

$$\mathbf{V}_P = \mathbf{V} + (\tilde{\mathbf{r}} + \tilde{\mathbf{u}})^T\boldsymbol{\Omega} + \dot{\mathbf{u}}, \tag{3.84}$$

where

$$\tilde{\mathbf{r}} = \begin{bmatrix} 0 & -z & y \\ z & 0 & -x \\ -y & x & 0 \end{bmatrix}, \tag{3.85}$$

and $\tilde{\mathbf{u}}$ is defined in a similar way.

Velocities $\dot{\mathbf{X}}$ and $\boldsymbol{\Omega}$ can be written as

$$\dot{\mathbf{X}} = P_1(\boldsymbol{\theta}^*)\dot{\boldsymbol{\theta}}^*, \qquad \boldsymbol{\Omega} = \mathbf{P}_2(\boldsymbol{\theta}^*)\dot{\boldsymbol{\theta}}^*. \tag{3.86}$$

The velocity can be expressed in terms of $\dot{\boldsymbol{\theta}}^*$ as

$$\mathbf{V}_P = \mathbf{R}^T\mathbf{P}_1\dot{\boldsymbol{\theta}}^* + (\tilde{\mathbf{r}} + \tilde{\mathbf{u}})^T\mathbf{P}_2(\boldsymbol{\theta}^*)\dot{\boldsymbol{\theta}}^* + \dot{\mathbf{u}}. \tag{3.87}$$

The kinetic energy of the infinitesimal volume dv about point P is

$$\begin{aligned} dT &= \frac{1}{2}\rho\mathbf{V}_P^T\mathbf{V}_P\,dv \\ &= \frac{1}{2}\rho\big[\dot{\boldsymbol{\theta}}^{*T}\mathbf{P}_1^T\mathbf{P}_1\dot{\boldsymbol{\theta}}^* + \dot{\boldsymbol{\theta}}^{*T}\mathbf{P}_2^T(\tilde{\mathbf{r}}\tilde{\mathbf{r}}^T + 2\tilde{\mathbf{r}}\tilde{\mathbf{u}}^T + \tilde{\mathbf{u}}\tilde{\mathbf{u}}^T)\mathbf{P}_2\dot{\boldsymbol{\theta}}^* \end{aligned}$$

$$+\dot{\mathbf{u}}^T\dot{\mathbf{u}}+2\dot{\boldsymbol{\theta}}^{*T}\mathbf{P}_1^T\mathbf{R}(\tilde{\mathbf{r}}^T+\tilde{\mathbf{u}}^T)\mathbf{P}_2\dot{\boldsymbol{\theta}}^*+2\dot{\boldsymbol{\theta}}^{*T}\mathbf{P}_1^T\mathbf{R}\dot{\mathbf{u}}$$

$$+2\dot{\boldsymbol{\theta}}^{*T}\mathbf{P}_2^T(\tilde{\mathbf{r}}+\tilde{\mathbf{u}})\dot{\mathbf{u}}\big]\,dv. \tag{3.88}$$

The kinetic energy of the link is thus

$$\mathcal{T}=\frac{1}{2}m\dot{\boldsymbol{\theta}}^{*T}\mathbf{P}_1^T\mathbf{P}_1\dot{\boldsymbol{\theta}}^*+\frac{1}{2}\dot{\boldsymbol{\theta}}^{*T}\mathbf{P}_2^T\mathbf{J}\mathbf{P}_2\dot{\boldsymbol{\theta}}^*+\frac{1}{2}\int_v\rho\dot{\mathbf{u}}^T\dot{\mathbf{u}}\,dv$$

$$+\frac{1}{2}\dot{\boldsymbol{\theta}}^{*T}\mathbf{P}_2^T\left(\int_v\rho(2\tilde{\mathbf{r}}+\tilde{\mathbf{u}})\tilde{\mathbf{u}}^T\,dv\right)\mathbf{P}_2\dot{\boldsymbol{\theta}}^*+\dot{\boldsymbol{\theta}}^{*T}\mathbf{P}_1^T\mathbf{R}\left(\int_v\rho\tilde{\mathbf{u}}^T\,dv\right)\mathbf{P}_2(\boldsymbol{\theta}^*)\dot{\boldsymbol{\theta}}$$

$$+\dot{\boldsymbol{\theta}}^{*T}\mathbf{P}_1^T\mathbf{R}\int_v\rho\dot{\mathbf{u}}\,dv+\dot{\boldsymbol{\theta}}^{*T}\mathbf{P}_2^T\int_v\rho(\tilde{\mathbf{r}}+\tilde{\mathbf{u}})\dot{\mathbf{u}}\,dv. \tag{3.89}$$

The global inertia properties of the link in the reference (undeformed) configuration are

$$m=\int_v\rho\,dv,\qquad \mathbf{J}=\int_v\tilde{\mathbf{r}}_c\tilde{\mathbf{r}}_c^T\rho\,dv.$$

Moreover, since point G is the mass center of the link,

$$\int_v\rho\tilde{\mathbf{r}}^T\,dv=0.$$

Given a set of virtual displacements $\delta\mathbf{X}$, $\delta\theta$ and $\delta\mathbf{u}$, the virtual displacements δx_{P}, expressed in the body-fixed frame, is

$$\delta x_{\mathrm{P}}=\mathbf{R}^T\mathbf{P}_1\delta\boldsymbol{\theta}^*+(\tilde{\mathbf{r}}+\tilde{\mathbf{u}})^T\mathbf{P}_2\delta\boldsymbol{\theta}^*+\delta\mathbf{u}. \tag{3.90}$$

The virtual work of a distributed force $\mathbf{f}(x,y,z,t)$ applied to the flexible link in the direction of the axes of the body-fixed frame is

$$\delta\mathcal{L}-\int_v\delta x_{\mathrm{P}}^T\mathbf{f}\,dv=\int_v\left(\delta\boldsymbol{\theta}^{*T}\mathbf{P}_1^T\mathbf{R}\mathbf{f}+\delta\boldsymbol{\theta}^{*T}\mathbf{P}_2^T(\mathbf{r}+\mathbf{u})\mathbf{f}+\delta\mathbf{u}^T\mathbf{f}\right)dv, \tag{3.91}$$

where v is the volume occupied by the compliant part of the system.

If a concentrated force $\mathbf{F}$ and a moment $\mathbf{M}$ act on the link (the force is applied in its center of mass G), the total virtual work acting on the system is thus

$$\delta\mathcal{L}=\delta\boldsymbol{\theta}^{*T}\mathbf{P}_1^T\mathbf{R}\mathbf{F}+\delta\boldsymbol{\theta}^{*T}\mathbf{P}_2^T(\tilde{\mathbf{r}}_c\mathbf{F}+\mathbf{M})$$

$$+\int_v\left[\boldsymbol{\theta}^{*T}\mathbf{P}_1^T\mathbf{R}\mathbf{f}+\delta\boldsymbol{\theta}^{*T}\mathbf{P}_2^T(\tilde{\mathbf{r}}+\tilde{\mathbf{u}})\mathbf{f}+\delta\mathbf{u}^T\mathbf{f}\right]dv. \tag{3.92}$$

The gravitational potential energy of the infinitesimal volume dv about point P is

$$d\mathcal{U}_g=-\mathbf{g}^T\left[\mathbf{X}+\mathbf{R}(\mathbf{r}+\mathbf{u})\right]dm. \tag{3.93}$$

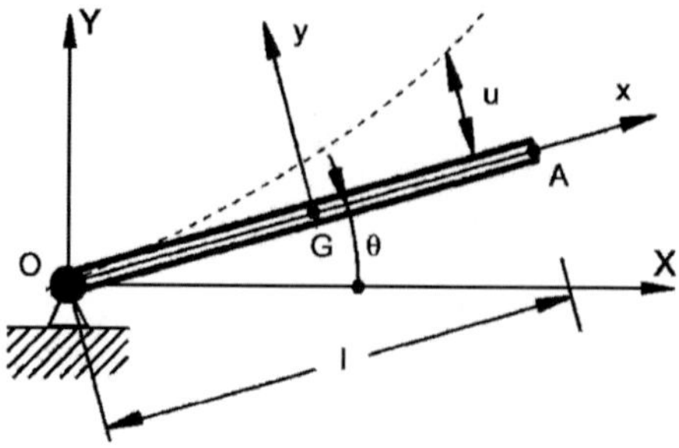

Fig. 3.25 Rotating elastic beam

Since point G is the center of mass of the link,

$$\int_v \rho \mathbf{r}\, dv = 0 \tag{3.94}$$

and, by integrating on the whole body, it follows that

$$\mathcal{U}_g = -m\mathbf{g}^T\mathbf{X} - \mathbf{g}^T\mathbf{R}\int_v \rho \mathbf{u}\, dv. \tag{3.95}$$

The elastic potential energy due to the deformation of the rigid body is not affected by generalized coordinates $\boldsymbol{\theta}^*$, but only by the coordinates $\mathbf{u}$ and their derivatives with respect to the spatial coordinates.

Remark 3.11 Usually the first and the second derivative $\mathbf{u}'$ and $\mathbf{u}''$ are included, but there may be cases in which also higher order derivatives are present

$$\mathcal{U}_e = \mathcal{U}_e(\mathbf{u}, \mathbf{u}', \mathbf{u}''). \tag{3.96}$$

It may be necessary to take into account also nonlinear terms in the strains to include effects like the influence of inertia forces on the elastic behavior of some parts of the system. This may be accounted for easily through a geometric matrix, but in general this matrix may be a function of the accelerations acting on the system.

Also a Rayleigh dissipation function, which is independent from generalized coordinates $\boldsymbol{\theta}^*$ and from velocities $\dot{\boldsymbol{\theta}}^*$, can be defined. It is a function only of $\dot{\mathbf{u}}$ (and possibly of $\mathbf{u}$) and of their derivatives with respect to space coordinates.

Example 3.10 Consider the same beam studied in Example 3.6. The beam is now a structural member moved by an actuator that exerts a torque at the hinged end (Point O in Fig. 3.16 and Fig. 3.25).

Compute its kinetic energy and the gravitational potential energy.

The system has just one rigid-body degree of freedom (θ), plus a single deformation degree of freedom ($u(x)$).

The various parameters of the system are

$$\mathbf{X}(\boldsymbol{\theta}^*) = \frac{l}{2}\begin{Bmatrix} \cos(\theta) \\ \sin(\theta) \\ 0 \end{Bmatrix}, \qquad \mathbf{R} = \begin{bmatrix} \cos(\theta) & -\sin(\theta) \\ \sin(\theta) & \cos(\theta) \end{bmatrix}, \qquad \mathbf{r} = \begin{Bmatrix} x \\ 0 \\ 0 \end{Bmatrix},$$

$$\mathbf{u} = \begin{Bmatrix} 0 \\ u \\ 0 \end{Bmatrix}, \qquad \tilde{\mathbf{r}} = \begin{bmatrix} 0 & 0 & 0 \\ 0 & 0 & -x \\ 0 & x & 0 \end{bmatrix}, \qquad \tilde{\mathbf{u}} = \begin{bmatrix} 0 & 0 & u \\ 0 & 0 & 0 \\ -u & 0 & 0 \end{bmatrix},$$

$$\dot{\mathbf{X}} = \frac{l}{2}\begin{Bmatrix} -\sin(\theta) \\ \cos(\theta) \\ 0 \end{Bmatrix}\dot{\theta}, \qquad \boldsymbol{\Omega} = \begin{Bmatrix} 0 \\ 0 \\ 1 \end{Bmatrix}\dot{\boldsymbol{\theta}},$$

and hence

$$\mathbf{P}_1 = \frac{l}{2}\begin{Bmatrix} -\sin(\theta) \\ \cos(\theta) \\ 0 \end{Bmatrix}, \qquad \mathbf{P}_2 = \begin{Bmatrix} 0 \\ 0 \\ 1 \end{Bmatrix}.$$

Since the beam has a constant cross section A, the kinetic energy expressed by (3.89) reduces to

$$\begin{aligned} \mathcal{T} &= \frac{1}{2}\frac{ml^2}{3}\dot{\theta}^2 + \frac{1}{2}\rho A\int_{-l/2}^{l/2} u^2\,dx + \frac{1}{2}\dot{\boldsymbol{\theta}}^2\rho A\int_{-l/2}^{l/2} u^2\,dx \\ &\quad + \dot{\theta}\rho A\frac{l}{2}\int_{-l/2}^{l/2} \dot{u}\,dx + \dot{\theta}\rho A\int_{-l/2}^{l/2} \dot{u}x\,dx. \end{aligned}$$

The same result could be obtained directly by writing the position in the inertial (non-rotating) reference frame XY of a point of the beam, located at coordinate x:

$$\begin{Bmatrix} X \\ Y \end{Bmatrix} = \begin{Bmatrix} (x + \frac{l}{2})\cos(\theta) - u\sin(\theta) \\ (x + \frac{l}{2})\sin(\theta) + u\cos(\theta) \end{Bmatrix},$$

where $u(t)$ is the displacement due to the bending of the beam. By differentiating the position, the velocity is readily obtained

$$\begin{Bmatrix} \dot{X} \\ \dot{Y} \end{Bmatrix} = \begin{Bmatrix} -[\dot{\theta}(x + \frac{l}{2}) + \dot{u}]\sin(\theta) - \dot{\theta}u\cos(\theta) \\ [\dot{\theta}(x + \frac{l}{2}) + \dot{u}]\cos(\theta) - \dot{\theta}u\sin(\theta) \end{Bmatrix}.$$

The kinetic energy of a length dx of beam with material density ρ and area of the cross section A, is readily obtained

$$\begin{aligned} d\mathcal{T} &= \frac{1}{2}\rho A\,d\xi\left(\dot{X}^2 + \dot{Y}^2\right) \\ &= \frac{1}{2}\rho A\left[\dot{\theta}^2\left(x + \frac{l}{2}\right)^2 + \dot{u}^2 + u^2\dot{\theta}^2 + 2\left(x + \frac{l}{2}\right)\dot{\theta}\dot{u}\right]dx. \end{aligned}$$

Since

$$\int_0^l \rho A\left(x+\frac{l}{2}\right)^2 dx = \frac{ml^2}{3},$$

the kinetic energy of the beam is then

$$\mathcal{T} = \frac{1}{2}\frac{ml^2}{3}\dot{\theta}^2 + \frac{1}{2}\rho A\int_{-l/2}^{l/2} \dot{u}^2\,dx + \frac{1}{2}\dot{\theta}^2\rho A\int_{-l/2}^{l/2} u^2\,dx + \dot{\theta}\rho A\int_{-l/2}^{l/2}\left(x+\frac{l}{2}\right)\dot{u}\,dx.$$

The first term is that expressing the rigid-body dynamics, while in structural dynamics only the second term is present.

The gravitational acceleration vector is

$$\mathbf{g} = [0\ -g\ 0]^T$$

and thus the gravitational potential energy of the length dx of beam is

$$d\mathcal{U}_g = \rho g A\left[\left(x+\frac{l}{2}\right)\sin(\theta) + u\cos(\theta)\right]dx.$$

The gravitational potential energy is thus

$$\mathcal{U}_g = \frac{mgl}{2}\sin(\theta) + \frac{mg}{l}\cos(\theta)\int_{-l/2}^{l/2} u\,dx.$$

Since the beam is an elastic body, an elastic potential energy must be added to the gravitational potential energy. If the damping of the structure is taken into account, also a Rayleigh dissipation function can be introduced.

The flexible body dynamics is often studied by using a modal approach. First the free vibration problem of the various links is studied, obtaining the eigenvalues and the eigenfunctions, if the system is modeled as a continuous system, or the eigenvalues and the eigenvectors if a discretization technique is employed. In the first case there is an infinity of eigenvalues and eigenfunctions, in the second case there are n eigenvalues and n eigenvectors, where n is the number of deformation degrees of freedom.

The displacement $\mathbf{u}(x, y, z, t)$ (in the general tridimensional case $\mathbf{u}$ is a vector with three components) of any point of each link can be expressed as a linear combination of the eigenfunctions $q_i(x, y, z)$

$$\mathbf{u}(x, y, z, t) = \sum_{i=1}^{\infty} \eta_i(t) q_i(x, y, z), \tag{3.97}$$

or, by truncating the modal expansion after a number of terms,

$$\mathbf{u}(x, y, z, t) = \boldsymbol{\phi}(x, y, z)\boldsymbol{\eta}(t),$$

where $\eta(t)$ is a column matrix containing the n modal coordinates retained and $\boldsymbol{\phi}(x, y, z)$ is a matrix containing n columns (in each column an eigenfunction) and three rows (the three components of the eigenfunction) in the case of a continuous system.

In the case of discrete systems, the displacement vector $\mathbf{u}(t)$ can be expressed as a linear combination of the eigenvectors

$$\mathbf{u}(t) = \sum_{i=1}^{n} \eta_i(t)\mathbf{q}_i = \boldsymbol{\Phi}\eta(t). \tag{3.98}$$

The functions of time $\eta_i(t)$ are the modal coordinates and the square matrix $\boldsymbol{\Phi}$ is the eigenvector matrix, a matrix whose columns are the eigenvectors of the system.

Remark 3.12 If all modes are accounted for, (3.97) and (3.98) are exact even if the system is damped and nonlinear. If, on the contrary just a limited number of eigenfunctions or eigenvectors are considered, as it is usually the case, they are approximated. In the case of a moving structure, this approximation is worse than usual, and a larger number of modes may be needed.

Once the number of eigenfunctions (eigenvectors) to be used has been stated, the generalized coordinates of the system are those corresponding to the rigid-body coordinates used for studying the rigid-body dynamics, plus the modal coordinates corresponding to the deformation of the system.

By introducing the eigenfunctions in the expression of the kinetic energy, the latter becomes

$$\begin{aligned}\mathcal{T} = {} & \frac{1}{2}m\dot{\boldsymbol{\theta}}^{*T}\mathbf{P}_1^T\mathbf{P}_1\dot{\boldsymbol{\theta}}^* + \frac{1}{2}\dot{\boldsymbol{\theta}}^{*T}\mathbf{P}_2^T\mathbf{J}\mathbf{P}_2\dot{\boldsymbol{\theta}}^* + \frac{1}{2}\dot{\eta}^T\left(\int_v \rho\boldsymbol{\phi}^T\boldsymbol{\phi}\,dv\right)\dot{\eta} \\ & + \dot{\boldsymbol{\theta}}^{*T}\mathbf{P}_2^T\left(\int_v \rho 2\tilde{\mathbf{r}}\tilde{\mathbf{u}}^T\,dv\right)\mathbf{P}_2\dot{\boldsymbol{\theta}}^* + \frac{1}{2}\dot{\boldsymbol{\theta}}^{*T}\mathbf{P}_2^T\left(\int_v \rho\tilde{\mathbf{u}}\tilde{\mathbf{u}}^T\,dv\right)\mathbf{P}_2\dot{\boldsymbol{\theta}}^* \\ & + \dot{\boldsymbol{\theta}}^{*T}\mathbf{P}_1^T\mathbf{R}\left(\int_v \rho\tilde{\mathbf{u}}^T\,dv\right)\mathbf{P}_2(\boldsymbol{\theta}^*)\dot{\boldsymbol{\theta}} + \dot{\boldsymbol{\theta}}^{*T}\mathbf{P}_1^T\mathbf{R}\left(\int_v \rho\boldsymbol{\phi}\,dv\right)\dot{\eta} \\ & + \dot{\boldsymbol{\theta}}^{*T}\mathbf{P}_2^T\left(\int_v \rho\tilde{\mathbf{r}}\boldsymbol{\phi}\,dv\right)\dot{\eta} + \dot{\boldsymbol{\theta}}^{*T}\mathbf{P}_2^T\int_v \rho\tilde{\mathbf{u}}\dot{\mathbf{u}}\,dv.\end{aligned} \tag{3.99}$$

One of the integrals is immediately computed:

$$\overline{\mathbf{M}} = \int_v \rho\boldsymbol{\phi}^T\boldsymbol{\phi}\,dv \tag{3.100}$$

is the (diagonal) modal mass matrix of the compliant system.

Other integrals are straightforward:

$$\overline{\mathbf{M}}_1 = \int_v \rho\boldsymbol{\phi}\,dv, \qquad \overline{\mathbf{M}}_2 = \int_v \rho\tilde{\mathbf{r}}\boldsymbol{\phi}\,dv. \tag{3.101}$$

It thus follows that

$$\mathcal{T} = \frac{1}{2} m \dot{\boldsymbol{\theta}}^{*T} \mathbf{P}_1^T \mathbf{P}_1 \dot{\boldsymbol{\theta}}^* + \frac{1}{2} \dot{\boldsymbol{\theta}}^{*T} \mathbf{P}_2^T \mathbf{J} \mathbf{P}_2 \dot{\boldsymbol{\theta}}^* + \frac{1}{2} \dot{\boldsymbol{\eta}}^T \overline{\mathbf{M}} \dot{\boldsymbol{\eta}}$$
$$+ \frac{1}{2} \dot{\boldsymbol{\theta}}^{*T} \mathbf{P}_2^T \left(\int_v \rho (2\tilde{\mathbf{r}} + \tilde{\mathbf{u}}) \tilde{\mathbf{u}}^T \, dv \right) \mathbf{P}_2 \dot{\boldsymbol{\theta}}^* + \dot{\boldsymbol{\theta}}^{*T} \mathbf{P}_1^T \mathbf{R} \left(\int_v \rho \tilde{\mathbf{u}}^T \, dv \right) \mathbf{P}_2 (\boldsymbol{\theta}^*) \dot{\boldsymbol{\theta}}$$
$$+ \dot{\boldsymbol{\theta}}^{*T} \mathbf{P}_1^T \mathbf{R} \overline{\mathbf{M}}_1 \dot{\boldsymbol{\eta}} + \dot{\boldsymbol{\theta}}^{*T} \mathbf{P}_2^T \overline{\mathbf{M}}_2 \dot{\boldsymbol{\eta}} + \dot{\boldsymbol{\theta}}^{*T} \mathbf{P}_2^T \int_v \rho \tilde{\mathbf{u}} \dot{\mathbf{u}} \, dv. \qquad (3.102)$$

The other integrals containing $\tilde{\mathbf{u}}$ must be solved for the various cases, so that to obtain an expression of the kinetic energy in terms of the rigid-body coordinates $\boldsymbol{\theta}^*$ and the modal coordinates $\dot{\boldsymbol{\eta}}$.

By integrating over the whole body, it follows that

$$\mathcal{U}_g = -m \mathbf{g}^T \mathbf{X} - \mathbf{g}^T \mathbf{R} \left(\int_v \rho \boldsymbol{\phi} \, dv \right) \boldsymbol{\eta}. \qquad (3.103)$$

The elastic potential energy in terms of modal coordinates is simply

$$\mathcal{U}_e = \frac{1}{2} \boldsymbol{\eta}^T \overline{\mathbf{K}} \boldsymbol{\eta}, \qquad (3.104)$$

where $\overline{\mathbf{K}}$ is the (diagonal) modal stiffness matrix of the compliant system.

If the links can be modeled as prismatic homogeneous Euler–Bernoulli beams, clamped on one side (to the previous joint) and free at the other end, the modal approximation is particularly simple. The eigenfunctions are

$$q_i(\zeta) = \frac{1}{N_2} \left\{ \sin(\beta_i \zeta) - \sinh(\beta_i \zeta) - N_1 \left[\cos(\beta_i \zeta) - \cosh(\beta_i \zeta) \right] \right\}, \qquad (3.105)$$

where

$$N_1 = \frac{\sin(\beta_i) + \sinh(\beta_i)}{\cos(\beta_i) + \cosh(\beta_i)},$$
$$N_2 = \sin(\beta_i) - \sinh(\beta_i) - N_1 \left[\cos(\beta_i) - \cosh(\beta_i) \right],$$

the nondimensional variable ζ is

$$\zeta = \frac{x}{l} + \frac{1}{2}$$

and parameters β_i for he various modes are summarized in the following table:

Mode number k	1	2	3	4	>4
β_i	1.875	4.694	7.855	10.996	$(k - 0.5)\pi$

These numerical values were obtained by solving numerically the characteristic equation.

Example 3.11 Write the equations of motion for the beam of the previous example, using a modal approximation and retaining n bending modes.

Since the displacement u has only one component, that in y direction, matrix $\boldsymbol{\phi}$ has the first and last row all full of zeros, and n columns to account for the n modes.

The kinetic and gravitational potential energy are easily computed by simply introducing the modal approximation into the expressions obtained in the previous example and here reported:

$$\mathcal{T} = \frac{1}{2}\frac{ml^2}{3}\dot{\theta}^2 + \frac{1}{2}\rho A \int_{-l/2}^{l/2} \dot{u}^2\,dx + \frac{1}{2}\dot{\theta}^2 \rho A \int_{-l/2}^{l/2} u^2\,dx$$
$$+ \dot{\theta}\rho A \int_{-l/2}^{l/2} \left(x + \frac{l}{2}\right)\dot{u}\,dx,$$
$$\mathcal{U}_g = \frac{mgl}{2}\sin(\theta) + \frac{mg}{l}\cos(\theta)\int_{-l/2}^{l/2} u\,dx.$$

The first yields

$$\mathcal{T} = \frac{1}{2}\frac{ml^2}{3}\dot{\theta}^2 + \frac{1}{2}\dot{\boldsymbol{\eta}}^T\left(\rho A \int_{-l/2}^{l/2} \mathbf{q}\mathbf{q}^T\,dx\right)\dot{\boldsymbol{\eta}}$$
$$+ \frac{1}{2}\dot{\theta}^2\boldsymbol{\eta}^T\rho A\left(\int_{-l/2}^{l/2} \mathbf{q}\mathbf{q}^T\,dx\right)\boldsymbol{\eta} + \dot{\theta}\rho A\left(\int_{-l/2}^{l/2} \mathbf{q}^T\left(x + \frac{l}{2}\right)dx\right)\dot{\boldsymbol{\eta}}.$$

Once the eigenfunctions are known, the integrals

$$\overline{\mathbf{M}} = \rho A \int_{-l/2}^{l/2} \mathbf{q}\mathbf{q}^T\,dx, \qquad \mathbf{M}_1 = \rho A \int_0^l \mathbf{q}\left(x + \frac{l}{2}\right)dx$$

are known constants. They are, respectively, the modal mass matrix (a diagonal matrix owing to the m-orthogonality properties of the eigenfunctions) and a vector. Their size is theoretically infinite, but practically is equal to the number of deformation modes considered.

The kinetic energy is thus

$$\mathcal{T} = \frac{1}{2}\frac{ml^2}{3}\dot{\theta}^2 + \frac{1}{2}\dot{\boldsymbol{\eta}}^T\overline{\mathbf{M}}\dot{\boldsymbol{\eta}} + \frac{1}{2}\dot{\theta}^2\boldsymbol{\eta}^T\overline{\mathbf{M}}\boldsymbol{\eta} + \dot{\theta}\mathbf{M}_1^T\dot{\boldsymbol{\eta}},$$

or, better,

$$\mathcal{T} = \frac{1}{2}\overline{\mathbf{M}}\begin{Bmatrix}\dot{\theta}\\ \dot{\boldsymbol{\eta}}\end{Bmatrix}^T\begin{bmatrix} J & \mathbf{M}_1^T \\ \mathbf{M}_1 & \overline{\mathbf{M}}\end{bmatrix}\begin{Bmatrix}\dot{\theta}\\ \dot{\boldsymbol{\eta}}\end{Bmatrix} + \frac{1}{2}\dot{\theta}^2\boldsymbol{\eta}^T\overline{\mathbf{M}}\boldsymbol{\eta},$$

where

$$J = \frac{ml^2}{3}$$

is the moment of inertia of the beam about its rotation axis.

The gravitational potential energy becomes

$$\mathcal{U}_g = \frac{mgl}{2}\sin(\theta) + \frac{mg}{l}\cos(\theta)\left(\int_{-l/2}^{l/2} \mathbf{q}^T\, dx\right)\boldsymbol{\eta},$$

or

$$\mathcal{U}_g = \frac{mgl}{2} g\sin(\theta) + g\cos(\theta)\mathbf{M}_2^T\boldsymbol{\eta},$$

where

$$\mathbf{M}_2 = \frac{m}{l}\int_{-l/2}^{l/2} \mathbf{q}\, d\xi.$$

The elastic potential energy is the same as that of the stationary vibrating beam, i.e.

$$\mathcal{U}_e = \frac{1}{2}\boldsymbol{\eta}^T\overline{\mathbf{K}}\boldsymbol{\eta},$$

where $\overline{\mathbf{K}}$ is the modal stiffness matrix. Owing to the k-orthogonality properties of the eigenfunctions, it is a diagonal matrix.

Since the beam is clamped at the left end, no rotation is possible there due to flexibility, and the generalized torque appearing there in the first equation is the motor torque M_θ. No generalized force is present in the other equations.

The first equation is thus

$$J\ddot{\theta} + \ddot{\theta}\boldsymbol{\eta}^T\overline{\mathbf{M}}\boldsymbol{\eta} + \mathbf{M}_1^T\ddot{\boldsymbol{\eta}} + 2\dot{\theta}\boldsymbol{\eta}^T\overline{\mathbf{M}}\dot{\boldsymbol{\eta}} + \frac{mgl}{2}\cos(\theta) - g\sin(\theta)\mathbf{M}_2^T\boldsymbol{\eta} = \mathbf{M}_\theta.$$

The derivatives appearing in the other equations are

$$\frac{\partial\mathcal{T}}{\partial\dot{\boldsymbol{\eta}}} = \overline{\mathbf{M}}\dot{\boldsymbol{\eta}} + \dot{\theta}\mathbf{M}_1,$$

$$\frac{d}{dt}\left(\frac{\partial\mathcal{T}}{\partial\dot{\boldsymbol{\eta}}}\right) = \overline{\mathbf{M}}\ddot{\boldsymbol{\eta}} + \ddot{\theta}\mathbf{M}_1,$$

$$\frac{\partial(\mathcal{T}-\mathcal{U})}{\partial\boldsymbol{\eta}} = \dot{\theta}^2\overline{\mathbf{M}}\boldsymbol{\eta} - g\cos(\theta)\mathbf{M}_2 - \overline{\mathbf{K}}\boldsymbol{\eta}.$$

The relevant equations are thus

$$\overline{\mathbf{M}}\ddot{\boldsymbol{\eta}} + \ddot{\theta}\mathbf{M}_1 - \dot{\theta}^2\overline{\mathbf{M}}\boldsymbol{\eta} + g\cos(\theta)\mathbf{M}_2 + \overline{\mathbf{K}}\boldsymbol{\eta} = \mathbf{0},$$

i.e. a set of nonlinear equations in the variables θ and $\boldsymbol{\eta}$.

If the term $\ddot{\theta}\eta^T\overline{\mathbf{M}}\eta$ in the first equation is neglected, by separating the linear from the nonlinear part, it follows that

$$\mathbf{M}\ddot{\mathbf{x}} + \mathbf{K}\mathbf{x} = \mathbf{f} + \mathbf{T},$$

where the mass matrix is constant

$$\mathbf{M} = \begin{bmatrix} J & \mathbf{M}_1^T \\ \mathbf{M}_1 & \overline{\mathbf{M}} \end{bmatrix}, \qquad \mathbf{K} = \begin{bmatrix} \mathbf{0} & \mathbf{0} \\ \mathbf{0} & \overline{\mathbf{K}} \end{bmatrix}, \qquad \mathbf{x} = \begin{Bmatrix} \theta \\ \eta \end{Bmatrix},$$

$$\mathbf{f} = \begin{Bmatrix} -2\dot{\theta}\eta^T\overline{\mathbf{M}}\dot{\eta} - Sg\cos(\theta) + g\sin(\theta)\mathbf{M}_2^T\eta \\ +\dot{\theta}^2\overline{\mathbf{M}}\eta - g\cos(\theta)\mathbf{M}_2 \end{Bmatrix}, \qquad \mathbf{T} = \begin{Bmatrix} M_\theta \\ \mathbf{0} \end{Bmatrix}.$$

If on the contrary the term $\ddot{\theta}\eta^T\overline{\mathbf{M}}\eta$ is not neglected, the mass matrix is not constant,

$$\mathbf{M} = \begin{bmatrix} J + \eta^T\overline{\mathbf{M}}\eta & \mathbf{M}_1^T \\ \mathbf{M}_1 & \overline{\mathbf{M}} \end{bmatrix},$$

a thing that, however, is not very problematic since the equation is at any rate nonlinear and numerical integration is required also for the simplified version.

The eigenfunctions are expressed by (3.105) and following. The value of N_2 is such that the maximum value of the eigenfunctions, occurring at the free end (at $\zeta = 1$), is equal to unity. In this way each modal coordinate is the contribution of the relevant mode to the displacement at the end of the beam. Remembering that the beam is prismatic and homogeneous, it follows that

$$\overline{\mathbf{M}}_{ii} = \frac{m}{l}\int_{-l/2}^{l/2} q_i^2\,dx, \qquad \overline{\mathbf{K}}_{ii} = \frac{\beta_i^4}{l^4}\frac{EI_y}{\rho A}\overline{\mathbf{M}}_{ii},$$

$$\mathbf{M}_{1_i} = m\int_{l/2}^{1/2} q_i\left(x + \frac{l}{2}\right)dx, \qquad \mathbf{M}_{2_i} = \frac{m}{l}\int_{l/2}^{1/2} q_i\,dx.$$

To avoid integrating complex harmonic and hyperbolic functions, the integration can be easily performed numerically. The modal masses and stiffness are simply

$$\overline{\mathbf{M}}_{ii} = \frac{m}{4}, \qquad \overline{\mathbf{K}}_{ii} = \frac{\beta_i^4 EI}{4l^3}.$$

Example 3.12 Compute the time history of the beam of the previous example, using a PID controller to reach and maintain the final position. Assume that at time $t = 0$ the beam is undeflected and lies on the x-axis.

The data are the same as in Example 3.6 ($l = 1$ m, $m = 5$ kg, $g = 9.81$ m/s^2, $K_p = 263$ Nm/rad, $K_d = 41.9$ Nm s/rad, $K_i = 200$ Nm/rad s) The data related to the material are $\rho = 2{,}700$ kg/m^3, $E = 7.2 \times 10^{10}$ N/m^2. The area of the cross section is $A = 1{,}851$ mm^2 and the area moment of inertia of the cross section is assumed to be $I = 7{,}410$ mm^4. With these data a very slender and flexible beam is obtained. Its slenderness, i.e. the ratio between the length and the radius of gyration of the cross section, is 500, a very high value that justifies the use of the Euler–Bernoulli beam model.

As a first attempt assume that the sensor measures angle θ. Clearly in this way there will be a positioning error, due to inflection of the beam. Using a PID controller, the motor torque is

$$M_\theta = -K_p(\theta - \theta_0) - K_d\dot{\theta} - K_i \int_0^t (\theta - \theta_0)\, du.$$

The auxiliary variables that can be introduced into the equation of motion are

$$\mathbf{v} = \dot{\mathbf{x}}, \qquad \mathbf{r} = \int_0^t \mathbf{x}\, du,$$

obtaining a set of $3(n+1)$ equations, where n is the number of modes considered. Actually, the variables $\mathbf{r}$ linked with the integrals of the coordinates need not to be n: since the integral control applies only to one coordinate, vector $\mathbf{r}$ needs to have just one element. Operating in this way, however, a simpler formulation of the equations is obtained.

The vector $\mathbf{T}$ to be introduced into the equations of motion is

$$\mathbf{T} = -\mathbf{K}_p\mathbf{x} - \mathbf{K}_d\mathbf{v} - \mathbf{K}_i\mathbf{r} + \mathbf{K}_p\mathbf{x}_0 + \mathbf{K}_i\mathbf{x}_0 t,$$

where the gain matrices, all with $n+1$ rows and columns, are

$$\mathbf{K}_p = \begin{bmatrix} K_p & \mathbf{0} \\ \mathbf{0} & \mathbf{0} \end{bmatrix}, \qquad \mathbf{K}_d = \begin{bmatrix} K_d & \mathbf{0} \\ \mathbf{0} & \mathbf{0} \end{bmatrix}, \qquad \mathbf{K}_i = \begin{bmatrix} K_i & \mathbf{0} \\ \mathbf{0} & \mathbf{0} \end{bmatrix}.$$

The state space equation of the system is thus

$$\begin{Bmatrix} \dot{\mathbf{v}} \\ \dot{\mathbf{x}} \\ \dot{\mathbf{r}} \end{Bmatrix} = \begin{bmatrix} -\mathbf{M}^{-1}\mathbf{K}_d & -\mathbf{M}^{-1}(\mathbf{K}_p + \mathbf{K}) & -\mathbf{M}^{-1}\mathbf{K}_i \\ \mathbf{I} & \mathbf{0} & \mathbf{0} \\ \mathbf{0} & \mathbf{I} & \mathbf{0} \end{bmatrix} \begin{Bmatrix} \mathbf{v} \\ \mathbf{x} \\ \mathbf{r} \end{Bmatrix} + \begin{Bmatrix} \mathbf{M}^{-1}\mathbf{f} \\ 0 \\ 0 \end{Bmatrix} + \begin{Bmatrix} \mathbf{M}^{-1}(\mathbf{K}_p + \mathbf{K}_i t)\mathbf{x}_0 \\ 0 \\ 0 \end{Bmatrix}.$$

The results of the numerical integration performed taking into account the first 4 flexible modes are reported in Fig. 3.26. Angle θ is reported in degrees, while instead of plotting the modal coordinates η_i, ratios η_i/l are reported. Their meaning is the angle under which the tip displacement due to the ith mode is seen from point O.

The vibrations of the beam as a flexible body are limited: the amplitudes of the flexible modes are small, and die out very quickly. The final value of θ_0 after 20 s is 30°, however, taking into account all four modes, the angle at which the tip is seen from point O is 29.43° corresponding to an error at the tip of the beam of 0.57°, i.e. of 8.6 mm.

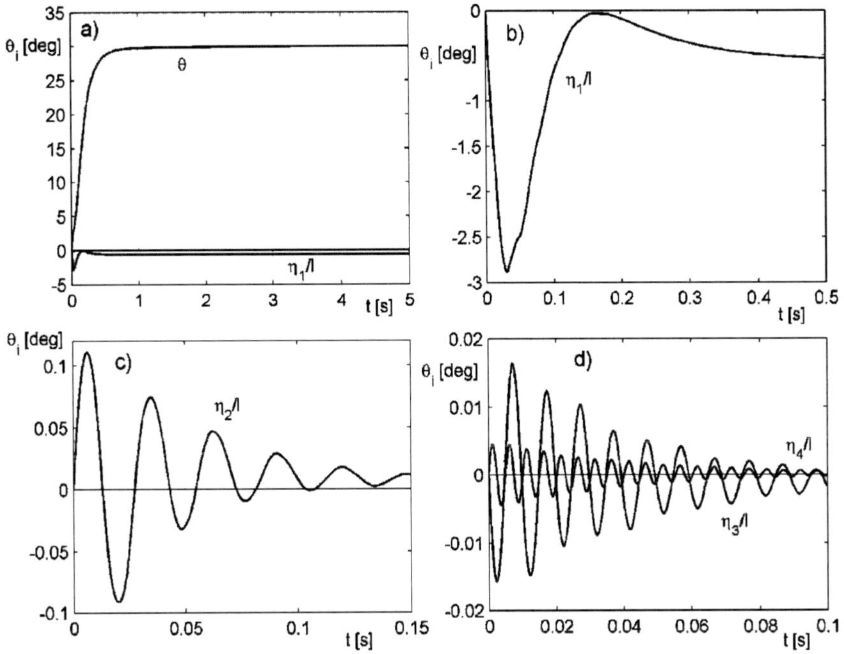

Fig. 3.26 Results of the numerical integration: angle θ and η_i/l for 4 modes. In (**b**), (**c**), and (**d**) enlargements related to the deformation modes just after $t = 0$ are reported

Example 3.13 Repeat the computations of the previous example, using a sensor that measures the displacement at the tip of the beam.

The angle between line OA and x-axis is the reference θ_0. The error is thus

$$e = \theta + \frac{1}{l}\sum_{i=1}^{n} \eta_i - \theta_0.$$

Assuming that the derivative control, which has no effect on the final value of the position, is done only on $\dot{\theta}$, the motor torque is

$$M_\theta = -K_p\left(\theta + \frac{1}{l}\sum_{i=1}^{n} \eta_i - \theta_0\right) - K_d\dot{\theta} - K_i\int_0^t \left(\theta + \frac{1}{l}\sum_{i=1}^{n} \eta_i - \theta_0\right) du.$$

Proceeding in the same way as above, and introducing the same auxiliary variables, it follows that the vector $\mathbf{T}$ to be introduced into the equations of motion is

$$\mathbf{T} = -\mathbf{K}_p x - \mathbf{K}_d \mathbf{v} - \mathbf{K}_i \mathbf{r} + \mathbf{K}_p \boldsymbol{\theta}_0 + \mathbf{K}_i \boldsymbol{\theta}_0 t.$$

where the gain matrices, all with $n+1$ rows and columns, are

$$\mathbf{K}_p = K_p \begin{bmatrix} 1 & \boldsymbol{\beta} \\ \mathbf{0} & \mathbf{0} \end{bmatrix}, \qquad \mathbf{K}_d = \begin{bmatrix} K_d & \mathbf{0} \\ \mathbf{0} & \mathbf{0} \end{bmatrix}, \qquad \mathbf{K}_i = K_i \begin{bmatrix} 1 & \boldsymbol{\beta} \\ \mathbf{0} & \mathbf{0} \end{bmatrix},$$

where

$$\boldsymbol{\beta} = \frac{1}{l}[1 \quad 1 \quad 1 \quad 1 \quad \dots \quad 1].$$

The state space equation of the system is thus

$$\begin{Bmatrix} \dot{\mathbf{v}} \\ \dot{\mathbf{x}} \\ \dot{\mathbf{r}} \end{Bmatrix} = \begin{bmatrix} -\mathbf{M}^{-1}\mathbf{K}_d & -\mathbf{M}^{-1}(\mathbf{K}_p + \mathbf{K}) & -\mathbf{M}^{-1}\mathbf{K}_i \\ \mathbf{I} & \mathbf{0} & \mathbf{0} \\ \mathbf{0} & \mathbf{I} & \mathbf{0} \end{bmatrix} \begin{Bmatrix} \mathbf{v} \\ \mathbf{x} \\ \mathbf{r} \end{Bmatrix} + \begin{Bmatrix} \mathbf{M}^{-1}\mathbf{f} \\ \mathbf{0} \\ \mathbf{0} \end{Bmatrix} + \begin{bmatrix} \mathbf{M}^{-1} \\ \mathbf{0} \\ \mathbf{0} \end{bmatrix} \begin{Bmatrix} (K_p + K_i t)\theta_0 \\ 0 \end{Bmatrix}.$$

The results of the numerical integration are not reported in detail, since the difference from those reported in Fig. 3.26 is very small. However, the final value of θ_0 after 20 s is 30.57°, which is slightly larger than the required value in order to compensate for the flexible displacements due to the static force (weight). Taking into account all 4 modes, the angle at which the tip is seen from point O is 30.00° showing that this kind of control can compensate completely for the error due to the compliance of the beam.

Example 3.14 Consider the same beam of the previous examples, but control it using an open-loop system that governs the actuator torque following a predetermined pattern.

Assuming that the torque is controlled following either a square-wave pattern (bang–bang control) (Fig. 3.27a) or a more elaborate double-versine time history (Fig. 3.27b) and neglecting weight, compute the maximum torque needed to achieve a rotation of 45° in 1 s and the time history of the tip of the beam during and after the maneuver.

In particular, study the vibration of the beam occurring after it has stopped in the end position.

The problem will be split into two separate problems related to rigid-body dynamics and beam vibration.

Rigid-Body Dynamics

The equation of motion of the beam as a rigid body can be obtained from the first equation of motion in the previous examples, by simply neglecting the terms in η:

$$J\ddot{\theta} + \frac{mgl}{2}\cos(\theta) = \mathbf{M}_\theta.$$

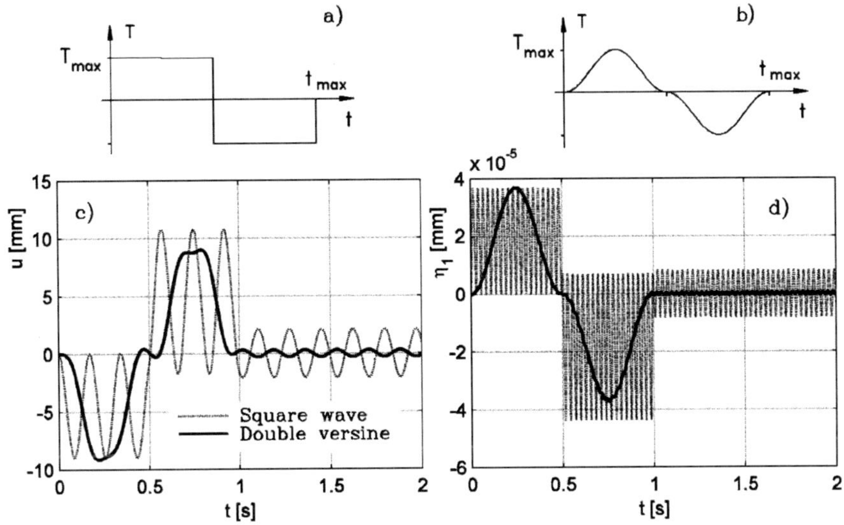

Fig. 3.27 Beam rotating about one of its ends under the effect of the driving torque $T(t)$: (**a**) and (**b**) time histories of the control torque during the maneuver: square-wave and double-versine patterns; (**c**) time history of the displacement of the free end: displacement from the rigid-body position, computed by numerically integrating the equations of the first five modes; (**d**) modal response: second mode

Neglecting the weight, in the case of the bang–bang control the time history of the torque is

$$M_\theta = M_{\theta\,\max} \quad \text{for } 0 \le t \le \frac{t_{\max}}{2},$$

$$M_\theta = -M_{\theta\,\max} \quad \text{for } \frac{t_{\max}}{2} < t \le t_{\max}.$$

Note that the square-wave law is that which allows the minimum travel time for a given value of the maximum driving torque. The control law is symmetrical in time and the speed of the beam at time $t_{\max}$ reduces to zero. By integrating the equation of motion, the time history of the displacement is

$$\theta = \frac{3T_{\max}}{2ml^2}t^2 \quad \text{for } 0 \le t \le \frac{t_{\max}}{2},$$

$$\theta = \frac{3T_{\max}}{4ml^2}\left(-t_{\max}^2 + 4t_{\max}t - 2t^2\right) \quad \text{for } \frac{t_{\max}}{2} \le t \le t_{\max}.$$

The relationship linking the torque with the displacement $\theta_{\max}$ and the time needed for rotation is

$$M_{\theta\,\max} = \frac{4ml^2\theta_{\max}}{3t_{\max}^2}.$$

In the case of the double-versine control, the time history of the torque is

$$M_\theta = \frac{M_{\theta\max}}{2}\left[1-\cos\left(\frac{4\pi}{t_{\max}}t\right)\right] \quad \text{for } 0 \le t \le \frac{t_{\max}}{2},$$

$$M_\theta = -\frac{M_{\theta\max}}{2}\left[1-\cos\left(\frac{4\pi}{t_{\max}}t\right)\right] \quad \text{for } \frac{t_{\max}}{2} \le t \le t_{\max}.$$

Also in this case the control law is symmetrical in time and the speed of the beam at time $t_{\max}$ reduces to zero. By integrating the equation of motion, the time history of the displacement is, for $0 \le t \le t_{\max}/2$ and $t_{\max}/2 < t \le t_{\max}$, respectively,

$$\theta = \frac{3T_{\max}}{4ml^2}\left\{t^2 + \frac{t_{\max}^2}{8\pi^2}\left[\cos\left(\frac{4\pi}{t_{\max}}t\right)-1\right]\right\},$$

$$\theta = \frac{3T_{\max}}{8ml^2}\left\{-t_{\max}^2 + 4t_{\max}t - 2t^2 - \frac{t_{\max}^2}{4\pi^2}\left[\cos\left(\frac{4\pi}{t_{\max}}t\right)-1\right]\right\}.$$

The relationship linking the torque with the displacement $\theta_{\max}$ and the time needed for rotation is

$$M_{\theta\max} = \frac{8ml^2\theta_{\max}}{3t_{\max}^2}.$$

The values of the maximum torque are thus $M_{\theta\max} = 13.09$ Nm for the square wave and $M_{\theta\max} = 26.17$ Nm for the double versine.

Vibration

By assuming that the law $\theta(t)$ is stated, the previously computed equation of motion reduces to

$$\overline{\mathbf{M}}\ddot{\eta} + \left(\overline{\mathbf{K}} - \dot{\theta}^2\overline{\mathbf{M}} + g\cos(\theta)\mathbf{M}_2\right)\eta = \ddot{\theta}\mathbf{M}_1.$$

The first and the second terms are the usual ones that appear in the equation of motion of the stationary beam. The third term is a centrifugal stiffening due to the component of the centrifugal force due to rotation acting in a direction perpendicular to the y-axis.

This effect can be neglected, since the stiffening effect due to the centrifugal field is small, if the angular velocity is small enough. Actually, in the small movement studied, the acceleration is high but the angular velocity maintains a low value. Moreover, the aim of the study is mainly to predict the free behavior of the beam after the required position is reached and this effect stops acting. Clearly the neglected term is not noninfluent, because the behavior after the beam has stopped depends on what happens during the motion, but the assumption that its effect is small can be accepted, at least in a first-approximation study.

Then there is a term due to weight, which is here neglected, and a term due to the acceleration of the beam that can be accounted for as an external excitation, because it does not contain the deformation of the beam.

The beam can thus be studied as a beam at standstill, under the effect of the inertia force $\mathbf{M}_1$

$$\overline{\mathbf{M}}\ddot{\eta} + \overline{\mathbf{K}}\eta = \ddot{\theta}\mathbf{M}_1 .$$

The solutions of the modal equations for the two cases obtained through the numerical integration of the first five modes are shown in Fig. 3.27c (total displacement at the end of the beam) and Fig. 3.27d (displacement at the end of the beam due to the second mode).

From the figure, it is clear that the double-versine control succeeds in positioning the beam without causing long-lasting vibrations as in the case of the bang–bang control pattern. The latter, however, strongly excites the first mode, which is little damped. The results are linked with the particular application, because a control input of the versine type can also excite some modes; however, the fact that the square-wave control input is more prone to exciting vibrations than more smooth control laws is a general feature.

In many cases, particular control laws that are much better than both the square-wave and versine ones are used, and much theoretical and experimental work has been devoted to identifying optimal control laws. Note that no damping of the beam material has been considered. In practice, the vibration of the beam is more damped than what has been computed.

In a practical case, a closed-loop control must be associated with the open-loop control to achieve a sufficient positioning accuracy.

3.16 High Level Control

Up to now, only the lowest level of control has been shown. The feedback control loop that uses the position and velocity signal from the sensors of the arm to drive the actuators to a given position, perhaps complemented by a feedforward action, is simple and does not require sophisticated control algorithms.

If the arm is a telemanipulator, the human in the control loop provides the high level control, setting the targets that the arm must reach and stating the position toward which the motion must be directed.

Industrial robots are often thought by a human master: this is particularly the case when the arm must move autonomously, but has just to perform repetitive tasks. The operator moves the arm along a given trajectory, often through a keyboard, memorizing a number of keypoints and then the controller repeats the movements autonomously without any change.

This strategy is, however, unsuitable when the various working cycles are not identical or the robot has to face unexpected situations. This is what usually occurs in the case of space robots.

If the robot must move in an autonomous way without repeating a fixed scheme, the difficulties quickly increase. Very complex control algorithms are required for the higher level tasks and the whole matter is still object of research.

Another difficulty comes from the need of understanding the presence of the object on which the arm must operate, its location and position. A camera (or better two cameras for stereoscopic vision) can be used in conjunction with a computer that elaborates the images but the whole field of robot vision is still a research topics.

As already stated, when the arm must follow a surface or must remain in touch with an object, the control is based on the force it exerts and not on the position. In this case, more than being based on vision, the control is based on touch.

Remark 3.13 The definition of the trajectory of a robotic arm is quite similar to that of the trajectory of a moving robot or an automatic vehicle.

3.17 Parallel Manipulators

Up to this point manipulators were assumed to be open kinematic chains. This is not always the case, and also closed kinematic chains can be used as manipulators, particularly when large loads must be handled with precision.

Parallel manipulators have a larger load capacity, stiffness, accuracy and speed than corresponding open chain manipulators, but their workspace, both in term of position and orientation, is usually much more limited. The Stewart platform is the better known parallel manipulator and is made of a rigid body (the platform) connected to a base through six links whose length can be changed at will. It has six joint coordinates (the lengths of the links) that determine the pose of the platform and thus it has as many joint coordinates as space coordinates of the end effector.

The Stewart platform was first used by Gough for a tire testing machine he built in the 1960s at Dunlop to position the tire and keep it in the required orientation. It has since used for many applications, also in the aerospace field. Although the name Hexapod (not to be confused with the same term used in walking machines technology to define a machine with six legs) is a trademark of Geodetic Technology for a Stewart platform, this name has been widely used.

As already stated, in a Stewart platform six linear actuators connect the base to the platform, whose position is determined by the length of the former (Fig. 3.28a). The points in which the actuators are connected to the base and to the platform can lie in two planes, like in the figure. In this case the base and the platform can have the shape of hexagons, but it is possible to demonstrate that if they are regular hexagons the platform is singular.

In parallel manipulators, the inverse kinematics is easier than the direct kinematics, which involves the solution of a complex set of nonlinear equations. The Stewart platform is no exception.

Chose point O on the base and a point G on the platform and state a base reference frame $OXYZ$ and a platform reference frame $Gxyz$. Defining vectors $\overline{(G-O)} = \mathbf{X}$, $\overline{(P_{bi}-O)} = \mathbf{r}_{Pbi}$, $\overline{(P_{pi}-O)} = \mathbf{r}_{Ppi}$, the latter expressed in the platform reference frame, as in Fig. 3.28a, the length of the generic ith actuator can be expressed as

$$l_i = |\mathbf{X} + \mathbf{R}\mathbf{r}_{Ppi} - \mathbf{r}_{Pbi}|. \tag{3.106}$$

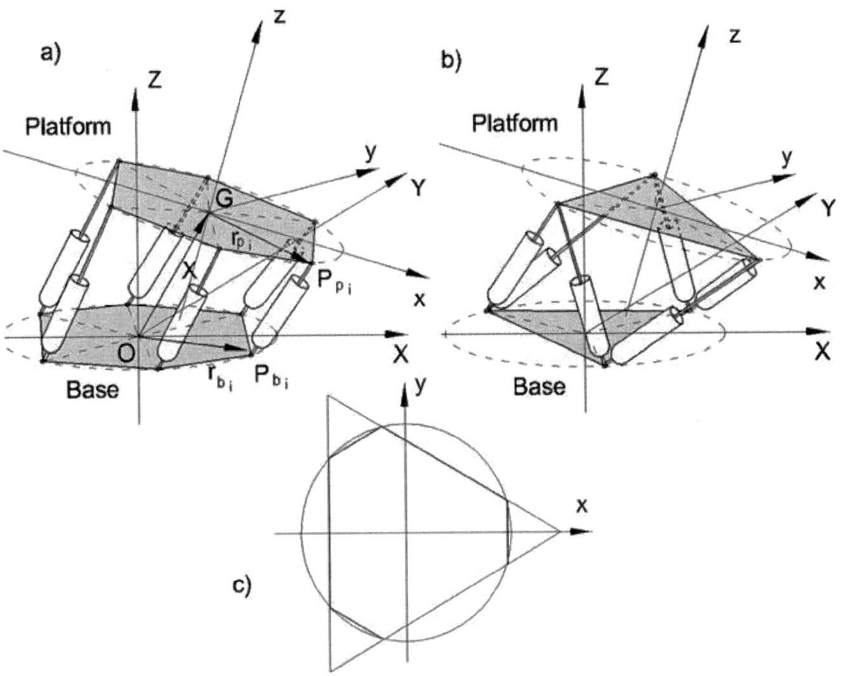

Fig. 3.28 (**a**) Generic Stewart platform. (**b**) Platform in which the base and the platform have the shape of triangles. (**c**) An irregular hexagon that can be inscribed in a circle

If the base is planar and lies in the XY plane, vector $\mathbf{r}_{Pbi}$ can be expressed as

$$\mathbf{r}_{Pbi} = \begin{bmatrix} R_{bi}\cos(\theta_{bi}) & R_{bi}\sin(\theta_{bi}) & 0 \end{bmatrix}^T. \tag{3.107}$$

In the same way, if the platform is planar and lies in the xy plane, vector $\mathbf{r}_{Ppi}$ can be expressed as

$$\mathbf{r}_{Ppi} = \begin{bmatrix} R_{pi}\cos(\theta_{pi}) & R_{pi}\sin(\theta_{pi}) & 0 \end{bmatrix}^T. \tag{3.108}$$

The pose of the platform is defined by the coordinates of point G (vector **X**) and by the roll, pitch and yaw angles that are included in the expression of matrix **R** (3.7) If the points in which the actuators are connected to the base and to the platform lie on two circles centered in O and G, respectively, (Fig. 3.28c), the distances R_{bi} and R_{pi} are all equal.

By introducing the expression of the rotation matrix into (3.28) and performing the relevant computations, the six equations defining the inverse kinematics are

$$\begin{aligned} l_i = \big\{ & X^2 + Y^2 + Z^2 + R_{pi}{}^2 + R_{bi}{}^2 - 2Y R_{bi}\sin(\theta_{bi}) - 2X R_{bi}\cos(\theta_{bi}) \\ & + 2X R_{pi}\big[\cos(\psi)\cos(\theta)\cos(\theta_{pi}) - \sin(\theta_{pi})\sin(\psi)\cos(\phi) \end{aligned}$$

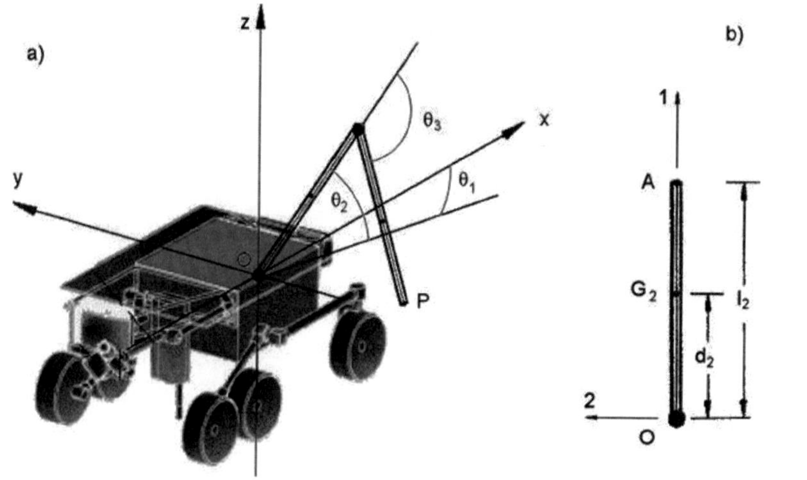

Fig. 3.29 (**a**) Sketch of the arm on board of the rover. (**b**) Sketch of the first part of the arm and its principal axes of inertia

$$
\begin{aligned}
&+\sin(\theta_{pi})\cos(\psi)\sin(\theta)\sin(\phi)\big]+2YR_{pi}\big[\sin(\psi)\cos(\theta)\cos(\theta_{pi})\\
&+\sin(\theta_{pi})\cos(\psi)\cos(\phi)+\sin(\theta_{pi})\sin(\psi)\sin(\theta)\sin(\phi)\big]\\
&+2YR_{pi}\big[\cos(\theta)\sin(\phi)\sin(\theta_{pi})-\sin(\theta)\cos(\theta_{pi})\big]\\
&-2R_{pi}R_{bi}\sin(\theta_{bi})\sin(\theta_{pi})\big[\sin(\psi)\sin(\theta)\sin(\phi)+\cos(\psi)\cos(\phi)\big]\\
&-2R_{pi}R_{bi}\cos(\theta_{pi})\cos(\theta)\big[\cos(\psi)\cos(\theta_{bi})+\sin(\psi)\sin(\theta_{bi})\big]\\
&+2R_{pi}R_{bi}\cos(\theta_{bi})\sin(\theta_{pi})\big[\sin(\psi)\cos(\phi)-\cos(\psi)\sin(\theta)\sin(\phi)\big]\big\}^{1/2},
\end{aligned}
\tag{3.109}
$$

for $i=1,\dots,6$.

These six equations must be solved in X, Y, Z, ϕ, θ and ψ to yield the direct kinematic, a thing that must be done numerically.

The inverse of the Jacobian can be computed in closed form, although not easily, by computing the derivatives of the expressions for l_i, and this allows to compute directly the velocity in the joint space from those in the physical space. The Jacobian for the opposite transformation can be obtained by numerical inversion.

Example 3.15 As a final example consider the arm of a rover operating on the surface of Mars ($g=3.77\ \text{m/s}^2$) that from its storage position must pick up a rock on the ground and then put it in the specimen basket on board (Fig. 3.29a).

The arm is an anthropomorphic arm with the first joint a spherical joint ($l_1=0$) and the following data: $l_2=1.5$ m, $d_2=0.9$ m, $m_2=5$ kg, $l_3=1.3$ m, $d_3=0.75$ m, $m_3=3$ kg. Axes 1, 2 and 3 of each section of the arm are assumed to be principal axes of inertia. The moment of inertia about axis 1 is assumed to be so small to be neglected. The other two moments of inertia are equal and their values are $J_2 = 1.25$ kg m^2 and $J_3 = 0.81$ kg m^2 for the two parts.

The mass of the specimen is $m_s = 10$ kg.

The arm is controlled by a PID controller, but the motors have a limited torque.

The rest position of the arm is characterized by the following values of the angles:

$$\boldsymbol{\theta}_0 = [0 \quad 90^\circ \quad -150^\circ]^T.$$

Each maneuver is performed assuming two waypoints and the final point and by stating a straight trajectory with constant acceleration and deceleration between them. They are:

- pick up run:
  waypoint 1 (get clear of the rover) $[200\ -400\ 600]^T$ mm, reached in 3 s;
  waypoint 2 (get close to the target) $[-1000\ -1000\ -700]^T$ mm, reached in further 5 s;
  target (specimen) $[-1000\ -1000\ -800]^T$ mm, reached in further 2 s;
- return run:
  waypoint 3 (get over the rover) $[-200\ -400\ 600]^T$ mm, reached in 3 s;
  waypoint 4 (get close to the target) $[-400\ 0\ 10]^T$ mm, reached in further 5 s;
  target (specimen basket) $[-400\ 0\ 0]^T$ mm, reached in further 2 s.

The gains of the controller are $K_p = 2000$ Nm/rad, $K_d = 4000$ Nm s/rad, $K_i = 600$ Nm/rad s for all joints. The saturation torques of the electric motors are

$$\mathbf{M}_{\theta\,\max} = [80 \quad 30 \quad 15]^T \text{ Nm}.$$

The arm stops when the target is reached with an error of 10 mm.

The kinematics and the inverse kinematics of the arm were dealt with in Example 3.1, while its open loop dynamics was studied in Example 3.5.

Since the motors are assumed to be limited in torque, the motor torques must be computed at each integration step and then corrected if the maximum toque is reached.

By introducing the usual auxiliary variables, the closed-loop equation of motion is

$$\begin{Bmatrix} \dot{\mathbf{v}} \\ \dot{\boldsymbol{\theta}} \\ \dot{\mathbf{r}} \end{Bmatrix} = \begin{bmatrix} \mathbf{0} & \mathbf{0} & \mathbf{0} \\ \mathbf{I} & \mathbf{0} & \mathbf{0} \\ \mathbf{0} & \mathbf{I} & \mathbf{0} \end{bmatrix} \begin{Bmatrix} \mathbf{v} \\ \boldsymbol{\theta} \\ \mathbf{r} \end{Bmatrix} + \begin{Bmatrix} \mathbf{M}^{-1}\mathbf{f} \\ \mathbf{0} \\ \mathbf{0} \end{Bmatrix} + \begin{Bmatrix} \mathbf{M}^{-1}\mathbf{M}_\theta \\ \mathbf{0} \\ \mathbf{0} \end{Bmatrix},$$

where $\mathbf{M}$ and $\mathbf{f}$ are the mass matrix and the vector obtained in Example 3.5, and the motor torques are

$$\mathbf{M}_\theta = -\mathbf{K}_p(\theta - \theta_0) - \mathbf{K}_d(\mathbf{v} - \mathbf{v}_0) - \mathbf{K}_i\left(\mathbf{r} - \int_0^t \theta_0\, du\right),$$

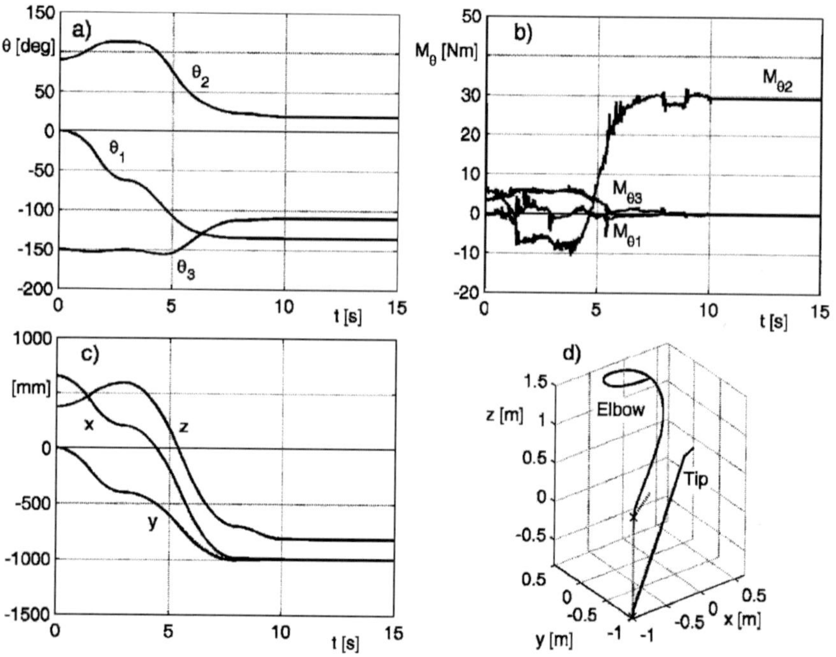

Fig. 3.30 Time history of the angles (**a**), motor torques (**b**) and coordinates of the tip of the arm (**c**) during the maneuver to get the specimen. (**d**) Trajectories of the elbow and the tip of the arm and sketch of the arm in the final position

if their absolute value does not exceed the saturation level, otherwise they are

$$M_{\theta i} = M_{\theta i\,\max}\,\mathrm{sign}(M_{\theta ic}),$$

where $M_{\theta ic}$ stands for the value computed using the PID algorithm.

First Part of the Manoeuvre

The time required to reach the target within the required tolerance is 9.33 s. The time histories of the angles, motor torques and coordinates of the tip of the arm are reported in Figs. 3.30a, b and c. The trajectories of the elbow and the tip of the arm are reported in Fig. 3.30d. The theoretical straight trajectories, plotted with dashed lines, are practically overwritten by the actual trajectories.

It can be clearly seen that the motion of the arm is smooth, without oscillations and that the torque required from the motors is not large. An exception is actuator for angle θ_2 that must bear the weight of the arm. There are some torque spikes due to the fact that the reference signals defining the trajectory are supplied every 0.1 s and not continuously.

Second Part of the Manoeuvre

After picking up the specimen, the inertial properties of the second part of the arm increase. Now $m_2 = 13$ kg, $d_2 = 1.173$ m and $J_2 = 1.51$ kg m^2. Note that the torque at the first joint to keep the arm in horizontal position is 148 Nm, greater

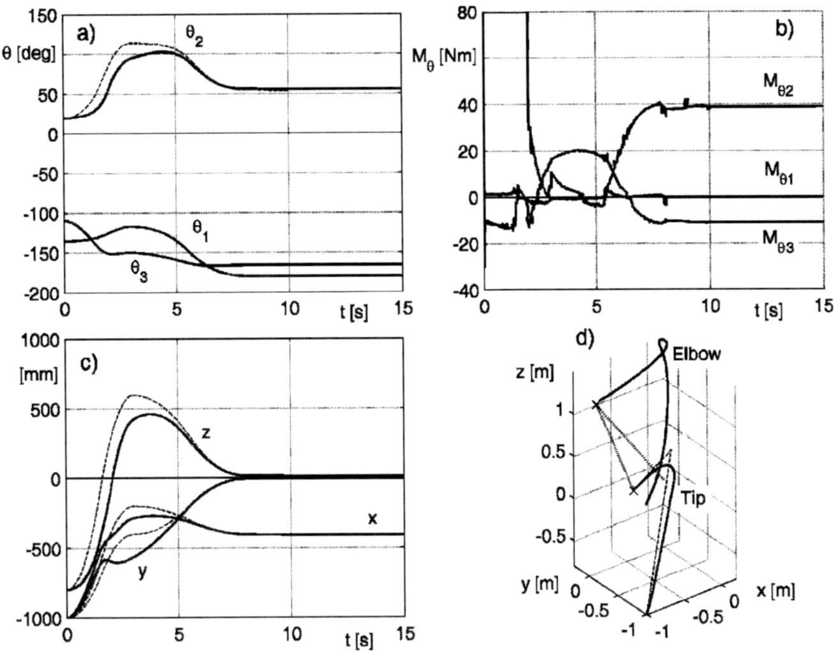

Fig. 3.31 Same as Fig. 3.30, but for the return run

than the saturation torque. It means that the arm cannot pick up such an heavy object at full elongation, a reasonable result. Owing to the saturation of the motor actuating angle θ_2, the error is now larger in the first part of the manoeuver, as seen by the fact that the dashed lines are not completely covered by the full lines.

The time required to reach the target is 12.70 s. The time history of the angles, the motor torques and the coordinates of the tip of the arm are plotted in Fig. 3.31 together with the trajectories of the tip and the elbow

Now strong saturation occurs at the beginning, when the arm must lift the specimen. However, the manoeuver goes on and is completed successfully in a reasonable time.

Chapter 4
Mobility on Planetary Surfaces

4.1 Mobility

Exploration vehicles, either manned or automatic, can use a variety of means of locomotion to achieve their goal. A first distinction must be made between ground vehicles, i.e. vehicles supported by a solid surface, atmospheric or sea vehicles, i.e. vehicles that move in a fluid without contact with the surface, be it a gas or a liquid, and space vehicles that move in the vacuum of space close to the surface.

The performance of ground vehicles is usually defined in terms of

- trafficability, defined as its ability to traverse difficult soil without loss of traction or even complete loss of mobility,
- maneuverability, which defines the ability of the vehicle to navigate through the environment, and
- terrainability, which defines the ability to negotiate terrain irregularities.

Ground vehicles may be supported and propelled by

- wheels,
- tracks,
- legs, or
- snakelike devices.

Other means of locomotion, often referred to as unconventional, may be used. They include, but do not reduce to, magnetic suspension, air-cushions, hopping devices and balloon (spherical) wheels. Often the devices that insure mobility (wheels, tracks etc.) are referred to as the running gear of the vehicle.

Vehicles moving in fluids may be supported by aerostatic (in general fluidostatic) or aerodynamic (in general fluid-dynamic) forces. Also here there are other alternatives like jet sustentation.

Finally, vehicles moving in space close to the surface of an airless body like the Moon are often referred to as hoppers, since they take off under rocket propulsion, perform a parabolic flight and then land braked by the same rocket used to take off. Hoppers can also be propelled by springs or electromagnetic actuators, particularly

G. Genta, *Introduction to the Mechanics of Space Robots*, Space Technology Library 26,
DOI 10.1007/978-94-007-1796-1_4,

in the case of low gravity bodies. Other possibilities like electromagnetic propulsion have been proposed but seldom studied in detail.

4.2 Vehicle–Ground Contact

All wheeled, tracked or legged vehicles moving on a solid surface use the contact with the ground as support mechanism, but often also to generate the forces in a direction parallel to the ground needed for propulsion. Normal and tangential forces are thus equally important to insure ground locomotion.

As stated in Chap. 2, most of the bodies to be explored in the solar system are characterized by a gravitational acceleration lower, in many cases much lower, than that of Earth. As a consequence the normal forces at the vehicle–ground contact are much lower than those encountered in traditional vehicle technology, at least with comparable vehicle mass. Tangential forces due to friction are correspondingly lower; but also in the case they are due to other mechanisms, like the so-called bulldozing force or the adhesion force, tangential forces are expected to be low in low gravity.

Low gravity simplifies the design of vehicles, since all contact forces and all stresses decrease with decreasing gravitational acceleration, but also sets limits to vehicle performance to correspondingly low levels. The limiting case is that of asteroids and comets, where locomotion on the ground may be problematic and where some sort of anchoring may be needed to avoid being projected into space.

Another difference between moving on Earth and on other celestial bodies is that most of the ground travel on Earth is done on prepared, often truly artificial, surface or at least on semi-prepared surface, while elsewhere vehicles and robots will have to manage unprepared ground.

Remark 4.1 In a sense, planetary mobility is similar to off-road locomotion on Earth, but with a number of differences.

Apart from the already mentioned lower gravitational acceleration, the main difference is due to the absence of humidity. The Moon, Mars and asteroid surface soil is completely free from liquid water. Since the presence of water is one of the main parameters in determining the characteristics of the soil, this makes a large difference between locomotion on Earth and on other bodies.

Cometary surface is very rich in water ice and little is known on the soil of the satellites of outer planets, except that there should be much ice too. However, ice is of hindrance to locomotion only at temperatures close to melting: the unique characteristic of water of decreasing its density when freezing causes the formation of a film of liquid water at the surface of the frozen soil under the running gear of the vehicle. The contact is then similar to a lubricated contact between solids and friction and adhesion forces are low. The higher the pressure, the more marked is this phenomenon. At low temperatures (e.g. at −40°C at Earth atmospheric pressure) this phenomenon does not take place at the contact pressures exerted by wheels and tracks and water ice is a very good surface to move onto.

Ice due to other substances, like frozen carbon dioxide common on Mars, does not melt under the running gear as a result of pressure, but it may do so due to heating: a film of liquid or vapor can form, reducing friction, if the part of the vehicle contacting the frozen ground is warm enough. This is, however, a different phenomenon and can be cared of simply by thermally insulating the parts of the vehicle in contact with the ground.

Titan soil may contain liquid hydrocarbons, producing effects like those of water on Earth, and surely contains a mixture of water and ice (slush) in the proximity of cold volcanoes. How this affects locomotion is still unknown.

Another point is the absence of products of biological origin, very important in giving peculiar characteristics to the wide variety of soils that can be found on Earth. The Moon and Mars are covered with regolith and so should be the surface of Mercury, of most asteroids and many satellites. Other satellites, comets and transneptunian objects should be covered with ice of different types.

Remark 4.2 It is likely that the variety of ground properties that may be found on a given body is much more limited than that found on our planet.

A further difference between the Earth and most other bodies is that the pressure and density of the atmosphere is much lower on the second ones: little atmosphere means low aerodynamic drag, and no constraint to the outer shape of the vehicle. However, this may be a negligible advantage, since all planetary exploration vehicles built up to now, and those that will be built in the predictable future, operate at a speed at which aerodynamic drag would at any rate be negligible.

Moving on unprepared ground means dealing with rough terrain and obstacles of different types. On the Moon and Mars there are stretches of fairly flat country, with many stones and boulders scattered on the ground, which proved to be manageable with confidence by wheeled vehicles. Lunar highlands and other zones on Mars are much more cratered and there are steeper slopes and more frequent obstacles, including ditches and cliffs.

As already stated, planetary soil is mostly made of regolith, an incoherent material, and its properties are mainly influenced by its granulometry and apparent density. The *apparent density* is the density of a given volume of soil taking into account also the voids between particles, while the *effective density* is the average density of the grains constituting the soil.

An incoherent soil deforms when the running gear of a vehicle exerts a pressure on it, owing to the sliding of the grains along their contact surface. The forces between grains that oppose to deformation are due to cohesion and increase with decreasing grain size: the smaller are the grains, the larger their surface/volume ratio is and the smaller the deformation under load.

Apart from cohesion, also friction forces between grains oppose to soil deformation.

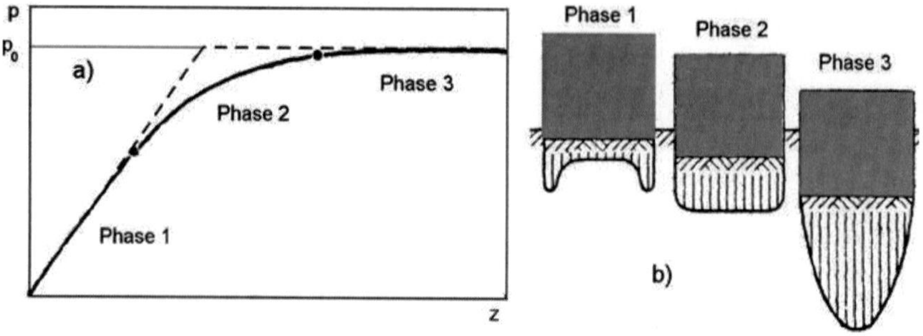

Fig. 4.1 Sinking of a pad into a homogeneous ground. (**a**) Pressure as a function of the sinking. (**b**) Pressure distribution in the various phases (from G. Genta, L. Morello, *The Automotive Chassis*, Springer, New York, 2009)

The study of the interaction between the running gear of the vehicle and the terrain is usually referred to as *terramechanics*[1] and has been developed for the rational study and design of off-road vehicles, in particular in the last 50 years.

Remark 4.3 Terramechanics relies mostly on empirical models, and different formulations can be found in the literature.

4.2.1 Contact Pressure

Consider a pad with area A that applies on the ground an average pressure p. A situation of this kind can occur when a rigid foot or a low pressure pneumatic tire are in contact with the ground. The experimental relationship between the pressure p and the sinking z is reported in Fig. 4.1a.

The sinking process can be divided into 3 phases. When the pressure on the ground is low, the soil is sheared at the periphery of the contact zone and is compressed. The pressure concentration at the periphery of the contact (Fig. 4.1b) increases at increasing cohesivity of the ground. A zone of compacted ground is formed at the center of the contact, and it compresses the ground at greater depths. The sinking is more or less proportional to the pressure, i.e. the behavior of the ground is linear.

The ground then starts behaving in a plastic way and larger and larger volumes of soil are interested by plastic deformations (phase 2). When all the ground in the contact zone is in plastic conditions (phase 3), an almost hydrostatic behavior is reached and the pressure does not increase any more with sinking. The pressure

[1]Perhaps the better known contributions to terramechanics are due to Bekker (e.g. M.G. Bekker, *Theory of Land Locomotion*, The University of Michigan Press, Ann Arbor, 1956; M.G. Bekker, *Off-the Road Locomotion*, The University of Michigan Press, Ann Arbor, 1960), who dealt also with the problem of ground locomotion on the lunar surface.

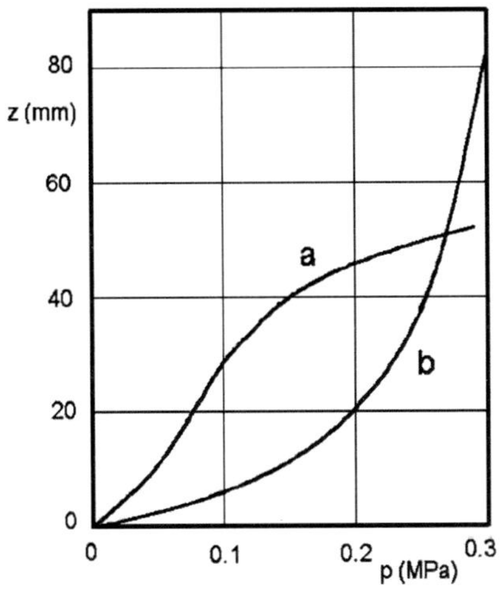

Fig. 4.2 Sinking as a function of the contact pressure, for a sandy ground. (**a**) Loose sand, with a depth of 200 mm; (**b**) compacted sand (from G. Genta, L. Morello, *The Automotive Chassis*, Springer, New York, 2009)

at which the phase 2 starts depends largely on the condition of the ground at the beginning of the contact: this explains why it is convenient to put a foot in the footprint of an earlier step.

A simple idealization that neglects phase 2 is that of a perfectly elastic–perfectly plastic ground: the ground is assumed to yield in a linear way up to a certain pressure p_0 and then the pad sinks without any increase in the reaction force. The maximum pressure p_0 the ground can exert is defined as the *bearing capacity* of the soil. Such an idealized behavior is shown by the dashed line in Fig. 4.1a.

The $z(p)$ curves for actual soils may deviate in different ways from the ones shown in the figure: as an example the curves for sandy soil (loose and compacted) are shown in Fig. 4.2.

In the first part of the curve in Fig. 4.1a (phase 1), where the soil behaves as an elastic material, it is possible to define a coefficient of proportionality, sometimes called the stiffness of the soil or the modulus of soil deformation

$$k = \frac{p}{z}. \tag{4.1}$$

Remark 4.4 The stiffness k does not depend only on the elastic characteristics of the ground (its Young's modulus E and its Poisson's ratio ν), but also on the area and the shape of the loading pad.

The stiffness k can be subdivided into k_c and k_ϕ, the cohesive and the frictional moduli, obtaining

$$k = \frac{k_c}{b} + k_\phi, \tag{4.2}$$

where b is the width (i.e., the smallest of the dimensions) of the pad. Operating in this way, k_c and k_ϕ are practically independent of the size of the pad, for small values of z.

Consider a pad exerting a constant pressure p on the soil. In Phase 1 the sinking z reduces to the elastic sinking z_e defined as

$$z_e = \frac{p}{k}. \tag{4.3}$$

The sinking z in the subsequent phases 2 and 3 may be expressed as

$$z = \frac{z_e p_0}{p_0 - p} = z_e \frac{1}{1 - p/p_0}, \tag{4.4}$$

where z_e is the elastic part of the soil deformation computed above, while p_0 is the *bearing capacity* of the soil. Obviously this formula is valid for $p \leq p_0$ only.

A more common empirical nonlinear expression used for approximating the law linking the pressure with the sinking into the ground is

$$p = \left(\frac{k_c}{b} + k_\phi\right) z^n, \tag{4.5}$$

that can be inverted, yielding the sinking of the pad as a function of the pressure

$$z = \left(\frac{p}{\frac{k_c}{b} + k_\phi}\right)^{1/n}. \tag{4.6}$$

The equation written in this way distinguishes between the reaction to the load due to cohesion and that due to the internal friction of the material. If the former is more important than the second the ground is said to be cohesive, otherwise it is frictional. While clay, particularly if wet, is a cohesive terrain, dry sand and regolith are mainly frictional.

Remark 4.5 The moduli k_c and k_ϕ have dimensions that depend on the exponent n.

Cohesive soils, like clay, have a low value of exponent n, in many cases as low as $\frac{1}{2}$. Frictional soils have larger values ad lunar regolith is characterized by a value of n close to 1.

The parameters entering into (4.5) are reported in Table 4.1 for dry sand and lunar regolith.

Remark 4.6 The values reported in the table are just typical values, and patches where the characteristics of the ground are quite different can be found both on Earth and on the Moon.

A model different from that shown in Fig. 4.1 and that of (4.5) is based on the assumption that when a pad is pressed on the ground the deformation is initially limited to small elastic deformation, which can be neglected. Then at a certain point, when a certain value of the pressure is reached, the soil starts yielding. This value

Table 4.1 Characteristics of dry sand and lunar regolith

	ρ kg/m^3	n	k_c N/m^{n+1}	k_ϕ N/m^{n+2}	c Pa	ϕ deg	K mm
Regolith	1,500–1,700	1	1,400	820,000	170	35–40	18
Dry sand	1,540	1.1	990	1,528,000	1,040	28–38	10–25

of the pressure under a plate with width (the smaller dimension) b and the ground can be referred to as the bearing capacity with no sinking

$$p_s = cJ_1N_c + \frac{1}{2}\rho g b J_2 N_\gamma, \tag{4.7}$$

where N_c and N_γ are nondimensional coefficients that are functions of the friction angle of the soil

$$\phi = \text{artg}(\mu^*), \tag{4.8}$$

and μ^* is the internal friction coefficient.[2] J_1 and J_2 are two other nondimensional coefficient that depend on the shape of the pad: for a long plate, like a track, $J_1 = J_2 = 1$; for a square plate $J_1 = 1.3$ and $J_2 = 0.8$; for a circular shape, like the contact area of a circular footplate or, approximately, a tire, $J_1 = 1.3$ and $J_2 = 0.6$.

In general, for a rectangular plate with length l and width b,

$$J_1 = \frac{l+b}{l+0.5b}, \qquad J_2 = \frac{l}{l+0.4b}.$$

The value of *bearing capacity factor* N_c was computed by Terzaghi[3] as

$$N_c = \cot(\phi)\left[e^{\pi\tan(\phi)}\tan^2\left(45+\frac{\phi}{2}\right)-1\right], \tag{4.9}$$

while there is no explicit formula for computing coefficient N_γ. N_c and N_γ are tabulated in Table 4.2.[4]

Regolith is a frictional soil but, in spite of the low value of c, on the Moon the second term in (4.7) may be comparable with the first (or even smaller) owing to the low gravitational acceleration. On asteroids the only term that does not vanish may be the first.

In cohesive soils the only thing that matters is the area of the pad, since the limitation to the supported load comes only from the pressure. On the contrary, on frictional soils the shape of the contact area matters, since the bearing capacity is

---

[2]The symbol μ^* is here used for the internal friction coefficient to avoid confusion with the traction coefficient μ defined later.

[3]K. Terzaghi, *Theoretical Soil Mechanics*, Wiley, New York, 1943; K. Terzaghi, R.B. Peck, G. Mesri, *Soil Mechanics in Engineering Practice*, Wiley, New York, 1996.

[4]D.P. Coduto, *Geotechnical Engineering*, Prentice Hall, Upper Saddle River, 1998.

Table 4.2 Values of the coefficients N_c, N_γ and N_q for different values of the friction angle ϕ

ϕ (deg)	N_c	N_γ	N_q	ϕ (deg)	N_c	N_γ	N_q
0	5.7	0	1	20	17.7	4.4	7.4
1	6	0.1	1.1	21	18.9	5.1	8.3
2	6.3	0.1	1.2	22	20.3	5.9	9.2
3	6.6	0.2	1.3	23	21.7	6.8	10.2
4	7.0	0.3	1.5	24	23.4	7.9	11.4
5	7.3	0.4	1.6	25	25.1	9.2	12.7
6	7.7	0.5	1.8	26	27.1	10.7	14.2
7	8.2	0.6	2.0	27	29.2	12.5	15.9
8	8.6	0.7	2.2	28	31.6	14.6	17.8
9	9.1	0.9	2.4	29	34.2	17.1	20.0
10	9.6	1.0	2.7	30	37.2	20.1	22.5
11	10.2	1.2	3.0	31	40.4	23.7	25.3
12	10.8	1.4	3.3	32	44.0	28.0	28.5
13	11.4	1.6	3.6	33	48.1	33.3	32.2
14	12.1	1.9	4.0	34	52.6	39.6	36.5
15	12.6	2.2	4.4	35	57.8	47.3	41.4
16	13.7	2.5	4.9	36	63.5	56.7	47.2
17	14.6	2.9	5.5	37	70.1	68.1	53.8
18	15.5	3.3	6.0	38	77.5	82.3	61.5
19	16.6	3.8	6.7	39	86.0	99.8	70.6

proportional to the width (the minimum dimension) of the pad: a square pad has a much higher carrying capacity than a long and narrow one.

The working condition of a vehicle that does not sink, i.e. if the bearing capacity with no sinking of the soil is not exceeded, is called *surface crossing*: this condition is very convenient, since it reduces the energy needed for motion and generally improves the performance of the vehicle.

When the pressure becomes larger, the pad starts sinking, while the pressure still increases: this condition is referred to as *subsurface crossing*. The relationship between the pressure and the plastic yield z is

$$p = cJ_1N_c + \frac{1}{2}\rho g b J_2 N_\gamma + \rho g z N_q, \tag{4.10}$$

where N_q is a third coefficient depending on ϕ (see Table 4.2). An expression for N_q is

$$N_q = e^{\pi \tan(\phi)} \tan^2\left(45 + \frac{\phi}{2}\right). \tag{4.11}$$

Other expressions for these coefficients can be found in the literature.[5]

[5] Ia.S. Ageikin, *Off-the-Road Mobility of Automobiles*, Balkema, Amsterdam, 1987.

Remark 4.7 While (4.7) can be used in this context without doubts, the use of (4.10) is questionable. The equation was developed for evaluating the bearing capacity of a foundation located a certain depth z, and not the bearing capacity of a pad set on the ground and then sinking to a depth z, which is clearly another thing.

Remark 4.8 Some doubts can also be cast on the applicability of these equations to situations in which the gravitational acceleration is very low, and certainly to the case where $g \to 0$, even if Terzaghi quotes as an example a weightless soil. In his book this reference is made for a soil in which the cohesive term dominates, and not to an actual very low gravity condition.

The first term of (4.10) depends on the cohesion of the soil, the second one on the density and the gravitational acceleration and the third one on the depth of penetration of the pad and also on the density and gravitation.

Equation (4.10) and the related coefficients are just approximations, and other formulae are also available. Moreover, in the case of low gravity bodies the apparent density and the cohesion of the material increase with the depth, even starting from the first few centimeters below the surface. An approach like the present one that assumes that the properties of the soil are constant is bound to yield conservative results. This is true not only for the third term, yielding the increase of carrying capacity with the depth of sinking, but also the other two that are anyway obtained integrating the characteristics of the material along the vertical direction.

Although newer, less empirical, methods are required to improve this situation, this approach allows to evaluate a conservative estimate of the carrying capacity of the soil.

The empirical equation (4.6) is usually applied instead of (4.10) to compute sinking.

Example 4.1 Consider a vehicle on the surface of the Moon, The running gear contacts the ground with 4 pads 200 mm wide and 200 mm long. Compute the maximum pressure on the soil that can be exerted without any plastic sinking.

If the mass of the vehicle is 1 ton, what is the safety factor against reaching the condition for starting sinking? Compute the depth of sinking using (4.6).

Assume $g = 1.62$ m/s^2 and the following data for the lunar soil: $n = 1$, $\rho = 1600$ kg/m^3, $c = 170$ Pa, $\phi = 37°$, $k_c = 1{,}400$ N/m^2, $k_\phi = 820{,}000$ N/m^3. Since the plates are square, $J_1 = 1.3$ and $J_2 = 0.8$.

From Table 4.2 it follows that $N_\gamma = 68.1$, $N_c = 70.1$.

The contribution to the bearing capacity with no sinking due to friction is 14,210 Pa, that due to cohesion is 15,490 Pa, for a total bearing capacity of 29,600 Pa. The force that can be withstood by the four plates is thus 4,738 N. Since the weight of the vehicle on the Moon is 1,620 N, the safety factor against sinking is 2.92.

The sinkage obtained from (4.6) is $z = 12$ mm.

4.2.2 Traction

Assuming no sinkage in the ground, i.e. neglecting the bulldozing component of the tractive force, the tangential force F_t that can be exerted by a wheel, track or foot with area A pressed on the ground with normal force F_n is given by the equation[6]

$$F_t = cA + F_n \mu^*. \tag{4.12}$$

The tangential (or shear) pressure

$$\tau = \frac{F_t}{A} \tag{4.13}$$

has then a maximum value, usually referred to as specific shear resistance of the soil:

$$\tau_0 = c + p\tan(\phi). \tag{4.14}$$

Remark 4.9 The traction that can be exerted at the vehicle–ground interface has thus two components. The first, due to cohesion, depends only on the supporting surface but not on the pressure exerted on the ground. The second one, due to friction, can be considered as a Coulomb friction and then does not depend on the contact area.

In case of purely frictional soil, like sand, the first term vanishes and the available traction depends only on the weight of the vehicle. In purely cohesive soils the effect of the normal force is nil and an heavy vehicle may experience large difficulties in proceeding; on the contrary what matters is the contact area of the vehicle with the ground. In the case of dry sand, the friction angle is about 35° and the corresponding friction coefficient is 0.7. In low gravity, the effect of friction can be small owing to the low load, and the cohesion forces can become important even in soils which would otherwise show low cohesion.

The ratio between the tangential and the normal force is the traction coefficient

$$\mu = \frac{F_t}{F_n} = \frac{cA}{F_n} + \mu^*. \tag{4.15}$$

Due to cohesion, the traction coefficient can increase in low gravity when the normal force is low even on soils whose cohesion is not high.

The value of the traction coefficient mentioned above is the maximum possible value, which is obtained when the shear resistance of the soil is reached simultaneously in the whole contact area.

The actual traction depends on the slip of the running gear (be it a foot, a track or a wheel) on the ground. To define the slip it must be remembered that the ground is a compliant body, and thus the part in contact with the running gear can move with

[6]M.G. Bekker, *Off the Road Locomotion*, The Univ. of Michigan Press, Ann Arbor, 1960.

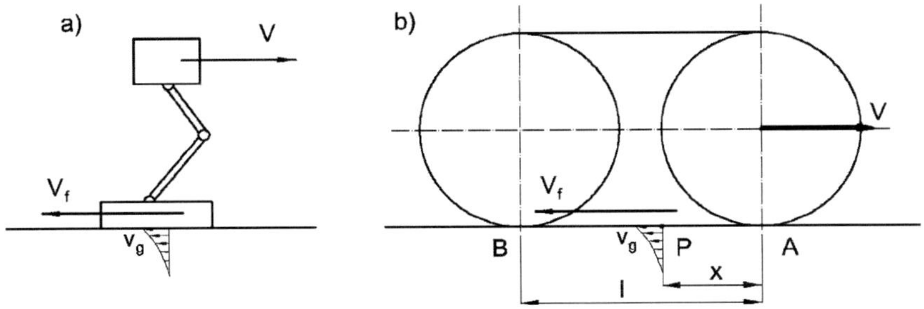

Fig. 4.3 Longitudinal slip in the case of a foot (**a**) and a track (**b**)

respect to a frame fixed to the ground and thus the presence a slip does not mean that the running gear slides on the surface of the soil.

Consider a foot of a vehicle traveling at a speed V (Fig. 4.3a). The foot is moving backward at a velocity V_f with respect to the body.[7] If $V = V_f$ there is no slip, but in general the part of the ground in contact with the foot moves backwards at a speed

$$v_g = V_f - V. \tag{4.16}$$

The speed of the contact area is an absolute velocity, and is possible only because the ground under the foot is compliant. The longitudinal slip can be defined as

$$\sigma = \frac{v_g}{V_f} = 1 - \frac{V}{V_f}. \tag{4.17}$$

The distance traveled backward by the contact area, i.e. the soil deformation, at time t is

$$d = v_g t = \sigma V_f t, \tag{4.18}$$

where t is the time elapsed from the instant the foot is pressed against the ground.

The shear stress of the ground can be expressed as a function of the soil deformation by the empirical formula due to Bekker,

$$\tau = \frac{c + p\tan(\phi)}{y_{\max}}\left[e^{(-K_2+\sqrt{K_2^2-1})K_1 d} - e^{(-K_2-\sqrt{K_2^2-1})K_1 d}\right], \tag{4.19}$$

where $y_{\max}$ is the maximum value of the function within braces.

The values of the two coefficients K_1 and K_2 must be measured on the particular soil: values $K_1 = 1$ and $K_2 = 1.1$ are mentioned for undisturbed, firmly settled silt and $K_1 = 0.32$ and $K_2 = 0.76$ for sandy loam.

---

[7] V is the absolute velocity of the vehicle (or better, the velocity of the vehicle with respect to the ground, assumed as fixed), while V_f is the relative velocity of the foot with respect to the vehicle. It is assumed to be positive when the foot moves backwards with respect to the body.

The tangential force thus starts from 0 at the instant when the foot enters contact with the ground (i.e. when $d = 0$), to increase until a maximum is reached and then decreases.

A simpler expression mentioned by Wong[8] is

$$\tau = \left[c + p\tan(\phi)\right]\left[1 - e^{-\frac{d}{K}}\right], \tag{4.20}$$

where K is the modulus of shear deformation (see Table 4.1 for the values for sand and lunar regolith). With this expression the tangential force starts from zero to increase asymptotically until the maximum value is reached for $d \rightarrow \infty$.

In the case of a foot, the pressure can be assumed to be constant and d increases linearly in time from the instant the foot touches the ground. In this case, the traction of a foot with area A on which a normal force F_z acts is

$$F_x = \left[Ac + F_z\tan(\phi)\right]\left[1 - e^{-\frac{\sigma V_f t}{K}}\right]. \tag{4.21}$$

In case of a track or a wheel the situation is different. The soil deformation d is zero in the point the running gear touches the ground (point A in Fig. 4.3b), and grows along the contact area, reaching a maximum in point B. Equation (4.18) still holds, but now t is the time elapsed from the instant a portion of the track touches the ground and that it reaches point P:

$$t = \frac{x}{V_f}. \tag{4.22}$$

The deformation in point P is thus

$$d = v_g t = \sigma x. \tag{4.23}$$

Using the formula suggested by Wong, the shear stress at point P is

$$\tau = \left[c + p\tan(\phi)\right]\left[1 - e^{-\frac{\sigma x}{K}}\right], \tag{4.24}$$

and the expression of the longitudinal force is

$$F_x = b\int_0^l \left[c + p\tan(\phi)\right]\left[1 - e^{-\frac{\sigma x}{K}}\right]dx. \tag{4.25}$$

This integral can be performed once the pressure distribution under the track is known.

If the pressure is constant, it follows that

$$F_x = \left[cA + F_z\tan(\phi)\right]\left[1 - \frac{K}{\sigma l}\left(1 - e^{-\frac{\sigma l}{K}}\right)\right]. \tag{4.26}$$

[8]J.Y. Wong, *Theory of Ground Vehicles*, Wiley, New York, 2001.

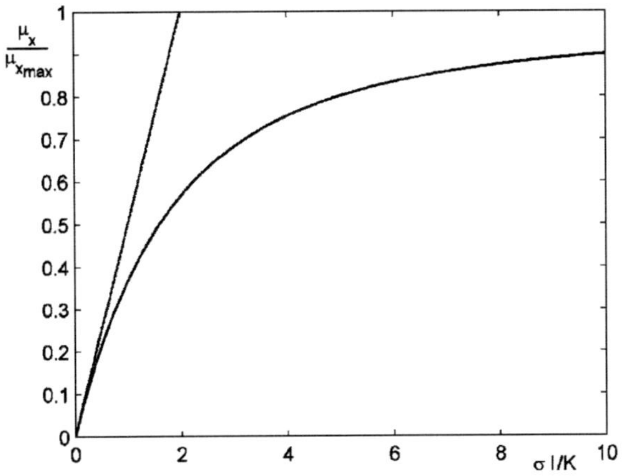

Fig. 4.4 Nondimensional traction coefficient $\mu_x/\mu_{x_{\max}}$ as a function of the nondimensional slip $\sigma l/K$

The traction coefficient is thus

$$\mu_x = \mu_{x_{\max}}\left[1 - \frac{K}{\sigma l}\left(1 - e^{-\frac{\sigma l}{K}}\right)\right], \tag{4.27}$$

where

$$\mu_{x_{\max}} = \left[\frac{cA}{F_z} + \tan(\phi)\right]. \tag{4.28}$$

A nondimensional plot $\mu_x/\mu_{x_{\max}}$ as a function of $\sigma l/K$ is reported in Fig. 4.4. For low values of the longitudinal slip the traction coefficient is almost linear with the slip and can be approximated as

$$\mu_x = C_\sigma \sigma, \tag{4.29}$$

where the *longitudinal force stiffness* is

$$C_\sigma = \left(\frac{\partial \mu_x}{\partial \sigma}\right)_{\sigma=0} = \left[\frac{cA}{F_z} + \tan(\phi)\right]\frac{l}{2K}. \tag{4.30}$$

For large values of the slip, the traction coefficient tends to a maximum value for $\sigma \to \infty$, since (4.20) was used. If the law $\tau(d)$ were expressed by (4.19), the law $\mu_x(\sigma)$ would have reached a maximum for a finite value of σ, to decrease in a more or less pronounced way.

Example 4.2 Compute the maximum value of the traction coefficient $\mu = F_t/F_n$ as a function of the normal pressure p for 3 different values of ϕ (25°, 30° and 35°) on lunar regolith with $c = 170$ Pa.

The results are reported in Fig. 4.5. The maximum traction coefficient increases strongly with decreasing pressure on the ground owing to the cohesive term that would lead to an infinitely large traction coefficient when the normal force tends to zero.

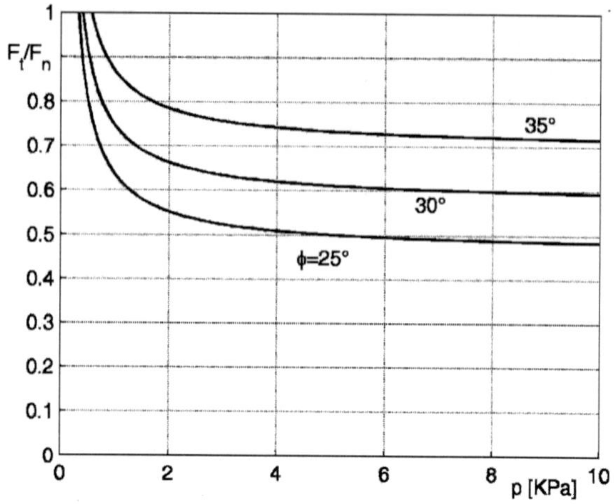

Fig. 4.5 Ratio between tractive and normal force F_t/F_n as functions of the normal pressure on the ground for 3 different values of the friction coefficient

When the pad sinks in the ground or is provided with spuds or threads that sink in the ground, another form of traction force is present, the so-called bulldozing force.

A formula proposed to take into account also the bulldozing force due to ridges or treads protruding into the ground for a depth h under a plate with width b is

$$F_b = 2cA\frac{h}{b} + 0.64F_n\mu^*\frac{h}{b}\left[\frac{\pi}{2} - \text{artg}\left(\frac{h}{b}\right)\right]. \tag{4.31}$$

Also this expression is made by a component due to cohesion, proportional to c and to the contact area but independent from the load, and a friction component, proportional to the normal force and to μ^* but independent from the area.

The cohesion term is proportional to the relative depth of the ridges h/b while the friction term grows initially almost linearly, but then much more slowly. The expressions

$$\Delta F_c = 2\frac{h}{b}, \qquad \Delta F_f = 0.64\frac{h}{b}\left[\frac{\pi}{2} - \text{artg}\left(\frac{h}{b}\right)\right] \tag{4.32}$$

are the factors by which the components of the traction force due, respectively, to cohesion and friction must be multiplied to take into account the effect the ridges. They are plotted as functions of h/b in Fig. 4.6.

Remark 4.10 The presence of spuds is much more effective on cohesive soil than on frictional soil, where they can produce only a moderate increase in traction due to internal friction.

If the contact area is loaded not only by a normal force but also by a tangential force, the sinking increases, as shown by some experimental curves reported in Fig. 4.7. The figure is referred to a sandy soil and curves a, b and c refer to values of the normal pressure p of 0.02, 0.03 and 0.04 MPa, respectively.

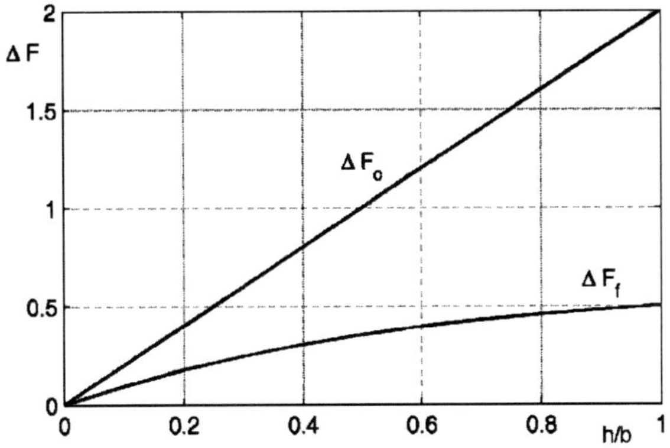

Fig. 4.6 Relative increase of the components due to cohesion and to friction of the tangential force at the vehicle–ground interface, due to the presence of ridges as function of the depth/width ratio h/b

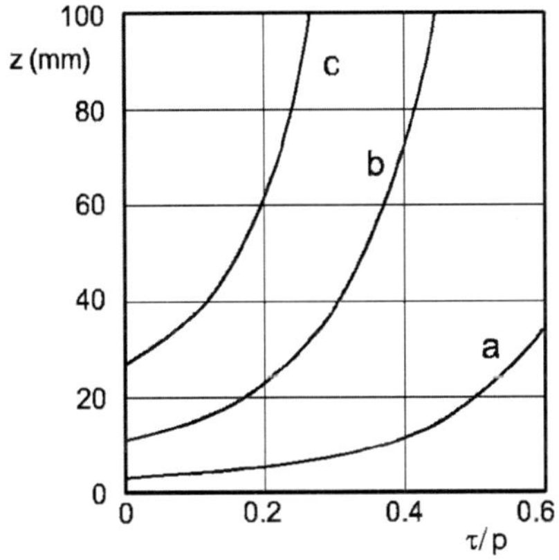

Fig. 4.7 Sinking of a plate on sand under the simultaneous effect of a normal pressure p and a tangential pressure τ. Curves a, b and c refer to values of the pressure p of 0.02 MPa, 0.03 MPa and 0.04 MPa, respectively

The bearing capacity of the soil decreases with increasing tangential force and (4.10) may be modified by introducing two correction factors[9] $K_{\beta 1}$ and $K_{\beta 2}$

$$p_0 = cJ_1 N_c K_{\beta 1} + \frac{1}{2}\rho g b J_2 N_\gamma K_{\beta 2} + \rho g z N_q. \tag{4.33}$$

Their values depend on the angle β

$$\beta = \text{artg}\left(\frac{F_t}{F_n}\right) \tag{4.34}$$

the resultant force makes with the perpendicular to the ground:

$$K_{\beta 1} = \frac{3\pi - 2\beta}{3\pi + 2\beta}, \qquad K_{\beta 2} = \frac{\pi - 4\beta \tan(\phi)}{\pi + 4\beta \tan(\phi)}. \tag{4.35}$$

[9]Ia.S. Ageikin, *Off-the-Road Mobility of Automobiles*, Balkema, Amsterdam, 1987.

4.3 Wheeled Locomotion

Most planetary rovers and vehicles, from the Russian Lunokhod and the American *Apollo* Roving Lunar Vehicle (RLV) of the 1970s to the recent robotic rovers such as *Spirit* and *Opportunity* and those still in the design stage are wheeled. Basically wheels are best suitable for prepared ground, but their simple mechanical design and control makes them a good choice also for off-road locomotion, in particular for dry ground.

4.3.1 Stiff Wheels Rolling on Stiff Ground

If a rigid wheel, which can be thought as a short cylinder, rolls on a rigid flat ground, the contact occurs along a line and the pressure is infinitely large, an impossible result. The high contact pressure at the wheel–ground interface leads to deformations of both bodies that reduce substantially the contact pressure.

The vehicle–soil contact must thus be considered as the interaction between compliant bodies. Even in the case their compliance is low, like in the contact between steel wheels and steel rails, the relevant phenomena can be understood only taking into account their compliance.

The Hertz theory for the contact between elastic bodies can be used for a cylindrical object pressed against a flat surface without rolling on it if

- the materials behave in a linear way,
- the deformations are small when compared with the size of the objects, and
- the bodies exchange no tangential force.

The contact area between two general bodies in contact, once projected on a plane perpendicular to the line connecting the centers of curvature in the contact point, has the shape of an ellipse. If the contact surfaces are two cylinders whose axes are parallel (or a cylinder and a plane, as a limiting case of a cylinder with a vanishing curvature) the ellipse degenerates into a rectangle.

The wheel–ground contact may thus be modeled as that of a cylinder with radius R and width b contacting, under the action of a force F_n, a planar strip with width b: clearly this is a further approximation, since in the real world the width of the plane simulating the ground is much larger than that of the wheel, but in this way it is possible to assume that in all planes perpendicular to the cylinder axis the situation is the same, and is equal to that occurring with an infinite cylinder contacting a semi-space.

The length of the contact zone a (Fig. 4.8a) is given by

$$a = 2\sqrt{\frac{F_n R(\theta_1 + \theta_2)}{\pi b}}, \tag{4.36}$$

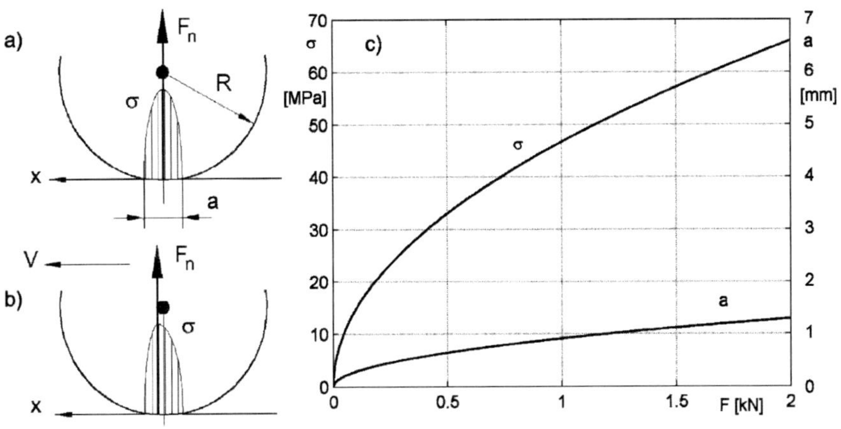

Fig. 4.8 Contact zone and pressure distribution for a cylinder pressed against a plane. (**a**) Elastic bodies, no rolling; (**b**) rolling with energy dissipation; (**c**) length of a contact zone and maximum pressure for an aluminum wheel pressed against a stone flat surface

where θ_i is a coefficient depending on the Young's modulus E_i and the Poisson's ratio ν_i of the material constituting the ith body:

$$\theta_i = 4\frac{1-\nu_i^2}{E_i}. \tag{4.37}$$

The average pressure on the ground is

$$\overline{\sigma} = \frac{F_n}{ab} = \frac{1}{2}\sqrt{\frac{\pi F_n}{Rb(\theta_1+\theta_2)}}. \tag{4.38}$$

The pressure distribution along the longitudinal direction is

$$\sigma = \frac{4F_n}{\pi ab}\sqrt{1-\frac{4x^2}{a^2}} = 2\sqrt{\frac{F_n}{\pi Rb(\theta_1+\theta_2)}}\sqrt{1-\frac{4x^2}{a^2}} \tag{4.39}$$

and its maximum value is

$$\sigma_{\max} = \frac{4}{\pi}\overline{\sigma}. \tag{4.40}$$

In case the two bodies are made by the same material, the expressions simplify as

$$a = 4\sqrt{\frac{4F_nR(1-\nu^2)}{\pi bE}}, \tag{4.41}$$

$$\sigma_{\max} = \sqrt{\frac{F_nE}{2\pi rh(1-\nu^2)}}. \tag{4.42}$$

This leads to the obvious consideration that the stiffer are the wheel and the ground, the smaller is the contact area and the greater is the contact pressure. Another, less obvious, consideration is that the contact pressure grows, for a given pair of bodies in contact, as the square root of the load.

The displacement in a direction perpendicular to the ground cannot be easily computed in the case of contact between a cylinder and a plane. From purely geometrical considerations, the displacement corresponding to a contact length a is

$$\Delta z = R\left\{1 - \cos\left[\text{arsin}\left(\frac{a}{2R}\right)\right]\right\}. \tag{4.43}$$

Since $a/2R$ is a small quantity,

$$\Delta z \approx \frac{a^2}{8R} = \frac{F_n(\theta_1 + \theta_2)}{2\pi b}. \tag{4.44}$$

This result is just a rough order of magnitude evaluation.

Example 4.3 Consider a solid aluminum ($E = 72$ GPa, $\nu = 0.33$) wheel with a diameter of 200 mm and a width of 30 mm, supported on a flat volcanic stone with $E = 25$ GPa, $\nu = 0.3$. Compute the length of the contact area and the pressure on the ground with increasing load up to 2,000 N.

The results are shown in Fig. 4.8c. Note that the length of the contact zone is quite small (about 1 mm, leading to a 30 mm^2 contact area, when the load is 1,000 N) and the contact pressure is correspondingly quite high. For the highest values of the pressure, plasticity of the materials starts entering the picture and the elastic approximation becomes less accurate. The assumption that the deformations are small if compared with the size of the objects is on the contrary applicable.

In case of a rigid wheel rolling on a rigid surface, the center of rotation is located at the wheel–ground contact point and the rolling radius coincides with the wheel radius R. Taking into account the deformation of the two bodies, the rolling radius decreases slightly and the center of rotation is located slightly below the ground surface.

If the wheel is rolling and the material is perfectly elastic, the situation is essentially the same and very little energy is lost, only due to the small sliding occurring at the contact. However, no material is exactly elastic and some energy is lost in the deformation and release cycles occurring in both the ground and the wheel. This energy dissipation is the main cause of rolling resistance, with sliding at the interface contributing only to a lesser extent.

As it will be seen in greater detail below, owing to energy dissipation, the pressure distribution, which in case of elastic, non-rolling, cylinders had an elliptical pattern (4.39), is no more symmetrical and its resultant displaces forward in the direction of motion (Fig. 4.8b). The resultant is no more passing for the center of the wheel, and this displacement causes a torque that hampers motion. This torque is seen on the vehicle as rolling resistance.

The rolling resistance is usually expressed in terms of the *rolling coefficient* defined as the ratio between the rolling resistance and the load acting on the wheel

$$f = \frac{F_x}{F_z}. \tag{4.45}$$

In general the rolling coefficient depends on many parameters, and also on the load acting on the wheel.

The value of the rolling coefficient for a steel wheel rolling on a steel rail is between 0.001 and 0.005.

4.3.2 Compliance of the Wheel and of the Ground

The Hertz theory is based on the assumption that the deformations are much smaller than the dimensions of the bodies in contact. This occurs only when a stiff wheel rolls on stiff ground as in rail transportation or in the case of a metal wheel rolling on a slab of stone. In all other cases, and above all in motion on unprepared ground, the deformations of the soil are such that the Hertz theory cannot be applied.

Three cases can be added to that of a stiff wheel rolling on a stiff ground studied above:

1. a stiff wheel rolling on compliant ground;
2. a compliant, possibly elastic, wheel rolling on stiff ground, and
3. a compliant wheel rolling on a compliant ground.

As a general rule, the resistance to motion encountered by the wheel is caused by the energy dissipation in both the wheel and the ground. Since the wheel can be designed in such a way that energy dissipations are as small as possible, while there is little to do to change the properties of natural ground, it is expedient that the latter deforms as little as possible, while all deformations are concentrated in the wheels (case 2).

Case 2 is typical of automotive technology, where low stiffness, mostly pneumatic, wheels are used on prepared road covered with tarmac or concrete that is quite stiff.

Case 3 occurs in off road locomotion, where the ground is compliant, but the wheels are as compliant as possible.

Case 1 is regarded as the worst situation, since the rigid wheel causes much permanent deformation in the ground and resistance to motion is large.

As already stated, the wheel must be compliant to prevent the ground from deforming too much. Usually this is achieved by building the wheel in two parts: a stiff, usually metallic, hub inset into a tire that supplies the required elasticity. The importance of the compliance of the wheel was realized long time ago and was at the base of the patent of the *aerial wheel* by Robert William Thomson, the first pneumatic tire. In 1849 he published some experimental results showing a decrease of the rolling resistance with respect to a wooden wheel with steel tire by 60% on hard macadam and 310% on broken flint, a type of ground not much different from some stretches of regolith with larger stones.

Thomson's pneumatic tire was, however, unpractical since it was made of a belt of leather riveted to the rim of the wheel, with an inner airtight, rubberized, fabric, tube and had no success. In 1888 the pneumatic tire was re-invented by John Boyd Dunlop and was a success, even if it was hardly more practical than that invented by Thomson. Since then, it has become the standard tire for vehicles of all kinds. At present the production of pneumatic tires is of about one billion per year.

In the beginning, however, pneumatic tires were not the only design and two other design trends were started in the nineteenth century: solid rubber tires and elastic, mainly metallic, tires in which deformations were due to spring-like mechanisms. Around this elastic structure there was at any rate either a compliant layer of solid rubber, or a pneumatic tire, usually thinner than the pneumatic tires used directly on the rigid hub. Pneumatic tires were initially used on bicycles, which were enjoying a growing popularity. Only later they could be used on heavier vehicles.

When a wheel with a pneumatic or elastic tire rolls on a prepared surface, like concrete of tarmac, the deformation of the wheel is large and the ground can be considered as a rigid surface. When the same wheel rolls on a natural surface, both objects in contact must be considered as compliant. On some robotic rovers wheels with no compliant tire are used: in this case the deformations are localized only in the ground.

As already stated, the wheel–ground contact is the contact between compliant bodies and this feature is essential to understand how a wheel works. The larger is the compliance of the bodies in contact, the larger is the contact area and thus the smaller is the contact pressure, at a given value of the contact force.

Remark 4.11 Not only the rolling resistance reduces when the tire is a compliant body, but also the generation of tangential forces in longitudinal and transversal direction improves.

In standard vehicles used on Earth, both on and off road, non-pneumatic tires have been abandoned since the 1920s, except in those cases where the vulnerability of tires may present an unacceptable disadvantage, like in some military vehicles. The rigid structure of the wheel, made by the disc and the rim, is thus surrounded by a compliant element, made by the tire and the tube. The latter can be absent in tubeless wheels, in which the tire fits airtight on the rim and the carcass contains directly the air. The tire is a complex structure, made by several layers of rubberized fabric, with a large number of cords running in the direction of the warp and only a few in that of the weft. The number of plies, their orientation, the formulation of the rubber and the material of the cords are widely variable: these are the parameters giving its peculiar characteristics to each tire.

Independently from their use, structurally tires belong mainly to two types: bias tires and radial tires, although a sort of in-between type is constituted by belted tires. In the older bias tires the carcass was made by a number of plies whose reinforcement runs at an angle of 35°–40° with respect to the circumferential direction. In

Fig. 4.9 (**a**) Picture of the wheel of the Lunar Roving Vehicle on display at the U.S. Space & Rocket Center in Huntsville; (**b**) an elastic motorized wheel designed by the Author for a small rover

the case of belted tires, a number of plies run under the tread with a small angle, about 15° from the circumference. This belt gives the tire a higher circumferential stiffness. Radial tires are made by plies oriented in direction perpendicular to the circumference and by belt plies. This structure leads to more compliant sidewalls and to a tread band which is stiffer in circumferential direction.

Remark 4.12 Presently radial tires have almost completely substituted the other types, owing to their superior comfort and performance.

Apart from the already mentioned main function of the tire, i.e. distributing the vertical load on a large enough area, a secondary function is that of insuring an adequate compliance, needed to absorb the irregularities of the road. It is essential that the compliance in different directions is suitably distributed: a tire must be compliant in vertical direction, while being stiffer in circumferential and lateral directions. This second function of the compliance of the tires is increasingly important with increasing speed: at the speed all present robotic rovers operate rigid wheels are adequate from this point of view.

When designing the LRV (Lunar Roving Vehicle) for the *Apollo* missions, some sort of elastic tires were needed but pneumatic or solid rubber tires were discarded mainly for reducing the vehicle mass. Designers resorted to metal elastic tires, of the kind that were widely tested at the end of the nineteenth century, when alternatives to pneumatic wheels were sought.

Tires made by an open steel wire mesh, with a number of titanium alloy plates acting as tread in the ground contact zone, were then built. Inside the tire a second smaller more rigid frame acted as a stop to avoid excessive deformation under high impact loads (Fig. 4.9a).

Fig. 4.10 The 'twheel' by Michelin, an airless tire being developed for automotive applications

Pneumatic tires designed to operate at Earth gravity may prove to be too stiff for the Lunar or Mars environment and, above all, elastomeric material used in tires are not suitable for the much more harsh space environment. It is still to be understood whether it is possible to obtain rubber formulations suitable for prolonged use in lunar or martian environment: if this can be done the long experience available in the field of tire design and construction will allow to build pneumatic wheels for planetary rovers and robots, a much desirable solution. Such pneumatic tires would be an enabling technology for planetary exploration.

In the recent years there was a revival of the research on non-pneumatic tires. Michelin presented a non-pneumatic tire, named *twheel* (tire-wheel, Fig. 4.10), whose design is based on a very flexible rim, carrying a thin solid rubber tire, attached to the hub through springs. The rim and the springs may be made from metal or composite material (CRP, carbon reinforced plastics, or GRP, glass reinforced plastics). The twheel, or similar structures, may be good solutions for planetary vehicles and robots.

4.3.3 Contact Between Rigid Wheel and Compliant Ground

Consider a rigid wheel, rolling on compliant soil (Fig. 4.11): the wheel is free to roll and is pulled in x direction by the force F_x while being pressed on the ground by force F_z. If the deformation is partially anelastic, there is some elastic return of the ground, as in Fig. 4.11a, and z_r is not 0. If the deformation of the soil is fully anelastic, its deformation is permanent and the contact is restricted to arc AB ($z_r = \theta_r = 0$).

The wheel exchanges with the ground a force per unit area σ_r directed radially.

Assuming that the wheel is a cylinder with radius R and width b, the equilibrium equations of the wheel in x and z directions are

$$F_x = bR \int_{\theta_r}^{\theta_0} \sigma_r \sin(\theta)\, d\theta, \tag{4.46}$$

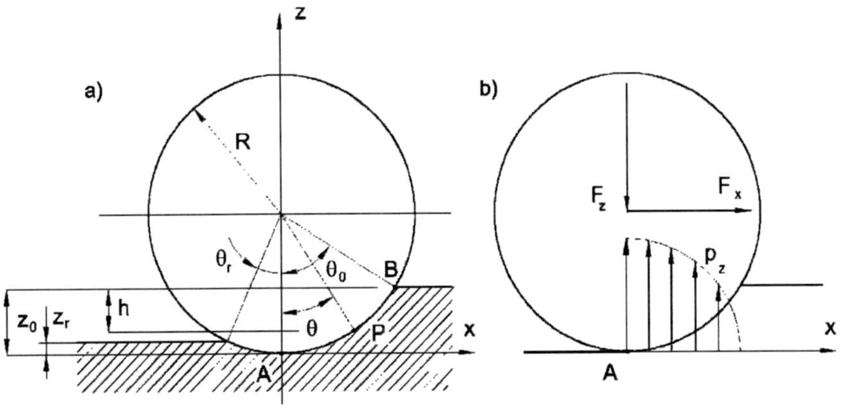

Fig. 4.11 Sinking of a rigid wheel into compliant, anelastic soil. (**a**) Geometry of the contact; (**b**) forces acting on the wheel, in the case of no elastic return of the ground

$$F_z = bR \int_{\theta_r}^{\theta_0} \sigma_r \cos(\theta)\, d\theta. \tag{4.47}$$

The equilibrium for rotation is always satisfied, since both forces F_x and F_z and pressures σ_r act through the center of the wheel O.

Assuming that the pressure is expressed in each point by (4.5),

$$\sigma_r = \left(\frac{k_c}{b} + k_\phi\right) h^n = R^n \left(\frac{k_c}{b} + k_\phi\right) \left[\cos(\theta) - \cos(\theta_0)\right]^n, \tag{4.48}$$

the forces are

$$F_x = R^{n+1}(k_c + bk_\phi) \int_{\theta_r}^{\theta_0} \left[\cos(\theta) - \cos(\theta_0)\right]^n \sin(\theta)\, dz, \tag{4.49}$$

$$F_z = R^{n+1}(k_c + bk_\phi) \int_{\theta_r}^{\theta_0} \left[\cos(\theta) - \cos(\theta_0)\right]^n \cos(\theta)\, dz. \tag{4.50}$$

Force F_x is the resistance to motion due to the anelastic deformation of the soil. It is referred to as *compaction resistance*.

The elastic return of the ground must be stated to compute the forces. Three cases are possible:

- Perfectly elastic ground: the elastic return is complete and $z_r = z_0$ or $\theta_r = -\theta_0$. This is just an ideal case that cannot occur in practice when the wheel is moving.
- Perfectly anelastic ground: the elastic return is null and $z_r = 0$ or $\theta_r = 0$. This condition may be close to the actual situation, and often the ground is assumed to be anelastic since the computations are much easier in this way and anyway there are no data about the elastic return.
- Partially elastic ground: the elastic return is incomplete. It is usually assumed that $z_r = \lambda z_0$, λ being a parameter that depend on the soil characteristics, but also on

the wheel pattern and, for braking or driving wheels, on the slip ratio (see below). In the latter case, if the wheel digs in the ground and accumulates material behind itself, it may be even larger than 1 (but in this case it is not a result of elasticity).

Angles θ_0 and θ_r can be written as functions of z_0:

$$\theta_0 = \arcos\left(1 - \frac{z_0}{R}\right), \qquad \theta_r = -\arcos\left(1 - \frac{\lambda z_0}{R}\right). \tag{4.51}$$

The vertical force

$$F_z = R^{n+1}(k_c + bk_\phi) \int_{-\arcos(1-\frac{\lambda z_0}{R})}^{\arcos(1-\frac{z_0}{R})} \left[\cos(\theta) - \cos(\theta_0)\right]^n \cos(\theta)\, dz \tag{4.52}$$

acting on the wheel is known and this equation allows to compute the sinking of the wheel z_0. Its solution must be performed numerically.

This approach has, however, an inconsistency: the contact pressure does not go to zero at the point the wheel parts contact with the ground. Some modifications, based on empirical results, have been proposed to overcome this difficulty.

Following Ishigami et al.,[10] it is possible to define an angle

$$\theta_m = (a_0 + a_1\sigma)\theta_0, \tag{4.53}$$

where a_0 and a_1 are parameters depending on the ground-wheel interaction (suggested values are $a_0 \approx 0.4$, $0 \le a_1 \le 0.3$) and σ is the longitudinal slip, at which the pressure on the ground reaches its maximum value. The pressure distribution, to be substituted to (4.48), is

$$\begin{aligned} \sigma_r &= R^n\left(\frac{k_c}{b} + k_\phi\right)\left[\cos(\theta) - \cos(\theta_0)\right]^n \quad \text{for } \theta_m \le \theta < \theta_0, \\ \sigma_r &= R^n\left(\frac{k_c}{b} + k_\phi\right)\left[\cos\left[\theta_0 - \frac{(\theta - \theta_r)(\theta_0 - \theta_m)}{\theta_m - \theta_r}\right] - \cos(\theta_0)\right]^n \\ &\quad \text{for } \theta_r < \theta \le \theta_m. \end{aligned} \tag{4.54}$$

Since the forces must be anyway computed numerically, the use of this more complex relationship does not complicate the study.

If the soil is considered as perfectly anelastic (Fig. 4.11b), things get much simpler. The forces can be written as

$$F_x = bR \int_0^{\theta_0} \sigma_r \sin(\theta)\, d\theta = b \int_0^{z_0} \sigma_r\, dz, \tag{4.55}$$

[10] G. Ishigami, A. Miwa, K. Nagatani, K. Yoshida, *Terramechanics-Based Model for Steering Maneuver of Planetary Exploration Rovers on Loose Soil*, Journal of Field Robotics, Vol. 24, No. 3, pp. 233–250, 2007.

$$F_z = bR \int_0^{\theta_0} \sigma_r \cos(\theta)\, d\theta = b \int_0^{x_0} \sigma_r\, dx. \tag{4.56}$$

Using (4.48) for the pressure, the horizontal force is

$$F_x = b\left(\frac{k_c}{b} + k_\phi\right) \int_0^{z_0} (z_0 - z)^n\, dz, \tag{4.57}$$

i.e.:

$$F_x = \frac{b}{n+1}\left(\frac{k_c}{b} + k_\phi\right) {z_0}^{n+1}. \tag{4.58}$$

The vertical force distribution is

$$dF_z = b\sigma_r\, dx = b\left(\frac{k_c}{b} + k_\phi\right)(z_0 - z)^n\, dx. \tag{4.59}$$

The relationship between x and z is

$$x^2 + (R - z)^2 = R^2, \tag{4.60}$$

which, in case of small sinking, can be simplified as

$$x^2 \approx 2Rz. \tag{4.61}$$

The maximum vertical pressure occurs for $x = 0$ and has the value

$$p_{\max} = \left(\frac{k_c}{b} + k_\phi\right) z_0^n. \tag{4.62}$$

The force distribution in vertical direction is thus

$$dF_z = p_{\max} b\left(1 - \frac{x^2}{2Rz_0}\right)^n dx. \tag{4.63}$$

This is just an approximation as is shown by the fact that the vertical force vanishes for

$$x = \sqrt{2Rz_0}, \tag{4.64}$$

while it should vanish in point B, i.e. for

$$x = \sqrt{2Rz_0 - z_0^2}. \tag{4.65}$$

For instance, if the sinkage is 20%, i.e. $z_0/R = 0.2$, the correct value of x/R is 0.49, while (4.64) yields 0.63. If the sinkage is 10%, the error becomes almost negligible: 0.44 against 0.45.

The approximated vertical pressure distribution is shown in Fig. 4.11b.

The expression of the vertical force is thus

$$F_z = p_{\max} b \int_0^{\sqrt{2Rz_0}} \left(1 - \frac{x^2}{2Rz_0}\right)^n dx. \tag{4.66}$$

In case $n = 1$, as often occurs for regolith, the vertical force is

$$F_z = \frac{2b\sqrt{2Rz_0}}{3} z_0 \left(\frac{k_c}{b} + k_\phi\right). \tag{4.67}$$

If $n \neq 1$, it is possible to write the series for $(1 - \frac{x^2}{2Rz_0})^n$:

$$\left(1 - \frac{x^2}{2Rz_0}\right)^n = 1 - n\frac{x^2}{2Rz_0} + \frac{n(n-1)}{2}\left(\frac{x^2}{2Rz_0}\right)^2 + \cdots, \tag{4.68}$$

and to retain only the first two terms, obtaining

$$F_z = \frac{(3-n)b\sqrt{2Rz_0}}{3} z_0^n \left(\frac{k_c}{b} + k_\phi\right). \tag{4.69}$$

By solving this equation in z_0, the sinkage of the wheel is obtained as a function of the load

$$z_0 = \left[\frac{3F_z}{(3-n)(\frac{k_c}{b} + k_\phi)b\sqrt{2R}}\right]^{\frac{2}{2n+1}}. \tag{4.70}$$

The resistance to motion, in this case referred to as *compaction resistance*, is obtained by introducing (4.70) into (4.58)

$$F_x = \frac{1}{n+1}\left[\frac{1}{b(\frac{k_c}{b} + k_\phi)}\right]^{\frac{1}{2n+1}} \left[\frac{3F_z}{(3-n)\sqrt{2R}}\right]^{\frac{2(n+1)}{2n+1}}. \tag{4.71}$$

If $n = 1$ the compaction resistance is

$$F_x = \frac{1}{2}\sqrt[3]{\frac{1}{b(\frac{k_c}{b} + k_\phi)}\left(\frac{3F_z}{2\sqrt{2R}}\right)^4}. \tag{4.72}$$

The rolling coefficient

$$f = \frac{F_x}{F_z} = \frac{1}{n+1}\left[\frac{F_z}{b(\frac{k_c}{b} + k_\phi)}\right]^{\frac{1}{2n+1}} \left[\frac{3}{(3-n)\sqrt{2R}}\right]^{\frac{2(n+1)}{2n+1}} \tag{4.73}$$

is thus a function of the vertical force and, in case of $n = 1$, is proportional to $\sqrt[3]{F_z}$.

Since the rolling coefficient is a function of the vertical force, from the viewpoint of the rolling resistance it is expedient to subdivide the load on a larger number of less loaded wheels: in this way the wheels sink less and the deformation of the ground is smaller. This is, however, true only at equal wheel diameter.

Example 4.4 Consider a Mars rover with 6 rigid wheels operating on a ground whose characteristics are similar to lunar regolith. Assume the following data:

General: $m = 150$ kg, $g = 3.77$ m/s^2; wheels: $R = 150$ mm; $b = 80$ mm; soil: $n = 1$, $c = 200$ Pa, $\phi = 35°$, $k_c = 1{,}400$ N/m^2, $k_\phi = 820{,}000$ N/m^3.

Compare the compaction resistance of the rover with 6 wheels with that of a rover with 4 wheels having the same characteristics.

Assuming that the load is equally subdivided on the six wheels, the force on each of them is $F_z = 94.25$ N. The maximum sinkage of the wheels in the ground is $z_0 = 25$ mm, corresponding to $0.164\ R$. The approximation of the formulae seen above is thus not too bad.

The maximum pressure on the ground is $p_{\max} = 20.6$ kPa, which should be compared with the bearing capacity of the ground.

The compaction resistance is $F_x = 20.2$ N per wheel, i.e. 121.2 N for the whole rover.

The rolling coefficient is $f = 0.21$.

If the rover has 4 wheels, the same values are: $z_0 = 32$ mm, $p_{\max} = 27$ kPa, $F_x = 34.74$ N, $f = 0.25$.

If the wheel sinks in the ground another form of resistance to motion is present: *bulldozing resistance*. Bulldozing resistance is particularly important if the ground is pushed forward by the wheel and is mitigated by the lateral flow that occurs if the wheel is not too wide. As a consequence, it is not very important in case of large diameter, narrow wheels, while can become a dominating factor if a wide wheel with a small diameter travels on loose soil, in particular if a layer of loose soil rests on a harder surface.

A formula for bulldozing resistance reported by Bekker is

$$F_{xb} = \frac{b\sin(\alpha+\phi)}{2\sin(\alpha)\cos(\phi)}\left[2z_0cK_c + \rho g z_0^2 K_\gamma\right] + \frac{\pi t^2 \rho g(90° - \phi)}{540} + \frac{\pi c t^2}{180} + ct^2\tan\left(45° + \frac{\phi}{2}\right), \tag{4.74}$$

where

$$\alpha = \arccos\left(1 - \frac{z_0}{R}\right) \tag{4.75}$$

is the angle of approach. This expression holds for rigid wheels only. Constants K_c and K_γ depend on the coefficients N_c and N_γ defined in Table 4.2

$$K_c = \left[N_c - \tan(\phi)\right]\cos^2(\phi), \tag{4.76}$$

$$K_{\gamma c} = \left[\frac{2N_\gamma}{\tan(\phi)} + 1\right]\cos^2(\phi), \tag{4.77}$$

$$t = z_0\tan^2\left(45° - \frac{\phi}{2}\right). \tag{4.78}$$

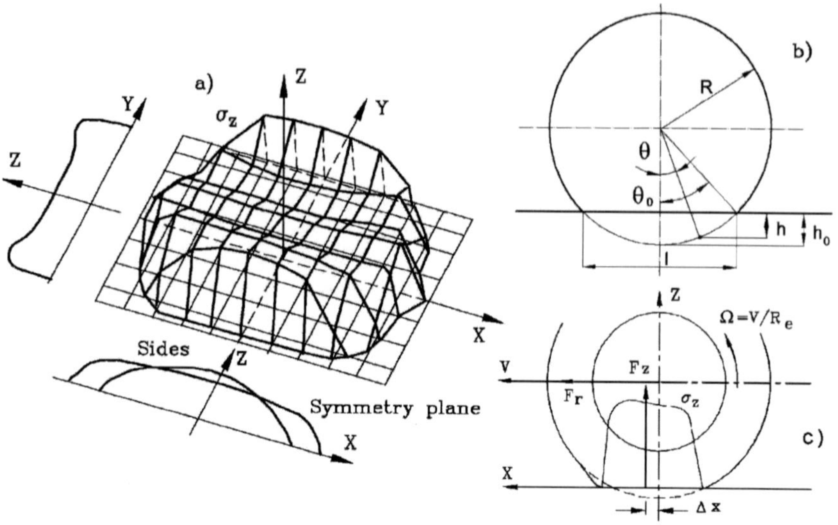

Fig. 4.12 (**a**) Pneumatic tire: pressure distribution at the wheel–road contact. (**b**) Solid rubber elastic wheel: geometrical definitions. (**c**) Pressure σ_z in the contact zone for a pneumatic rolling wheel (from G. Genta, L. Morello, *The Automotive Chassis*, Springer, New York, 2009)

4.3.4 Contact Between Compliant Wheel and Rigid Ground

The pressure that a compliant wheel exerts on a rigid ground depends on the wheel's structure and its computation can be quite complex. If the wheel is not rolling, the pressure distribution is symmetrical with respect to the YZ plane (Fig. 4.12a). If the wheel is perpendicular to the ground and symmetrical with respect to XZ plane, the pressure distribution is symmetrical with respect to this plane as well.

An ideal pneumatic wheel is made by a toroidal membrane with negligible bending stiffness filled with air under pressure. In this case the pressure it exerts on the ground is constant and is equal to the inflation pressure p_i.

The force F_z corresponding to this pressure depends only on the area A of the contact surface with the ground and thus on the deflection h_0 of the tire.

Remark 4.13 The relationship between A and h_0 is quite complex and has little physical significance, since the tire structure has its own stiffness and this deeply affects the pressure distribution on the ground.

As seen in Fig. 4.12a, the pressure is constant and close to the inflation pressure only at the center of the contact area, while on the sides the stiffness of the carcass deeply affects the pressure on the ground.

While the contact between a pneumatic tire and the ground is difficult to be studied, the pressure distribution on the ground of a solid tire made of low stiffness material (in terrestrial applications solid rubber) is much easier to study (Fig. 4.12b).

Assuming that the pressure is proportional to the vertical deformation of the material

$$p = c_r h, \tag{4.79}$$

where c_r is a constant of the material, the maximum pressure occurs at the center of the contact

$$p_{\max} = c_r h_0. \tag{4.80}$$

The relationships between the local deformation h, the total deformation h_0 and angles θ and θ_0 are

$$h_0 = R\big[1 - \cos(\theta_0)\big], \qquad h = R\big[\cos(\theta) - \cos(\theta_0)\big]. \tag{4.81}$$

The *loaded radius* R_l is defined as the height of the center of the wheel on the ground:

$$R_l = R - h_0. \tag{4.82}$$

Assuming that angles θ and θ_0 are small, it follows that

$$p \approx \frac{c_r R}{2}\left(\theta_0^2 - \theta^2\right), \qquad p_{\max} \approx \frac{c_r R}{2}\theta_0^2. \tag{4.83}$$

Under the same assumption, the vertical force acting on the ground is linked with angle θ_0 by the relationship

$$F_z \approx b \int_{-l/2}^{l/2} p\, dx = \frac{bR}{2} \int_{-l/2}^{l/2} c_r \left(\theta_0^2 - \theta^2\right) dx. \tag{4.84}$$

Since

$$l = 2R \sin(\theta_0) \approx 2R\theta_0, \qquad x = 2R \sin(\theta) \approx 2R\theta, \tag{4.85}$$

the force can be written as

$$F_z = \frac{bR^2 c_r}{2} \int_{-\theta_0}^{\theta_0} \left(\theta_0^2 - \theta^2\right) d\theta, \tag{4.86}$$

i.e.

$$F_z = \frac{2}{3} bR^2 c_r \theta_0^3. \tag{4.87}$$

Since

$$h_0 = R\frac{\theta_0^2}{2}, \tag{4.88}$$

the relationship linking the force with the deformation of the tire is

$$F_z = \frac{2}{3} b\sqrt{8R} c_r \sqrt{h_0^3}. \tag{4.89}$$

This formula can be written in the usual form linking the maximum pressure with the force and the deformation of the tire

$$p_{\max} = \sqrt{\frac{9}{32}} \frac{F_z}{b\sqrt{h_0 R}} = \frac{0.53 F_z}{b\sqrt{h_0 R}}. \tag{4.90}$$

In the case of an actual pneumatic tire the situation is intermediate between that of an ideal pneumatic tire and a solid rubber tire: the pressure on the ground is due to both the inflation pressure and the stiffness of the structure. An empirical formula is

$$F_z = (p_i + p_c)\frac{h_0^2}{h_0 + 1}\sqrt{4Rr - 2h_0(R + r)}, \tag{4.91}$$

where r is the transversal radius of the tire and p_c is the mean vertical pressure of the uninflated carcass.

In general, the force the tire receives from the ground is assumed to be located at the center of the contact area and can be decomposed along the axes of the XYZ frame of Fig. 4.12a, where axes X and Y lie on the ground, respectively, in longitudinal and transversal direction (i.e. in the midplane of the wheel and perpendicular to it) and Z has a direction perpendicular to the ground. The longitudinal force F_x, the lateral force F_y and the normal force F_z are so obtained. Similarly, the moment the tire receives from the road in the contact area can be decomposed along the same directions, yielding the overturning moment M_x, the rolling resistance moment M_y and the aligning torque M_z. The moment applied to the tire from the vehicle about the spin axis is referred to as wheel torque T.

The situation described above is related to a stationary wheel. If the wheel rolls on a level road with no braking or tractive moment applied to it and its mean plane perpendicular to the road, the distribution of the normal force is no more symmetrical with respect to the YZ plane, and a rolling resistance moment is originated (Fig. 4.12c).

While the relationship between the angular velocity Ω and the forward speed V of a rolling rigid wheel of radius R is simply

$$V = \Omega R,$$

for a compliant wheel an effective rolling radius R_e can be defined as the ratio between V and Ω

$$R_e = V/\Omega. \tag{4.92}$$

This amounts to define as effective rolling radius the radius of a rigid wheel which travels and rotates at the same speed as the compliant wheel.

The wheel–road contact is far from being a point-contact and the tread band is compliant also in circumferential direction; as a consequence radius R_e coincides neither with the loaded radius R_l nor with its unloaded radius R and the center of instantaneous rotation is not coincident with the center of contact A (Fig. 4.13).

Owing to the longitudinal deformations of the tread band, the peripheral velocity of any point of the tread varies periodically: when it gets close to the point in

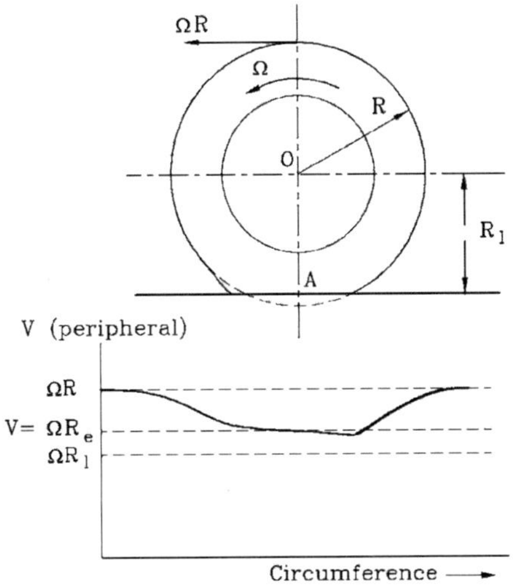

Fig. 4.13 Compliant wheel rolling on a flat road: geometrical configuration and peripheral speed in the contact zone (from G. Genta, L. Morello, *The Automotive Chassis*, Springer, New York, 2009)

which it enters the contact zone it slows down and consequently a circumferential compression results. In the contact zone there is no or limited sliding between tire and road.

Remark 4.14 The peripheral velocity of the tread (relative to the wheel center) in that zone coincides with the velocity of the center of the wheel V.

After leaving the contact zone, the tread regains its initial length and its peripheral velocity ΩR is restored. As a consequence of this mechanism, the spin speed of a compliant wheel tire is smaller than that of a rigid wheel with the same loaded radius R_l (and faster than that of a rigid wheel with radius R) and traveling at the same speed

$$R_l < R_e < R.$$

The center of rotation of the wheel lies then under the surface of the road, at a short distance from it.

The stiffer is the thread band circumferentially, the closest is R_e to R. Owing to their lower vertical stiffness, radial tires have a lower loaded radius R_l than bias-ply tires with equal radius R but their effective rolling radius R_e is closer to the unloaded radius, as the tread is circumferentially stiffer. For instance, in a bias-ply tire R_e can be about 96% of R while R_l is 94% of it, while in a radial tire R_e and R_l can be, respectively, 98% and 92% of R. This effect can be even larger in elastic wheels like the twheel: if the external rim is circumferentially stiff, R_e can be equal to R, while R_l depends on the stiffness of the radial springs.

Remark 4.15 The effective rolling radius depends on many factors, some of which are determined by the tire as the type of structure, the wear of the tread, and others by the working condition as load, speed and, in pneumatic tires, inflation pressure.

An increase of the vertical load F_z and a decrease of the inflation pressure p lead to similar results: a decrease of both R_l and R_e. With increasing speed, the tire expands under centrifugal forces, and consequently R, R_l and R_e all increase. This effect is larger in bias-ply tires while, owing to the greater stiffness of the tread band, radial tires expand to a very limited, and usually negligible, extent.

As will be shown in the following sections, any tractive or braking torque applied to the wheel will cause strong variations of the effective rolling radius.

The rolling resistance of an elastic wheel rolling on a hard surface is mostly due to the energy dissipated in the tire. Other mechanisms, like small sliding between road and wheel, aerodynamic drag on the disc and friction in the hub are responsible for a small contribution to the overall resistance, of the order of a few percent. The resultant F_z of the contact pressures moves forward (Fig. 4.12c) producing a torque

$$M_y = -F_z \Delta x$$

with respect to the rotation axis that is seen as rolling resistance.

To maintain a free wheel spinning, a force at the wheel–ground contact is required and then some of the available traction is used: on the free wheel, to supply a torque which counteracts the total moment M_y, and on the driving wheels which must supply a tractive force against the rolling resistance of the former. On driving wheels the driving torque is directly applied through the driving shafts to overcome rolling resistance moment. Rolling resistance of driving wheels thus does not involve forces acting at the road-wheel contact and does not use any of the available traction.

The rolling resistance is thus

$$F_r = \frac{-F_z \Delta x}{R_l}. \tag{4.93}$$

Equation (4.93) is of limited practical use, since Δx is not easily determined.

For practical purposes, rolling resistance is usually expressed as

$$F_r = -f F_z, \tag{4.94}$$

where the *rolling resistance coefficient* (or simply *rolling coefficient*) f must be determined experimentally. The minus sign in (4.94) comes from the fact that traditionally the rolling resistance coefficient is expressed by a positive number.

Coefficient f depends on many parameters, like the traveling speed V, the inflation pressure p (in pneumatic tires), the normal force F_z, the size of the wheel and of the contact zone, the structure and the material of the tire, the working temperature, the road conditions and, last but not least, the forces F_x and F_y exerted by the wheel.

The rolling resistance coefficient f generally increases with the speed V of the vehicle, at the beginning very slowly and then at an increased rate (Fig. 4.14).

The law $f(V)$ can be approximated by a polynomial expression of the type

$$f = f_0 + KV \quad \text{or} \quad f = f_0 + KV^2, \tag{4.95}$$

the second being generally preferred.

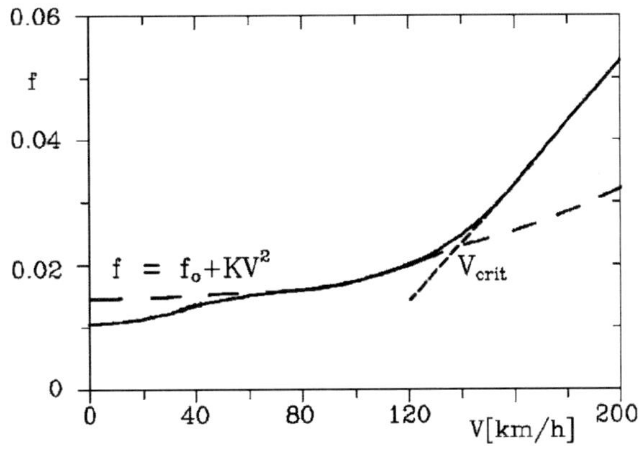

Fig. 4.14 Rolling coefficient as a function of speed for a wheel with pneumatic tire. Experimental curve (radial tire 5.20-14 inflated at 190 kPa, with a load of 3.40 kN rolling on tarmac) compared with (4.95) (from G. Genta, L. Morello, *The Automotive Chassis*, Springer, New York, 2009)

The values of f_0 and K must be measured on any particular tire; as an example, the tire of Fig. 4.14 is characterized, in the test conditions reported, by the values: $f_0 = 0.013$, $K = 6.5 \times 10^{-6}$ s^2/m^2.

A semi-empirical expression of the same kind is suggested by the Society of Automotive Engineers (SAE) to take into account the influence of both load and pressure on the rolling resistance coefficient:

$$f = \frac{K'}{1,000}\left(5.1 + \frac{5.5 \times 10^5 + 90F_z}{p} + \frac{1,100 + 0.0388F_z}{p}V^2\right), \tag{4.96}$$

where coefficient K' takes the value 1 for conventional tires and 0.8 for radial tires. The normal force F_z, the pressure p and the speed V must be expressed respectively in N, N/m^2 (Pa) and m/s.

The speed at which the curve $f(V)$ shows a sharp bend upward is generally said to be the *critical speed* of the wheel (not to be confused with the critical speed of the vehicle). Its presence can be easily explained by vibratory phenomena which take place in the tire at high speed. Non-pneumatic elastic wheels may have a low damping and thus be even more prone to vibrate: in the design of such wheels care must be given to assess their vibrational characteristics. However, the critical speed is usually much higher than the speed that can be reached by rovers and moving robots, so it may be of little relevance in the present context.

The type of structure and the material used for the construction of the tire play an important role in determining the rolling resistance. Even small differences like ply orientation or exact rubber composition can cause strong differences. However, most existing data are for standard pneumatic tires and specific designs departing from the usual approach must be tested to assess their rolling resistance.

At low speed, typical values for the rolling coefficient of standard tires are included between

$$f_0 = 0.008\text{–}0.02$$

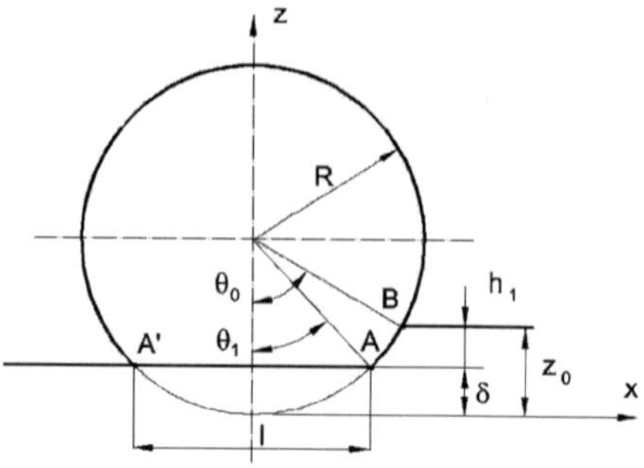

Fig. 4.15 Contact between compliant wheel and compliant terrain. Geometrical definitions

for good concrete or tarmac surfaces and

$$f_0 = 0.04\text{–}0.1$$

for hard and flat natural surfaces.

Wear, temperature, inflation pressure (in pneumatic tires), load and other operating conditions cause changes in the rolling resistance: tires must be tested in conditions simulating as closely as possible the operating ones.

4.3.5 Contact Between Compliant Wheel and Compliant Ground

If both the wheel and the ground are compliant both may undergo deformation. This can, however, occur only if the maximum pressure exerted by the ground exceeds the pressure needed to deform the tire, which in (4.91) was expressed as $p_i + p_c$. If this pressure is not exceeded the situation is that shown in Fig. 4.11. If it is exceeded in point A of the contact zone, the situation is that shown in Fig. 4.15.

Assuming that the deformation is linked to the pressure by (4.5) and that angle θ_1 is small, so that its cosine is close to 1, the pressure in point A is

$$\sigma_{r\mathrm{A}} = \left(\frac{k_c}{b} + k_\phi\right) h_1^n = p_i + p_c. \tag{4.97}$$

The sinkage of the wheel is thus

$$h_1 = \left(b\frac{p_i + p_c}{k_c + bk_\phi}\right)^{1/n}. \tag{4.98}$$

If $n = 1$ the sinkage is linear in the total pressure $p_i + p_c$ exerted by the tire.

Remark 4.16 When traveling on soft ground the inflation pressure p_i is usually reduced to decrease sinking.

The coordinate x_A of point A is linked with the vertical deformation of the tire by the relationship

$$x_A^2 + \left[R - (z_0 - h_1)\right]^2 = R^2. \tag{4.99}$$

If the deformation of the wheel is small when compared with its radius, it follows that

$$x_A \approx \sqrt{2R(z_0 - h_1)}. \tag{4.100}$$

Assume that the wheel is free to roll and is pulled in x direction by the force F_x while being pressed on the ground by force F_z. Assume again that the deformation of the soil is anelastic and thus its deformation is permanent. In this situation the contact occurs only along arc A′AB. The wheel exchanges with the ground a force per unit area

$$\sigma_r = \left(\frac{k_c}{b} + k_\phi\right) h^n \tag{4.101}$$

directed radially along arc AB, plus a force

$$F_{1z} = bl(p_i + p_c) \tag{4.102}$$

along line A′A. The latter is directed in z direction.

The equilibrium equations of the wheel in x and z directions are

$$F_x = bR \int_{\theta_1}^{\theta_0} \sigma_r \sin(\theta)\, d\theta = b \int_{\delta}^{z_0} \sigma_r\, dz, \tag{4.103}$$

$$F_z = F_{1z} + bR \int_{\theta_1}^{\theta_0} \sigma_r \cos(\theta)\, d\theta = bl(p_i + p_c) + b \int_{x_A}^{x_B} \sigma_r\, dx. \tag{4.104}$$

In the latter equation it has been explicitly assumed that the pressure acting on the whole line A′A is constant, which is consistent with the simplified model of the tire here used.

The equilibrium for rotation is always satisfied, since both forces F_x and F_z and pressures σ_r act through the center of the wheel O, while the pressure acting on line A′A is symmetric about the center of the wheel.

The horizontal force is

$$F_x = (k_c + bk_\phi) \int_{\delta}^{z_0} h^n\, dz, \tag{4.105}$$

i.e.:

$$F_x = (k_c + bk_\phi) \int_{z_0 - h_1}^{z_0} (z_0 - z)^n\, dz. \tag{4.106}$$

By performing the integration, it follows that the compaction resistance is

$$F_x = (k_c + bk_\phi) \frac{h_1^{n+1}}{n+1}, \tag{4.107}$$

i.e.

$$F_x = \frac{[b(p_i + p_c)]^{n+1/n}}{(n+1)(k_c + bk_\phi)^{1/n}}. \tag{4.108}$$

If $n = 1$ the expression for the compaction resistance is

$$F_x = \frac{b^2(p_i + p_c)^2}{2(k_c + bk_\phi)}. \tag{4.109}$$

The compaction resistance can thus be computed without obtaining explicitly the sinkage of the wheel.

The explicit computation of the vertical force is not reported here, since in the present case the sinkage depends only on the pressure, which is known, and the computation of the force is not needed.

The resistance to motion given by (4.108) takes into account only the compaction of the ground; to it the rolling resistance due to the deformation of the tire must be added. When the deformation of the ground is small, the value obtained for the elastic wheel on rigid ground can be assumed, while it reduces with increasing ground deformation, until it vanishes when the deformation of the wheel δ (see Fig. 4.11) becomes negligible.

To take into account that the resistance of the wheel increases with decreasing inflation pressure, Bekker in the mentioned books suggests to add a term

$$F_{xi} = \frac{F_z u}{p_i^{a^*}}, \tag{4.110}$$

where u and a^* are empirical coefficients related to the internal structure of the tire.

Remark 4.17 The application of this formula, with the coefficients obtained from a test in which the tire is rolling against a hard surface, is questionable.

While for small values of the pressure the deformations of the wheel are large (although less than when rolling against a hard surface), when the pressure increases the deformations are small and the formula gives a rolling resistance which can be larger than the correct value. In particular, when the pressure approaches the value at which the wheel is not deformed, this component of the rolling resistance should tend to zero.

For a 7.00-16 tire the values suggested are $u = 0.12$ and $a^* = 0.64$, when the pressure is measured in psi. The corresponding values for pressure measured in kPa are $u = 0.413$ and $a^* = 0.64$. Tires designed for low gravity applications should have much lower values of u, which must be obtained experimentally for each case.

The total rolling resistance is thus

$$F_x = b\frac{(p_i + p_c)^{n+1/n}}{(n+1)(\frac{k_c}{b} + k_\phi)^{1/n}} + \frac{F_z u}{p_i^{a^*}}. \tag{4.111}$$

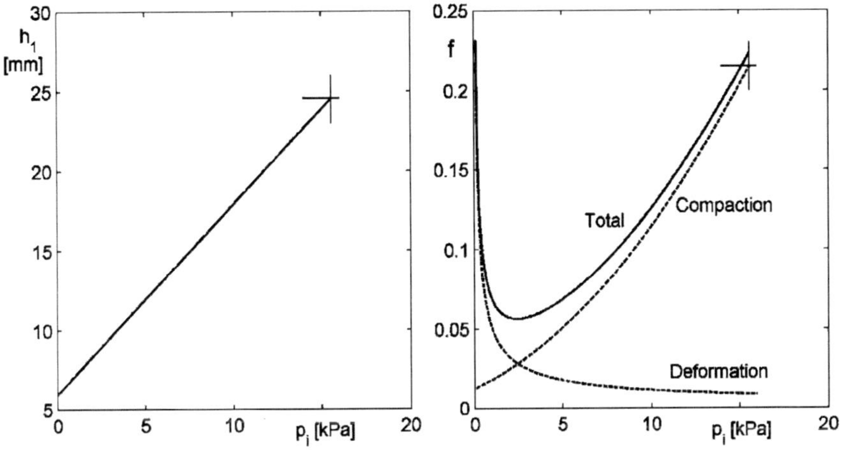

Fig. 4.16 Sinking and rolling coefficient as functions of the inflation pressure for a pneumatic wheel

Example 4.5 Consider the same Mars rover with 6 wheels of the previous example, but now assume that the wheels are elastic. The characteristics of the ground and all the other data are the same:

General: $m = 150$ kg, $g = 3.77$ m/s^2; wheels: $R = 150$ mm; $b = 80$ mm; soil: $n = 1$, $c = 200$ Pa, $\phi = 35°$, $k_c = 1{,}400$ N/m^2, $k_\phi = 820{,}000$ N/m^3.

Assume that the tire has been specifically designed for the application, and the values of p_c, a^* and u are, respectively, $p_c = 5$ kPa, $a^* = 0.64$ and $u = 0.05$ kPa. The inflation pressure varies from 0 to 15.6 kPa (the latter value corresponds to the maximum pressure computed in the previous example and thus if the inflation pressure exceeds this value the wheel remains rigid).

Plot the sinking of the wheels and the rolling coefficient as functions of the inflation pressure p_i.

Assume that $p_i = 5$ kPa. Compute the sinkage of the wheels and the rolling coefficient.

Assuming that the load is equally subdivided on the six wheels, the force on each of them is $F_z = 94.25$ N. The maximum sinkage of the rigid wheels was $z_0 = 25$ mm, the compaction resistance was $F_x = 20.2$ N per wheel, and the corresponding rolling coefficient was $f = 0.21$.

The results for various values of p_i are reported in Fig. 4.16. Note that for values of the pressure approaching 15.6 kPa, the deformation component of the rolling coefficient should approach zero.

At a pressure $p_i = 5$ kPa the sinkage is 11.9 mm and the compaction component of the rolling coefficient is 0.05. The total rolling coefficient is 0.069.

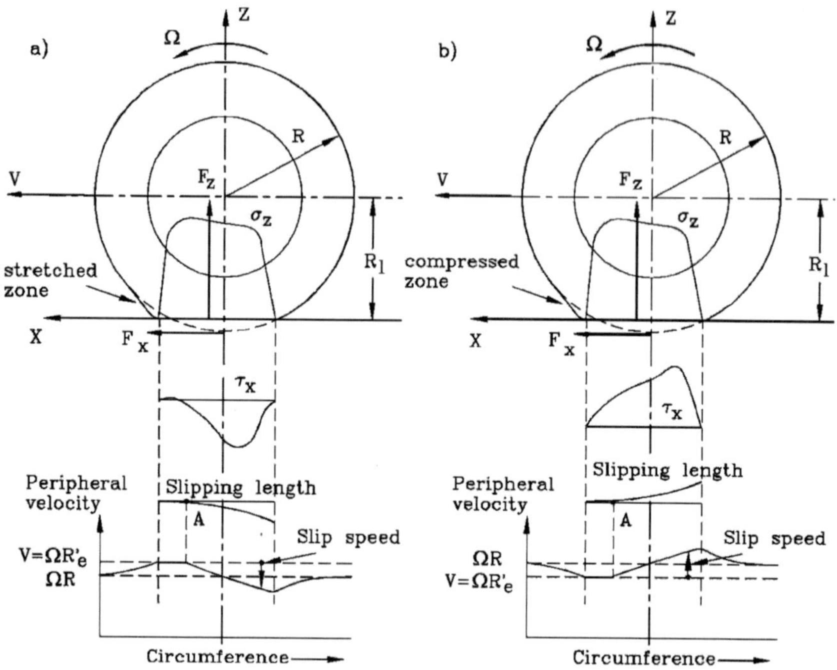

Fig. 4.17 Force distributions and peripheral velocity in a braking (**a**) and in a driving (**b**) wheel operating on a hard surface. Note that the equivalent rolling radius R'_e differs from R_e defined for free straight rolling conditions (force F_x lies on the ground) (from G. Genta, L. Morello, *The Automotive Chassis*, Springer, New York, 2009)

4.3.6 Tangential Forces: Elastic Wheels on Rigid Ground

Longitudinal Forces

A wheel can produce tractive or braking forces only if a longitudinal slip is present, i.e. if the wheel rotates (slightly) faster (for tractive forces) or slower (for braking forces) than a wheel when in pure rolling.

Consider a compliant wheel rolling on a level hard surface. If a braking moment M_b is applied to it, the distributions of normal pressure and longitudinal forces resulting from that application are qualitatively sketched in Fig. 4.17a.

The tread band is circumferentially stretched in the zone in front of the contact with the ground, while in free rolling the same part of the tire was compressed. The peripheral velocity of the tread band in the leading zone of the contact $\Omega R'_e$ is consequently higher than that the peripheral velocity ΩR of the undeformed wheel. The effective rolling radius R'_e, whose value R_e in free rolling was between R_l and R, grows toward R and, if M_b is large enough, becomes greater than R.

The instantaneous center of rotation is consequently located under the road surface (Fig. 4.18). The angular velocity Ω of the wheel is lower than that characterizing free rolling in the same conditions ($\Omega_0 = V/R_e$).

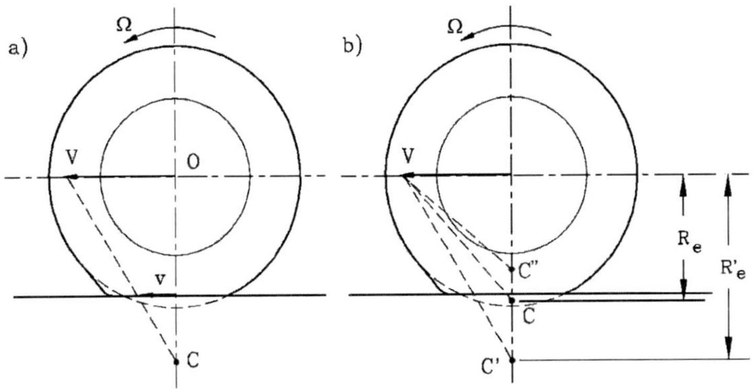

Fig. 4.18 (**a**) Braking wheel, center of instantaneous rotation and slip speed. (**b**) Position of the center of instantaneous rotation in free rolling C, braking C′ and driving C″ (from G. Genta, L. Morello, *The Automotive Chassis*, Springer, New York, 2009)

In such conditions it is possible to define a longitudinal slip as

$$\sigma = \frac{R\Omega - V}{V} = \frac{v}{V}, \tag{4.112}$$

where v is the linear speed at which the contact zone moves on the ground. The longitudinal slip is often expressed as a percentage; in the present book, however, the definition of (4.112) will be strictly adhered to.

Remark 4.18 The slip here defined is due to the deformation of the wheel, while that defined in the case of a rigid wheel is due to the deformation of the soil. The two definitions are slightly different, but their meaning is essentially the same.[11]

The limiting case is that of a wheel that slides on the ground without rotating: the longitudinal slip is $\sigma = -1$ and the center of rotation is at infinity below the road.

If instead of braking, the wheel is driving, the leading part of the contact zone is compressed instead of being stretched (Fig. 4.17b). The value of the effective rolling radius R'_e is smaller than that characterizing free rolling and is usually smaller than R_l; the angular velocity of the wheel is greater than Ω_0.

[11]This definition, suggested by SAE, is different from that of (4.17), which in this case is

$$\sigma = \frac{R\Omega - V}{R\Omega}.$$

The two formulations are essentially equivalent for small values of the slip.

Often they are both used for braking and driving, respectively:

$$\sigma = \frac{R\Omega - V}{R\Omega} \quad \text{for driving} \quad \text{and} \quad \sigma = \frac{R\Omega - V}{V} \quad \text{for braking.}$$

Here the limiting case is that of a wheel that spins without moving forward: the longitudinal slip is $\sigma = \infty$ and the center of rotation is in the center of the wheel.

The slip defined by (4.112) is positive for driving conditions and negative for braking. The presence of the slip velocity[12] v does not mean, however, that there is an actual sliding of the contact zone as a whole. The peripheral velocity of the leading part of that zone is actually

$$V = \Omega R'_e,$$

and consequently in that zone no sliding can occur. The speed of the tread band starts to decrease (in braking, increase in driving) and sliding begins only at the point indicated in Fig. 4.17 as point A. The slip zone, which extends only to a limited part of the contact zone for small values of σ, gets larger with increasing slip and, at a certain value of that parameter, reaches the leading part of the contact zone and global sliding of the tire occurs (Fig. 4.19a).

The longitudinal force F_x the wheel exchanges with the ground is a function of σ. It vanishes when $\sigma = 0$ (free rolling conditions)[13] to increase almost linearly for values of σ from (−0.15)–(−0.30) to 0.15–0.30.

The longitudinal force can be approximated in this range by a linear expression

$$F_x = C_\sigma \sigma, \tag{4.113}$$

where C_σ is usually referred to as the *longitudinal force stiffness* of the wheel.

Outside this range, which depends on many factors, the absolute value of the force increases less sharply, then reaches a maximum and eventually decreases. In braking the maximum value of $|\sigma|$ occurs for $\sigma = -1$, characterizing free sliding (locking of the wheel), while in driving σ can have any positive value, up to infinity when the wheel spins while the vehicle is not moving.

As a first approximation, force F_x can be considered as roughly proportional to the load F_z, at equal value of σ. It is consequently useful to define a longitudinal force coefficient

$$\mu_x = \frac{F_x}{F_z}. \tag{4.114}$$

The qualitative trend of such coefficient is reported against σ in Fig. 4.19b.

Two important values of μ can be identified on the curve both in braking and in driving: the peak value μ_p and the value μ_s characterizing pure sliding. The first is referred to as *driving traction coefficient* when the wheel is exerting a positive longitudinal force and as *braking traction coefficient*, usually reported in absolute value, in the opposite case. The second is the *sliding driving traction coefficient* or the *sliding braking traction coefficient*.

---

[12]The slip velocity is defined by SAE Document J670 as $\Omega - \Omega_0$, i.e. the difference between the actual angular velocity and the angular velocity of a free rolling tire. Here a definition based on a linear velocity rather than an angular velocity is preferred: $v = R_e(\Omega - \Omega_0)$.

[13]Actually free rolling is characterized by a very small negative slip, corresponding to the rolling resistance. This is, however, usually neglected when plotting curves $F_x(\sigma)$.

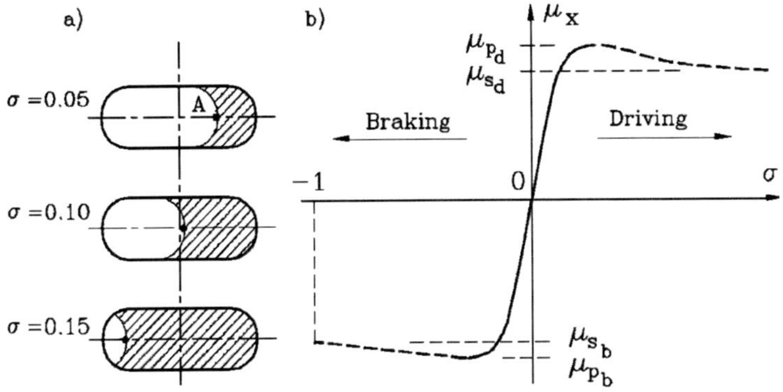

Fig. 4.19 (**a**) Slip zone at the wheel–road contact with different values of the longitudinal slip σ. (**b**) Qualitative trend of the longitudinal force coefficient μ_x as a function of the longitudinal slip σ (from G. Genta, L. Morello, *The Automotive Chassis*, Springer, New York, 2009)

The part of the curve $\mu(\sigma)$ which lies beyond the range included by the two peak values, represented by a dashed line in Fig. 4.19b, is a zone of instability. When the peak value of σ, characterized by μ_{p_b}, is exceeded, the wheel locks in a very short time. In order to prevent the locking of wheels, devices generally defined as antilock or antiskid systems are widely used in the automotive field. The traction and braking control device of robots must also take into account this phenomenon. To do so, it is possible to detect the deceleration of the wheel and, when it reaches a predetermined value, to decrease the braking moment avoiding the locking of the wheel. Antilock devices can operate on each wheel separately or, more often, on both wheels of an axle.

Similarly, to avoid that a wheel slips under the effect of a driving torque applied to it, antispin devices limit the driving moment (or apply a braking torque) when the acceleration of the wheels exceeds a stated value.

The curves usually show a certain symmetry between the braking and driving conditions and often the maximum braking and driving forces are assumed to be equal. The values of function $\mu_x(\sigma)$ depend on a number of parameters, such as the type of wheel, ground conditions, speed, magnitude of the side force F_y exerted by the tire and many others. Moreover, there is a significant difference between the curves obtained by different experimenters in conditions not exactly comparable.

The maximum value of the longitudinal force decreases with increasing speed but this reduction is much influenced by operating conditions. Generally speaking, it is not very marked on dry hard surfaces, while it is greater on wet or dirty roads. Also the difference between the maximum value and the value related to sliding is more notable in these cases.

Remark 4.19 High performance tires show peak values of μ_x that can be as high as 1.5–1.8 on hard surfaces, but even these tires do not reach very high values of longitudinal force coefficient in sliding condition.

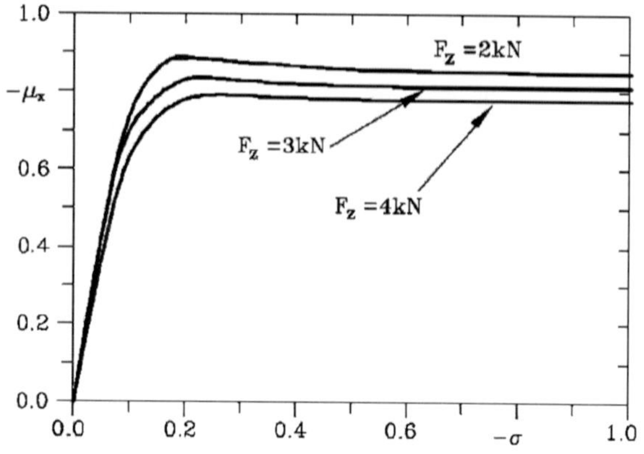

Fig. 4.20 Influence of normal force F_z on the curve $\mu_x(\sigma)$. Pneumatic tire 6.00-15, $p = 170$ kPa, $V = 100$ km/h (from G. Genta, L. Morello, *The Automotive Chassis*, Springer, New York, 2009)

Remark 4.20 On hard surfaces tread wear has a great influence on the longitudinal forces, particularly at high speed. Worn tires have a larger traction coefficient on dry roads.

The presence of a film of liquid can change drastically these results. If the liquid film is thick, the tire can lift from the road surface as a result of hydrodynamic lift (hydroplaning). A liquid film can slip between the tire and the ground thus reducing the contact area. With increasing speed the area of the contact zone further reduces, until a complete lifting of the tire takes place. True hydrodynamic lubrication conditions can be said to exist in this case and consequently the force coefficient or, better, the friction coefficient as in this condition sliding usually occurs, reduces to very low values, of the order of 0.05.

Remark 4.21 This is why traction on ice at a temperature high enough to cause melting under pressure is so low.

Remark 4.22 The only place in the solar system (apart from Earth) where there might be danger of hydroplaning is Titan, owing to the presence of liquid hydrocarbons.

The assumption, in a way implicit in the definition of the longitudinal force coefficient, that longitudinal forces are proportional to the normal force acting on the wheel is only a crude approximation. Actually the longitudinal force coefficient decreases with increasing load as shown in Fig. 4.20.

The curves $\mu_x(\sigma)$ can be approximated by analytical expressions. One of the formulas which can be used in the range $-1 < \sigma < 1$ is

$$\mu_x = A\left(1 - e^{-B\sigma}\right) + C\sigma^2 - D\sigma, \tag{4.115}$$

where

$$B = \left(\frac{K}{\alpha + d}\right)^{1/n}$$

is a factor which takes into account the interaction between the longitudinal slip σ and the sideslip α (see Sect. 4.3.6). The derivative in the origin is

$$\left(\frac{\partial \mu_x}{\partial \sigma}\right)_{\sigma=0} = AB - D. \tag{4.116}$$

Coefficients A, C, D, K, d and n must be obtained from the experimental curves and have no physical meaning. They depend not only on the ground conditions but also on the load. Two curves $\mu_x(\sigma)$ obtained through (4.115) are reported in Fig. 4.21. Curve B refers to a racing tire, with high traction.

A very good approximation of longitudinal force F_x as a function of the slip σ can be obtained through the empirical equation introduced by Pacejka[14] and known as the *magic formula*. Such mathematical expression allows one to express not only force F_x as function of the normal force F_z and the longitudinal slip σ, but also the side force F_y and the aligning torque M_z as functions of various parameters (see below).

The equation yielding the longitudinal force F_x, as a function of the slip σ, is

$$F_x = D\sin\bigl(C\arctan\bigl\{B(1-E)(\sigma+S_h)+E\arctan\bigl[B(\sigma+S_h)\bigr]\bigr\}\bigr)+S_v, \tag{4.117}$$

where B, C, D, E, S_v and S_h are six coefficients which depend on the load F_z. They must be obtained from experimental testing and do not have any direct physical meaning. In particular, S_v and S_h have been introduced to allow nonvanishing values of F_x when $\sigma = 0$.

Coefficient D yields directly the maximum value of F_x, apart from the effect of S_v. The product BCD gives the slope C_σ of the curve for $\sigma + S_h = 0$. The values of the coefficients are expressed as functions of a number of coefficients b_i which can be considered as characteristic of any specific tire, but depend also on ground conditions and speed

$$C = b_0, \qquad D = \mu_p F_z,$$

where for b_0 a value of 1.65 is suggested and

$$\mu_p = b_1 F_z + b_2,$$
$$BCD = \bigl(b_3 F_z^2 + b_4 F_z\bigr)e^{-b_5 F_z},$$
$$E = b_6 F_z^2 + b_7 F_z + b_8,$$
$$S_h = b_9 F_z + b_{10}, \qquad S_v = 0.$$

If a symmetrical behavior for positive and negative values of force X is accepted, this model can be used for both braking and driving. The curve is usually extended to braking beyond the point where $\sigma = -1$, to simulate a wheel rotating in backward direction while moving forward.

---

[14] E. Bakker, L. Lidner, H.B. Pacejka, *Tire Modelling for Use in Vehicle Dynamics Studies*, SAE Paper 870421; E. Bakker, H.B. Pacejka, L. Lidner, *A New Tire Model with an Application in Vehicle Dynamics Studies*, SAE Paper 890087.

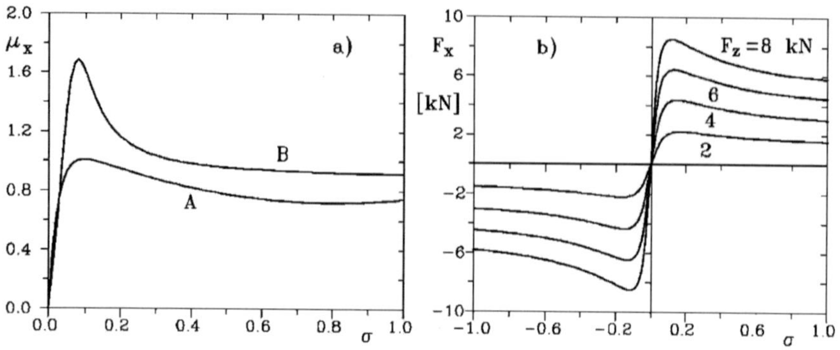

Fig. 4.21 (**a**) Curves $\mu_x(\sigma)$ for a pneumatic tire 145/80 R 13 4.5J obtained through (4.115) (curve A) and a tire 245/65 R 22.5 obtained through (4.117) (curve B). (**b**) Curves $F_x(\sigma)$ for different values of the load obtained using (4.117) for a radial tire 205/60 VR 15 6J (from G. Genta, L. Morello, *The Automotive Chassis*, Springer, New York, 2009)

A set of curves $F_x(\sigma)$ obtained for vertical loads $F_z = 2$, 4, 6 and 8 kN for a radial tire 205/60 VR 15 6J is shown in Fig. 4.21b.

Remark 4.23 The coefficients introduced in (4.117) and the results obtained from it are usually expressed in non consistent units: force F_z is in kN, longitudinal slip is expressed as a percentage and force F_x is in N.

Remark 4.24 The importance of the model expressed by (4.117) is mainly linked to the fact that tire manufacturers are increasingly giving the performances of their tires in terms of the coefficients to be introduced into it and in the similar expressions for the cornering force and the aligning torque. The magic formula is a simple and accurate model for tire behavior and, which is even more important, one for which the data are readily available.

Rolling resistance can also be defined when tractive or braking moments are applied to the wheel. In this case the power dissipated by rolling resistance $F_r V$ can be expressed as

$$|F_r|V = \begin{cases} |F_b|V - |M_b|\Omega & \text{(braking)} \\ |M_t|\Omega - |F_t|V & \text{(traction)} \end{cases} \tag{4.118}$$

where F_b, F_t, M_b and M_t are, respectively, the braking and tractive forces and moments. Equations (4.118) should be applied only in constant speed motion, since they do not include tractive (braking) moments needed to accelerate (decelerate) rotating parts.

In general rolling resistance increases with both tractive or braking longitudinal force F_x and this effect is not negligible in case of strong longitudinal forces, particularly in the case of braking. This is due mainly to the fact that the generation of longitudinal forces is always accompanied by the presence of sliding in at least a part of the contact zone. The minimum rolling resistance may, however, occur not when the wheel exerts no driving force but when it exerts a very small tractive force. This is, however, a small effect and depends on the structure of the wheel.

Lateral (or Cornering) Forces

Pure rolling condition is characterized by no longitudinal slip and by the velocity V of the center of the wheel being contained in the midplane of the wheel. In this case there is rolling without sliding. It has already been stated that to produce longitudinal force a longitudinal slip must be present. In the same way, to produce a lateral force a lateral slip must be present, i.e. the velocity of the center of the wheel must make an angle with respect to the midplane of the wheel, the XZ plane of Fig. 4.12. The angle between XZ plane and the direction of the velocity of the center of the wheel is the *sideslip angle* α of the wheel. Another characteristic angle is that between XZ plane and the mean plane of the wheel: it is called the *inclination* (or *camber*[15]) *angle* of the wheel and symbol γ is used for it.

The fact that the wheel has a sideslip angle, i.e. is not in pure rolling, does not mean that in the contact zone the tire slips on the ground: also in this case, as seen for longitudinal forces, the compliance of the tire allows the tread to move, relatively to the center of the wheel, with the same velocity as the ground. The generation of tangential forces in the rigid ground–wheel contact is thus directly linked with the compliance of the tire. However, some localized sliding between the wheel and the road can be present and, with increasing sideslip angle, they become more and more important, until the whole wheel is in actual, macroscopic sliding.

If the velocity of the center of the wheel does not lie in its midplane, i.e. if the wheel travels with a sideslip angle, the shape of the contact zone is distorted (Fig. 4.22). Consider a point belonging to the midplane on the tread band. Upon approaching the contact zone it tends to move in a direction parallel to the velocity V, relatively to the center of the wheel, and consequently goes out of the midplane.

After touching the ground at point A, it continues following the direction of the velocity V (for an observer fixed to the ground, it remains still) until it reaches point B. At that point, the elastic forces pulling it toward the midplane are strong enough to overcome those due to friction, forcing it to slide on the ground and to deviate from its path. This sliding continues for the remaining part of the contact zone, until point C is reached. The contact zone can thus be divided into two parts: a leading zone where no sliding occurs and a trailing one where the tread slips toward the midplane. This second zone grows with the sideslip angle (Fig. 4.22b), until it includes the whole contact zone and the wheel actually slips on the ground.

The lateral deformations of the tire are plotted in a qualitative way in Fig. 4.23, together with the distribution of the normal and tangential forces per unit area σ_z and τ_y, and of the lateral velocity. The resultant F_y of the distribution of side forces

[15] Often the sign of the inclination angle is defined with reference to frame XYZ, while the sign of the camber angle depends on whether the wheel is at the right or left side of the vehicle. Here reference to frame XYZ will always be made.

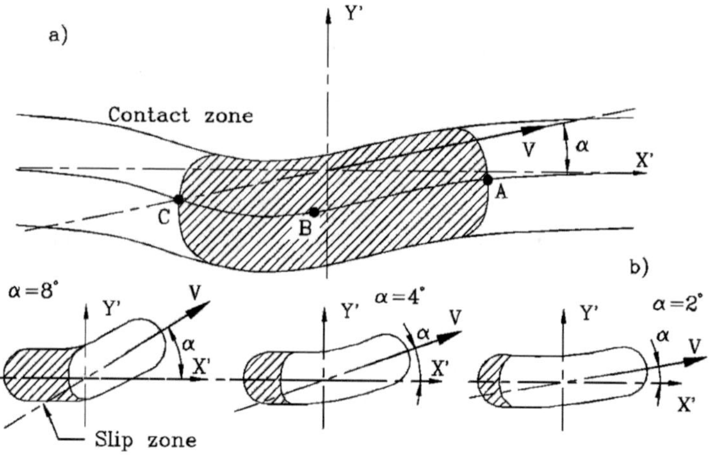

Fig. 4.22 Wheel–ground contact when a sideslip angle is present. (**a**) Contact zone and path of a point belonging to the midplane of the tread band. (**b**) Contact and slip zones at various sideslip angles α (from G. Genta, L. Morello, *The Automotive Chassis*, Springer, New York, 2009)

Fig. 4.23 Lateral deformations, distribution of σ_z and τ_y, sliding and lateral velocities in a wheel rolling with a slip angle α (from G. Genta, L. Morello, *The Automotive Chassis*, Springer, New York, 2009)

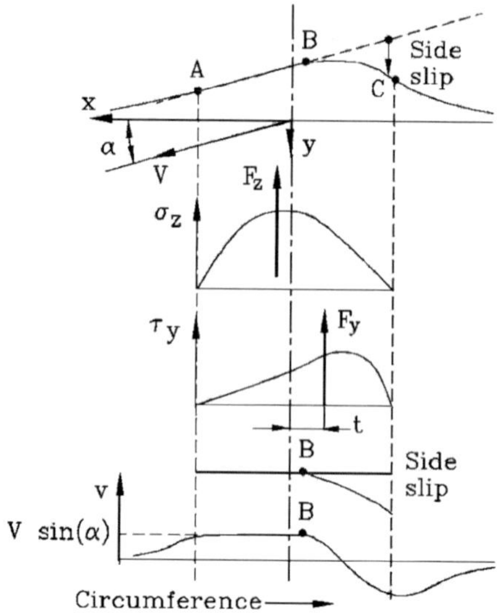

is not applied at the center of the contact zone but at a point which is located behind it at a distance t. Such a distance is defined as the *pneumatic trail*.

The moment

$$M_z = F_y t$$

is the aligning moment as it tends to force the mean plane of the wheel toward the direction of the velocity V. At first the absolute value of the side force F_y grows

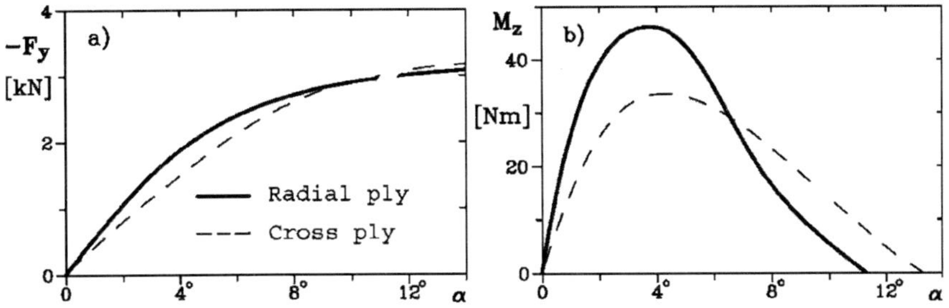

Fig. 4.24 Side force F_y and aligning moment M_z for pneumatic tires of the same size but different type. Tire 5.60-13; $F_z = 3$ kN, $p = 170$ kPa; $V = 40$ km/h (from G. Genta, L. Morello, *The Automotive Chassis*, Springer, New York, 2009)

almost linearly with α, then, when the limit conditions of sliding are approached, in a slower way. Eventually it remains constant, or decreases slightly, when sliding conditions are reached.

The side force F_y is plotted as a function of α for the cases of a radial and a bias-ply pneumatic tire in Fig. 4.24a. Radial tires show a "stiffer" behavior than bias-ply ones for what side forces are concerned, as they require smaller sideslip angles to produce the same side force.

With increasing sideslip angle, τ_y is more evenly distributed and the pneumatic trail decreases. The aligning moment is thus the product of a force which increases with α and a distance which decreases; its trend is consequently of the type shown in Fig. 4.24b. At high values of α, M_z can change direction, as is shown in the figure.

The side force coefficient

$$\mu_y = \frac{F_y}{F_z}$$

is often used for the side force. Its maximum value μ_{y_p} is usually defined as the *lateral traction coefficient*, while the value of the lateral traction coefficient in sliding conditions is μ_{y_s}.

The cornering force increases linearly with α for low values of the sideslip angle. The slope $\partial F_y / \partial \alpha$ of the curve in the origin is usually defined as *cornering stiffness* or *cornering power* and written as C. Since the cornering stiffness is expressed as a positive number while, at least in the initial part of the curve $F_y(\alpha)$ the derivative $\partial F_y / \partial \alpha$ is always negative, the cornering force can be expressed, for low values of α, as

$$F_y = -C\alpha. \tag{4.119}$$

Expression (4.119) is quite useful to study the dynamic behavior of vehicles under the assumption of small sideslip angles, as it actually occurs in normal driving conditions. In particular, it is essential in the study of the stability of linearized models.

Also the aligning moment can be expressed by a linear law

$$M_z = (M_z)_{,\alpha}\alpha, \tag{4.120}$$

where $(M_z)_{,\alpha}$ is the derivative $\partial M_z/\partial\alpha$ computed for vanishingly small α and is defined as *aligning stiffness coefficient* or simply *aligning coefficient*. This linear relationship holds for a much more restrict range of sideslip angles than that regarding the cornering force.

Remark 4.25 Both force F_y and moment M_z depend on many factors, besides the angle α, as normal force F_z, speed, pressure p, ground conditions etc.

At increasing speed, the curve $F_y(\alpha)$ lowers, mainly in the part corresponding to the higher values of the sideslip angle. The linear part remains almost unchanged. Also the pneumatic trail t decreases with increasing speed and consequently the aligning torque shows a decrease which is more marked than that of the side force.

The decrease of F_y, t and M_z is more pronounced in the case of bad ground conditions. As far as hydrodynamic lifting (hydroplaning) is concerned, the same considerations seen for the longitudinal force F_x can be repeated for the side force F_y.

If the mean plane of the wheel is not perpendicular to the ground, i.e. if an *inclination* or *camber* angle γ (Fig. 4.25a) is present, the wheel produces a lateral force, even if no slip angle is present. It is usually said camber thrust or camber force, as distinct from the cornering force, due to sideslip angle alone. The total side or lateral force is given by the camber force added to the cornering force. The camber force is usually far smaller than the cornering force, at least at equal values of angles α and γ. It depends on the load F_z, is practically linear with it (Fig. 4.25), and is strongly dependent on the type of tire considered.

Remark 4.26 The camber and sideslip force act in the same direction (i.e. the camber force helps in producing the lateral force needed to bend the trajectory) if the wheel leans toward the inside of the bend, like in motorcycles. If the wheel leans outwards, the camber thrust detracts from the cornering force.

The camber thrust is usually applied in a point leading the center of the contact zone, producing a small moment M_{z_γ} that is usually neglected, due to its small value. Bias-ply tires usually produce greater camber thrusts and moments than radial ones.

Usually both sideslip and camber are simultaneously present. Ideally, when both sideslip and camber angle are equal to zero the lateral force and the aligning torque should be vanishingly small. In pneumatic tires, this is in practice not true for a number of reasons. Firstly, the lateral behavior of tires exhibits a hysteresis, in such a way that when the zero sideslip angle condition is reached from a condition in which a force was exerted in a certain direction, a small residual force in the same direction remains. This can give a feeling of lack of precision of the steering system and compels the driver or the control system to make continuous corrections.

Moreover, the center of the hysteresis cycle is not at the point in which both angle and force are equal to zero: owing to lack of geometrical symmetry, a tire working in symmetrical conditions may produce a side force. A first effect is due to a possible conicity of the outer surface of the tire: a conical drum would roll on a circular

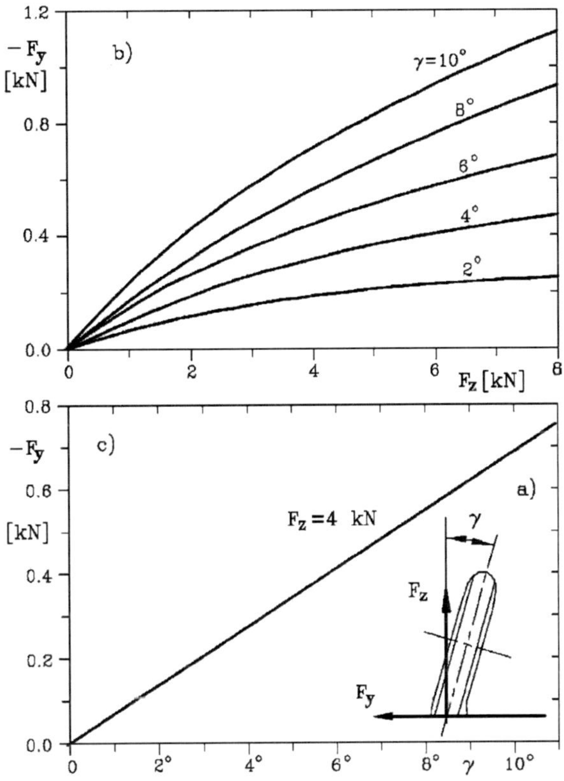

Fig. 4.25 Camber thrust. (**a**) Sketch; note that the force F_y is negative and hence is directed in a direction opposite to that shown. Force as a function of the normal load (**b**) and of the camber angle (**c**). Tire 6.40-13, $p = 200$ kPa (from G. Genta, L. Morello, *The Automotive Chassis*, Springer, New York, 2009)

path whose center coincides with the apex of the cone. Conicity is due to lack of precision during the manufacturing process and hence is linked with manufacturing quality control; its direction is random and its amount changes from tire to tire of the same model. If a tire is turned on the rim of the wheel, the direction of the conicity is reversed, as is the force it causes when the tire rolls along a straight path.

In pneumatic tires, another unavoidable lack of symmetry is linked with the angles of the various plies and their stacking order; the effect it causes is called ply steer. If the wheel rolls free, ply steer causes it to roll along a straight line angled with respect to the plane of symmetry; if the wheel rolls with no sideslip angle the generation of a side force results. If a tire is turned on the rim the direction of the force due to ply steer is not reversed. As it is caused by a factor included in the tire design, unlikely the effect of conicity, that of ply steer is consistent between tires of the same model.

Generally speaking, the lateral force offset is subdivided into two parts: the part which does not change sign when the wheel is turned on the rim is said to be ply-steer force, while the part that changes sign is said to be conicity force.

Conicity effects may be present also in non-pneumatic tires, while ply steer is too dependent on the structure of the wheel to say anything before the exact design is defined. While conicity can be included into the models of the tire only in a statistical way, ply steer is one of the peculiarities of each tire and can be accounted for with precision.

Remark 4.27 While these effects are usually considered as a nuisance, opposite ply steer of the wheels of a given axle can be used as a substitute for toe-in or toe-out. While the latter two increase the rolling resistance the former has no effect on it.

The ratio between the cornering stiffness and the normal force is usually referred to as *cornering stiffness coefficient* (the term *cornering coefficient* is also used but SAE recommendation J670 suggests to avoid it for clarity). For bias-ply tires it is of the order of $0.12\ \text{deg}^{-1} = 6.9\ \text{rad}^{-1}$ and for radial tires of the order of $0.15\ \text{deg}^{-1} = 8.6\ \text{rad}^{-1}$.

In the same way the camber stiffness can be defined as the slope of the curve $F_y(\gamma)$ for $\gamma = 0$:

$$C_\gamma = \frac{\partial F_y}{\partial \gamma}. \tag{4.121}$$

The camber thrust produced by a positive camber angle is negative and hence the camber stiffness is negative. The ratio between the camber stiffness and the normal force is usually referred to as camber stiffness coefficient. This coefficient is higher for bias-ply tires than for radial tires: in the first case an average value is of the order of $0.021\ \text{deg}^{-1} = 1.2\ \text{rad}^{-1}$ and in the second is of the order of $0.01\ \text{deg}^{-1} = 0.6\ \text{rad}^{-1}$.

The value of the camber stiffness is important in the case a wheel rolls on a road with a transversal slope with its midplane remaining vertical: in this case there is a component of the weight which is directed downhill and the camber thrust which is directed uphill. The net effect can be in one direction or the other depending on the magnitude of the camber stiffness coefficient: the downhill component of the weight is

$$W \sin(\alpha_t) \approx W \alpha_t,$$

where α_t is the transversal inclination of the road while the camber thrust is equal to the weight multiplied by the camber stiffness coefficient and the angle. It is clear that if the value of the camber stiffness coefficient is larger than one (measured in rad^{-1}), as it occurs for bias ply tires, the net force is directed uphill; the opposite occurs for radial tires. This situation occurs when a rut is present in the road: a radial tire tends to track in the bottom while a bias-ply tire tends to climb out of the rut.

To include camber thrust into the linearized model, (4.119) can be modified as

$$F_y = -C\alpha + C_\gamma \gamma. \tag{4.122}$$

It can be used with confidence for values of α up to about 4°and of γ up to 10°.

The effect of the camber angle can be included in the linearized expression of the aligning torque by modifying (4.120) as

$$M_z = (M_z)_{,\alpha}\alpha + (M_z)_{,\gamma}\gamma, \tag{4.123}$$

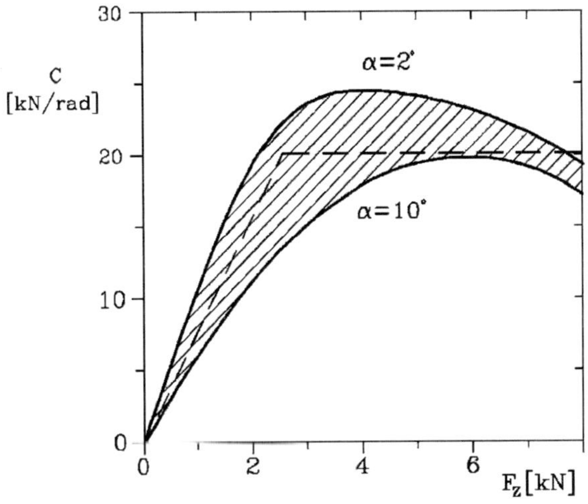

Fig. 4.26 Cornering stiffness as a function of the load F_z (the curve labeled $\alpha = 10°$ is related to a sort of 'secant' stiffness) (from G. Genta, L. Morello, *The Automotive Chassis*, Springer, New York, 2009)

where $(M_z)_{,\gamma}$ is the derivative $\partial M_z/\partial\gamma$ computed for vanishingly small α and γ but, as already stated, the second effect is so small that is usually neglected.

Equation (4.123) supplies a good approximation of the aligning torque for a range of α far more limited than that in which (4.119) holds. It must, however, be noted that the importance of the aligning torque in the study of the behavior of the vehicle is limited and consequently a precision lower than that required for side forces can be accepted. Practically, a good approximation of the aligning torque is important only when studying the steering mechanism.

The *aligning stiffness coefficient* due to sideslip angle is of about 0.01 m/deg (Nm/N deg) for bias ply tires and of 0.013 m/deg for radial tires while that due to camber (aligning camber stiffness coefficient) is approximately of 0.001 m/deg for the first ones and of 0.0003 m/deg for the latter.

A small aligning moment is due to the curvature of the path even if the sideslip angle is equal to zero; however, this effect is not negligible only if the radius of the trajectory is very small, of the order of a few meters, and consequently it is present only in low speed manoeuvres. It may be important for the dimensioning the steering system for the mentioned conditions.

The definition of the cornering coefficient implies that the cornering stiffness is linear with the normal load F_z; actually the cornering stiffness behaves in this way only for low values of force F_z and then increases to a lesser extent (Fig. 4.26). When the limit value has been reached it remains constant or slightly decreases. It is often expedient to approximate the cornering stiffness as a function of the load with two straight lines, the second of which is horizontal. Note that in the figure the line corresponding to a sideslip angle of 2° refers to the true cornering stiffness while the other curve ($\alpha = 10°$) is related to a sort of 'secant' stiffness.

When the need for a more detailed numerical description of the lateral behavior of a tire arises, there is no difficulty, at least in theory, to approximate the experimental law $F_y(\alpha, \gamma, F_z, p, V, \ldots)$ and the similar relationship for the aligning torque, using the algorithms which are common in numerical analysis. This approach can be used with success in the numerical simulations of the behavior of the vehicle, even if it is often quite expensive in terms of time needed for data preparation and computation. A problem which is common to many numerical approaches like this is that of requiring a great amount of experimental data, which are often difficult, or costly, to obtain.

Polynomial approximations, with terms including the third power of the slip angle α, can be used.

As already stated, (4.117) can also be used to express the cornering force and the aligning moment as function of the various parameters.

In the case of the side force, the *magic formula* is

$$F_y = D\sin\bigl(C\arctan\bigl\{B(1-E)(\alpha+S_h)+E\arctan\bigl[B(\alpha+S_h)\bigr]\bigr\}\bigr)+S_v, \quad (4.124)$$

where the product of coefficients B, C and D yields directly the cornering stiffness. The values of the coefficients are

$$C = a_0, \qquad D = \mu_{y_p} F_z,$$

where a value of 1.30 is suggested for a_0 and $\mu_{y_p} = a_1 F_z + a_2$,

$$E = a_6 F_z + a_7,$$

$$BCD = a_3 \sin\left[2\arctan\left(\frac{F_z}{a_4}\right)\right]\bigl(1 - a_5|\gamma|\bigr),$$

$$S_h = a_8\gamma + a_9 F_z + a_{10},$$

$$S_v = a_{11}\gamma F_z + a_{12} F_z + a_{13}.$$

To obtain a better description of the camber thrust, the constant a_{11} is often substituted by the linear law

$$a_{11} = a_{111} F_z + a_{112}.$$

Coefficients S_h and S_v account for ply steer and conicity forces.

Similarly, in the case of the aligning torque the formula is

$$M_z = D\sin\bigl(C\arctan\bigl\{B(1-E)(\alpha+S_h)+E\arctan\bigl[B(\alpha+S_h)\bigr]\bigr\}\bigr)+S_v, \quad (4.125)$$

$$C = c_0, \qquad D = c_1 F_z^2 + c_2 F_z,$$

where a value of 2.40 is suggested for c_0,

$$E = \bigl(c_7 F_z^2 + c_8 F_z + c_9\bigr)\bigl(1 - c_{10}|\gamma|\bigr),$$

$$BCD = \bigl(c_3 F_z^2 + c_4 F_z\bigr)\bigl(1 - c_6|\gamma|\bigr)e^{-c_5 F_z},$$

$$S_h = c_{11}\gamma + c_{12} F_z + c_{13},$$

$$S_v = \bigl(c_{14} F_z^2 + c_{15} F_z\bigr)\gamma + c_{16} F_z + c_{17}.$$

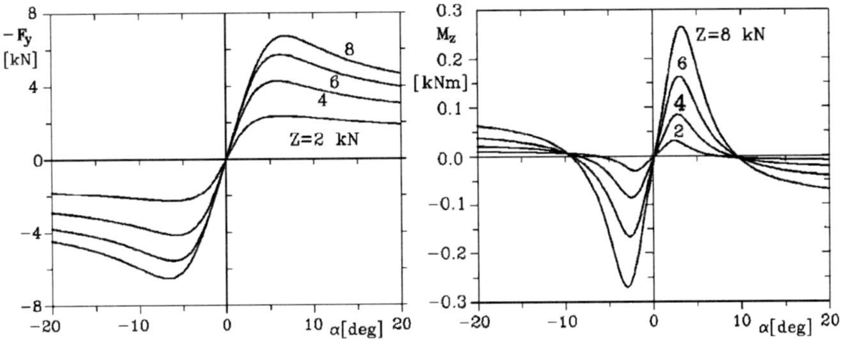

Fig. 4.27 Curves $F_y(\alpha)$ and $M_z(\alpha)$ obtained by using the "magic formula" (4.124) and (4.125). Radial tire 205/60 VR 15 6J (from G. Genta, L. Morello, *The Automotive Chassis*, Springer, New York, 2009)

Also in this case the units introduced into the *magic formula* (4.124) and (4.125) are usually not consistent: the load F_z is often expressed in kN, angles α and γ are in degrees, F_y and M_z are obtained in N and Nm, respectively.

The curves $Y(\alpha)$ and $N_a(\alpha)$ for values of the vertical load F_z equal to 2, 4, 6 and 8 kN for a radial tire 205/60 VR 15 6J are shown in Fig. 4.27.

It is also possible to build structural models of the tire to express the forces it exerts by taking into account the deformations and stresses their structure is subjected to. Apart from very complex numerical models, mainly based on the finite element method, which allow one to compute the required characteristics but are so complex that they are of little use in vehicle dynamics computations, it is possible to resort to simplified models, dealing with the tread band as a beam or as a string on elastic foundations.[16] These models allow one to obtain interesting results, particularly from a qualitative viewpoint, as they link the performance of the tire with its structural parameters, but their quantitative precision is usually smaller than that of empirical models, in particular of those based on the *magic formula* which is now a standard in tire modeling.

If the wheel travels with a sideslip angle α, as it is the case any time it exerts a side force F_y, a strong increase of rolling resistance can be expected. The force in the mean plane of the wheel increases but above all the transversal force F_y has a component which adds to the rolling resistance (Fig. 4.28). The rolling resistance is by definition the component of the force due to the road-tire contact directed as the velocity V; it can thus be expressed as

$$F_r = F_x \cos(\alpha) + F_y \sin(\alpha). \tag{4.126}$$

[16] See, for instance, J.R. Ellis, *Vehicle Dynamics*, Business Books Ltd., London, 1969; G. Genta, *Meccanica dell'autoveicolo*, Levrotto & Bella, Torino, 1993. In the case of elastic non-pneumatic wheels such models may be more accurate than for pneumatic tires.

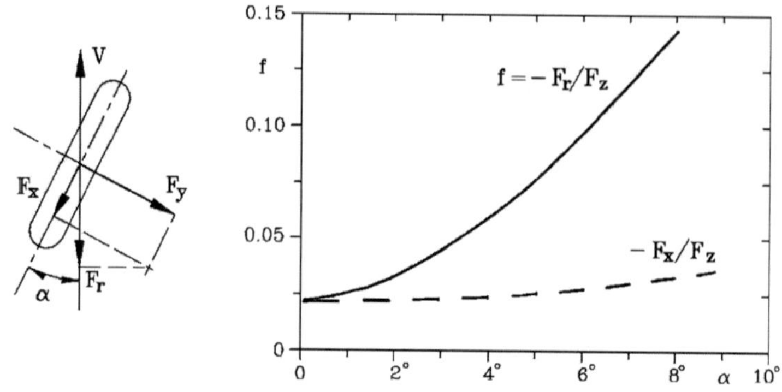

Fig. 4.28 Rolling resistance coefficient as a function of the slip angle α. Pneumatic tire 7.50-14, $F_z = 4$ kN, $p = 170$ kPa (from G. Genta, L. Morello, *The Automotive Chassis*, Springer, New York, 2009)

If the component in the plane of symmetry of the wheel F_x were independent of the sideslip angle and the cornering force F_y were linear with it, (4.119), for small values of α the rolling resistance would follow a quadratic law

$$F_r = F_x - C\alpha^2. \tag{4.127}$$

If the mean plane of the wheel is not perpendicular to the ground, a component of the aligning torque M_z contributes to rolling resistance. Equation (4.93) becomes

$$F_r = \frac{-F_z \Delta x \cos(\gamma) - M_z \sin(\gamma)}{R_l}. \tag{4.128}$$

Remark 4.28 This effect is usually small, due to the fact that γ is usually small. It is, however, dependent on the sideslip angle α through the aligning torque M_z.

Interaction Between Longitudinal and Side Forces

The considerations seen in the preceding sections apply only in the case in which longitudinal and side forces are generated separately. If the tire produces simultaneously forces in longitudinal and lateral direction the situation can be different as the traction used in one direction limits that available in the other.

By applying a driving or braking force to a tire that is rolling with a sideslip angle, the cornering force reduces and the same applies to the longitudinal force a tire can exert if it is called to exert also a lateral force.

It is possible to obtain a polar diagram of the type shown in Fig. 4.29a in which the force in Y direction is plotted versus the force in X direction for any given value of the sideslip angle α. Each point of the curves is characterized by a different value of the longitudinal slip σ. In a similar way it is possible to plot a curve $F_y(F_x)$ at constant σ.

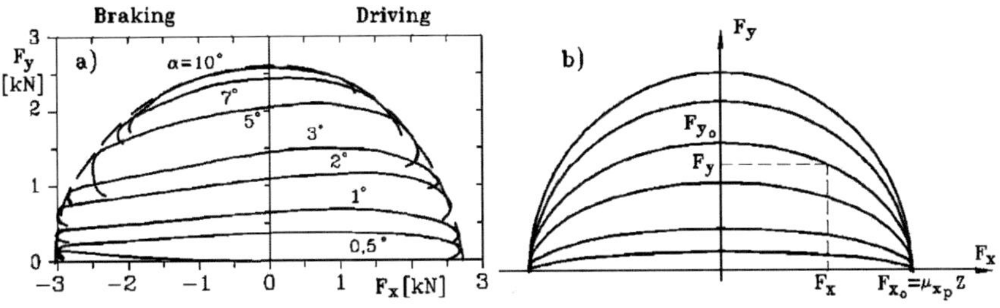

Fig. 4.29 Polar diagrams of the force exerted on the wheel with constant sideslip angle. (**a**) Experimental plots; (**b**) elliptic approximation (from G. Genta, L. Morello, *The Automotive Chassis*, Springer, New York, 2009)

Remark 4.29 The curves are not exactly symmetrical with respect to the F_y-axis: Usually tires develop the maximum value of the force F_y when they exert a small longitudinal force, particularly at a moderate sideslip angle.

If F is the total force exerted on the wheel by the road while F_x and F_y are its components, the resultant force coefficient is

$$\mu = \frac{F}{F_z} = \sqrt{\mu_x^2 + \mu_y^2}. \tag{4.129}$$

The various curves plotted for different values of α are enveloped by the polar diagram of the maximum force the tire can exert. If it were a circle, the so-called *traction circle*, as in simple models it can be assumed to be, the maximum force coefficient would be independent of the direction.

Actually, not only the value of μ_x is greater than that of μ_y but, as already stated, there is some difference in longitudinal direction between driving and braking conditions. The envelope, as well as the whole diagram, is a function of many parameters. Apart from the already mentioned dependence on the type of tire and road conditions, there is a strong reduction of the maximum value of force F with the speed, which is particularly strong in conditions of low traction.

A model allowing one to approximate the curves $F_y(F_x)$ at constant α with simple functions can be quite useful. This can be obtained by using the elliptical approximation (Fig. 4.29b)

$$\left(\frac{F_y}{F_{y0}}\right)^2 + \left(\frac{F_x}{F_{x0}}\right)^2 = 1, \tag{4.130}$$

where forces F_{y0} and F_{x0} are, respectively, the force F_y exerted, at the given sideslip angle, when no force F_x is exerted and the maximum longitudinal force exerted at zero sideslip angle. The envelope curve is then elliptical, the *traction ellipse*.

If (4.130) is used in order to express function $F_y(F_x)$, the cornering stiffness of a tire which is exerting a longitudinal force F_x can be expressed as a function of the cornering stiffness C_0 (i.e. the cornering stiffness when no longitudinal force is produced) by the expression

$$C = C_0\sqrt{1 - \left(\frac{F_x}{\mu_p F_z}\right)^2}, \tag{4.131}$$

where force F_{x_0} has been substituted by $\mu_p F_z$.

Although a rough approximation, particularly for the case in which the longitudinal force approaches its maximum value (the differences between the curves of Fig. 4.29a and those of Fig. 4.29b are evident), the elliptical approximation is often used for all the cases where the concept of cornering stiffness is useful.

The empirical model expressed by (4.117) and (4.124) can be modified to allow for the interaction between longitudinal and lateral forces in a better way than that of computing separately the two forces and then using the elliptic approximation.

4.3.7 Tangential Forces: Rigid Wheel on Compliant Ground

The situation of a wheel rolling on unprepared ground is similar to that described above, but the longitudinal and lateral slip is larger and the maximum force that can be obtained is usually smaller.

Consider a rigid wheel rolling on compliant ground or, as it is sometimes the case, a compliant wheel rolling on ground that is so soft that the wheel remains undeformed. In this case the deformation and the slip occur in the ground instead of occurring in the wheel and the situation is that shown in Fig. 4.11a.

If a driving or braking torque is applied to the wheel, in the contact area the wheel exerts on the ground a longitudinal shear stress τ_x, while a lateral shear stress τ_y is exerted in presence of a sideslip angle α.

The maximum shear stress that the wheel can withstand is expressed by the specific shear resistance of the soil (4.14):

$$\tau_0 = c + \sigma_z \tan(\phi),$$

where the pressure σ_z is in general not constant in the contact area. To the thrust supplied by the resultant of such shear stresses a further force due to the spuds that may be present at the periphery of the wheel, expressed by (4.31), must be added.

A longitudinal slip can be defined as seen in (4.17)[17]

$$\sigma = \frac{\Omega R - V}{\Omega R} = 1 - \frac{V}{\Omega R}. \tag{4.132}$$

[17] As already stated, this definition is equivalent but not identical to that suggested by SAE. However, at the denominator both the velocity of the wheel V or the peripheral velocity ΩR can be used.

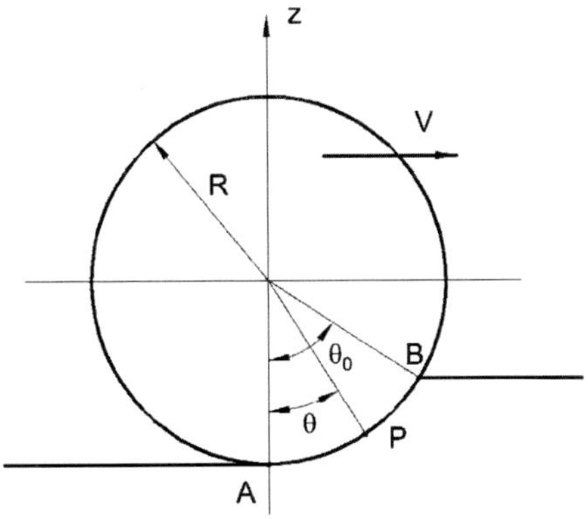

Fig. 4.30 Rigid wheel on compliant ground with no elastic return: geometrical definitions

The slip velocity at point P (see Fig. 4.30, in which the elastic return has been neglected) is

$$V = \Omega R - V\cos(\theta) = \Omega R\big[1 - (1-\sigma)\cos(\theta)\big]. \tag{4.133}$$

The distance traveled backward by point P, i.e. the soil deformation is

$$d = \int_0^t v_g\,dt = R\int_\theta^{\theta_0}\big[1-(1-\sigma)\cos(\theta)\big]\,d\theta, \tag{4.134}$$

i.e.

$$d = R(\theta_0 - \theta) - R(1-\sigma)\big[\sin(\theta_0) - \sin(\theta)\big]. \tag{4.135}$$

Using the formula suggested by Wong (4.20), the shear stress at point P is

$$\tau_x(\theta) = \big[c + \sigma_z\tan(\phi)\big]\big[1 - e^{-\frac{R}{K_x}\{(\theta_0-\theta)-(1-\sigma)[\sin(\theta_0)-\sin(\theta)]\}}\big]. \tag{4.136}$$

The longitudinal modulus of shear deformation K_x has been used instead of K, to take into account the fact that the modulus of shear deformation in x direction may be different from that in y direction.

The pressure is a function of θ, for instance expressed by (4.48)

$$\sigma_r = R^n\left(\frac{k_c}{b} + k_\phi\right)\big[\cos(\theta) - \cos(\theta_0)\big]^n. \tag{4.137}$$

The more complex expression (4.54) can be used as an alternative.

The forces in x and z directions are thus obtained by performing the integrals

$$F_z = Rb\int_{\theta_r}^{\theta_0}\big[\sigma_z(\theta)\cos(\theta) + \tau_x(\theta)\sin(\theta)\big]\,d\theta, \tag{4.138}$$

$$F_x = Rb\int_{\theta_r}^{\theta_0}\big[-\sigma_z(\theta)\sin(\theta) + \tau_x(\theta)\cos(\theta)\big]\,d\theta. \tag{4.139}$$

The wheel torque is

$$M_w = R^2 b \int_{\theta_r}^{\theta_0} \tau(\theta)\, d\theta. \tag{4.140}$$

These integrals must be performed numerically. The normal force F_z is usually known and (4.138) can be used to compute the sinkage of the wheel, i.e. to compute θ_0 as a function of F_z and of the longitudinal slip σ. This implies to compute the integral for various values of z_0, i.e. of θ_0 and θ_r, and then to find the value that corresponds to the given force F_z. Once the sinking of the wheel has been obtained, the other two equations allow one to compute the longitudinal force and the moment as functions of the slip.

Remark 4.30 The longitudinal force so obtained is the total longitudinal force, exerted by the driving wheel with the compaction resistance already accounted for. The net longitudinal force is usually referred to as the *drawbar pull.*

The computation has been performed assuming that the wheel is driving, and the conditions are different from those of a towed wheel: in particular, in the case of a towed wheel the shear stresses are not directed in the same direction in the whole contact zone.

If angle θ_0 is small, the normal component of the tangential forces can be neglected, and the vertical force can be assumed not to be influenced by the longitudinal slip. The sinkage can thus be computed as seen in Sect. 4.3.3.

For the evaluation of the longitudinal force the pressure can be considered a constant in the contact zone, obtained by dividing the normal force by the contact area. Assuming that the ground is perfectly anelastic ($\theta_r = 0$), the expression of F_x reduces to

$$F_x \approx Rb\left[c + \frac{F_z}{A}\tan(\phi)\right] \int_0^{\theta_0} \left[1 - e^{-\frac{R\sigma}{K}(\theta_0 - \theta)}\right] d\theta. \tag{4.141}$$

This yields

$$F_x \approx bR\theta_0\left[c + \frac{F_z}{A}\tan(\phi)\right]\left[1 - \frac{K}{R\theta_0\sigma}\left(1 - e^{-\frac{R\sigma}{K}\theta_0}\right)\right], \tag{4.142}$$

which coincides with (4.27).

Remark 4.31 To obtain the drawbar pull, the compaction resistance should be subtracted from the force so computed, but this approach introduces further errors, since the former was obtained assuming that the wheel is towed.

On compliant ground, usually characterized by high rolling drag and low available traction, in most cases motion is possible only if all wheels are driving wheels. If some of the wheels are not driving, the traction the driving wheels can supply may not be sufficient to overcome the drag (mostly compaction, but also bulldozing) of the free wheels and motion may be impossible, even on level road.

If the wheel works with a non negligible sideslip angle, a lateral force results. Since the wheel sinks for a certain depth in the ground, a bulldozing force acting on the side of the wheel must be added to the force exerted on the cylindrical surface of the wheel owing to the tangential stresses in lateral direction τ_y.

The total lateral force is thus

$$F_y = Rb \int_{\theta_r}^{\theta_0} \tau_y(\theta)\, d\theta + \int_{\theta_r}^{\theta_0} F_{yb}\big[R - h\cos(\theta)\big]\, d\theta. \tag{4.143}$$

The lateral tangential stresses τ_y are expressed by a relationship similar to (4.136) in which the lateral displacement

$$d = R(1-\sigma)(\theta_0 - \theta)\tan(\alpha) \tag{4.144}$$

is introduced in place of the longitudinal displacement:

$$\tau_y(\theta) = \big[c + \sigma_z \tan(\phi)\big]\big[1 - e^{-\frac{R}{K_y}(1-\sigma)(\theta_0-\theta)\tan(\alpha)}\big]. \tag{4.145}$$

The second part of the expression of the lateral force is the bulldozing force, expressed following Ishigami et al.,[18] where h is the depth of sinking

$$h(\theta) = R\big[\cos(\theta) - \cos(\theta_0)\big]$$

and F_{yb} is the bulldozing force per unit width

$$F_{yb} = D_1\left[ch(\theta) + D_2\frac{\rho h^2(\theta)}{2}\right], \tag{4.146}$$

where ρ is the ground density and

$$D_1 = \cot\left(45^\circ - \frac{\phi}{2}\right) + \tan\left(45^\circ + \frac{\phi}{2}\right), \tag{4.147}$$

$$D_2 = \cot\left(45^\circ - \frac{\phi}{2}\right) + \tan(\phi)\cot^2\left(45^\circ - \frac{\phi}{2}\right). \tag{4.148}$$

This formulation relies on the same approximation for the destructive angle given by Bekker as the expression of the bulldozing resistance in (4.74).

The bulldozing component of the side force does not depend on the sideslip angle: it is zero if the latter vanishes, and has the mentioned value if $\alpha \neq 0$. A smoother behavior is to be expected in actual conditions, but this drawback is not very important: the bulldozing component is quite small anyway and could even be neglected except in the case of large sinking, particularly in frictional soil.

[18] G. Ishigami, A. Miwa, K. Nagatani, K. Yoshida, *Terramechanics-Based Model for Steering Maneuver of Planetary Exploration Rovers on Loose Soil*, Journal of Field Robotics, Vol. 24, No. 3, pp. 233–250, 2007.

Another point must be considered: while the lateral force depends on the longitudinal slip, the longitudinal and vertical forces do not depend on the sideslip angle. This type of interaction between the longitudinal and lateral forces is questionable.

Example 4.6 Consider the Mars rover with 6 rigid wheels already studied in Example 4.4. It operates on the same ground whose characteristics are similar to lunar regolith. The data are:

General: $m = 150$ kg, $g = 3.77$ m/s^2; wheels: $R = 150$ mm; $b = 80$ mm; soil: $n = 1$, $c = 200$ Pa, $\phi = 35°$, $k_c = 1{,}400$ N/m^2, $k_\phi = 820{,}000$ N/m^3, $K_x = K_y = 18$ mm, $a_0 = 0.4$, $a_1 = 0.1$ and $\rho = 1{,}700$ kg/m^3. Assume that the soil is either perfectly anelastic ($\lambda = 0$) or has a weak elastic return ($\lambda = 0.2$).

Plot the longitudinal force and driving moment as functions of the longitudinal slip and compare them with those obtained using the simplified models. Plot also the lateral force as a function of the sideslip angle.

Assuming that the load is equally subdivided on the six wheels, the results obtained in the previous example for rigid wheels on anelastic ground were $F_z = 94.25$ N, $z_0 = 25$ mm, $p_{\max} = 20.6$ kPa. The compaction resistance was 20.2 N per wheel. The value of θ_0 corresponding to a sinkage of 25 mm is

$$\theta_0 = \arccos\left(1 - \frac{z_0}{R}\right) = 0.586 \text{ rad} = 33.26°.$$

The vertical force can be obtained by integrating numerically equation (4.138). Since the value of the vertical force is known, for each value of the slip it is thus possible to obtain the value of θ_0, i.e. of the sinking of the wheel.

The results are reported in Fig. 4.31a: in the present case the sinking is almost independent from σ, while in most cases it decreases with increasing longitudinal slip showing that the wheel floats better when exerts a driving force.

This result depends much on the pressure distribution assumed, and in particular on the value of coefficient a_1. The sinking computed using the present formulas (for $\lambda = 0$) is larger than that obtained in Example 4.4, since there the pressure was assumed to be distributed on the ground in a different way.

Equations (4.139) and (4.140) are integrated for different values of the slip to yield the longitudinal force and the moment. The results are plotted as functions of the slip in Fig. 4.31b. When the slip is small, the longitudinal force is the drawbar pull and is negative, i.e. the wheel is exerting a resisting force, the compaction resistance, and not an actual traction.

The net traction coefficient, obtained by dividing the drawbar pull by the normal force, is reported as a function of the slip in Fig. 4.31c. In spite of the fairly large sinking, the net traction coefficient is not too low.

The pressure and the tangential stress are reported as functions of angle θ in Fig. 4.31d: the computation has been performed for $\sigma = 0.2$.

The lateral force was computed by integrating (4.143) for different values of σ and α. The results are reported in Fig. 4.32. From the figure it is clear that in the present case the influence of the bulldozing force on the results is marginal and can be neglected.

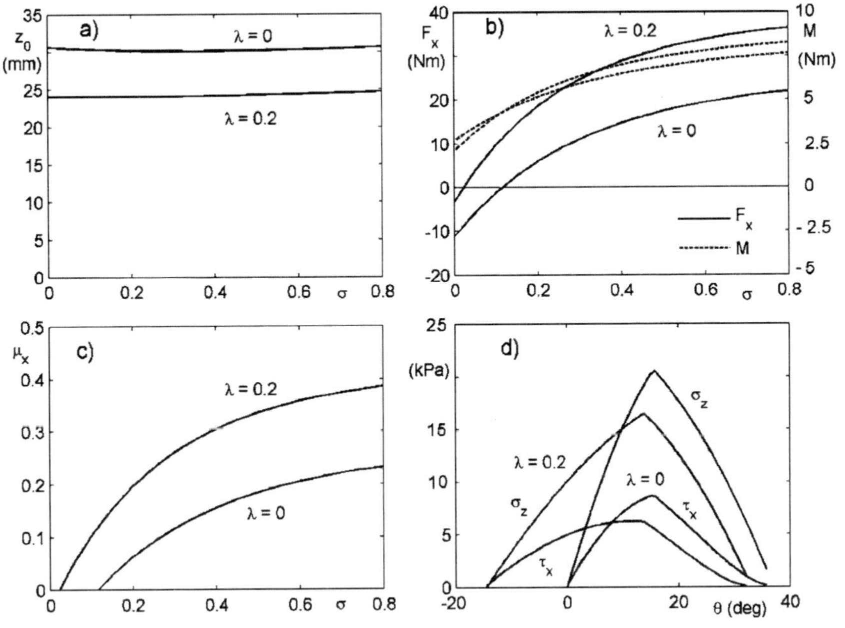

Fig. 4.31 Rigid wheel on a compliant ground. (**a**) Sinking for two different values of the elastic return of the ground. (**b**) Longitudinal force and moment as functions of the longitudinal slip. (**c**) Longitudinal traction coefficient as a function of the longitudinal slip. (**d**) Pressure distribution at the contact with the ground for a slip $\sigma = 0.2$

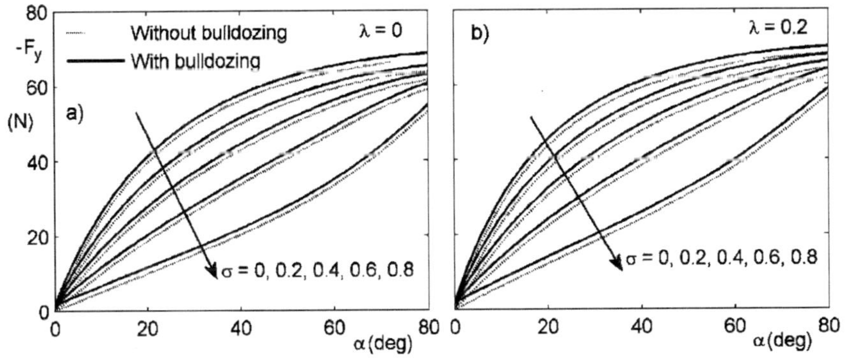

Fig. 4.32 Cornering force as a function of the side slip angle for different values of coefficient λ and of the longitudinal slip σ

4.3.8 *Tangential Forces: Compliant Wheel on Compliant Ground*

The two cases here studied, compliant wheel on rigid ground and rigid wheel on compliant ground, are limiting cases. If both are compliant some slip occurs in both and the situation is more complicated. A much simplified model in which the slip is assumed to occur only in the ground is, however, reported here on the grounds that the stiffness in longitudinal and lateral direction of the tire is much higher than the stiffness in radial direction.

As a consequence, the outer part of the wheel, i.e. the thread band, behaves as a rigid body in circumferential and axial direction, while bending in the plane of the wheel, and each one of its points travel at a constant speed equal to ΩR.

The situation may be modeled as shown in Fig. 4.15, and the pressure can be assumed to be constant and equal to the pressure $p_i + p_c$ causing the elastic wheel to yield in the second part of the contact. In the first part the pressure is given by the usual expression

$$\sigma_r = \left(\frac{k_c}{b} + k_\phi\right)h^n = R^n\left(\frac{k_c}{b} + k_\phi\right)\left[\cos(\theta) - \cos(\theta_0)\right]^n.$$

As already stated, if angle θ_1 is small enough to assume that its cosine is nearly equal to 1, point A can be obtained from the equation

$$\sigma_{r\text{A}} = \left(\frac{k_c}{b} + k_\phi\right)h_1^n = p_i + p_c$$

yielding

$$h_1 = \left(b\frac{p_i + p_c}{k_c + bk_\phi}\right)^{1/n}.$$

This approach has the disadvantage of yielding a pressure that does not vanish in the point where the contact is released. The x coordinate of point A is

$$x_\text{A} = R\sin(\theta_1). \tag{4.149}$$

The longitudinal slip length is

$$d = (\theta_0 - \theta) - (1 - \sigma)\left[\sin(\theta_0) - \sin(\theta)\right], \tag{4.150}$$

along arc AB (i.e. for $\theta_1 \le \theta \le \theta_0$) and

$$d = \sigma(\theta_1 - \theta) + (\theta_0 - \theta_1) - (1 - \sigma)\left[\sin(\theta_0) - \sin(\theta_1)\right], \tag{4.151}$$

along line A′A (i.e. for $-\theta_1 \le \theta \le \theta_1$).

If angle θ is small enough to linearize its trigonometric functions ($\sin(\theta) \approx \theta$), the expression of the longitudinal shear stresses at a generic point of the contact zone is

$$\tau_x(\theta) = \left[c + \sigma_z\tan(\phi)\right]\left[1 - e^{-\frac{R\sigma}{K_x}(\theta_0 - \theta)}\right]. \tag{4.152}$$

In a similar way, the lateral shear stress is

$$\tau_y(\theta) = \left[c + \sigma_z\tan(\phi)\right]\left[1 - e^{-\frac{R}{K_y}(1-\sigma)(\theta_0-\theta)\tan(\alpha)}\right]. \tag{4.153}$$

The forces can thus be obtained by integrating the pressures on the contact zone:

$$F_z = Rb\int_{\theta_1}^{\theta_0}\big[\sigma_r(\theta)\cos(\theta) + \tau_x(\theta)\sin(\theta)\big]\,d\theta + 2Rbx_A(p_i + p_c), \tag{4.154}$$

$$F_x = Rb\int_{\theta_1}^{\theta_0}\big[-\sigma_r(\theta)\sin(\theta) + \tau_x(\theta)\cos(\theta)\big]\,d\theta + Rb\int_{-\theta_1}^{\theta_1}\tau_x(\theta)\,d\theta, \tag{4.155}$$

$$F_y = Rb\int_{-\theta_1}^{\theta_0}\tau_y(\theta)\,d\theta. \tag{4.156}$$

In the last equation the bulldozing force has been neglected.

The first equation allows to compute the value of θ_0 from the known value of the normal force F_z, while the other two equations allow to compute the longitudinal and transversal forces $F_x(\sigma)$ and $F_x(\alpha, \sigma)$.

In this case the sinking of the wheel h_1 is almost independent of σ and may be considered as a constant, while θ_0, although depending on the load, is little affected by the slip.

This model, however, is just a first approximation and its results must be considered as indicative.

Remark 4.32 In general, when a compliant wheel travels on soft ground, the deformation of the soil contributes to the sideslip and the cornering stiffness can be expected to decrease. The picture is made more complex by the possibility of the presence of a bulldozing component of the side force. The latter consideration holds particularly in the case of a rigid wheel on very soft ground.

Example 4.7 Consider the same elastic wheel for a Mars rover seen in Example 4.5 and assume the same data for vehicle, wheels and soil: number of wheels $n = 6$, $m = 150$ kg, $g = 3.77$ m/s^2, $R = 150$ mm; $b = 80$ mm, $p_c = 5$ kPa, $n = 1$, $c = 200$ Pa, $\phi = 35°$, $k_c = 1{,}400$ N/m^2, $k_\phi = 820{,}000$ N/m^3. Assume an inflation pressure $p_i = 5$ kPa and moduli of shear deformation $K_x = K_y = 18$ mm.

Compute the longitudinal force as a function of the longitudinal slip and the lateral force as a function of the sideslip angle.

Since the force acting on each wheel is $F_z = 94.25$ N, the sinkage of a rigid wheel (i.e. a wheel inflated at more than 15.6 kPa) was computed in Example 4.5 as $z_0 = 25$ mm. At a pressure $p_i = 5$ kPa a sinkage of 11.9 mm was obtained.

The results for rigid wheels ($p_i > 15.6$ kPa) are different from those obtained in Example 4.6 since in that example a different distribution of the vertical pressure was assumed.

The results for $p_i = 5$ kPa are plotted in Fig. 4.33.

4.3.9 Tangential Forces: Empirical Models

The models seen above for compliant ground rely on the terramechanics approach and are based on many crude assumptions. Models of this kind require experimental

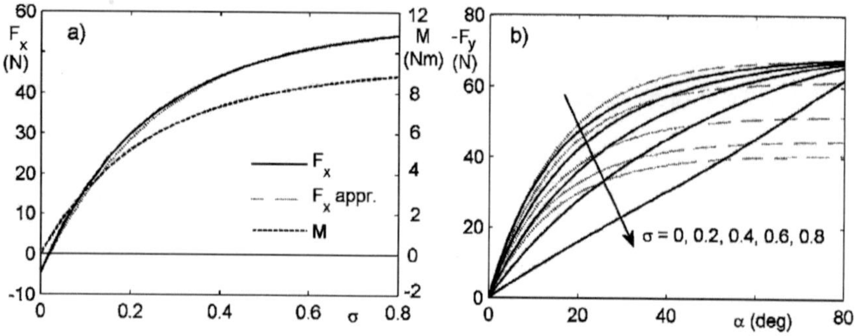

Fig. 4.33 Tangential force and aligning moment on a compliant wheel rolling on compliant ground. (**a**) Longitudinal force and aligning moment as functions of the longitudinal slip; (**b**) cornering force as a function of the side slip angle for different values of the longitudinal slip σ. The results of the approximated empirical formulae are also reported (*dotted lines*)

validation performed in conditions simulating as closely as possible the actual situation.

Once the forces acting between the wheel and the ground have been measured, it is possible to compute the coefficients to be introduced into equations of the type of the magic formula to summarize the experimental results in a few equations that can be included in the model of the vehicle.

The magic formula is known to yield accurate results, but requires many experimental data from which to compute the tens of coefficients involved.

A simple, although much less accurate approach is that of using the following exponential expressions:

- Longitudinal force (drawbar pull):

$$F_x(\sigma) = \text{sgn}(\sigma)\mu_{x_p} F_z\left[1 - e^{-\frac{C_\sigma}{\mu_{xp} F_z}|\sigma|}\right] - |F_{x_r}|, \tag{4.157}$$

where C_σ is the longitudinal force stiffness, which can be computed from the slope of the curve in the point for $\sigma = 0$, $|F_{x_r}|$ is the resistance to motion at $\sigma = 0$ and $\mu_{x_p} F_z$ is the force at the horizontal asymptote. It can be computed from an experimental value of the longitudinal force F_{x1} and the corresponding value of the slip σ_1 from the equation

$$\text{sgn}(\sigma_1)\mu_{x_p} F_z\left[1 - e^{-\frac{C_\sigma}{\mu_{xp} F_z}|\sigma_1|}\right] - F_{x1} - |F_{x_r}| = 0. \tag{4.158}$$

- Lateral force (at $\sigma = 0$):

$$F_y(\alpha) = -\text{sgn}(\alpha)\mu_{y_p} F_z\left[1 - e^{-\frac{C}{\mu_{yp} F_z}|\alpha|}\right], \tag{4.159}$$

where C is the cornering stiffness, computed from the slope of the curve in the point for $\alpha = 0$ (changed in sign) and $\mu_{y_p} F_z$ is the force at the horizontal asymptote. It can be computed from an experimental value of the cornering force F_{y1}

and the corresponding value of the sideslip angle α_1 from the equation

$$-\mathrm{sgn}(\alpha_1)\mu_{y_p}F_z\left[1-e^{-\frac{C}{\mu_{yp}F_z}|\alpha_1|}\right]-F_{y1}=0. \tag{4.160}$$

- Aligning moment

$$M_z(\alpha)=\mathrm{sgn}(\alpha_1)M_{z0}e^{-C_1|\alpha_1|}\left[1-e^{-\frac{C}{\mu_{yp}F_z}|\alpha_1|}\right], \tag{4.161}$$

where C is the cornering stiffness computed above, parameter M_{z0} can be computed from the slope in the origin from the equation

$$M_{z0}=\frac{\mu_{y_p}F_z}{C}\left(\frac{dM_z}{d\alpha}\right) \tag{4.162}$$

and C_1 can be computed from a pair of experimental values M_{z0} and α.

The aligning moment is, however, more difficult to predict and the last expression can give only an indication. In case of compliant wheel on rigid road M_{z0} can be approximated as

$$M_{z0}=\frac{1}{6}\mu_{y_p}F_z a \tag{4.163}$$

where a is the length of the contact area. This expression comes from the assumption that at small sideslip angles the lateral stress distribution is triangular. If the ground is compliant, the contact area displaces forward (in some simplified models it lies all forward of the intersection of the z axis with the ground) and the aligning torque is negative even for small sideslip angles.

- Longitudinal–lateral forces interaction.

  A simple way to account for the interaction is the already mentioned elliptic approximation, yielding

$$(F_y)_{\sigma\neq 0}=(F_y)_{\sigma=0}\sqrt{1-\left(\frac{F_x}{\mu_{x_p}F_z}\right)^2}. \tag{4.164}$$

  This approach yields only a first approximation evaluation.

Example 4.8 Approximate with exponential functions the longitudinal and lateral forces of the elastic wheel seen in Example 4.7 as functions of the longitudinal slip and the sideslip angle.

From the plots obtained in the previous example, it is possible to obtain

Drawbar pull: $C_\sigma=240$ N, $\mu_{x_p}F_z=61.8$ N (i.e. $\mu_{x_p}=0.66$), $|F_{x_r}|=4.68$ N.
Cornering force: $C=254$ N/rad, $\mu_{y_p}F_z=68.0$ N (i.e. $\mu_{y_p}=0.72$).

The results obtained through these approximated formulae are plotted in Fig. 4.33, dotted lines. The results are quite close to the ones previously obtained, particularly for the longitudinal force.

In the same figure the interaction between longitudinal and lateral forces obtained using the elliptical approximation are also reported. Here the results are clearly less accurate, in particular for values of σ larger than 0.2.

4.3.10 Dynamic Behavior of Tires

Elastic wheels are prone to vibrate and their dynamic behavior is important in determining the forces transferred to the vehicle.

In particular, the forces the wheel exchanges with the ground in dynamic conditions are different from those characterizing steady-state running. If the geometrical parameters (slip, slip angle and camber angles) or the forces in X and Z directions are variable during motion, the values of the longitudinal and side force and of the aligning moment at any instant are usually lower than those which would characterize stationary conditions with the same values of all parameters. As an example, if a tire is tilted about the vertical axis at standstill and then it is allowed to roll, the side force reaches the steady-state value only after a certain time, after rolling for a certain distance, usually referred to as relaxation length. This effect is usually not noticeable in normal driving as the time delay is very small, but the fact that there is a delay between the setting of the sideslip angle and the force generation is very important in dynamic conditions.

If the sideslip angle is changed with harmonic law in time, the side force and the aligning torque follow the sideslip angle with a certain delay, function of the frequency, and their value is lower than that obtained in quasi-static conditions, i.e. with very low frequency.

If the frequency is not very high, at the speeds encountered in normal driving the average values are not much lower than those characterizing static conditions, but a certain phase lag between the sideslip angle and the F_y force remains. More important for what practical applications are concerned is the case in which the load F_z applied by the wheel on the ground is variable, as is the case of rolling on uneven ground (Fig. 4.34). The frequency may be high, if the speed is high enough, and the decrease of lateral force due to dynamic effects may be large. In the figure the law $z(t)$ of the vertical displacement of the hub of the wheel is harmonic with a frequency of about 7 Hz while the response $F_y(t)$ has a more complicated time history, with even an inversion of sign occurring at each cycle. The decrease of the average value of the lateral force at increasing frequency is shown in Fig. 4.34b.

Strictly linked with the dynamic behavior of the tire are the self-excited vibrations of the wheel and the whole steering mechanism known as *wheel shimmy*. Such vibrations are today mainly of historical interest, as modern vehicles are free from this problem which was very important about half a century ago when it represented an actual danger in the automotive and aeronautical fields.

Remark 4.33 If elastic non-pneumatic wheels are used at high speed on future rovers, wheel shimmy may become again an actual danger.

4.3.11 Omni-Directional Wheels

It is possible to arrange wheels that have the possibility to roll freely in two directions, i.e. omni-directional wheels. The simplest way this can be accomplished

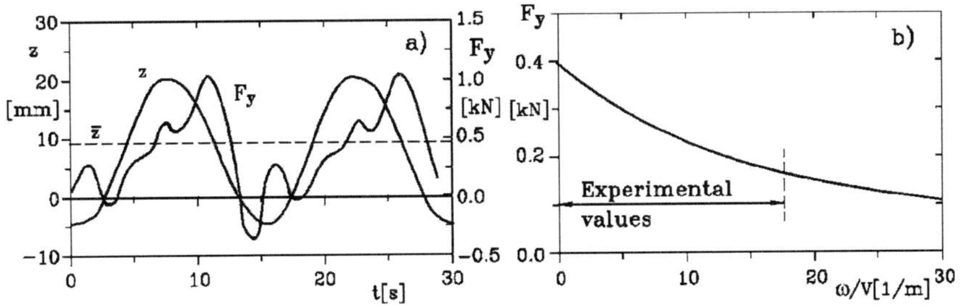

Fig. 4.34 (**a**) Lateral force generated by a pneumatic tire working with constant slip angle but with the hub moving vertically with harmonic law $z(t)$. (**b**) Average value of the lateral force F_y as a function of the ratio between the circular frequency ω of the law $\alpha(t)$ and the speed V (from G. Genta, L. Morello, *The Automotive Chassis*, Springer, New York, 2009)

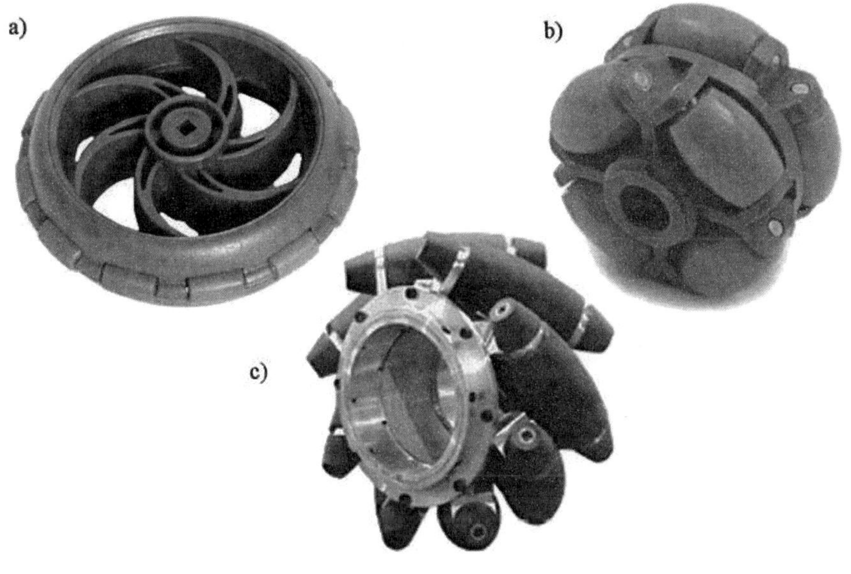

Fig. 4.35 Omni-directional wheels. (**a**) Wheel with a row of rollers (VEX Robotics); (**b**) wheel with two rows of rollers (North American Roller Products). In both cases the axes of the rollers lie in the rotation plane of the wheel; (**c**) wheel with skew rollers (Airtrax)

is by locating rollers at the periphery of the wheel like in the examples shown in Fig. 4.35 so that the motion in the plane of rotation is allowed by the rotation of the wheel, while the motion in axial direction is allowed by the rollers. Omni-directional wheels may then be considered as wheels with vanishing cornering stiffening.

The wheel may have a single row of rollers (Fig. 4.35a) or two rows (Fig. 4.35b) so that at least one roller is always in contact with the ground in any position.

Usually omnidirectional wheels are propelled only in the longitudinally direction, while the rollers are free-wheeling. If the wheels of a vehicle are of this kind, they must be set with their midplanes not parallel to each other, otherwise the motion in a direction perpendicular to their midplanes is not controllable. Similarly, the perpendicular to the midplanes of the wheels must not converge in a point, otherwise rotation about that point cannot be controlled.

Another possibility is orienting the axes of the roller in a direction at an angle with the symmetry plane of the wheel (Fig. 4.35c). In this case the midplanes of the wheels may be parallel and it is possible to produce lateral motion by differential rotation of the wheels.

Omni-directional wheels have severe limitations. First, the rollers have a small diameter (at least if compared with the diameter of the wheel), causing and increase of the pressure on the ground and limiting their mobility on uneven ground. Moreover, it is very difficult to protect the mechanisms of the roller from dust and dirt. Omni-directional wheels are suitable only for motion on hard, flat and possibly horizontal surfaces and thus they are seldom considered for planetary vehicles and robots. They will no more be dealt with here.

4.4 Tracks

The main disadvantage of wheels in off road locomotion is the relatively high pressure they exert on the ground. To decrease the pressure on the ground it is possible to increase the size of the wheels, both in diameter and axial thickness, or to use some unconventional layouts, like spherical wheels. However, this is a disadvantage more felt in terrestrial applications than in planetary exploration vehicles, mainly for two reasons:

- As already stated, all celestial bodies of the solar system that are possible targets for exploration are characterized by fairly low gravity. The weight of even large vehicles is much lower than on Earth and consequently the pressure the running gear exerts on the ground is lower.
- The worst conditions encountered on Earth occur when the soil is very wet and rich in organic material. But there is no mud on the surface of other planet and the bearing capacity of regolith is much higher than that encountered in bad conditions on our planet.

To reduce the pressure on the ground it is possible to use tracks instead of wheels: a tracked vehicle is essentially a vehicle that lays a hard surface under its own wheels that distributes the pressure on the ground on a larger area, and then removes it.

The pressure distribution on the ground depends on many parameters, like the tightness and the flexibility of the track, the distance between the wheels or rollers and the characteristics of the soil. The distribution of the tractive force depends on the same parameters and also on the presence of treads or plates in a direction perpendicular to the ground.

With respect to wheels, tracks are usually heavier, mechanically more complex and produce a greater resistance to motion, at least on a soil with good carrying capacity. They become more convenient than wheels only when the carrying capacity is low, causing the wheels to sink.

Fig. 4.36 The tracked microrover Nanokhod

For all these reasons tracks are seldom considered for planetary exploration vehicles except when the particular characteristics of the vehicle would lead to unacceptably high contact pressure if wheels where used.

Remark 4.34 Patches of very fine regolith with a low carrying capacity were anyway found both on the Moon and on Mars, in particular close to the rim of some craters. They resulted in severe mobility problems for wheeled vehicles. Tracks may prove to be more suitable than wheels on these patches.

A further disadvantage of tracks is the difficulty of protecting the mechanisms, and particularly the zone where the wheels are supported by the track, from sand and dust. This is already a serious problems on Earth, but becomes much more severe on planets where in general the dust is finer and the environment is much drier.

Tracks may be made of rigid plates hinged to each other or by a flexible element, similar to a belt. In the first case the tracks are heavier, cause a larger energy consumption and are subjected to a quicker wear in dusty environment.

One or the few examples of tracked rovers designed and built (but never used in actual missions) is the Nanokhod microrover (Fig. 4.36).

Since tracks seem to be little suited for planetary exploration and exploitation machines, they will not be further dealt with. The required details can be obtained from the many books devoted to off-road locomotion, agricultural machinery and military vehicles.

4.5 Legged Locomotion

Legged locomotion is the most common way of moving on the surface of our planet used by animals of all kind. Continuous rotational motion is unknown in nature, except for some microscopic animals which are propelled by the continuous rotation of cilia or flagella,[19] and nothing similar to wheels was developed by evolution. The number of legs on which animals support their weight during walking has continuously reduced during evolution, while their layout assumed the configuration of a chain of rigid segments, connected by cylindrical or spherical hinges—from the filaments (parapods) of Annelida (e.g. the millipedes) to the articulated legs of the arthropods. In the latter a continuous reduction of their number (ten in the crustacea, eight in the Arachnida, six in the insects) has occurred. With terrestrial vertebrates the number of legs reduced to four.

A high number of legs, together with a low position of the center of mass (i.e. a small height of the center of mass if compared with the 'track' of the legs) allows the animal to remain easily in static equilibrium conditions during all phases of walking. A quadruped, particularly if its center of mass is high, usually goes through positions that are not of equilibrium and therefore must coordinate its movements with a greater precision and have quicker reactions than a hexapod or an octopod. Besides, the larger is the animal and the lower is the gravity of the planet, the easier is to remain in equilibrium on fewer legs, in the sense that the response of the nervous system to avoid tipping over may be less quick. From this point of view low gravity simplify the operations linked with motion.

On the other hand, however, a maximum walking speed exists for any animal; to go faster the animal must change its gait and perform a transition from walking to running or jumping. This speed depends on the size of the animal and on the gravity of the planet on which it moves. This dependence can be expressed by the Froude number[20]

$$\mathcal{F}_r = \frac{V}{\sqrt{gL}}, \tag{4.165}$$

where V, g and L are, respectively, the speed, the gravitational acceleration and a characteristic length, in this case the length of the legs.[21] When the Froude number reaches a value of about 0.7, a change of gait occurs: the animal starts running (in some cases jumping), indicating that running becomes more efficient than walking. Another important value is 1: when the Froude number is equal to unity walking is no more possible. These considerations come from modeling the legs as pendulums or inverted pendulums: when the speed reaches a value equal to $\sqrt{gL}$, i.e. $\mathcal{F}_r = 1$, the centrifugal acceleration of the inverted pendulum of length L, V^2/L, equals the gravitational acceleration and the pendulum lifts off the ground, which is incompatible with walking. This means that the transition and the maximum walking speed for a human with a leg length of 0.8 m are, on the surface of the Earth, 2 and 2.8 m/s (7.1 and 10.1 km/h), respectively.

---

[19] A. Azuma, *The Biokinetics of Flying and Swimming*, Springer, Tokyo, 1992.

[20] The Froude number can be also defined as $\mathcal{F}_r = \frac{V^2}{gL}$, i.e. the square of that defined above. With this definition it can be interpreted as the ratio between inertial and gravitational forces.

[21] G.A. Cavagna, P.A. Willems, N.C. Heglund, *Walking on Mars*, Nature, Vol. 393, p. 636, June 1998; A.E. Minetti, *Invariant Aspects of Human Locomotion in Different Gravitational Environments*, Acta Astronautica, Vol. 39, No 3–10, pp. 191–198, 2001.

The energy requirements for motion is an important parameter, and from this viewpoint locomotion on a solid surface is more expensive than swimming in water or flying. The energy needed to move on level ground ideally should compensate for the aerodynamic drag and energy losses at the foot–ground contact, mainly due to irreversible deformations on the bodies in contact. Both these sources of losses are small, at least in low speed walking on a hard surface. Most of the energy is lost by internal friction and by the need to accelerate and decelerate continuously some of the parts of the system (mainly the legs, but also the body, since usually a walking animal or machine does not move at constant speed). Recovery of the kinetic energy of the legs can thus be important from the energetic viewpoint and can be achieved through their pendular motion. The optimal speed from this viewpoint can be shown to correspond to $\mathcal{F}_r = 0.5$. For a human on Earth this speed is about 1.4 m/s (5 km/h).

The energy dissipated for motion in the case of a walking vehicle is linked with the irreversible deformation of the ground, compressed by the feet. If the soil is very compliant there is a definite advantage with respect to wheeled vehicles: a walking machine compresses just some small zones of soil where it puts its feet, while a wheeled machine compresses a continuous strip of ground. Another advantage is that the tractive force exerted by feet may be larger than that due to wheels in soft and slippery ground. The bulldozing force due to the sinking of the feet may be much larger than the traction force exerted by a wheel.

Another advantage is the possibility of adequately choosing where to put the feet: just a number of 'good' spots are required and not a continuous strip. Similarly, also the capability of overcoming obstacles is much better.

These advantages are often compensated by the much greater complexity of both the mechanical design and the control system, which depend strictly on the exact leg kinematics. These topics will be dealt with when dealing with the architecture of walking machines.

Another weak point of legged locomotion is the need that the actuators supply forces to support the weight of the machine even when non moving, or moving very slowly. Natural actuators (muscles) do this with high efficiency, while most artificial actuators, like electric motors, have a very low (even zero) efficiency in this case.

Legged locomotion was often considered for planetary rovers, but seldom used for prototypes and never in actual missions. This is justified considering that wheeled vehicles have a tradition that cannot be matched by other configurations and the designer can rely on a well consolidated technology, without the need of resorting to simulations, experimental tests and other studies, slowing down the design process and increasing costs. But it is not just a matter of a consolidated design practice: legged vehicles are usually highly stressed, have reciprocating parts undergoing a large number of fatigue cycles, require complex control systems and in some cases have a higher energy consumption in actual working, in spite of a greater theoretical efficiency.

4.6 Fluidostatic Support

If the solid surface of a celestial body is covered by a layer of fluid, be it a liquid or a gas, it is possible to exploit fluidostatic forces to support a vehicle. On our planet ships, balloons and blimps are all examples of fluidostatic vehicles.

The force a fluid exerts on any body immersed in it is the so-called Archimedes' force:

$$F = \rho g V, \tag{4.166}$$

where ρ is the density of the fluid and V is the volume of the object, or better, of the displaced fluid. The force is directed vertically upwards and (4.166) holds only if the gravitational acceleration is constant in all the zone occupied by the body.

In such a case, the force is applied in the geometric center of the volume of displaced fluid or center of buoyancy. If the body is only partially immersed in the fluid, like in the case of surface ships, the position of the center of buoyancy changes when the object rolls. Assume that the body has a longitudinal plane of symmetry, and that in normal conditions it floats with the plane of symmetry vertical. To assess the stability of this equilibrium position for small roll motions a point, which in ship technology is referred to as the *metacenter*, is defined as the intersection of the vertical passing through the center of buoyancy in a position characterized by a small roll angle and the symmetry plane. The hull is stable if the metacenter is located above the center of mass.

Remark 4.35 From (4.166) it is clear that, while in liquids it is possible to obtain large forces even with relatively small volumes, owing to their high density, in gases aerostatic support requires to displace large volumes of fluid.

The bodies of the solar system where large surfaces covered with a liquid exist are very few, so hydrostatic support is seldom considered. Two possible exceptions are Titan and Europa. On Titan there are lakes of liquid hydrocarbons (methane and ethane), and it is possible to consider robotic boats or submarines to explore them. But it is on Europa that exploration submarines will be perhaps more useful, if under the ice surface there is a water ocean, where there are chances to find life.

Aerostatic robots have been proposed for Mars and Titan. If robots will be sent to explore the upper layers of the atmosphere of gas giants it is likely that they will be aerostatic vehicles.

Aerostatic vehicles are usually subdivided in balloons and blimps (airships). The former have no propulsion and are just carried by atmospheric winds, while the second have a propulsion device and can maintain a given course.

Usually the atmosphere is assumed to be made by a mixture of perfect gases. The pressure p and the density ρ are linked by the relationship

$$\frac{p}{\rho} = R^* T, \tag{4.167}$$

where T is the absolute temperature and R^* is a constant characterizing the given gas or mixture of gases. It is the ratio between the universal gas constant $R =$

8,314 J/(mol K) and the average molecular mass of the gas.The average molecular mass for Earth's atmosphere is 29 and thus $R^* = 287$ m^2/s^2 K. On Mars, for instance, the atmosphere is made mainly by carbon dioxide with molecular mass 44 and hence $R^* = 188$ m^2/s^2 K.

Consider an aerostat filled with a gas with molecular mass $\mathcal{M}_1$ flying in an atmosphere made by gases with an average molecular mass $\mathcal{M}_2$. Assuming that the pressure of the gas in the aerostat and outside it are equal and that also the temperatures are equal, the lift is the difference between the aerostatic force and the weight of the gas:

$$F = \rho g V(\rho_o - \rho_i) = \frac{pgV}{RT}(\mathcal{M}_o - \mathcal{M}_i), \tag{4.168}$$

where subscripts o and i refer to the gases outside and inside the aerostat.

Remark 4.36 It may seem strange that aerostats have been proposed for planets with a very thin atmosphere, but the low temperature and the atmospheric composition may produce a density high enough to sustain a vehicle: the colder is the planet and the higher is the molecular mass and the pressure of the atmosphere, the larger is the lift produced.

Example 4.9 Compute the volume of a balloon filled with hydrogen or helium able to lift a payload of 100 kg (including the structure of the balloon) on Mars. Assume a pressure of 600 Pa and a temperature of $-50°\text{C} = 223$ K.

Since $g = 3.77$ m/s^2, the weight to be lifted is 377 N. The volume is then

$$V = \frac{FRT}{pg(\mathcal{M}_o - \mathcal{M}_i)},$$

i.e. 7,721 m^3 if the balloon is filled with helium (molecular mass 4) or 7,354 if it is filled with hydrogen. If the balloon is spherical, its diameter is 24.5 m for helium and 24.1 m for hydrogen.

The difference between the two cases is marginal.

The formulas above assume that the balloon is in equilibrium of pressure and temperature with the surrounding atmosphere. The lift increases if the gas is heated so that it expands: either the balloon can increase its volume or some of the gas is vented out and the balloon gets lighter. An interesting possibility is a balloon that heats strongly and expands by day, rising in the atmosphere, to land by night when it cools.

It is unlikely that hot air balloons are used on Mars or on other planets: a hot air balloon uses the same gas of the atmosphere that is heated and thus has a lower density. On Earth this solution is easy, since air is heated by combustion (early balloons used straw as a fuel, modern ones use propane), but in a non oxidizing atmosphere everything is more difficult. A possibility on Titan might be to burn the methane existing in the atmosphere by mixing oxygen to it. The very low outside temperature might make possible to reach a good temperature difference, producing a good lift, but the amount of methane in the atmosphere is low and a way to enrich the gas must be found.

Both balloons and airships have a long history of technological developments. This old technology may prove to be useful for planetary exploration.

4.7 Fluid-Dynamics Support

On Earth both hydrodynamic and aerodynamic forces are used for transportation.

Generally speaking, fluid-dynamic forces exerted on an object moving in the fluid are proportional to the square of the relative speed, the density of the fluid and the square of the linear dimension of the object

$$F = \frac{1}{2}\rho V^2 S C_f, \tag{4.169}$$

where coefficient $1/2$ is included just for historical reasons, surface S is a reference surface and C_f is a coefficient depending on the shape of the body and its position with respect to the direction of the relative velocity. However, C_f depends also on two nondimensional parameters, the Reynolds and the Mach numbers:

$$\mathcal{R}_e = \frac{VL}{\nu}, \qquad \mathcal{M}_a = \frac{V}{V_s},$$

where ν is the kinematic viscosity of the fluid, L is a reference length and V_s is the speed of sound in the fluid. The first one is a parameter showing the relative importance of the viscous and inertial effects in determining the aerodynamic forces. If its value is low the former are of great importance while if it is high aerodynamic forces are mainly due to the inertia of the fluid.

The Mach number shows the importance of the effects due to fluid compressibility.

The reference surface S and length L are arbitrary, to the point that in some cases a surface not existing physically is used, like in the case of airships where S is the power $2/3$ of the displacement. It is, however, clear that the numerical values of the coefficients depend on the choice of S and L, which must be clearly stated. In the case of airplane wings, S is the wing surface and length L is the mean chord c, i.e. the average width of the wing. The wing area is then

$$S = bc,$$

where b is the wingspan.

The study of the aerodynamic forces is performed using a reference frame Gxyz fixed to the body and moving with it (Fig. 4.37). It is centered in the center of mass G, the x-axis has the direction of the resultant air velocity vector V_r of the object with respect to the atmosphere. The z-axis is contained in the symmetry plane of the body (if it exists), perpendicular to the x-axis and the y-axis is perpendicular

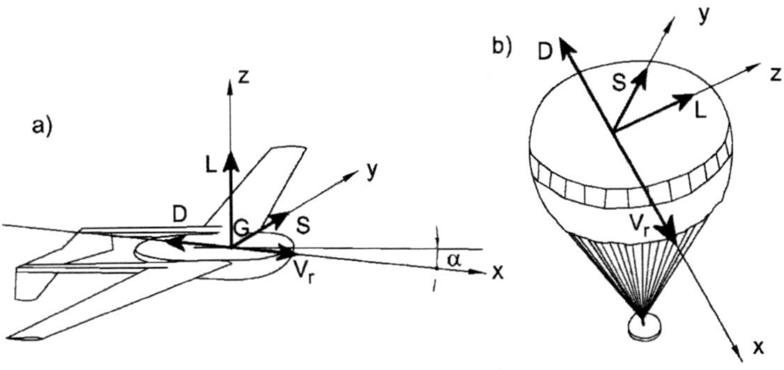

Fig. 4.37 Aerodynamic forces: reference frame and components. (**a**) Aircraft (the sketch is that of the UAV NASA planned to launch on Mars in 2003); (**b**) parachute

to the other two. Angle α, referred to as the *angle of attack*, is the angle between the x-axis and a reference direction in the symmetry plane, usually located so that if $\alpha = 0$ the lift vanishes. Another angle is often defined, the *sideslip angle* δ, between the x-axis and the symmetry plane; if there is no symmetry plane it is defined so that if $\delta = 0$ the side force vanishes.

If the aerodynamic force is decomposed along the axes of frame xyz, the components are referred to as *drag* D, *side force* S and *lift* L. In the figure the drag points backwards, as physically does; however, it would have been more consistent with sign conventions to plot it pointing forward and stating it is negative.

The expressions of the components of the aerodynamic force is the same equation (4.169), where instead of the generic force coefficient C_f, the drag coefficient C_D, the lift coefficient C_L and the side force coefficient C_S are introduced. As a first approximation, for small angles α and δ, the lift can be considered as proportional to α and the side force to δ.

Remark 4.37 The reference situation is often that occurring in the a wind tunnel, with the object stationary and the air rushing against it; the velocity of the air relative to the body is then usually displayed, instead of the velocity of the body.

The aerodynamic force is the resultant of the forces the fluid exerts in a direction perpendicular to the surface (pressure forces) and those exerted in a direction tangential to it. The latter are nil if there is no relative velocity between the body and the fluid.

If the fluid were inviscid, i.e. if its viscosity were nil, no tangential forces could act on the surface of the body and it can be demonstrated that no force could be exchanged between the body and the fluid, apart from aerostatic forces, at any relative speed, since also the resultant of the pressure distribution always vanishes. This

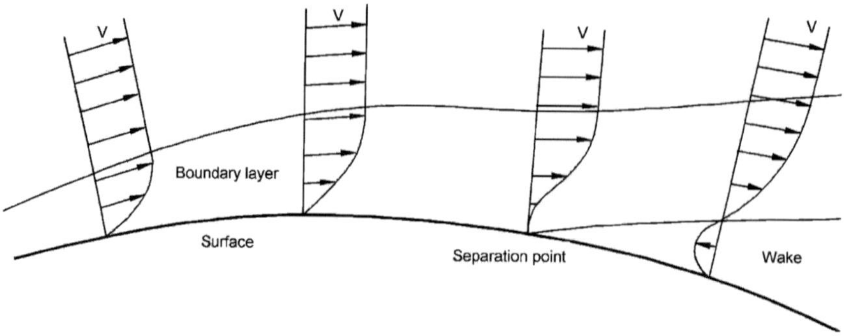

Fig. 4.38 Boundary layer: velocity distribution in direction perpendicular to the surface. The separation point is also represented (from G. Genta, L. Morello, *The Automotive Chassis*, Springer, New York, 2009)

result, due to D'Alembert, was formulated in 1744[22] and then again in 1768.[23] It is since known as the D'Alembert Paradox.

In the case of a fluid with no viscosity, the pressure p and the velocity V can be linked by the Bernoulli equation,

$$p + \frac{1}{2}\rho V^2 = \text{constant} = p_0 + \frac{1}{2}\rho V_0^2, \tag{4.170}$$

where p_0 and V_0 are the values of the ambient pressure and of the velocity far enough upstream from the body.

The Bernoulli equation, which holds along any streamline, has been written without the gravitational term, the one linked with aerostatic forces. It states simply that the total energy is conserved along any streamline.

No fluid has actually zero viscosity and the paradox is not applicable to any real fluid. Viscosity has a twofold effect: it causes the tangential forces giving way to the so-called *friction drag* and modifies the pressure distribution, whose resultant is no longer equal to zero. The latter effect, which for fluids with low viscosity is generally more important than the former, generates the lift, the side force and the *pressure drag*. The direct effects of viscosity (i.e. the tangential forces) can usually be neglected while its modifications on the aerodynamic field must be accounted for.

Owing to viscosity, the layer of fluid in immediate contact with the surface tends to adhere to it, i.e. its relative velocity vanishes, and the body is surrounded by a zone in which there are strong velocity gradients. This zone is usually referred to as the *boundary layer* (Fig. 4.38) and all viscous effects are concentrated in it. The viscosity of the fluid outside the boundary layer is usually neglected and Bernoulli equation can be used in this region.

[22] D'Alembert, *Traité de l'équilibre et du moment des fluides pour servir de suite un traité de dynamique*, 1774.

[23] D'Alembert, *Paradoxe proposé aux geometres sur la résistance des fluides*, 1768.

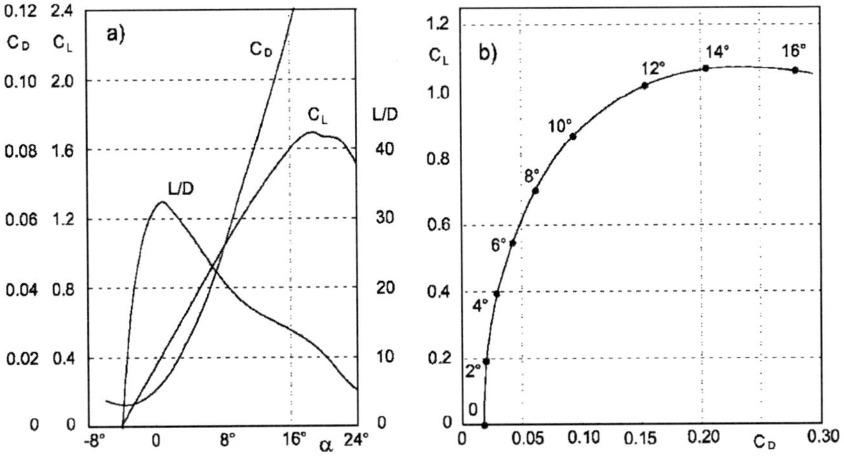

Fig. 4.39 (**a**) Lift and drag coefficients and efficiency of a wing as functions of the angle of attack. (**b**) Polar diagram of an airplane

The thickness of the boundary layer increases as the fluid in it loses energy owing to viscosity and slows down. If the fluid outside the boundary layer increases its velocity, a negative pressure gradient along the separation line between the external flow and the boundary layer is created, and this decrease of pressure in a way helps the flow within the boundary layer contrasting its slowing down. On the contrary, if the outer flow slows down, the pressure gradient is positive and the airflow in the boundary layer is hampered.

At a certain point, the flow in the boundary layer may stop causing a zone of stagnant air to form in the vicinity of the body: the flow separates from the surface possibly starting the formation of a wake. This is particularly important when takes place on the wing: at increasing angle of attack the lift increases initially in a linear way and this is accompanied by a moderate, although increasingly important, increase in drag. Then the lift increase becomes slower, and the drag grows more substantially. Finally, when a critical value of the angle of attack is reached, the flow detaches from the upper surface and the wing *stalls*. The lift abruptly decreases and an even more marked increase of drag occurs.

The lift and drag coefficients of a wing are plotted as functions of the angle of attack in Fig. 4.39a. On the same plot also the efficiency of the wing

$$E = \frac{L}{D} = \frac{C_L}{C_D} \tag{4.171}$$

is reported. The curves are for a given wing, but are typical.

Remark 4.38 The curves $C_L(\alpha)$ and $C_D(\alpha)$ are influenced by many characteristics of the wing, like the airfoil and the planform. In particular, the drop of the lift after the stall is reached can be more or less abrupt.

It is possible to increase the lift coefficient, although at the expense of an increase of the drag coefficient, by using suitable moving surfaces located at the trailing edge (flaps) or at the leading edge (slats), which change the airfoil characteristics. These high lift devices, used in all modern aircraft for take-off and landing, may be even more important in case of planets with low atmospheric density like Mars.

The dependence of the aerodynamic characteristics of a body on the angle of attack can be summarized in the polar diagram: a plot of the lift coefficient as a function of the drag coefficient. The polar diagram for an aircraft is reported in Fig. 4.39b.

There is some difference between aerodynamic vehicles operating in a fluid at a certain distance from the ground and vehicle operating just above the surface. Close to a surface the lift strongly increases and the overall aerodynamic performance changes (ground effect). Close to the ground it is also possible to produce a gas cushion, using suitable fans, which can be used for hovering vehicles (hovercraft).

The most common types of aircraft are fixed wing (aeroplanes) and rotary wing (autogyros and helicopters) craft. Both have been considered for planetary exploration robots and vehicles.

The speed at which a fixed wing aircraft must fly to sustain itself can be easily computed by equating the weight with the aerodynamic lift

$$mg = \frac{1}{2}\rho V^2 S C_L. \tag{4.172}$$

This yields the flying speed

$$V = \sqrt{\frac{2mg}{\rho S C_L}}. \tag{4.173}$$

The minimum take-off speed can thus be computed by introducing the maximum value of the lift coefficient into (4.173). At higher speeds the aircraft can fly with a lower lift coefficient.

The drag at the flight speed is

$$D = \frac{1}{2}\rho V^2 S C_D = mg\frac{C_D}{C_L} = \frac{mg}{E}, \tag{4.174}$$

i.e. is equal to weight divided by the aerodynamic efficiency.

Remark 4.39 The reciprocal of the efficiency is thus a sort of a friction coefficient, i.e. a number that multiplied by the weight gives the force that opposes to motion.

Remark 4.40 The attitude (i.e. the angle of attack) of the aircraft that minimizes the drag is that characterized by the maximum aerodynamic efficiency.

The power required for flight is the product of the drag by the speed

$$P = DV = mg\frac{C_D}{C_L}\sqrt{\frac{2mg}{\rho S C_L}} = \sqrt{\frac{2m^3g^3C_D^2}{\rho S C_L^3}}. \tag{4.175}$$

Remark 4.41 The attitude of the aircraft that minimizes the power required for motion is that at which the product

$$\sqrt{\frac{C_L^3}{C_D^2}} = \sqrt{C_L}E$$

is maximum. Such attitude is also that allowing the maximum flight duration for a given quantity of energy stored on board.

The time a glider (an unpowered aircraft) can fly losing an altitude Δz (Δz must be small enough to consider the density ρ as a constant) can be easily computed by equating the loss of potential energy with the energy required for flying for a time t

$$mg\Delta z = Pt, \tag{4.176}$$

i.e.

$$t = \frac{mg\Delta z}{P} = \Delta z\sqrt{\frac{\rho S C_L^3}{2mgC_D^2}}. \tag{4.177}$$

The attitude maximizing the flight time is the same as the one that minimizes the power required for flight. If a glider is released in the high atmosphere of a planet, the time it takes to reach the surface can be easily computed by integrating (4.177), taking into account that the density of the atmosphere changes with the altitude.

Other aerodynamic devices used in planetary atmospheres are parachutes. The drag coefficient of a domed parachute is about 1.5, but there are other types of parachute with different shape. Inflatable devices that combine the ways of working of parachutes and balloons have been suggested; they are usually referred to as ballutes.

Seldom rotary wing aircraft (helicopters or autogyros) are considered as planetary exploration rovers. This can be justified by the difficulty of flying such machines in the low density atmosphere of Mars, but the possibility of taking off and landing with a very short run (autogyros) or even vertically and flying without moving forward (helicopters) is an important advantage. They are, however, fully adequate to the thick atmosphere of Venus and low gravity and dense air of Titan. A configuration for unmanned rotorcraft that is now quite popular is the so-called quadrotor or quadrocopter, generally consisting of a cruciform structure with a rotor at the end of each one of the four arms. Quadcopter UAVs (Unmanned Aerial Vehicles) or drones have usually fixed pitch rotors, that in the smallest models usually reduce to four propellers rotating about vertical axes. Their control is much simpler than that of single or twin rotors helicopters, whose rotors have a variable pitch with both

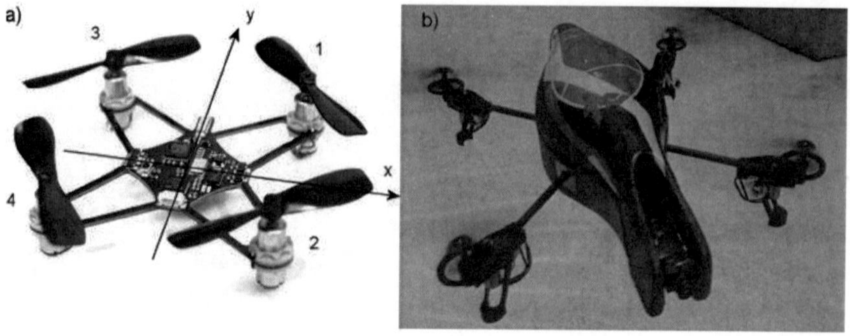

Fig. 4.40 (**a**) A miniature quadcopter UAV; (**b**) the Parrot AR.Drone, a commercial quadcopter UAV

collective and cyclic pitch control. A picture of a miniature quadcopter is shown in Fig. 4.40a (the size of the machine is just a few centimeters across), while a larger machine (the Parrot AR.Drone) is shown in Fig. 4.40b.

The control of a quadcopter is achieved by varying the relative speed, and thus the thrust, of each rotor, a thing which is easy if electric motors are used. With reference to Fig. 4.40a, rotors 1 and 4 rotate in one direction while rotors 2 and 3 rotate in the opposite direction, so that the reaction torques are balanced and no tail rotor is needed, as in single rotor helicopters. By

- reducing the speed of rotors 1 and 3 and increasing that of rotors 2 and 4, a roll rotation (rotation about x axis) to the left is obtained, while the torques are still balanced;
- reducing the speed of rotors 1 and 2 increasing that of rotors 3 and 4, a pitch rotation (rotation about y axis) to dive is obtained, while the torques are still balanced;
- reducing the speed of rotors 1 and 4 increasing that of rotors 2 and 3, a yaw rotation (rotation about z axis) is obtained. The direction of the yaw rotation depends on the direction of the rotation of the rotors.

By using a simple control electronics and sensors (generally rate gyros) a quadcopter can be easily controlled, achieving a good maneuverability, being able to fly in any direction and turn on the spot. They are scalable from miniature to large machines. Even in the thin air on Mars, it is conceivable to build a small quadcopter carried by a wheeled rover, which can take off vertically and remain a few meters over the machine on the ground, powered and controlled through an umbilical. The batteries and the controller need not to be carried on the flying machine, which can be light enough to fly. The quadcopter can lift a camera and an antenna, so that the rover can see much farther and remain in contact with a fixed antenna at a much larger distance.

Example 4.10 Consider a UAV (unmanned aerial vehicle) designed for flying at low altitude in the Mars atmosphere.

As in the previous example, assume a pressure of 600 Pa and a temperature of $-50°\text{C} = 223$ K.

The aircraft has a wingspan of 10 m, a mean wing chord of 1.5 m and a ready-to-fly mass of 150 kg. Assuming a lift coefficient at take-off (with extended flaps) $C_L = 1.4$ and a corresponding drag coefficient $C_D = 0.15$, compute the minimum take-off speed and the power needed to take off.

The atmospheric density, computed as in the previous example, is

$$\rho = \frac{p}{R^*T} = 0.0142 \text{ kg/m}^3.$$

The wing area is $S = 15$ m^2. The minimum take-off speed

$$V = \sqrt{\frac{2mg}{\rho S C_L}}$$

is thus $V_{\min} = 61.6$ m/s $= 221$ km/h.

The drag at this speed is

$$D = \frac{1}{2}\rho V^2 S C_D = 60.58 \text{ N}.$$

The power is thus

$$P = VD = 3.73 \text{ kW}.$$

This value does not include the power needed to accelerate and at any rate the power needed to fly at a safely higher speed is larger.

Example 4.11 Compute the minimum diameter of a parachute allowing to land a mass of 100 kg on Mars with a vertical speed of 5 m/s. Assume the same data for Mars atmosphere as before.

The system descends at constant speed when the drag is equal to the weight

$$mg = \frac{1}{2}\rho V^2 S C_D.$$

The surface of the parachute is thus

$$S = \frac{2mg}{\rho V^2 C_D} = 1{,}416 \text{ m}^2.$$

The diameter is 42.4 m. The speed so computed is the asymptotic speed reached in steady-state conditions; during the descent the speed is higher, since the atmospheric density decreases with increasing altitude.

4.8 Other Types of Support

Many other types of vehicles have been suggested for planetary exploration, even if they have seldom been seriously considered for actual projects. Many of them work on principles similar to those described above: for instance, jumping vehicles that use ground–vehicle contact forces are not dissimilar to walking machines, the difference being not in the type of contact forces, but in the mechanism that provides propulsion and controls the trajectory. Another case is that of snake robots: the contact area with the ground is similar to that of tracks, the difference being that in this case the motion is obtained by deforming the body and changing the pressure in selected points of the contact area.

Hydrostatic vehicles, ships and submarines, normally move under the action of propellers or hydrojets, but swimming vehicles, which are propelled by waving motions of the body like fish have been designed and tested.

It is, however, possible to conceive vehicles that work on principles that are altogether different.

Magnetic levitation for instance is usually considered suitable only for guided vehicles, and requires the construction of fixed infrastructures. It may well be the best way to support vehicles in low gravity: the strength of the magnetic fields is lower and above all it is possible to exert traction, braking and cornering forces much higher than those available using other supporting principles in low gravity. However, it is conceivable that the required infrastructures may be built only after the colonization of other celestial bodies is quite advanced.

Other devices use jets. Jet hopping vehicles can be used on the Moon or low gravity bodies, in spite of the need of using rockets: low gravity can make this way of moving convenient, and progress in magnetic materials may make possible to use also some sort of electromagnetic launchers to propel hopping vehicles.

Chapter 5
Wheeled Vehicles and Rovers

5.1 Introduction

At present there is a limited experience in operating rovers on Mars and an even smaller experience on robotic Moon rovers, while the experience regarding man-carrying vehicles is limited to the Moon and a single case (although with some differences between a mission and another): the LRV (Lunar Roving Vehicle) of the last *Apollo* Missions.

Remark 5.1 All the successful machines used up to now in planetary exploration are wheeled devices.

The rovers used in Mars exploration are three: the *Sojourner* rover (Microrover Flight Experiment, MFEX, Fig. 5.1a) and the two MERs (Mars Exploration Rovers) *Spirit* (Fig. 5.1b) and *Opportunity*. The latter two machines are still operating at the time of writing (2011; actually one of them is bogged down and unable to move), and their expected operating life has been exceeded by several times. Earlier devices, based on skis, carried by the Russian *Mars 2* and *3* probes in 1971 had a limited mobility (about 15 meters from the lander) and could not be tested in operation since *Mars 2* crash-landed on the planet and *Mars 3* ceased transmissions 20 seconds after landing.

These wheeled machines have a similar architecture: they are all based on rigid wheels and articulated non elastic suspensions, of the rocker bogie type (patented by NASA). Their size is, however, different, the Sojourner being much smaller than the MERs. Some of their main characteristics are summarized in Table 5.1.

The only robotic rover used on the Moon is the Russian *Lunokhod* (Fig. 5.2).

The only man-carrying vehicle ever used outside our planet is the *Lunar Rover* (Lunar Roving Vehicle, LRV) used by the astronauts of the last three Apollo missions to move on the lunar surface. It incorporated the top automotive technology of its times, blended with aerospace technology. It was an outstanding success and constitutes a reference for any kind of human rated vehicle that in the future will be used on celestial bodies, but it is now outdated in the light of the recent advances of current motor vehicles. It will be dealt with in detail in Sect. 5.7

G. Genta, *Introduction to the Mechanics of Space Robots*, Space Technology Library 26, DOI 10.1007/978-94-007-1796-1_5, © Springer Science+Business Media B.V. 2012

Fig. 5.1 Rovers for Mars: (**a**) *Sojourner*; (**b**) *Spirit* (NASA images)

Table 5.1 Main characteristics of the robotic rovers actually used in Mars and Moon exploration

	Mars		Moon
	Sojourner	Spirit & Opportunity	Lunokhod 1 & 2
Year	1997	2004	1970–1973
Mass (kg)	11.5	185	840
Size (L × W × H) (m)	0.68 × 0.48 × 0.28	1.6 × 2.3 × 1.5	1.7 × 1.6 × 1.35
Max. speed (km/h)	0.0036	0.018	2
Locomotion devices	Wheels (6)	Wheels (6)	Wheels (8)
Suspensions	Rocker bogie	Rocker bogie	Independent
Power source	GaAs/Ge solar cells	GaAs/Ge solar cells	Solar panels
Power (W on Mars)	16.5	140	–
Batteries	Lithium non-rechargeable	Lithium ion rechargeable	Rechargeable

While the examples of rovers actually used in space exploration are few, the number of designs is quite large. Many of them remained at the level of paper study, while a certain number were implemented in the form of demonstrators or engineering models. Here the variety of types is large and almost every kind of possible locomotion device has been used, particularly in the case of the designs that remained on paper. Demonstrators and engineering models are mostly based on wheels and legs.

Two basic functions can be identified in all types of vehicles: propulsion and trajectory control. In case of wheeled devices they are usually implemented through the wheel–ground contact, i.e. through the forces applied by the wheels on the ground, which, as seen in the previous chapter, are caused by the deformation of the wheel and of the ground. As a consequence, the wheels of a vehicle are always operating with some sideslip and longitudinal slip and are never in pure rolling.

Fig. 5.2 The Russian *Lunokhod* rover

5.2 Uncoupling of the Equations of Motion of Wheeled Vehicles

Consider the body of a vehicle[1] as a rigid body. Assume at first that also the wheels and the ground are rigid bodies and that the ground is flat. If the wheels are attached to the vehicle in a rigid way, the motion is planar and the vehicle can be considered as a system with three degrees of freedom.

Let Gxy be a reference frame fixed to the vehicle with x and y axes parallel to the ground and centered in its center of mass G. By using the inertial reference frame[2] XY shown in Fig. 5.3, it is possible to use the coordinates X and Y of the center of mass G of the vehicle and the yaw angle ψ between X and x axes as generalized coordinates.

The components of the velocity in the body-fixed frame u and v can be expressed as functions of the components of the absolute velocity as

$$\left\{ \begin{array}{c} u \\ v \end{array} \right\} = \begin{bmatrix} \cos(\psi) & -\sin(\psi) \\ \sin(\psi) & \cos(\psi) \end{bmatrix} \left\{ \begin{array}{c} \dot{X} \\ \dot{Y} \end{array} \right\}. \tag{5.1}$$

Instead of writing the equations of motion in the inertial reference frame, it is possible to write them in the body-fixed frame: velocities u and v are not the derivatives of actual coordinates, but the equations can be written using pseudo-coordinates without any problem (see Appendix A).

In most working conditions of vehicles, particularly in high speed motion, the sideslip angles of the wheels α and of the vehicle β are small, and it is possible to

[1] In the following the term vehicle is always used for any moving machine, be it a human-carrying vehicle, a rover, a moving robot, etc.

[2] Strictly speaking such reference frame is not inertial, as it is fixed to the ground and hence it follows the motion of the planet. It is, however, "enough" inertial for the problems here studied and this issue will not be dealt with any further.

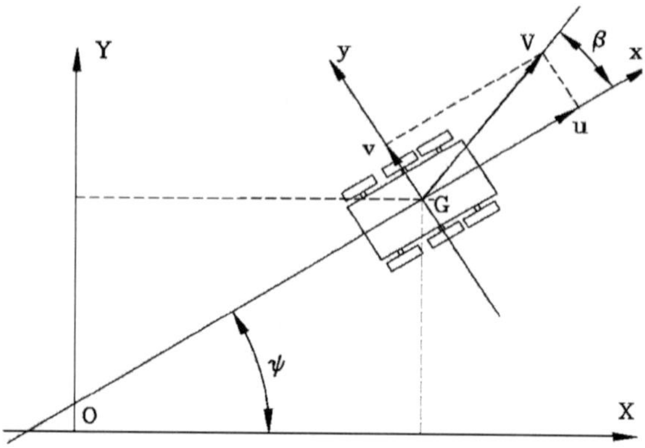

Fig. 5.3 Reference frame for the study of the motion of a rigid vehicle. The vehicle has three degrees of freedom and the coordinates X and Y of the center of mass G and the yaw angle ψ can be used as generalized coordinates

linearize their trigonometric functions, while clearly the yaw angle ψ can assume any value from 0 to 360°.

In these conditions it is well known[3] that the three equations of motion uncouple into two separate sets:

- A single equation in the forward velocity u describing the longitudinal behavior.
- A set of two equations in the lateral velocity v and the yaw angle ψ describing the lateral behavior or, as it is usually referred to, the handling.

A similar situation occurs if the presence of compliant suspensions and possibly the compliance of the wheels is accounted for. In this case the body of the vehicle is assumed to be a rigid body moving in three dimensions and its number of degrees of freedom is 6. Three of them can be considered as translational and the corresponding generalized coordinates can be the coordinates of its center of mass in any suitable inertial reference frame. For the three rotational degrees of freedom a set of three Tait–Brian angles can be chosen (see Sect. 3.6).

As seen in Sect. 3.6, the yaw angle ψ, the pitch angle θ and the roll angle ϕ can be defined.

Again, instead of using the coordinates X, Y and Z of the center of mass of the vehicle together with the three angles ψ, θ and ϕ, it is possible to write the equations of motion with reference to the non-inertial frame x^*y^*Z (Fig. 3.10b). The velocities u and v directed along axes x^* and y^* are the derivatives of pseudo-coordinates.

Often, not only the sideslip angles are small, but also the pitch and roll angles θ and ϕ can be considered as small angles. If this occurs, and if the vehicle is symmetrical with respect to xz plane, the six equations of motion can be subdivided into three uncoupled sets of equations:

- A single equation in the forward velocity u describing the longitudinal behavior.
- A set of three equations in the lateral velocity v, the yaw angle ψ and the roll angle ϕ describing the lateral behavior or, as it is usually referred to, the handling.

[3] G. Genta, *Motor Vehicle Dynamics, Modelling and Simulation*, World Scientific, Singapore, 2005; G. Genta, L. Morello, *The Automotive Chassis*, Vol. 2, Springer, New York, 2009.

- A set of two equations in the vertical displacement Z and the pitch angle θ describing the suspension motion of the vehicle, usually referred to as its ride (or, for human-carrying vehicles, comfort) behavior.

This uncoupling is strictly linked with a number of assumptions and, as a consequence, becomes inapplicable if one of them is dropped. As already stated, the first is the existence of a plane of symmetry, namely xz plane. Usually the lack of inertial symmetry of the structure and the differences between the characteristics of the individual components located at opposite sides of the vehicle are small enough to be neglected.

A second assumption is that of a perfect linearity of all compliant and damping components. The linearity of the elastic behavior of springs and wheels is an acceptable assumption in the motion about any equilibrium position, provided that its amplitude is small enough. On the contrary, the nonlinearity of the shock absorbers used in most vehicles can be a factor that cannot be neglected even in the motion in the small if their force–velocity characteristic is unsymmetrical, since in the jounce and rebound movements they act with different damping coefficients even if the amplitude of the motion tends to zero.

A third assumption regards all angles except the yaw angle ψ, which must be small enough to allow the linearization of their trigonometric functions. This assumption holds only for small displacements from the equilibrium position and depends also on the characteristics of the vehicle: the harder the suspensions, the more extended is the range in which the uncoupling assumption holds. However, in general the mentioned angles are small enough, except for vehicles with two wheels which can work with large roll angles.

On the contrary, the linearization of the wheel–ground contact is not strictly required for uncoupling: even if the forces exchanged by the wheels and the ground are not linear in the sideslip and inclination angles, the three sets of equations would remain uncoupled, although nonlinear. This last statement is important, since the linear model for the behavior of the tires holds only for values of angles α and γ that are far smaller than those for which the trigonometric functions can be linearized.

This uncoupling is, however, more general and can be extended to vehicles that cannot be considered as rigid bodies. If the vehicle has a symmetry plane, the vibration modes can be subdivided into symmetrical and skew symmetrical modes. The dynamics of a compliant vehicle can be expressed in terms of its vibration modes: the dynamics involving symmetrical modes couples with the ride behavior, while that involving skew symmetrical modes couples with lateral dynamics. In a similar way, the dynamics of the driveline, if it exists, and of the traction control couples with the longitudinal behavior.

But these considerations can be applied to vehicles of altogether different type: since no particular assumption on the nature of the forces supporting the vehicle has been done, the same uncoupling holds also for vehicles supported by hydrostatic, aerostatic or aerodynamic forces. Even the presence of aerodynamic forces due to the deformation of the structure does not change the overall picture, provided that they can be assumed to depend linearly on the modal coordinates.

Remark 5.2 The main cause of coupling between lateral and ride behavior is a large roll angle: on vehicles with two wheels, for instance, or on aircraft, the roll angle may be large enough to prevent from linearizing its trigonometric functions.

5.3 Longitudinal Behavior

5.3.1 Forces on the Ground

A rigid vehicle with more than three wheels is like a rigid body supported on a plane on more than three points and thus it is statically indeterminate unless the wheels are connected to the body through elastic suspensions or a suitable articulated device allowing to distribute the load on the ground is a predictable way. However, a four-wheeled vehicle with two axles that is symmetrical with respect to xz plane,[4] can be considered as a beam on two supports and the normal forces F_{z_1} and F_{z_2} on the axles can be easily determined.

Owing to symmetry, the loads on each wheel are respectively $F_{z_1}/2$ and $F_{z_2}/2$ for the front and the rear wheels. To simplify the equations, z-axis is assumed to be perpendicular to the ground (and x-axis is parallel to it).

With the vehicle at standstill on level road the normal forces are

$$\begin{cases} F_{z_1} = mg\epsilon_{0_1}, \\ F_{z_2} = mg\epsilon_{0_2}, \end{cases} \quad \text{where} \quad \begin{cases} \epsilon_{0_1} = \dfrac{b}{l}, \\ \epsilon_{0_2} = \dfrac{a}{l}. \end{cases} \tag{5.2}$$

Equation (5.2) can also be used to find the position of the center of mass of the vehicle by simply measuring the load on the ground on the two axles:

$$a = \frac{l F_{z_2}}{mg}.$$

The forces acting on a two-axle vehicle moving on flat ground with longitudinal grade angle α (positive when moving uphill) are sketched in Fig. 5.4.

Taking into account also the aerodynamic forces[5] and moments and the inertia force $-m\dot{V}$ acting in x direction on the center of mass, the dynamic equilibrium

[4]In the present section, where the longitudinal dynamics is studied, a complete symmetry with respect to xz plane is assumed.

[5]All resistances (aerodynamic drag and rolling resistance) are directed in the figure in forward direction (the direction of the positive x-axis) but their value is negative, so they are, as they must obviously be, directed backwards.

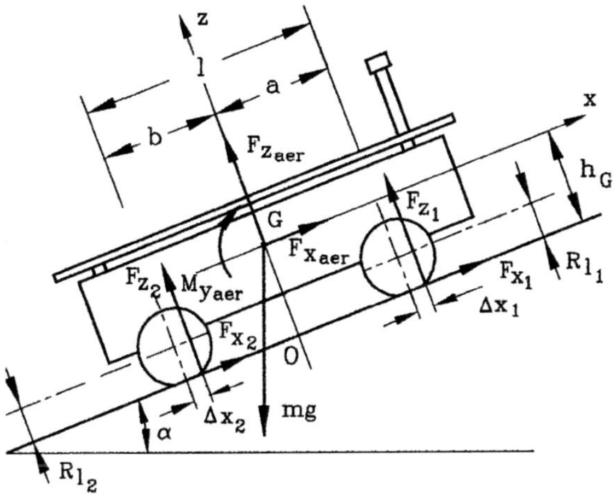

Fig. 5.4 Forces acting on a vehicle moving on inclined ground

equations for translations in x and z direction and rotations about point O are

$$\begin{cases} F_{x_1} + F_{x_2} + F_{x_{\text{aer}}} - mg\sin(\alpha) = m\dot{V}, \\ F_{z_1} + F_{z_2} + F_{z_{\text{aer}}} - mg\cos(\alpha) = 0, \\ F_{z_1}(a + \Delta x_1) - F_{z_2}(b - \Delta x_2) + mgh_G\sin(\alpha) - M_{\text{aer}} + |F_{x_{\text{aer}}}|h_G = -mh_G\dot{V}. \end{cases} \tag{5.3}$$

If the vehicle moves on rigid ground and all the rolling resistance is ascribed completely to the forward displacement of the resultant F_{z_i} of contact pressures σ_z, distances Δx_i can be easily computed by using (4.95) for the rolling coefficient

$$\Delta x_i = R_{l_i} f = R_{l_i}\left(f_0 + KV^2\right). \tag{5.4}$$

Except for the case of vehicles with different wheels on the various axles, the values of Δx_i are all equal.

In general forces F_{x_1} and F_{x_2} are the differences between the tractive and resistive forces acting on the wheels and so are the drawbar pulls of the various axles. Their sum is the drawbar pull of the whole vehicle.

The second and third equations (5.3) can be solved in the normal forces acting on the axles, yielding

$$\begin{cases} F_{z_1} = mg\dfrac{(b - \Delta x_2)\cos(\alpha) - h_G\sin(\alpha) - K_1V^2 - \dfrac{h_G}{g}\dot{V}}{l + \Delta x_1 - \Delta x_2}, \\ \\ F_{z_2} = mg\dfrac{(a + \Delta x_1)\cos(\alpha) + h_G\sin(\alpha) - K_2V^2 + \dfrac{h_G}{g}\dot{V}}{l + \Delta x_1 - \Delta x_2}, \end{cases} \tag{5.5}$$

where

$$\begin{cases} K_1 = \dfrac{\rho S}{2mg}\left[C_x h_G - l C_{M_y} + (b - \Delta x_2) C_z\right], \\ K_2 = \dfrac{\rho S}{2mg}\left[-C_x h_G + l C_{M_y} + (a + \Delta x_1) C_z\right]. \end{cases}$$

The values of Δx_i are usually quite small (in particular their difference is usually equal to zero) and can be neglected. If considered, they introduce a further weak dependence of the vertical loads on the square of the speed, owing to the term KV^2 in the rolling resistance.

If aerodynamic forces can be neglected, either because the speed is small or because there is no atmosphere, and Δx_i are negligible, the forces on the ground simplify as

$$\begin{cases} F_{z_1} = \dfrac{mg}{l}\left[b\cos(\alpha) - h_G \sin(\alpha) - \dfrac{h_G}{g}\dot{V}\right], \\ F_{z_2} = \dfrac{mg}{l}\left[a\cos(\alpha) + h_G \sin(\alpha) + \dfrac{h_G}{g}\dot{V}\right]. \end{cases} \tag{5.6}$$

If more than two axles are present, even in symmetrical conditions the system remains statically indeterminate and it is necessary to take into account the exact geometrical and elastic parameters of the wheel suspensions.

5.3.2 Resistance to Motion

Consider a vehicle moving at constant speed on a straight trajectory on level ground. The forces that must be overcome to maintain its speed constant are rolling resistance and, if there is an atmosphere, aerodynamic drag. The former is usually the most important form of drag at low speed, while the other one is important only at high speed, except if the fluid is very thick. In the solar system this can happen only on Venus (if wheeled vehicles will ever be used there) or on the bottom of some sea.

If the ground is not level, the component of weight acting in a direction parallel to the velocity V, i.e. the grade force, must be added to the resistance to motion: It may become far more important than all other forms of drag even for moderate values of the grade (Fig. 5.4).

The total resistance to motion, or road load, as it is commonly referred to, can be written in the form

$$R = F_{xr} + \frac{1}{2}\rho V^2 S C_x + mg\sin(\alpha), \tag{5.7}$$

where F_{xr} is the rolling resistance, which includes also compaction and bulldozing resistance. On compliant ground, however, this equation must be used with care because it is often impossible to separate the traction from rolling resistance and it is better to reason in terms of drawbar pull. The condition for constant speed motion is that the drawbar pull F_{DB} equals the other forms of resistance:

$$F_{DB} = \frac{1}{2}\rho V^2 S C_x + mg \sin(\alpha). \tag{5.8}$$

If the expression (4.95) for the rolling coefficient can be used, and taking into account also aerodynamic lift, the road load is

$$R = \left[mg\cos(\alpha) - \frac{1}{2}\rho V^2 S C_L\right]\left(f_0 + K V^2\right) + \frac{1}{2}\rho V^2 S C_x + mg\sin(\alpha), \tag{5.9}$$

i.e.

$$R = A + B V^2 + C V^4, \tag{5.10}$$

where

$$A = mg\left[f_0\cos(\alpha) + \sin(\alpha)\right],$$

$$B = mgK\cos(\alpha) + \frac{1}{2}\rho S[C_D - C_L f_0],$$

$$C = -\frac{1}{2}\rho S K C_L.$$

Note that the usual formulae seen in the section on aerodynamics have been used for aerodynamic drag and lift

$$F_{x_{aer}} = \frac{1}{2}\rho V^2 S C_D, \qquad F_{z_{aer}} = \frac{1}{2}\rho V^2 S C_L;$$

this is inconsistent with the usual praxis in road vehicle dynamics, where the components of the aerodynamic force along the body axes and not along the 'wind axes' (a frame where x-axis has the direction of the relative velocity) are used. Owing to the generally limited importance of the aerodynamic forces in this context, this topics will not be dealt with in detail.

If the grade angle of the ground is small, it is possible to assume that $\cos(\alpha) \approx 1$ and $\sin(\alpha) \approx \tan(\alpha) = i$, where i is the grade of the road. In this case coefficient B is independent of the grade of the road and

$$A \approx mg(f_0 + i)$$

depends linearly on it.

The power needed to move at constant speed V is simply obtained by multiplying the road load given by (5.10) by the value of the velocity,

$$P_r = VR = AV + BV^3 + CV^5. \tag{5.11}$$

Usually the last term is dropped, even if there is an atmosphere. For slow vehicles, like most rovers, only the first term is present, and the power needed for motion

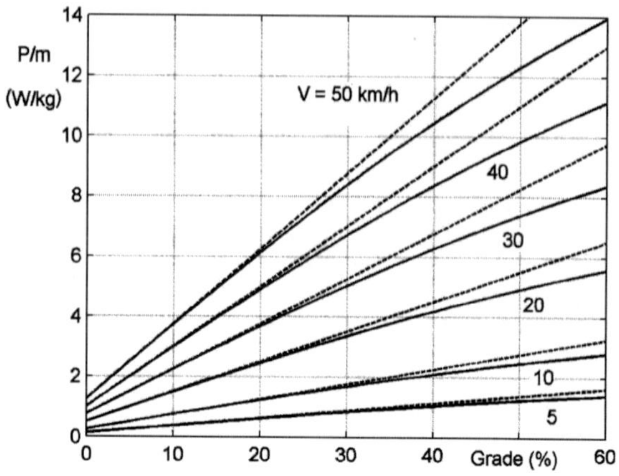

Fig. 5.5 Specific power needed to travel on a slope on the Moon at various speeds, with $f_0 = 0.05$ and $\eta_t = 0.9$ (*full lines*: non simplified expression; *dashed lines*: simplified expression)

reduces to

$$P_r = mgV\left[f_0\cos(\alpha) + \sin(\alpha)\right] \approx mgV(f_0 + i). \tag{5.12}$$

Motion at constant speed is possible only if the power available at the wheels at least equals the required power given by (5.11). This means that the engine must supply a sufficient power, taking into account also the losses in the transmission, and that the road–wheel contact must be able to transmit it.

In most cases the motor is connected to the wheels through a transmission system including a number of gear wheels and possibly joints and other mechanical or electromechanical devices. The efficiency η_t of the transmission must be accounted for and the power the motor must supply is

$$P_m = \frac{P_r}{\eta_t}. \tag{5.13}$$

Example 5.1 Compute the specific power needed for a wheeled vehicle to travel on a slope on the Moon, assuming that the wheels do not sink appreciably in the ground and the concept of rolling coefficient can be used. Assume also that $f_0 = 0.05$ and $\eta_t = 0.9$. Remembering that on the Moon $g = 1.62\ \text{m/s}^2$, the results obtained from (5.12) are plotted in Fig. 5.5.

The simplified expression can be used for slopes up to about 20%.

From the plot it is clear that, for instance, a power of just 2 kW allows a 1 ton vehicle to travel on level ground at more than 50 km/h and to overcome a 35% grade at 10 km/h.

5.3.3 *Model of the Driveline*

The longitudinal behavior of the vehicle is strictly linked with the behavior of the engine and the driveline.

The most common type of motors considered for planetary vehicles and rovers are electric motors. A very interesting layout is that or using a motor in each wheel, without the use of transmission shafts and differential gears, which are needed when a motor drives the two wheels of an axle or, even more, all the wheels of a vehicle. However, in most cases it is not possible to connect directly the wheel to the motor shaft, since the speed at which the wheels must turn is much lower than that at which the motor works at its best. A reduction gear is thus usually interposed between the motor and the wheels. If the speed range of the vehicle is large, it may be impossible to have the motor working in good conditions through the whole speed range and a variable ratio transmission, either continuously variable (CVT) or with a number of different ratios, is needed. Low speed, high torque, motors (the so-called torque motors) are a very interesting alternative to using reduction gears.

The simplest way of modeling the inertia of the motor(s) and of the vehicle is considering them as moments of inertia (flywheels), rotating at a given speed. If, for instance, the angular velocity Ω_m of the motor(s) is taken as a reference, the transmission ratio between the velocity of the wheels and of the motors is defined as

$$\tau = \frac{\Omega_w}{\Omega_m}$$

and no longitudinal slip is considered, the kinetic energy due to the vehicle mass can be expressed as

$$\mathcal{T}_v = \frac{1}{2} m V^2 = \frac{1}{2} m R_e^2 \tau^2 \Omega_m^2, \tag{5.14}$$

where R_e is the rolling radius of the wheels.

The mass of the vehicle can thus be modeled as an equivalent moment of inertia rotating at the motor speed

$$J_{\text{eq}} = m R_e^? \tau^?. \tag{5.15}$$

If the torsional compliance of the transmission is neglected, the total inertia of the n_m motors, the n_w wheels and the vehicle mass is

$$J_{\text{eq}_t} = m R_e^2 \tau^2 + n_w J_w \tau^2 + n_w J_m. \tag{5.16}$$

In the unlikely case the radii of the wheels, the transmissions ratios and the moments of inertia of the motors are different, the equivalent moment of inertia can be written as

$$J_{\text{eq}_t} = m R_e^2 \tau^2 + \sum_{j=1}^{n_w} J_{w_j} \left(\frac{R_e}{R_{e_j}} \right)^2 \tau^2 + \sum_{j=1}^{n_m} J_{m_j} \left(\frac{R_e \tau}{R_{e_j} \tau_j} \right)^2, \tag{5.17}$$

where R_e and τ are reference values, related to the wheel and engine whose speed is considered.

The total resistance to motion (road load) F_r can be modeled as a drag torque M_r applied to the moment of inertia simulating the vehicle

$$M_r = F_r R_e \tau. \tag{5.18}$$

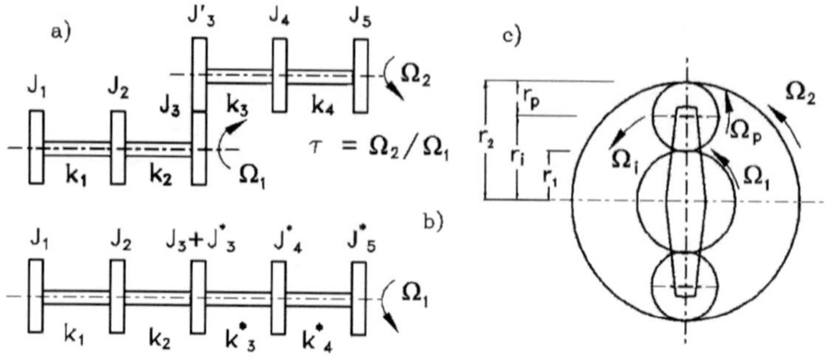

Fig. 5.6 Geared system: sketch of the (**a**) actual system and (**b**) equivalent system; (**c**) planetary gear train. Sketch of the system and notation

The dynamic model of the system is thus

$$\sum_{j=1}^{n_m} M_{m_j} - M_r = J_{\text{eq}_t} \ddot{\Omega}_m, \tag{5.19}$$

where the motor torques M_{m_j} can be computed from the characteristics of the motor, their rotational speed and above all the control parameters through a suitable model of the motor (see Sect. 7.3.1).

This model can be made more realistic by adding the torsional compliance of all the element that are present in the system, like the motor shaft, the joints and possibly the gear wheels, as well as the rotational inertia of the various elements of the driveline. A model of the whole driveline is thus obtained, with the motors and vehicle modeled as flywheels located at its ends. Usually, it is a lumped parameters system made by massless shafts where the elastic properties of the system are concentrated, with lumped moments of inertia modeling its inertial properties.

The damping of the system may be neglected altogether, or modeled by introducing suitable viscous dampers in parallel to the springs modeling the various parts of the shaft and joints.

As said above, the fact that the various elements of the driveline rotate at different speeds must be accounted for. Consider the system sketched in Fig. 5.6a, in which the two shafts are linked by a pair of gear wheels, with transmission ratio τ. For the study of the torsional vibrations of the system, it is possible to replace the actual system with a suitable equivalent system, in which one of the two shafts is replaced by an expansion of the other (Fig. 5.6b).

Assuming as well that the deformation of gear wheels is negligible, the equivalent rotations ϕ_i^* may be obtained from the actual rotations ϕ_i simply by dividing the latter by the transmission ratio $\tau = \Omega_2/\Omega_1$,

$$\phi_i^* = \frac{\phi_i}{\tau}. \tag{5.20}$$

The kinetic energy of the ith flywheel, whose moment of inertia is J_i, and the elastic potential energy of the ith span of the shaft are, respectively,

$$\mathcal{T} = \frac{1}{2} J_i \dot{\phi}_i^2 = \frac{1}{2} J_i^* \dot{\phi}_i^{*2},$$
$$\mathcal{U} = \frac{1}{2} k_i \left(\phi_{i+1}^2 - \phi_i^2\right) = \frac{1}{2} k_i^* \left(\phi_{i+1}^{*2} - \phi_i^{*2}\right), \tag{5.21}$$

where the equivalent moment of inertia and stiffness are, respectively,

$$J_i^* = \tau^2 J_i, \qquad k_i^* = \tau^2 k_i. \tag{5.22}$$

The moments of inertia and the torsional stiffness of the various elements of the geared system can thus be reduced to the main system simply by multiplying them by the square of the gear ratio.

In the same way, if damping of the shafts is accounted for by introducing dampers in parallel to the springs, the damping coefficient must be multiplied by the square of the gear ratio.

$$c_i^* = \tau^2 c_i. \tag{5.23}$$

If the system includes a planetary gear train, the computation can be performed without difficulties: the equivalent stiffness can be computed simply from the overall transmission ratio and the total kinetic energy of the rotating parts must be taken into account. The angular velocities of the sun gear Ω_1, of the ring gear Ω_2, of the revolving carrier Ω_i, and of the intermediate pinions Ω_p of the planetary gear shown in Fig. 5.6c are linked by the equation

$$\frac{\Omega_1 - \Omega_i}{\Omega_2 - \Omega_i} = -\frac{r_2}{r_1}, \qquad \Omega_p = (\Omega_1 - \Omega_i)\frac{r_1}{r_p} - \Omega_i. \tag{5.24}$$

The equivalent moment of inertia of the system made of the internal gear, with moment of inertia J_1, the ring gear, with moment of inertia J_2, the revolving carrier, with moment of inertia J_i, and n intermediate pinions, each with mass m_p and moment of inertia J_p, referred to the shaft of the internal gear is

$$J_{\text{eq}} = J_1 + J_2\left(\frac{\Omega_2}{\Omega_1}\right)^2 + \left(J_i + n m_p r_i^2\right)\left(\frac{\Omega_i}{\Omega_1}\right)^2 + n J_p\left(\frac{\Omega_p}{\Omega_1}\right)^2. \tag{5.25}$$

If the deformation of the meshing teeth must be accounted for, it is possible to introduce two separate degrees of freedom for the two meshing gear wheels into the model, modeled as two different inertias, and to introduce a shaft between them whose compliance simulates the compliance of the transmission. This is particularly important when a belt or flexible transmission of some kind is used instead of the stiffer gear wheels. In a driveline there may be several shafts connected to each other, in series or in parallel, by gear wheels with different transmission ratios.

The equivalent system is referred to one of the shafts and the equivalent inertias and stiffness of the elements of the others are all computed using the ratios between

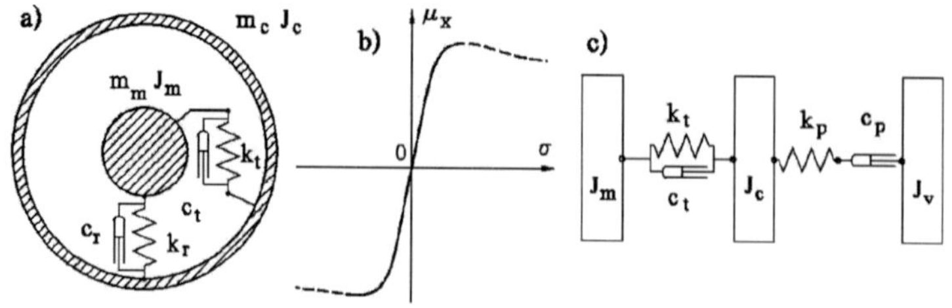

Fig. 5.7 Model of the wheel and the wheel–ground contact. (**a**) Dynamic model of the wheel; (**b**) force-longitudinal slip characteristic for the wheel; (**c**) dynamic model of the wheel–ground contact (the moments of inertia and the torsional characteristics are drawn as masses and translational characteristics) (from G. Genta, L. Morello, *The Automotive Chassis*, Springer, New York, 2009)

the speeds of the relevant element and the reference shaft. The equivalent system will then be made of a set of elements, in series or in parallel, following the scheme of the actual system, but with rotations that are all consistent.

Remark 5.3 If the compliance of the gears is to be accounted for in detail, the nonlinearities due to the contacts between the meshing teeth and backlash must be considered.

5.3.4 Model Including the Longitudinal Slip

In the previous section the longitudinal slip of the wheels was neglected, as is usually done when performing a first approximation study. For a more detailed analysis, both the compliance of the wheels and their longitudinal slip at the wheel–ground contact must be accounted for. The simplest way to model the former is by simulating the tire as a rigid ring, with mass m_c and moment of inertia J_c. This is connected to the wheel hub, whose mass and moment of inertia are m_m and J_m through an elastic system having a radial and torsional stiffness equal to k_r and k_t respectively.

The rim and hub are also assumed to be rigid bodies. Viscous dampers with coefficients c_r and c_t (Fig. 5.7a) may be added in parallel to the springs. The masses and the radial stiffness and damping are included in the ride comfort models (see following sections), while the moments of inertia, the torsional stiffness and the torsional damping are included in the driveline and longitudinal models.

The wheel–ground contact may be characterized by the plot of the longitudinal force coefficient versus the longitudinal slip $\mu_x(\sigma)$ (Fig. 5.7b). Usually only the first part of the curve, approximated as a straight line, is used in the study of the driveline dynamics. The slope C_σ of the line may be easily obtained from the models seen in the previous chapter or, in case of pneumatic tires, from the coefficients of the magic formula and is given by product BCD.

The longitudinal slip σ as defined by S.A.E (4.112)

$$\sigma = \frac{R\Omega - V}{V},$$

can be written in terms of the speed Ω_v of the moment of inertia simulating the vehicle. By identifying the radius R_e with R, it follows that

$$\sigma = \frac{\Omega_c - \Omega_v}{\Omega_v} = \frac{\Omega_c}{\Omega_v} - 1. \tag{5.26}$$

The longitudinal force the wheel exerts is thus

$$F_x = F_z \mu_x = C_\sigma F_z \sigma = C_\sigma F_z \left(\frac{\Omega_c - \Omega_v}{\Omega_v} \right). \tag{5.27}$$

The moment exerted on the wheel due to the longitudinal slip is

$$M = R_e F_x = C_\sigma F_z R_e \frac{(\Omega_c - \Omega_v)}{\Omega_v}. \tag{5.28}$$

The wheel–ground contact may thus be modeled as a torsional viscous damper with damping coefficient

$$c_p = \frac{C_\sigma F_z R_e}{\Omega_v} = \frac{C_\sigma F_z R_e^2}{V}. \tag{5.29}$$

The coefficient c_p depends first upon the vehicle speed and then upon the variable of motion Ω_v: the equation of motion is thus nonlinear. For small velocity variations it is, however, possible to linearize the equations by using an average value of the speed in the expression of c_p. When the speed tends to zero, the damping coefficient tends to infinity: this linearized model cannot be used in the first instants of a take-off manoeuvre, when the vehicle is still stationary, because in these conditions the longitudinal slip is high, or better tends to infinity.

The tire model is usually complemented by adding a spring in series with the damper. Its stiffness is

$$k_p = \frac{2C_\sigma F_z R_e^2}{a}, \tag{5.30}$$

where a is the length of the contact zone.

Stiffness k_p can be suitably modified to take into account the longitudinal compliance of the suspension of the driving wheels.

The wheel is thus modeled with two moments of inertia connected to each other with a spring and a damper in parallel, and connected to the flywheel simulating the vehicle with a spring and a damper in series. The model is sketched in Fig. 5.7c, where the torsional springs and the moments of inertia are drawn as springs and masses.

Example 5.2 Build a simple linearized torsional model (in the configuration and the state space) of the driveline of the Apollo LRV. The LRV had four elastic wheels, each one with an electric motor in the hub connected to the wheel through a harmonic drive.

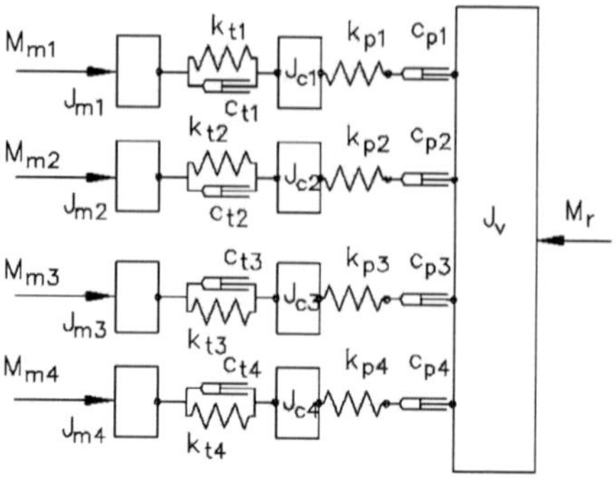

Fig. 5.8 Linearized model of the driveline of the *Apollo* LRV

If the compliance of the transmission is negligible, each wheel can be modeled as shown in Fig. 5.7c, where J_c is the moment of inertia of the rim of the wheel, reduced to motor shaft and J_m is the moment of inertia of the rotor of the motor and of the gearing, plus the inner part of the wheel. A single moment of inertia is used because the transmission is assumed to be torsionally stiff. Note that the transmission ratio is a very small number (about $1/80$) and as a consequence the inertia of the rotor of the motor may be much larger than the inertia of the wheels and even of the equivalent inertia of the vehicle.

The four driving torques M_{mi} act on the rotors of the electric motors and the drag torque M_v acts on the flywheel simulating the vehicle.

The model is shown in Fig. 5.8. It has a has 13 degrees of freedom. Because four of them have a vanishing associated mass, it has only 22 state variables. The generalized coordinates are the rotations of the various moments of inertia. It is possible to order them by separating the nodes where there is a mass from those that are massless. Thus

$$\mathbf{x} = \begin{bmatrix} \mathbf{x}_1^T & \mathbf{x}_2^T \end{bmatrix}^T,$$

where

$$\mathbf{x}_1 = \begin{bmatrix} \theta_{m1} & \theta_{c1} & \theta_{m2} & \theta_{c2} & \theta_{m3} & \theta_{c3} & \theta_{m4} & \theta_{c4} & \theta_v \end{bmatrix}^T,$$
$$\mathbf{x}_2 = \begin{bmatrix} \theta_{p1} & \theta_{p2} & \theta_{p3} & \theta_{p4} \end{bmatrix}^T,$$

where subscripts m, v, c and p designate respectively the moments of inertia simulating the motors and the wheel hubs, the moments of inertia simulating the vehicle, the wheel rims and the massless points located between the dampers and the springs simulating the wheel–ground contact.

The mass matrix of the system may be partitioned in four parts as

$$\mathbf{M} = \begin{bmatrix} \mathbf{M}_{11} & \mathbf{M}_{12} \\ \mathbf{M}_{21} & \mathbf{M}_{22} \end{bmatrix},$$

where

$$\mathbf{M}_{11} = \mathrm{diag}\begin{bmatrix} J_{m1} & J_{c1} & J_{m2} & J_{c2} & J_{m3} & J_{c3} & J_{m4} & J_{c4} & J_v \end{bmatrix}$$

and all other sub-matrices are null.

The stiffness matrix may be partitioned in the same way and

$$\mathbf{K}_{11} = \begin{bmatrix} k_{t1} & -k_{t1} & 0 & 0 & 0 & 0 & 0 & 0 & 0 \\ & k_{t1}+k_{p1} & 0 & 0 & 0 & 0 & 0 & 0 & 0 \\ & & k_{t2} & -k_{t2} & 0 & 0 & 0 & 0 & 0 \\ & & & k_{t2}+k_{p2} & 0 & 0 & 0 & 0 & 0 \\ & & & & k_{t3} & -k_{t3} & 0 & 0 & 0 \\ & & & & & k_{t3}+k_{p3} & 0 & 0 & 0 \\ & & & & & & k_{t4} & -k_{t4} & 0 \\ & & & & & & & k_{t4}+k_{p4} & 0 \\ & \text{symm} & & & & & & & 0 \end{bmatrix},$$

$$\mathbf{K}_{21} = \begin{bmatrix} 0 & -k_{p1} & 0 & 0 & 0 & 0 & 0 & 0 & 0 \\ 0 & 0 & 0 & -k_{p2} & 0 & 0 & 0 & 0 & 0 \\ 0 & 0 & 0 & 0 & 0 & -k_{p3} & 0 & 0 & 0 \\ 0 & 0 & 0 & 0 & 0 & 0 & 0 & -k_{p4} & 0 \end{bmatrix},$$

$$\mathbf{K}_{22} = \text{diag}\begin{bmatrix} k_{p1} & k_{p2} & k_{p3} & k_{p4} \end{bmatrix}, \qquad \mathbf{K}_{12} = \mathbf{K}_{21}^T.$$

In a similar way the submatrices of the damping matrix are

$$\mathbf{C}_{11} = \begin{bmatrix} c_{t1} & -c_{t1} & 0 & 0 & 0 & 0 & 0 & 0 & 0 \\ & c_{t1} & 0 & 0 & 0 & 0 & 0 & 0 & 0 \\ & & c_{t2} & -c_{t2} & 0 & 0 & 0 & 0 & 0 \\ & & & c_{t2} & 0 & 0 & 0 & 0 & 0 \\ & & & & c_{t3} & -c_{t3} & 0 & 0 & 0 \\ & & & & & c_{t3} & 0 & 0 & 0 \\ & & & & & & c_{t4} & -c_{t4} & 0 \\ & & & & & & & c_{t4} & 0 \\ & \text{symm} & & & & & & & c_{vt} \end{bmatrix},$$

where

$$c_{vt} = c_{p1} + c_{p2} + c_{p3} + c_{p4},$$

$$\mathbf{C}_{21} = \begin{bmatrix} \mathbf{0}_{4\times 9} & \begin{bmatrix} -c_{p1} \\ -c_{p2} \\ -c_{p3} \\ -c_{p4} \end{bmatrix} \end{bmatrix},$$

$$\mathbf{C}_{22} = \text{diag}\begin{bmatrix} c_{p1} & c_{p2} & c_{p3} & c_{p4} \end{bmatrix}, \qquad \mathbf{C}_{12} = \mathbf{C}_{21}^T.$$

The force vector can be partitioned as well:

$$\mathbf{F}_1 = \begin{bmatrix} M_{m1} & 0 & M_{m2} & 0 & M_{m3} & 0 & M_{m4} & 0 & M_v \end{bmatrix}^T$$

and $\mathbf{F}_2 = 0$.

The state vector can be written in the form

$$\mathbf{z} = \begin{bmatrix} \mathbf{v}_1^T & \mathbf{x}_1^T & \mathbf{x}_2^T \end{bmatrix}^T, \tag{5.31}$$

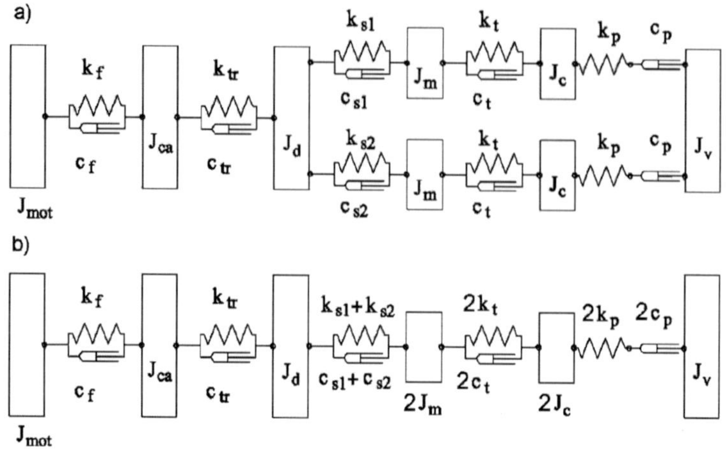

Fig. 5.9 Model of a driveline of a vehicle with a single motor and two driving wheels, like those common in automobiles, for the study of low frequency dynamics (**a**), and model in which the presence of two separate wheel shafts is neglected (**b**) (from G. Genta, L. Morello, *The Automotive Chassis*, Springer, New York, 2009)

where $\mathbf{v}_1$ contains the derivatives of coordinates $\mathbf{x}_1$.

The state equation is thus

$$\begin{bmatrix} \mathbf{M}_{11} & \mathbf{0} & \mathbf{C}_{12} \\ \mathbf{0} & \mathbf{0} & \mathbf{C}_{22} \\ \mathbf{0} & \mathbf{I} & \mathbf{0} \end{bmatrix} \dot{\mathbf{z}} = - \begin{bmatrix} \mathbf{C}_{11} & \mathbf{K}_{11} & \mathbf{K}_{12} \\ \mathbf{C}_{21} & \mathbf{K}_{21} & \mathbf{K}_{22} \\ -\mathbf{I} & \mathbf{0} & \mathbf{0} \end{bmatrix} \mathbf{z} + \begin{Bmatrix} \mathbf{F}_1 \\ \mathbf{0} \\ \mathbf{0} \end{Bmatrix}. \tag{5.32}$$

The dynamic matrix of the system is thus

$$\mathbf{A} = - \begin{bmatrix} \mathbf{M}_{11} & \mathbf{0} & \mathbf{C}_{12} \\ \mathbf{0} & \mathbf{0} & \mathbf{C}_{22} \\ \mathbf{0} & \mathbf{I} & \mathbf{0} \end{bmatrix}^{-1} \begin{bmatrix} \mathbf{C}_{11} & \mathbf{K}_{11} & \mathbf{K}_{12} \\ \mathbf{C}_{21} & \mathbf{K}_{21} & \mathbf{K}_{22} \\ -\mathbf{I} & \mathbf{0} & \mathbf{0} \end{bmatrix}. \tag{5.33}$$

A complete driveline model can thus be assembled using the partial models seen above. The model shown in the example is typical for a four-wheeled rover with motors in the wheel hubs and torsionally rigid transmission. A simple model for a typical automotive manual transmission, with a friction clutch, a gearbox and a differential gear driving two of the vehicle's wheels is shown in Fig. 5.9a. If low frequency oscillations are to be studied, the engine can be modeled as a single moment of inertia. The two wheel shafts are modeled separately, but the two branches of the driveline can be joined if only a first approximation study of the low frequency dynamics is required (Fig. 5.9b). This can be done by introducing inertias and stuffiness equal to the sum of those of the single branches.

The model shown in Fig. 5.9a has 10 degrees of freedom. Because two of them have a vanishing associated mass, it has only 18 state variables.

The assumption that the damping of the various components of the driveline can be modeled as viscous is only approximate, but it cannot anyway be modeled as hysteretic damping (which is not much better for elements like the clutch damper springs) because that would not allow numerical time-domain simulations. However, if the phenomena under study occur at a well determined frequency, it is possible to approximate hysteretic damping with an equivalent viscous damping

$$c_{eq} = \frac{\eta k}{\omega}, \tag{5.34}$$

where η and k are the loss factor and the stiffness of the relevant elements and ω is the frequency of the oscillations of the driveline. It is possible to perform a first computation with no damping (except that used to simulate tire slip) to compute a value for the frequency of the free oscillations, and then to proceed with calculations that include an equivalent damping.

5.3.5 Maximum Torque that Can Be Transferred to the Ground

The driving torque needed to overcome the resistance to motion must be transferred through the ground–wheel contact. As the road load increases with increasing speed and grade of the road, there is a limit to the maximum speed that can be reached and the maximum grade that can be managed due to the maximum driving force the vehicle can exert, even if no limit would exist to the power supplied by the motor.

Remark 5.4 Such a speed can even be zero if the resistance to motion at a vanishingly small speed is larger than the maximum available traction. This means that the vehicle cannot move.

The difference between the traction and the resistance to motion is the drawbar pull F_{DB}. Motion is thus possible if

$$F_{DB} > 0. \tag{5.35}$$

If this condition is realized, the drawbar pull can be used for accelerating the vehicle or for climbing. If the ground has a longitudinal slope α (i.e. a longitudinal grade $i = \tan(\alpha)$) the maximum grade that can be overcome is that for which the grade force equals the drawbar pull

$$F_{DB} = mg \sin(\alpha_{max}), \tag{5.36}$$

i.e.

$$\alpha_{max} = \text{arsin}\left(\frac{F_{DB}}{mg}\right). \tag{5.37}$$

Assume that all wheels are driving and that the rolling resistance is due only to the forward displacement of the normal force F_z at the ground–wheel contact (negligible compaction and bulldozing resistance) and hence is overcome directly by the driving torque exerted by the engine, the only road load that must be overcome at the wheel–road contact at low speed is the grade force. The maximum grade that can be overcome is thus computed by equating the maximum traction the wheels can exert to the grade force:

$$\sum_{\forall i} F_{z_i} \mu_{i_p} = mg \sin(\alpha_{\max}). \tag{5.38}$$

Assuming that the maximum value of the traction coefficient of all wheels is the same, it follows that

$$\mu_{x_p} \sum_{\forall i} F_{z_i} = \mu_{x_p} mg \cos(\alpha_{\max}) = mg \sin(\alpha_{\max}), \tag{5.39}$$

i.e.

$$\tan(\alpha_{\max}) = i_{\max} = \mu_{x_p}. \tag{5.40}$$

Remark 5.5 The gravitational acceleration is not present in the expression of the maximum slope: at low gravity, both the available traction force and the resistance to motion are smaller, and these two effects compensate each other. Low gravity thus does not affect, at least as a first approximation, mobility on slopes. On the contrary, on soft ground low gravity may be an advantage, since it reduces sinking.

If the ground can support well the vehicle allowing to reach high speeds, it is possible to state what is the maximum speed that can be reached for traction reasons. The maximum power that can be transferred by the vehicle–ground interaction is

$$P_{\max} = V \sum_{\forall i} F_{z_i} \mu_{i_p}, \tag{5.41}$$

where the sum is extended to all the driving wheels.

If the maximum longitudinal force coefficient μ_{i_p} and the load acting on the driving wheels were independent of the speed, the maximum power would increase linearly with V. The optimum motor characteristic $P_m(\Omega_m)$ for a vehicle with a transmission with fixed ratio would be a linear characteristic. This is, however, not the case as the situation is far more complicated.

Consider first the case of a vehicle with all wheels driving and assume that all wheels work with the same longitudinal slip, i.e. that the values of μ_i are all equal. This situation will be referred here to as "ideal driving force".

Taking into account also aerodynamic lift, the maximum power that can be transferred to the road is

$$P_{\max} = V \mu_p \left[mg \cos(\alpha) - \frac{1}{2} \rho V^2 S C_L \right]. \tag{5.42}$$

The last term, due to aerodynamic lift, may usually be neglected. To model in a simple way the decrease of driving force occurring with increasing speed, it is possible to use a linear law,

$$\mu_{i_p} = c_1 - c_2 V. \tag{5.43}$$

The maximum speed can be obtained by equating the power required for motion at constant speed (except that due to rolling resistance) to (5.42) and using (5.43) for expressing the decrease of the available driving force with the speed. This results in the cubic equation

$$C_L c_2 V^3 + (C_L c_1 + C_D) V^2 - \frac{2mg}{\rho S}\big[(c_1 - c_2 V)\cos(\alpha) - \sin(\alpha)\big] = 0. \tag{5.44}$$

Neglecting aerodynamic effects, it follows that

$$V_{\max} = \frac{c_1 - \tan(\alpha)}{c_2}. \tag{5.45}$$

Remark 5.6 The maximum speed is again independent from the gravitational acceleration, and is dominated by the decrease of the available traction with the speed. Since expression (5.43) is approximated, it can be expected that the value of the maximum speed so obtained is just a rough approximation.

The values of the slope and speed so obtained are only ideal reference values, since they were obtained assuming that the longitudinal slip of all wheels were equal. However, if each wheel has its own motor, it is possible to control the motor torques in such a way that the traction is distributed on the various wheels so that conditions close to the optimal ones are obtained.

5.3.6 Maximum Performances Allowed by the Motors

The maximum speed that can be reached on level ground with a given installed power can be found by equating the expressions for the available power at the wheels to the required power to travel at constant speed

$$AV + BV^3 + CV^5 = P_{m_{\max}} \eta_t, \tag{5.46}$$

whose solution yields directly the maximum value of the speed.

If aerodynamic lift is neglected (actually it is sufficient to neglect the contribution to rolling resistance proportional to the square of the speed due to lift), the equation is cubic and its solution can be obtained in closed form:

$$V_{\max} = A^*\big(\sqrt[3]{B^* + 1} - \sqrt[3]{B^* - 1}\big), \tag{5.47}$$

where

$$A^* = \sqrt[3]{\frac{P_{e_{\max}} \eta_t}{2mgK + \rho S C_X}},$$

$$B^* = \sqrt{1 + \frac{8m^3 g^3 f_0^3}{27 P_{e_{\max}}^2 \eta_t^2 (2mgK + \rho S C_X)}}.$$

Remark 5.7 The speed so obtained is, however, just a potential maximum speed, since to actually obtain it the motor (or motors) must be allowed to develop its maximum power at the maximum speed of the vehicle.

This implies a suitable gear ratio of the transmission

$$\tau = \frac{V_{\max}}{R_e (\Omega_m)_{P_{\max}}}, \tag{5.48}$$

where $(\Omega_m)_{P_{\max}}$ is the speed of the motor at which the peak power is obtained.

This value of the maximum speed can, however, be reached only in one given condition, since the load, but also the rolling resistance coefficient and even the air density, affect the resistance to motion.

The maximum slope that can be managed with a given gear ratio can be obtained by plotting the curves of the required power at various values of the slope and looking for the curve that is tangent to the curve of the available power. The slope so obtained is, however, only a theoretical result, since it can be managed only at a single value of the speed. This condition may be unstable depending of the shape of the power curve of the motor.

To be able to manage with safety a given slope, the curve of the available power must be above that of the required power in a whole range of speeds, starting from a value low enough to assure that also starting on that slope is possible. Assuming that the motor torque M_m at start (or at the lowest speed, if the motor cannot start under load and some sort of clutch must be used to start the vehicle) is known and remains constant for a small speed range, it is possible to equate the power of the engine at low speed,

$$P_m = M_m \Omega_m,$$

with the required power, which at low speed can be computed by neglecting the terms in B and C of the resistance to motion:

$$P_r = \frac{mgV}{\eta_t} \big[f_0 \cos(\alpha) + \sin(\alpha) \big].$$

Remembering that the speed of the motor is linked to the speed of the vehicle by the relationship

$$\Omega_m = \frac{V}{R_e \tau}, \tag{5.49}$$

it follows that

$$M_m = \frac{mgR_e\tau}{\eta_t}\big[f_0\cos(\alpha) + \sin(\alpha)\big]. \tag{5.50}$$

This equation can be used to compute the maximum slope that can be managed with a given motor and gear ratio at very low speed, or the gear ratio needed to start on a given slope.

In general, at least two different gear ratios are needed to optimize the performance using a given motor: one to obtain the maximum speed and another one to manage the required slope. To simplify the layout of the drive system, in exploration rovers speed performance is usually sacrificed and a fixed ratio, allowing to obtain the required *gradeability* (performance on a sloping ground), is used

$$\tau = \frac{M_m\eta_t}{mgR_e[f_0\cos(\alpha) + \sin(\alpha)]}. \tag{5.51}$$

5.3.7 Energy Consumption at Constant Speed

The energy needed to travel at constant speed for a time t can be immediately computed by multiplying the power required for constant speed driving by the time

$$E = P_r t = \frac{P_r d}{V}, \tag{5.52}$$

where d is the distance traveled. Note that (5.52) gives the energy required at the wheels: to obtain the energy actually required for motion, this expression must be divided by the various efficiencies (transmission, motor, etc.):

$$E = d\frac{A + BV^2 + CV^4}{\eta_t\eta_m}. \tag{5.53}$$

If the aerodynamic lift is neglected and the efficiency of the motor can be considered as a constant, the energy consumption obtained from (5.53) would be a quadratic function of the speed. Actually the efficiency of all kinds of motors depend on the speed and can be quite low at low speed. This may cause the energy required for motion to be high at low speed, to decrease, to reach a minimum at a given speed to increase again.

Remark 5.8 When traveling at a very low speed, the efficiency or the motor can reduce to almost zero, with a resulting increase of energy consumption. This compels to use a low transmission ratio, so that the motors can rotate at a speed at which their efficiency is not too low even when the speed of the vehicle is low.

5.3.8 Acceleration

If the required power is, at a certain speed, lower than the power available at the wheels, the difference between the two is the power which is available to accelerate the vehicle.

During an acceleration a number of rotating elements (wheels, transmission, the motor itself) must increase their angular velocity. It is expedient to write an equation linking the power supplied by the motor with the kinetic energy $\mathcal{T}$ of the vehicle

$$\eta_t P_m - P_r = \frac{d\mathcal{T}}{dt}. \tag{5.54}$$

The power P_m should be that provided in non steady-state running; but owing to the different time scale between the phenomena generating the mechanical power in the motor and the acceleration of the vehicle, the error introduced by using the values obtained from the steady-state motor characteristics is negligible. Moreover, the efficiency of the transmission should not be considered when dealing with the power needed to accelerate the inertia of the motor, which is accelerated directly by the engine torque. However, the error introduced in this way is usually negligible.

If the transmission has a fixed ratio, the speed of the motor and that of the other rotating parts of the vehicle is proportional to the speed of the vehicle through the gear ratio and the rolling radius. Once that the transmission ratio has been chosen, (5.49) gives the relationship between the speed of the vehicle and the rotational speed of the motor. Similar relationships can be used for the other rotating elements which may be present and must be accelerated when the vehicle speeds up.

The kinetic energy of the vehicle can then be expressed as

$$\mathcal{T} = \frac{1}{2} m V^2 + \frac{1}{2} \sum_{\forall i} J_i \Omega_i^2 = \frac{1}{2} m_e V^2, \tag{5.55}$$

where the sum extends to all rotating elements which must be accelerated when the vehicle speeds up. The term m_e is the equivalent or apparent mass of the vehicle, i.e. the mass of an object that, when moving at the same speed of the vehicle, has the same total kinetic energy. Considering only the rotational inertia of the wheels and the motor, it can be written in the form

$$m_e = m + n_w \frac{J_w}{R_e^2} + n_m \frac{J_m}{R_e^2 \tau^2}, \tag{5.56}$$

where J_w is the moment of inertia of each wheel, which are assumed to have the same radius and hence to rotate at the same speed, and of all elements rotating at its speed and J_m is the moment of inertia of each motor and all the elements rotating together with it. To account for the fact that the motor is accelerated directly, at least in an approximated way, the last term is sometimes multiplied by η_t. The modifications to (5.56) to take into account the presence of different wheels, different motors and transmission ratios are obvious.

Of the two last terms, the first is usually small, while second can be important, if the reduction gear has a low transmission ratio.

As the equivalent mass is a constant, once that the gear ratio has been chosen, (5.54) yields

$$\eta_t P_e - P_r = m_e \frac{dV}{dt}. \tag{5.57}$$

Equation (5.57) holds only in the case of constant equivalent mass. If a continuously variable transmission (CVT) is used, the overall transmission ratio, and hence the equivalent mass, changes in time and the equation should be modified as

$$\eta_t P_e - P_r = m_e V \frac{dV}{dt} + \frac{1}{2} V^2 \frac{dm_e}{dt}. \tag{5.58}$$

The correction present in (5.58) is, however, usually very small, since the equivalent mass does not change quickly.

From (5.57) the maximum acceleration of the vehicle is immediately obtained as a function of the speed

$$\left(\frac{dV}{dt}\right)_{\max} = \frac{\eta_t P_m - P_r}{m_e V}, \tag{5.59}$$

where the power P_m is the maximum power the motor can deliver at the speed Ω_m, corresponding to speed V.

The minimum time needed to accelerate from speed V_1 to speed V_2 can be computed by separating the variables in (5.59) and integrating

$$t_{V_1 \to V_2} = \int_{V_1}^{V_2} \frac{m_e}{\eta_t P_e - P_r} V \, dV. \tag{5.60}$$

If there are different transmission ratios, the integral must be performed separately for each velocity range in which the equivalent mass is constant, i.e. a given transmission ratio is used.

By further integration it is possible to obtain the distance needed to accelerate to any value of the speed

$$s_{V_1 \to V_2} = \int_{t_1}^{t_2} V \, dt. \tag{5.61}$$

Sometimes instead of modeling the vehicle as an equivalent mass which is accelerated along the road, it is modeled as an equivalent moment of inertia attached to the motor

$$J_e = m R_e^2 \tau^2 + J_w \tau^2 + J_m. \tag{5.62}$$

In low gravity conditions the worst limitation to the available acceleration does not came from the available power of the motor but from the available traction at the wheel–ground contact.

If all wheels are accelerated by their own motor, and neglecting compaction and bulldozing resistance, only the force needed to accelerate the mass of the vehicle is transferred to the ground by the wheel–ground contact. Using (5.42) to express the maximum power that can be transferred and neglecting aerodynamic forces, the maximum value of the acceleration that can be obtained is

$$a_{\max} = \mu_{x_p} g. \tag{5.63}$$

The motion of an accelerating vehicle can also be studied using a model of the driveline, in which the vehicle is modeled as a moment of inertia connected to the motors by the driveline (see Sect. 5.3.3 and following). The response to a given time history of the engine torque is computed and the time history of the vehicle speed is obtained.

5.3.9 Braking

Braking is an important function in all vehicles, but its importance increases with increasing vehicle speed. Very slow rovers may have little need of braking and it is possible to avoid altogether installing a braking system by using a non-reversible transmission. Brakes are not only used to slow down the vehicle, but also to keep the vehicle stationary on sloping ground and most vehicles have what is usually referred to as station brakes.

The braking functions can be performed by different kind of devices:

- non-reversible transmission,
- regenerative braking systems,
- dissipative braking systems.

As already stated, in case of very slow vehicles it is possible to avoid using brakes by resorting to a non reversible transmission. This approach has, however, some drawbacks. Firstly the efficiency of the transmission is low: a transmission is non reversible when its efficiency is approximately lower than 50%. Since the efficiency depends on many parameters, to ensure that the transmission is non reversible, a lower efficiency must be achieved. Low efficiency may cause also cooling problems, particularly in cases when a cooling fluid is unavailable. Another disadvantage is linked to the tendency of the transmission to lock if the motor exerts no torque. For this reason a non reversible transmission can be used only on slow vehicles where the inertia forces are low and the locking of the wheels does not cause dangers.

A typical example of nonreversible gearing is the worm drive: high reduction ratios are obtained and the device cannot be backdriven if the worm has a single thread.

If the vehicle is driven by electric motors and the transmission is reversible, it is possible to brake by recovering the kinetic energy of the vehicle by operating the motors in the generator mode. Some sort of energy storage system must be present to accept this energy and, if the vehicle is fast and a strong deceleration is needed, the energy storage system must be able to accept large electric power. The advantages of this approach are self-evident: the same driving system acts also as a braking device (although the size of the drive motors can be not large enough to act as brakes) and some of the energy is recovered—how much depends on the efficiency of the system. Regenerative braking is best applied if all wheels have drive motors, since braking is best performed on all wheels. A disadvantages is that a separate station brake is needed and possibly also an emergency braking system if the vehicle is fast.

Dissipative brakes convert the kinetic energy of the vehicle into heat, usually through friction. All the kinetic energy of the system is lost, but the system is simple and works also at zero speed.

When there is no cooling fluid, like on an airless world, there may be problems in dissipating the heat produced.

All the mentioned braking devices act on the wheels and so rely on the wheel–ground contact to exert the forces needed to slow down the vehicle. In low gravity the deceleration that can be produced in this way is limited.

The total braking force F_x is

$$F_x = \sum_{\forall i} \mu_{x_i} F_{z_i}, \tag{5.64}$$

where the sum is extended to all the wheels. The longitudinal equation of motion of the vehicle is

$$\frac{dV}{dt} = \frac{\sum_{\forall i} \mu_{x_i} F_{z_i} - \frac{1}{2}\rho V^2 S C_D - f \sum_{\forall i} F_{z_i} - mg\sin(\alpha)}{m}, \tag{5.65}$$

where m is the actual mass of the vehicle and not the equivalent mass, and α is positive for uphill grades. The rotating parts of the vehicle are directly slowed down by the brakes and hence do not enter the evaluation of the forces exchanged between vehicle and road. They must be accounted for when assessing the required braking power of the brakes and the energy that must be dissipated.

In a simplified study of braking, aerodynamic drag and rolling resistance can be neglected, since they are usually far smaller than braking forces. Also, in case of limited sinking of the wheels, rolling resistance can be considered as causing a braking moment on the wheel more than a braking force directly on the ground.

Ideal braking can be defined as the condition in which all wheels brake with the same longitudinal force coefficient μ_x. As in ideal braking all force coefficients μ_{x_i} are assumed to be equal, and the acceleration is

$$\frac{dV}{dt} = \mu_x \left[g\cos(\alpha) - \frac{1}{2m}\rho V^2 S C_Z \right] - g\sin(\alpha). \tag{5.66}$$

In case of level road, for a vehicle with no aerodynamic lift, (5.66) reduces to

$$\frac{dV}{dt} = \mu_x g. \tag{5.67}$$

The maximum deceleration in ideal conditions can be obtained by introducing the maximum negative value of μ_x into (5.66) or (5.67).

The assumption of ideal braking implies that the braking torques applied to the various wheels are proportional to the forces F_z, if the radii of the wheels are all equal. In case of regenerative braking operated using the electric motors, it is possible to do so through the motor controllers.

To compute the forces F_x the wheels must exert to perform an ideal braking manoeuvre, forces F_z on the wheels must be computed first. This can be done using the formulae seen in Sect. 5.3.1. In case of a two-axle vehicle, neglecting aerodynamic forces, the equations reduce to

$$F_{z_1} = \frac{m}{l}\left[gb\cos(\alpha) - gh_G\sin(\alpha) - h_G\frac{dV}{dt}\right], \tag{5.68}$$

$$F_{z_2} = \frac{m}{l}\left[ga\cos(\alpha) + gh_G\sin(\alpha) + h_G\frac{dV}{dt}\right]. \tag{5.69}$$

From (5.65)

$$\frac{dV}{dt} = \frac{\mu_{x_1}F_{z_1} + \mu_{x_2}F_{z_2}}{m} - g\sin(\alpha), \tag{5.70}$$

since the values of μ_x are all equal in ideal braking, the values of longitudinal forces F_x are

$$F_{x_1} = \mu_x F_{z_1} = \mu_x\frac{mg}{l}\left[b\cos(\alpha) - h_G\mu_x\right], \tag{5.71}$$

$$F_{x_2} = \mu_x F_{z_2} = \mu_x\frac{mg}{l}\left[a\cos(\alpha) + h_G\mu_x\right]. \tag{5.72}$$

By eliminating μ_x using (5.71) and (5.72), the following relationship between F_{x_1} and F_{x_2} is readily obtained:

$$(F_{x_1} + F_{x_2})^2 + mg\cos^2(\alpha)\left(F_{x_1}\frac{a}{h_G} - F_{x_2}\frac{b}{h_G}\right) = 0. \tag{5.73}$$

The plot of (5.73) in F_{x_1},F_{x_2} plane is a parabola whose axis is parallel to the bisector of the second and fourth quadrants if $a = b$ (Fig. 5.10). The parabola is thus the locus of all pairs of values of F_{x_1} and F_{x_2} leading to ideal braking.

Actually, only a part of the plot is of interest: that with negative values of the forces (braking in forward motion) and with braking forces actually achievable, i.e. with reasonable values of μ_x (Fig. 5.11).

On the same plot it is possible to draw the lines with constant μ_{x_1}, μ_{x_2} and acceleration. On level ground, the first two are straight lines passing respectively through points B and A, while the lines with constant acceleration are straight lines parallel to the bisector of the second quadrant.

Remark 5.9 The forces so obtained are related to each axle and not to each wheel: in the case of axles with two wheels their values are thus twice the values referred to the wheel.

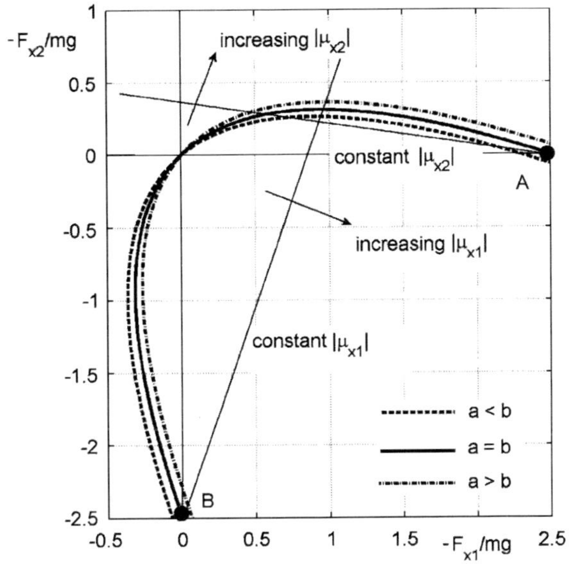

Fig. 5.10 Braking in ideal conditions. Nondimensional relationship between F_{x_1} and F_{x_2} for vehicles with the center of mass at mid-wheelbase, forward and backward of that point. Plots obtained with $l/h_G = 5$, level road

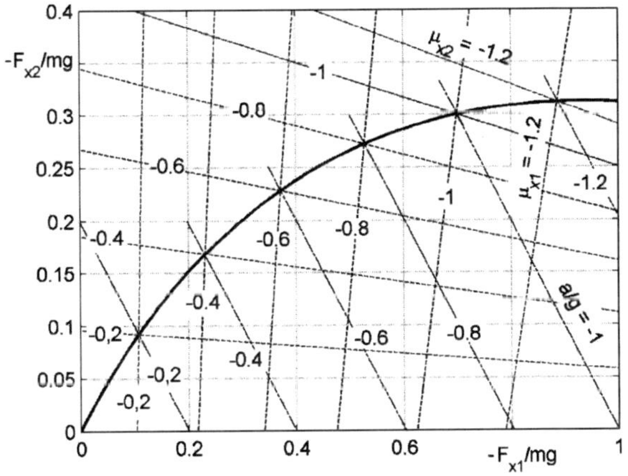

Fig. 5.11 Enlargement of the useful zone of the plot of Fig. 5.10. Also the lines with constant μ_{x_1}, μ_{x_2} and acceleration are reported (from G. Genta, L. Morello, *The Automotive Chassis*, Springer, New York, 2009)

The moment to be applied to each wheel is approximately equal to the braking force multiplied by the loaded radius of the wheel: if the wheels have equal radii, the same plot holds also for the braking torques. If this condition does not apply the scales are just multiplied by two different factors and the plot is distorted, but remains essentially unchanged.

To perform a more precise computation, the rolling resistance should be accounted for, which is a small correction, and the torque needed for decelerating the rotating inertias should be added. This correction may be important in the case of high transmission ratios and motors with large rotor inertia.

If the relationship between the braking moments at the rear and front wheels is different from that stated to comply with the conditions to obtain ideal braking, the

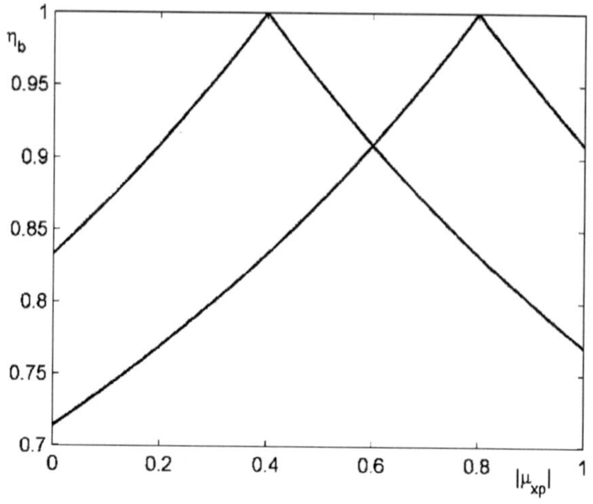

Fig. 5.12 Braking efficiency η_b as a function of the limit value of μ_x for two different values of constant K_B. Plot obtained with $a = b$ and $a/h_G = 2$ (after G. Genta, L. Morello, *The Automotive Chassis*, Springer, New York, 2009)

deceleration made possible by a given traction coefficient decreases. This is the case when friction brakes are controlled by a simple hydraulic, pneumatic or mechanical system—like in the case of the *Apollo* LRV—and is imposed by the parameters of the actual braking system of the vehicle. It is thus possible to define an efficiency of braking as the ratio between the (negative) acceleration obtained in actual conditions and that occurring in ideal conditions, obviously at equal value of the coefficient μ_x of the wheels whose longitudinal force coefficient is higher

$$\eta_b = \frac{(dV/dt)_{\text{actual}}}{(dV/dt)_{\text{ideal}}} = \frac{(dV/dt)_{\text{actual}}}{\mu_x g}, \tag{5.74}$$

where the last expression holds only on level road for a vehicle with negligible aerodynamic loading. With simple computations it is possible to demonstrate that in the latter case the braking efficiency is

$$\eta_b = \min\left\{\frac{a(K_B+1)}{l - \mu_p h_G(K_B+1)}, \frac{b(K_B+1)}{lK_B + \mu_p h_G(K_B+1)}\right\}. \tag{5.75}$$

where K_B is the ratio between the braking force at the front axle and that at the rear axle. The first value holds when the rear wheels lock first, the second one when the limit conditions are reached at the front wheels first.

A typical plot of the braking efficiency versus the peak braking force coefficient is plotted in Fig. 5.12.

The value of the maximum longitudinal force coefficient μ_p at which the condition $\eta_b = 1$ holds can be stated and the value of ratio K_B can be easily computed. For values of $|\mu_p|$ lower than the chosen one, the rear wheels lock first while for higher values locking occur at the front wheels. Once K_B is known, the braking system can easily be designed.

The curve $\eta_b(\mu_x)$ can be plotted by stating increasing values of the braking forces and then computing the values of μ_x and η_b referred to the front and rear wheels. The result is of the type shown in Fig. 5.12.

Operating in this way the rear wheels lock when the road is in good conditions. To postpone the locking of the rear wheels a value of K_B causing the condition $\eta_b = 1$ to occur for higher values of the traction coefficient can be used, but this reduces the efficiency when the road is in poor conditions.

To avoid locking of the rear wheels without lowering the efficiency at low values of μ_x, devices which reduce the braking force of the rear wheels when the deceleration increases above a certain value can be used.

This type of control of braking can be considered a feedforward control: the braking force is controlled through a predetermined law of variation of parameter K_B to achieve a required value of the braking efficiency. Antilock systems (ABS) are on the contrary based on a feedback approach and act directly to reduce the braking force when a wheel approaches slipping conditions, i.e. when the longitudinal slip σ approaches the value characterized by the maximum value of $|\mu_x|$. They are normally based on wheel speed sensors which allow to compare the instantaneous speed of the wheels and the speed corresponding to the velocity of the vehicle. If a slip that exceeds the allowable limits is detected, the device acts to reduce the braking torque, restoring appropriate working conditions. However, once the wheel has resumed low slip conditions, the device allows the braking torque to increase again to the previous value and incipient locking can occur.

The brakes may thus operate in a cyclic way, with subsequent interventions of the ABS system, thus maintaining the longitudinal force coefficient near its maximum value (Fig. 4.19). Different devices, however, can operate in different ways, both for what the hardware characteristics and the control algorithms are concerned.[6]

The instantaneous power the brakes must dissipate is

$$|P| = |F_x|V = V\left|\frac{dV}{dt}m_e + mg\sin(\alpha)\right|, \qquad (5.76)$$

where all forms of drag have been neglected.

The brakes cannot dissipate it directly; they usually work as a heat sink, storing some of the energy in the form of thermal energy and dissipating it in due time. Obviously care must be exerted to design the brakes in such a way that they can store the required energy without reaching too high temperatures and to ensure adequate cooling. The average value of the braking power must at any rate be lower than the thermal power the brakes can dissipate.

Remark 5.10 When a vehicle operates on a planet without atmosphere, brake cooling may become an important issue. Perhaps it could be better to speak of brake thermal control instead of cooling, since there are cases in which the brakes must be prevented from cooling too much.

[6]See, for instance, G. Genta, L. Morello, *The Automotive Chassis*, Springer, New York, 2009.

In this case operating in low gravity may help, but only in an indirect way. The energy involved in slowing down a vehicle from a given speed is the same, independent from the gravitational acceleration, but low gravity usually compels to limit the maximum speed of the vehicle and compels to brake more gradually, since the maximum deceleration is lower, reducing the braking power and making it easier to recover it, if regenerative braking is planned, or to dissipate it if no energy is recovered.

5.4 Lateral Behavior

5.4.1 Trajectory Control

The second main function encountered in all vehicles is trajectory control. From this viewpoint, all vehicles can be divided into two categories:

(a) Guided vehicles, or better, kinematically guided vehicles, whose trajectory is fixed by a set of kinematic constraints;
(b) Piloted vehicles, in which the trajectory, a tridimensional or a planar curve, is determined by a guidance system, controlled by a human pilot or by a device, usually electromechanical. The guidance system acts by exerting forces on the vehicle that are able to change its trajectory.

In the first case the kinematic constraint exerts all forces needed to modify the trajectory without any deformation, i.e. is assumed to be infinitely stiff and infinitely strong. A perfect kinematic guidance is therefore an abstraction, although it is well approximated in many actual cases.

In the second case the forces are due to the changes of the attitude of the vehicle which in turn are caused by forces and moments due to the guidance devices. These vehicles can be said to be dynamically guided.

Apart from the cases where the forces needed to change the trajectory are directly exerted by thrusters (usually rockets), dynamic guidance can be due to attitude changes large enough to be directly felt by the pilot or driver, or small enough to go unnoticed. The first case is that of aerodynamically or hydrodynamically controlled vehicles, in which the pilot acts on a control surface, causing the changes of attitude needed to generate the forces that modify the trajectory. Usually there is also a certain delay between the changes of attitude and the actual generation of forces and consequently the drivers feels clearly that a dynamic control, i.e. a control through the application of forces, takes place.

In the case of wheeled vehicles the situation is similar but the driver has a completely different impression: The driver, or the automatic trajectory control system, operates the steering causing some wheels to work with a sideslip and to generate lateral forces. These forces cause a change of attitude of the vehicle (the attitude or sideslip angle β is defined as the angle between the longitudinal direction of the vehicle and the velocity vector) and then a sideslip of all wheels: the resulting forces bend the trajectory. However, the linearity of the behavior of the tires, at least for small sideslip angles, the high value of the cornering stiffness and the short time delay with which the wheels respond to changes in the sideslip and camber angles, give the driver the impression of a kinematic, not dynamic, driving. The wheels seem to be in pure rolling and the trajectory seems to be determined by the directions of the midplanes of the wheels.

This impression has influenced the study of the handling of wheeled vehicles and above all of wheeled robots for a long time, originating the very concept of kinematic steering and in a sense hiding the true meaning of the phenomena. Actually, the concept of kinematic steering was applied first to horse drawn wagons: Ackermann patented in 1818 the idea that the lines perpendicular to the midplanes of the wheels passing through the contact points on the ground should converge in the instant center of rotation of the vehicle.[7] This concept worked well in connection with carriage wheels provided with steel tires and later was applied to motor vehicles. Only in the 1930s some experiments did show the importance of the sideslip angle to generate lateral forces and this concept was finally formalized by Olley in 1937.

The impressions received by the driver are in good accordance with the kinematic approach, at least for all the linear part of the behavior of the tire. When high values of the sideslip angles are reached, the average driver has the impression of losing control of the vehicle, much more so if this occurs abruptly. This impression is confirmed by the fact that in normal road conditions, particularly if radial tires are used, the sideslip angles become large only when approaching the limit lateral forces.

This way of controlling a wheeled vehicle is typical of motor vehicles used on hard surfaced roads, but there is another possibility. The torque about z-axis causing the vehicle to rotate and to assume the required attitude can be produced by differential traction on the wheels of the same axle instead of steering some of the wheels. In this case, usually referred to as *slip steering*, no steering wheel in the classical sense may be present.

This way of controlling the trajectory may be implemented together with the more usual way, as in the case of many VDC (Vehicle Dynamics Control) systems where the driver chooses the trajectory by steering the front wheels while the device maintains the vehicle in the required trajectory by differentially braking the left and right wheels with suitable control algorithms. In other cases this may be the main way of controlling the trajectory like in some wheeled earth moving machines and above all in all tracked vehicles. Many three-wheeled small robots work in this way too, but the third wheel is either an omnidirectional wheel or a swiveling wheel.

Remark 5.11 If there is an omnidirectional wheel the vehicle is not able to withstand lateral forces and, when rolling on an inclined surface, it will tend to go downhill, if the omnidirectional wheel is at the front axle.

[7] Erasmus Darwin had actually this idea in 1758 but the one who applied it first was a German carriage builder, who had his agent in England, Rudolph Ackermann, to patent in 1818 what is now called the Ackerman steering geometry.

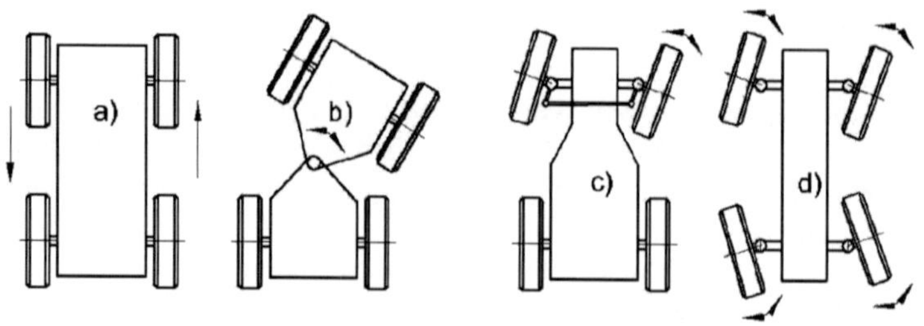

Fig. 5.13 (**a**) Slip steering; (**b**) articulated steering; (**c**) coordinated steering; (**d**) independent steering

The distinction between kinematically and dynamically guided vehicles is only a rough distinction, since there are cases in between like kinematic guidance with deformable constraints or magnetic levitation vehicles. The difference turns out to be more quantitative than qualitative, and depends mostly on the greater or smaller "stiffness" with which the vehicle responds to the variations of attitude due to the guidance devices.

Propulsion of wheeled vehicles and rovers is usually implemented through the wheel–ground contact. Both propulsion and trajectory control functions are thus implemented through the forces applied by the wheels on the ground, which are caused by the deformation of the wheels and of the ground. As a consequence, the wheels of a vehicle are always operating with some sideslip and longitudinal slip and are never in pure rolling.

The steering function can be implemented in the following ways:

- Slip steering (Fig. 5.13a).[8] There is no particular steering mechanism, but the driving wheels can produce differential traction on the two sides of the vehicle. This can be achieved by using independent motors in the wheels, two motors actuating the wheels of the two sides independently, or by using a differential gear. The brakes can also be used in differential mode to achieve some slip steering. Very small trajectory curvature radius can be achieved by rotating backwards the wheels on one side, and turning on the spot is possible by using equal and opposite rotation speed. No kinematic steering (see below) is possible.
- Articulated steering (Fig. 5.13b). One or more whole axles, carrying each the wheels of both sides, are articulated and can steer under the control of one or more actuators. The body can be made of two or more subunits, each carrying one or more axles, or the wheels can be carried by bogies, pivoted under the body. With proper control kinematic steering (see below) is possible. As an alternative, instead of using actuators to rotate one of the parts of the vehicle with respect to the other, it is possible to resort to differential longitudinal forces, like in slip steering. In this case, however, lower sideslip angles of the wheels usually occur.

[8]Sometimes the term *skid steering* is used instead of slip steering. It should be avoided, since this way of steering is due to longitudinal and lateral slip of the wheels, but only in extreme cases it produces actual skidding (i.e. global slipping) of the wheels on the ground.

- Coordinated steering (Fig. 5.13c). The two wheels of the same axle steer about two different steering axes (kingpin axes), but their steering angles are coordinated by a mechanical linkage. If the linkage satisfies the Ackermann condition (see below), an Ackermann steering, which makes kinematic steering possible, is realized. No practical steering linkage, like the much used Jeantaud linkage, realizes an exact Ackermann steering. If more than one axle is steering, the steering angles of the various axles can be independent or (seldom) coordinated by another mechanical linkage. Most automotive vehicles have a steering of this kind.
- Independent steering (Fig. 5.13d). Each wheel of one axle or of more axles can steer independently about a steering (kingpin) axis, under the action of an actuator. The possibility of realizing the Ackermann conditions (and thus making kinematic steering possible) depends on the control systems of the actuators. If large steering angles are possible, and the control system is flexible enough, any kind of trajectory is possible, including turning on the spot, moving sideways, etc. A steering of this kind was used on the *Apollo* LRV.

Other steering strategies are possible too, like when using omnidirectional wheels that allow the vehicle to move even sidewise by counter-rotating the wheels of a given axle.

5.4.2 Low-Speed or Kinematic Steering

Low speed or *kinematic steering* is defined as the motion of a wheeled vehicle determined by pure rolling of the wheels. The velocities of the centers of all the wheels lie in their midplane, i.e. the sideslip angles α are vanishingly small. In these conditions the wheels can exert no cornering force to balance the centrifugal force due to the curvature of the trajectory.

Remark 5.12 Kinematic steering is possible only if the velocity is vanishingly small.

Consider a vehicle with four wheels, two of which (the front wheels) can steer (Fig. 5.14). The relationship that must be verified to allow kinematic steering is easily found by imposing that the perpendiculars to the midplanes of the front wheels meet the one of the rear wheels at the same point

$$\tan(\delta_1) = \frac{l}{R_1 - \frac{t}{2}}, \qquad \tan(\delta_2) = \frac{l}{R_1 + \frac{t}{2}}. \tag{5.77}$$

Instead of the track t, (5.77) should contain the distance between the kingpin axes of the wheels, or better, between their intersections with the ground. By eliminating R_1 between the two equations, a direct relationship between δ_1 and δ_2 is readily found

$$\cot(\delta_1) - \cot(\delta_2) = \frac{t}{l}. \tag{5.78}$$

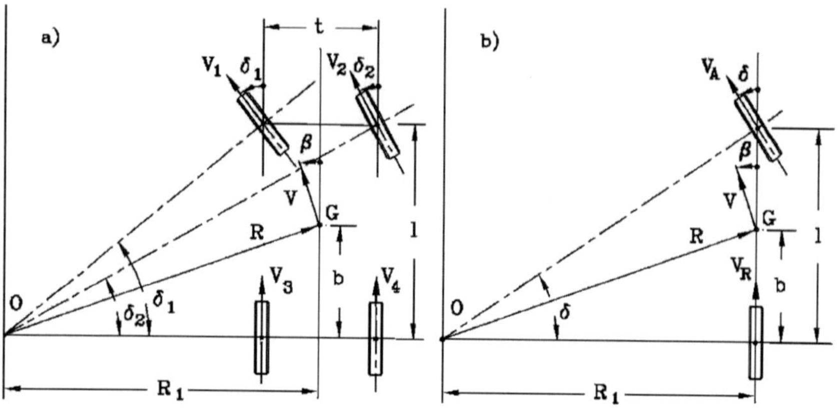

Fig. 5.14 Kinematic steering of a four-wheeled and a two-wheeled vehicle. The sideslip angle of the vehicle β is also shown (from G. Genta, L. Morello, *The Automotive Chassis*, Springer, New York, 2009)

A device allowing one to steer the wheels complying exactly with (5.78) is usually referred to as *Ackermann steering* or *Ackermann geometry*. No actual steering mechanism allows to follow exactly such law and a *steering error*, defined as the difference between the actual value of δ_2 and that obtained from (5.78) can be obtained as a function of δ_1.

Consider for instance the device based on an articulated quadrilateral (Jeantaud linkage) shown in Fig. 5.15a. The relationship linking angle δ_1 to angle δ_2 is[9]

$$\sin(\gamma - \delta_2) + \sin(\gamma + \delta_1) = \sqrt{\left[\frac{l_1}{l_2} - 2\sin(\gamma)\right]^2 - \left[\cos(\gamma - \delta_2) - \cos(\gamma - \delta_1)\right]^2}. \qquad (5.79)$$

A steering error

$$\Delta\delta_2 = \delta_2 - \delta_{2_c},$$

i.e. the difference between the actual value of δ_2 and the kinematically correct one can be computed for each value of δ_1, as shown in Fig. 5.15b. Three values of angle γ have been considered: 16°, 18° and 20°; the higher the value of γ the lower the error is for small values of the steering angle. However, low values of the error at low steering angles are accompanied by large errors at large steering angles and a trade-off is needed: In the case of the figure a value of 18° can be a good starting point.

Much effort has been devoted to design devices allowing to minimize this error; the importance of this issue has, however, been overstated from the viewpoint of the directional response of the vehicle: the facts that

- a sideslip angle is always present,
- most suspension mechanisms allow a certain amount of roll steer,

[9] G. Genta, L. Morello, *The Automotive Chassis*, Springer, New York, 2009.

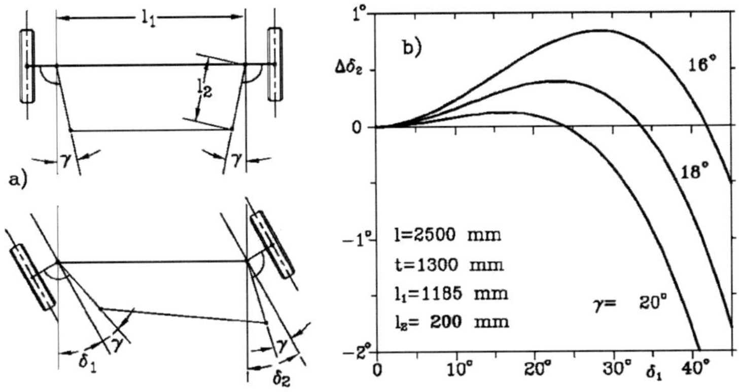

Fig. 5.15 Steering mechanism based on an articulated quadrilateral. (**a**) Sketch; (**b**) steering error $\Delta\delta_2 = \delta_2 - \delta_{2_c}$ as a function of δ_1 (from G. Genta, L. Morello, *The Automotive Chassis*, Springer, New York, 2009)

- in most cases the steering wheels are intentionally not exactly parallel but have a certain toe-in, and
- the deformations of the suspensions induce small steering angles depending on the forces exerted by the wheel on the road,

reduce the importance of small steering errors.

In case of articulated steering, on the contrary, kinematic steering is always possible for a four-wheeled vehicle. When independent steering is used, the realization of the Ackermann condition depends only on the control of the steering actuators, and is possible for a vehicle with any number of wheels. The rovers used on Mars, for instance, are provided with six wheels: the central ones are either fixed or steer only in particular cases, like when rolling sideways, while the front and rear ones have independent steering programmed to satisfy kinematic steering conditions.

The radius of the trajectory of the center of mass of a four-wheeled vehicle is

$$R = \sqrt{b^2 + R_1^2} = \sqrt{b^2 + l^2 \cot^2(\delta)}, \tag{5.80}$$

where δ is the steering angle of the *equivalent* two-wheeled vehicle (Fig. 5.14b). Although it should be computed by averaging the cotangents of the angles of the two wheels,

$$\cot(\delta) = \frac{R_1}{l} = \frac{\cot(\delta_1) + \cot(\delta_2)}{2}, \tag{5.81}$$

it is very close to the direct average of the angles. Consider for example the same vehicle of Fig. 5.15 with center of mass at mid wheelbase on a curve with a radius as small as $R = 10$ m. The correct values of the steering angles are $\delta_1 = 15.090°$, $\delta_2 = 13.305°$ and $\delta = 14.142°$. By direct averaging the steering angles of the wheels, it would follow that $\delta = 14.197°$, with an error of only 0.36%.

In case the radius of the trajectory is large if compared with the wheelbase of the vehicle, (5.80) reduces to

$$R \approx l \cot(\delta) \approx \frac{l}{\delta}. \tag{5.82}$$

Equation (5.82) can be rewritten in the form

$$\frac{1}{R\delta} \approx \frac{1}{l}. \tag{5.83}$$

The expression $1/R\delta$ has an important physical meaning: it is the ratio between the response of the vehicle, in terms of curvature $1/R$ of the trajectory, and the input causing the vehicle to steer. It is therefore a sort of transfer function for the directional control and can be referred to as *trajectory curvature gain*. In kinematic steering it is approximately equal to the reciprocal of the wheelbase.

If all the wheels of the vehicle can steer and are controlled to satisfy kinematic steering conditions, a larger trajectory curvature gain can be obtained. Its maximum value is

$$\left(\frac{1}{R\delta}\right)_{\max} \approx \frac{1}{2l}. \tag{5.84}$$

The optimum condition for low-speed steering of a vehicle with four steering wheels (4WS) is that the steering angles are equal and opposite, realizing the above defined maximum gain.

Another important transfer function is the ratio β/δ. The sideslip angle of the vehicle at its center of mass can be expressed as a function of the radius of the trajectory R as

$$\beta = \arctan\left(\frac{b}{\sqrt{R^2 + b^2}}\right). \tag{5.85}$$

By linearizing (5.85) and introducing expression (5.83) linking R with δ, it follows that

$$\frac{\beta}{\delta} = \frac{b}{l}. \tag{5.86}$$

Ratio β/δ can be referred to as *sideslip angle gain*.

In case of a vehicle with more than two axles, a true kinematic steering is possible only if the wheels of several axles (all except one) can steer and if the steering angles comply with conditions similar to (5.77).

Particularly in the case of long vehicles, the off-tracking distance, i.e. the difference of the radii of the trajectories of the front and the rear wheels is an important parameter. If R_f is the radius of the trajectory of the front wheels, the off-tracking distance is

$$R_f - R_1 = R_f\left\{1 - \cos\left[\arctan\left(\frac{l}{R_1}\right)\right]\right\}. \tag{5.87}$$

If the radius of the trajectory is large when compared to the wheelbase, (5.87) reduces to

$$R_f - R_1 \approx R\left[1 - \cos\left(\frac{l}{R}\right)\right] \approx \frac{l^2}{2R}. \tag{5.88}$$

Example 5.3 Consider a six-wheel mars exploration rover, with the central wheels located at the center of the wheelbase. Compute the steering angles of the wheels to perform Ackerman steering as functions of the radius of the trajectory and compare the results obtained from the complete equations with those obtained from the simplified linearized model.

Data: wheelbase $l = 1.4$ m, track $t = 1.9$ m.

The angles δ_{ij} (with $i = 1$: front; $i = 2$: middle; $i = 3$: rear; $j = 1$: left; $j = 2$: right) when steering to the right are

$$\delta_{11} = -\delta_{31} = \operatorname{atan}\left(\frac{l}{2R - t}\right),$$

$$\delta_{21} = \delta_{22} = 0,$$

$$\delta_{12} = -\delta_{32} = \operatorname{atan}\left(\frac{l}{2R + t}\right).$$

The results for the linearized computation are

$$\delta_{11} = \delta_{12} = -\delta_{31} = -\delta_{32} = \frac{l}{2R}.$$

The results are reported in Fig. 5.16a for radii between 1.8 and 10 m and Fig. 5.16b for radii between 10 and 100 m.

5.4.3 *Ideal Steering*

If the speed is not vanishingly small, the wheels must move with suitable sideslip angles to generate cornering forces. A simple evaluation of the steady state steering of a vehicle in high-speed or *dynamic*[10] *steering* conditions can be performed as follows. Consider a rigid vehicle moving on level ground with transversal slope angle α_t and neglect aerodynamic forces. Define a η-axis parallel to the ground, passing through the center of mass of the vehicle and intersecting the vertical for the center of the trajectory, which in steady-state condition is circular (Fig. 5.17). In

[10]The term *dynamic steering* is used here to denote a condition in which the trajectory is determined by the balance of forces acting on the vehicle, as opposed to *kinematic steering* in which the trajectory is determined by the directions of the midplanes of the wheels. Note that dynamic steering applies to both steady state and unstationary turning.

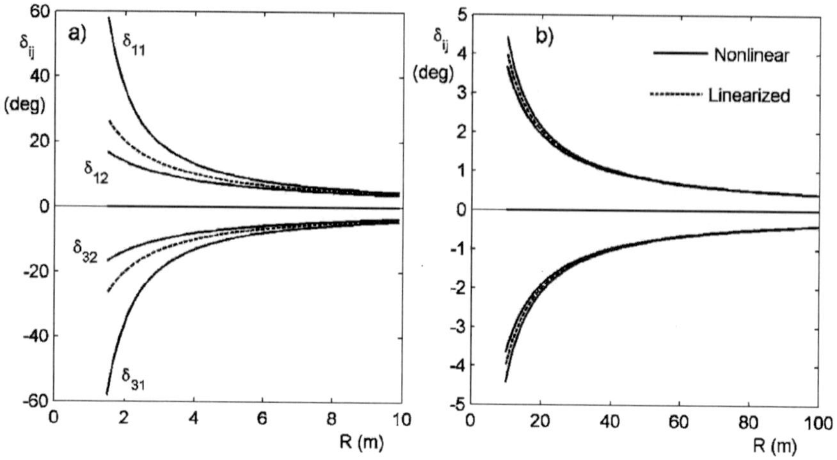

Fig. 5.16 Steering angles of the six wheels as functions of the trajectory curvature radius (the central wheels do not steer)

the figure a four-wheeled vehicle has been sketched, but this simplified model holds for any vehicle with four or more wheels. It can be used also for vehicles with three wheels, provided that some changes to rollover conditions are introduced.

Axis η does not coincide with the y axis, except at one particular speed.

The equilibrium equation in η direction is immediately written by equating the components of weight mg, of centrifugal force mV^2/R and of the forces P_η due to the wheels

$$\frac{mV^2}{R}\cos(\alpha_t) - mg\sin(\alpha_t) = \sum_{\forall i} P_{\eta_i}. \tag{5.89}$$

For a first approximation study, forces P_η can be confused with the cornering forces F_y of the tires and all wheels can be assumed to work with the same side force coefficient μ_y. As the last assumption is similar to that seen for braking in ideal conditions, this approach will be referred to as *ideal steering*. These two assumptions lead to substituting the expression $\sum_{\forall i} P_{\eta_i}$ with $\mu_y F_z$.

Again neglecting aerodynamic forces, force $F_z = \sum F_{z_i}$ exerted by the vehicle on the road is

$$F_z = mg\cos(\alpha_t) + \frac{mV^2}{R}\sin(\alpha_t). \tag{5.90}$$

By introducing (5.90) into (5.89), the ratio between the lateral acceleration and the gravitational acceleration g is

$$\frac{V^2}{Rg} = \frac{\tan(\alpha_t) + \mu_y}{1 - \mu_y\tan(\alpha_t)}. \tag{5.91}$$

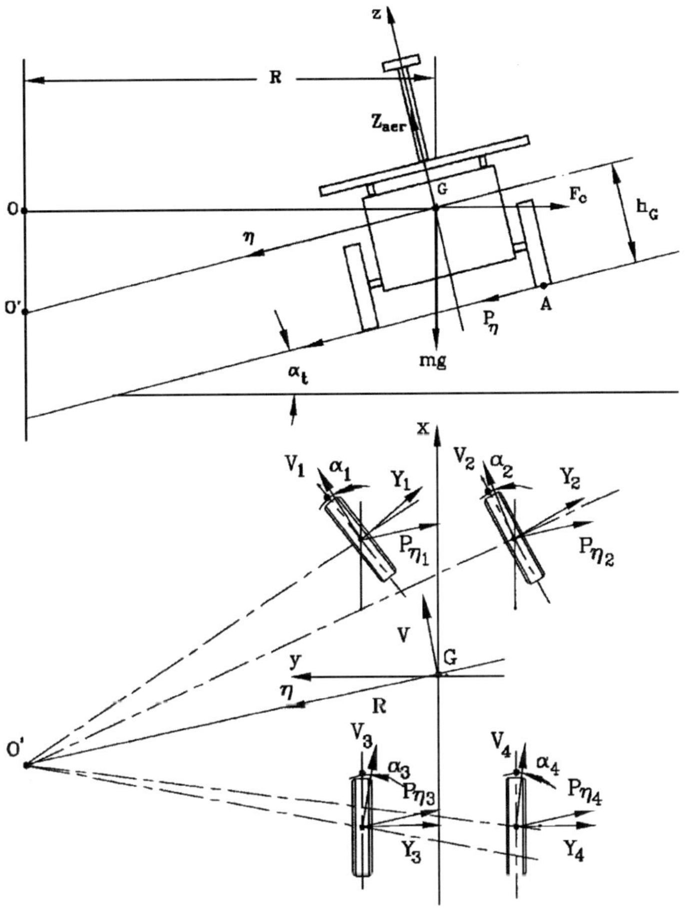

Fig. 5.17 Simplified model for dynamic steering

By introducing the maximum value of the side force coefficient μ_{y_p} into (5.91) it is possible to obtain the maximum value of the lateral acceleration

$$\left(\frac{V^2}{R}\right)_{\max} = g f_s, \tag{5.92}$$

where the so-called sliding factor f_s can be defined as[11]

$$f_s = \frac{\tan(\alpha_t) + \mu_{y_p}}{1 - \mu_{y_p}\tan(\alpha_t)}. \tag{5.93}$$

[11]The sliding factor is more commonly defined as the square root of the same quantity considered here. The present definition, which refers directly to the lateral acceleration instead of the speed at which a given radius can be obtained, is here preferred, as in particular conditions it reduces to the side force coefficient.

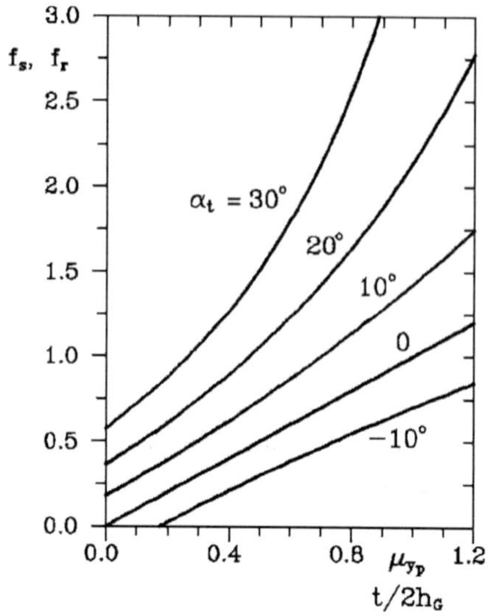

Fig. 5.18 Sliding and rollover factors as functions of μ_{y_p} and of $t/2h_G$ for roads with different transversal slope

The sliding factor is reported as a function of μ_{y_p} for different values of the transversal slope of the ground in Fig. 5.18. Note that if the ground is flat it reduces to the maximum value of the side force coefficient μ_{y_p}.

The maximum speed at which a bend with radius R can be negotiated is

$$V_{\max} = \sqrt{Rg}\sqrt{\frac{\tan(\alpha_t) + \mu_{y_p}}{1 - \mu_{y_p}\tan(\alpha_t)}}. \tag{5.94}$$

The limitation to the maximum lateral acceleration due to the cornering force the tires can exert is, however, not the only one, at least theoretically. A further one can come from the danger of rollover, occurring if the resultant of forces in yz plane crosses the road surface outside point A (Fig. 5.17).

The limit condition for rollover is

$$\left(\frac{V^2}{R}\right)_{\max} = gf_r, \tag{5.95}$$

where the rollover factor can be defined as

$$f_r = \frac{\tan(\alpha_t) + \frac{t}{2h_G}}{1 - \frac{t}{2h_G}\tan(\alpha_t)}. \tag{5.96}$$

Its expression is identical to that of the sliding factor, once ratio $t/2h_G$ has been substituted for μ_{y_p} (Fig. 5.18).

The maximum lateral acceleration is thus

$$\left(\frac{V^2}{R}\right)_{\max} = g\min\{f_s, f_r\}. \tag{5.97}$$

Whether the limit condition first reached is that related to sliding, with subsequent spin out of the vehicle, or related to rolling over depends on the relative magnitude of f_s and f_r. If $f_s < f_r$, as it often occurs, the vehicle spins out. This condition can be written in the form

$$\mu_{y_p} < \frac{t}{2h_G}.$$

The present model is only a rough approximation of the actual situation, as is based on the assumption that the side force coefficients μ_y of all wheels are equal, which implies that all wheels work with the same sideslip angle α. Also it ignores the effect of the different directions of the cornering forces of the various wheels which should be considered as perpendicular to the midplanes of the wheels and not directed along η axis. The load transfer between the wheels of the same axle and the presence of the suspensions have also been neglected, other assumptions which contribute to the lack of precision of this model.

If the maximum speed at which a circular path can be negotiated is measured during a steering pad test and the value of the lateral force coefficient is computed through (5.93), a value of μ_{y_p} which is well below that obtained from tests on the tires is obtained. The cornering force coefficient obtained in this way is that of the vehicle as a whole and the difference between its value and that referred to the tires gives a measure of how well the vehicle is able to exploit the cornering characteristics of its wheels.

The side force coefficient measured on the whole vehicle depends also on the radius of the trajectory, with a notable decrease on narrow bends. The majority of vehicles are able to use only a fraction from 50 to 80% of the potential cornering force of the tires. This reduction of the lateral forces makes the danger of rollover a more remote one. Actually rolling over in quasi-static condition is impossible for most vehicles; rollover may, however, occur owing to dynamic phenomena in non-stationary conditions or to lateral forces due to side contacts ruling out the possibility of side slipping and then causing far stronger lateral forces to be exerted on the wheels. Also the presence of the suspensions modifies this picture, making rollover a possibility.

The cornering dynamics of a vehicle with two wheels is radically different from that of vehicles with three or more wheels, but it will not be dealt here because this is an unlikely configuration for robots or planetary exploration vehicles.

Example 5.4 Compute the cornering performances of a vehicle operating on the Moon surface. Assume that the maximum value of the side force coefficient of the wheels varies between 0.1 and 0.5, and that the sliding factor is 70% of the side force coefficient.

The results are reported in Fig. 5.19. in the form of a plot of the minimum radius of the trajectory as a function of the speed.

Remark 5.13 The transversal load transfer is independent from the gravitational acceleration. On the Moon, in spite of the poor performance, is not less than on Earth at equal value of the force coefficient.

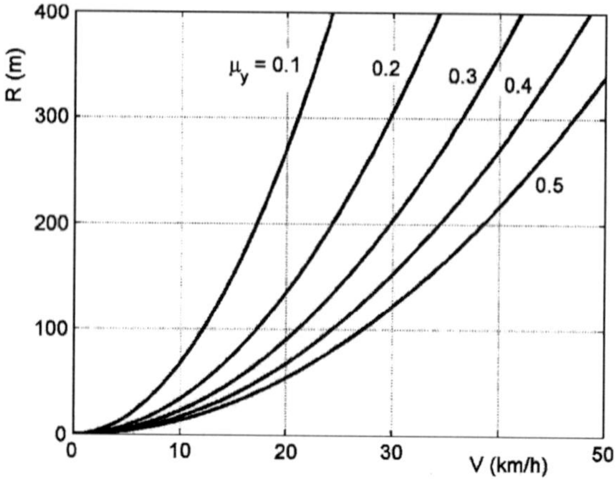

Fig. 5.19 Relationship between the minimum radius of the trajectory and the maximum speed for a vehicle on the Moon with different values of the maximum value of the side force coefficient. The sliding factor is assumed as 70% of the side force coefficient of the wheel

The main effect of low gravity is that the tires are likely to be oversized owing to the low load, and thus they may work in the part of the cornering stiffness versus load curve where load transfer is less important, at least using linearized models (see below). The very low cornering forces available may make worth while to use the modern stability enhancement (ESP, VDC, etc.) systems; they may be even more useful in low gravity environments than on Earth. If electric steering control is used, in a global by-wire architecture, they can be easily integrated in the vehicle control system.

5.4.4 Ground–Wheel Contact as a Non-holonomic Constraint

A traditional way to study the motion of wheeled robots is to model the wheel–ground interface as a non-holonomic constraint. This approach has a number of drawbacks, as it will be clear in this section.

Consider a robot with a number n of axles, each with two wheels, and controlled by independent steering. The rover is considered as a rigid body with mass m and moment of inertia J_z about the z axis, moving on a flat and terrain sloping at an angle α with respect to the horizontal (Fig. 5.20a and b). The inertial reference frame lies on the ground, with its X axis horizontal and Y axis sloping upwards. Without considering the constraints due to the trajectory control, the rover has 3 degrees of freedom and the coordinates X and Y of the center of mass G and the yaw angle ψ can be taken as generalized coordinates.

To define the trajectory two constraint equations must be defined that express the condition that the velocity of all wheels are contained in their symmetry planes.

The generic velocity of the center P_i of the contact area of the jth wheel ($j = 1, 2$) of the ith axle ($i = 1, \dots, n$) located in a point whose coordinates are x_i and y_{ij} in the reference frame of the vehicle, is

$$\mathbf{V}_{P_{ij}} = \mathbf{V}_G + \dot{\psi} \Lambda \overline{(P_{i,j} - G)}, \tag{5.98}$$

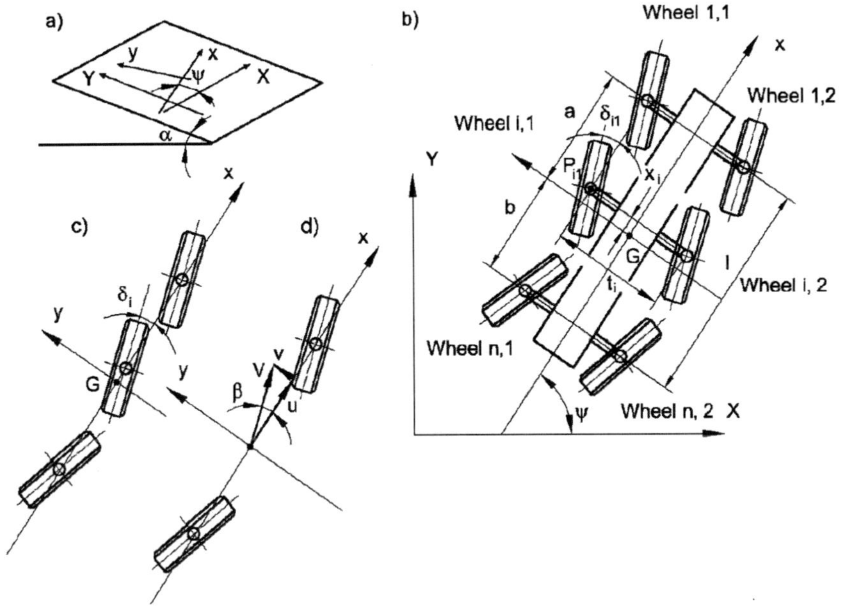

Fig. 5.20 Rover with n axles, each with two wheels, moving on a plane inclines with respect to a horizontal plane. (**a**) Reference frames; (**b**) sketch of the rover; (**c**) monotrack model, used to compute the conditions for kinematic steering; (**d**) monotrack model with two axles, used to define the non-holonomic constraints, once the kinematic steering conditions are satisfied

i.e.

$$\mathbf{V}_{P_i} = \begin{Bmatrix} u - \dot{\psi} y_{ij} \\ v + \dot{\psi} x_i \end{Bmatrix}. \tag{5.99}$$

If the ith wheel has a steering angle δ_i, the condition for pure rolling is

$$\tan(\delta_{ij}) = \frac{v + \dot{\psi} x_i}{u - \dot{\psi} y_{ij}}. \tag{5.100}$$

The constraint equations are thus

$$v + \dot{\psi} x_i - \tan(\delta_{ij})(u - \dot{\psi} y_{ij}) = 0 \quad \text{for } i = 1, \dots, n; \; j = 1, 2. \tag{5.101}$$

The $2n$ angles δ_{ij} appearing in the $2n$ equations (5.101) are linked to each other by the conditions that allow to perform kinematic steering.

Remembering that

$$y_{i1} = \frac{t_i}{2}, \qquad y_{i2} = -\frac{t_i}{2}, \tag{5.102}$$

where t_i is the track of the ith axle, the constraint equations of the two wheels of each axle are

$$\begin{cases} v + \dot{\psi} x_i - \tan(\delta_{i1})\left(u - \dot{\psi}\dfrac{t}{2}\right) = 0, \\ v + \dot{\psi} x_i - \tan(\delta_{i2})\left(u + \dot{\psi}\dfrac{t}{2}\right) = 0. \end{cases} \tag{5.103}$$

A first set of relationships to grant the possibility of kinematic steering is thus

$$\tan(\delta_{i1})\left(u - \dot{\psi}\frac{t}{2}\right) = \tan(\delta_{i2})\left(u + \dot{\psi}\frac{t}{2}\right) = u\tan(\delta_i), \tag{5.104}$$

where δ_i is an 'average steering angle' of the ith axle.

The steering angles of the wheels are thus linked to the average steering angles of the axles by the relationships

$$\tan(\delta_{ij}) = \frac{u}{u \mp \dot{\psi}\frac{t}{2}}\tan(\delta_i), \tag{5.105}$$

where the upper sign (−) holds for $j = 1$ (left wheel) and the lower one for $j = 2$ (right wheel).

The relationships for kinematic steering are

$$v + \dot{\psi} x_i - u\tan(\delta_i) = 0 \quad \text{for } i = 1, \dots, n. \tag{5.106}$$

Only two of the angles δ_i are independent, say the first δ_1 and the last δ_n. The first, the last and a generic intermediate axles are thus linked by the relationships

$$v = -\dot{\psi} a + u\tan(\delta_1) = \dot{\psi} b + u\tan(\delta_n) = -\dot{\psi} x_i + u\tan(\delta_i), \tag{5.107}$$

where $a = x_1$ and $b = -x_n$.

From (5.107) it follows that

$$\tan(\delta_i) = \frac{1}{2}\left[\tan(\delta_1) + \tan(\delta_n) + \frac{\dot{\psi}}{u}(b - a + 2x_i)\right] \quad \text{for } i = 2, \dots, n-1. \tag{5.108}$$

All steering angles δ_{ij} can thus be computed, once δ_1 and δ_n have been stated. In this way, the model of the vehicle reduces to a monotrack model, with single wheels instead of axles having two (or more) wheels each (Fig. 5.20c).

As already stated, the condition allowing one to minimize the radius of the trajectory is

$$\delta_n = -\delta_1 \tag{5.109}$$

and thus

$$\tan(\delta_i) = \frac{\dot{\psi}}{2u}(b - a + 2x_i) \quad \text{for } i = 2, \dots, n-1. \tag{5.110}$$

The two constraint equations are thus

$$\begin{cases} f_1(\mathbf{x}, \dot{\mathbf{x}}) = v + \dot{\psi} a - u \tan(\delta_1) = 0, \\ f_2(\mathbf{x}, \dot{\mathbf{x}}) = v - \dot{\psi} b - u \tan(\delta_n) = 0, \end{cases} \tag{5.111}$$

that correspond to the non-holonomic constraints for the two-wheeled monotrack vehicle of Fig. 5.20d.

The potential energy due to gravity is

$$\mathcal{U} = mgY \sin(\alpha). \tag{5.112}$$

To write the equations of motion, the Lagrangian of the system can be written in terms of velocities in the body-fixed frame as

$$\mathcal{L} = \frac{1}{2} m\left(u^2 + v^2\right) + \frac{1}{2} J_z \dot{\psi}^2 - mgY \sin(\alpha). \tag{5.113}$$

The velocities in the body-fixed frame are linked to the derivatives of the generalized coordinates by the relationship

$$\begin{Bmatrix} u \\ v \\ \dot{\psi} \end{Bmatrix} = \begin{bmatrix} \cos(\psi) & \sin(\psi) & 0 \\ -\sin(\psi) & \cos(\psi) & 0 \\ 0 & 0 & 1 \end{bmatrix} \begin{Bmatrix} \dot{X} \\ \dot{Y} \\ \dot{\psi} \end{Bmatrix}. \tag{5.114}$$

Since the constraints are non-holonomic and the Jacobian matrix of functions f_1 and f_2 has rank 2, i.e. all constraints are non-holonomic and are independent, the equation of motion is (see Appendix A)

$$\frac{d}{dt}\left(\frac{\partial \mathcal{L}}{\partial \dot{x}_i}\right) - \frac{\partial \mathcal{L}}{\partial x_i} + \sum_{j=1}^{2} \lambda_j \frac{\partial f_j}{\partial \dot{x}_i} = Q_i, \tag{5.115}$$

where λ_i are Lagrange multipliers.

The generalized forces Q_j can be computed easily as the derivatives of the virtual work of the forces applied to the system $\delta\mathcal{W}$ with respect to the virtual displacements δx_i.

The forces applied to the system are the net tractions, i.e. the 'drawbar pulls' referred to the single wheels, which are directed in the direction of the midplanes. For the generic jth wheel of the ith axle it can be decomposed along the body axes x and y as

$$\mathbf{F}_{xij} = \begin{Bmatrix} F_{xij} \cos(\delta_{ij}) \\ F_{xij} \sin(\delta_{ij}) \end{Bmatrix}. \tag{5.116}$$

The virtual displacement of the center of the contact area of the ijth wheel is

$$\delta x_{\mathrm{P}_{ij}} = \begin{Bmatrix} \delta x - y_{ij}\delta\psi \\ \delta y + x_i\delta\psi \end{Bmatrix} = \begin{Bmatrix} \delta X\cos(\psi) + \delta Y\sin(\psi) - y_{ij}\delta\psi \\ -\delta X\sin(\psi) + \delta Y\cos(\psi) + x_i\delta\psi \end{Bmatrix} \tag{5.117}$$

and thus the virtual work is

$$\begin{aligned} \delta\mathcal{W}_{\mathrm{P}_{ij}} &= F_{xij}\cos(\delta_{ij})\big[\delta X\cos(\psi) + \delta Y\sin(\psi) - y_{ij}\delta\psi\big] \\ &\quad + F_{xij}\sin(\delta_{ij})\big[-\delta X\sin(\psi) + \delta Y\cos(\psi) + x_i\delta\psi\big]. \end{aligned} \tag{5.118}$$

By performing the relevant derivatives, and remembering, for instance, that

$$\frac{\partial\mathcal{L}}{\partial\dot{X}} = \frac{\partial\mathcal{L}}{\partial u}\frac{\partial u}{\partial\dot{X}} + \frac{\partial\mathcal{L}}{\partial v}\frac{\partial v}{\partial\dot{X}} + \frac{\partial\mathcal{L}}{\partial\dot{\psi}}\frac{\partial\dot{\psi}}{\partial\dot{X}} = \frac{\partial\mathcal{L}}{\partial u}\cos(\psi) + \frac{\partial\mathcal{L}}{\partial v}\cos(\psi)$$

and that the same holds for the functions expressing the constraints, e.g.

$$\frac{\partial f_1(\mathbf{x},\dot{\mathbf{x}})}{\partial\dot{X}} = \frac{\partial f_1}{\partial u}\frac{\partial u}{\partial\dot{X}} + \frac{\partial f_1}{\partial v}\frac{\partial v}{\partial\dot{X}} + \frac{\partial f_1}{\partial\dot{\psi}}\frac{\partial\dot{\psi}}{\partial\dot{X}} = \frac{\partial f_1}{\partial u}\cos(\psi) + \frac{\partial f_1}{\partial v}\cos(\psi),$$

the equations of motion are

$$\begin{cases} \big[m(\dot{u} - v\dot{\psi}) + \lambda_1\sin(\delta_1) + \lambda_2\sin(\delta_n)\big]\cos(\psi) \\ \qquad - \big[m(\dot{v} + u\dot{\psi}) + \lambda_1\cos(\delta_1) + \lambda_2\cos(\delta_n)\big]\sin(\psi) \\ \quad = \displaystyle\sum_{i=1}^{n}\sum_{j=1}^{2} F_{xij}\cos(\psi + \delta_{ij}), \\ \big[m(\dot{u} - v\dot{\psi}) + \lambda_1\sin(\delta_1) + \lambda_2\sin(\delta_n)\big]\sin(\psi) \\ \qquad + \big[m(\dot{v} + u\dot{\psi}) + \lambda_1\cos(\delta_1) + \lambda_2\cos(\delta_n)\big]\cos(\psi) \\ \quad = \displaystyle\sum_{i=1}^{n}\sum_{j=1}^{2} F_{xij}\sin(\delta_{ij} + \psi) - mg\sin(\alpha), \\ J_z\ddot{\psi} + \lambda_1 a\cos(\delta_1) - \lambda_2 b\cos(\delta_n) \\ \quad = \displaystyle\sum_{i=1}^{n}\sum_{j=1}^{2} F_{xij}\big[-y_{ij}\cos(\delta_{ij}) + x_i\sin(\delta_{ij})\big]. \end{cases} \tag{5.119}$$

By premultiplying the right and left sides of this equation by the matrix in (5.114), and adding the constraint equations, the following set of five equations is obtained:

$$\begin{cases} m(\dot{u} - v\dot{\psi}) + \lambda_1 \sin(\delta_1) + \lambda_2 \sin(\delta_n) \\ \quad = \sum_{i=1}^{n}\sum_{j=1}^{2} F_{xij}\cos(\delta_{ij}) - mg\sin(\alpha)\sin(\psi), \\ m(\dot{v} + u\dot{\psi}) + \lambda_1 \cos(\delta_1) + \lambda_2 \cos(\delta_n) \\ \quad = \sum_{i=1}^{n}\sum_{j=1}^{2} F_{xij}\sin(\delta_{ij}) - mg\sin(\alpha)\cos(\psi), \\ J_z\ddot{\psi} + \lambda_1 a\cos(\delta_1) - \lambda_2 b\cos(\delta_n) \\ \quad = \sum_{i=1}^{n}\sum_{j=1}^{2} F_{xij}\left[-y_{ij}\cos(\delta_{ij}) + x_i\sin(\delta_{ij})\right], \\ v = \frac{u}{l}\left[b\tan(\delta_1) + a\tan(\delta_n)\right], \\ \dot{\psi} = \frac{u}{l}\left[\tan(\delta_1) - \tan(\delta_n)\right]. \end{cases} \tag{5.120}$$

A solution in the state space can be obtained by introducing the yaw velocity as an auxiliary variable $r = \dot{\psi}$

$$\begin{cases} \dot{u} = vr - \frac{\lambda_1}{m}\sin(\delta_1) - \frac{\lambda_2}{m}\sin(\delta_n) - g\sin(\alpha)\sin(\psi) \\ \quad + \sum_{i=1}^{n}\sum_{j=1}^{2} \frac{F_{xij}}{m}\cos(\delta_{ij}), \\ \dot{\psi} = r, \\ \dot{v} = -ur - \frac{\lambda_1}{m}\cos(\delta_1) - \frac{\lambda_2}{m}\cos(\delta_n) - g\sin(\alpha)\cos(\psi) \\ \quad + \sum_{i=1}^{n}\sum_{j=1}^{2} \frac{F_{xij}}{m}\sin(\delta_{ij}), \\ \dot{r} = -\lambda_1\frac{a}{J_z}\cos(\delta_1) + \lambda_2\frac{b}{J_z}\cos(\delta_n) \\ \quad + \sum_{i=1}^{n}\sum_{j=1}^{2} \frac{F_{xij}}{J_z}\left[-y_{ij}\cos(\delta_{ij}) + x_i\sin(\delta_{ij})\right], \\ v = \frac{u}{l}\left[b\tan(\delta_1) + a\tan(\delta_n)\right], \\ r = \frac{u}{l}\left[\tan(\delta_1) - \tan(\delta_n)\right]. \end{cases} \tag{5.121}$$

The trajectory is easily computed by integrating the equation

$$\begin{Bmatrix} \dot{X} \\ \dot{Y} \end{Bmatrix} = \begin{bmatrix} \cos(\psi) & -\sin(\psi) \\ \sin(\psi) & \cos(\psi) \end{bmatrix} \begin{Bmatrix} u \\ v \end{Bmatrix}. \tag{5.122}$$

It is easy to verify that in steady state conditions $\dot{u} = \dot{v} = \ddot{\psi} = 0$ and the velocity of the vehicle is

$$V = R\dot{\psi},$$

where R is the radius of the curvature of the (circular, since in steady state the radius is constant) trajectory.

From the fifth equation, remembering that

$$u^2 + v^2 = V^2, \tag{5.123}$$

the steady state conditions are readily obtained: once that the velocity on the trajectory V and the steering angles δ_1 and δ_2 are stated, the components of the velocity along the x and y axes are

$$u = V\frac{l}{\sqrt{l^2 + [b\tan(\delta_1) + a\tan(\delta_n)]^2}}, \tag{5.124}$$

$$v = V\frac{b\tan(\delta_1) + a\tan(\delta_n)}{\sqrt{l^2 + [b\tan(\delta_1) + a\tan(\delta_n)]^2}}. \tag{5.125}$$

From the last equation the angular velocity r can thus be obtained

$$r = V\frac{\tan(\delta_1) - \tan(\delta_n)}{\sqrt{l^2 + [b\tan(\delta_1) + a\tan(\delta_n)]^2}}. \tag{5.126}$$

The radius of the trajectory is thus

$$R = \frac{V}{r} = \frac{\sqrt{l^2 + [b\tan(\delta_1) + a\tan(\delta_n)]^2}}{\tan(\delta_1) - \tan(\delta_n)}. \tag{5.127}$$

These values coincide with those obtained in Sect. 5.4.2.

The values of the Lagrange multipliers and of the longitudinal forces can be obtained from the first three equations. The steady state trajectory is a circle, and thus angle ψ is a linear function of time: a steady state solution is thus possible only if the slope α is equal to zero.

This approach, as already stated, neglects the sideslip angle (and also the longitudinal slip, which is, however, of little importance in the lateral behavior), leading to large inaccuracies in predicting the trajectory at high speed, since it neglects completely the over- or understeering behavior of the vehicle. However, it is inaccurate also at low speed (as a limit, even when the speed tends to zero) when the ground has a lateral slope. The models based on assuming that the wheels are non-holonomic constraints can thus be used only in structured environments, where the robot travels at very low speed on perfectly smooth and flat terrain, while in unstructured environments, like those found in planetary exploration, they are hardly applicable.

5.4.5 Model for High-Speed Cornering

Equations of Motion

The difference between high-speed and kinematic steering is that in the former the sideslip angles of the wheels are accounted for.

Using the same model shown in Fig. 5.20b, from (5.99) the angle between the velocity of the ijth wheel and the x-axis is

$$\beta_{ij} = \arctan\left(\frac{v_{ij}}{u_{ij}}\right) = \arctan\left(\frac{v + \dot{\psi} x_i}{u - \dot{\psi} y_{ij}}\right). \tag{5.128}$$

If the ijth wheel has a steering angle δ_{ij}, its sideslip angle is

$$\alpha_{ij} = \beta_{ij} - \delta_{ij} = \arctan\left(\frac{v + \dot{\psi} x_i}{u - \dot{\psi} y_{ij}}\right) - \delta_{ij}. \tag{5.129}$$

Neglecting the effects due to the camber angle,[12] the lateral force exerted on the ijth wheel depends only on the sideslip angle:

$$F_{yij} = f_1(\alpha_{ij}). \tag{5.130}$$

The forces applied to the system are now the sum of the drawbar pulls and the side forces applied to the wheels. When decomposed along the body axes x and y, the forces acting on the ijth wheel are

$$\mathbf{F}_{xij} = \left\{ \begin{array}{c} F_{xij} \cos(\delta_{ij}) - F_{yij} \sin(\delta_{ij}) \\ F_{xij} \sin(\delta_{ij}) + F_{yij} \cos(\delta_{ij}) \end{array} \right\}. \tag{5.131}$$

The virtual displacement of the center of the contact area of the ijth wheel is expressed by (5.117).

On each wheel also a moment, the aligning torque, is applied by the ground. With the same assumptions seen above, it is a function of the sideslip angle:

$$M_{zij} = f_2(\alpha_{ij}). \tag{5.132}$$

The virtual work applied on each wheel is thus

$$\begin{aligned} \delta \mathcal{W}_{\mathrm{P}_{ij}} = {} & \left[F_{xij} \cos(\delta_{ij}) - F_{yij} \sin(\delta_{ij})\right]\left[\delta X \cos(\psi) + \delta Y \sin(\psi) - y_{ij} \delta \psi\right] \\ & + \left[F_{xij} \sin(\delta_{ij}) + F_{yij} \cos(\delta_{ij})\right]\left[-\delta X \sin(\psi) + \delta Y \cos(\psi) + x_i \delta \psi\right] \\ & + M_{zij} \delta \psi. \end{aligned} \tag{5.133}$$

---

[12] In the present model no account is taken for the presence of the suspensions and the vehicle is assumed to be a rigid body. For symmetry reasons, the camber angles of the two wheels of the same axle can be assumed to be equal and opposite, so that no camber force acts on each axle: this is just an approximation, but holds quite well for small camber angles.

The equations of motion can be obtained in the same way seen in the previous case, the only differences being that now there is no non-holonomic constraint and that also the lateral forces and the aligning torques must be accounted for.

The equations of motion are thus (5.120), without the terms (at left hand side)

$$\begin{cases} \lambda_1 \sin(\delta_1) + \lambda_2 \sin(\delta_n), \\ \lambda_1 \cos(\delta_1) + \lambda_2 \cos(\delta_n), \\ \lambda_1 a \cos(\delta_1) - \lambda_2 b \cos(\delta_n) \end{cases}$$

and with the additional terms (at right hand side)

$$\begin{cases} -\sum_{i=1}^{n}\sum_{j=1}^{2} F_{yij} \sin(\delta_{ij}), \\ \sum_{i=1}^{n}\sum_{j=1}^{2} F_{yij} \sin(\delta_{ij}), \\ \sum_{i=1}^{n}\sum_{j=1}^{2} \{F_{yij}[y_{ij} \sin(\delta_{ij}) + x_i \cos(\delta_{ij})] + M_{zij}\}. \end{cases}$$

Obviously, the constraint equations must now be omitted.

The lateral forces acting on the wheels now substitute the Lagrange multipliers, whose physical meaning is the forces due to the non-holonomic constraints. This is fairly obvious.

The state space model expressed by (5.121) now becomes

$$\begin{cases} \dot{u} = vr - g\sin(\alpha)\sin(\psi) + \frac{1}{m}\sum_{i=1}^{n}\sum_{j=1}^{2}[F_{xij}\cos(\delta_{ij}) - F_{yij}\sin(\delta_{ij})], \\ \dot{\psi} = r, \\ \dot{v} = -ur - g\sin(\alpha)\cos(\psi) + \frac{1}{m}\sum_{i=1}^{n}\sum_{j=1}^{2}[F_{xij}\sin(\delta_{ij}) + F_{yij}\cos(\delta_{ij})], \\ \dot{r} = \frac{1}{J_z}\sum_{i=1}^{n}\sum_{j=1}^{2}\{F_{xij}[-y_{ij}\cos(\delta_{ij}) + x_i\sin(\delta_{ij})] \\ \qquad + F_{yij}[y_{ij}\sin(\delta_{ij}) + x_i\cos(\delta_{ij})] + M_{zij}\}. \end{cases} \tag{5.134}$$

Remark 5.14 The state space model is made of four first order equations instead of six, as it would be expected for a system with 3 degrees of freedom. The point is that the configuration space model is made by a second order differential equation (that regarding the yaw rotation) plus two first order equations (those regarding the displacement degrees of freedom).

Two further first order equations must anyway be added for computing the trajectory

$$\begin{Bmatrix} \dot{X} \\ \dot{Y} \end{Bmatrix} = \begin{bmatrix} \cos(\psi) & -\sin(\psi) \\ \sin(\psi) & \cos(\psi) \end{bmatrix} \begin{Bmatrix} u \\ v \end{Bmatrix}. \tag{5.135}$$

A further set of $6n$ algebraic equations must be added. They are equation (5.129) yielding the sideslip angles α_{ij} as functions of u, v, $\dot{\psi}$ and δ_{ij} and (5.130) and (5.132) yielding the forces and moments as functions of the sideslip angles α_{ij}. The steering angles and the longitudinal forces are assumed as know functions of time, but they are outputs of the equations modeling the control system of the rover.

Steady-State Solution

A steady-state solution is characterized by constant values of the velocities u, v and $\dot{\psi}$, the steering angles δ_{ij}, the sideslip angles α_{ij} and the forces. The radius of the trajectory is constant, and thus the trajectory is circular. These conditions cannot be met if the lateral slope of the ground is accounted for, so α must be assumed to be equal to zero.

The equations of motion reduce to

$$\begin{cases} mvr + \sum_{i=1}^{n}\sum_{j=1}^{2}\left[F_{xij}\cos(\delta_{ij}) - F_{yij}\sin(\delta_{ij})\right] = 0, \\ -mur + \sum_{i=1}^{n}\sum_{j=1}^{2}\left[F_{xij}\sin(\delta_{ij}) + F_{yij}\cos(\delta_{ij})\right] = 0, \\ \sum_{i=1}^{n}\sum_{j=1}^{2}\left\{F_{xij}\left[-y_{ij}\cos(\delta_{ij}) + x_i\sin(\delta_{ij})\right]\right. \\ \quad \left. + F_{yij}\left[y_{ij}\sin(\delta_{ij}) + x_i\cos(\delta_{ij})\right] + M_{zij}\right\} = 0. \end{cases} \tag{5.136}$$

These three equations, plus the $6n$ equations (5.129), (5.130) and (5.132) are all nonlinear.

A simple procedure based on the Newton–Raphson method to solve the steady-state dynamics of the vehicle is by writing a set of three equations in the form

$$p_k\,(u, v, \dot{\psi}) = 0, \tag{5.137}$$

where the functions p_k are the left-hand sides of (5.136)

The sideslip angles, the side forces and aligning torques to be introduced in these functions can be computed from (5.129), (5.130) and (5.132) and there is no need to introduce specific unknowns and equations in the set (5.137).

A set of values of the unknown is assumed, for instance those obtained from the kinematic steering model. The values $\mathbf{x}^{(l+1)}$ of the unknown at the $(l+1)$th iteration from that at the lth iteration are obtained from the usual formula

$$\mathbf{x}^{(l+1)} = \mathbf{x}^{(l)} - \mathbf{S}^{-1}\mathbf{p}. \tag{5.138}$$

Matrix **S** is the Jacobian matrix of functions p_k with respect to unknowns $\mathbf{x}_m$. Since the expressions of the various functions may be very complicated (some of them may be known only experimentally), the Jacobian matrix can be obtained only numerically, computing the various functions three times at each iteration with small increments to each unknown (one by one). The incremental ratios

$$\frac{\Delta p_k}{\Delta x_m}$$

thus obtained are used instead of the derivatives.

Instead of using the component of the velocity u as an unknown, it is possible to state the speed along the trajectory V, assuming as an unknown a parameter linked with the longitudinal forces.

5.4.6 Linearized Model for High-Speed Cornering

Equations of Motion

Equations (5.134) are nonlinear in the velocities u, v and $\dot{\psi}$, but can be easily linearized, to obtain closed form solutions allowing one to perform a general study of the handling of the vehicle and particularly to study its stability in the small. This linearization can be performed only if the lateral slope of the ground is neglected ($\alpha = 0$), since otherwise the nonlinearities due to the trigonometric functions of angle ψ, which is generally not a small angle, must be considered.

As a first consideration, angle β (Fig. 5.3) can be considered small and its trigonometric functions can be linearized

$$\begin{cases} u = V\cos(\beta) \approx V, \\ v = V\sin(\beta) \approx V\beta. \end{cases} \tag{5.139}$$

Also the steering angles and the yaw velocity can be considered, in most driving conditions, as small quantities.

Since angle ψ does not appear explicitly in the equations of motion, it needs not to be a small angle. The equations reduce to

$$\begin{cases} \dot{V} = vr + \dfrac{1}{m}\displaystyle\sum_{i=1}^{n}\sum_{j=1}^{2}(F_{xij} - F_{yij}\delta_{ij}), \\ \dot{v} = -Vr + \dfrac{1}{m}\displaystyle\sum_{i=1}^{n}\sum_{j=1}^{2}(F_{xij}\delta_{ij} + F_{yij}), \\ \dot{r} = \dfrac{1}{J_z}\displaystyle\sum_{i=1}^{n}\sum_{j=1}^{2}\left[F_{xij}(-y_{ij} + x_i\delta_{ij}) + F_{yij}(y_{ij}\delta_{ij} + x_i) + M_{zij}\right]. \end{cases} \tag{5.140}$$

If the speed V is a known function of time (often the dynamic behavior at constant speed is studied, in other cases a law $V(t)$ is stated), the last two equations uncouple from the first one, allowing to study the lateral behavior of the vehicle.

The expression of the sideslip angle can be easily linearized. By noting that $y_i\dot{\psi}$ is far smaller than the speed V, it follows that

$$\alpha_i = \frac{v + rx_i}{V} - \delta_i = \beta + \frac{x_i}{V}r - \delta_i. \tag{5.141}$$

The coordinate y_i of the center of the contact area of the wheel does not appear in the expression for the sideslip angle α_i. If the differences between the steering angles δ_i of the wheels of the same axle are neglected, the values of their sideslip angles are thus equal. This allows one to work in terms of axles instead of single wheels and to substitute a model of the type of that of Fig. 5.20c to that of Fig. 5.20d.

Cornering forces can be linearized by expressing them as the product of the cornering stiffness by the sideslip angle

$$F_{yi} = -C_i\alpha_i = -C_i\left(\beta + \frac{x_i}{V}r - \delta_i\right). \tag{5.142}$$

Remark 5.15 Equation (5.142) is written in terms of axles. The cornering stiffness is then that of the axle and not of the single wheel. As already stated, in this way no allowance is taken for the camber force as, owing to the assumption of rigid vehicle, no roll is considered and the wheels of any axle have opposite camber. The camber forces thus cancel each other.

A similar expression can be used for the aligning torques

$$M_{zi} = (M_{zi})_{,\alpha}\alpha_i = (M_{zi})_{,\alpha}\left(\beta + \frac{x_i}{V}r - \delta_i\right). \tag{5.143}$$

The steering angles of the various axles can be expressed as

$$\delta_i = K_i'\delta, \tag{5.144}$$

where parameters K_i' are usually considered as constants and δ is the control input from the directional control system. In the case of vehicles with only one steering axle (the front axle), all K_i' vanish except $K_1' = 1$; in other cases, they may be functions of many parameters. If also the variables of motion β or r enter the equations defining the various K_i', the model is no longer linear.

The total lateral force to be introduced in the second equation (5.140) can be reduced to the linear expression

$$F_y = Y_\beta\beta + Y_r r + Y_\delta\delta + F_{y_e}, \tag{5.145}$$

where

$$\begin{cases} Y_\beta = -\sum_{\forall i} C_i + \frac{1}{2}\rho V_r^2 S(C_y)_{,\beta}, \\ Y_r = -\frac{1}{V}\sum_{\forall i} x_i C_i, \\ Y_\delta = \sum_{\forall i}(K_i' C_i + F_{xi}). \end{cases} \tag{5.146}$$

In the first equation (5.146) also an aerodynamic term has been added. It is due to the lateral force and contains the derivative $(C_y)_{,\beta}$ with respect to the sideslip angle of the vehicle β of the side force coefficient C_y. This aerodynamic term becomes important only at high speed in an environment in which the atmospheric density is not too low. Usually it is neglected when studying planetary vehicles.

The linearized expression of the total yawing moment about the center of mass to be introduced in the third equation (5.140) is

$$M_z = N_\beta \beta + N_r r + N_\delta \delta + M_{z_e}, \tag{5.147}$$

where

$$\begin{cases} N_\beta = \sum_{\forall i}\left[-x_i C_i + (M_{z_i})_{,\alpha}\right] + \frac{1}{2}\rho V_r^2 S l (C_{M_z})_{,\beta}, \\ N_r = \frac{1}{V}\sum_{\forall i}\left[-x_i^2 C_i + (M_{z_i})_{,\alpha} x_i\right], \\ N_\delta = \sum_{\forall i} K_i'\left[C_i x_i - (M_{z_i})_{,\alpha} + F_{xi} x_i\right]. \end{cases} \tag{5.148}$$

Again an aerodynamic term (the yaw aerodynamic moment) has been included. It is usually not considered in the study of motion of planetary vehicles.

The terms Y_β, Y_r, Y_δ, N_β, N_r and N_δ are the derivatives $\partial F_y/\partial\beta$, $\partial F_y/\partial r$, etc. They are usually referred to as derivatives of stability. N_r is sometimes referred to as yaw damping, as it is a factor that multiplied by an angular velocity yields a moment, like a damping coefficient.

Remark 5.16 If aerodynamic forces are neglected, Y_β, Y_δ, N_β and N_δ are constant while Y_r and N_r are proportional to $1/V$. They are strongly influenced by the load and ground conditions through the cornering stiffness of the tires.

In many cases also self-aligning torques and the longitudinal forces are neglected, and the derivatives of stability depend only on the cornering stiffness of the wheels

$$\begin{cases} Y_\beta = -\sum_{\forall i} C_i, \\ Y_r = -\frac{1}{V}\sum_{\forall i} x_i C_i, \\ Y_\delta = \sum_{\forall i} K_i' C_i, \end{cases} \qquad \begin{cases} N_\beta = -\sum_{\forall i} -x_i C_i = V Y_r, \\ N_r = -\frac{1}{V}\sum_{\forall i} -x_i^2 C_i, \\ N_\delta = \sum_{\forall i} K_i' C_i x_i. \end{cases} \tag{5.149}$$

By adding also an external lateral force and an external yawing moment acting on the vehicle, the final expression of the linearized equations of motion for the handling model is thus

$$\begin{cases} m\dot{v} + mVr = Y_\beta \beta + Y_r r + Y_\delta \delta + F_{y_e}, \\ J_z \dot{r} = N_\beta \beta + N_r r + N_\delta \delta + M_{z_e}. \end{cases} \tag{5.150}$$

They are often written in terms of the sideslip angle β, instead of the lateral velocity v:

$$\begin{cases} mV(\dot{\beta} + r) + m\dot{V}\beta = Y_\beta \beta + Y_r r + Y_\delta \delta + F_{y_e}, \\ J_z \dot{r} = N_\beta \beta + N_r r + N_\delta \delta + M_{z_e}. \end{cases} \tag{5.151}$$

They are two first order differential equations in the two unknown v and r or β and r.

The steering angle δ can be considered as an input to the system, together with the external force and moment F_{y_e} and M_{z_e}. This way of proceeding is usually referred to as *locked controls* behavior. Alternatively it is possible to study the *free controls* behavior, in which the steering angle δ is one of the variables of the motion and a further equation expressing the dynamics of the steering system is added.

The equations of motion can be written in terms of state equations in the form

$$\{\dot{z}\} = [A]\{z\} + [B_c]\{u_c\} + [B_e]\{u_e\}, \tag{5.152}$$

where the state and input (control and external) vectors $\{z\}$ and $\{u\}$ are

$$\{z\} = \begin{Bmatrix} \beta \\ r \end{Bmatrix}, \qquad \{u_c\} = \delta, \qquad \{u_e\} = \begin{Bmatrix} F_{y_e} \\ M_{z_e} \end{Bmatrix},$$

the dynamic matrix is

$$[A] = \begin{bmatrix} \frac{Y_\beta}{mV} - \frac{\dot{V}}{V} & \frac{Y_r}{mV} - 1 \\ \frac{N_\beta}{J_z} & \frac{N_r}{J_z} \end{bmatrix}$$

and the input gain matrices are

$$[B_c] = \begin{bmatrix} \frac{Y_\delta}{mV} \\ \frac{N_\delta}{J_z} \end{bmatrix}, \qquad [B_e] = \begin{bmatrix} \frac{1}{mV} & 0 \\ 0 & \frac{1}{J_z} \end{bmatrix}.$$

The study of the system is straightforward: the eigenvalues of the dynamic matrix allow one to see immediately whether the behavior is stable or not and the study of the solution to given constant inputs yields the steady state response to a steering input or to external forces and moments.

Spring–Mass–Damper Analogy

There is an interesting analogy that can be used. If the speed is kept constant in such a way that the derivatives of stability are constant in time, there is no difficulty in obtaining r from the first equation (5.151) and substituting it into the second, which becomes a second order differential equation in β. Similarly, solving the second in β and substituting it in the first one, an equation in r is obtained. The result is

$$P\ddot{\beta} + Q\dot{\beta} + U\beta = S'\delta + T'\dot{\delta} - N_r F_{y_e} + J_z \dot{F}_{y_e} - (mV - Y_r)M_{z_e} \tag{5.153}$$

or

$$P\ddot{r} + Q\dot{r} + Ur = S''\delta + T''\dot{\delta} + N_\beta F_{y_e} - Y_\beta M_{z_e} + mV\dot{M}_{z_e}, \tag{5.154}$$

where

$$\begin{cases} P = J_z mV, \\ Q = -J_z Y_\beta - mVN_r, \\ U = N_\beta(mV - Y_r) + N_r Y_\beta, \end{cases} \qquad \begin{cases} S' = -N_\delta(mV - Y_r) - N_r Y_\delta, \\ S'' = Y_\delta N_\beta - N_\delta Y_\beta, \\ T' = J_z Y_\delta, \\ T'' = mVN_\delta. \end{cases}$$

If the simplified expressions of the derivatives of stability are used, the expressions for P, Q, etc. reduce to

$$\begin{cases} P = J_z mV, \\ Q = J_z \sum_{\forall i} C_i + m \sum_{\forall i} x_i^2 C_i, \\ U = \dfrac{1}{V}\left[\sum_{\forall i} C_i \sum_{\forall i} x_i^2 C_i - \sum_{\forall i} (x_i C_i)^2\right] - mV \sum_{\forall i} x_i C_i, \end{cases}$$

$$\begin{cases} S' = -mV \sum_{\forall i} K_i' C_i - \dfrac{1}{V}\left[\sum_{\forall i} K'_i x_i C_i \sum_{\forall i} x_i C_i - \sum_{\forall i} K_i' C_i \sum_{\forall i} x_i^2 C_i\right], \\ S'' = \sum_{\forall i} C_i \sum_{\forall i} K_i' x_i C_i - \sum_{\forall i} K_i' C_i \sum_{\forall i} x_i C_i, \\ T' = J_z \sum_{\forall i} K_i' C_i, \\ T'' = mV \sum_{\forall i} K_i' x_i C_i. \end{cases}$$

Moreover, if the vehicle has only two axles with the front one steering, they can be simplified even more

$$\begin{cases} P = J_z m V, \\ Q = J_z(C_1 + C_2) + m\left(a^2 C_1 + b^2 C_2\right), \\ U = mV(-aC_1 + bC_2) + C_1 C_2 \dfrac{l^2}{V}, \end{cases} \qquad \begin{cases} S' = C_1\left(-amV + C_2 \dfrac{bl}{V}\right), \\ S'' = l C_1 C_2, \\ T' = J_z C_1, \\ T'' = mVaC_1. \end{cases}$$

Each one of equations (5.153) and (5.154) is sufficient for the study of the dynamic behavior of the vehicle. They are formally identical to the equation of motion of a spring–mass–damper system.

Remark 5.17 The linearized behavior of a rigid vehicle at constant speed is identical to that of a mass P suspended to a spring with stiffness U and a damper with damping Q, excited by the different forcing functions stated above.

The analogy here suggested is only a formal one: the state variables β and r are dimensionally an angular velocity (r) or are related to velocities (β has been introduced to express the lateral velocity v) and not displacements and thus P, Q and U are dimensionally far from being a mass, a damping coefficient and a stiffness.

Steady-State Response to a Steering Input

In steady state driving the radius of the trajectory is constant, i.e. the path is circular. The relationship linking r to the radius R of the trajectory is thus

$$r = \frac{V}{R}. \tag{5.155}$$

To compute the steady state response is the same as computing the equilibrium position of the equivalent mass–spring–damper system under the effect of a constant force $S'\delta$ or $S''\delta$ since in steady state motion $\dot{\delta} = 0$ and we have

$$\begin{cases} \beta = \dfrac{S'}{U}\delta = \dfrac{-N_\delta(mV - Y_r) - N_r Y_\delta}{N_\beta(mV - Y_r) + N_r Y_\beta}\delta, \\ r = \dfrac{S''}{U}\delta = \dfrac{Y_\delta N_\beta - N_\delta Y_\beta}{N_\beta(mV - Y_r) + N_r Y_\beta}\delta. \end{cases} \tag{5.156}$$

The transfer functions of the vehicle are thus the *trajectory curvature gain*

$$\frac{1}{R\delta} = \frac{Y_\delta N_\beta - N_\delta Y_\beta}{V[N_\beta(mV - Y_r) + N_r Y_\beta]}, \tag{5.157}$$

expressing the ratio between the curvature of the trajectory and the steering input, the *lateral acceleration gain*

$$\frac{V}{R\delta} = \frac{V^2[Y_\delta N_\beta - N_\delta Y_\beta]}{N_\beta(mV - Y_r) + N_r Y_\beta}, \tag{5.158}$$

expressing the ratio between the centrifugal acceleration and the steering input, the *sideslip angle gain*

$$\frac{\beta}{\delta} = \frac{-N_\delta(mV - Y_r) - N_r Y_\delta}{N_\beta(mV - Y_r) + N_r Y_\beta}, \tag{5.159}$$

expressing the ratio between the sideslip angle and the steering angle and the *yaw velocity gain*

$$\frac{r}{\delta} = \frac{Y_\delta N_\beta - N_\delta Y_\beta}{N_\beta(mV - Y_r) + N_r Y_\beta}, \tag{5.160}$$

expressing the ratio between the yaw velocity and the steering angle.

Assume that

- the vehicle has only two axles and only the front wheels steer,
- a simplified expression of the derivatives of stability with only the lateral forces of the tires accounted for is used,
- no dependence of the cornering stiffness of the tires with the speed due to longitudinal load shift is considered.

In this case it is possible to define a constant *stability factor* K or an *understeer gradient* K^*:

$$K = \frac{m}{l^2}\left(\frac{b}{C_1} - \frac{a}{C_2}\right), \qquad K^* = \frac{mg}{l}\left(\frac{b}{C_1} - \frac{a}{C_2}\right). \tag{5.161}$$

The expressions of the above defined gains reduce to

- *trajectory curvature gain*

$$\frac{1}{R\delta} = \frac{1}{l}\frac{1}{1 + KV^2}, \tag{5.162}$$

- *lateral acceleration gain*

$$\frac{V^2}{R\delta} = \frac{V^2}{l}\frac{1}{1 + KV^2}, \tag{5.163}$$

- *sideslip angle gain*

$$\frac{\beta}{\delta} = \frac{b}{l}\left(1 - \frac{maV^2}{blC_2}\right)\frac{1}{1 + KV^2}, \tag{5.164}$$

- *yaw velocity gain*

$$\frac{r}{\delta} = \frac{V}{l}\frac{1}{1 + KV^2}. \tag{5.165}$$

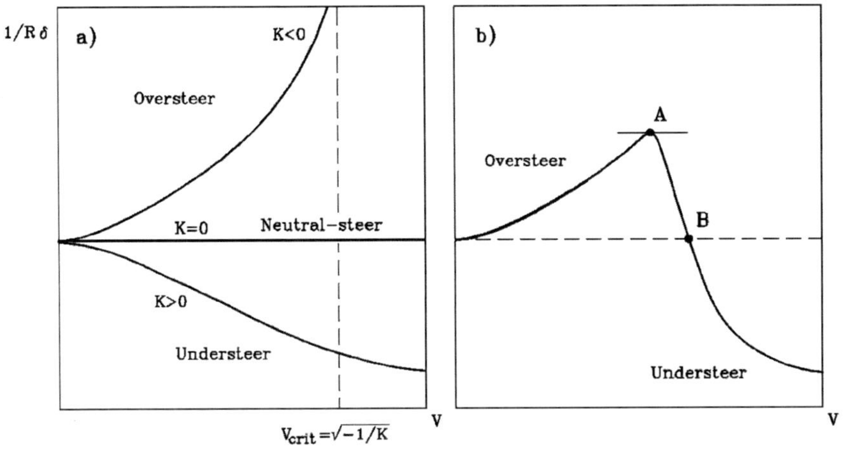

Fig. 5.21 Steady state response to a steering input (from G. Genta, L. Morello, *The Automotive Chassis*, Springer, New York, 2009)

Under the mentioned conditions, the kinematic trajectory curvature gain is

$$\left(\frac{1}{R\delta}\right)_c = \frac{1}{l}.$$

The expression $1 + KV^2$ can thus be considered as a correction factor giving the response of the vehicle in dynamic conditions from that in kinematic conditions. This is true for all the mentioned gains except for the sideslip angle gain which depends on the speed even if factor

$$1 + KV^2 = 1,$$

i.e., if $K = 0$. If (5.161) holds, i.e. if only the cornering forces of the tires are considered, K is a constant.

If $K = 0$ the value of $1/R\delta$ is constant and equal to the value characterizing kinematic steering, i.e. the response of the vehicle to a steering input is, at any speed, equal to that in kinematic conditions. This, however, does not mean that the vehicle is in kinematic conditions, since the value of the sideslip angle is not equal to its kinematic value and those of the sideslip angles of the wheels are not equal to zero. A vehicle behaving in this way is said to be *neutral-steer* (Fig. 5.21a).

If $K > 0$ the value of $1/R\delta$ decreases with increasing speed. The response of the vehicle is thus smaller than that in kinematic conditions and, at equal radius of the trajectory, the steering angle increases at increasing speed. A vehicle which behaves in this way is said to be *understeer*. A quantitative measure of the understeering of a vehicle is given by the *characteristic speed*, defined as the speed at which the steering angle needed to negotiate a turn is equal to twice the Ackermann angle, i.e.

the trajectory curvature gain is equal to

$$\frac{1}{2l}.$$

Using the simplified approach outlined above, the characteristic speed is

$$V_{\text{char}} = \sqrt{\frac{1}{K}}. \tag{5.166}$$

If $K < 0$ the value of $1/R\delta$ increases with increasing speed until, for a speed

$$V_{\text{crit}} = \sqrt{-\frac{1}{K}} \tag{5.167}$$

the response tends to infinity, i.e., the system develops an unstable behavior. A vehicle behaving in this way is said to be *oversteer* and the speed given by (5.167) is said *critical speed*. The critical speed of any oversteer vehicle must be well above the maximum speed it can reach, at least in normal road conditions.

The value of β, or better, of β/δ, decreases with the speed from the kinematic value up to the speed

$$(V)_{\beta=0} = \sqrt{\frac{blC_2}{am}} \tag{5.168}$$

at which it vanishes. At higher speed it becomes negative, tending to $-\infty$ when approaching the critical speed for oversteer vehicles and tending to

$$\frac{aC_1}{aC_1 - bC_2}$$

when the speed tends to infinity in the case of understeer vehicles.

The sideslip angles of the front and rear wheels are equal in the case of neutral-steer vehicles. In the case of oversteer vehicles the rear wheels have a greater sideslip angle (in absolute value, as the sideslip angles are negative when the radius of the trajectory is positive), while the opposite holds in the case of understeer vehicles. It follows that oversteer vehicles can be expected to reach the limit conditions at the rear wheels and understeer vehicles at the front wheels, even if the present model cannot be applied in conditions approaching any limit.

The above mentioned considerations hold only in the case of vehicles with two axles and in which all effects causing a dependence of the derivatives of stability from the speed (for Y_r and N_r a dependence different from that from $1/V$) are neglected. If this last assumption is dropped, the stability factor K is not constant and the vehicle can have a different behavior at different speeds.

A strong effect is due to the aerodynamic yawing moment. If

$$(C_{M_z})_{,\beta} = \frac{\partial C_{M_z}}{\partial \beta}$$

is negative (the side force F_y acts forward of the center of mass), the effect is to increase oversteer or to decrease understeer, at increasing speed. This effect, which increases with the absolute value of $(C_{M_z})_{,\beta}$, is destabilizing and causes a decrease of the critical speed, if it exists. The opposite occurs if $(C_{M_z})_{,\beta}$ is positive.

Another important effect is due to the longitudinal load shift. If the load on the rear axle increases more, or decreases less, than that on the front axle, understeer increases with increasing speed.

The case of a vehicle that is oversteer at low speed and understeer at high speed, as it can be caused by a positive value of $(C_{M_z})_{,\beta}$, is shown in Fig. 5.21b. Following the definition seen above, the speed at which neutral-steer is obtained is that of point B.

If the simplified expressions for the derivatives of stability are not used, a new definition of neutral-steer, and hence under- and over-steer, can be introduced. Instead of referring to the condition

$$\frac{1}{R\delta} = \frac{1}{l},$$

neutral-steer can be defined by the relationship

$$\frac{d}{dV}\left(\frac{1}{R\delta}\right) = 0. \tag{5.169}$$

It is obvious that in case the derivatives of stability are constant (Y_r and N_r are proportional to $1/V$) the first definition, which can be said to be *absolute*, and the second, which can be said to be *incremental*, coincide.

On the plot of Fig. 5.21b the speed at which neutral-steering (following the incremental definition) is obtained is point A, where the curve reaches its maximum. The incremental definition follows more closely the feeling of the driver, who feels the vehicle as oversteer if an increase of speed is accompanied by a decrease of radius of the trajectory and vice versa. The driver has clearly no reference to feel the kinematic value of the radius of the trajectory and hence the absolute definition has little meaning for him.

From the viewpoint of the equations of motion, on the contrary, the absolute definition is more significant.

Neutral-Steer Point and Static Margin

The neutral-steer point of the vehicle is usually defined as the point laying on the plane of symmetry in which is applied the resultant of the cornering forces due to the tires as a consequence of a sideslip angle β, obviously with $\delta = 0$ and $r = 0$. The cornering forces, computed through the linearized model, in these conditions are simply $-C_i\beta$ and the x coordinate of the neutral point is

$$x_N = \frac{\sum_{\forall i} x_i C_i}{\sum_{\forall i} C_i}. \tag{5.170}$$

A better definition of neutral-steer point can, however, be introduced. If all forces and moments due to a sideslip angle β, with $\delta = 0$ and $r = 0$ are considered, the resultant force and moment are simply $Y_\beta \beta$ and $N_\beta \beta$ respectively.[13] The x coordinate of the neutral-steer point, defined as the point of applications of the resultant of all lateral forces, is thus

$$x_N = \frac{N_\beta}{Y_\beta}. \tag{5.171}$$

The static margin $\mathcal{M}_s$ is the ratio between the x coordinate of the neutral-steer point and the wheelbase

$$\mathcal{M}_s = \frac{x_N}{l}. \tag{5.172}$$

As will be seen when dealing with the response to external forces and moments, if an external force is applied to the neutral-steer point it does not cause any steady-state yaw velocity. Owing to the mathematical model used in the present chapter, the position in height of the neutral-steer point cannot be defined.

Remark 5.18 The condition to obtain a neutral-steer response is that the neutral-steer point coincides with the center of mass, i.e. $x_N = 0$, $\mathcal{M}_s = 0$, $N_\beta = 0$. If they are positive the vehicle is oversteer[14] (center of gravity behind the neutral point); the opposite applies to understeer vehicles.

The signs of parameters K, K^*, $\mathcal{M}_s$, x_N, $|\alpha_1| - |\alpha_2|$ and N_β corresponding to an oversteer, understeer or neutral-steer behavior are

Behavior	K	K^*	$\mathcal{M}_s$	x_N	$\lvert\alpha_1\rvert - \lvert\alpha_2\rvert$	N_β
Understeer	>0	>0	<0	<0	>0	>0
Neutral-steer	0	0	0	0	0	0
Oversteer	<0	<0	>0	>0	<0	<0

Steady-State Response to External Forces and Moments

From the equivalent mass–spring–damper model the steady state response to an external force F_{y_e} or an external moment M_{z_e} is immediately obtained. The relevant gains are

[13] Y_β can be considered as a sort of cornering stiffness of the vehicle.

[14] Sometimes the position of the neutral-steer point and the static margin are defined with different sign conventions: instead of referring to the position of the neutral-steer point with respect to the center of mass, the position of the latter with respect to the former is given. In this case the signs of x_N and $\mathcal{M}_s$ are changed and an understeer vehicle has a positive static margin.

$$\begin{cases} \dfrac{1}{RF_{y_e}} = \dfrac{N_\beta}{VU}, \\ \dfrac{V^2}{RF_{y_e}} = \dfrac{VN_\beta}{U}, \\ \dfrac{\beta}{F_{y_e}} = \dfrac{-N_r}{U}, \end{cases} \qquad \begin{cases} \dfrac{1}{RM_{z_e}} = \dfrac{-Y_\beta}{VU}, \\ \dfrac{V^2}{RM_{z_e}} = \dfrac{-VY_\beta}{U}, \\ \dfrac{\beta}{M_{z_e}} = \dfrac{-mV + Y_r}{U}. \end{cases} \tag{5.173}$$

If the vehicle is neutral-steer, $N_\beta = 0$ and consequently

$$\frac{1}{RF_{y_e}} = 0.$$

The trajectory remains straight under the effect of an external force. The trajectory is, however, changed from the one preceding the application of force F_{y_e}: the deviation is equal to angle β, i.e. to $-F_{y_e}/Y_\beta$. The lateral velocity of the vehicle is simply

$$v = V\beta = -\frac{VF_{y_e}}{Y_\beta}.$$

Y_β must be as large as possible, in particular in the case of fast vehicles, in order to avoid large lateral velocities. The fact that the trajectory remains straight can be easily understood considering that the neutral-steer point lies in the center of mass, i.e. in the point of application of the external force.

Remark 5.19 This condition can be used to define the neutral-steer point as the point in which the application of an external force does not cause a yaw rotation of the vehicle. If the presence of the suspension is accounted for, instead of a neutral-steer point it is possible to define a neutral steer line as the locus of the points in xz plane in which an external force applied in y direction does not cause any yaw rotation.

The most common case of a side force applied to the center of mass is the component in the ground plane of the weight of the vehicle when traveling on laterally sloping ground.

Considerations on the Linearized Model

Linearized models are often used in the automotive field, since the dynamic study is performed with reference to high speed driving, usually involving motion on roads with large curvature radii, resulting in small steering and sideslip angles and low yaw velocities. Also the sideslip angles of the wheels are usually small, if the conditions are far from the limit conditions.

In case of planetary rovers often the speeds are low and the radius of the trajectory is small; as a result the applicability of the linearized model may be questionable. In low gravity conditions it is possible that even at low speed the limit conditions for lateral forces are approached, limiting the possibility of considering as small angles the sideslip angles of the wheels.

In the model here studied (even before linearization) the vehicle is assumed to be a rigid body. As a consequence, if the number of wheels is larger than 3, the forces it exerts on the road in normal direction are undetermined. However, this limitation it can be circumvented by taking into account the presence of the suspensions in the computation of the forces while neglecting them into the dynamic study of the vehicle.

The forces acting on the various axles in symmetrical conditions have already been obtained. When the motion occurs on a curved trajectory, the forces acting on each wheel cannot be simply obtained as half (or one quarter, for axles with four wheels) of the total force on the axle.

The forces F_{z_l} and F_{z_r} acting on the left and right wheels of the ith axle can be expressed as

$$\begin{cases} F_{z_{i_l}} = \dfrac{F_{z_i}}{2} + \Delta F_{z_i}, \\ F_{z_{i_r}} = \dfrac{F_{z_i}}{2} - \Delta F_{z_i}, \end{cases} \tag{5.174}$$

where F_{z_i} and ΔF_{z_i} are respectively the total load and the transversal load shift (or load transfer) of the relevant axle.

If the cornering stiffness depends on the load F_z in a linear way, the increase of cornering stiffness of the more loaded wheel would exactly compensate for the decrease of cornering stiffness of the other one, and transversal load shift would have no effect whatsoever. Since this is not exactly the case, the load transfer causes a decrease of cornering stiffness of each axle that becomes more relevant with increasing lateral acceleration. In vehicle dynamics it is a common practice to neglect transversal load transfer if the lateral acceleration is not high, usually is lower than $0.5\ g$.[15]

With reference to the dynamics of planetary rovers and vehicles, this condition may be critical too, since the gravitational acceleration can be small.

Once the second and third equation (5.140) have been solved, the first equation (5.140) can be used to study the longitudinal behavior. Since the velocity V has been assumed to be a known quantity, the first equation can be solved in the longitudinal forces.

This uncoupling is only approximated, because the longitudinal forces are included in the expressions of the derivatives of stability. This dependence is, however, weak, and can be neglected, at least in a first approximation study of the lateral behavior.

Influence of Longitudinal Forces and Load Transfer

As stated above, the directional behavior can be strongly influenced by the presence of longitudinal forces between the wheels and the ground. Any longitudinal force

---

[15] L. Segel, *Theoretical Prediction and Experimental Substantiation of the Response of the Automobile to Steering Control*, Cornell Aer. Lab., Buffalo, N.Y.

causes a reduction of the cornering stiffness: in a two axles vehicle, if it is applied to the front axle it reduces the value of C_1 and consequently makes the vehicle more understeer or less oversteer. Opposite effect is due to a longitudinal force applied to the rear axle.

In the linearized model, this can be easily accounted for by using the elliptical approximation (4.131) which, if a complete linearization of the behavior of the tires is assumed, can be applied directly to each axle

$$C_i = C_{0_i}\sqrt{1-\left(\frac{F_{x_i}}{\mu_p F_{z_i}}\right)^2}.$$

Note that the forces and the cornering stiffness are referred to the whole axle.

The driving force needed to maintain a constant speed increases with the latter and, as a consequence, the cornering stiffness of the tires of the driving axles decreases. This effect is felt particularly if the conditions of the ground are poor or the gravity is low, since the ratio between the actual and the maximum value of the driving force is present.

In the case of rear wheel drive vehicles, the driving forces increase the oversteer behavior or decrease the understeer one. The critical speed, if it exists, decreases or a critical speed may appear. In bad road conditions a rear wheel drive vehicle may have a low critical speed and the driver may be required to limit the speed for stability reasons, to avoid spinout. Starting and accelerating the vehicle may be difficult and great care must be exerted when operating the accelerator control; antispin devices are useful in these conditions.

Front wheel drive vehicles on the contrary become tendentially understeer and more stable with increasing speed or decreasing μ_p and an increasingly large steering angle is needed to maintain the vehicle on a given trajectory. The limit condition is that of an infinitely stable vehicle, i.e. a vehicle that can only move on a straight line.

In the case the vehicle has more than one driving axle and in the case of braking, the effect on handling depends on how the longitudinal forces are shared between the axles. If the front axle is working with a larger longitudinal force coefficient μ_x than the rear axle, which does not necessarily imply that force F_x is larger but that the ratio F_x/F_z of the front wheels is larger than that of the rear wheels, the vehicle becomes more understeer and is in a sense more stable. When the limit conditions are reached and the front wheels slip (lock in braking or spin in traction) the vehicle cannot be steered and follows a straight trajectory. A larger ratio F_x/F_z at the rear wheels makes the vehicle more oversteer and easily introduces a critical speed.

No allowance has yet been taken for the transversal load shift. If the dependence of the cornering stiffness of a single wheel from the load is of the type shown in Fig. 5.22, this does not introduce errors if the load transfer ΔF_z is small, lower than $(\Delta F_z)_{\lim}$ in the figure (condition a).

If the load transfer is larger, as in the case of ΔF_{zb}, the increase of the cornering stiffness of the more loaded wheel cannot compensate for the decrease of the cornering stiffness of the other one and the cornering stiffness of the axle is reduced. This effect introduces a nonlinearity in the behavior of the vehicle.

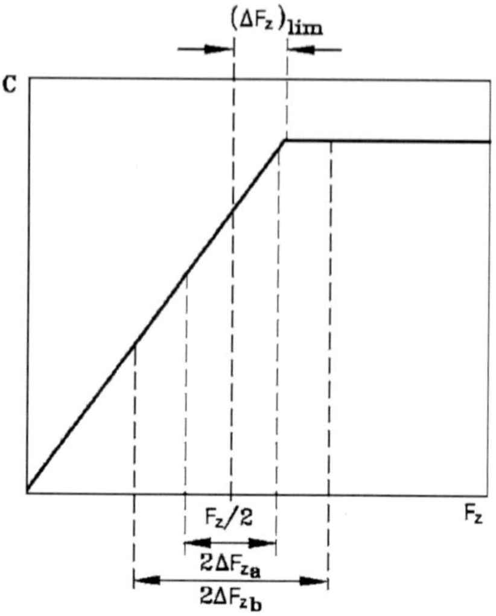

Fig. 5.22 Effect of load transfer on the cornering stiffness (from G. Genta, L. Morello, *The Automotive Chassis*, Springer, New York, 2009)

The simultaneous presence of longitudinal forces and load transfer makes things more complicated. Even if the cornering stiffness is still in the linear part of the plot of Fig. 5.22, i.e. the load transfer is smaller than $(\Delta F_z)_{\text{lim}}$, the combined effect yields a nonlinear behavior. Assuming that the longitudinal force splits equally on its two wheels, the cornering stiffness of the axle, computed using the elliptical approximation, is

$$C = \frac{1}{2}\left(C_0 + \Delta F_z \frac{\partial C}{\partial F_z}\right)\sqrt{1 - \left[\frac{F_x}{\mu_p(F_z + 2\Delta F_z)}\right]^2} + \frac{1}{2}\left(C_0 - \Delta F_z \frac{\partial C}{\partial F_z}\right)\sqrt{1 - \left[\frac{F_x}{\mu_p(F_z - 2\Delta F_z)}\right]^2}, \qquad (5.175)$$

where forces F_x and F_z are referred to the whole axle.

Owing to the presence of the square root, the decrease of the cornering stiffness of the less loaded wheel is greater, particularly if μ_x is low, than the increase at the other wheel.

Load transfer on the driving axle increases the effect of longitudinal forces; this combined action can be reduced by introducing an anti-roll bar on the other axle. Operating in this way, the increased load transfer on the non-driving axle reduces also its cornering stiffness, reducing the overall effect of longitudinal forces on handling.

Open-Loop Stability

A dynamically guided vehicle must be guided by a control system, be it a human driver onboard the vehicle, a human driver teleoperating the vehicle or an electromechanical control device. What actually matters for correct operation is the closed-loop stability of the controlled vehicle.

However, it is customary to verify the open loop stability of the vehicle alone, since if the vehicle is stable in this way the control system must perform the task of choosing the trajectory and controlling the vehicle so that it follows it (high level control) but not that of ensuring that the attitude of the vehicle on the trajectory is the correct one (low level control).

Vehicles are intrinsically nonlinear systems, and the stability that can be usually reached is just a "stability in the small", i.e. for small variations of the values of the state variables about each equilibrium point in the state space. The linearized model studied here can be considered as a linearization well suited for this study in the small.

The above mentioned definition of stability refers to the state of the system and in the case of the handling model with two degrees of freedom the state variables are β and r (or v and r). A vehicle is thus stable if, when in motion with given values β_0 and r_0 of β and r, after a small external perturbation, it follows that

$$\beta(t) \to \beta_0, \qquad r(t) \to r_0.$$

No reference is made to the trajectory: after a perturbation the vehicle cannot return to the previous trajectory and a correction by the driver or by an automatic control system is required in order to maintain the vehicle on the required path.

If the steering angle δ is kept at a given value or follows a determined law, the motion is said to occur in locked control condition. Only locked control stability will be studied here.

Stability can be studied by using the homogeneous equation of motion

$$\{\dot{z}\} = [A]\{z\}.$$

The eigenvalues of the dynamic matrix $[A]$ are readily found and the stability is assessed from the sign of their real part, which must be negative. If the imaginary part is nonzero the behavior is oscillatory; which does not necessarily imply that the trajectory is oscillatory but only that the time histories $\beta(t)$ and $r(t)$ are such.

The analogy with the spring–mass–damper system allows a simpler approach to the study of the stability at constant speed:

- To ensure static stability the stiffness U must be positive;
- To ensure dynamic stability the damping coefficient Q must be positive;
- To allow an oscillatory free behavior Q must be lower than the critical damping $2\sqrt{PU}$.

By inspecting the relevant mathematical expressions, it is easy to verify that U is always positive for understeer and neutral-steer vehicles, and in the latter case it tends to zero when the speed tends to infinity. In the case of oversteer vehicles it is positive up to the critical speed, where it vanishes to become negative at higher speed. The critical speed is thus the threshold of instability for oversteer vehicles. Similar results hold also if the complete expressions for the derivatives of stability are used.

It is also easy to verify that Q is always positive: if the vehicle is statically stable it is also dynamically stable. As a first approximation, neutral-steer vehicles are critically damped, while understeer and oversteer vehicles are respectively underdamped and overdamped: the free behavior of the former can thus be expected to be oscillatory. It must, however, be noted that the issue whether a given vehicle has an oscillatory behavior or not cannot be satisfactorily solved using the present rigid body model as the presence of rolling motions, which are neglected here and are almost always underdamped and then oscillatory, can induce an oscillatory behavior also for what β and r are concerned. This is particularly true for vehicles whose suspensions exhibit roll steer.

Nonstationary Motion

There is no difficulty in integrating numerically the equation of motion (5.152) or even the complete nonlinear equation (5.134) once laws $\delta(t)$, $F_{y_e}(t)$, $M_{z_e}(t)$ and $V(t)$ are stated. Once the law $r(t)$ has been obtained, it is possible to integrate it to yield the yaw angle

$$\psi(t) = \int_0^t r(u)\,du. \tag{5.176}$$

The trajectory can then be obtained directly in the inertial coordinates X, Y. The velocities $\dot{X}$ and $\dot{Y}$ can be expressed in terms of angles β and ψ

$$\begin{Bmatrix} \dot{X} \\ \dot{Y} \end{Bmatrix} = V \begin{bmatrix} \cos(\psi) & -\sin(\psi) \\ \sin(\psi) & \cos(\psi) \end{bmatrix} \begin{Bmatrix} \cos(\beta) \\ \sin(\beta) \end{Bmatrix}. \tag{5.177}$$

By integrating equations (5.177) the trajectory is readily obtained

$$\begin{cases} X = \displaystyle\int_0^t V\big[\cos(\beta)\cos(\psi) - \sin(\beta)\sin(\psi)\big]\,du, \\ Y = \displaystyle\int_0^t V\big[\cos(\beta)\sin(\psi) + \sin(\beta)\cos(\psi)\big]\,du. \end{cases} \tag{5.178}$$

Remark 5.20 The integration must be performed numerically even in the case of the linearized model since angle ψ may be too large to linearize its trigonometric functions.

Vehicles with a Number of Steering Axles

In the majority of the vehicles with two axles only the front wheels are provided with a steering system. However, steering can be performed on several, or even all, wheels. The main aim has been an increase of manoeuvrability and in general of the handling characteristics both in low-speed and high-speed steering.

With reference to two axle vehicles, to reduce the radius of the trajectory in low-speed (kinematic) conditions, the rear axle must steer in opposite direction from the front one; if the absolute values of the steering angles are equal, the radius is halved and the off-tracking of the rear axle is reduced to zero. Using the notation introduced in the preceding sections, this situation is characterized by

$$K_1' = 1, \qquad K_2' = -1$$

(in the following it will always be assumed that $K_1' = 1$). Practically this value is too high as, if starting the motion with the wheels in a steered position, the rear axle would initially be displaced too much outwards the line connecting the centers of the wheels in the initial position.

The Lunar Roving Vehicle had a 4WS system that allowed only opposite steering, with a value of K_2' close to -1. This is a clear indication that the designers had in mind kinematic steering, even if the vehicle in actual use worked with large sideslip angles, at least to judge from the movies taken on the Moon.

Assuming that $K_1' = 1$ and K_2' is constant, in kinematic steering the trajectory curvature gain and the off-tracking distance are

$$\frac{1}{R\delta} \approx \frac{1 - K_2'}{l}, \qquad R_f - R_1 \approx \frac{l^2(1 - K_2')}{2R(1 + K_2')}. \tag{5.179}$$

In high-speed cornering the situation is different: the possibility of putting all wheels with a sideslip angle without waiting for a rotation of the whole vehicle makes the response to a steering input far quicker. In this case rear wheels must exert cornering forces with the same direction as front ones and consequently the steering angles must be in the same direction. The limiting case, for a vehicle with neutral-steer point at the center of the wheelbase, will be that of equal steering angles $K_1' = K_2' = 1$. This is again an unpractical result, as the vehicle would react very quickly in a lane change, simply moving sideways, but would never be able to negotiate a road bend: Instead of turning it would accelerate laterally.

The steering mechanism must adapt the value of K_2' to the external conditions and the requests of the driver. The simplest strategy is that of using a device, possibly mechanical, linking the steering boxes of the axles with a variable gear ratio: when angle δ is small, as typically occurs in high speed driving, K_2' is positive and the steering angles have the same directions while when δ is large, as occurs when manoeuvring at low speed, K_2' is negative. However, to exploit fully the potential advantages of multiple steering, more complicated control laws must be implemented. The parameters which can enter such law are plenty, e.g. the speed V, the lateral acceleration, the sideslip angles α_i, etc.

From the viewpoint of the mathematical modeling, the situation is, at least in principle, simple. There is no difficulty in introducing a suitable function $K_2'(V, \delta, \dots)$ into the equations (actually it would appear only in the derivatives of stability Y_δ and N_δ) and to modify accordingly the equations of the rigid-body model seen above. If function K_2' includes some of the state variables, the modifications can be larger but no conceptual difficulty arises.

Remark 5.21 Except for the latter case, the locked control stability is not affected by the introduction of 4WS, while the stability with free controls can be affected by it.

Generally speaking, the advantages are mainly linked with a quicker response of the vehicle to a steering input, but this cannot be true for all types of manoeuvres: steering all axles in the same direction can make the vehicle faster in lane change manoeuvres but slower in acquiring a given yaw velocity. The feeling of the driver can be strange and, at least at the beginning, unpleasant. A solution may be initially steering the rear wheels in opposite direction for a very short time, to initiate a yaw rotation, and then in the same direction as the front wheels, to generate cornering forces. This requires a more complicated control logic, possibly based on microprocessors.

Actually, in dynamic steering the trajectory is controlled by forces perpendicular to the trajectory and then, if the sideslip angle of the vehicle β is small, almost perpendicular to the xz plane. However, dynamic driving occurs in two phases: initially the driver, operating some control device, exerts forces (mainly moments) changing the attitude of the vehicle and then, owing to this change of attitude, the required lateral forces are produced. In both two-wheel and multi-wheel steering the forces which change the attitude of the vehicle are lateral forces due to the steering of some wheels, but also longitudinal forces can generate the yaw moment used to modify the attitude of the vehicle.

The inputs to the increasingly common vehicle dynamics control (VDC) system are some state variables of the vehicle, like the yaw velocity r and the sideslip angle β. A reasonable control philosophy would be to keep the former at the value required by the driver by stating a steering wheel angle (yaw-rate control); as a consequence a third variable which must be acquired is the steering wheel angle i.e., neglecting the steering gear ratio, angle δ. The quantities r, β and δ are the parameters entering the rigid-body handling models described in the previous sections, so it is fairly easy to introduce VDC in the models seen above.

While the yaw velocity can be easily acquired using a rate gyro, the sideslip angle is difficult to measure. Instead, the lateral acceleration $\dot{v}$ can be easily measured and is even more directly linked with the reaching of critical conditions for handling: while in steady state cornering the lateral acceleration and the yaw velocity are linked by the relationship $\dot{v} = Vr$ (confusing the component of the acceleration perpendicular to the trajectory with that along the y-axis of the vehicle), in a spinout the yaw rate grows without a matching increase of the lateral acceleration.

The use of the longitudinal (braking or driving) forces for implementing VDC systems is particularly interesting both for the quickness of the control action and for the possibility of integrating the relevant hardware with that already present for anti-lock and anti-spin systems. The main cornering control is still performed by steering some of the wheels and is operated directly by the driver, while quick corrections can be performed by the VDC system using differential application of longitudinal forces.

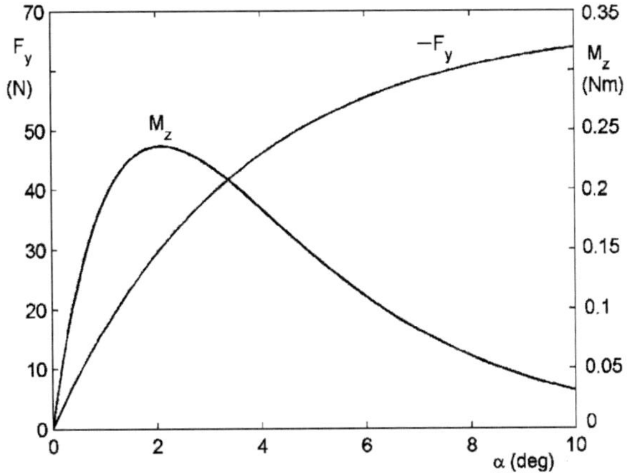

Fig. 5.23 Cornering forces and aligning torques as functions of the sideslip angle

Example 5.5 Consider the six-wheels Mars exploration rover already studied in Example 5.3.

Assume that the suspensions distribute equally the vertical forces on the ground on all wheels, when there is no load shift due to inertia forces.

The data are: mass $m = 180$ kg, gravitational. acceleration $g = 3.77\ \text{m/s}^2$, wheelbase $l = 1.4$ m, position of axles $x = [0.6\ -0.1\ -0.8]$ m (the center of mass is located slightly forward the central wheels), track (all axles) $t = 1.9$ m.

Simple expressions are assumed for the lateral characteristics of the wheels: (4.159) for the side force

$$F_y(\alpha) = -\text{sgn}(\alpha)\mu_{y_p} F_z\left[1 - e^{-\frac{C}{\mu_{yp} F_z}|\alpha|}\right],$$

and (4.161) for the aligning torque

$$M_z(\alpha) = \text{sgn}(\alpha_1) M_{z0} e^{-C_1|\alpha_1|}\left[1 - e^{-\frac{C}{\mu_{yp} F_z}|\alpha_1|}\right].$$

The following values of the parameters are assumed: $\mu_{y_p} = 0.6$, $C = 1{,}100$ N/rad, $M_{z0} = 1.1$ Nm and $C_1 = 20$. The forces and moments are reported as functions of the sideslip angle in Fig. 5.23.

Since the center of mass does not lie on the central axis, the radius of the trajectory of the former is

$$R = \sqrt{R_2^2 + \left(\frac{a-b}{2}\right)^2}.$$

The average steering angles of the first and last axles are

$$\delta_1 = -\delta_2 = \text{atan}\left[\frac{l}{\sqrt{4R^2 - (a-b)^2}}\right].$$

A linearized study is straightforward. The results in terms of the trajectory curvature gain as a function of the speed is reported in Fig. 5.24a. Both the dynamic

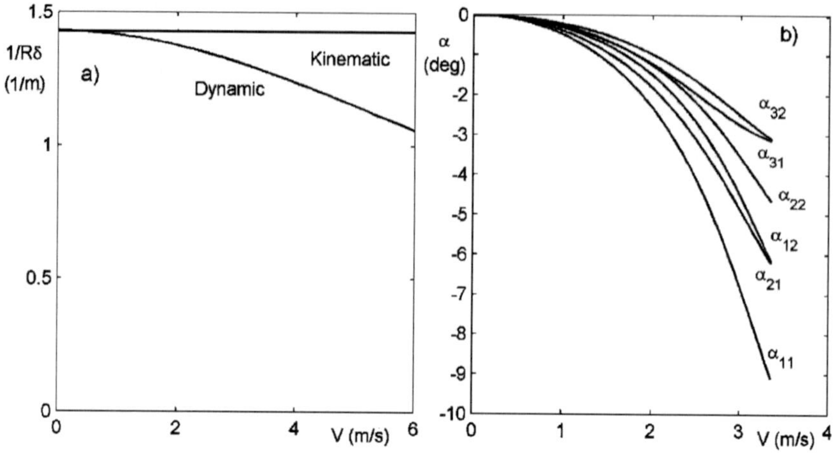

Fig. 5.24 (**a**) Linearized solution: trajectory curvature gain as a function of the speed. (**b**) Nonlinear solution: sideslip angles of the wheels on a trajectory with a radius of 5 m as functions of the speed

and the kinematic values are reported: the vehicle is clearly understeer, as could be expected from the position of the center of mass. The dynamic effects are already felt at speeds as low as 2 m/s.

The sideslip angles of the wheels when the rover travels on a trajectory with a curvature radius of 5 m are reported as functions of the speed in Fig. 5.24b. Owing to the low value of the radius, the nonlinear steady-state solution has been used. The sideslip angles are quite large even when the speed is low (a sideslip angle of 2° is already considered as large): this is due to a combined effect of the low trajectory radius and low gravity.

The difference between kinematic and dynamic steering in a speed range up to 6 m/s is shown in Fig. 5.25. The curves were computed for speeds giving way to a centrifugal acceleration lower than the ideal maximum value $\mu_{yp}g$.

From the radius of curvature of the trajectory it is clear that the rover is understeer also taking into account the nonlinear terms: the dynamic radius increases with the speed. The lateral velocity v and the yaw angular velocity r increase linearly with the speed in case of kinematic steering, while varying in a nonlinear way for dynamic steering.

The sideslip angle β and the curvature radius of the trajectory do not depend on the speed in kinematic steering, while being a function of the latter in dynamic conditions. The sideslip angle, which in the present case is positive in kinematic conditions, decreases with the speed, to become negative at higher speed. When the speed tends to 0 the conditions tend to the kinematic conditions.

Example 5.6 Consider a light human-carrying unpressurized rover similar to the LRV (Lunar Roving Vehicle) of the last three *Apollo* Missions.

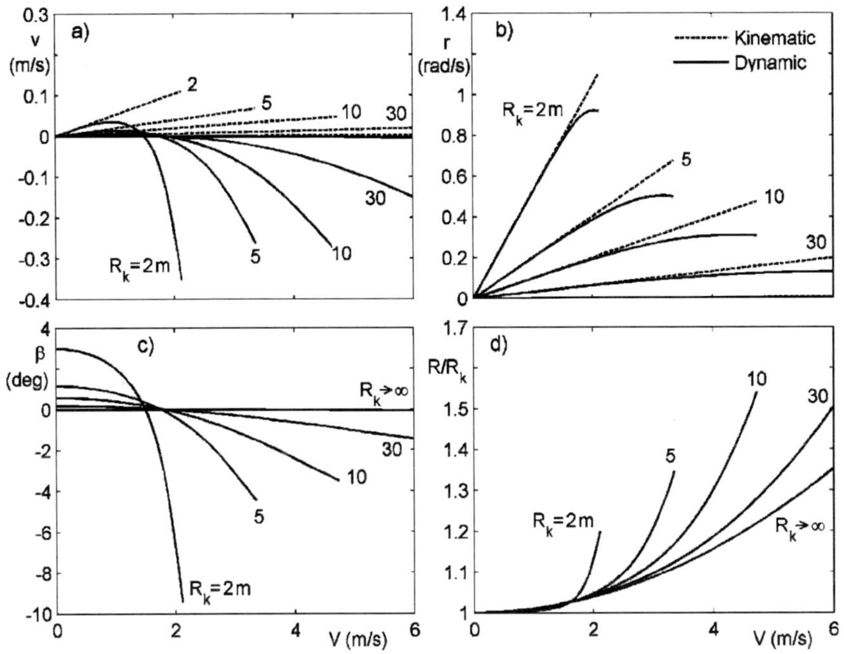

Fig. 5.25 (**a**) and (**b**) Lateral velocity v and yaw velocity r as functions of the speed for different values of the kinematic curvature radius of the trajectory. (**c**) and (**d**) Sideslip angle β and ratio between the dynamic and kinematic radius of the trajectory as functions of the speed

Assume that the suspensions distribute equally the vertical forces on the ground on all wheels, when there is no load shift due to inertia forces.

The data are: mass (fully loaded) $m = 660$ kg, gravitational acceleration $g = 1.62\ \text{m/s}^2$, wheelbase $l = 2.3$ m, position of axles $x = [\,1.2\ -1.1\,]$ m (the center of mass is located slightly behind the mid wheelbase point), track (all axles) $t = 1.8$ m.

For the lateral characteristics of the wheels the same equations seen in the previous example (4.157) and (4.161) are used. The data for the wheels are: $\mu_{y_p} = 0.6$, $C = 5000$ N/rad, $M_{z0} = 20$ Nm and $C_1 = 20$.

The average steering angles of the first and last axles are assumed as

$$\delta_1 = -\delta_2 = \text{atan}\left(\frac{l}{2R}\right).$$

The trajectory curvature gain obtained from the linearized study is plotted as a function of the speed in Fig. 5.26a. Both the dynamic and the kinematic values are reported: the vehicle is clearly oversteer, as it could be expected from the position of the center of mass.

The sideslip angles of the wheels when the rover travels on a trajectory with a radius of 5 m are reported as functions of the speed in Fig. 5.26b.

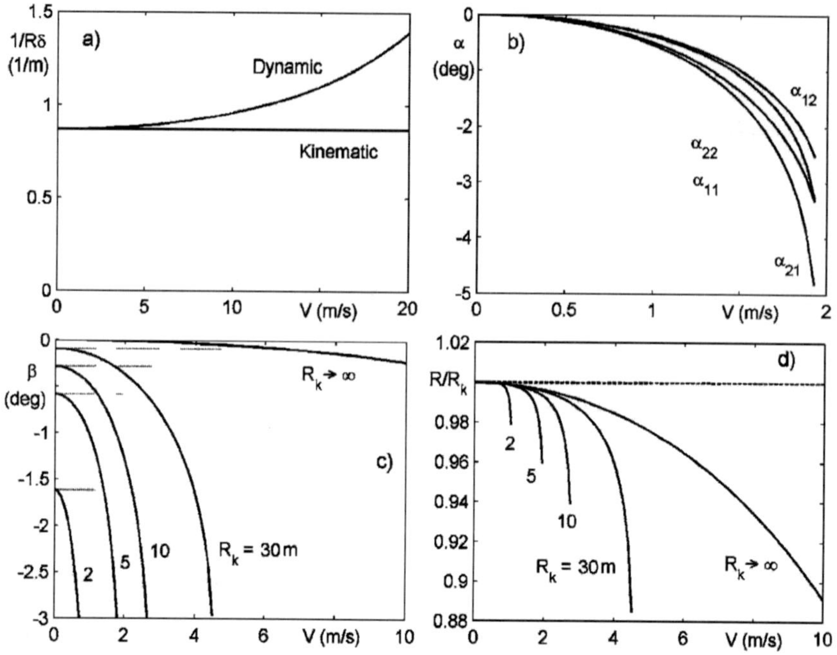

Fig. 5.26 (**a**) Linearized solution: trajectory curvature gain as a function of the speed. (**b**), (**c**) and (**d**) Nonlinear solution. Steering angles of the four wheels as functions of the trajectory curvature radius. Sideslip angle β and ratio between the dynamic and kinematic radius of the trajectory as functions of the speed

The difference between kinematic and dynamic steering in terms of sideslip angle β and curvature radius of the trajectory in a speed range up to 10 m/s is shown in Fig. 5.26c and d. The curves were computed for speeds giving way to a centrifugal acceleration lower than the ideal maximum value $\mu_{yp} g$.

5.4.7 *Slip Steering*

Equations of Motion

A possible architecture, sometimes used in both wheeled earth moving machines and robotics, is a vehicle with no steering wheels, in which the trajectory is controlled only by differential braking and driving of right and left wheels.

One axle, for instance the rear one, may control the trajectory in this way, while the other axis may have a single swiveling wheel: this is used in aircraft (differential braking of the main wheels during takeoff and landing) and on some robots. However, this approach results in poor ability to withstand lateral forces.

A mathematical model for a multi-axle vehicle with slip steering is still (5.134) in which the steering angles δ_{ij} are set to zero:

$$\begin{cases} \dot{u} = vr - g\sin(\alpha)\sin(\psi) + \displaystyle\sum_{i=1}^{n}\sum_{j=1}^{2}\frac{F_{xij}}{m}, \\ \dot{\psi} = r, \\ \dot{v} = -ur - g\sin(\alpha)\cos(\psi) + \displaystyle\sum_{i=1}^{n}\sum_{j=1}^{2}\frac{F_{yij}}{m}, \\ \dot{r} = \displaystyle\sum_{i=1}^{n}\sum_{j=1}^{2}\left(-\frac{F_{xij}}{J_z}y_{ij} + \frac{F_{yij}}{J_z}x_i + M_{zij}\right). \end{cases} \tag{5.180}$$

The steering control is now provided by the longitudinal forces, which can be written as

$$F_{xij} = F_{xi} \mp \Delta F_{xi}, \tag{5.181}$$

where the upper sign holds for $j = 1$. The simplest control law is stating that all axles supply the same average and differential longitudinal forces, i.e. all F_{xi} and ΔF_{xi} are equal. In this case the equations of motion reduce to

$$\begin{cases} \dot{u} = vr - g\sin(\alpha)\sin(\psi) + \dfrac{2nF_x}{m}, \\ \dot{\psi} = r, \\ \dot{v} = -ur - g\sin(\alpha)\cos(\psi) + \displaystyle\sum_{i=1}^{n}\sum_{j=1}^{2}\frac{F_{yij}}{m}, \\ \dot{r} = \displaystyle\sum_{i=1}^{n} t_i\frac{\Delta F_x}{J_z} + \frac{1}{J_z}\sum_{i=1}^{n}\sum_{j=1}^{2}(F_{yij}x_i + M_{zij}). \end{cases} \tag{5.182}$$

The unknown F_x appears only in the first equation, which can be separated from the other ones and solved in F_x by itself. If the interaction of the forces in x and y directions is accounted for, this uncoupling is not complete, but can be nevertheless used inside each iteration loop.

Steady-State Response

The steady-state equations of motion, which do not include the term linked with the slope of the ground, can be written in the usual form

$$p_k(u, v, r) = 0, \tag{5.183}$$

where functions p_k are

$$\mathbf{p}=\left\{\begin{array}{l} mvr+2nF_x \\ -mur+\sum_{i=1}^{n}\sum_{j=1}^{2}F_{yij} \\ \sum_{i=1}^{n}t_i\Delta F_x+\sum_{i=1}^{n}\sum_{j=1}^{2}(F_{yij}x_i+M_{zij}) \end{array}\right\}. \tag{5.184}$$

Equations (5.129), yielding the sideslip angles α_{ij} as functions of u, v, and ψ, and (5.130) and (5.132), yielding the forces and moments as functions of the sideslip angles α_{ij}, are also used.

A simpler approach is to assume that the speed of the vehicle V and the radius of the trajectory (and then the yaw velocity $r = V/R$) are known. The second and the third equation can thus be solved separately,

$$p_k(v, \Delta F_x)=0, \tag{5.185}$$

where

$$\mathbf{p}=\left\{\begin{array}{l} -mr\sqrt{V^2-v^2}+\sum_{i=1}^{n}\sum_{j=1}^{2}F_{yij} \\ \sum_{i=1}^{n}t_i\Delta F_x+\sum_{i=1}^{n}\sum_{j=1}^{2}(F_{yij}x_i+M_{zij}) \end{array}\right\}. \tag{5.186}$$

In the computation of the forces F_{yij} at each iteration of the Newton–Raphson procedure it is possible to take into account the interaction between longitudinal and lateral forces by computing the longitudinal force from the first equation

$$F_x=-\frac{mvr}{2n}. \tag{5.187}$$

Linearized Solution

Also in this case it is possible to obtain a linearized solution, based on the assumption that the wheelbase is much smaller than the radius of the trajectory and all angles are small. Since in slip steering the sideslip angles are usually larger than in the case of conventional steering, the linearized solution holds only for very large radii.

The equations of motion of the vehicle are the usual ones

$$\left\{\begin{array}{l} mV(\dot{\beta}+r)+m\dot{V}\beta=Y_\beta\beta+Y_r r+Y_e, \\ J_z\dot{r}=N_\beta\beta+N_r r+M_{z_e}+M_{z_c}, \end{array}\right. \tag{5.188}$$

where the yawing control moment M_{z_c} is added to the yawing moment M_{z_e} applied to the vehicle. The control moment is due to the differential braking or driving forces applied to the right and left wheels.

The directional behavior in terms of the various gains can be expressed by (5.173) when the control moment M_{z_c} is substituted for the external moment M_{z_e}, which is now set to zero. The relevant equations are here reported again:

- *Trajectory curvature gain*

$$\frac{1}{RM_{z_c}} = \frac{-Y_\beta}{VU}; \tag{5.189}$$

- *Lateral acceleration gain*

$$\frac{V^2}{RM_{z_c}} = \frac{-VY_\beta}{U}; \tag{5.190}$$

- *Sideslip angle gain*

$$\frac{\beta}{M_{z_c}} = \frac{-mV + Y_r}{U}. \tag{5.191}$$

Example 5.7 The Mars exploration rover studied in Example 5.6 is controlled through slip steering instead of conventional steering.

Since the differential longitudinal forces needed to perform slip steering may be quite high on sharp corners, the interaction between longitudinal and cornering forces must be accounted for. As no detailed characteristics of the wheel were available, the simple elliptical approximation

$$F^*_{yij} = F_{yij}\sqrt{1 - \left(\frac{F_{xij}}{\mu_{x_p} F_z}\right)^2}$$

can be used to compute the actual side force F^*_y from the value F_y obtained neglecting the interaction. No interaction was considered for the aligning torque, owing to its small effect on the results.

The results of the steady-state solution of the nonlinear equations of motion are reported in Fig. 5.27 (lateral and yaw velocity v and r, sideslip angle β and control differential force ΔF as functions of the speed) and Fig. 5.28 (sideslip angles as functions of the speed for different values of the radius of the trajectory).

The differential force increases with increasing curvature of the trajectory, as clear from Fig. 5.27d. However, when the curve becomes very sharp, the sideslip angles become quite large and the wheels work in a zone of its characteristics where the sideforce is almost constant with the sideslip angle. Due to this fact and also to the interaction between longitudinal and cornering forces, the differential longitudinal force needed to keep the vehicle on its trajectory does not increase much further.

As shown in Fig. 5.28, on a radius of 2 m the sideslip angles of the front wheel remain positive, i.e. they generate a force which are opposite to those of the other wheels. While on a curve with a radius of 2 m the sideslip angles reach values of more than 50°, on radii of 30 m or more their values are less than 2° at low speed.

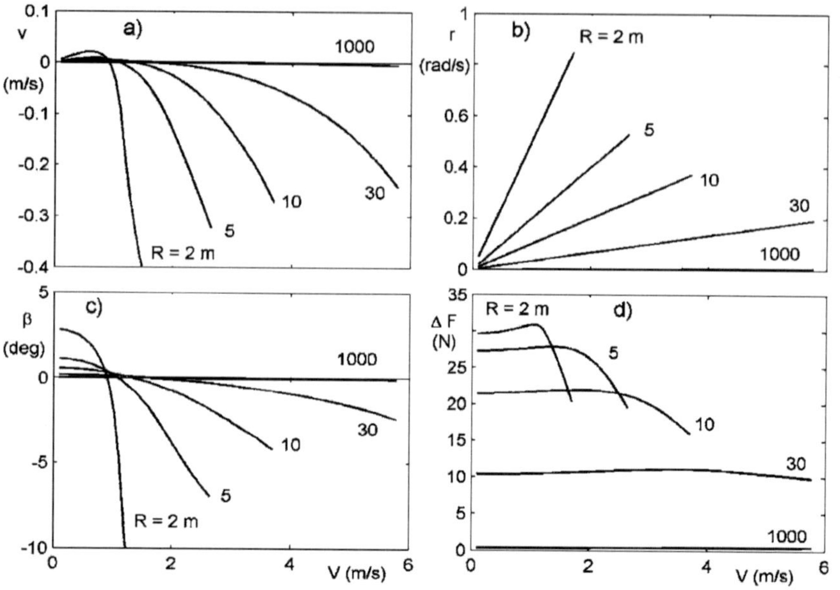

Fig. 5.27 (**a**) and (**b**) Lateral velocity v and yaw velocity r as functions of the speed for different values of the radius of the trajectory. (**c**) and (**d**) Sideslip angle β and differential longitudinal force exerted by each wheel as functions of the speed

A parameter that has a particular importance in vehicles controlled by slip steering is the ratio between the track and the wheelbase t/l. In standard vehicles this ratio is usually smaller than unity, and can be quite small in large industrial vehicles (down to 0.2 or less). Slip steering requires that ratio t/l is larger than usual, possibly close to 1.[16] As a consequence, slip steering results in wide vehicles, which become very wide in case of large vehicles. This can prevent using this type of control for large vehicles on Earth, were a limitation to the width of vehicles is imposed by existing infrastructures. On planetary surfaces there are no existing infrastructures, and nothing prevents from using very wide pressurized rovers, a thing that allows to consider slip steering also for this category of vehicles.

5.4.8 Articulated Steering

A vehicle with articulated steering (Fig. 5.13) is made by two or more subunits, or modules, each carrying one or more axles, so that steering can be performed by rotating these subunits with respect to each other. Alternatively, the wheels can be

[16] G. Genta, *Study of the Lateral Dynamics of a Large Pressurized Lunar Rover: Comparison Between Conventional and Slip-Steering*, 61st Int. Astronautical Congress, Prague, Sept. 2010.

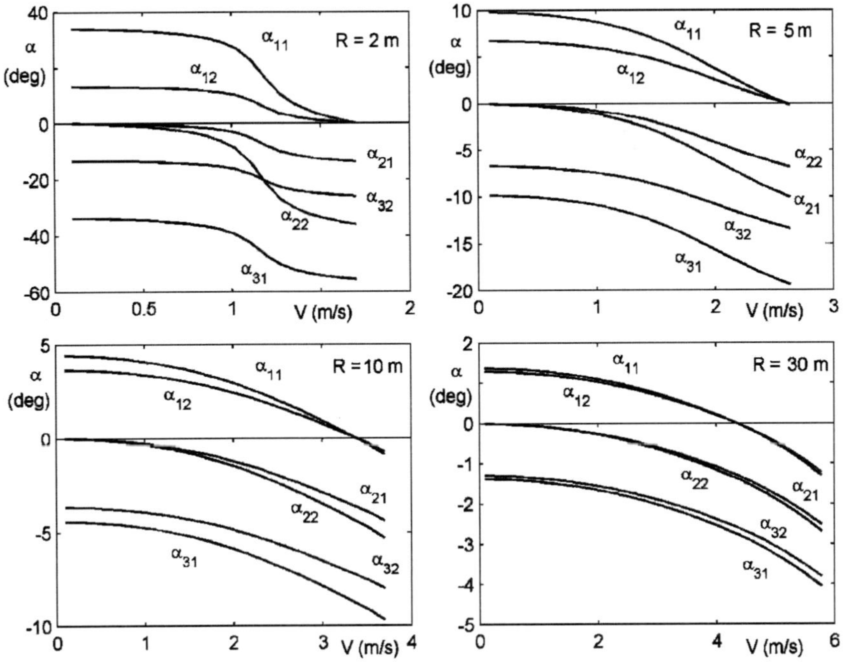

Fig. 5.28 Sideslip angles of the wheels as functions of the speed on trajectories with different radii

carried by bogies, pivoted under the main body. If each subunit carries no more than a single axle, kinematic steering is possible without any additional constraint on the steering angles being needed.

Articulated steering has a number of advantages, in particular linked with the possibilities of adapting the body to the ground configuration, if all the joints between the various segments except one are spherical hinges, and of maneuvering in narrow spaces. As an example, a rover proposed by NASA for a Mars Sample Return Mission is shown in Fig. 5.29a. The body consists in three rigid segments connected by two joints. Another example is the rover built by NASA and now on display at the U.S. Space & Rocket Center in Huntsville (Fig. 5.29b).

In automotive applications, articulated steering is common in trailers, where the front axle is attached to the drawbar assembly, which causes also the axle to steer. Two axles full trailers can be thought as two bodies vehicles, the first being the drawbar assembly with the front axle, and the second the vehicle body with the rear axle. In this case there is no steering actuator and the system is fully passive. On the contrary, it is little used on isolated vehicles; only some construction machines are based on this architecture.

Although never used in actual space missions, many designs based on articulated steering were proposed and in some cases demonstrators were built. These designs go from a simple two-bodies articulated vehicle, to the three body configuration of

Fig. 5.29 (**a**) Artist impression of an articulated rover on Mars (NASA image). (**b**) Articulated rover demonstrator built by NASA and now on display at the U.S. Space & Rocket Center in Huntsville

Fig. 5.29a, to snake-like multibody systems, like the Genbu robots developed at the Tokyo Institute of Technology.[17]

Remark 5.22 These multibody devices, although defined snake-like machines, have nothing to do with true snake (apodal) robots, which are propelled by the body–ground friction forces and not by wheels.

Multibody vehicles can be subdivided into four classes, depending whether the wheels and the joints between the bodies are actuated or not:

- Passive wheels–Passive joints (PW–PJ) vehicles are unpowered vehicles that cannot move autonomously, but only be pulled by a tractor. Full trailers are of this type, as well as multiple trailers forming road trains that are not road legal in Europe, but can be used in some extra-European country.
- Passive wheels–Active joints (PW–AJ) vehicles are propelled by the motion of the body in a way that reminds true snakes, even if snakes have obviously no wheels. Since they resort to the snaking movement of the body for propulsion, they cannot travel in straight line and require a large number of articulations. They have many degrees of freedom controlled by actuators.
- Active wheels–Passive joints (AW–PJ) vehicles are propelled and steered by the action of the wheels. The steering action is similar to slip steering, since differential tractive forces on at least one of the modules produce a yawing moment that makes that module to steer with respect to the others, but the wheels operate at sideslip angles much smaller that those of slip steering vehicles. As a limiting case, they can perform kinematic steering, a thing impossible for vehicles with slip steering. The number of bodies can be small, down to a minimum of 2.

[17]H. Kimura, K. Shimizu, S. Hirose, *Development of Genbu: Active-Wheel Passive- Joint Snake-Like Mobile Robot*, Journal of Robotics and Mechatronics, Vol. 16, No. 3, 2004.

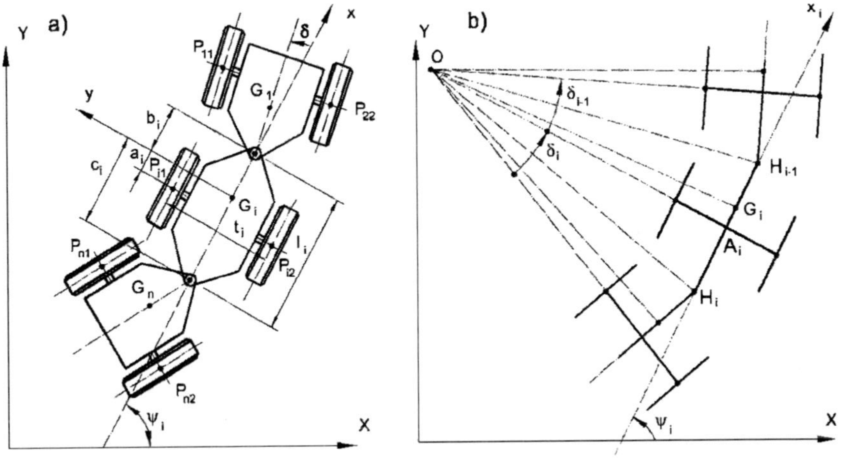

Fig. 5.30 (**a**) Sketch of a multibody rover for the case with $n = 3$. a_i, b_i and c_i are the x coordinates of the center of the wheels and of the hinges. a_i may be either positive (the axle is in front of the center of mass), negative (the axle is behind the center of mass) or 0 (the center of mass is on the axle) and in the figure is negative. b_i is usually positive and c_i is usually negative. (**b**) Steady-state kinematic steering

- Active wheels–Active joints (AW–AJ) vehicles are propelled by the actuators of the wheels and steered by the actuators of the joints. They can travel on a straight line and be made by a small number (down to 2) modules.

When traveling on a flat surface, the joints can be thought as cylindrical joints with their axis perpendicular to the ground. If they are active, their single degree of freedom is controlled by a single actuator. When moving on non-flat terrain, the joints must allow out-of-plane motions, and this implies that the joints have at least another degree of freedom about the pitch axis. If this degree of freedom is controlled, the capability of managing obstacles is greatly increased. If all the modules have both yaw and pitch mobility, at least one of the joints must either be locked or controlled about its pitch axis, to ensure stable operation. A third degree of freedom about the longitudinal (roll) axis can be allowed at the joints, to allow relative rolling of the various modules so that the vehicle can follow an irregular ground profile.

In any case, ultimately trajectory control is performed by exploiting the lateral forces (cornering forces) exerted by the wheels that bend the trajectory while also compensating for any lateral force exerted on the vehicle, as typical for all wheeled vehicles and robots.

A sketch of a multibody rover is shown in Fig. 5.30a. It is made of n bodies, each with one axle with two wheels, a mass m_i and a moment of inertia J_{zi} about the z axis. The rover, controlled either by the steering angle δ or by differential longitudinal forces on the wheels, moves on a flat terrain sloping at an angle α with respect to the horizontal (Fig. 5.20a). The inertial reference frame has its X axis horizontal and Y axis sloping upwards. The total number of the degrees of freedom of the rigid bodies is thus $3n$, and the coordinates X_i and Y_i of the centers of mass G_i and the yaw angles ψ_i can be taken as generalized coordinates.

Steady-State Kinematic Steering

In steady-state kinematic steering the perpendiculars to the midplanes of the wheel converge in a single point, the center of curvature of the trajectory (Fig. 5.30b). The radii of the trajectories of the various points may differ from each other, but are constant, i.e. the trajectory is circular. Since the configuration of the system does not change, the angular velocities of the various segments are all equal: $\dot{\psi} = r = r_i$ $\forall i$.

The velocities and the radii of the trajectory of the centers of mass G_i of the various bodies are linked by the relationships

$$\frac{V_i}{R_i} = r. \tag{5.192}$$

In the following, the velocity V and radius of curvature of the trajectory R of the vehicle are those of the center of mass of the second body.

By introducing the sideslip angles of the various bodies β_i, the components of the velocity are

$$u_i = V_i \cos(\beta_i), \qquad v_i = V_i \sin(\beta_i). \tag{5.193}$$

The radii R_i^* and R_{hi} of the trajectories of the center of the ith axle and of the ith hinge (the rear hinge of the ith body, i.e. the hinge between the ith and the $(i+1)$th body) are

$$R_i^* = \sqrt{R_i^2 - a_i^2}, \qquad R_{hi} = \sqrt{R_i^{*2} + c_i^2} = \sqrt{R_i^2 + c_i^2 - a_i^2}. \tag{5.194}$$

The radius $R_{h(i-1)}$ of the trajectories of the forward hinge of the ith body is linked with R_i by the relationship

$$R_{h(i-1)} = \sqrt{R_i^{*2} + b_i^2} = \sqrt{R_i^2 + b_i^2 - a_i^2}. \tag{5.195}$$

The radii of the trajectories of the various hinges are thus linked by the relationship

$$R_{hi} = \sqrt{R_{h(i-1)}^2 + c_i^2 - b_i^2} \tag{5.196}$$

that allows to compute all the radii of the trajectories of the various bodies, once that of one of the bodies is known. If the axle is at the same distance from the two hinges ($c_i = -b_i$), the radii of the trajectories of the hinges are equal.

The sideslip angles of the various bodies are

$$\beta_i = \operatorname{atan}\left(\frac{-a_i}{R_i^*}\right). \tag{5.197}$$

When a_i is positive (the axle is in front of the center of mass), β_i is negative (the velocity points outside the trajectory).

The relationship linking the radius of curvature with the steering angle δ_i is

$$\delta_i = \psi_i - \psi_{(i+1)} = \text{asin}\left(\frac{b_{(i+1)}}{R_{hi}}\right) - \text{asin}\left(\frac{c_i}{R_{hi}}\right). \tag{5.198}$$

If the radii of the trajectories of the various points are large with respect to the dimensions of the vehicle, the steering angle δ_i and the sideslip angles β_i are small and their trigonometric functions can be linearized. The radii of the trajectories of the various points and their velocities are almost equal to each other. It then follows

$$\delta_i = \psi_i - \psi_{(i+1)} \approx \frac{b_{(i+1)} - c_i}{R_{hi}} \approx \frac{b_{(i+1)} - c_i}{R}, \tag{5.199}$$

$$\beta_i \approx \frac{-a_i}{R}. \tag{5.200}$$

The trajectory curvature gain in kinematic conditions, written with reference to the steering angle between the first and second body is thus

$$\frac{1}{R\delta_1} \approx \frac{1}{b_2 - c_1}. \tag{5.201}$$

The components of the velocities of the various bodies are thus

$$u_i \approx V, \qquad v_i \approx V\beta_i. \tag{5.202}$$

Equations of Motion

If the assumption of kinematic steering is done, the wheels are modeled as non-holonomic constraints. If, on the contrary, dynamic steering is assumed, the cornering forces acting on the wheels are functions of the sideslip angles.

The components u_i and v_i of the velocity of the center of mass G_i expressed in the frame fixed to the ith body are

$$\begin{Bmatrix} u_i \\ v_i \end{Bmatrix} = \begin{bmatrix} \cos(\psi_i) & \sin(\psi_i) \\ -\sin(\psi_i) & \cos(\psi_i) \end{bmatrix} \begin{Bmatrix} \dot{X}_i \\ \dot{Y}_i \end{Bmatrix}. \tag{5.203}$$

The velocity of the of the center P_{ij} of the contact area of the jth wheel of the ith axle is

$$V_{Pij} = V_{Gi} + \dot{\psi}_i \wedge \overline{(P_{ij} - G_i)} = \begin{Bmatrix} u_i \mp \dot{\psi}_i \frac{t_i}{2} \\ v_i + \dot{\psi}_i a_i \end{Bmatrix} \tag{5.204}$$

for $i = 1, \dots, n$. The upper sign holds for $j = 1$ (left wheel).

In kinematic steering, the condition for pure rolling is

$$f_{ni}(v_i, \dot{\psi}_i) = v_i + \dot{\psi}_i a_i = 0. \tag{5.205}$$

The constraints on the velocities of the two wheels of the same axle coincide, which means that there are just n non-holonomic constraint equations.

In case of dynamic steering, the sideslip angle of the wheels of the ith body are

$$\alpha_{ij} = \text{atan}\left(\frac{v_i + \dot{\psi}_i a_i}{u_i \mp \dot{\psi}_i \frac{t_i}{2}}\right). \tag{5.206}$$

In both cases, there are other $2(n-1)$ constraint equations stating that each hinge point is in common between two subsequent bodies:

$$\begin{Bmatrix} X_i + c_i \cos(\psi_i) \\ Y_i + c_i \sin(\psi_i) \end{Bmatrix} = \begin{Bmatrix} X_{i+1} + b_{i+1} \cos(\psi_{i+1}) \\ Y_{i+1} + b_{i+1} \sin(\psi_{i+1}) \end{Bmatrix} \tag{5.207}$$

for $i = 1, \dots, n-1$. These constraints are holonomic, and can be written in the following form

$$\begin{aligned} &f_{hxi}(X_i, X_{i+1}, \psi_i, \psi_{i+1}) \\ &\quad = X_i + c_i \cos(\psi_i) - X_{i+1} - b_{i+1} \cos(\psi_{i+1}) = 0, \\ &f_{hyi}(Y_i, Y_{i+1}, \psi_i, \psi_{i+1}) \\ &\quad = Y_i + c_i \sin(\psi_i) - Y_{i+1} - b_{i+1} \sin(\psi_{i+1}) = 0. \end{aligned} \tag{5.208}$$

In case of kinematic steering there are $3n$ dynamic equations, $2n-2$ holonomic constraint equations plus n non-holonomic constraint equations: the system has thus 2 degrees of freedom. There is only one way of controlling the system in a way consistent with the kinematic steering assumption: imposing the steering angle δ. This can be defined as a locked control strategy of the Active Wheels–Active Joints (AW–AJ) type (δ may be either constant or an imposed function of time) and a further relationship can be added:

$$\psi_1 - \psi_2 = \delta. \tag{5.209}$$

The steering angle δ can be considered as a control variable, and be obtained from the equation describing the behavior of the control system. There is just a single free generalized coordinate remaining, that expressing the motion of the vehicle along its trajectory that is univocally determined by the steering angle.

Kinematic steering will not be dealt with any further.

In case of dynamic steering there are $3n$ dynamic equations and $2n-2$ holonomic constraint equations: the system has thus $n+2$ degrees of freedom. In this case AW–AJ or AW–PJ strategies can be used. In the first case δ may be either constant or an imposed function of time and, adding (5.209), the number of degrees of freedom reduces to $n+1$.

To write the equations of motion, the Lagrangian of the system can be written in terms of velocities in the body-fixed frame as

$$\mathcal{L} = \sum_{i=1}^{n}\left[\frac{1}{2}m_i\left(u_i^2 + v_i^2\right) + \frac{1}{2}J_{zi}\dot{\psi}_i^2 - m_i g Y_i \sin(\alpha)\right]. \tag{5.210}$$

The rotational kinetic energy of the wheels has been neglected: no gyroscopic effect of the wheels will be obtained in this way.

Since the constraints are holonomic, they can be introduced into the Lagrangian function, yielding

$$\mathcal{L}^* = \sum_{i=1}^{n}\left[\frac{1}{2}m_i\left(u_i^2+v_i^2\right)+\frac{1}{2}J_{zi}\dot{\psi}_i^2 - m_i g Y_i \sin(\alpha)\right] + \sum_{i=1}^{n-1}(\lambda_{xi} f_{hxi} + \lambda_{yi} f_{hyi}). \tag{5.211}$$

The virtual displacement of the center of the contact area of the ijth wheel in the body-fixed frame is

$$\delta q^*_{\mathrm{P}_{ij}} = \left\{\begin{array}{c} \delta X_i \cos(\psi_i) + \delta Y_i \sin(\psi_i) \mp \delta\psi \frac{t_i}{2} \\ -\delta X_i \sin(\psi_i) + \delta Y_i \cos(\psi_i) + \delta\psi a_i \end{array}\right\} \tag{5.212}$$

for $i = 1, \dots, n$. As usual, the upper sign holds for $j = 1$.

Forces F_{xij} and F_{yij} act in the direction of axes x_i and y_i, on the centers of the contact areas. Moreover, since the wheels now work with a sideslip angle, also the aligning torques M_{zij} should be added. The virtual work is thus the product of the forces and moments by the virtual displacements and rotations

$$\begin{aligned} \delta\mathcal{W}_{\mathrm{P}_{ij}} &= F_{xij}\left[\delta X_i \cos(\psi_i) + \delta Y_i \sin(\psi_i) - y_{ij}\delta\psi_i\right] \\ &\quad + F_{yij}\left[-\delta X_i \sin(\psi_i) + \delta Y_i \cos(\psi_i) + x_i\delta\psi_i\right] + M_{zij}\delta\psi_i. \end{aligned} \tag{5.213}$$

The $3n$ differential equations obtained from the Lagrange equations can thus easily be written in the form

$$\left\{\begin{array}{l} m_i(\dot{u}_i - v_i\dot{\psi}_i) + (\lambda_{x(i-1)} - \lambda_{xi})\cos(\psi_i) \\ \quad + \left[m_i g \sin(\alpha) + \lambda_{y(i-1)} - \lambda_{yi}\right]\sin(\psi_i) = F_{xi1} + F_{xi2}, \\ m_i(\dot{v}_i + u_i\dot{\psi}_i) - (\lambda_{x(i-1)} - \lambda_{xi})\sin(\psi_i) \\ \quad + \left[m_i g \sin(\alpha) + \lambda_{y(i-1)} - \lambda_{yi}\right]\cos(\psi_i) = F_{yi1} + F_{yi2}, \\ J_{zi}\ddot{\psi}_i - (b_i\lambda_{x(i-1)} - c_i\lambda_{xi})\sin(\psi_i) + (b_i\lambda_{y(i-1)} - c_i\lambda_{yi})\cos(\psi_i) \\ \quad = a_i(F_{yi1} + F_{yi2}) + \frac{t_i}{2}(F_{xi2} - F_{xi1}) + M_{zi1} + M_{zi2} \end{array}\right. \tag{5.214}$$

in which the multipliers with subscript 0 and n must be set to zero.

The $2(n-1)$ constraint equations (5.208) can be differentiated with respect to time to write a relationship between the velocities in the reference frames of the various bodies. The result is

$$\begin{aligned}
&u_i\cos(\psi_i)-(v_i+c_i\dot{\psi}_i)\sin(\psi_i)-u_{i+1}\cos(\psi_{i+1})\\
&\quad+(v_{i+1}+b_{i+1}\dot{\psi}_{i+1})\sin(\psi_{i+1})=0,\\
&u_i\sin(\psi_i)+(v_i+c_i\dot{\psi}_i)\cos(\psi_i)-u_{i+1}\sin(\psi_{i+1})\\
&\quad-(v_{i+1}+b_{i+1}\dot{\psi}_{i+1})\cos(\psi_{i+1})=0.
\end{aligned}\tag{5.215}$$

If the control parameters are the longitudinal forces, the longitudinal forces are stated, while the lateral forces can be computed from the sideslip angles, which are in turn obtained through (5.206) from the velocities.

If the vehicle is controlled by imposing a torque at some of the joints using actuators, the virtual work due to the torque T_i at the ith actuator is located between the $(i-1)$th and the ith body

$$\delta\mathcal{W}_{T_i}=T_i(\delta\psi_i-\delta\psi_{i-1}),$$

for $i=1,\dots,n-1$, where obviously $\psi_0=0$.

The third equation (5.214) thus becomes

$$\begin{aligned}
&J_{zi}\ddot{\psi}_i-(b_i\lambda_{x(i-1)}-c_i\lambda_{xi})\sin(\psi_i)+(b_i\lambda_{y(i-1)}-c_i\lambda_{yi})\cos(\psi_i)\\
&\quad=a_i(F_{yi1}+F_{yi2})+\frac{t_i}{2}(F_{xi2}-F_{xi1})+M_{zi1}+M_{zi2}+T_i-T_{i-1}.
\end{aligned}\tag{5.216}$$

Steady State Solution

In steady state operation u_i, v_i and $r_i=\dot{\psi}_i$ are constant. Since angles ψ_i change in time, also the forces due to the component of weight along the slope change in time. Only the case of motion on horizontal ground is consistent with the steady-state assumption and thus it will be assumed that $\alpha=0$.

Moreover, the curvatures of the trajectories of the various points are constant, meaning that the trajectories are circular. Since the configuration of the vehicle rotates about the center of curvature of the trajectories but does not change in time, all angular velocities are equal and it is possible to define a single angular velocity. The yaw angles are thus

$$\psi_i=\psi_{i0}+rt\tag{5.217}$$

where $r=r_i\ \forall i$.

The configuration of the vehicle can be computed at any instant, for instance for $t=0$, since in steady state conditions it remains always the same. Assuming that the vehicle is controlled by the longitudinal forces, at that instant the equations of motion reduce to

$$\left\{\begin{array}{l} -m_i v_i r + (\lambda_{x(i-1)} - \lambda_{xi}) \cos(\psi_{i0}) + (\lambda_{y(i-1)} - \lambda_{yi}) \sin(\psi_{i0}) \\ \quad = F_{xi1} + F_{xi2}, \\ m_i u_i r - (\lambda_{x(i-1)} - \lambda_{xi}) \sin(\psi_{i0}) + (\lambda_{y(i-1)} - \lambda_{yi}) \cos(\psi_{i0}) \\ \quad = F_{yi1} + F_{yi2}, \\ -(b_i \lambda_{x(i-1)} - c_i \lambda_{xi}) \sin(\psi_{i0}) + (b_i \lambda_{y(i-1)} - c_i \lambda_{yi}) \cos(\psi_{i0}) \\ \quad = a_i (F_{yi1} + F_{yi2}) + \frac{t_i}{2}(F_{xi2} - F_{xi1}) + M_{zi1} + M_{zi2}. \end{array}\right. \tag{5.218}$$

To obtain the relationship between the input variable, i.e. the differential traction of the wheels in case of AW–PJ machines or the torque of the actuators in case of AW–AJ machines, and the output, the curvature of the trajectory, it is possible to state the radius R_i and the velocity V_i of the center of mass of any one of the bodies, for example the second one.

A relationship linking the driving forces F_{xij} with each other, i.e. a control law for the various power drivers must be stated. The forces at the wheels of the same axle can be written as

$$F_{xij} = F_{xi} \pm \Delta F_{xi}, \tag{5.219}$$

where F_{xi} is the average force and ΔF_{xi} is the differential force causing steering.

In the following a control law stating that all F_{xi} are equal to a force F_x and that all differential force vanish except the first that is equal to ΔF_x. Other strategies are possible, for instance to state that the average force is proportional to the load on the axle, and the differential force is suitably distributed, to obtain a required shape of the vehicle.

In the case of AW–AJ machines, a law linking together the control torques can be added. For instance, only the first torque acting between the first and the second body can be different from zero.

The unknowns thus are $5n$ in number: the values of v_i, u_i, ψ_{i0}, λ_{yxi}, λ_{yxi} (which are only $n-2$), F_x and ΔF_x.

The equations also number $5n$: the $3n$ equations (5.218), the $2n-2$ equations (5.215) yielding the constraint conditions, plus the relationships

$$u_2^2 + v_2^2 = V^2 \quad \text{and} \quad \psi_{20} = 0, \tag{5.220}$$

that are written assuming, arbitrarily, that the second body is taken as a reference. Further equations must also be stated to compute forces F_{yi1}, F_{yi2} and the aligning torques. They are the $2n$ equations (5.206) yielding the sideslip angles, and the characteristics of the wheels yielding the lateral forces and the aligning torques as functions of the sideslip angles. Actually, as already stated, only $5n$ equations can be used since those yielding the sideslip angles, the forces and the aligning torques can be used 'off line'. A Newton–Raphson scheme can thus be devised, where the Jacobian matrix is computed numerically. Starting the computation from the kinematic values of the parameters, the convergence on the solution is straightforward.

In case of a AW–AJ vehicle, the differential force ΔF_x vanishes, and the torque(s) produced by the actuators on the joint(s) take its place as unknowns.

If a linearized solution is acceptable, a thing that is possible if one of the angles ψ_{i0} is set to zero and the others are small, as it occurs when traveling on a trajectory with a radius much larger than the length of the vehicle, the sideslip angles of the two wheel of the same axle are (almost) equal and (5.206) reduce to

$$\alpha_{ij} = \frac{v_i + ra_i}{V}. \tag{5.221}$$

The second and third sets of n equations uncouple and, by introducing the cornering and the aligning stiffness C_i and $M_{zi,\alpha}$, they reduce to

$$\begin{cases} 2C_i v_i + V\lambda_{y(i-1)} - V\lambda_{yi} = -2C_i ra_i - m_i V^2 r, \\ 2(C_i a_i - M_{zi,\alpha})v_i + b_i V\lambda_{y(i-1)} - c_i V\lambda_{yi} + Vt_i \Delta F_{xi} \\ \quad = 2(-C_i a_i + M_{zi,\alpha})ra_i, \end{cases} \tag{5.222}$$

allowing to compute v_i, λ_{yi} and ΔF_x (or the torque T_1).

The first n equations

$$-m_i v_i r + \lambda_{x(i-1)} - \lambda_{xi} - 2F_{xi} = 0 \tag{5.223}$$

can be used to compute λ_{xi} and F_{xi}.

Since ψ_{20} was assumed to be equal to zero, the second constraint equation allows us to compute ψ_{10}:

$$\psi_{10} = \frac{v_2 - v_1 + (b_2 - c_1)r}{V}. \tag{5.224}$$

All other angles ψ_{i0} can be easily computed

$$\psi_{(i+1)0} = \psi_{i0}\frac{v_i - v_{i+1} + (c_i - b_{i+1})r}{V} \tag{5.225}$$

for $i = 2, \dots, n-2$

It is thus possible to compute the trajectory curvature gain and the sideslip angles gains

$$\frac{1}{R\Delta F_x} \quad \text{and} \quad \frac{\beta_i}{\Delta F_x} = \frac{v_i}{V\Delta F_x}$$

in case of AW–PJ, or

$$\frac{1}{R\delta} \quad \text{and} \quad \frac{\beta_i}{\delta} = \frac{v_i}{V\delta}$$

in case of AW–AJ machines controlled with a locked-control strategy. The gains, as obvious in a linearized system, are constant, although being functions of the speed.

If the centers of mass are on the axles ($a_i = 0$) and the aligning torques are neglected, the vehicle is neutral steer, i.e. the trajectory curvature gain does not

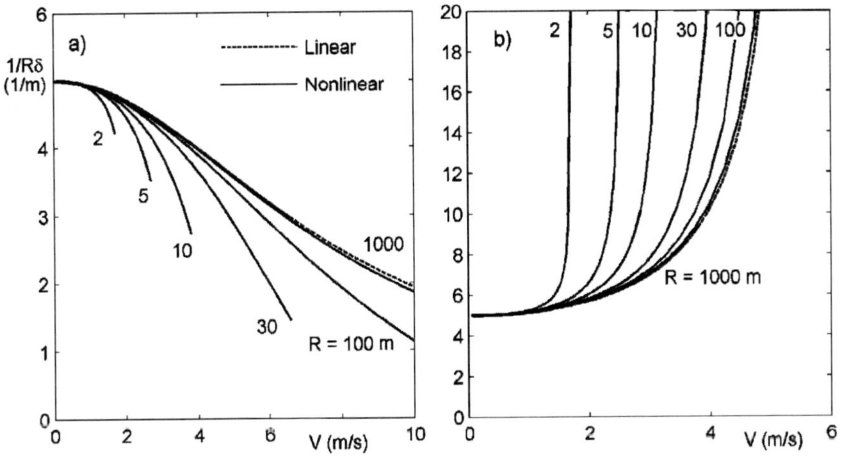

Fig. 5.31 Trajectory curvature gain of the multibody vehicle as function of the speed. (**a**) Vehicle with nine segments; (**b**) vehicle with 10 segments

depend on the speed. The equations yielding the lateral behavior reduce to

$$\begin{cases} 2C_i v_i + V\lambda_{y(i-1)} - V\lambda_{yi} = -m_i V^2 r, \\ b_i \lambda_{y(i-1)} - c_i \lambda_{yi} + t_i \Delta F_{xi} = 0. \end{cases} \tag{5.226}$$

Example 5.8 Consider a 'snake' robot made of nine identical segments with two wheels each operating on Mars. The data are the following: $m_i = 3$ kg, $g = 3.77$ m/s^2, $a_i = 10$ mm, $b_i = 100$ mm, $c_i = -100$ mm, track (all axles) $t = 300$ mm.

For the lateral characteristics of the wheels the same equations (4.157) and (4.161) used in the previous examples are employed, without accounting for the interaction between longitudinal and cornering forces. The data for the wheels are: $\mu_{y_p} = 0.6$, $C = 100$ N/rad, $M_{z0} = 0.005$ Nm and $C_1 = 20$.

The control is performed by applying a control torque at the joint between the first and second segment.

A linearized steady-state solution is first performed. A value of the radius $R =$ 1,000 m is assumed and the angular velocity is computed for each value of the speed. The trajectory curvature gain is reported in Fig. 5.31a, together with the solution obtained for the nonlinear model on different trajectories.

From the plot it is clear that the vehicle is understeer, which is an apparent contradiction with the fact that the center of mass of each segment is behind its wheels.

Actually, using the same data for each segment and reducing the vehicle to two segments, a strongly oversteer response is obtained. Different models with a different number of segments show that by assembling an odd number of bodies an understeer response is obtained, while an even number cause an oversteer response. The trajectory curvature gain of the vehicle with 10 segments is reported in Fig. 5.31b: the response is oversteer, with a low value of the critical speed.

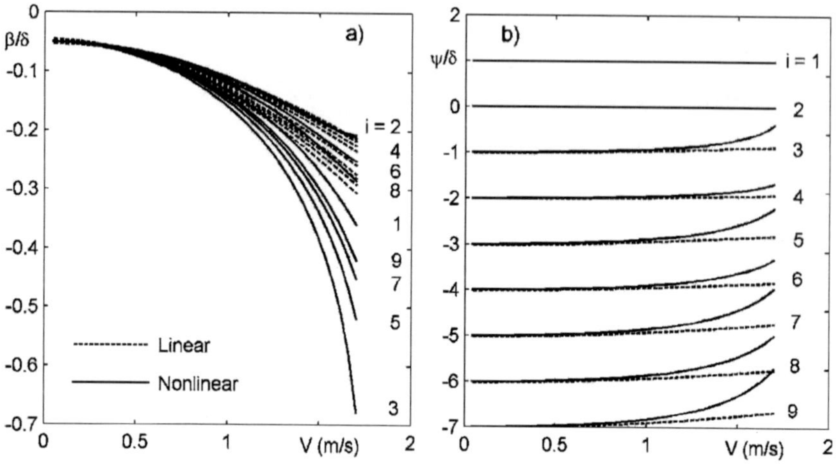

Fig. 5.32 Sideslip angle gains β_i/δ (**a**) and gains of the yaw angles ψ_i/δ (**b**) form the multibody vehicle with nine segments

The sideslip angle gains β_i/δ and the gains of the yaw angles ψ_i/δ are reported in Fig. 5.32.

5.4.9 Trajectory Definition

Up to this point the problem has been stated in these terms: given a trajectory, how to ensure that the robot follows it as faithfully as possible? In the case of vehicles at present used on our planet or on an extraterrestrial body (the LRV) and of teleoperated devices, this task is entrusted to a human controller.

An ideal trajectory must first be defined. When a human operator is in charge, this is can be done almost without the operator realizes it consciously: usually it is enough to give a look to the terrain or the road to define an acceptable trajectory and optimal trajectories are found by trying a few times the same road. A careful study of the maps may be needed to plan trajectories beyond the range of human eyes.

In case of semi-autonomous devices, the trajectory can be defined by a supervising human in the same way seen for robotic arms, perhaps by defining waypoints and then approximating the path through interpolations of different kinds.

A well known approach to define a path is based on the generation of a potential attracting the rover toward its destination point and repelling it from obstacles.[18] It usually requires that a map of the zone surrounding the rover is known so that the guidance system has a world model to refer to, but can work also if the guidance is

[18] A. Ellery, *An Introduction to Space Robotics*, Springer Praxis, Chichester, 2000.

made using only reflexes. In the latter case the potential function at the beginning is built from the knowledge of the destination point and is then modified by adding the obstacles as they are discovered while the rover is proceeding along the path.

A possible potential function may be:[19]

$$U = U_0 + \sum_{\forall i} U_i, \quad \text{or better,} \quad U = U_0 + \max\{U_i\}, \tag{5.227}$$

where

- U_0 is the potential of the destination point, defined as

$$U_0 = G_0 d_0, \tag{5.228}$$

  where G_0 is a gain and d_0 is the distance of the rover from the destination point.
- U_i is the potential of the ith obstacle, having a 'sphere of influence' with radius S_i and a radius, augmented of the radius of the rover and possibly by a safety distance, r_i

$$\begin{cases} U_i = \infty & \text{if } d_i \leq r_i, \\ U_i = G_i \frac{S_i - d_i}{S_i - r_i} & \text{if } r_i < d_i \leq S_i, \\ U_i = 0 & \text{if } d_i > S_i. \end{cases} \tag{5.229}$$

  In the first case (actually U_i is a very large number), the rover collides with the obstacle, while in the last one the rover is outside the sphere of influence of the obstacle.

The potential may have local minima, which attract the rover. A random perturbation can be added to the potential to avoid this problem, so that the rover will get out of local minima in case it falls in it.

Once the trajectory has been defined, the simplest possible guidance algorithm is to set a steering angle proportional to the difference between the actual and the required yaw angle

$$\delta = -K(\psi - \psi_0). \tag{5.230}$$

The latter (ψ_0) is computed from the potential function and is the angle the projection on the xy plane of the steepest descent of the potential surface makes with the x-axis.

A variety of more complex and effective guidance algorithms have been used, but this goes beyond the scope of the present book.

Example 5.9 Consider a four-wheeled rover traveling on the surface of Titan[20] that has to reach a destination point having coordinates (850, 900) m from a starting

[19] Y.K. Huang, N. Ahuja, *A Potential Field Approach to Path Planning*, IEEE Trans. on Robotics and Automation, Vol. 8, No. 1, 1992.

[20] G. Genta, A. Genta, *Preliminary Assessment of a Small Robotic Rover for Titan Exploration*, Acta Astronautica, Vol. 68, No. 5–6, 556–566, March–April 2011.

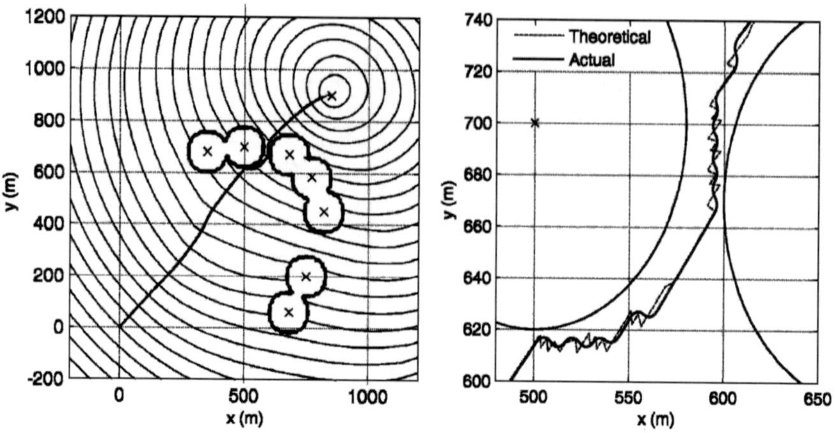

Fig. 5.33 (**a**) Potential field about a set of seven obstacles and a destination point and trajectory from the origin to the destination point. (**b**) Enlargement of the zone of the trajectory between the obstacles

point located in the origin of the reference frame, traveling on a flat surface. Seven round obstacles are located in points having x coordinates 350, 500, 680, 770, 820, 750 and 680 m and y coordinates 680, 700, 670, 580, 450, 200 and 60 m. The parameters for computing the potential are $G_0 = -500$ 1/m, $G_i = 2$, $r_i = 80$ m, $S_i = 500$ m. A limit value equal to 10 is imposed to the potential of each obstacle.

Compute the trajectory to reach the destination point, and the actual trajectory obtained by simulating the motion of the rover at a constant speed of 1 m/s. The data of the rover, modeled using the linearized model of (5.152) and following, are: mass $m = 40$ kg, moment of inertia $J_z = 12$ kg m^2, longitudinal positions of the axles $x_1 = 0.4255$ m and $x_2 = -0.4255$ m, equal and opposite steering angles as in the case of optimal kinematic steering ($K'_1 = 1$ and $K'_2 = -1$), cornering stiffness of the axles $C_1 = C_2 = 194$ N/rad, gain for the proportional trajectory control $K = 0.5$, maximum absolute value of the steering angle 5°, initial yaw angle equal to 0.

The potential obtained using (5.227) is shown in Fig. 5.33a. A nominal trajectory, computed assuming straight segments 5 m long, is also shown. Owing to the way it has been obtained, it is a serrated line, as shown in the enlargement of the zone close to the obstacles shown in Fig. 5.33b.

The simulation was performed by numerically integrating the equation of motion. Initially the rover moves in X direction and the first maneuver is a turn to the left of about 45°. The results of the numerical integration is also reported on the same figure, showing that the control strategy used allows the rover to follow the path with good precision. The low value imposed to the maximum steering angle produces a smoothing the path with respect to the theoretical one.

5.4.10 Steering Activity

To evaluate how much the steering control is actuated during the rover's motion, an index called Steering Activity was introduced. It can be defined as the ratio between the difference of the distance travelled by the wheels on the two sides and the distance travelled by the center of the rover

$$S_a = \frac{\int_{t_1}^{t_2} |V_L - V_R|\, dt}{\int_{t_1}^{t_2} |V|\, dt} = d\frac{\int_{t_1}^{t_2} |\dot{\psi}|\, dt}{\int_{t_1}^{t_2} |V|\, dt}, \tag{5.231}$$

where subscripts R and L indicate the left and right wheels and d is the track of the vehicle. In the second expression the speed of the wheels are substituted by the speed of two points located on the y-axis at the same lateral position as the wheels.

In case of a circular trajectory with radius R gone through at constant speed, it follows

$$S_a = \frac{d}{R}. \tag{5.232}$$

When turning on the spot the steering activity tends to infinity, while it tends to zero when travelling on a straight line.

If the steering activity is computed on different trajectories connecting the same points, the smoother is the trajectory the lower is the value of S_a.

5.5 Suspension Dynamics

If the number of wheels is greater than three, a rigid vehicle is a statically undetermined system, as already stated in the previous sections. The structure of the vehicle must either be compliant enough or include a compliant connection between the wheels and the body.

The suspensions have two main functions:

- Distributing the load on the ground in a way corresponding to the design goals and allowing the vehicle to sit in the proper attitude, under the effect of the static and quasi-static loads in stationary conditions;
- Allowing the wheels to follow an uneven road profile without transferring excessive loads and accelerations to the vehicle body.

When the vehicle is designed to work on unprepared ground, with obstacles of a size that may be not negligible with respect to the radius of the wheels, a further task, namely obstacle management, is associated to the suspensions.

For the first task the suspensions may be an articulated system with no compliance at all or may be an elastic system, linear or nonlinear, while for the second they must also incorporate damping, at least to avoid the onset of resonant oscillations. The second task is so important that suspensions are also used in the case of vehicles with two or three wheels, which do not need them in order to distribute the loads on the ground in a predictable way.

Remark 5.23 Theoretically both tasks could be performed by the wheels, if they are compliant enough, also in the case of rigid vehicles, but their compliance is not usually sufficient and their damping is too low to do it in a satisfactory way. Anyway, damping should be associated with non rotating elements, otherwise it increases rolling resistance.

Ideally, the suspensions should allow the wheels to move with respect to the body of the vehicle in a direction perpendicular to the ground (z direction), maintaining the plane of the wheel parallel to itself and constraining all motions in x and y directions: A suspension of a single wheel should be a system with a single degree of freedom, the displacement in z direction.

However, none of the systems which are used at present are able to perform in that way and each one of them has a peculiar behavior; the approximations with which the suspensions perform their task of constraining the five other degrees of freedom of the wheel hub are important in giving any particular vehicle its own character. As when the body of the vehicle is moved in vertical direction (displacement z) or rotates about its x axis (roll angle ϕ) the position of the wheel changes, it is possible to plot the camber angle γ, the track t, the characteristic angles of the steering system, the steering angle δ, etc. as functions of z and ϕ. These functions are generally strongly nonlinear but can be linearized about any equilibrium position and the derivatives $\partial t/\partial z$, $\partial t/\partial \phi$, $\partial \gamma/\partial z$, $\partial \gamma/\partial \phi$, $\partial \delta/\partial z$, $\partial \delta/\partial \phi$, etc.[21] can be easily defined. They can be considered as constants in the small motions about an equilibrium position and define the behavior of the suspension.

Apart from these kinematic characteristics of the suspension, the position of the wheel with respect to both the ground and the body can be influenced by the compliance of the joints, which are often not spherical or cylindrical hinges but compliant links, and that of other parts of the suspension. If the effect of compliance is accounted for, the displacements due to deformations are not univocally determined by the position of the body and for them it is impossible to introduce the relevant derivatives. It must be stated that the trend in the design of automotive suspensions is towards a replacement of joints working in a kinematically correct way by elastic hinges or compliant elements and towards an integration in a single element of the guiding functions of the linkages and the elastic functions of the springs. It is thus increasingly more difficult to define the kinematic parameters of the suspension.

A vehicle on elastic suspensions can be modeled as a multibody system with a rigid body, the *sprung mass*, connected to a number of masses which include the wheels, the *unsprung masses*, through massless springs and dampers simulating the suspensions. The unsprung masses are connected to the ground through massless springs and dampers simulating the tires. This model is clearly an approximation, as the various links, the springs and the tires have their own mass and hence also their own natural frequencies and the body and the linkages are not rigid bodies. Models of this type, whose complexity and precision can be increased by modeling

[21] In the following the notation $(t)_{,z}$, $(t)_{,\phi}$, etc. will be used.

the vehicle body as an elastic body, are, however, very useful in the study of the dynamic behavior of vehicles.

A vehicle with four wheels can be modeled as a system with 10 degrees of freedom, six for the body and one for each wheel. This holds for any type of suspension, as the wheels of each axle can be suspended separately (independent suspensions) or together (solid axle suspensions) but the total number of degrees of freedom is the same. Additional degrees of freedom as the rotation of the wheels about their axis or about the kingpin axis[22] can be inserted into the model to allow one to take into account the longitudinal slip or the compliance of the steering system.

Each suspension is thus characterized by its mass, the stiffness and damping parameters of the suspension and of the tire. The latter can be strongly nonlinear but can often be linearized, particularly for what the stiffness is concerned, if small motions about an equilibrium position are studied.

5.5.1 Non Compliant Suspensions

Wheels can manage at low speed obstacles whose height is only slightly smaller than their radius and can traverse ditches whose width is no more than 70% of their diameter, unless the vehicle layout allows working with some wheels off the ground.

Robotic rovers used on Mars have no elastic and damping suspensions and sometimes their wheels are rigid. It is well known that elastic suspensions are useful only at speeds higher than a given value, depending on many factors and in particular on the natural frequency of the whole vehicle, on the stiffness of the tires and on the size of the ground irregularities. Since no elastic suspensions are present, a linkage allowing to distribute evenly the load on the ground is used: the experimental NASA rover Rocky III and all the rovers actually used on Mars up to now use a *rocker bogie* linkage, which distributes evenly the load on the six wheels of the machine. Linkages of other types, allowing a greater travel of the wheels, but always without compliant elements were used on the Russian lunar robotic rovers (Lunokhod). The maximum speed of the latter was 2 km/h while that of Mars rovers was lower, less than 0.02 km/h. At those speeds no elastic suspensions are needed. As an added consideration, all devices on board of robotic rovers must be suited to withstand high acceleration, in particular if the airbag landing technique is used. Suspensions aimed at reducing accelerations during driving on uneven ground are so even less useful.

Multibody Vehicles

The simplest layout allowing a four-wheeled vehicle to be statically determinate is by subdividing the vehicle body into two segments and allowing one segment to

[22]The kingpin axis is the axis about which the wheel hub rotates when steering.

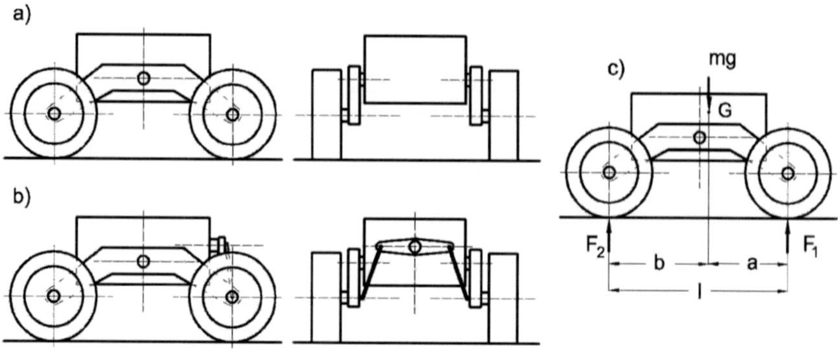

Fig. 5.34 Rocker arm suspension for a four-wheeled rover. (**a**) The body is supported by a differential gear; (**b**) the body is suspended with a transversal arm; (**c**) forces on the ground

rotate with respect to the other about a longitudinal axis. Articulated body devices can be designed with any number of wheels: for the first two axles a single internal degree of freedom is required; for any added axles two further degrees of freedom must be included. As an example, the already mentioned rover proposed by NASA for a Mars Sample Return Mission, shown in Fig. 5.29a, can be mentioned. The body consists of three rigid segments connected with two joints. The joints may be controlled by actuators to improve the ability of overcoming obstacles.

Rocker Arms Suspensions: Four Wheels

A possible solution for a two-axle vehicle is the one shown in Fig. 5.34: the two wheels on each side are attached to a rocker arm that is in turn attached to the vehicle body through a cylindrical hinge. The body can be kept at an angle with the horizontal that is an average of the angles of the two rocker arms either by using a differential gear between the trunnions or by leaving the two hinges free and using a further rocker arm in transversal direction as shown in Fig. 5.34b.

When standing on horizontal flat ground the force acting on the front and rear wheels are

$$F_{z_1} = mg\frac{b}{2l}, \qquad F_{z_2} = mg\frac{a}{2l}. \tag{5.233}$$

Remark 5.24 The position of the hinge is immaterial: if the hinge is on the center of mass the trunnions do not carry any moment or the transversal rocker arm is not loaded, but the above mentioned load distribution holds also if the center of mass is displaced with respect to the hinge.

The rockers carrying the wheels can be attached to the body by motorized hinges, in such a way that the position of the body can be controlled; for instance the body can be kept horizontal even when managing a longitudinal slope, as far as the geometry of the system allows it.

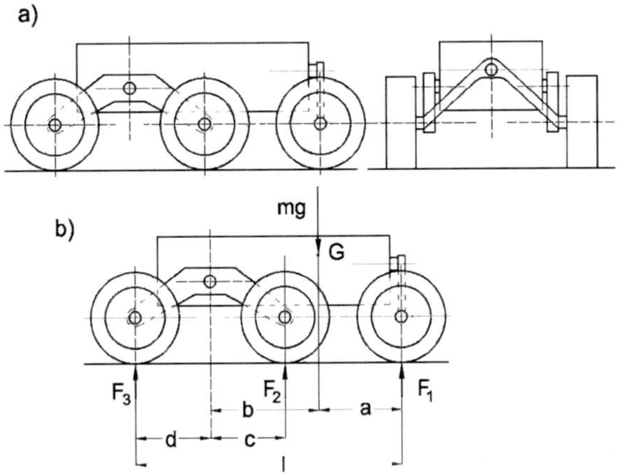

Fig. 5.35 Rocker arm suspension for a six-wheeled rover. (**a**) Sketch of the suspension system; (**b**) forces on the ground

In a solution of this kind the best arrangement is to have each wheel propelling itself with its own motor. The direction control may be performed by differential traction, i.e. slip steering, a thing that gives the capability of turning on the spot by having the wheels on the two sides turning in different directions, or by using actuators that steer two or all four wheels. To turn the vehicle on the spot the steer actuators must be able to steer the wheels by almost 90°.

If more than two wheels are steering, a strategy based on kinematic steering is used, and wheels steer always in opposite direction. This is well justified by the low speed of this kind of vehicles.

The capability of overcoming obstacles is limited, but the system is very simple, both from the mechanical and control viewpoint.

Rocker Arms Suspensions: Six Wheels

Rocker arm suspensions are also applicable to three-axle vehicles, as shown in Fig. 5.35: two wheels on each side are attached to a rocker arm, while the remaining two wheels, one on each side, are attached to a transversal rocker arm. The configuration is statically determined and the wheels can follow an irregular ground surface, at least within the limits imposed by the geometry of the rocker arms.

The forces on the ground are easily computed: when standing on horizontal flat ground the force acting on each front wheel (if the front of the vehicle is the end characterized by the transversal rocker arm) and on the hinges connecting the body to the rear rocker arms are

$$F_{z_1} = mg\frac{b}{2(a+b)}, \qquad F_r = mg\frac{a}{2(a+b)}. \tag{5.234}$$

The forces acting on the rear wheels are

$$F_{z_2} = mg\frac{ad}{2(a+b)(c+d)}, \qquad F_{z_3} = mg\frac{ac}{2(a+b)(c+d)}. \tag{5.235}$$

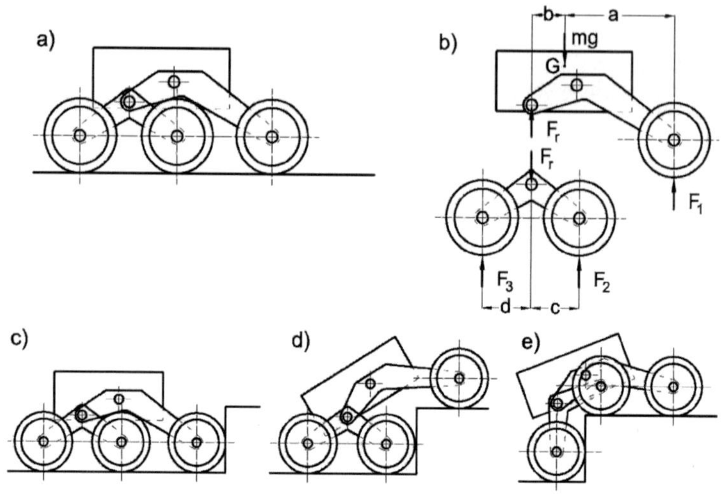

Fig. 5.36 Six-wheeled rocker bogie device. (**a**) Sketch of the vehicle; (**b**) forces when standing on level ground; (**c**)–(**e**) overcoming a step higher than the wheels' diameters

If

$$a = 2b, \qquad c = d, \tag{5.236}$$

each wheel carries $1/6$ of the weight of the rover.

Remark 5.25 This computation holds not only for the vehicle at standstill on level ground, but also for constant speed running on level ground, provided that all wheels are motorized: if the torque at each wheel balances exactly rolling resistance, the situation is the same.

Rocker Bogie Suspensions: Six Wheels

Rocker-Bogie (RB) suspensions were used by all NASA Mars Exploration Rovers (MER) like Spirit and Opportunity and, in 1997, the Mars Pathfinder Rover. The RB system permits each wheel to independently conform to uneven terrain, theoretically allowing the rover to traverse obstacles higher than the diameter of the rover's wheels. The RB system also provides good stability when the rover is operating on steep sloping surfaces. The RB suspension system was invented by Don Bickler, and patented by NASA/JPL in 2000 and 2001.

Six wheels rocker bogie suspensions appear to be a development of the rocker arm concept. The three wheels on each side are attached to a separate linkage made of a front rocker arm carrying the front wheel, the trunnion that connects the wheels to the body and a rear cylindrical hinge. The latter carries a rear rocker arm, on which the middle and the rear wheels are attached (Fig. 5.36a).

Provisions for folding or reducing the size of the system (telescopic arms, for instance) are often included for stowage in the lander, but are not important in understanding how the device works.

The way the body is attached to the two rocker bogies is the same as seen for the rocker arms of the arrangement shown in Fig. 5.34.

When standing on horizontal flat ground the force acting on each front wheel and on the hinge connecting the front arm with the rear one are

$$F_{z_1} = mg\frac{b}{2(a+b)}, \qquad F_r = mg\frac{a}{2(a+b)}. \tag{5.237}$$

The forces acting on the rear wheels are

$$F_{z_2} = mg\frac{ad}{2(a+b)(c+d)}, \qquad F_{z_3} = mg\frac{ac}{2(a+b)(c+d)}. \tag{5.238}$$

Again, the position of the main hinge is immaterial like in four-wheels rocker arms suspensions.

As in the previous case, if

$$a = 2b, \qquad c = d,$$

each wheel carries $1/6$ of the weight of the rover. If a is greater than $2b$ the front wheel is less loaded than the other ones, and this may be useful in overcoming obstacles.

The sequence of the rover crossing a step is shown in Fig. 5.36c–e. The front wheels are first forced against the step by the rear wheels and the rotation of the front wheel then lifts the front of the vehicle up and over the obstacle.

Assuming that the vehicle is on level ground, the force the rear wheels exert to push the front wheel against the obstacle is

$$F_x = 2\mu_x(F_{z2} + F_{z3}). \tag{5.239}$$

Assume that the front wheels has just parted contact with the ground to climb the obstacle. The force in vertical direction the front wheels exert at the contact with the latter is

$$F_x\mu_{x1} = 2\mu_{x_1}\mu_x(F_{z2} + F_{z3}),$$

where μ_{x1} is the traction coefficient against the vertical surface of the step, which is not said to be equal to μ_x. This force must be greater than the force $2F_{z1}$ acting on the front wheels for being able to climbing over the obstacle. This means that the product of the friction coefficients must be high enough:

$$\mu_{x_1}\mu_x > \frac{F_{z1}}{F_{z2} + F_{z3}}, \tag{5.240}$$

or, if $\mu_{x_1} = \mu_x$,

$$\mu_x > \sqrt{\frac{F_{z1}}{F_{z2} + F_{z3}}}. \tag{5.241}$$

If the load is evenly distributed on the wheels, it follows that for being able to climb over the obstacle

$$\mu_x > \frac{1}{\sqrt{2}} \approx 0.7. \tag{5.242}$$

After the wheel has started climbing, the conditions become easier, since some load transfer toward the rear wheels occurs, particularly if the center of mass is high.

From Fig. 5.36d it is clear that the second wheel finds a more difficult situation: the backward displacement of the center of mass unloads the front wheel, while loading the two rear ones, so that the required traction coefficient increases. The same holds for the last wheel.

During the overcoming of the obstacle, the vehicle slows down or stops altogether. This is not an issue for the operational speeds at which these vehicles operate, but requires a good control of the motors, to avoid that the wheels dig in loose ground.

As a conclusion, kinematically rocker bogies vehicles can climb over obstacles higher than the diameter of the wheels, but this is limited by the available traction and in practice it may be difficult to do so. Steps are, however, the most difficult type of obstacles and inclines allow a much better mobility.

If the center of mass can be displaced, which is a difficult issue, or there are actuators applying torques at the hinges of the rocker arms to change the load distribution on the ground, obstacle overcoming may be easier.

The Mars Exploration Rovers built with this configuration are able to withstand a tilt of 45° in any direction without tipping over and their low maximum speed justifies this choice.

Usually steering is implemented by rotating each wheel about a vertical axis using a suitable steer actuator. Normally both the front and rear wheels steer, while the central wheels are fixed and a kinematic steering strategy is implemented. To turn on the spot, a steering angle of almost 90° must be reached. If also the central wheels can steer, moving sideways becomes possible, a feature that increases the possibility to maneuver in tight places.

Rocker Bogie Suspensions: Eight Wheels

Rocker-Bogie suspensions can be used also for vehicles with a larger number of wheels. An example of a layout for an 8-wheeler is shown in Fig. 5.37. In this case the vehicle may be symmetrical, so that it can run in both direction without any difference.

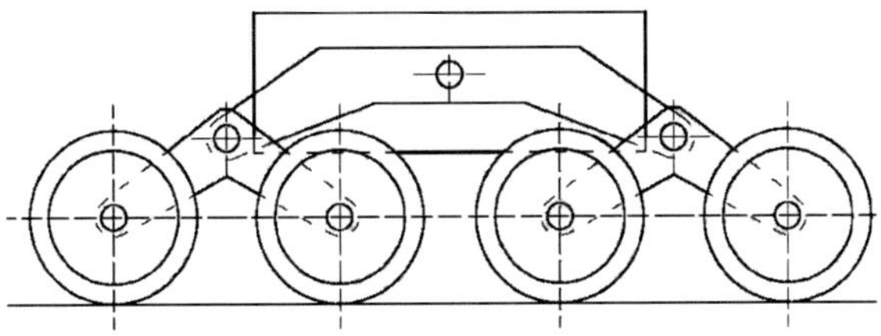

Fig. 5.37 Rocker bogey suspension for a rover with eight wheels

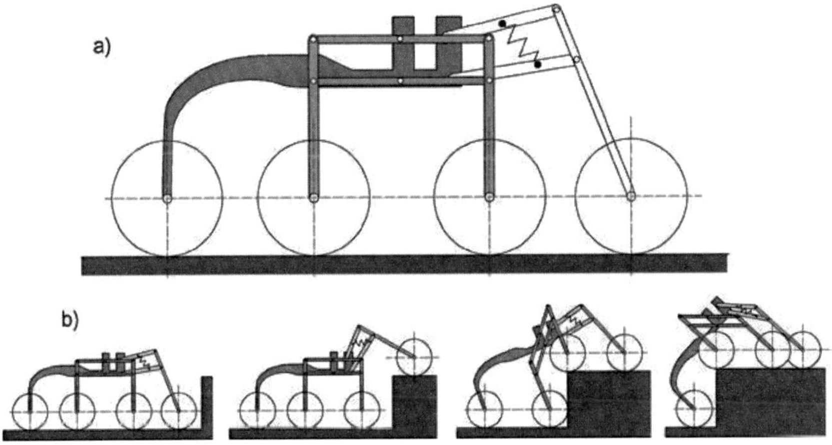

Fig. 5.38 The Shrimp, a six-wheeled rover (the front and rear axles have a single wheel) built by the Technical University of Lausanne (EPFL)

Other Non Elastic Suspensions

There is a large variety of possible kinematic layouts for non elastic suspensions. As an example, in Fig. 5.38 a six-wheeled rover built by the Technical University of Lausanne (EPFL) is shown. It has a rhombus configuration (single wheel at the first and last axle, two wheels at the other axles), and the first wheels is carried by a spring-loaded fork. The lateral wheels are carried by two bogies with a parallelogram configuration, while the rear wheel is carried directly by the body.

As shown in Fig. 5.38b it has a good capability of managing obstacles (it is shown when climbing a step whose height is 1.5 times the diameter of the wheels) both in structured and unstructured environment. The front wheel fork has a configuration causing the front wheel to raise when it is pushed backward, so making it easier to climb steps.

There are many cases where the articulated linkages connecting the wheels to the body are driven by actuators and may be considered as legs: these cases will be considered when dealing with wheels-legs hybrid vehicles.

5.5.2 Elastic Suspensions

The speeds that can be reached by machines provided with rocker bogie suspensions are too low for giving the required mobility to astronauts and consequently man-carrying planetary exploration vehicles must be much faster. Since the need to reduce vibration is greater due to the presence of humans on board, these vehicles must be endowed with elastic and damping suspensions. For instance the maximum speed of the Lunar Roving Vehicle used in the *Apollo* missions was 18 km/h and its four wheels were carried by transversal quadrilaterals independent suspensions with springs and dampers in a fairly conventional automotive layout.

It is likely that also robotic rovers will be supplied with elastic suspensions when higher speeds will be reached, for instance in case of robotic or teleoperated devices used to help astronauts in their exploration tasks. These devices must be able of reaching a speed at least equal to that of a human walking, taking also into account that in low gravity astronauts will perhaps walk faster than in 1 g conditions. A speed of at least 3 m/s (about 10 km/h) should be obtained and this implies the use of true suspensions, like those applied to a number of fast exploration microrobots built for military applications.

Solid-Axle

The simplest class of suspensions is that in which both wheels of the same axle are connected by a rigid beam. A solid axle of this kind must have two degrees of freedom, namely a translation in vertical direction and a roll rotation.

The simplest layout is that used in the early motor vehicles and often in modern industrial vehicles: it is based on a rigid beam, which includes also the final drive (solid drives or live axles, Fig. 5.39a), and is guided directly by the springs. Usually the latter are semi-elliptic leaf springs as shown in Fig. 5.39, in which the shock absorbers are not represented.

The solution of Fig. 5.39a, often referred to as Hotchkiss axle, has the disadvantages of approximating the correct kinematic behavior in which the axle can move only along z coordinate and rotate about the roll axis, in a poor way. The stiffness in x and y directions, although much higher than that in z direction, is not high enough, as is the stiffness for rotations about the y axis and, above all, any rolling motion is linked to a steering of the whole axis (roll steer). In other words, the derivative $\partial\delta/\partial\phi$ can be quite high. The latter characteristic is due to the fact that the motion of the points in which the axle is connected to the springs is not exactly vertical: while one of the wheels moves towards the body and the other one away from it, they move longitudinally in opposite directions causing the axle to steer.

The axle shown in Fig. 5.39b is usually referred to as a De Dion axle. It has been widely used also in slightly different versions, in which guiding elements are present and helical spring are used instead of leaf springs.

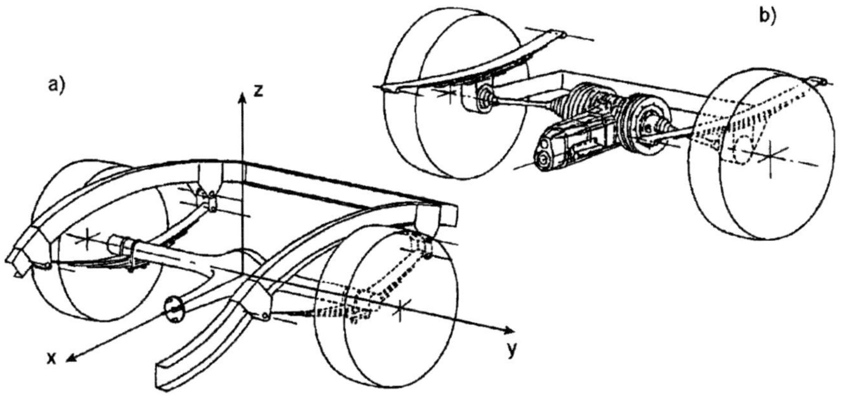

Fig. 5.39 Typical rigid-axle suspensions used on rear axles: the leaf springs also act as constraints to guide the axle in the suspension motions (from G. Genta, L. Morello, *The Automotive Chassis*, Springer, New York, 2009)

Some solutions in which different types of linkages are used to control the motion of the axle are shown in Fig. 5.40. The transversal guide can be supplied by a Watt quadrilateral (a), by a reaction bar (g), by a triangle hinged with two joints to the body (c) or to the axle (e), or by a straight guide (f). In other cases (b, c) the bending stiffness in xy plane of the leaf springs or of the longitudinal links act as transversal constraints. Only in the cases (a) and (f) the axle is guided in transversal direction in an almost kinematically correct way.

The longitudinal guide is supplied by an articulated quadrilateral (a, f) which constrains also rotations about y-axis, by a Watt's quadrilateral (c), which links rotations about y-axis with translations along z-axis, or by the longitudinal stiffness of the springs. Solutions (b), which contains also a compliant element, (d) and (e), kinematically similar to (f), can be assimilated to a quadrilateral. Only solution (c) uncouples exactly displacements z and roll rotations with displacements in x direction and rotations about z-axis, avoiding completely roll steering.

In the architecture (c) the torsional stiffness of the axle must be low, or better, a cylindrical hinge must uncouple the rotations of the two parts of the axle which are different from each other in any rolling motion. If this uncoupling is obtained through the deformation of the axle, a correct kinematic working of the suspension is impossible, as this layout relies on the deformation of some elements to allow the required displacements.

Devices which rely on the compliance of some elements allow one to reduce the number of parts and the cost of the system. The same can be said for the traditional suspensions based on leaf springs acting as guide elements, but the use of springs of a different type allows one to obtain better performances, particularly if the friction between metal parts (as occurs between the leaves of the spring) is avoided.

The choice of the type of guide elements is dictated by the stressing and deformation of the various elements, trying to avoid contact forces which can lock the motion in certain cases, as can occur in solution (f).

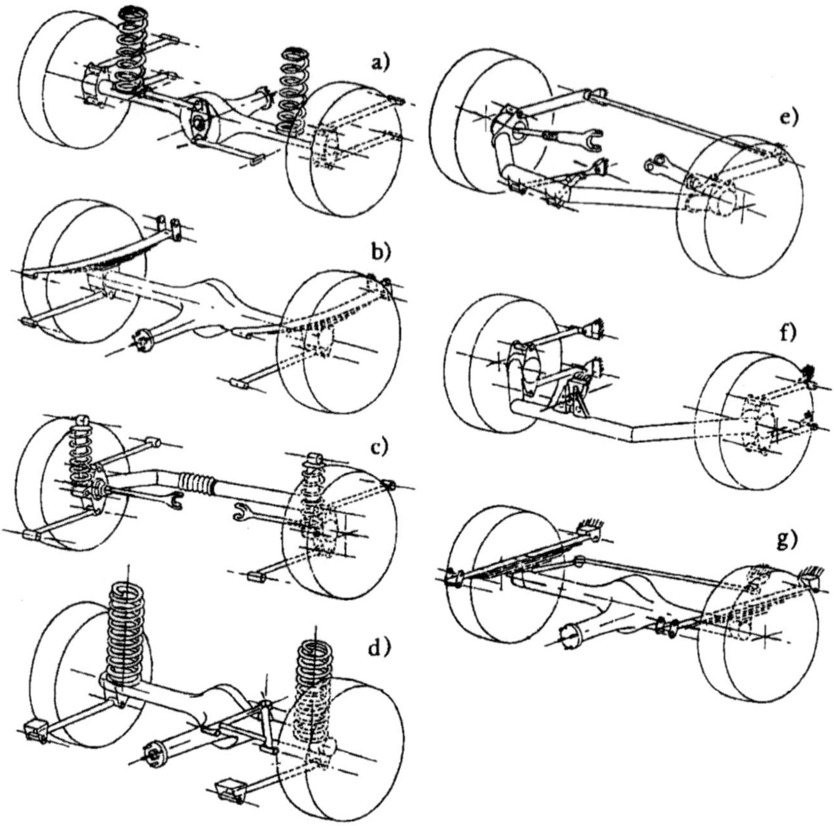

Fig. 5.40 Solid-axle suspensions with different geometry of the various linkages (from G. Genta, L. Morello, *The Automotive Chassis*, Springer, New York, 2009)

In all cases the track is constant as is the camber angle, at least if the compliance of the wheels is neglected. It follows that $(t)_{,z} = (t)_{,\phi} = (\gamma)_{,z} = (\gamma)_{,\phi} = 0$.

An important parameter in the study of suspensions is the position of the roll center RC of the suspension. The roll axis of the vehicle is the instantaneous axis for roll rotations of the vehicle when it is in symmetrical conditions (i.e., with the roll angle $\phi = 0$).

Remark 5.26 The roll axis is an instantaneous axis of rotation as the rolling motion with a large angle is not a pure rotation about a well defined axis and can be defined only for small rotations about a given position, namely the symmetrical equilibrium position.

In two axles vehicles, the point in which the roll axis crosses the plane perpendicular to the ground through the centers of the wheels of a given suspension is the roll center of that suspension. For symmetry reasons, the roll axis must lie in

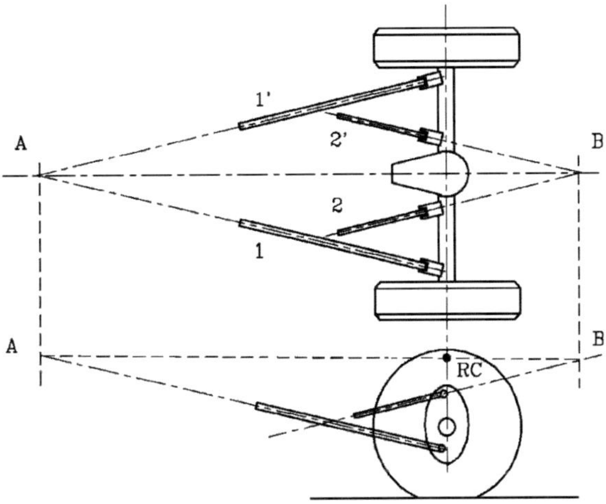

Fig. 5.41 Four-link solid-axle suspension. Position of point RC (from G. Genta, L. Morello, *The Automotive Chassis*, Springer, New York, 2009)

the symmetry plane of the vehicle (xz plane) and therefore also the roll centers of the suspensions must lie in it. Note that in case of a two-axle vehicle the roll center of each suspension can be determined from the characteristics of the relevant suspension only and that the roll axis can be defined as a line connecting the roll centers of the two suspensions. If the vehicle has more than two axles, the roll centers of the suspensions need not be aligned: a roll axis still exists, but it does not pass through the roll centers of the single suspensions, considered as insulated.

The roll center of each suspension can also be defined as the point on a plane perpendicular to the ground and to the symmetry plane in which the application of a lateral force F_y to the vehicle body does not cause any roll. The two definitions obviously coincide.

A four-link suspension is shown in Fig. 5.41. To obtain the position of the roll center, the intersections A and B of the axes of links 1–1′ and 2–2′ must be found first. They lie in the midplane of the vehicle. The roll center is found as the intersection of line AB with the plane perpendicular to the ground containing the centers of the wheels. If two links are parallel (say links 1 and 1′) the intersection is at infinity and line AB is parallel to the projection of the relevant links on the symmetry plane.

Independent-Wheels Suspensions

If the wheels are suspended independently, the linkages must constrain five out of the six degrees of freedom of the wheel (or better, of the wheel hub, because the wheel is also free to rotate about its axis). The unconstrained degree of freedom should be the translation in a direction perpendicular to the ground. As already stated, none of the many devices currently used fulfills exactly this requirement.

Since the suspension must restrain five degrees of freedom, it can be materialized as a system made of five bars with spherical hinges at their ends (Fig. 5.42). This

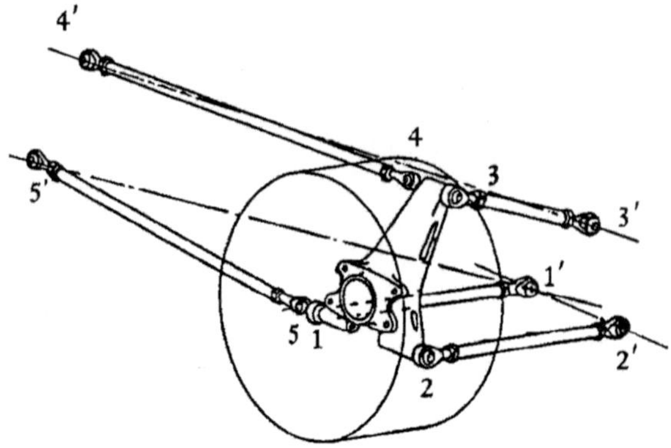

Fig. 5.42 General suspension based on five linkages to constrain the five degrees of freedom of the wheel (from G. Genta, L. Morello, *The Automotive Chassis*, Springer, New York, 2009)

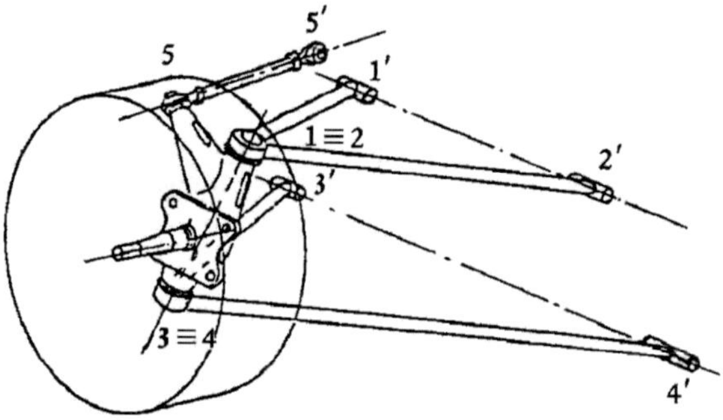

Fig. 5.43 Suspension based on transversal articulated quadrilaterals (SLA or A-arm suspension) (from G. Genta, L. Morello, *The Automotive Chassis*, Springer, New York, 2009)

layout, often referred to as multilink suspensions, has the advantage of allowing a large freedom of adjustment by changing the length of the bars by screwing in or out the joints It has little application outside the field of racing cars for its complexity, even if simpler multilink suspensions are now more widespread.

The suspensions of recent racing cars are similar to the multilink suspensions here described, but elastic hinges are used instead of the spherical hinges. This is a solution that can well be suitable for space applications, since metal elastic hinges are more tolerant to the harsh space environment than spherical joints or elastomeric elements.

Almost all independent suspensions can be obtained by grouping the bars of 5 bars multilink suspensions in different ways.

In general the motion of the wheel is not planar and as a consequence the study of the kinematic behavior is not easy. Nowadays this difficulty is easily overcome by using computer generated trajectories.

If points 1 and 2 and points 3 and 4 coincide with each other the corresponding bars become triangular elements: the suspension obtained is a transversal quadrilaterals suspension, often referred to as SLA (short-long arm) or A-arm suspension (Fig. 5.43). If lines $1'2'$ and $3'4'$ are parallel, the motion of the wheel is contained in a plane perpendicular to the line $1'2'$ and the projection of the mechanism in such plane is an articulated quadrilateral, whose side $1'3'$ is made by the vehicle body.

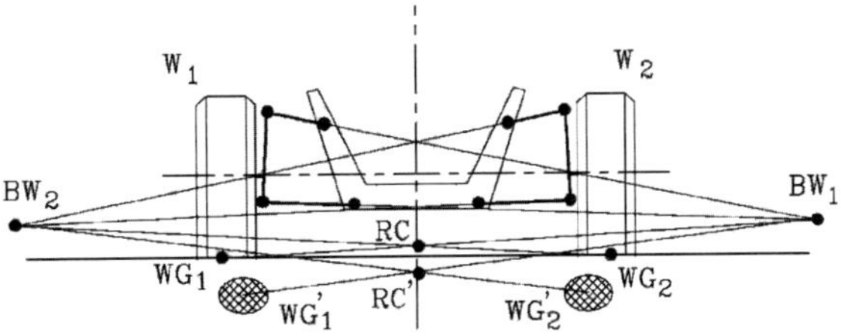

Fig. 5.44 Position of the roll center RC for a front suspension based on transversal articulated quadrilaterals with axes $1'2'$ and $3'4'$ parallel to x-axis. RC: Position with rigid tires; RC′: Position obtained taking into account the compliance of the tires (from G. Genta, L. Morello, *The Automotive Chassis*, Springer, New York, 2009)

Remark 5.27 The only vehicle used on the Moon, the *Apollo* RLV had all four wheels suspended by a linkage of this kind.

One of the main reasons of this choice is that it leaves much free space for the mechanical parts and the payload, as shown in Fig. 5.44.

In the same figure the construction to obtain the roll center is also shown. At first the center of rotation of the wheels with respect of the ground (in the suspension motion) WG_1 and WG_2 are located. If the wheels were rigid discs they would be at the center of the contact zone (WG_1 and WG_2). If the compliance of the wheels is also accounted for, they are below the ground in the shaded zones (WG'_1 and WG'_2), located slightly inboard with respect to the centers of the contact areas. The centers of rotation of the wheels with respect to the vehicle body BW_1 and BW_2 are thus located at the intersection of the directions of the upper and lower links, which can converge towards the outside of the vehicle (Fig. 5.45a, negative swing arm suspension) or towards the midplane (Fig. 5.44, positive swing arm suspension). It is possible to obtain

$$\frac{\partial t}{\partial z} = 0$$

by stating that points BW_1 and BW_2 lie on the ground (Fig. 5.45b), but this condition can be obtained only for one or two values of the load. If $\partial\gamma/\partial\phi$ must also vanish, points BW_1, BW_2 and RC must be located on the ground in the symmetry plane (Fig. 5.45c).

By connecting points BW1 and WG1 and points BW2 and WG2 and intersecting such lines the roll center RC, which lies in the symmetry plane, can be located. In the case of transversal articulated quadrilaterals, it is usually close to the ground or, if the deformation of the tires is considered, even below it. If the axes of the hinges of the two triangular linkages are not horizontal or not parallel (Fig. 5.46) the determination of the roll center and of the motion of the latter is far more complicated.

If the upper triangle is substituted by a prismatic guide, a MacPherson suspension is obtained (Fig. 5.47). It is a very simple solution that leaves much free space for the mechanical parts of the vehicle, quite common for the front axle of cars, particularly small ones. However, the fact that the shock absorber is a structural element acting as

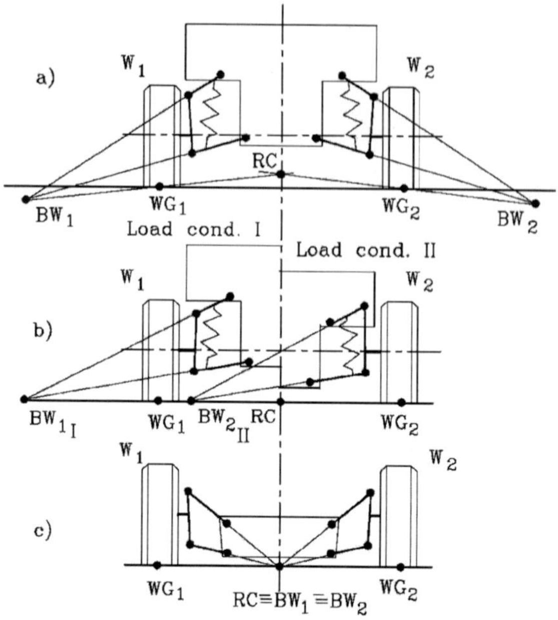

Fig. 5.45 Different schemes for transversal articulated quadrilateral suspensions (from G. Genta, L. Morello, *The Automotive Chassis*, Springer, New York, 2009)

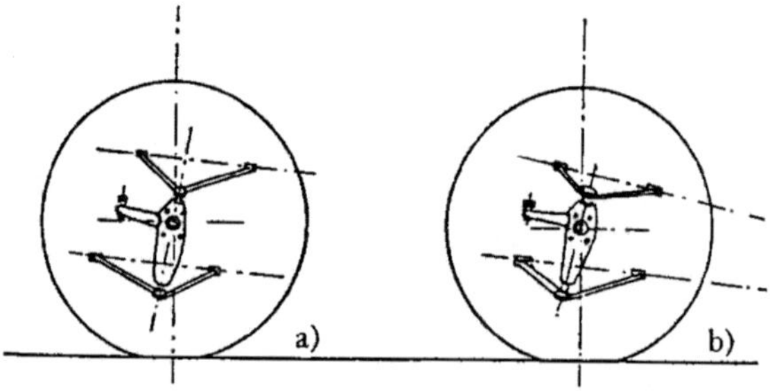

Fig. 5.46 Suspension based on articulated quadrilaterals with hinge axes not horizontal (**a**) and not parallel (**b**) (from G. Genta, L. Morello, *The Automotive Chassis*, Springer, New York, 2009)

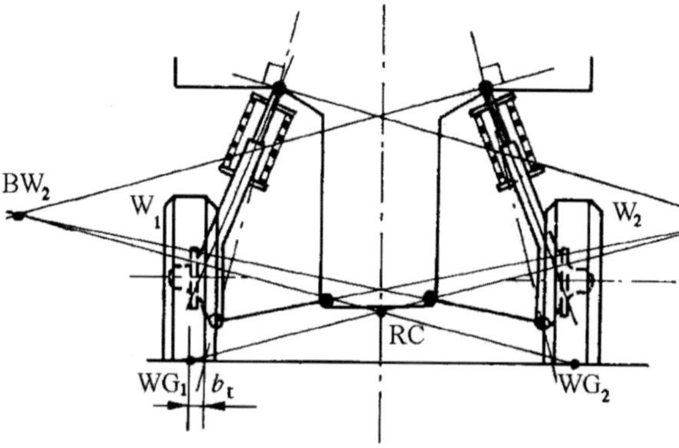

Fig. 5.47 MacPherson suspension (from G. Genta, L. Morello, *The Automotive Chassis*, Springer, New York, 2009)

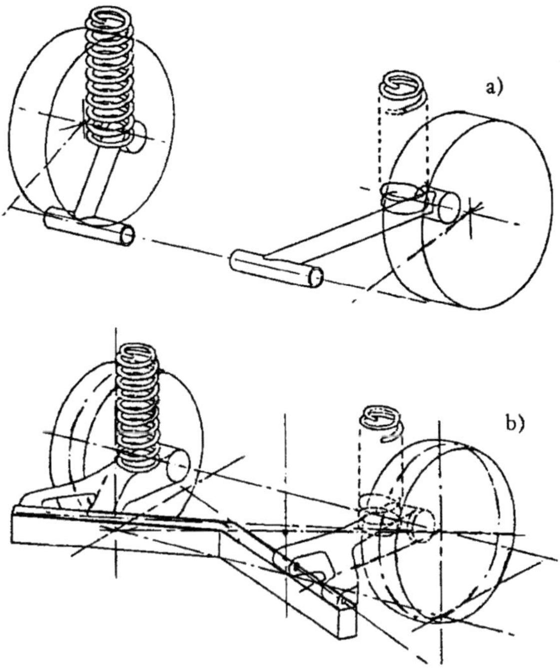

Fig. 5.48 Suspension based on trailing arms, with hinge axis parallel to y-axis (**a**) and inclined (**b**) (from G. Genta, L. Morello, *The Automotive Chassis*, Springer, New York, 2009)

a prismatic guide may be a limiting factor for demanding applications like planetary rovers and vehicles.

A different approach, which is promising for the use on planetary vehicles and rovers, is that of trailing arm suspensions (Fig. 5.48). The arms can be hinged to an axis which is perpendicular to the symmetry plane of the vehicle but this is not always the case. In the first case the track remains constant,

$$\frac{\partial t}{\partial z} = \frac{\partial t}{\partial \phi} = 0$$

and the camber angle does not change in the vertical motions and is equal to the roll angle

$$\frac{\partial \gamma}{\partial z} = 0, \qquad \frac{\partial \gamma}{\partial \phi} = 1.$$

If the compliance of the tires is neglected, the roll center is on the ground (Fig. 5.48a) or slightly below (Fig. 5.48b).

When a suspension of this type is used for a steering axle, the orientation of the kingpin axis (the axis about which the steering motion occurs) changes in both vertical and rolling motions.

Trailing arms are mostly used for rear axles. When a solution of this kind is used for a front axle, the swing arms are directed forward and not backwards and the term pushed arms is often used. While in the past there were cars with pushed arms at the

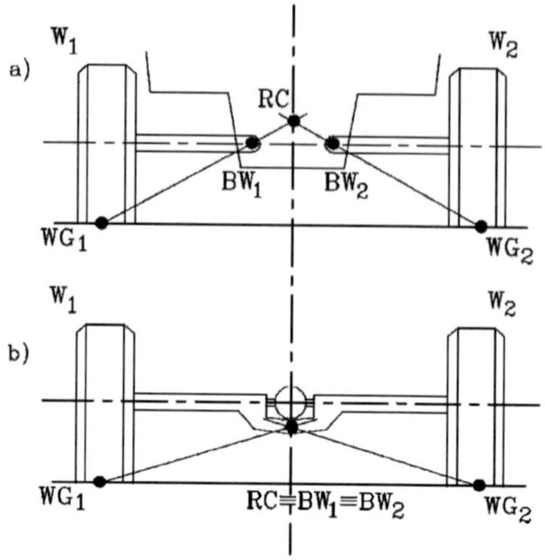

Fig. 5.49 Suspension based on transversal swing arms, with hinges located in two different points (**a**) and in the plane of symmetry (**b**) (from G. Genta, L. Morello, *The Automotive Chassis*, Springer, New York, 2009)

front axle and trailing arms at the rear wheels, this solution is now quite uncommon. An advantage of trailing arms is that when the wheel encounter an obstacle and is forced backwards, it raises, allowing to get easily over the obstacle. On the contrary, when a pushed arm suspension encounters an obstacle and the wheel is pushed backwards, the wheel goes down, against the obstacle, transferring a large force to the body. In low speed planetary rovers this drawback has little importance, and the combination of trailing and pushed arms is a viable and interesting solution.

Another solution is that based on transversal swing arms. The hinges of the arms can be located at different points (Fig. 5.49a) or be coincident (Fig. 5.49b). The roll center can be quite high on the road and the values of $\partial t/\partial z$ and of $\partial t/\partial \phi$ may be not small.

Another possible alternative, very seldom considered, is the use of vertical prismatic guides. Though it seems to be the simplest and more kinematically correct solution, friction due to transversal forces, which in some cases are large, may cause the suspension to lock or at least not to move freely. All solutions based on prismatic guides have this problem, which in MacPherson suspensions is often solved by setting the spring in the direction of the load and not of the strut, in order to relieve from it all transversal forces.

The two suspensions of the same axle can be interconnected using mechanical springs (as the antiroll bars), or pneumatic or hydraulic devices. Apart from this "transversal" interconnection, sometimes the two wheels of the same side are connected to each other, leading to a "longitudinal" interconnection.

The condition

$$\frac{\partial \gamma}{\partial \phi} = 1$$

is generally regarded as an unwanted characteristic of some suspensions, like transversal parallelograms or longitudinal swing arms. When managing a turn, the vehicle body rolls outwards (e.g., in curves to the left the body rolls to the right) and, if the wheels remain parallel to the body, they produce a camber thrust with a direction opposite to that of the cornering forces due to the sideslip angles. This reduces the handling performance of the vehicle.

The opposite situation occurs in motorcycles, which roll inwards with respect to the direction of the bend, producing camber thrusts that add to the cornering forces. Since this is impossible in vehicles with four or more wheels, the usual approach is to strive to have a derivative $(\gamma)_{,\phi}$ as small as possible. In transversal quadrilaterals suspensions, this is done having an upper arm shorter than the lower one (from which the definition of short and long arms suspension). The difference between the camber and the roll angles is often referred to as *camber recovery*.

Remark 5.28 Apparently, this may appear to have little relevance in space robots and rovers, owing to their low speed. The low gravitational acceleration, however, makes it important to exploit as fully as possible the low lateral forces of the wheels, and a good camber recovery may be an important requirement of the suspension.

5.5.3 Anti-dive and Anti-squat Designs

When the vehicle accelerates or brakes a load transfer between front and rear wheels occurs. This causes the body to pitch up (lift or squat) or down (dive). Apparently, in the case of two-axle vehicles, the forces acting on the front and rear axles F_{z_1} and F_{z_2} can be approximated as

$$\begin{cases} F_{z_1} = F^*_{z_1} - m\dfrac{h_G}{l}\dot{V}, \\ F_{z_2} = F^*_{z_2} + m\dfrac{h_G}{l}\dot{V}, \end{cases} \tag{5.243}$$

where forces $F^*_{z_i}$ are those occurring when the vehicle does not accelerate. The lift of the front and the rear of the body are thus, respectively,

$$\Delta z_1 = m\frac{h_G}{lK_f}\dot{V}; \qquad \Delta z_2 = -m\frac{h_G}{lK_r}\dot{V},$$

where K_f and K_r are the vertical stiffness of the front and rear suspensions. The pitch angle due to an acceleration is thus

$$\theta = \frac{1}{l}(-\Delta z_1 + \Delta z_2) = -m\frac{h_G}{l^2}\dot{V}\left(\frac{1}{K_f} + \frac{1}{K_r}\right). \tag{5.244}$$

A positive value of θ occurs when the vehicle dives (pitches down) as it occurs with a negative acceleration, hence the minus sign in the formula.

This expression is, however, an oversimplification, for two reasons: firstly the longitudinal forces due to the driving or braking wheels can cause themselves a pitching moment due to the coupling of the suspensions and, secondly, the driving and braking torque reactions can be applied, at least partly, to the suspensions instead of the body, inducing further pitching. Both effects cause pitching even in constant speed driving. For driving torques this effect is much stronger when each wheel has its own motor in the hub, like in the case of the Lunar Roving Vehicle or in multi-wheeled rovers

If the suspension allows the wheels to move also in x direction, i.e. if the characteristic $\partial x/\partial z$ is not vanishingly small, a fraction

$$F_x \frac{\partial x}{\partial z}$$

of force F_x acting between the road and the wheel acts on the suspension and causes pitching. Equation (5.243) thus becomes

$$\begin{cases} F_{z_1} = F^*_{z_1} - m\dfrac{h_G}{l}\dot{V} - \left(\dfrac{\partial x}{\partial z}\right)_1 F_{x1}, \\ F_{z_2} = F^*_{z_2} + m\dfrac{h_G}{l}\dot{V} - \left(\dfrac{\partial x}{\partial z}\right)_2 F_{x2} \end{cases} \tag{5.245}$$

and (5.244) transforms into

$$\theta = -m\frac{h_G}{l^2}\dot{V}\left(\frac{1}{K_f} + \frac{1}{K_r}\right) - \left(\frac{\partial x}{\partial z}\right)_1 \frac{F_{x1}}{lK_f} + \left(\frac{\partial x}{\partial z}\right)_2 \frac{F_{x2}}{lK_r}. \tag{5.246}$$

If only longitudinal forces needed to accelerate or to brake the vehicle are considered and the percentage of the longitudinal force assigned to the front axle is k_l, it follows that

$$F_{x_1} = k_l m\dot{V}, \qquad F_{x_2} = (1 - k_l)m\dot{V}.$$

Equation (5.244) thus yields

$$\theta = -m\frac{\dot{V}}{l}\left[\frac{h_G}{lK_f} + \frac{h_G}{lK_r} + \frac{k_l}{K_f}\left(\frac{\partial x}{\partial z}\right)_1 - \frac{(1 - k_l)}{K_r}\left(\frac{\partial x}{\partial z}\right)_2\right]. \tag{5.247}$$

Obviously (5.247) holds in the case of acceleration and braking alike, provided that the sign of $\dot{V}$ is correct and a suitable value for k_l is used.

Consider for example the trailing arm suspension of Fig. 5.50a. With simple geometrical reasoning it is easy to assess that

$$\left(\frac{\partial x}{\partial z}\right) = \frac{e}{d}. \tag{5.248}$$

A similar equation holds also for the suspension of Fig. 5.50b. If a torque M_y is applied to the sprung mass, it causes an increase of the force acting on the spring equal to

$$\frac{M_y}{d}.$$

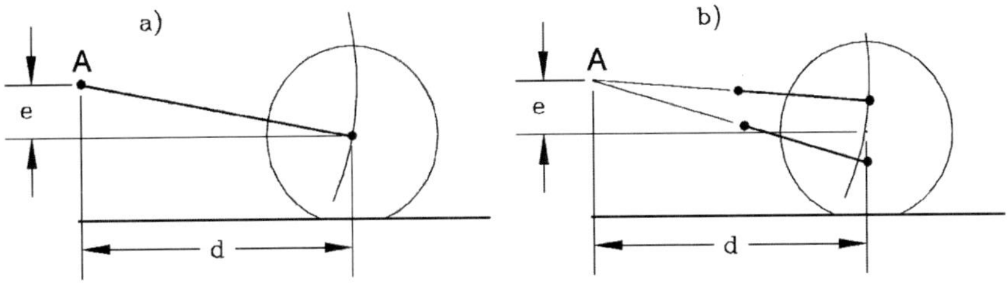

Fig. 5.50 Relationship between $\partial x/\partial z$ and geometry in suspensions with a single (**a**) and two (**b**) trailing arms. Note that if the arms are parallel $d \to \infty$ (from G. Genta, L. Morello, *The Automotive Chassis*, Springer, New York, 2009)

As the torque linked to the generation of driving or braking forces is equal to $-F_x R_l$, the result is that the application of the braking torque to the suspension can be accounted for by substituting

$$\left(\frac{\partial x}{\partial z}\right)_i + \left(\frac{R_l}{d}\right)_i \quad \text{for} \left(\frac{\partial x}{\partial z}\right)_i .$$

Note that d is positive when point A is in front of the wheel and negative otherwise.

The driving torque is applied to the unsprung mass in the case of live axles and, above all, when the motors are located in the wheel hubs, while in De Dion axles and independent suspensions it is applied directly to the vehicle body and this correction does not apply. Braking torques are on the contrary applied usually to the unsprung masses, so the term in R_l/d must always be accounted for. However, if the torque transmission between the sprung and the unsprung masses is supplied by linkages which prevent any relative rotation about y axis, these effects are minimized, since d tends to infinity.

The above relationships allow one to design the suspensions to compensate, usually partially, for squat or dive. A total compensation occurs when (5.246) yields $\theta = 0$. If

$$\frac{h_G}{lK_f} + \frac{k_l}{K_f}\left(\frac{\partial x}{\partial z}\right)_1 = 0 \tag{5.249}$$

the front of the car does not lift in acceleration or dive in braking, while if

$$\frac{h_G}{lK_r} - \frac{(1-k_l)}{K_r}\left(\frac{\partial x}{\partial z}\right)_2 = 0 \tag{5.250}$$

the rear does not squat in acceleration or lift in braking.

Note that in case of a single driving axle either $k_l = 0$ or $k_l = 1$ and both front and rear compensations cannot be performed together. To obtain a complete compensation the term in square brackets in (5.246) must vanish and the front of the vehicle must dive to compensate for the squat of the rear axle in front drives or the rear must lift in rear drives.

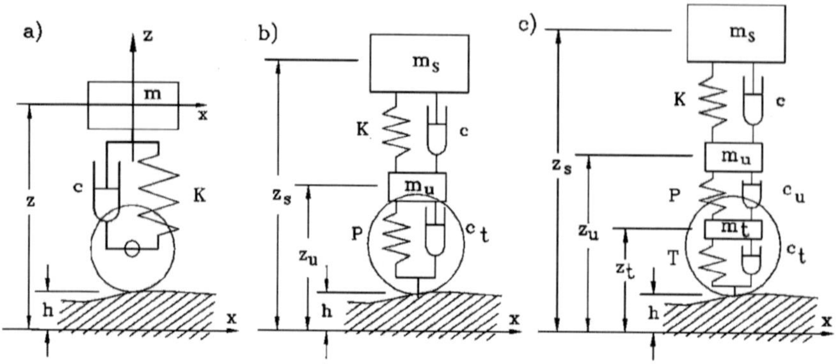

Fig. 5.51 Quarter-car models with one (**a**), two (**b**) and three (**c**) degrees of freedom (from G. Genta, L. Morello, *The Automotive Chassis*, Springer, New York, 2009)

In case of braking a total compensation of the front axle leads to the condition

$$\frac{k_l}{K_f}\left[\left(\frac{\partial x}{\partial z}\right)_1+\left(\frac{R_l}{d}\right)_1\right]=-\frac{h_G}{lK_f},\quad \text{i.e.}\quad \frac{k_l}{K_f}\left(\frac{e+R_l}{d}\right)_1=-\frac{h_G}{lK_f} \tag{5.251}$$

and that of the rear axle to

$$\frac{(1-k_l)}{K_r}\left[\left(\frac{\partial x}{\partial z}\right)_2+\left(\frac{R_l}{d}\right)_2\right]=\frac{h_G}{lK_r},\quad \text{i.e.}\quad \frac{(1-k_l)}{K_r}\left(\frac{e+R_l}{d}\right)_2=\frac{h_G}{lK_r}. \tag{5.252}$$

5.5.4 Quarter-Car Models

The simplest model to study the suspension dynamics is the so-called *quarter car model*, consisting of a single wheel, its suspension and the part of the vehicle body insisting on it. In the case of a rigid-axle suspension, its vertical dynamic can be studied with the same model, but referred to the axle (i.e. all masses, stiffness and damping coefficients are double, being referred to the axle). Three of the possible quarter car models are shown in Fig. 5.51.

The first model has a single degree of freedom. The wheels are considered as rigid bodies and the only mass considered is the sprung mass. This model holds well for motions taking place at low frequency, in a range extending to frequencies slightly in excess of the natural frequency of the sprung mass (in most cases, for human-carrying vehicles, up to 3–5 Hz).

The second model has two degrees of freedom. The wheels are considered as massless springs and both the unsprung and the sprung masses are considered. This model holds well for frequencies up to the natural frequency of the unsprung mass and slightly over (in most cases, up to 30–50 Hz).

The third model has three degrees of freedom. The wheels are modeled as a spring–mass–damper system, which represent their dynamic characteristics in their lowest mode. This model allows to study motions taking place at frequencies in excess to the first natural frequency of the wheels (120–150 Hz for pneumatic tires).

If higher frequencies must be accounted for, it is possible to introduce a number of modes of the wheel by inserting other masses. These models, which are essentially based on the modal analysis of the suspension-wheel system, are clearly approximated since the behavior of the wheel is usually nonlinear.

Quarter-Car with a Single Degree of Freedom

Consider the simplest quarter-car model shown in Fig. 5.51a. Using the symbols shown in the figure, the equation of motion of the system is

$$m\ddot{z} + c\dot{z} + Kz = c\dot{h} + Kh, \tag{5.253}$$

where z is the displacement from the static equilibrium position.

The frequency response of the quarter car is shown in Fig. 5.52a; it can be obtained simply by stating a harmonic input of the type

$$h = h_0 e^{i\omega t}.$$

The output is itself harmonic and can be expressed as

$$z = z_0 e^{i\omega t},$$

where both amplitudes h_0 and z_0 are complex numbers to account for the different phasing of response and excitation. The amplification factor, i.e. the ratio between the absolute values of the amplitudes of the response and the excitation and the phase of the first with respect to the second, can be easily shown to be

$$\begin{cases} \dfrac{|z_0|}{|h_0|} = \sqrt{\dfrac{K^2 + c^2\omega^2}{(K - m\omega^2)^2 + c^2\omega^2}}, \\ \Phi = \arctan\left(\dfrac{-cm\omega^3}{K(K - m\omega^2) + c^2\omega^2}\right). \end{cases} \tag{5.254}$$

More than the frequency response expressing the ratio between the amplitudes of response and excitation, what matters in vehicle suspensions is the inertance, i.e. the ratio between the amplitudes of the acceleration of the sprung mass and that of the displacement of the supporting point. Since in harmonic motion the amplitude of the acceleration is equal to the amplitude of the displacement multiplied by the square of the frequency, the inertance is

$$\frac{|(\ddot{z})_0|}{|h_0|} = \omega^2 \frac{|z_0|}{|h_0|}.$$

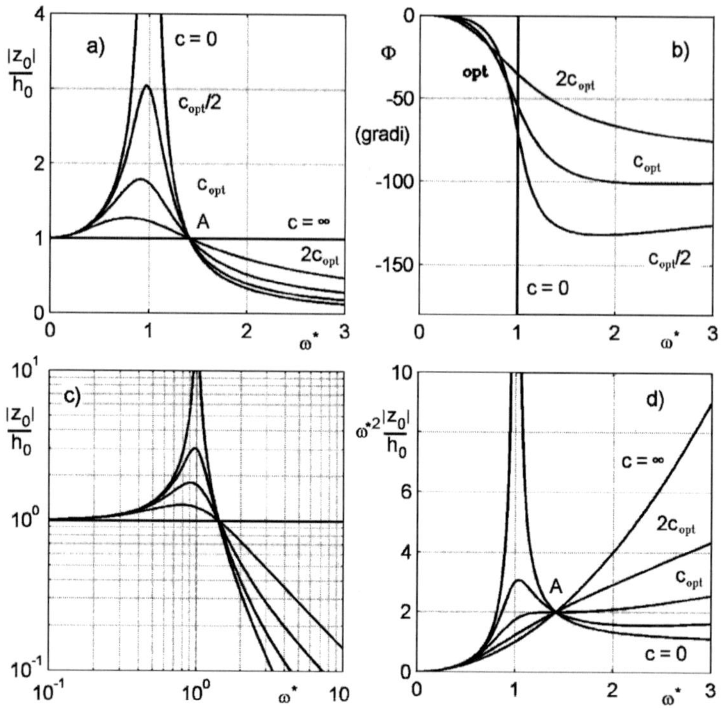

Fig. 5.52 Quarter car with a single degree of freedom, response to harmonic excitation. (**a**, **c**) Ratios between the amplitudes of the displacement and (**d**) of the acceleration of the sprung mass and the amplitude of the displacement the ground and (**b**) phase, for different values of the damping of the shock absorber. The responses are plotted as functions of the nondimensional frequency $\omega^* = \omega\sqrt{m/K}$ (from G. Genta, L. Morello, *The Automotive Chassis*, Springer, New York, 2009)

Both frequency responses are plotted in Fig. 5.52 for different values of the damping of the shock absorber, together with the phase Φ. The responses are plotted as functions of the nondimensional frequency,

$$\omega^* = \omega\sqrt{\frac{m}{K}}.$$

All curves pass through point A, located at a frequency equal to $\sqrt{2K/m}$. Since to obtain a good riding comfort the acceleration of the sprung mass must be kept to a minimum, a reasonable way to optimize the suspension is to choose a value of the damping of the shock absorber leading to a relative maximum, or at least point of stationarity, at point A on the curve related to the acceleration. By differentiating the expression of

$$\omega^2\frac{|z_0|}{|h_0|}$$

with respect to ω and equating the derivative to zero at point A, the following value of the damping is obtained:

$$c_{opt} = \sqrt{\frac{Km}{2}} = c_{cr}\frac{1}{2\sqrt{2}}, \tag{5.255}$$

where

$$c_{cr} = 2\sqrt{Km}$$

is the critical damping of the suspension.

For manned vehicles ride comfort is an important criterion, but also in the case of robots or other unmanned vehicles the vertical acceleration must be kept low, at least with the goal of avoiding damages to the rover itself or the payload, and to allow the latter to perform its tasks in the best way.

Although this way of optimizing the suspension can be easily criticized, since the comfort of a suspension is a far more complex concept that the simple reduction of the vertical acceleration (the so-called *jerk*, i.e. the third derivative of the acceleration with respect to time d^3z/dt^3 also plays an important role), it already gives important indications.

The dynamic component of the force the wheel exerts on the ground is

$$F_z = c(\dot{z} - \dot{h}) + K(z - h) = -m\ddot{z}. \tag{5.256}$$

To minimize the vertical acceleration leads to the minimization of the dynamic component of the vertical load on the wheel, which has a negative influence on its ability to exert longitudinal or cornering forces. The condition leading to the optimum comfort seems then to coincide with that leading to the optimum handling performances.

Equation (5.255) allows one to choose the value of the damping c. For the value of the stiffness K there is no such optimization: to minimize both the acceleration and the dynamic component of the force K should be as low as possible, the only limit to the softness of the springs coming from available space considerations: the softer the springs the larger the oscillations of the sprung mass.

Remark 5.29 This conclusion is drawn from an oversimplified model and applies only approximately to actual vehicles.

As a last consideration, the optimum damping expressed by (5.255) is lower than the critical damping. The quarter-car model is underdamped and can undergo free oscillations, although highly damped, since the damping ratio

$$\zeta = \frac{c}{c_{cr}}$$

is quite high, namely $1/2\sqrt{2} \approx 0.354$.

Example 5.10 Compute, using the quarter car model, the natural frequency of the suspensions of the Apollo LRV. As already stated, the suspensions had a fairly standard transversal quadrilateral layout, with the upper and lower arms almost parallel.

Using a one degree of freedom model, and assuming that the mass (210 kg + a payload of 450 kg) was evenly distributed on the wheels, from the measured values of the ground clearance (356 and 432 mm in full load and unloaded conditions) it is possible to find out the vertical stiffness of the suspension–tire assembly: $K = 2.40$ kN/m.

It is a very low value,yielding a natural frequency in bounce of just 0.6 Hz for the fully loaded vehicle or 1.1 Hz in empty conditions.

Quarter-Car with Two Degrees of Freedom

The following model is that shown in Fig. 5.51b: it is made by two masses, the sprung and the unsprung mass, connected by springs and dampers simulating the suspension and the tire. It is well suited for the study of the behavior of vehicle suspensions in a frequency range which goes beyond the natural frequency of the unsprung mass.

With reference to Fig. 5.51b, the equation of motion of the model is

$$\begin{bmatrix} m_s & 0 \\ 0 & m_u \end{bmatrix} \begin{Bmatrix} \ddot{z}_s \\ \ddot{z}_u \end{Bmatrix} + \begin{bmatrix} c & -c \\ -c & c + c_t \end{bmatrix} \begin{Bmatrix} \dot{z}_s \\ \dot{z}_u \end{Bmatrix} + \begin{bmatrix} K & -K \\ -K & K + P \end{bmatrix} \begin{Bmatrix} z_s \\ z_u \end{Bmatrix} = \begin{Bmatrix} 0 \\ c_t \dot{h} + Ph \end{Bmatrix}, \tag{5.257}$$

where z_s and z_u are the displacements from the static equilibrium position and are referred to an inertial frame.

The response to a harmonic excitation is readily obtained in the same way as seen for the previous model. By neglecting the damping of the tire c_t, which is usually small, it follows that

$$\begin{cases} \dfrac{|z_{s0}|}{|h_0|} = P\sqrt{\dfrac{K^2 + c^2\omega^2}{f^2(\omega) + c^2\omega^2 g^2(\omega)}}, \\ \dfrac{|z_{n0}|}{|h_0|} = P\sqrt{\dfrac{(K - m\omega^2)^2 + c^2\omega^2}{f^2(\omega) + c^2\omega^2 g^2(\omega)}}, \end{cases} \tag{5.258}$$

where

$$\begin{cases} f(\omega) = m_s m_u \omega^4 - \left[P m_s + K(m_s + m_u)\right]\omega^2 + KP, \\ g(\omega) = (m_s + m_u)\omega^2 - P. \end{cases}$$

The dynamic component of the force exerted in z direction by the tire on the ground can be easily computed in the same way as seen for the model with a single

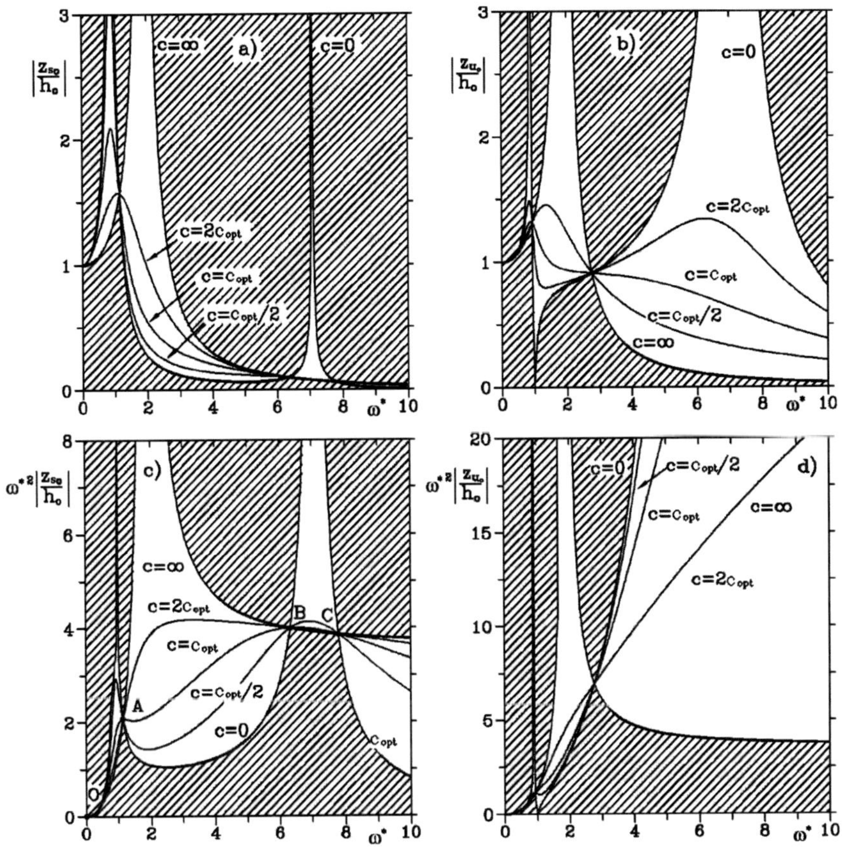

Fig. 5.53 Quarter car with two degrees of freedom, response to harmonic excitation. Ratios between the amplitudes of the displacements of the sprung and the unsprung masses (**a**, **b**) and of the accelerations (**c**, **d**) to the amplitude of the displacement of the ground, for different values of the damping of the shock absorber. The responses are plotted as functions of the nondimensional frequency $\omega^* = \omega\sqrt{m/K}$ (from G. Genta, L. Morello, *The Automotive Chassis*, Springer, New York, 2009)

degree of freedom, obtaining

$$\frac{|F_{z0}|}{|h_0|} = P\omega^2\sqrt{\frac{[K(m_s+m_u)-m_s m_u\omega^2]^2+c^2(m_s+m_u)\omega^2}{f^2(\omega)+c^2\omega^2g^2(\omega)}}. \tag{5.259}$$

The frequency responses related to both the sprung and the unsprung masses are plotted in Figs. 5.53a and b for a system with $P = 4K$ and $m_s = 10m_u$. The plots, shown using the nondimensional frequency

$$\omega^* = \omega\sqrt{\frac{m}{K}}, \tag{5.260}$$

show curves obtained with different values of damping c; all curves lie in the non-shaded region of the graph.

If $c = 0$ the natural frequencies are two and the peaks are infinitely high. Also for $c \to \infty$ the peak, corresponding to the natural frequency of the whole system, which is now rigid, over the spring simulating the tire, goes to infinity.

The frequency responses of Figs. 5.53a (sprung mass) and 5.53b (unsprung mass) multiplied by ${\omega^*}^2$ are shown in Figs. 5.53c and 5.53d; they give the nondimensional ratio between the accelerations of the two masses and the displacement of the supporting point. All curves pass through points O, A, B and C. Between O and A and between B and C the maximum acceleration of the sprung mass increases with decreasing damping, while between A and B and from C upwards it increases with the damping.

An optimum value of the damping can be found by trying to keep the acceleration as low as possible in a large field which goes up to the natural frequency of the unsprung mass, i.e. by looking for a curve which has a relative maximum or a stationarity point in A. Operating as seen for the previous model the following value is obtained:

$$c_{\mathrm{opt}} = \sqrt{\frac{Km}{2}}\sqrt{\frac{P+2K}{P}}. \tag{5.261}$$

Since P is much larger than K, the value of $\sqrt{(P+2K)/P}$ is close to unity and the optimum damping is only slightly larger than the one computed for the model with a single degree of freedom (5.255). In the case of Fig. 5.53

$$\sqrt{\frac{P+2K}{P}} = 1.22.$$

From Fig. 5.53c it is clear that this value of the damping is effective in maintaining the acceleration low in wide frequency range.

The amplitude of the dynamic component of force F_z expressed by (5.261) is plotted in nondimensional form (divided by $P|h_0|$) as a function of the nondimensional frequency in Fig. 5.54. The value of the optimum damping expressed by (5.261) is effective in keeping as low as possible also the maximum value of the dynamic component of force F_z at least at low frequencies. At higher frequencies a slightly higher value of damping could be effective, even if it would result in a larger acceleration of the sprung mass.

Example 5.11 Consider a quarter car model with the following data: sprung mass 240 kg, unsprung mass 38 kg, wheel stiffness 135 kN/m, spring stiffness 15.7 kN/m, damping coefficient equal to optimum damping (1.52 kNs/m). Compute the response in terms of minimum force on the ground $F_{\mathrm{st}} - F$ to a harmonic input in the frequency range between 0 to 12 Hz on the Moon and compare the results with those obtainable on Earth.

The static force is 450 N while on Earth it is 2.73 kN. The response for an amplitude of the excitation of 2, 3 and 10 mm is reported in Fig. 5.55.

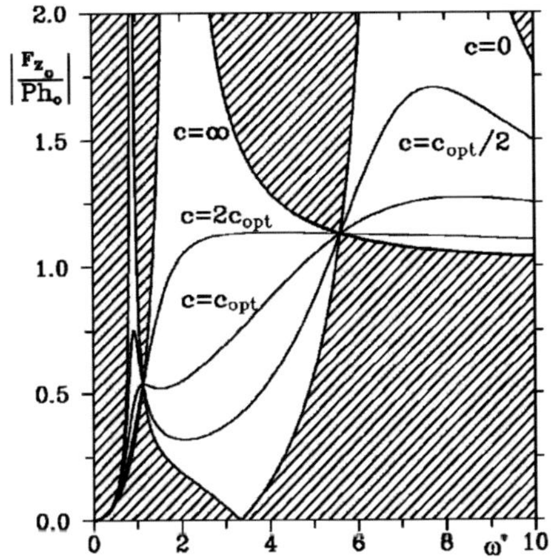

Fig. 5.54 Quarter car with two degrees of freedom, response to harmonic excitation. Ratio between the amplitude of the dynamic component of force F_z between tire and road and the displacement of the ground, made nondimensional by dividing it by the stiffness of the tire P, for different values of the damping of the shock absorber. The response is plotted as a function of the nondimensional frequency $\omega^* = \omega\sqrt{m/K}$ (from G. Genta, L. Morello, *The Automotive Chassis*, Springer, New York, 2009)

Fig. 5.55 Ratio between the minimum vertical force on the ground and the static force for a quarter car model as a function of the frequency when driving on Earth (*dashed lines*) and on the Moon (*full lines*) on a road with harmonic profile with a given amplitude h_0

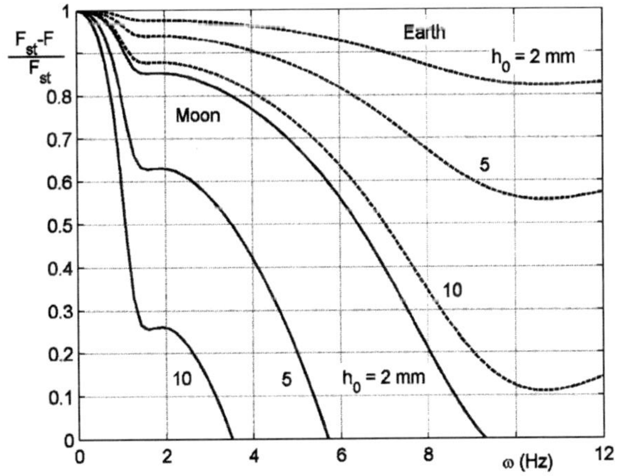

The curves stop when the minimum vertical force vanishes, since the linearized model loses any meaning when the wheel loses contact with the ground. On the Moon this occurs with an amplitude of the excitation as small as $h_0 = 2$ mm, and that causes the wheel to bounce at a frequency of about 9 Hz. For larger amplitudes, the contact of the wheel on the ground is quite uncertain, with the ensuing reduction of the already poor traction and cornering forces. On the contrary, on Earth even at an amplitude of 10 mm contact is assured.

Owing to low gravity, the wheels tend to lift from the ground as shown in the movies taken in the *Apollo* missions.

From the considerations seen above it is possible to draw the conclusion that the value of the damping coefficient expressed in (5.261) is optimal both from the viewpoint of comfort and that of handling, since it leads to low variations of the forces on the ground. A slightly higher damping can, however, improve slightly handling performances. This conclusion, obtained from a highly simplified model, is not in good accordance with the experimental evidence, which states that the value of damping optimizing riding comfort is lower than that optimizing handling.

A plot obtained considering a random velocity input whose power spectrum is a white noise in a frequency range between 0.1 and 100 Hz can shed some light on this issue. The power spectral density of the output of the system S_o is obtained from that of the input S_i by simply multiplying the latter by the square of the frequency response $H(\omega)$

$$S_o = H^2(\omega) S_i. \tag{5.262}$$

If the input is a white noise acting in a frequency range between ω_1 and ω_2, the r.m.s. (root mean square) value of the output is simply

$$O_{\text{r.m.s.}} = \sqrt{\int_{\omega_1}^{\omega_2} S_o \, d\omega} = \sqrt{S_i} \sqrt{\int_{\omega_1}^{\omega_2} H^2(\omega) \, d\omega}. \tag{5.263}$$

The r.m.s. value of the acceleration is easily computed: as the output is the acceleration of the sprung mass and the input is the vertical velocity of the contact point, the frequency response is the first equation (5.258) multiplied by ω. Also the r.m.s. value of the dynamic component of force F_z is readily obtained, using the frequency response (5.259) divided by ω. An interesting result is the graph obtained by plotting the r.m.s. value of the acceleration versus that of the force for various values of the damping of the shock absorber (Fig. 5.56).

The conditions leading to the optimum comfort (in the sense of minimum acceleration) and to the optimum handling (in the sense of minimum force variations) are readily identified: The first is obtained with a damping which is lower than the optimum damping defined above while the second for a value which is higher. This result is in better accordance with the experimental results than the previous one.

As already stated, the compliance of the wheel has a negligible effect on the frequency response at low frequency, while at higher frequencies it must be accounted for. A comparison between the results obtained using the quarter car models with one and two degrees of freedom is shown in Fig. 5.57.

5.5.5 *Bounce and Pitch Motions*

To study the suspension dynamics of a vehicle with two axles the model shown in Fig. 5.58a can be used. Note that the same model can be extended to a vehicle with any number of axles, just by adding a further unsprung mass for each axle beyond the second one.

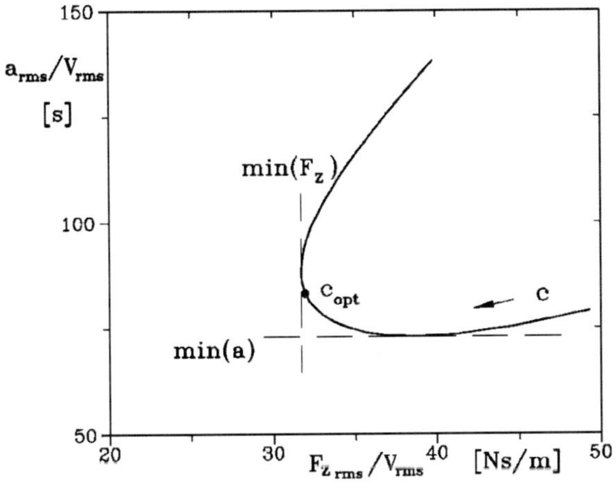

Fig. 5.56 Ratio of the r.m.s. value of the acceleration of the sprung mass and the r.m.s. value of the input velocity versus the ratio between the r.m.s. value of the dynamic component of force F_z and the r.m.s. value of the input velocity for various values of the damping c of the shock absorber. White noise velocity input in a range between 0.1 and 100 Hz. Quarter car model with two degrees of freedom. Data: $m_s = 238$ kg; $m_u = 38$ kg; $K = 15.7$ kN/m; $P = 135$ kN/m (from G. Genta, L. Morello, *The Automotive Chassis*, Springer, New York, 2009)

Fig. 5.57 Acceleration of the sprung mass as a function of the frequency for a unit displacement input. Comparison between the quarter car model with one and two degrees of freedom (in the latter case $P = 4K$, $m_s = 10m_u$). (1) 2 d.o.f.; (2) 1 d.o.f., damping defined by (5.261); (3) 1 d.o.f., damping defined by (5.255) (from G. Genta, L. Morello, *The Automotive Chassis*, Springer, New York, 2009)

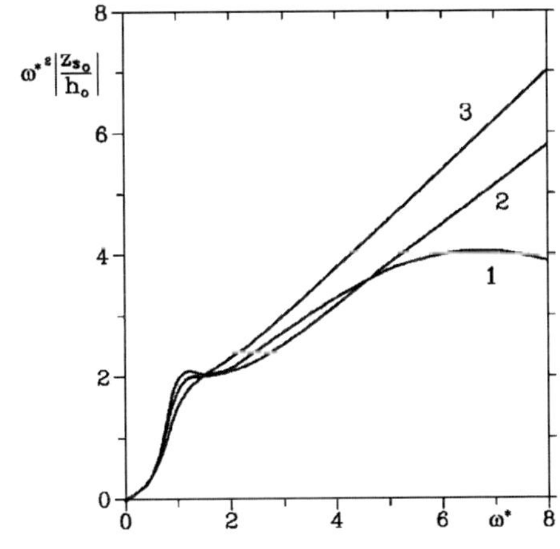

Note that:

- the model does not include some aerodynamic effects, which are marginal on Earth and nil on airless worlds,
- the moment due to weight is also neglected, since usually is small and to introduce it the position of the pitch center must be known. It can be introduced in the study of any particular vehicle, once the geometry of the suspensions is fully defined,
- the uncoupling between longitudinal, lateral and suspension dynamics was introduced speaking of rigid-body vehicles, but it holds also in the case of vehicles

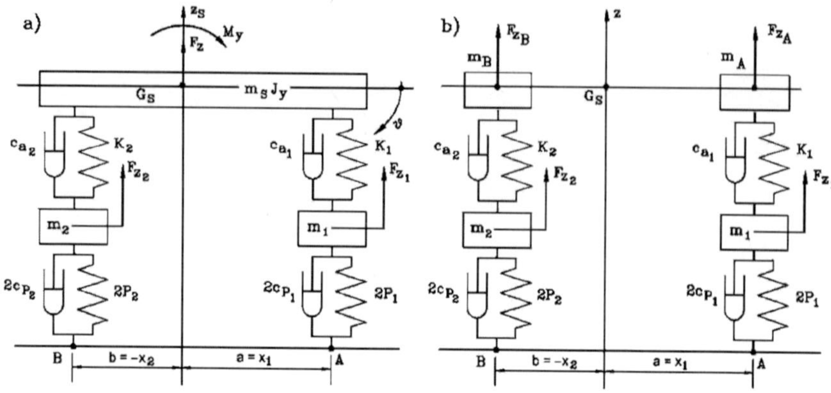

Fig. 5.58 (**a**) Model for the suspension dynamics of a vehicle with two axles. (**b**) Case in which the model splits in that of two quarter car models (after G. Genta, L. Morello, *The Automotive Chassis*, Springer, New York, 2009)

with elastic suspensions. If the suspensions are based on rigid axles, it is straightforward to include the degrees of freedom for the vertical displacements of the axles (here defined as Z_1 and Z_2) in the suspension dynamics and those related to the rolling of the axles (angles ϕ_1 and ϕ_2) in the lateral dynamics, together with body rolling ϕ. If one or more suspensions are of the independent type it is still possible to uncouple the equations, introducing a mean vertical displacement of the ith suspension

$$Z_i = \frac{Z_{ir} + Z_{il}}{2},$$

where subscript r and l refer to the right and left wheel, and a sort of rolling of the axle

$$\phi_i = \frac{Z_{il} - Z_{ir}}{d},$$

where d is a suitable length. The first variable enters suspension dynamics, while the second one enters lateral dynamics.

Obviously the mass m_i and the other characteristics of the suspension are the characteristics of the whole axle and not those of the single independent suspension. The equation of motion can be written in the form

$$\begin{bmatrix} m_S & 0 & 0 & 0 \\ 0 & J_y & 0 & 0 \\ 0 & 0 & m_1 & 0 \\ 0 & 0 & 0 & m_2 \end{bmatrix} \begin{Bmatrix} \ddot{Z}_s \\ \ddot{\theta} \\ \ddot{Z}_1 \\ \ddot{Z}_2 \end{Bmatrix}$$

$$+ \begin{bmatrix} c_1 + c_2 & -ac_1 + bc_2 & -c_1 & -c_2 \\ & a^2c_1 + b^2c_2 & ac_1 & -bc_2 \\ & & c_1 + 2c_{t_1} & 0 \\ \text{symm} & & & c_2 + 2c_{t_2} \end{bmatrix} \begin{Bmatrix} \dot{Z}_s \\ \dot{\theta} \\ \dot{Z}_1 \\ \dot{Z}_2 \end{Bmatrix}$$

$$+\begin{bmatrix} K_1+K_2 & -aK_1+bK_2 & -K_1 & -K_2 \\ & a^2K_1+b^2K_2 & aK_1 & -bK_2 \\ & & K_1+2P_1 & 0 \\ \text{symm} & & & K_2+2P_2 \end{bmatrix} \begin{Bmatrix} Z_s \\ \theta \\ Z_1 \\ Z_2 \end{Bmatrix}$$

$$= 2\begin{Bmatrix} 0 \\ 0 \\ c_{t_1}\dot{h}_{\mathrm{A}} + P_1 h_{\mathrm{A}} \\ c_{t_2}\dot{h}_{\mathrm{B}} + P_2 h_{\mathrm{B}} \end{Bmatrix}. \tag{5.264}$$

The excitation vector has been written considering only the forcing functions due to the vertical motions of points A and B, neglecting longitudinal forces at the road–wheel interface. If a coupling between vertical and horizontal motions of the suspensions is present, the effects of the inertia of the wheels and driveline on the ride comfort are also neglected.

Equation (5.264) allows one to study bounce and pitching motions of the suspended mass of the vehicle and the corresponding motions of the unsprung masses.

In the case of two axles vehicles, an important parameter is the dynamic index

$$I_d = \frac{J_y}{mab} = \frac{r^2}{ab}, \tag{5.265}$$

where r is the radius of gyration of the sprung mass about y-axis.

If the dynamic index is equal to one, the free oscillations of the sprung mass are rotations about the points of attachment of the suspensions: the model of Fig. 5.58a coincides with that of Fig. 5.58b and it is not possible to identify a bounce and a pitch mode but the free motion is better described in terms of front oscillation and rear oscillation.

To demonstrate this feature, the equations of motion can be written in terms of coordinates Z_A and Z_{B} instead of Z_s and θ. The coordinate transformation can be expressed as

$$\begin{Bmatrix} Z_s \\ \theta \\ Z_1 \\ Z_2 \end{Bmatrix} = \frac{1}{l}\begin{bmatrix} b & 0 & a & 0 \\ -1 & 0 & 1 & 0 \\ 0 & l & 0 & 0 \\ 0 & 0 & 0 & l \end{bmatrix} \begin{Bmatrix} Z_{\mathrm{A}} \\ Z_1 \\ Z_{\mathrm{B}} \\ Z_2 \end{Bmatrix}. \tag{5.266}$$

The inertia matrix with the new coordinates can be obtained as

$$\mathbf{M}' = \mathbf{T}^T\mathbf{M}\mathbf{T},$$

where $\mathbf{T}$ is the transformation matrix defined by (5.266). All other matrices are obtained in the same way. Equation (5.264) becomes

$$\begin{bmatrix} m_S\dfrac{b^2+r^2}{l^2} & 0 & m_S\dfrac{ab-r^2}{l^2} & 0 \\ & m_1 & 0 & 0 \\ & & m_S\dfrac{a^2+r^2}{l^2} & 0 \\ \text{symm} & & & m_2 \end{bmatrix} \begin{Bmatrix} \ddot{Z}_A \\ \ddot{Z}_1 \\ \ddot{Z}_B \\ \ddot{Z}_2 \end{Bmatrix}$$

$$+ \begin{bmatrix} c_1 & -c_1 & 0 & 0 \\ & c_1+2c_{t_1} & 0 & 0 \\ & & c_2 & -c_2 \\ \text{symm} & & & c_2+2c_{t_2} \end{bmatrix} \begin{Bmatrix} \dot{Z}_A \\ \dot{Z}_1 \\ \dot{Z}_B \\ \dot{Z}_2 \end{Bmatrix}$$

$$+ \begin{bmatrix} K_1 & -K_1 & 0 & 0 \\ & K_1+2P_1 & 0 & 0 \\ & & K_2 & -K_2 \\ \text{symm} & & & K_2+2P_2 \end{bmatrix} \begin{Bmatrix} Z_A \\ Z_1 \\ Z_B \\ Z_2 \end{Bmatrix}$$

$$= 2 \begin{Bmatrix} 0 \\ c_{t_1}\dot{h}_A + P_1 h_A \\ 0 \\ c_{t_2}\dot{h}_B + P_2 h_B \end{Bmatrix}. \tag{5.267}$$

It is thus clear that if

$$r^2 = ab$$

the first two equations uncouple from the other two, yielding two equations of motion of independent quarter cars with sprung masses

$$m_S\frac{b}{l} \quad \text{and} \quad m_S\frac{a}{l}$$

respectively.

In the study of the low frequency modes of the sprung mass the wheels can be considered as rigid. A model in which the sprung mass is a beam directly suspended on the ground by the suspension springs and dampers can be used. Its equation of motion is simply

$$\begin{bmatrix} m_S & 0 \\ 0 & J_y \end{bmatrix} \begin{Bmatrix} \ddot{Z}_s \\ \ddot{\theta} \end{Bmatrix} + \begin{bmatrix} c_1+c_2 & -ac_1+bc_2 \\ -ac_1+bc_2 & a^2c_1+b^2c_2 \end{bmatrix} \begin{Bmatrix} \dot{Z}_s \\ \dot{\theta} \end{Bmatrix}$$

$$+ \begin{bmatrix} K_1+K_2 & -aK_1+bK_2 \\ -aK_1+bK_2 & a^2K_1+b^2K_2 \end{bmatrix} \begin{Bmatrix} Z_s \\ \theta \end{Bmatrix}$$

$$= \frac{1}{l} \begin{Bmatrix} bc_1\dot{h}_A + ac_2\dot{h}_B + bK_1h_A + aK_2h_B \\ -c_1\dot{h}_A + c_2\dot{h}_B - K_1h_A + K_2h_B \end{Bmatrix}. \tag{5.268}$$

The modes of oscillation of the undamped system can be obtained by studying the homogeneous equation without the damping terms. It is thus clear that if

$$aK_1 = bK_2$$

the bounce mode uncouples from the pitch mode: the first is a vertical translation of the sprung mass while the second one is a rotation about its center of mass.

In general such condition is not met and both modes involve translation of the center of mass and pitch rotation, i.e. both modes are rotations about two centers none of them coinciding with the center of mass. If the dynamic index is equal to unity they coincide with the suspension points. Otherwise the positions of the centers depend on both the value of

$$-aK_1 + bK_2$$

and on the dynamic index. In general the two points are located one in front and the other behind the center of mass and one lies inside and one outside the wheelbase. The mode whose center of rotation lies outside the wheelbase is mainly translational and is considered as a bounce mode, while the other, being mainly rotational, is considered as a pitch mode.

The natural frequencies ω of the undamped system can be obtained from the equation

$$\omega^4 - \omega^2 \frac{K_1(r^2 + d_1^2) + K_2(r^2 + d_2^2)}{m_s r^2} + K_1 K_2 \frac{l^2}{m_s^2 r^2} = 0, \tag{5.269}$$

where again the term due to weight has been neglected.

If bounce and pitch uncouple, $aK_1 = bK_2$, the natural frequencies are

$$\begin{cases} \omega_1 = \sqrt{\dfrac{lK_1}{bm_s}} & \text{bounce,} \\ \omega_2 = \sqrt{\dfrac{laK_1}{r^2 m_s}} = \omega_1 \sqrt{\dfrac{ab}{r^2}} & \text{pitch.} \end{cases} \tag{5.270}$$

From (5.270) it follows that when the dynamic index has a unit value the two natural frequencies coincide. This solves an apparent inconsistency: if $aK_1 = bK_2$ the centers of rotations are one in the center of mass (pitch mode) and one at infinity (bounce mode) while when the dynamic index is equal to one the rotation centers are at the suspension points. When both conditions apply at the same time the natural frequencies coincide, and when two coincident eigenvalues are present any linear combination of the eigenvectors is itself an eigenvector. This means that, when dealing with a rigid beam, any point of the beam can be considered as rotation center.

The bounce and pitch dynamics of the suspended mass are strictly related to each other. Some empirical criteria, dating back from the 1930s, for the choice of the relevant parameters are here reported:[23]

[23] T.D. Gillespie, *Fundamentals of Vehicle Dynamics*, SAE, Warrendale, 1992.

- The vertical stiffness of the front suspension must be about 30% lower than that of the rear suspension;
- The pitch and bounce frequencies must be close to each other; the bounce frequency should be less than 1.2 times the pitch frequency;
- Neither frequency should be greater than 1.3 Hz;
- The roll frequency should be approximately equal to the bounce and pitch frequency.

The first rule states that the natural frequency of the rear suspension is higher than that of the front one, at least if the weight distribution is not such that the rear wheels are far more loaded than the front ones. The importance of having a lower natural frequency of the front suspension can be explained by observing that any road input reaches first the front suspension and only after a certain time the rear one. If the natural frequency of the latter is higher, when the vehicle rides over a bump the rear part "catches up" rapidly the motion of the front part and after the first oscillation the body of the vehicle moves in bounce rather than in pitch, which is considered good for ride comfort. Then the rear part should lead the motion, but in the meantime damping has caused the amplitude to decrease.

The second rule states that it must be avoided that the pitch frequency is much higher than the bounce one, as it may occur when the dynamic index is smaller than unity (vehicle with long wheelbase and small front/rear overhang). Generally speaking, a dynamic index near unity is considered as a desirable condition for good ride properties, while a complete bounce–ride uncoupling as occurs when $aK_1 = bK_2$ is considered a nuisance. Coupling between bounce and pitching is good as it tends to avoid strong pitch oscillations.

A low value of natural frequencies leads to soft suspensions and large travels. If the value of the pitch natural frequency is too high, compared with that of bounce motions, ride comfort can be affected. To control the pitch and bounce natural frequency independently, without changing the wheel positions and the inertial properties of the body, the suspensions can be interconnected. If the front and rear wheels are connected by a spring which opposes to pitching motions in a way similar to antiroll bars for rolling motions, the pitching frequency can be raised without increasing bouncing frequency and the damping of pitching motion is decreased. This is, however, the opposite of what is usually needed.

Various types of mechanical, hydraulic or pneumatic interconnections can be used, the latter particularly when air or hydraulic springs are used.

5.5.6 Wheelbase Filtering

The study of the free bounce and pitch oscillations does not, however, supply a complete picture of the riding behavior of the vehicle. The excitations applied to the front and rear wheels are not independent and the rear wheels are excited by the same forcing function as the front ones but with a delay

$$\tau = \frac{l}{V}$$

equal to the time needed to travel a distance equal to the wheelbase.

Consider a two-axle vehicle traveling on a ground with undulations having a harmonic shape. If the wavelength of the ground is long, i.e. the frequency of the forcing function is small, the excitation of the front and rear wheels occurs almost in phase, with the result of exciting mostly bounce modes. A road input with a wavelength equal to the wheelbase or to its whole submultiples will excite in phase the two axles and then will excite bouncing but not pitching modes. An input with a wavelength equal to twice the wheelbase (or an odd submultiple of it) will excite the two axles at 180° phasing, producing pitching but not bouncing. The last statement is true only if the center of mass is at mid-wheelbase; qualitatively, however, holds even if pitch and bounce motions are not uncoupled and *wheelbase filtering*, as this phenomenon is referred to, is present being immaterial the type of suspensions and the free response of the system.

Wheelbase filtering introduces a dependence of the response of the system from the speed. As an example, if the wheelbase is 2 m and the speed is 20 m/s, the time delay τ is 0.1 s. Maximum pitching and vanishing bouncing response occur with wavelengths of twice the wheelbase and its odd submultiples, $4, 4/3, 4/5, \dots$ m. At 20 m/s the corresponding frequencies at which the bouncing response is canceled are $5, 15, 25, \dots$ Hz. The lowest frequency at which wheelbase filtering occurs is quite high if compared with the bounce natural frequency of the sprung mass: the response of the vehicle is similar to that of the quarter car model, with some frequencies at which the response is filtered out in the high frequency range. The higher the speed the higher is the frequency range in which wheelbase filtering occurs.

In the same example maximum bouncing and cancellation of pitching response occur with wavelengths equal to the wheelbase and its submultiples, $2, 1, 0.5, \dots$ m. At 20 m/s the corresponding frequencies at which the pitching response is canceled are $10, 20, 30, \dots$ Hz, even higher than those for which bounce is canceled. However, no pitching excitation occurs at very low frequency, as stated above, and generally speaking very little pitching occurs when driving on a flat surface.

In the case of large vehicles, like large pressurized rovers, the long wheelbase and the low speeds, together with high spring stiffness can change this picture: here wheelbase filtering can lead to high pitching and low bouncing response. This is further aggravated by the fact that in tall vehicles pitching oscillations are felt, at locations above the center of gravity, as horizontal oscillations which can be a nuisance to riding comfort.

The subjective feeling of riding comfort is also affected by the position in which the passengers are located; while near the center of mass pitching oscillations are little felt, far from it they detract to comfort to a greater extent.

Remark 5.30 This may be a problem in large pressurized rovers, where the occupants may be located in an elevated position and quite far from the mass center.

The coupling between horizontal and vertical motions due to the suspension geometry can severely reduce comfort. For small oscillations about the equilibrium position it is possible to identify a pitch center a way similar to what has been done for the roll center. Its position allows to study the coupling between vertical and longitudinal dynamics and particularly that between pitching and longitudinal oscillations.

5.5.7 Roll Motions

The equations of motion governing roll motions are coupled with those governing handling and not with those related to ride comfort. However, it is also true that rolling can affect strongly the subjective feeling of riding comfort.

If the roll axis is assumed to be a baricentric principal axis of inertia and the aerodynamic forces and moments are neglected, roll motions can be studied through the equation

$$\begin{bmatrix} J_x & 0 & 0 \\ 0 & J_{x_1} & 0 \\ 0 & 0 & J_{x_2} \end{bmatrix} \begin{Bmatrix} \ddot{\phi} \\ \ddot{\phi}_1 \\ \ddot{\phi}_2 \end{Bmatrix} + \begin{bmatrix} \Gamma_1 + \Gamma_2 & -\Gamma_1 & \Gamma_2 \\ -\Gamma_1 & \Gamma_1 + \Gamma_{t_1} & 0 \\ -\Gamma_2 & 0 & \Gamma_2 + \Gamma_{t_2} \end{bmatrix} \begin{Bmatrix} \dot{\phi} \\ \dot{\phi}_1 \\ \dot{\phi}_2 \end{Bmatrix}$$
$$+ \begin{bmatrix} \chi_1 + \chi_2 & -\chi_1 & \chi_2 \\ -\chi_1 & \chi_1 + \chi_{t_1} & 0 \\ -\chi_2 & 0 & \chi_2 + \chi_{t_2} \end{bmatrix} \begin{Bmatrix} \phi \\ \phi_1 \\ \phi_2 \end{Bmatrix} = 2 \begin{Bmatrix} 0 \\ \Gamma_{t_1}\dot{\alpha}_{t_1} + \chi_{t_1}\alpha_{t_1} \\ \Gamma_{t_2}\dot{\alpha}_{t_2} + \chi_{t_2}\alpha_{t_2} \end{Bmatrix}, \quad (5.271)$$

where χ_i, χ_{t_i}, Γ_i, Γ_{t_i} are respectively the torsional stiffness and damping of the ith suspension and the corresponding tires. The excitation is given by the transversal slope of the road α_{t_1} and α_{t_2} at the front and rear suspensions.

Equation (5.271) can be solved numerically and allows the computation of the natural frequencies of roll oscillations. However, a further simplification allows to obtain some interesting information: if the moments of inertia of the unsprung masses are neglected and the excitations at the front and rear wheels are assumed to be equal, the two unsprung masses can be considered as a single body and the equation of motion reduces to

$$\begin{bmatrix} J_x & 0 \\ 0 & 0 \end{bmatrix} \begin{Bmatrix} \ddot{\phi} \\ \ddot{\phi}_u \end{Bmatrix} + \begin{bmatrix} \Gamma & -\Gamma \\ -\Gamma & \Gamma \end{bmatrix} \begin{Bmatrix} \dot{\phi} \\ \dot{\phi}_u \end{Bmatrix}$$
$$+ \begin{bmatrix} \chi & -\chi \\ -\chi & \chi + \chi_t \end{bmatrix} \begin{Bmatrix} \phi \\ \phi_u \end{Bmatrix} = 2 \begin{Bmatrix} 0 \\ \chi_t \alpha_t \end{Bmatrix}, \quad (5.272)$$

where

$$\chi = \chi_1 + \chi_2, \qquad \chi_t = \chi_{t_1} + \chi_{t_2}, \qquad \Gamma = \Gamma_1 + \Gamma_2,$$

ϕ_u is the rotation of the unsprung mass, which is modeled as a single body, and the damping of the tires has been neglected.

The equation of motion is formally identical to that of the quarter car model with two degrees of freedom, except for the fact that here the mass of the unsprung body is vanishingly small.

The optimum value of the damping can be obtained from (5.261)

$$\Gamma_{\text{opt}} = \sqrt{\frac{\chi}{J_x}}\sqrt{\frac{2\chi + \chi_t}{\chi_t}}. \tag{5.273}$$

This condition is usually not met, particularly when antiroll bars are present. The torsional damping of the suspensions is supplied by the same shock absorbers which are usually designed to optimize bounce behavior and the damping supplied in rolling is generally lower than optimum. If antiroll bars are present, the increase of stiffness is not accompanied by an increase of damping, which causes, together with an increase of the natural frequency, also a decrease of the damping ratio. The stiffer the suspensions are in torsion, the more underdamped is rolling behavior, if the stiffness increase has been obtained through antiroll bars. While they reduce rolling in steady state conditions, they can actually increase rolling in dynamic conditions and, in particular, they can cause a strong overshot in the response to a step input as it occurs when a rolling moment is applied suddenly, by a steering input or other causes. High rolling in dynamic conditions can well be a cause of rollover.

5.5.8 Ground Excitation

The excitation due to motion on uneven ground is important for the riding comfort, for the ability of the tires to exert forces in x and y direction, as it causes a variable normal load F_z, and for the stressing of the structural elements. Such excitation cannot be studied with a deterministic approach and the methods used for random vibrations must be applied.

When motion occurs on artificial surfaces one of the many models developed for road profiles can be used. In case of natural surfaces the characteristics are much more variable and need to be experimentally defined in each case. At any rate it is possible to express the ground profile as a law $h(x)$ and to obtain its power spectral density through harmonic analysis.

The profile is a function of space and not of time and the frequency referred to space $\overline{\omega}$ is expressed in rad/m or cycles/m and not in rad/s or in Hz. The power spectral density $\overline{S}$ of law $h(x)$ is thus expressed in $\text{m}^2/(\text{rad/m})$ or in $\text{m}^2/(\text{cycles/m})$.

The simplest approximation is to express the law $\overline{S}(\overline{\omega})$ as a straight line on a logarithmic plot, i.e. a law of the type

$$\overline{S} = c\overline{\omega}^{-n}, \tag{5.274}$$

where n is a nondimensional constant while the dimensions of c depend on n.

Often it is assumed that $n = 2$: this is for instance the case of ISO standards for road profiles. In this case c is expressed in $\text{m}^2(\text{cycles/m})$.

If the vehicle travels with velocity V, it is possible to transform the law $h(x)$ into a law $h(t)$ and compute a frequency ω and a power spectral density S (measured in $\mathrm{m}^2/(\mathrm{rad/s})$ or m^2/Hz) referred to time from $\overline{\omega}$ and $\overline{S}$ defined with respect to space

$$\begin{cases} \omega = V\overline{\omega}, \\ S = \dfrac{\overline{S}}{V}. \end{cases} \tag{5.275}$$

The dependence of S from ω is thus

$$S = cV^{n-1}\omega^{-n}. \tag{5.276}$$

Note that if $n = 2$, the power spectral density of the displacement is proportional to ω^{-2} and the power spectral density of the vertical velocity is constant: road excitation is thus equivalent to a white noise in terms of vertical velocity of the contact point. This justifies the use of an input of this type for the plot of Fig. 5.56. The power spectral density of the vertical acceleration is proportional to ω^2, showing that high frequency disturbances are strongly felt as acceleration.

Once that the power spectral density $S(\omega)$ of the excitation (namely of function $h(t)$) and the frequency response $H(\omega)$ of the vehicle are known, the power spectral density of the response $S_r(\omega)$ is easily computed as

$$S_r(\omega) = H^2(\omega)S(\omega). \tag{5.277}$$

If the frequency response $H(\omega)$ is the ratio between the amplitude of the acceleration of the sprung mass and that of the displacement of the contact point, the power spectral density $S_r(\omega)$ refers directly to the acceleration of the sprung mass. The root mean square value of the acceleration, for instance, computed with reference to a given frequency range, is simply

$$a_{\mathrm{r.m.s.}} = \sqrt{\int_{\omega_1}^{\omega_2} S_a(\omega)\,d\omega}. \tag{5.278}$$

An index, aimed to define the quality of road surfaces was introduced starting from the 1940s. It is named International *Roughness Index* (IRI) and since 1982 is used by the World Bank to compare road conditions in various countries. It has been shown that a good correlation exists between the index and both the vertical acceleration and the variation of the force on the ground, a property that allows to predict the comfort and the performance of vehicles on a given road.

A similar approach can be used to predict the suitability of a natural surface to be crossed by wheeled vehicles and rovers at various speeds.

The IRI is defined with reference to a particular quarter-car with two degrees of freedom moving at a specified speed, usually 80 km/h. The data of the quarter-car, often defined as golden car, are:

$$\frac{K}{m_s} = 63.3\ \mathrm{s}^{-2}, \qquad \frac{P}{m_s} = 653\ \mathrm{s}^{-2}, \qquad \frac{m_u}{m_s} = 60.15, \qquad \frac{c}{m_s} = 6\ \mathrm{s}^{-1}.$$

The value of the optimum damping is

$$\frac{c_{\text{opt}}}{m_s} = 6.147\ \text{s}^{-1}$$

and thus the model has a damping that is close to optimal.

To define the Roughness Index of a given road profile, the motion of the quarter-car is simulated and the cumulative travel of the sprung mass with respect to the unsprung mass is calculated over time. The index is the total value of the travel divided by the distance travelled by the vehicle

$$\text{IRI} = \frac{1}{VT}\int_{VT}^{T} |\dot{z}_s - \dot{z}_u|\, dt. \tag{5.279}$$

The index so defined is a non-dimensional quantity, but one that is often measured in non-consistent units, [m/km] or [in/mi]. If the speed tends to zero also the IRI tends to zero: in a way it is a measure of the dynamic effects in a direction perpendicular to the ground encountered when travelling of a given surface.

The Roughness Index may also be interpreted as the average value of the absolute value of the relative speed of the two masses divided by the vehicle speed.

By using numerical values of the quarter car model and of the speed similar to those typical of planetary exploration vehicles and rovers, it is possible to compare the surfaces encountered on different planetary environments, assessing the possibility of traveling on them at a given speed.

5.5.9 Effects of Vibration on the Human Body

The optimization of the suspension has been performed with the aim of reducing to a minimum the vertical acceleration of the sprung mass. This would be correct if the discomfort were caused directly by the acceleration, being immaterial the frequency at which the acceleration is applied.

The ability of the human body to withstand vibration and the related discomfort has been the object of countless studies and several standards on the subject have been stated. Only the plot of Fig. 5.59, taken from ISO 2631 standard, is reported here.

The standard states the r.m.s. value of the acceleration which causes, in a given time, a reduction of the physical efficiency. The exposure limits can be obtained by multiplying by 2 the values reported in the figure, while the "reduced comfort boundary" is obtained by dividing the same values by 3.15 (i.e., by decreasing the r.m.s. value by 10 dB). From the plot it is clear that the frequency field in which humans are more affected by vibration lies between 4 and 8 Hz.

Frequencies lower than 1 Hz produce sensations that can be assimilated to motion sickness. They depend on many parameters other than acceleration and vary from individual to individual. Above 80 Hz the effect of vibration depends on the

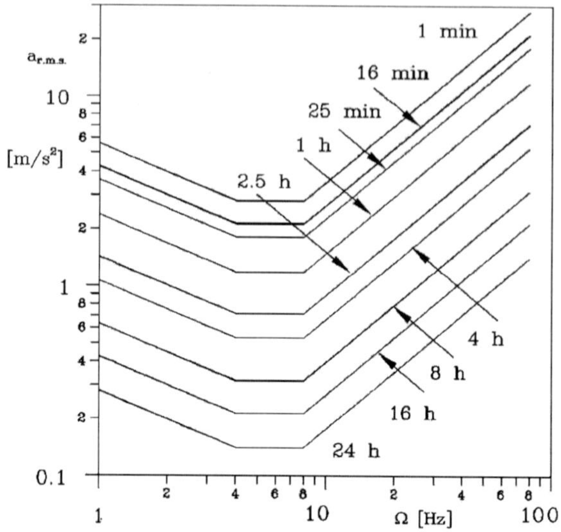

Fig. 5.59 r.m.s. value of the vertical acceleration causing reduced physical efficiency to a sitting subject as a function of the frequency. The curves for different exposure times have been reported (ISO 2631 standard) (from G. Genta, L. Morello, *The Automotive Chassis*, Springer, New York, 2009)

interested part of the body and on the skin conditions, as local vibrations become the governing factor and it is difficult to give general guidelines. There are also resonance fields at which some parts of the body vibrate with particularly large amplitudes. As an example, the thorax–abdomen system has a resonant frequency at about 3–6 Hz, although all resonant frequency values are much dependent on individual characteristics. The head-neck-shoulder system has a resonant frequency at about 20–30 Hz, and many other organs have more or less pronounced resonances at other frequencies (e.g., the eyeball at 60–90 Hz, the lower jaw-skull system at 100–220 Hz, etc.).

At low frequency the curves are, on a logarithmic plane, straight lines sloping downwards while at high frequency they are lines parallel to the bisector of the first quadrant: at low frequency the derivative of the acceleration with respect to time, i.e. the jerk d^3z/dt^3, has the same importance as the acceleration, while at high frequency the amplitude of the velocity is constant. This suggests that what actually causes discomfort is either the vertical jerk or the acceleration or the velocity depending on the frequency range and not the acceleration alone.

In pressurized rovers the habitat can be connected to the vehicle chassis through elastic suspensions to improve comfort. In small rovers on which astronauts ride using their space suits, like the *Apollo* LRV, the space suit can provide a further vibration isolation, which must be accounted for when studying the comfort of the vehicle.

5.5.10 Concluding Remarks on Ride Comfort

The linearized study of suspension motions, based mostly on quarter-car models, shows that the value of the damping of the shock absorbers optimizing comfort is the same as that reducing to a minimum the dynamic component of the force on the ground and hence optimizing handling. However, some results obtained considering the root mean square value of the acceleration and of the dynamic component of the force show that, even using a simplified linearized model, the value optimizing comfort is lower than that optimizing handling.

The last statement is also confirmed by other considerations. Firstly, to optimize handling the reduction of the force is not the only goal and the displacement of the sprung mass with respect to the unsprung ones is also important. Every type of suspension has some deviations from a perfect kinematic guide and thus causes the wheels to be set in a position which is different from the nominal one (e.g., changes of the camber angles, roll steer etc.); this affects negatively the handling characteristics of the vehicle. The larger is the displacement of the sprung mass, the worse is this problem.

Operating in the same way as for minimizing the acceleration, it can be shown that the value of the damping which minimizes the displacement is

$$c = \sqrt{\frac{m(P+K)(P+2K)}{2P}}, \tag{5.280}$$

which is higher than the optimum value computed above.

Remark 5.31 This also suggests the increase of the stiffness of the suspensions and goes against the criterion of 'the softer the better' deriving from the consideration of the vertical acceleration alone.

An important point is the reduction, to the lowest possible value, of the unsprung masses. This is particularly important in the case of fast vehicles and affects both handling and comfort performance. This consideration limits the possibility of directly mounting the motors in the wheels, in particular in the case of direct-drive torque motors, that are often heavier than a corresponding higher speed motor driving the wheel through a reduction gear.

On fast vehicles it is usually expedient to locate the motors on the vehicle body, driving the wheels through shafts connecting the body to the suspension, while the solution with sealed motorwheel units can be more suited to slow vehicles, in particular when operating in a dusty and difficult environment. The advances in low-weight torque motors may in the future widen the applicability of the latter solution.

Another point is linked to roll oscillations. The damping of the shock absorbers is usually chosen considering mainly bounce; this causes rolling motions to be in most cases excessively underdamped. When antiroll bars are used the situation becomes worse: by increasing the roll stiffness without increasing the corresponding damping they cause a more marked underdamped behavior and a decrease of the dynamic stability of roll motions. This not only increases the amplitude of rolling motions, while lowering the roll angle in steady state conditions, and the dynamic load transfer but also makes the rollover of the vehicle in dynamic conditions easier.

The increase of the damping of the shock absorbers above the value defined above as optimum is effective in reducing these effects, which affect more the handling characteristics than comfort.

On the contrary, the importance of reducing the jerk to increase comfort goes in the opposite direction. The value of damping minimizing jerk is lower than that minimizing the acceleration; this leads to a better comfort when the damping is decreased.

The effect of the stiffness of the springs on comfort is in a way contradictory: on one hand, as already stated, the need of reducing the acceleration suggests to reduce the stiffness as much as possible, but this would lead to very low natural frequencies which can in turn cause motion sickness and similar effects.

As already stated the behavior of the shock absorbers is usually far from being linear. Moreover, the force they exert is not only a function of the velocity but also of the displacement and of other parameters, like the temperature. Usually this last consideration is neglected and the force F is considered as a function of the vertical velocity alone ($F = F(v_z)$) but this function is almost always nonsymmetric, with a higher damping capacity in the rebound stroke than in the jounce one. If the force is proportional to the velocity, the law $F(v_z)$ is at best made by two straight lines through the origin. The linear model seen above would thus seem not to be applicable even to the case of small oscillations.

Any law $F(v_z)$ can be thought as the sum of an odd function and an even function. It is possible to demonstrate[24] that the even function causes a displacement of the center of the oscillations in dynamic conditions from the static equilibrium position but has little effect on the response of the system. If the even part of the law $F(v_z)$ is neglected, the characteristic of the shock absorber can be linearized in the origin and an equivalent linear damping can be used for the study of the small oscillations of the system: this explains why linearized models can be used with some confidence even in a case in which the effect of the nonlinearities would seem to be important in all working conditions.

Dry friction, like that occurring in leaf springs, introduces hysteresis and an apparent increase of stiffness in low amplitude motion. If small amplitude oscillations occur about the equilibrium position, the apparent stiffness is strongly dependent on the amplitude, with a value tending to infinity when the amplitude tends to zero. This behavior is typical of dry friction and causes the spring to lock when very small movements are required. The stiffness for the small oscillations typical of ride behavior can be far larger than the overall stiffness of the spring.

The presence of dry friction makes linear models not applicable, or at least makes their results quite inaccurate and causes a deterioration of the ride qualities of the suspension.

All the above mentioned considerations about comfort are little influenced by the gravitational acceleration of the planet on which the vehicle moves: the same criteria for bounce and pitch motions developed for vehicles on Earth hold in the same way as usual. There is even an advantage: a limit to the spring softness in vehicular suspensions usually comes from the need to limit suspension travel with changing load. In low gravity, if the spring are designed with dynamic considerations in mind, the

[24] G.Genta, P.Campanile, *An Approximated Approach to the Study of Motor Vehicle Suspensions with Nonlinear Shock Absorbers*, Meccanica, Vol. 24, pp. 47–57, 1989.

static deflection under load is small and there is no limit of this kind to suspension softness.

The only limitation in this area is the need of avoiding too low bounce and pitch frequencies, which can give motion sickness. The difference may be from the human side: we know very little on how may react to vibration a human body accustomed to no gravity after a few days of space travel and then to a period of low gravity. The usual guidelines may not apply so well as on Earth: for instance the LRV had a bounce frequency in full loaded conditions too low for comfort, but it is not known that astronauts suffered from motion sickness when driving on the Moon. Further studies are needed, but they must be conducted on site, since low gravity cannot be properly simulated on Earth.

However, low gravity causes an unwanted effect on bounce and pitch motion: as already stated the wheels tend to lift from the ground and this is apparent also in the movies taken in the *Apollo* missions.

This problem can be lessened by increasing the damping of the shock absorber or decreasing the stiffness of the spring, but this would in turn deteriorate comfort. This could suggest the use of at least semi-active suspensions, or even fully active ones.

An interesting possibility is the use of electromagnetic damping, also because of the difficulties of cooling standard shock absorbers in space vacuum (Moon) or in a very thin atmosphere (Mars).

The considerations about comfort hold mainly for man-carrying vehicles. Robots and telemanipulators need not to control so strictly vertical and pitch acceleration. However, even if there is no human on board, suspension dynamics must be controlled, since it strongly influences the forces the vehicle can exchange with the ground, affecting its longitudinal and lateral performance, and in extreme case can jeopardize the structural integrity of the rover itself and its payload.

5.6 Coupled Longitudinal, Lateral and Suspension Models

The mathematical models for the study of the dynamic behavior of wheeled vehicles and robots seen in the previous sections were all based on the uncoupling between the longitudinal, lateral and suspension dynamics. The limitations of this approach were stated in Sect. 5.2.

Mathematical models allowing the study of the dynamic behavior of wheeled vehicles without resorting to such an approximation are commonly used in automotive technology, and they are implemented in a number of commercial codes. These codes are basically of two types: general purpose multibody codes, with specific adaptation to deal with those components that are peculiar to the wheeled vehicle technology, and specifically designed codes for vehicle simulation. An example of a code of the first type is ADAMS® CAR, while one of the second kind is CarSim®. A discussion on the relative advantages and disadvantages of the two approaches, and a synthetic description on their foundations can be found in automotive textbooks, like for instance.[25]

There is no difficulty in using these codes for the dynamic simulation of wheeled rovers designed for planetary exploration, once the correct value of the gravitational acceleration, the different kinematics of the suspensions and the required changes related to a possibly different model of the wheel-ground interaction are introduced. For the latter point, the magic formula or Pacejka tire model is flexible enough to be extended to non-pneumatic wheels rolling on non-prepared ground.

However, the working conditions of planetary rovers are quite different from those typical of road vehicles operating on Earth: apart from the reduced gravity, that is anyway an important factor, the speeds are much lower and operation of unprepared and rough ground must be considered as a rule instead of being an exception. Moreover, when dealing with robotic rovers the speeds are usually so low that inertia forces due to the forward velocity may be small (even negligible) if compared with the inertia forces linked with the suspension motion due to ground irregularity.

Dedicated models for the simulation of the coupled dynamics of planetary wheeled rovers, implemented in the form of specialized codes, can be used. Their aims can be, for instance,

- to simulate the response of the vehicle to a given ground profile and to control inputs like the motor torques and, in case of conventional steering, the steering control,
- to evaluate the control strategy of the suspension, in case of active or semiactive suspensions, and to design the relevant control system,
- to simulate the high-level trajectory control strategy in case of robotic rovers, and
- to train the operators in case of teleoperated devices.

In the latter case, it may be expedient to use simplified models, since this tasks requires real time operation of the model, but can tolerate lower simulation precision.

If the mathematical model of the rover is based on the assumption that all its parts, except for the wheels, are rigid bodies and that the latter can be modeled as massless, possibly damped, springs, a small number of generalized coordinates can be used. The vehicle body has six degrees of freedom and the corresponding generalized coordinates can be the coordinates of its center of mass in an inertial reference frame and a set of three Tait-Brian angles, of the usual yaw-pitch-roll type. The number of generalized coordinates depends on the suspension type: if for instance a rocker arm scheme like the one of Fig. 5.34 is used, a single additional degree of freedom is required. In the case of the solutions shown in Figs. 5.35 and 5.36 the additional degrees of freedom required are three in number, while a number of degrees of freedom equal to the number of wheels is needed for independent suspensions.[26] Even if the number of degrees of freedom

[25] G. Genta, L. Morello, *The Automotive Chassis*, Springer, New York, 2009.

[26] G. Genta, *Dynamic Modelling of a Wheeled Lunar Microrover*, 61st International Astronautical Congress, Prague, Oct. 2010.

is small, the computer time for the simulation of the motion on rough terrain may be long.

5.7 The *Apollo* LRV

The Lunar Roving Vehicle (LRV, Fig. 5.60) is the only planetary exploration vehicle built and operated with humans of board and is also the only one operated at speeds allowing to cover significant distances in relatively short times. After about 40 years it is interesting to analyze its design and its details, since it represented at that time a concentrate of high automotive technology and in a way can be considered as a precursor of the all-electric, drive-by-wire approach today much discussed in the automotive industry.

In recent years, many other projects of planetary exploration vehicles were developed (e.g., the six-wheelers developed by LunaCorp and Carnegie Mellon University or the robotic rover presently studied by Tsinghua University, Beijing), but they remain at the design stage.

The basic constraints of the LRV were the mass and space available on board of the Lunar Excursion Module (LEM). They forced the designers to adopt a foldable architecture and dictated many design choices. It was a remarkable feat that a successful vehicle could be built within such strict constraints.

The main characteristics of the LRV were[27]

- Mass + payload: $210 + 450 = 660$ kg
- Length (overall): 3,099 mm
- Wheelbase: 2,286 mm
- Track: 1,829 mm
- Maximum speed: 18 km/h
- Maximum manageable slope: 25°
- Maximum manageable obstacle: 300 mm height
- Maximum manageable crevasse: 700 mm length
- Range: 120 km in four traverses
- Operating life: 78 h

A short analysis of the various subsystems and some considerations on what is feasible today is reported in the following sections.

5.7.1 Wheels and Tires

Pneumatic or solid rubber tires were discarded mainly for reducing the vehicle mass. Tires made by an open steel wire mesh, with a number of titanium alloy plates acting

[27] A. Ellery, *An Introduction to Space Robotics*, Springer Praxis, Chichester, 2000; A. Ellery, *Lunar Roving Vehicle Operations Handbook*, http://www.hq.nasa.gov/office/pao/History/alsj/lrvhand.html.

Fig. 5.60 Image of the Lunar Rover (RLV) with astronaut Eugene Cernan on board taken during the third EVA in Apollo 17 mission (NASA image)

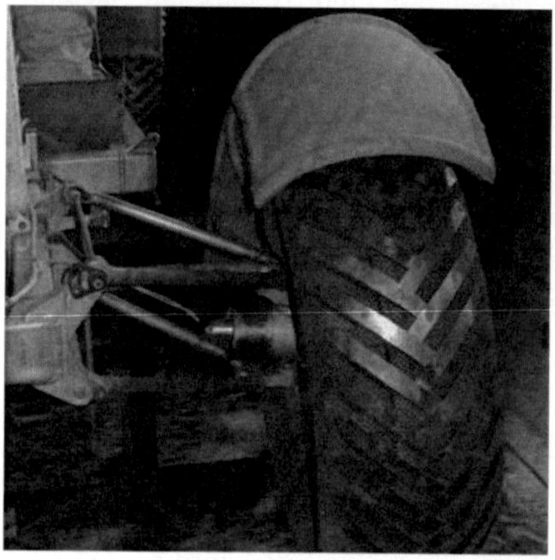

Fig. 5.61 Picture of the front suspension of the Lunar Roving Vehicle on display at the U.S. Space & Rocket Center in Huntsville

as tread in the ground contact zone, were built. Inside the tire a second smaller more rigid frame acted as a stop to avoid excessive deformation under high impact loads. The tire outer diameter was 818 mm (Figs. 4.9a and 5.61).

Tires are one of the largest deviation of the LRV from consolidated automotive technology.

5.7.2 Drive and Brake System

The LRV was a 4WDS electric vehicle, with four independent, series wound DC brush electric motors mounted in the wheel hubs. Each motor was rated 180 W, had a maximum speed of 17,000 rpm and was connected to the relevant wheel using a harmonic drive reduction unit with a gear ratio of 1/80. The nominal input voltage was 36 V, controlled by PWM from the Electronic Control Unit. The drive unit was sealed, maintaining inside a pressure of about 0.5 bar for proper lubrication and brush operation.

Four cable actuated drum brakes directly mounted on the wheels were used. Future planetary vehicles could still have drum brakes, since the braking torque is low enough not to require disc brakes and drums can solve potential problems caused by the lunar dust in disc brakes. Low speed and low deceleration capability makes thermal problems of the brakes less important, even if the absence of air makes cooling much more difficult. Modern brake by wire layouts, in which an electric motor located in the wheel hub actuates directly the brake are a very good alternative to cable control, and are worth to be considered in future designs. Also, the low braking power, linked with the low speed and low gravitational acceleration, makes regenerative braking using traction motors an interesting possibility, both for recovering energy (extending the range) and for making brake cooling easier. No regenerative braking was used on the LRV, mainly due to the use of primary batteries.

The driver interface for the longitudinal control was the same T handle actuating also the steering. Advancing the joystick actuated the motors forward, pulling it back actuated the motors in reverse, but only if the reverse switch was engaged. To actuate the brakes the handle had to be pivoted backward about the brake pivot point.

The wheels could be disengaged from the drive-brake system into a freewheeling condition in case of a failure of the drive system.

5.7.3 Suspensions

Suspensions had a fairly standard transversal quadrilateral layout, with the upper and lower arms almost parallel. Apparently they did not include anti-dive or anti-squat provisions. Springs were torsion bars applied to the two arms and a conventional shock absorber was located along a diagonal of the quadrilateral (Fig. 5.61). The ground clearance varied from full load and unloaded conditions between 356 and 432 mm. These values yield a vertical stiffness of the suspension–tire assembly of 2.40 kN/m, a very low value. Assuming that the tire was much harder than the suspension, the natural frequency in bounce was of just 0.6 Hz for the fully loaded vehicle or 1.1 Hz in empty conditions. Lower values would be obtained if the compliance of the tires was accounted for.

5.7.4 Steering

Steering was performed on all wheels and was electrically actuated (steer by wire). The geometry was designed with kinematic steering in mind: Ackerman steering on each axle and opposite steering of the rear axle with equal angles at front and rear wheels. The kinematic wall-to-wall steering radius was 3.1 m. Each steering mechanism was actuated by an electric motor through a reduction gear and a spur gear sector; in case of malfunctioning of one of the two steering devices, the steering of the interested axle could be centered and locked and the vehicle could be driven by steering one axle only.

The same T handle controlling the motors and the brakes operated also the steering, by lateral displacement. A feedback loop ensured that the wheels were steered by an angle proportional to the lateral displacement of the handle, but there was no force feedback except for a restoring force increasing linearly with the steering angle up to a 9° handle angle, then increasing with a step and then increasing again linearly with greater stiffness.

Perhaps steering control is the most outdated part of the LRV. In a way it was a forerunner of four-wheel steering (4WS) and steer by wire systems presently developed, but with important differences. The 4WS logic was suitable only for kinematic (low speed) steering. On a way, the low top speed justifies this choice, but only to a point: the low gravitational acceleration of the Moon implies that sideslip angles are much higher, for a given trajectory and a given speed, than on Earth so the very concept of kinematic steering is applicable only at speeds much lower than on Earth. Since centrifugal acceleration is proportional to the square of the speed, the top speed $V_{\max} = 18$ km/h is equivalent, from this viewpoint, to a speed at which dynamic effects start to be present.

No 4WS vehicle has such a large rear steering angle and above all a much more sophisticated rear steering strategy is present in even the simplest road vehicle with steering on all wheels.

A second difference is that today steer by wire systems are reversible, in the sense that the driver interface is haptic, i.e. there is an actuator supplying a feedback to allow the driver to feel the wheel reaction, like in conventional mechanical steering systems. The fact that it is possible to drive a vehicle without a feedback is proven by the fact that radio controlled model cars can be operated (the control of the RLV has surprising similarities with that of R/C model cars) but it is considered unsafe and difficult to operate a full size car in this way. *Apollo* astronauts needed much training to operate the LRV and a purposely designed trainer (using conventional pneumatic tires) simulating its performances had to be built, since it was impossible to operate the LRV on Earth.

In future planetary exploration vehicles the know-how gained in 4WS steer by wire applications has to be incorporated in the design.

5.7.5 Power System

Power was supplied by two primary silver-zinc batteries, with a nominal voltage of 36 V and a capacity of 115 Ah each (4.14 kWh). Except for the case of the LRV and other machines with a very short planned duration, available power is one of the main limitations of planetary rovers. The LRV was designed to be used in a single mission for a short time, so primary batteries could be used.

5.8 Conclusions on Wheeled Vehicles

Man-carrying vehicles for future lunar and planetary missions may, as already stated, exploit the recent advances in vehicular technology. A modular approach based on the "four active corners" philosophy may be the best choice. Each corner should contain:

- A wheel, with its tire. Likely purposely designed lightweight and low stiffness non pneumatic elastic tires are the best choice.
- An electric drive unit, made by a brushless electric motor mounted in the wheel hub. Torque motors connected directly to the wheels without reduction gears seem to be a good solution, in particular if the low speed of the vehicle makes the reduction of the unsprung masses less important.
- A by-wire brake unit, to be used for stopping the vehicle and for emergency braking, while normal slow-down is performed by the motor as regenerative braking. The simplest solution seems to be a light duty disc brake, with an electric caliper, possibly operated by an electric motor through a ball screw, even if a piezoelectric solution might be more suitable.
- A by-wire steering unit, powered by a small electric, possibly brushless, motor, turning the wheel hub about the kingpin axis through a reduction gear, possibly a worm gear.
- The wheel suspension, which may be of the SLA or Mc Pherson type, with the possibility of using other types like trailing arms depending on constraints and available space. The suspension will be provided with the relevant spring and damper system. The damper can be a standard shock absorber, but an electromagnetic device, possibly a MEMD (motional electromagnetic damper) or a TEMD (transformer electromagnetic damper) seems to be more suitable. It may work in a fully passive, semiactive or fully active way.
- All the sensors required to supply the control system the relevant information: basically the speed of the wheel and the steering angle, but many other information like steering torque, braking and driving torque, suspension travel and acceleration and many other are important for controlling adequately the vehicle.

The four corners can be identical (or better, some components may be specular) with the possible exception that, if regenerative braking is performed by the motors, brakes may be installed on the front axle only. Also steering may be installed on the front axle only, but four-wheel steering is highly advisable.

The vehicle body has no mechanical parts and may be just a platform carrying the astronauts with their individual life support system or a true habitat allowing a shirt sleeve environment.

The power system (possibly secondary electric accumulators or a more complex generating system), the control system and the man–machine interface may be part of the vehicle body or separate subsystems carried on board.

The interface between each corner and the body is made by

- A mechanical interface, consisting in just a number of attachment points where the corner is attached (likely bolted) to the body.
- A power interface, supplying the required power to the electric motors. Likely, the power amplifiers are installed in the body, together with the control system, so that the corner receives directly a modulated power input.
- A sensor interface, connecting the sensors on the corner with the control system.

The man–machine interface may be of different types, depending on how many functions are entrusted to the human controllers. In a not too far future, it is likely that the actual driving of the vehicle will be performed by the crew, so that a vehicular type interface is predictable. However, the drive-by-wire architecture allows the driver interface to be movable and even may be detached from the vehicle so that driving may be performed both by astronauts on board and by astronauts not on the vehicle, through cables or a radio (or laser) link. Likely, the main control will be performed using some sort of joystick or knob like those seen in some drive by wire prototypes rather than a standard steering wheel. At any rate a haptic interface is highly advisable.

While a modular structure like the one suggested above is suitable for people carrying vehicles, it may be less applicable to robots. Robotic rovers have a much wider spectrum of applications and it may be difficult to reduce their configurations to a number of standard components. Some robots may require true suspensions while for other ones simple linkages made by rigid bodies, like the rocker bogie configuration, may be suitable. While for fast rovers operating on not too rough terrain a four-wheels configuration may be the best ones, robots that must manage a very rough terrain, particularly if they are small and slow, may be best provided with six or eight wheels.

Slow machines, particularly in the case of devices aimed to perform construction works or other heavy operations, may have no steering wheels and control their trajectory by controlling the rotation speed of the right and left wheels. Slip steering allows for a much simpler and sturdy configuration and its lower energy efficiency may be not so important in reduced gravity.

Some effort may be devoted to identify a small number of basic configurations and to standardize components. With the exception of the rocker bogie configuration, also in this case it is possible to define a number of corners (they might number more than four if a higher number of wheels are used) and to separate the mobility systems (basically the corners) from the main vehicle body and possible other systems like manipulators, scientific payloads etc.

The modular approach may be useful not only to reduce costs and development time but also, when a human outpost has been established, can make maintenance and repair much easier. A philosophy similar to the 'plug and play' approach made so popular in the computer field may be developed, and maintenance and upgrading operations may be performed by the astronauts with a minimum of tooling and spare parts.

Chapter 6
Non-wheeled Vehicles and Rovers

The variety of configurations for vehicles and moving robots not based on wheeled locomotion that have been suggested for the exploration of planets with a solid surface is so wide that it is impossible to deal with, or even just to mention, them all.

Since the most common are those based on some form of legs, the present section is mainly devoted to walking machines. Apart from legged vehicles, only a few other approaches will be briefly described: wheels–legs hybrid, track–leg hybrids, jumping devices, vehicles on skis and apodal devices.

6.1 Walking Machines

6.1.1 General Layout

Several times legs have been suggested for vehicles moving on uneven ground, where wheeled and even tracked vehicles may have difficulties in managing obstacles and maintaining a good mobility. Also from the energy viewpoint legs are, at least theoretically, more efficient than wheels on rough ground. The configurations for walking machines suggested and tested in the past are so many that it is impossible to list all of them.

A selection of images of walking devices is shown in Fig. 6.1.

Most configurations for walking machines try to imitate at least some features of natural walking systems, a consideration leading to a high number of internal degrees of freedom and to a complex control system. Independently from the number of legs chosen, the leg configuration (Fig. 6.2) can be defined as *mammalian* (Figs. 6.1b, c, e, g, h), *reptilian* (Fig. 6.1i) or *insect-like* (Fig. 6.1f)—all terms showing the zoomorphic origin of these configurations. In the latter two the stroke of the leg is mostly due to a motion occurring in the horizontal plane, while in the first one the motion is mostly contained in a vertical (sagittal) plane. The pendular recovery of the kinetic energy of the legs is thus possible only if a mammalian configuration

G. Genta, *Introduction to the Mechanics of Space Robots*, Space Technology Library 26,
DOI 10.1007/978-94-007-1796-1_6,

Fig. 6.1 A selection of images of walking devices; (**a**) Walking Horse; (**b**) Iron Mule Train; (**c**) General Electric Walking Truck; (**d**) Odeics Functionoid; (**e**) Ohio State University Adaptive Suspension Vehicle; (**f**) University of Karlsruhe Lauron II; (**g**) Plustech; (**h**) University of Karlsruhe Bisam; (**i**) Tokyo Institute of Technology Titan VIII; (**j**) Martin Marietta Walking Beam; (**k**) Carnegie Mellon Hopper; (**l**) Honda humanoid robot

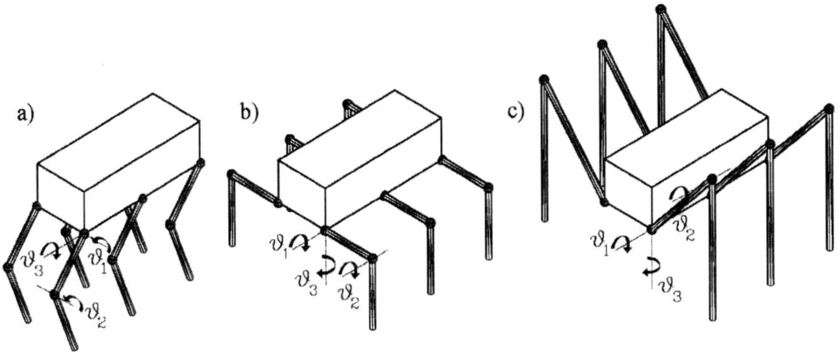

Fig. 6.2 Basic leg configurations for walking machines (**a**) Mammalian; (**b**) reptile; (**c**) insect. In the first configuration the motion of the foot is mostly due to angles θ_1 and θ_2; in the second one to θ_1 and θ_3

is chosen, and only for this configuration the considerations based on the Froude number apply (see Sect. 4.5).

For the vehicle to move in a kinematically correct way on an arbitrary trajectory on uneven ground, each foot must have, independently from the leg configuration chosen, a minimum of three degrees of freedom (Fig. 6.2). The number of degrees of freedom to be controlled is thus large, particularly if the machine has six or eight legs. In most cases the first (proximal) leg segment (thigh) has two degrees of freedom, while the second (distal) one (shank) has one. The knee is thus a cylindrical joint, while the hip is a spherical joint. Legged animals have usually a higher number of degrees of freedom, since in many cases the legs have more than two segments and above all the body is articulated and changes its shape during walking. Some walking machines with articulated body have been built, but this increases the mechanical and control complexity. A lower number of degrees of freedom has sometimes been suggested, but this usually leads to kinematic inaccuracies, with consequent slipping of the feet on the ground and generation of large forces, or to limitations to the ability of following the required trajectory or of walking on rough terrain.

On the other side, it is possible to use redundant degrees of freedom: for example, a zoomorphic leg must have a minimum of three degrees of freedom, but in some cases a fourth one, e.g. at the ankle, is added. One or even two degrees of freedom at the ankle allow the machine to adapt better to rough terrain but, again, adds much to the complexity of the system and, adding mass at the end of the leg, increases the energy requirements for motion. A possible solution is to give the ankle one or two passive degrees of freedom, for instance using spring loaded hinges. In this way the machine adapts better to ground irregularities, while adding little to the complexity and the mass of the system.

Another feature of all legged animals is bilateral symmetry, with the ensuing even number of legs, although an additional support may be offered by the tail, leading to an odd number of supports. Also here most walking machines are zoomorphic, even if some machines with a radial symmetry and even with an odd number of legs exist. Perhaps there is a good reason to this, since bilateral symmetry may be better suited than radial symmetry to move along a straight line (true radial symmetry implies that all legs are equal and the body has no 'front' or 'rear').

The mechanical configuration of walking machines has a deep impact on the requirements for the control system. Early attempts to built walking machines, like the automata designed, and sometimes built, before the nineteenth century[1] and machines dating back to the first half of the twentieth century (Figs. 6.1a and b) relied on mechanical devices, like gear wheels, cams, levers, etc. for generating the correct feet trajectories. Needless to say that little flexibility can be obtained in this way and that walking on rough ground is difficult.

Other solutions (Fig. 6.1c) relied on the human presence on board to control the motion of the feet: in the G.E. Walking Truck the legs were like manipulators which duplicated the motion of the legs and arms of the driver who could 'animate' the machine. The latter could thus perform complex movements. This strategy proved to be both successful and unworkable, since it allows the machine to perform satisfactorily, at the expense of large efforts from its human controller.

At present the trend is to entrust all low-level control tasks to the machine, leaving the human driver only the task of choosing the trajectory, like in conventional motor vehicles, or even entrusting all tasks to the machine. This can make a distinction between walking machines, or vehicles, which are in way controlled by a human, and walking robots, even if this distinction is not clearly cut.

Each configuration has peculiar advantages and disadvantages, which must be carefully balanced when designing any particular machine for a given mission. The performance of the proposed walking machine must be also weighted against that of its more conventional wheeled or tracked counterpart. The result of this trade-off has been in the past quite unequivocal: except for very few cases, wheeled vehicles have always been chosen or showed a better commercial success, to the point that walking machines are still experimental devices, with little chances of entering common use in a short time.

The usual choice of wheeled vehicles, in spite of the advantages of other solutions, is justified by a number of reasons. Wheeled vehicles have a tradition that cannot be matched by other configurations and the designer can rely on a well consolidated technology, without the need of resorting to simulations, experimental tests and other studies, slowing the design process and increasing costs. But it is not just a matter of a consolidated design practice: as already stated in Chap. 4, legged vehicles are usually highly stressed, have reciprocating parts undergoing a large number of fatigue cycles, require complex control systems and in some cases have a higher energy consumption in actual working, in spite of a greater theoretical efficiency. Other unconventional layouts suffer from the same problems to an even higher degree.

---

[1]See for example M.E. Roseheim, *Robot Evolution: The Development of Anthrorobotic*, Wiley, New York, 1994; F. Junko, *Enchanting Gadgets and Engaging Contraptions. Japanese Mechanical Dolls*, The East, Vol. VIII, No. 4, April 1972; C. Singer, E.J. Holmyard, A.R. Hall (editors), *A History of Technology*, Clarendon Press, Oxford, 1954.

6.1.2 Generation of Feet Trajectories

Legs are similar to arms, and a wide variety of layouts can be devised. Once the kinematical configuration has been stated, it is possible to define a workspace and to obtain the direct and inverse kinematics. In general, it must be remembered that the precision requirements for the motion of legs are usually lower than those for arms: the exact point in which the foot touches the ground is less critical than the exact positioning of an arm and, while the trajectory of the foot with respect to the moving vehicle must be accurate enough to avoid large slipping, some slipping may be accepted. When applying longitudinal or transversal forces, the foot has at any rate some slip, as seen in Sect. 4.2.2, and obtaining a very accurate kinematics is not more important than obtaining a perfect Ackermann steering for wheels.

Remark 6.1 Many walking machines have articulated mechanism to move the feet along straight lines during the support stroke without the need for the controller and the actuators to be involved in the trajectory generation (Figs. 6.1d, e).

However, in the past years the growing confidence in computer controlled devices, made many think that the mechanical layout could be made very simple, with the control system able to generate the correct feet trajectories with legs made of just two articulated beams (Figs. 6.1 from f to i).

The legs shown in Fig. 6.2 can be assimilated to revolute arms. The relationship between angles θ_i and the coordinates of the foot with respect to the hip are, for a mammalian configuration (Figs. 6.1a and 6.2a)

$$\left\{\begin{array}{c} x \\ y \\ z \end{array}\right\} = \left\{\begin{array}{c} l_1 \sin(\theta_1) - l_2 \sin(\theta_2) \\ [l_1 \cos(\theta_1) + l_2 \cos(\theta_2)] \sin(\theta_3) \\ [l_1 \cos(\theta_1) + l_2 \cos(\theta_2)] \cos(\theta_3) \end{array}\right\}, \tag{6.1}$$

where l_1 and l_2 are the lengths of the two leg segments. Relationships (6.1) can be easily inverted, yielding the values of the angles allowing to set the foot in the required position:

$$\begin{cases} \theta_3 = \text{artg}\left(\dfrac{y}{z}\right), \\ \theta_1 = \text{asin}\left[\dfrac{\alpha\delta + \gamma\sqrt{4\alpha^2 + 4\gamma^2 - \delta^2}}{2(\alpha^2 + \gamma^2)}\right], \\ \theta_2 = \text{asin}\left[\dfrac{\sin(\theta_1) - \alpha}{\epsilon}\right], \end{cases} \tag{6.2}$$

where

$$\alpha = \frac{x}{l_1}, \qquad \epsilon = \frac{l_1}{l_2}, \qquad \gamma = \frac{z}{l_1 \cos(\theta_3)}, \qquad \delta = 1 + \alpha^2 + \gamma^2 - \epsilon^2.$$

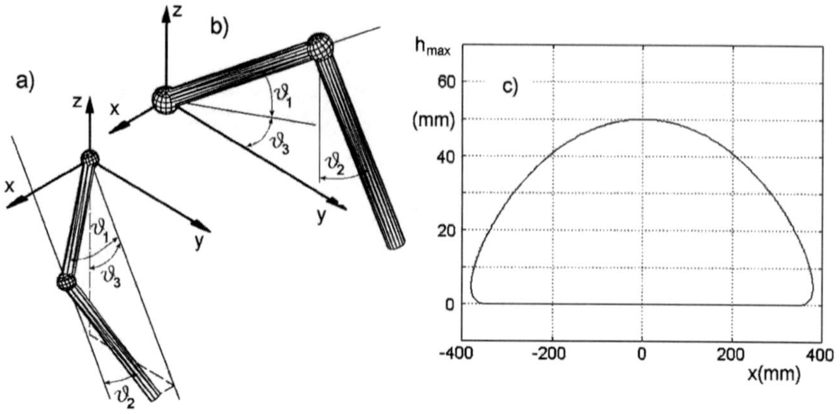

Fig. 6.3 (**a**) and (**b**) Definition of angles θ_1 and θ_2 ($\theta_3 = 0$) for a mammalian and a reptile leg, respectively. (**c**) Trajectory of the foot with respect to the body. Data: step length $L = 700$ mm, maximum foot clearance on level ground $h_{\text{max}} = 50$ mm

For a reptile or insect configuration (Figs. 6.2b, c and 6.3b) the same relationships are

$$\begin{Bmatrix} x \\ y \\ z \end{Bmatrix} = \begin{Bmatrix} [l_1 \cos(\theta_1) + l_2 \sin(\theta_2)] \sin(\theta_3) \\ [l_1 \cos(\theta_1) + l_2 \sin(\theta_2)] \cos(\theta_3) \\ l_1 \sin(\theta_1) - l_2 \cos(\theta_2) \end{Bmatrix}. \tag{6.3}$$

The inverse kinematic can be easily obtained also in this case

$$\begin{cases} \theta_3 = \text{artg}\left(\dfrac{x}{y}\right), \\ \theta_1 = \text{asin}\left[\dfrac{\alpha\delta + \gamma\sqrt{4\alpha^2 + 4\gamma^2 - \delta^2}}{2(\alpha^2 + \gamma^2)}\right], \\ \theta_2 = \text{asin}\left[\dfrac{\gamma - \cos(\theta_1)}{\epsilon}\right], \end{cases} \tag{6.4}$$

where

$$\alpha = \frac{z}{l_1}, \qquad \epsilon = \frac{l_1}{l_2}, \qquad \gamma = \frac{y}{l_1 \cos(\theta_3)}, \qquad \delta = 1 + \alpha^2 + \gamma^2 - \epsilon^2.$$

Assume a law of motion of the foot with respect to the body in straight walking made of a straight line in the *support phase*, i.e. when the foot is in contact with the ground (kinematically correct motion), with sinusoidal laws $x(t)$ and $h(t)$ for the *swing* or *return phase*. The trajectory of the foot, reported in Fig. 6.3c, is thus made of a straight line plus three arcs of ellipses. Assuming that the body moves with constant speed, the time histories of angles θ_1, θ_2 and θ_3 for two typical cases are shown in Figs. 6.4 and 6.5.

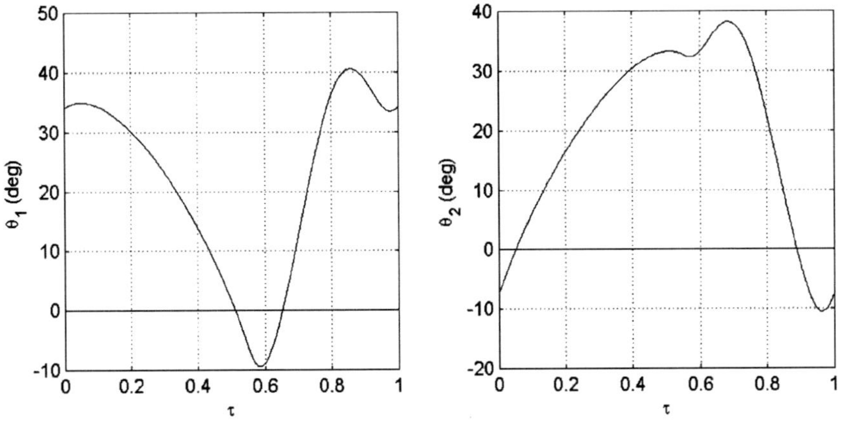

Fig. 6.4 Time histories of angles θ_1 and θ_3 ($\theta_2 = 0$) for a mammalian leg in straight walking with the following data: $l_1 = 500$ mm, $l_2 = 550$ mm, step length $L = 700$ mm. The foot remains on the ground for 0.55% of the time needed for each step. The nondimensional time t is defined with reference to the step time. The support phase lasts for $0 < \tau < 0.55$ while the swing phase for $0.55 < \tau < 1$

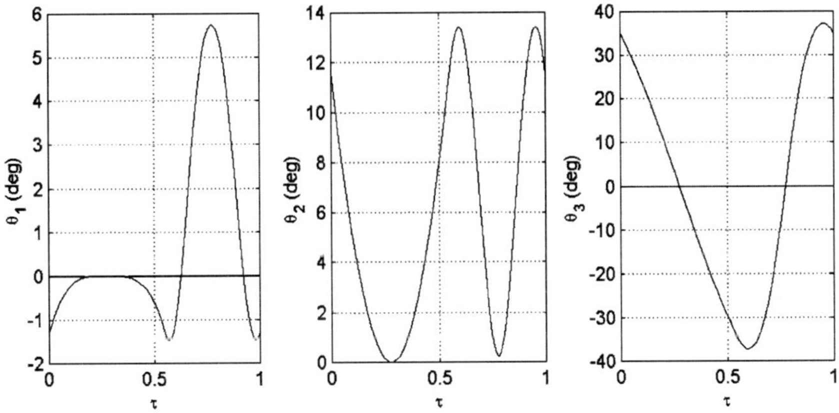

Fig. 6.5 Time histories of angles θ_1, θ_2 and θ_3 for a reptilian leg in straight walking. Same data as in Fig. 6.4, but $h = 550$ mm and lateral distance of the foot $y_0 = 500$ mm

From Fig. 6.4 it is clear that in the case of a mammalian leg the actuators must support the weight of the vehicle during the support phase while moving the joint. This implies that, if each joint is powered by a separate actuator, they perform work, even if the vehicle is walking on a level surface. Some actuators actually supply work, while others brake, but in practice it is very difficult to recover the energy from the latter. The behavior in the swing phase can be different, since there is no strict constraint on the trajectory of the legs which are raised from the ground, provided that they do not touch the ground or other obstacles.

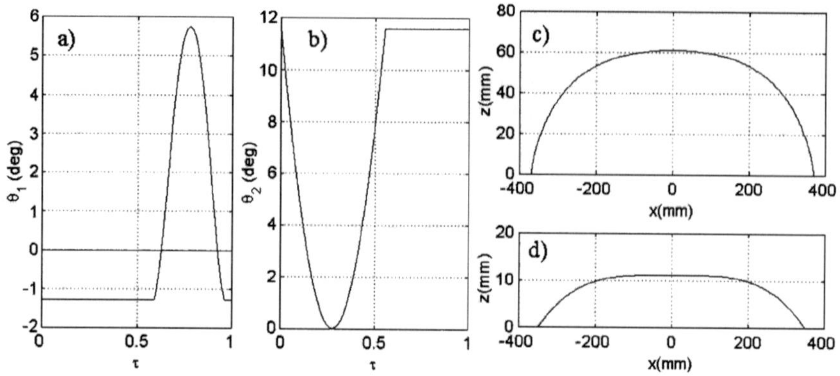

Fig. 6.6 Time histories of angles θ_1 (**a**) and θ_2 (**b**) (θ_3 is not represented since it is almost the same as in Fig. 6.5) for the same case of Fig. 6.5, but with the hip and knee joints locked. The trajectory of the foot and that of the hip are shown in (**c**) and (**d**)

Figure 6.5 shows that also in the case of a reptile (or insect) configuration the actuators continuously adjust their position, but in this case the actuator that supports most of the weight of the vehicle is the actuator for θ_1 (at least if, as in the example, the second segment is almost vertical, i.e. if θ_2 is close to 0), and its motion in the support phase is small.

Two strategies are possible for walking on a slope: the body may be maintained horizontal or parallel to the slope. In the first case the actuator for θ_3 does not perform any useful work, i.e. does not carry a part of the weight of the machine, and the actuators performing most of the work are those for θ_1 and θ_2 as in level walking.

The figures show what is the actual difficulty in building a really zoomorphic walking machine, i.e. one in which the various leg segments are powered by different actuators, either directly acting on the structural members (the 'bones') or operating through tendons. While the generation of the trajectories does not involve serious problems for modern control systems, also owing to the low frequencies involved, the actuators must supply a torque with very slow motion in certain phases of the step, while they must move quickly in other ones. This can be possible if pneumatic or hydraulic actuators are used (which in space applications is commonly ruled out for their bulk and mass—except for very large and heavy walking machines), but is very difficult with electric actuators. Both the motors and the power electronics must work with high currents for large fractions of the working cycle, with low efficiency and heating problems. Also the design of power amplifiers is a difficult task.

A possible solution in the case of reptilian or insect configurations is a modification of the trajectory of the feet, in which the actuator for θ_1 is locked by a brake as soon as the foot touches the ground and the other two angles are modified to compensate for this (Fig. 6.6). The compensation cannot, however, be complete, and a correct kinematics cannot be achieved. The weight of the vehicle is supported by the brake for the whole stance phase, and this reduces greatly the energy consumption. Some power is used by the actuator for θ_2, which cannot be locked in the stance phase but, if the shank is close to be vertical, the energy used is not large. The actuator can be locked during the swing phase, as shown in the figure, reducing the energy consumption due to inertia forces, if it is acceptable that the foot moves sideways when not in contact with the ground.

Since the foot trajectory is not a straight line with respect to the hip during the stance phase, the hip does not move in a horizontal straight line. The trajectory of the foot in the xz-plane referred to the intersection of the ground with z-axis is reported in Fig. 6.6c. The trajectory of the hip with respect to a reference frame moving horizontally at the height of the hip when the foot touches the ground is reported in Fig. 6.6d. The motion of the body is not exactly horizontal, and also slight roll and pitch motions may occur. However, the example shows that, even in the case of a very long stroke (almost a 800 mm stroke with a length l_1 of 500 mm), the kinematic inaccuracies are small and likely are less than the errors due to other effects like the compliance of the ground and the control dynamics not modeled here. Slight sideways motions may also occur at the beginning and end of the stance phase.

6.1.3 Non-zoomorphic Configurations

Up to this point, the trajectories of the feet have been computed with the aim of obtaining a kinematically correct motion. This has the advantage of allowing the body to move at constant speed, while the legs cycle supplying support and propulsion. This requires, however, control and mechanical performance beyond what is possible from machines, and up to now the high speed and low energy consumption permitted by zoomorphic configurations haves never been achieved.

Even animals often deviate from truly correct kinematic performance, particularly when moving at low speed.

The unavailability of electromechanical actuators able of supplying torque while moving very little like muscles is one of the worst limitations. While in the case of reptilian and insect configurations the mentioned approximations allow to avoid to use the actuators to carry the load during the whole stance phase, nothing of this kind can be done for mammalian configurations. In that case it is possible to use a mechanical linkage able to uncouple the vertical and the horizontal motion of the foot, in such a way that the actuator for the former can be locked in the stance phase. The most common linkage is the pantograph both in the planar (particularly suited for mammalian legs, see for example Fig. 6.7[2]) and three dimensional versions.

This solution, constituting a first deviation from the zoomorphic approach, and most of the other ones which have been proposed for the same purpose, are not free from drawbacks. The forces in the links and on the joints are usually large and the guides, which allow the linkage to generate the required trajectory, are much loaded in a direction perpendicular to that of the motion.

---

[2]S.M. Song, K.J. Waldron, *Machines that Walk: The Adaptive Suspension Vehicle*, MIT Press, Cambridge, 1989.

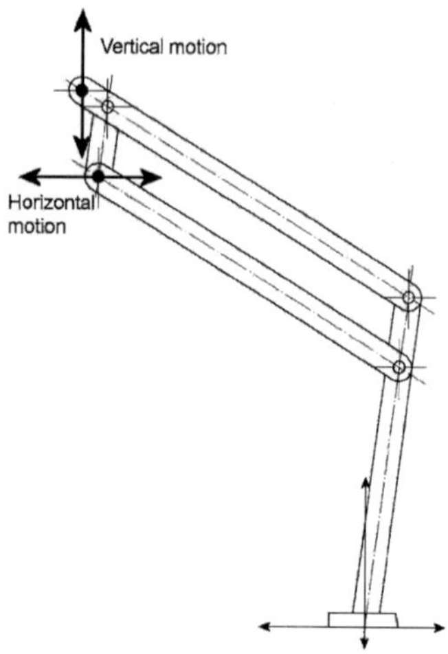

Fig. 6.7 Sketch of the pantograph legs of the Ohio State University Adaptive Suspension Vehicle

Another deviation from a zoomorphic configuration is the use of linear motion joints instead of revolute joints. A Cartesian leg, for instance can work in a kinematically correct way if has three degrees of freedom like the Cartesian arm of Fig. 3.1. A mixed solution, like the spherical arm shown in the same figure, can be a good compromise. The first leg segment (thigh) has just one degree of freedom, identical to θ_3 of the reptilian configuration, while the knee has its second degree of freedom. The third degree of freedom is provided by the second segment (shank) that is made by a linear actuator.

Apart from peculiar applications, zoomorphic configurations for walking machines have little advantages or, even if they are better on paper, they still cannot exploit in real working conditions the advantages they theoretically have. As an example, most of the advantages of zoomorphic configurations are linked with the ability of reaching speed higher than those achievable by alternative solutions, but in practice all walking machines built up to now are quite slow. Not only no walking machine able to run has yet been built, but few machines were able of getting close to the speed at which transition between walking and running occurs. Also the alternative of jumping instead of running has been explored, with some machines like the Carnegie Mellon Hopper (Figs. 6.1k and 6.25b), with some success—but again no machine of this type has reached the operational stage. Except for the Aibo dog, the only zoomorphic walking machine which up to now is operational is the hexapod forestry machine Plustech (Fig. 6.1g).

However, the possibility of achieving high speeds is of little concern in most robots for planetary exploration, where the velocity is strongly limited by other considerations, namely strict limitations to the available power, the impossibility of having a human in the control loop or the need of accurately scanning the ground with the sensors. Simplified, non-zoomorphic, layouts may be expedient in such applications.

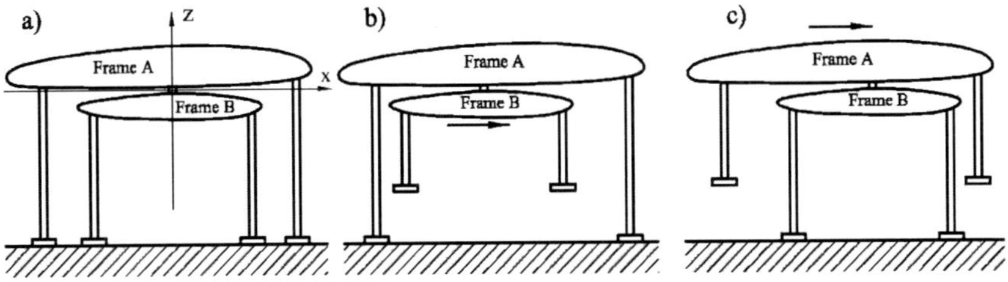

Fig. 6.8 (**a**) Scheme of a twin rigid-frames walking machine; (**b**) phase 2 of a step-motion of frame B; (**c**) phase 5 of a step-motion of frame A

Rigid-Frames Layout

Among non-zoomorphic layouts, one of the most interesting configurations is that based on two rigid frames (Fig. 6.8). The two frames (A and B in Fig. 6.8) can move with respect to each other in a single direction (taken as longitudinal direction of the machine) and rotate about a common vertical axis. Each frame may have any number of legs (at least three for equilibrium); since the two frames need not to have the same number of legs, the total number of legs may be odd. Each leg has a single degree of freedom, namely vertical translation. In this way, if the frames are kept horizontal, the motion of the feet with respect to the body is always kinematically correct and horizontal and vertical motions are completely uncoupled.

Each step (forward or backward) is performed in six phases:

1. rising the feet of the legs of frame B;
2. moving frame B forward (Fig. 6.8b);
3. lowering the feet of legs of frame B until each one touches the ground;
4. rising the feet of legs of frame A;
5. moving forward frame A (Fig. 6.8c);
6. lowering the feet of legs of frame A.

To steer, a step in which the frames rotate with respect to each other instead of moving in longitudinal direction during phases 2 and 5 is performed.

The actuators need not to be controlled in velocity or to be accurately synchronized: a simple on-off controller is adequate. Each leg needs only a touch sensor, to check when the foot touches the ground, and a position sensor, in such a way the controller knows at any instant the configuration of the machine. Also the actuators which move the body (longitudinal translation and rotation about the vertical axis) need just a touch sensor and a position sensor. In this way the machine realizes when it touches an obstacle. In case electric actuators are used, monitoring the actuation currents has proved to be an effective way for detecting when the feet or the body touch against an obstacle, at least in Earth's gravity.

The total number of degrees of freedom of a machine of this type is $2 + n_A + n_B$, where n_A and n_B are the number of legs, respectively, of frames A and B: the minimum number of degrees of freedom is thus equal to 8, for a hexapod machine. Some proposals to reduce the number of degrees of freedom to 4, by operating

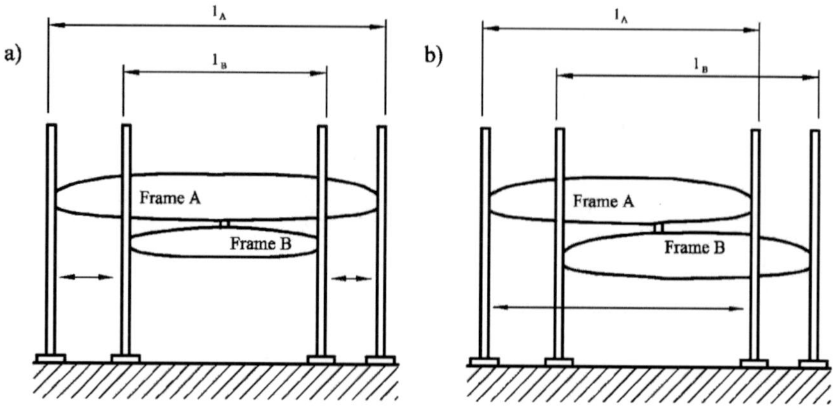

Fig. 6.9 Sketch of an internal (**a**) and an external (**b**) twin rigid-frames walking machine

all the legs of each frame using a single actuator, have been forwarded,[3] but this solution can work only on a perfectly flat and smooth terrain, and hence is of no use in practical applications (why using a walking machine where wheels are at their best?). Another disadvantage is that the body is not kept horizontal while walking on a slope, and the actuator for longitudinal motion is much loaded.

The operational mode described above has the disadvantage of requiring a start–stop cycle for each frame at each step. This is of little inconvenience in slow walking, but is a limiting factor if higher speeds are required. To increase the speed and to reduce the power required to accelerate the frames, their travel must be as long as possible: an important parameter of any twin-frames machine is the ratio t/l_{max} between the travel t and the overall length l_{max}.

The sketch of Fig. 6.8 is just a functional scheme. For practical reasons, legs made by a single element telescopic actuator are usually considered. To avoid mechanical complexities, it is usually unfeasible that a leg of a frame which in any phase of a step is internal to those of the other frame can become external in another phase of the step.

Twin rigid-frames machines can thus be subdivided in two types, which may be referred to as *internal* and *external* machines. In the former (Fig. 6.9a) the legs of one frame lie all inside the 'legbase' of the other frame, while in the latter (Fig. 6.9b) some of the legs of a frame are inside and some are outside the 'legbase' of the other one.

Remark 6.2 This distinction is not an absolute one, since it is possible to conceive different configurations, but there are no examples of that since they are unpractical.

[3] J. Peabody, H.B. Gurocak, *Design of a Robot that Walks in Any Direction*, Journal of Robotic System, pp. 75–83, 1998.

In the first case an upper limit to the travel is

$$t < l_A - l_B,$$

while in the second is

$$t < \min(l_A, l_B).$$

As a consequence an external machine can have a much longer travel, up to twice as large. The travel strongly affects the equilibrium of the machine and hence it results as a trade-off among various requirements, influenced by such characteristics as the mass distribution between the frames, the possible shifts of the center of mass, the number of the legs of each frame and many other, including the precision with which the control system keeps the body horizontal on uneven and sloping ground.[4]

The first examples of successful rigid-frames walking machines are the RECUS (Remote Controlled Underwater Surveyor) built by Komatsu[5] and the Martin Marietta Walking Beam (Fig. 6.1j).[6] The first machine is a very heavy and slow octopod, which has been used for underwater civil engineering works. The second one is an eptapod, designed by Martin Marietta for Mars exploration. They are both "internal" machines. More recently, several versions of a demonstrator of a microrover for planetary exploration based on the twin-frames hexapod configuration, initially designated as Algen and then WALKIE 6 (Fig. 6.10)[7] were built at the Mechatronics Lab. of the Politecnico di Torino. WALKIE 6 has an *external* configuration.

The experience accumulated in the design of several versions of WALKIE 6 and their operation for prolonged periods of time did show that this approach can overcome the reliability problems, reducing at the same time the control complexity, without penalizing too much performances. Actually WALKIE 6.2 is one of the few walking machines able to perform autonomously (the power and the control systems are on board and no umbilical cord is needed) for long periods without maintenance. Its low power requirements (less than 3 Watts at 52 m/h) remains unmatched, while its speed is similar to, and even higher than, that of wheeled planetary exploration microrovers.

Although it is clear that rigid frames machines are only suitable for low speed, the question of how fast they can operate is still open. In particular, it still debatable that they are actually slower in practice (no doubt they are in theory) than

---

[4] G. Genta, M. Chiaberge, N. Amati, *Non Zoomorphic Rigid Frame Walking Micro-Rover for Uneven Ground: From a Demonstrator to an Engineering Prototype*, International Conference on Smart Technology Demonstrators & Devices, Edimbourgh, Dec. 2001.

[5] D.J. Todd, *Walking Machines: An Introduction to Legged Robots*, Kogan Page Ltd., London, 1985.

[6] M.E. Roseheim, *Robot Evolution: The Development of Anthrorobotic*, Wiley, New York, 1994.

[7] L. Bussolino, D. Del Corso, G. Genta, M.A. Perino, R. Somma, *ALGEN—A Walking Robotic Rover for Planetary Exploration*, Int. Conf. on Mobile Planetary Robots & Rover Roundup, Santa Monica, 1997; N. Amati, M. Chiaberge, G. Genta, E. Miranda, L.M. Reyneri, *Twin Rigid-Frames Walking Microrovers: A Perspective for Miniaturization*, Journal of the British Interplanetary Society, Vol. 52, No. 7/8, pp. 301–304, 1999; N. Amati, M. Chiaberge, G. Genta, E. Miranda, L.M. Reyneri, *WALKIE 6—A Walking Demonstrator for Planetary Exploration*, Space Forum, Vol. 5, No. 4, pp. 259–277, 2000.

Fig. 6.10 Walkie 6.2 while being tested on a Mars analog surface on Mount Etna

zoomorphic machines. The average speed of a walking machine of this kind can be approximated as

$$V = \frac{V_f}{2(1+\alpha)}, \tag{6.5}$$

where

$$\alpha = \frac{2h_z V_f}{t V_z},$$

V_f and V_z are the average velocities of one frame with respect to the other and of the leg actuators and h_z is the vertical throw of the legs. The speed limitations come mostly from the fact that each frame must be accelerated and stopped once at each step and the maximum acceleration is limited by the available power, the ability of the payload to withstand the acceleration, the available traction on the ground and the gravitational acceleration. Any increase of the travel t of the frames play a double role: it decreases parameter α and allows to reach higher frame speeds, at equal acceleration. The power needed to move increases strongly with the speed, unless some ways for recovering the energy used to accelerate the frames is used.

To increase the speed it is possible to put as much as possible of the vehicle mass on a third frame, which can move independently from the two carrying the legs.[8] If the third frame is powered by the same actuator which moves the other frames, or at least is not controlled independently, the added complexity is very limited and the number of degrees of freedom is not affected. The advantage is that the acceleration of the largest part of the mass, and above all of the payload, is typically half than that of the frames, at equal speed.

A far cleverer strategy can be devised, if a separate control of the third frame is accepted. The third frame can move forward at constant speed, while the other

[8]G. Genta, N. Amati, L.M. Reyneri, *Three Rigid Frames Walking Planetary Rovers: A New Concept*, 50th Int. Astronautical Congress, Amsterdam, Oct. 1999.

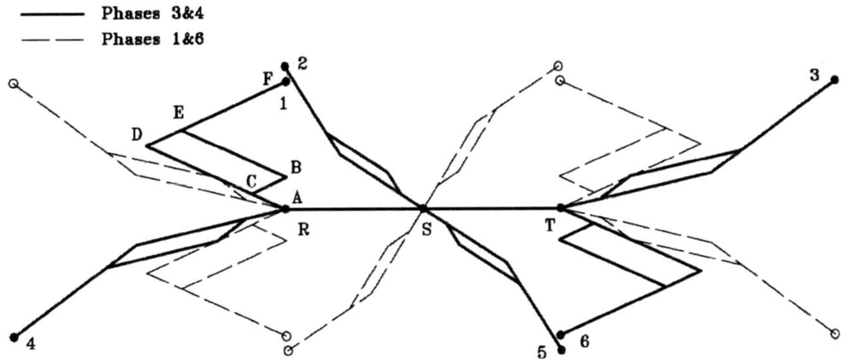

Fig. 6.11 Hexapod planar-motion walking machine, with the legs in their extreme positions

two frames cycle, so that a large part of the mass of the machine is not subjected to frequent start–stop cycles. One of the most interesting features of the zoomorphic layouts is thus introduced, while retaining a very simple mechanical layout. However, the control system must be much more complex: while the drivers of the motors can operate, in the simplest rigid-frames machines, in an on-off pattern, to achieve this more sophisticated control the speed of the actuators must be controlled and some form of predicting the ground conditions must be introduced—just as in zoomorphic machines.

Planar-Motion Walking Machines

Another non-zoomorphic layout can be conceived: the legs, similar to those of rigid-frames machines, can be moved in longitudinal direction by using some sort of kinematic linkage instead of being rigidly attached to the two frames which constitute the body of the machine. The simplest way of obtaining the longitudinal motion of the legs is to use pantographs, not in the vertical plane as in the Adaptive suspension vehicle of the Ohio State University (Figs. 6.1e and 6.7), but in the horizontal plane.

Each leg is thus made by a single-degree of freedom pantograph located in a horizontal plane (Fig. 6.11).[9]

Point A (referred to leg #1) is the cylindrical hinge in which the leg is articulated to the body, while point B moves in longitudinal direction, being guided along a straight line. The foot is carried by a vertical linear actuator located in point F, in a way that is identical to rigid-frames walking machines. Like in rigid-frames machines, the vertical and horizontal motions of each foot are uncoupled from each other, and if the body is always maintained in a horizontal position even when walking on a slope, the direction of the linear actuators is always vertical, while the

[9]G. Genta, N. Amati, *Planar Motion Hexapod Walking Machines: A New Configuration*, CLAWAR 2001, Professional Engineering Publishing, London, October 2001.

motion of all the pantographs occur in a horizontal plane (hence the definition of planar-motion walking machines).

Since no forces (except inertia forces which in low speed walking are negligible) act in the horizontal plane, the guide used to allow point B to move along a straight line has to withstand no force in a direction perpendicular to the motion. The actuator moving point B has to supply only the power needed to overcome friction (neglecting inertia forces). All loads act in a direction perpendicular to the plane of the pantograph, generating bending moments in the beams and torques in the joint, whose vector, however, is always perpendicular to the axis of the hinge.

It is possible to design the leg in such a way that all loads are carried by beams AD and DF (for instance by using spherical joints in points C and E) so that bars BC and BE are just push-pull rods, not loaded in bending. In this way the bending moment in D is simply the load on the foot multiplied by distance DF and that in A is the load multiplied by distance AF. These bending moments are not constant, since the load on the foot is not constant during walking, but their variation is neither large nor quick, as occurs for the loads in many elements of walking machines of different type.

If legs built in this way are used for a hexapod walking machine, the six pantographs may be moved by a single actuator, realizing a type of motion and a gait (alternate tripod gait, see below) which is exactly the same as for rigid-frames machines. As an alternative, six independent actuators may be used, allowing to walk with any type of gait like more complex zoomorphic machines and allowing the body to move at a constant speed.

A basic layout for a hexapod walking machine of this type is shown in Fig. 6.11. Assuming an alternating tripod gait, the same control architecture as for rigid-frames machines can be used. Each step can be subdivided in six phases. Assuming that at the beginning all feet are on the ground, legs 1, 3, 5 are in forward position and legs 2, 4, 6 are in backward position (dotted lines in Fig. 6.11):

1. Legs 2, 4, 6 are raised (the vehicle is supported by legs 1, 3, 5);
2. Legs 2, 4, 6 move forward and legs 1, 3, 5 move backward (with respect to the body). The body moves forward;
3. Legs 2, 4, 6 are lowered to the ground;
4. Legs 1, 3, 5 are raised (the vehicle is supported by legs 2, 4, 6);
5. Legs 1, 3, 5 move forward and legs 2, 4, 6 move backward (with respect to the body). The body moves forward;
6. Legs 1, 3, 5 are lowered to the ground.

The body moves forward only in phases 2 and 4, while in all other phases only the vertical actuators move.

Remark 6.3 This type of gait and control strategy is the worst choice for what static equilibrium is concerned, but allows to control the vehicle using simple on-off switches and requires no velocity control on the actuators.

The number of total degrees of freedom is 8 (the eighth degrees of freedom is used for rotation) since, even if the six pantographs are operated by six separate electric motors, they may be controlled by a single driver: the machine is exactly equivalent to a twin-frames walker, with the difference that only cylindrical hinges are used, no slider guide operating under load is present and the stroke may be longer and hence the machine can move faster. The maximum speed is roughly twice that of a two frames machine, at equal acceleration. If the electric motors are controlled separately, a machine with 13 degrees of freedom is obtained and it is possible to have a continuous motion of the body, with no acceleration–deceleration cycle and any of the gaits typical of hexapod machines may be used.

Steering may be implemented in different ways, the simplest being to move the legs at different sides with different strokes; a sort of slip steering. This is perhaps not advisable, since large slipping of the feet and large stresses in the legs may occur. The best way would be to have the feet moving on curved trajectories, which introduces mechanical and control complexities. A simple solution is making the body in two parts, one articulated on the other, with three legs each, which is identical to the strategy used in twin-frames machines and requires a single additional degree of freedom. It compels the machine to stop for changing direction while walking occurs only on a straight path.

The total number of degrees of freedom for a machine with n legs is thus $n+2$, if the strategy of moving all pantograph together is used, or $2n+1$ if each leg is controlled separately. Most of the advantages and drawback are similar to those of rigid-frames machines, but this configuration has much more flexibility and may be seen as an intermediate layout between simple rigid-frames machines and much more complicated zoomorphic machines.

6.1.4 Gait and Leg Coordination

The motion of a leg can be subdivided into two phases: the *support phase*, when the leg contributes to support the weight of the body, and the *transfer* or *swing* or *return phase*, when the leg, at the end of its stroke, lifts from the ground to go back and gets ready to start the following cycle.

The *cycle time* of a leg is the time a leg takes to perform both support and return phases.

In diagrams the legs are numbered from the most forward one going backward, with legs on the left being given an odd number (Fig. 6.12a). In a hexapod machine the legs are: 1: front left; 2: front right; 3: center left; 4: center right; 5: rear left; 6: rear right.

The most important parameter to define the gait is the *duty factor* β of each leg

$$\beta_i = \frac{t_{si}}{t_{ci}}, \tag{6.6}$$

where t_{si} and t_{ci} are, respectively, the duration of the support phase of leg i and the cycle time of leg i.

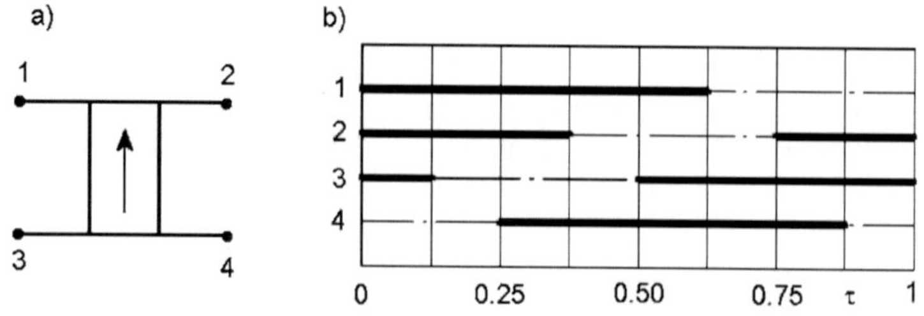

Fig. 6.12 (**a**) Leg numbering in a quadruped; (**b**) example of gait diagram for a quadruped

A gait characterized by the same value of the duty factor of all legs is defined as a *regular gait*.

The duty factor is by definition a number smaller than one; the higher the duty factor, the longer time a leg spends on the ground. The average velocity, relative to the vehicle body, of the foot is the same in the support and return phase when $\beta = 0.5$; otherwise the ratio between the relative velocities is

$$\frac{V_{ri}}{V_{si}} = \frac{\beta_i}{1-\beta_i}. \tag{6.7}$$

The nondimensional time

$$\tau = \frac{t}{t_{ci}} \tag{6.8}$$

is used instead of the standard time. Since leg #1 is usually assumed to touch the ground at $\tau = 0$, its support phase is characterized by

$$0 < \tau < \beta$$

and the return phase by

$$\beta < \tau < 1.$$

The *leg phase* is a measures of the delay with which the ith leg contacts the ground with the respect to leg #1.

A common representation of the gait is the *gait diagram*, like the one, referred to a quadruped, shown in Fig. 6.12b. The nondimensional time is reported on the horizontal axis, usually for values between 0 and 1, i.e. for a complete cycle. Each leg is represented by a horizontal line, dark when the leg in its support phase and light when is in its transfer phase.

The plot of the figure refers to a regular gait, with a duty factor $\beta = 5/8 = 0.625$, which means that each leg spends the 62.5% of its time on the ground. As an average, the speed the foot moves forward (V_r) must then be 5/3 of the speed the foot moves backward (V_s).

At time $t = 0$ leg #1 has just started its support phase, legs #3 and #2 are on the ground while leg #4 is returning. At $\tau = 0.125$ leg #2 has reached full backward position and starts returning. At $\tau = 0.25$ leg #4 has reached its full forward position

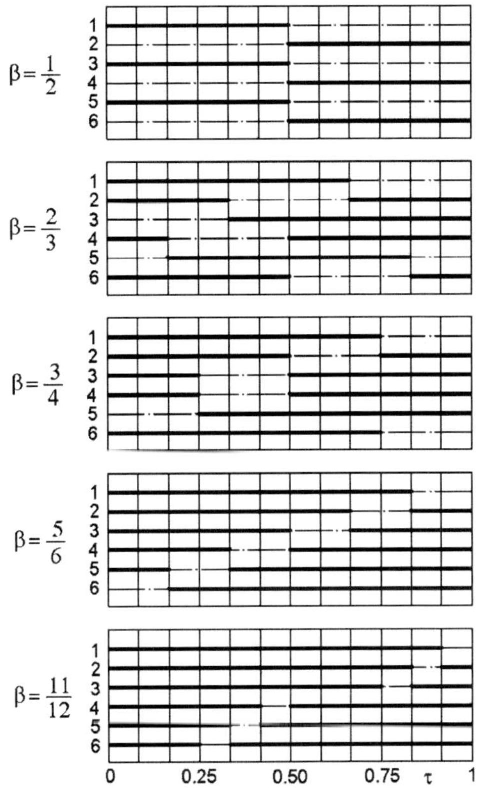

Fig. 6.13 Diagrams for wave gaits with different duty factors for hexapod walkers

and lowers on the ground, ready to start its support phase. Than in sequence leg #3 raises, leg #2 goes down, leg #1 raises, leg #3 goes down, and finally leg #4 raises.

The phases of the legs are equi-spaced: 0, 180°, 270°, 90°.

Other definition are the *leg stride*, distance the center of mass of the vehicle moves in complete locomotion cycle and the *leg stroke*, distance the foot moves relative to the body in the support phase.

A gait is *periodic* if the events occurring in the cycle time repeat themselves in the following cycles; otherwise it is non-periodic.

The variety of possible gaits is large. In general, when walking on a more or less regular surface it is convenient to use a periodic gait. A particular class of periodic gaits are *wave gaits*: the legs on each side are placed on the ground at regular intervals, starting from the rearmost ones. In backward wave gaits the sequence is reversed and the first leg to be placed on the ground is the front one.

The diagrams for wave gaits with different duty factor for hexapod machines are shown in Fig. 6.13.

The gait of the first diagram is usually referred to as alternate tripod gait, and is the simplest of the gaits for a hexapod: the legs are divided into two sets of three and touch the ground alternately. It is also the gait of a twin rigid-frames hexapod,

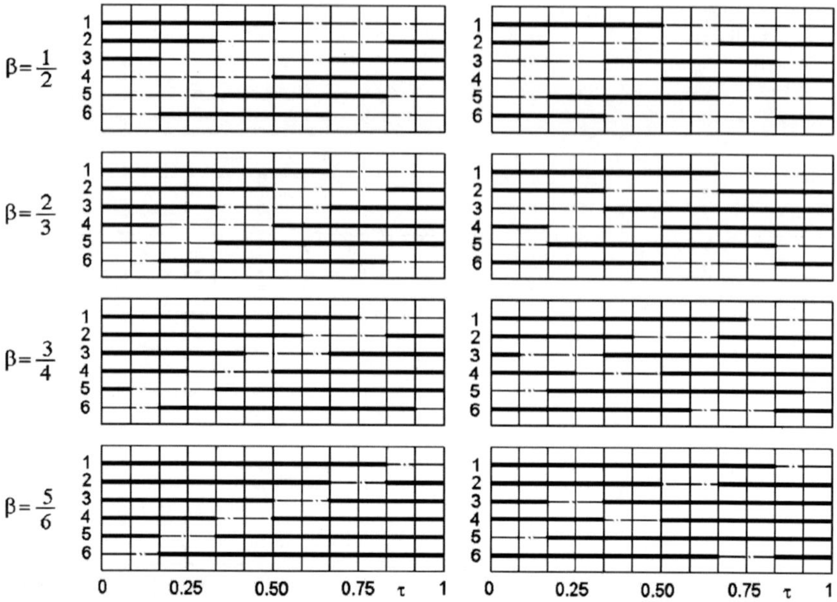

Fig. 6.14 Equal phase, half-cycle gaits (*on the left*) and equal phase full-cycle gaits (*on the right*) for a hexapod, with different values of the duty factor

as described in the previous section. At each time there are always three legs not on the ground.

In the second diagram at any time there are always two legs off the ground, while in the third one the number of legs not on the ground is sometimes one and sometimes two.

Starting from the fourth diagram the number of legs on the ground is never less than five. This improves stability, but compels to accelerate much the legs to ensure a quick return.

Another class of regular gaits are *equal phase gaits* (Fig. 6.14): the cycles of the various legs are equi-spaced in time. For a machine with n legs, this means that the phase angles are

$$0, \frac{2\pi}{n}, \frac{4\pi}{n}, \dots, \frac{2(n-1)\pi}{n}.$$

The legs on each side may be equi-spaced by $2\pi/n$ (*half-cycle gait*, like the ones on the left of the figure, where the phases are 0, 300°, 240° on the left side and 180°, 120° and 60° on the right) or by $4\pi/n$ (*full-cycle gait*, like the ones on the right of the figure, where the phases are 0, 240°, 120° on the left side and 180°, 60° and 300° on the right). Also equal phase gaits may be reversed.

Remark 6.4 For some values of the duty factor the equal phase gait coincides with a wave gait.

Table 6.1 A comparison of various periodic and non-periodic gaits

Gaits	Stability	Suitable terrain	Control	Power cons.	Smooth. of ride
Periodic					
Wave	good	perfect	easy	uneven	good
Equal phase	good	perfect	easy	even	good
BWD wave	fair	perfect	easy	uneven	good
BWB equal phase	fair	perfect	easy	even	good
Dexterous periodic	good/fair	fair	fair	even/uneven	good
Cont. f.-the-lead.	fair	fair/rough	fair	uneven	good
Non-periodic					
Discont. f.-the-lead.	very good	rough	hard	uneven	poor
Large obstacle	fair	obstacle	fair	uneven	poor
Precision footing	very good	rough with obst.	very hard	uneven	poor
Free	good	rough	hard	uneven	fair

The advantage of equal phase gaits is that of requiring a more constant input power than wave gaits, minimizing the need for short-term energy storage.

Periodic gaits are well suited for terrains where the feet can be placed in any point, or at lest the high-level controller can prevent the machine from stepping in a 'bad' spot just by steering the trajectory. If the places unsuitable to put a foot on are many, the high-level controller must state also the position of the footholds, and non-periodic gaits become advisable. A strategy, usually referred to as *follow the leader*, is to define the spots where to put the front feet and then the low-level controller puts the other feet in the same places, central feet after the front ones, rear feet after the middle ones.

In moderately bad conditions it is possible to define periodic follow the leader gaits. Other strategies are those called large obstacle gaits, where the high-level controller accurately set the feet clear of ditches and other large obstacles, or precision footing gaits. A comparison of a number of periodic and non-periodic gaits is shown in Table 6.1.

6.1.5 Equilibrium

The motion of any walking machine can be performed in static equilibrium or dynamic equilibrium conditions. In the first case the motion can be considered as a sequence of static equilibrium positions and inertia forces play a small role in the motion. Ideally, the motion can be stopped in any instant.

Remark 6.5 This condition holds only if the speed of the machine tends to zero, but most artificial walking devices operate at such a low speed that they are not far from this condition.

To reach stability in static equilibrium conditions the minimum number of legs that must be on their support phase at each moment is three.

Most animals are able to move in conditions that are dominated by inertia forces: in most instants the static equilibrium conditions are not satisfied and the motion cannot be stopped in any arbitrary instant. In this case there is no minimum number of legs that must be in support: when running, there are instants when no legs are on the ground.

Speaking in terms of duty factor, the duty factor during dynamic equilibrium running may become very low. Most animals walk with gaits with high values of duty factor when moving slowly, changing their gait and lowering the duty factor when increasing the speed, to transition from walking to running above a certain speed (or better, as already stated, when the Froude number goes beyond a given value).

In low gravity inertia forces can become a dominating factor at speeds much lower than usual (usual meaning in 1 g conditions), so the importance of dynamic equilibrium conditions may well be greater in planetary exploration (above all in asteroid and comet exploration) than on Earth.

It must be clearly stated that most walking machines that were successful in dealing with static equilibrium conditions never attempted working in dynamic conditions, and if machines able to run on flat ground were built, running on uneven ground is still a difficult achievement.

One of the few walking machines able to achieve a true dynamic stability is the BigDog, a quadruped robot, built in 2005 by Boston Dynamics with Foster-Miller, the NASA JPL and the Harvard University with DARPA funding (Fig. 6.15). It is a sort of small ‘artificial mule’, designed mostly for military applications, about 910 mm long, 760 mm tall, with a mass of about 110 kg, able to carry a mass of 150 kg. Powered by a two-stroke, one-cylinder, 11 kW internal combustion engine driving the 16 actuators of the legs through a hydraulic transmission, it is capable of managing difficult terrain at a speed of 8 km/h.

With its performance and impressive stability, the Big Dog could be the prototype of an astronaut assistant for planetary exploration. In low gravity its structural mass could be further lowered and its carrying capacity increased, even if a completely different power system would be required.

Only static stability will be dealt with here.

Consider a walking machine in a given instant of its motion. Since inertia force are not considered, there is no difference whether the machine is in that position because it is standing in that way or is going through that position during motion. Moreover, assume that the ground is horizontal.

At least three legs must be in support (in Fig. 6.16 the machine is supported by four legs). The feet are assumed to be located in points F_i; such points may be considered as the centers of pressure of the feet. The *support pattern* is the largest convex polygon whose vertices are points F_i. Point G is the projection on the ground of the position of the center of mass.

The *instant stability margin* is the minimum distance of the projection of the center of mass from the boundary of the support pattern

$$S_m = \min\{d_1, d_2, \dots, d_n\}. \tag{6.9}$$

Fig. 6.15 The Big Dog, one of the few successful walking machine able to achieve a true dynamic stability

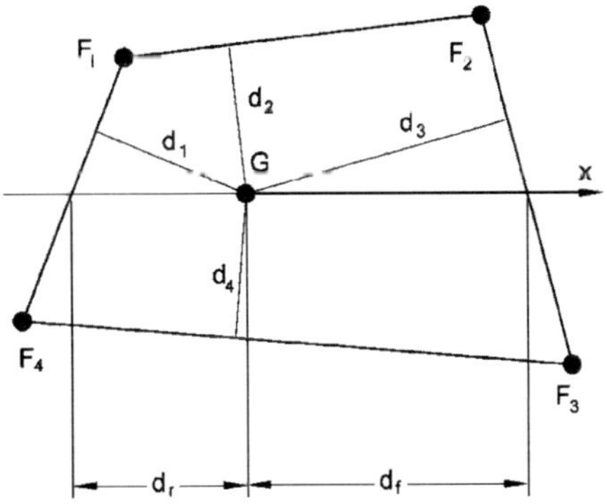

Fig. 6.16 Support pattern in an instant in which four legs are in their support phase. Note that the criterion for numbering the footprints is different from the criterion for numbering the legs seen above

Since the center of mass moves with respect to the support pattern, and the support pattern itself changes when a foot contacts the ground or a foot leaves the ground (if the trajectory of the feet is not kinematically correct a foot may also move on the ground, slipping on it), the stability margin is a function of time: $S_m = S_m(t)$.

Similarly, it is possible to define a *front* and *rear instant stability margin* (in the figure are indicated as d_f and d_r) and a *longitudinal instant stability margin*

$$S_l = \min\{d_f, d_r\}. \tag{6.10}$$

Since the instant stability margin is a function of time, the *stability margin* is defined as the minimum value in time taken by the instant stability margin in one cycle

$$S = \min\{S_m(t)\}. \tag{6.11}$$

If the stability margin is larger than 0, static equilibrium is always ensured. If not, in some position the walking machine is unstable and starts to fall down. Whether motion is possible in this condition must be stated considering dynamic equilibrium. In the same way a longitudinal stability margin can be defined as the minimum value taken by the instant longitudinal stability margin.

Remark 6.6 The higher is the number of legs and the value of the duty factor β, the more stable is a walking machine.[10]

6.1.6 Biped and Humanoid Robots

Biped walking machines have peculiar problems, particularly for what equilibrium is concerned. It is a fact that also in the animal world bipedal locomotion appeared much later than locomotion with a larger number of legs, and had to wait for the development of a much larger brain, able to control an equilibrium that is always on the verge of instability or is outright unstable.[11] Many anthropologists link the development of human brain and bipedal locomotion, saying that not only the latter had to wait for the development of the former, but also our brain was at least in part a result of the needs imposed by our way of walking.

Apart from a much more developed controller, walking on two legs requires also sophisticated sensors to identify the direction of the gravitational field: human inner ear equilibrium organs are unique in the animal world.

When walking, a biped needs to remain in equilibrium on a single foot for a long time (always, if $\beta = 0.5$) and thus the support pattern is the outer shape of a single feet. In all other cases the foot can be assumed to be a point, with a vanishing area, without this causing the support area to vanish. The feet of many multileg walking

---

[10]For a detailed analysis of the static equilibrium of walking machines, see for instance S.M. Song, K.J. Waldron, *Machines that Walk: The Adaptive Suspension Vehicle*, MIT Press, Cambridge, 1989.

[11]Birds are not considered here: for them walking is only a secondary form of locomotion, they can anyway use their wings to produce aerodynamic forces that help to maintain equilibrium, and above all their body is extended in horizontal direction and not vertical, yielding a low position of the center of mass.

machines reduce to simple small pads located at the end of the last segment. This is impossible for bipeds, whose feet must have a larger area and above all must be articulated on the last leg segment with an ankle articulation with usually two degrees of freedom.

If walking on rough terrain is expected, this articulation is usually a true active joint, so that all the foot is in contact with the ground and the joint can transfer to the body not only the supporting force but also the moment needed to keep the body in equilibrium. The human foot is a complex, delicate and highly loaded structure, with many additional degrees of freedom that help in properly distribute the load on the ground: the articulation of the toes, for instance, is important in achieving a correct walking.

The larger the foot area, the easier is to maintain the equilibrium. In particular, the early biped machines, which tried to mimic human shape and gait, had oversized feet, often provided with appendages to increase the area (the spurs of the *Steam Man* of the end of the nineteenth century and the lateral beams protruding from the inner side of the feet of many other similar devices).

Except in the case of feet that have a laterally interlocking shape, which limits severely the gait to prevent the robot from stepping on its own feet, the machine must displace laterally its body, so that the vertical through its center of mass falls always within the contour of the supporting foot. This can be done either by displacing laterally the hip, by inclining laterally the bust, by moving the arms, or, in a less humanoid way, by moving laterally a ballast mass mounted on a rail at the shoulder level. Often more than one of these approaches are followed.

The number of degrees of freedom of the leg of a biped is thus larger than the minimum number seen for other types of walking machines (three degrees of freedom), and is usually 5 (including the degrees of freedom of the ankle) or 6 (with an additional one at the hip). To them, possible internal degrees of freedom of the feet and above all the degrees of freedom needed to displace the center of mass, must be added.

The picture of an experimental biped robot, the Johnnie 2 built by the Technical University of Munich, is shown in Fig. 6.17a. In this case, more than in the case of more commercial and more human-like biped robots, it is possible to see the internal complexity of such machines. Being an experimental device, a safety line is attached to the head to prevent falling, and an umbilical to take out signals is provided.

Whether building humanoid robots is a worthwhile goal is still unclear. While it is certain that humanoids are useful for scientific purposes and can be widely accepted in specific markets, like the so-called edutainment (education and entertainment) sector, the need for robots with a human shape for general applications is uncertain. An often mentioned reason is that all artificial environments are designed with humans in mind and thus a robot (or better a machine, be it autonomous, teleoperated or even programmed to perform simple tasks) working in them is more efficient if it has a human shape and is able to walk on two legs, to use hands attached at the ends of two arms (but perhaps four hands with four arms would be even better), to have stereoscopic vision using two cameras located in the head, etc.

There is no doubt that a biped can manage stairs, enter small elevators, pass through small doors and go around in the often cluttered space in houses, offices,

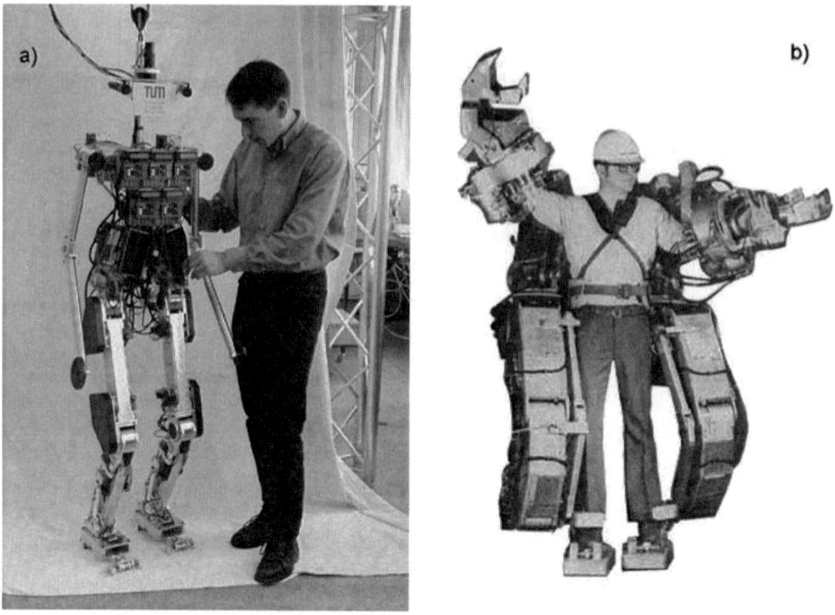

Fig. 6.17 (**a**) A research humanoid robot: Johnnie 2 built by the Technical University of Munich. (**b**) The Hardyman, a human-amplifying exoskeleton built by General Electrics at the end of the 1960s

hospitals, etc. better than a wheeled or a multileg machine. Communicating with voice and simple human gestures helps cooperating with human workers or receiving orders from customers who need not to be expert in using a human-machine interface (HMI) and humanoid arms with hands allow to operate tools that are designed to be used by humans. These machines can perform menial jobs in houses, hospitals, shops, offices and other places in which at present all these jobs are directly performed by humans. Other applications are supplying telepresence to meetings, performing simple telework in difficult or distant places and, with increasing performances, performing more complex tasks.

A more questionable, but often mentioned, application for humanoid teleoperators is driving machinery in dangerous or uncomfortable places, like construction or mining machines, underwater devices, etc. Here the question is whether it is worth while building a teleoperator that seats in the cockpit of these machines under the control of a human operator, while the latter can teleoperate directly the machine.

The feasibility of these applications will depend not only on the possibility of building humanoid robots with the operating characteristics, but also with the size and mass of a human, while being able to interact safely and efficiently with humans, but also on the cost of complex machines of this kind. Perhaps mass production will in the long run reduce costs, like it was the case in the recent past for personal computers, but at present such cost reduction may seem questionable.

Fig. 6.18 Astronaut Catherine Coleman poses with Robonaut 2 on the ISS. (NASA image)

It must be, however, remembered that statements saying that personal computers will never have a market not only because they will never be cheap enough, but above all because normal people will never be interested in buying them, were still written in the 1980s by well knows historians of technology. The slogan 'a personal robot in every house' might well become a reality in the future. And these personal robots will be humanoid robots.

What said above will have some application to space robotics too? As a first point, up to now there are not many artificial environments in space designed for humans to require humanoid robots. The only place where at present a humanoid robot may be useful is the International Space Station, where, however, the usefulness of legs is questionable, due to microgravity, and the cramped space suggests to build robotic assistants as small as possible. The other artificial environments, from lunar and Mars bases to large exploration spacecraft, have still to be designed, and while doing so they may be shaped so that they can accommodate centaur or other wheeled or multileg robots. Their design is far enough in the future to allow astronaut assistant of various types to be built in advance and to influence their design.

If personal (humanoid) robots will be common on Earth for various applications, their availability will influence the design of space environments, and we will see humanoid robots in space. But perhaps a lesson can be learned from science fiction: in the Star Wars saga there are many humanoid robots used on planets, but the robots to be used in space are not humanoid.

The NASA astronaut assistant Robonaut 2 is at present in operation inside the International Space station (Fig. 6.18). It was launched aboard Space Shuttle Discovery on the STS-133 mission in February 2011 and then installed permanently on the Unity node. As clearly seen in the figure, its upper half is humanoid, but it has no mobility devices and in particular no legs. It is an experimental device, aimed to test the possibility of running robotic astronaut assistants in actual space missions, and many improvements are planned to increase its autonomy and to give it mobility. At present it is not suited to operate outside the station, but also this feature is expected to be introduced in the future. Several mobility devices, in particular based on wheels, were tested on the ground, but also versions provided with humanoid legs have been studied.

Exoskeletons are biped machines with peculiar characteristics: they are a sort of wearable robots that can be put around the human body to enhance its strength, to protect it from a hostile environment or to compensate for motion handicaps. An interesting type of exoskeletons for space use are motorized space suits.

The first exoskeleton to be built was the *Hardyman* built by General Electric in the end of the 1960s (Fig. 6.17b). Although not reaching a fully functional stage, it demonstrated the possibility of building a human-amplifying device. Subsequent work on exoskeletons was mostly oriented toward military applications, like the *Pitman*, a Body Armor Powered (BAP), developed at Los Alamos National Lab since the early 1990s. The human operator inside the robot is fully insulated from the environment, protected from possible chemical and bacteriological attacks, more or less as he would be in a space suit.

Motorized space suits can be developed along these lines, with the goal of making EVA less demanding; this approach is in competition with the development of less stiff and more wearable space suits. A particularly interesting point is that a motorized space suit can be developed in a modular way: the first part of the suit that will be developed is likely the glove, since at present space suit gloves are quite stiff and hamper the ability of astronauts to perform delicate jobs with their fingers. Motorized space suits may in the future be used to perform heavy tasks on the Moon and on Mars, while it is possible to imagine that in a very distant future humans wearing motorized suits will be able to work on high gravity planets.

Exoskeletons are strictly speaking more telemanipulators than robots: they can be controlled by the human inside who simply moves as if his body were free: the device copies and amplifies the motion to produce the same outputs. The human-machine interface may thus consist in pressure sensors located between the human body and the inner surface of the robot. A more sophisticated approach is reading the nervous signals driving the muscles of the human body, or even the brain activity, so that the exoskeleton performs the movements the astronauts 'think' to perform. Much research is going on to implement these strategies.

In a way the control of an exoskeleton is easier than that of a biped robot, since the task of maintaining the equilibrium can be entrusted to the human being inside; it is, however, uncertain whether this can be done in practice, without putting too much burden on the human, who would play the role of a low-level controller.

6.1.7 Conclusions

The control of zoomorphic walking machines is much demanding. The complexity of generating the correct feet trajectories and choosing the most suitable gait are just two of the many problems the designer of the control system of a walking machine must solve.

If many legs are used, the number of degrees of freedom is high. The motion of the actuators must be coordinated and the control system must ensure that the legs follow well determined trajectories with a good precision. Even if the precision required is lower than that typical of industrial robots, the larger number of degrees of freedom makes the control task much more complicated than what is typical in robotics. The reduction of the number of legs is accompanied by an increase of the performance required from the control system which, although controlling a smaller number of degrees of freedom, must deal with the intrinsic instability of this type of layout. While a hexapod can use a gait in which all subsequent positions are static equilibrium positions, a quadruped can do so only if some additional degree of freedom is present, to deform the body or to shift masses in order to move the center of mass. The situation is even worse for a biped: the fact that the reduction of the number of legs during animal evolution was accompanied (or better, made possible) by an increase of the brain mass and complexity, shows clearly that the control difficulties greatly increase in this way.

The choice of the gait and the adjustment of the trajectories of the feet is much influenced by the nature of the terrain, and most animals rely on visual information to predict the ground irregularities and make adjustments before actually having to manage obstacles. This can be done by using radars, ultrasonic sensors or image recognition devices, but no available and proven technology at present exists. Some attempts to simplify the layout to avoid closed loop control gave promising results, but only in case of very small machines and fairly smooth ground. It is doubtful whether walking machines much larger than insects can be made without complex control systems.

Many elements of walking machines built following zoomorphic configurations are often highly stressed, a thing that in living beings can be accommodated for by the ability of bones to remodel themselves, becoming stronger in the most stressed zones. Biological materials, from bones to tendons to muscles, have properties that are still unmatched by materials used in machines, which explains why the very high stressing encountered in walking machines may be acceptable in animals and unacceptable in artificial walkers.

But it is not just a matter of materials and control: natural walking machines have a reliability which is unacceptably low for artificial machinery. All organisms shaped by evolution are expendable, since what matters is the species and not the individual. Reliability is thus a secondary factor. To this the possibility of self-repair of biological materials must be added: bones and muscles remodel and repair themselves in a way that is impossible for the materials used in machines.

Owing to these difficulties few of the many walking machines prototypes have on-board power and control systems. If power and control signals are supplied from the outside through an umbilical (for the control system a radio link could be used; to supply power through a microwave link has never been attempted, as far as the author knows, on walking machines) the mass of the power system and of a control computer can be well beyond the carrying capacity of the machine.

Supplying energy from outside makes the use of pneumatic or hydraulic actuators possible or, in the case of electric actuators, allows to support the weight of the machine using the motors. Some walking machines have been controlled by a mainframe, and so no strict limitations on the computational power is present. Obviously this is acceptable for a demonstrator or for some specialized applications, but if a workable machine, which can compete with standard wheeled or tracked vehicles or robots must be built, all systems must be on board.

Some doubts may be cast on the concepts of biomimetism and biomorphism in machines, at least until what is taken from nature is the very way in which materials are built and machines are designed. That metal is different from bones and electric motors are different from muscles is trivial, but perhaps the largest difference is not in their properties but in the way they are built. Living beings are built by bottom-up processes, assembling their molecular components, and this yields a machinery of outstanding complexity, but one which builds by itself, following encoded instructions. Trying to duplicate this complexity with a top-down procedure, by machining the various parts with the standard technologies used to build electromechanical devices may well be a losing strategy, leading to unbearable manufacturing costs and poor performances.

Micro- and nano-technologies may allow a bottom-up approach similar to that seen in nature, and perhaps in this way biomimetic and biomorphic machines may in the future be built. It is a perhaps commonplace to say that the twentieth century was the age of physics while the twenty first will be the age of biology, but this could well turn out to be true, also in the fields of mobile robotics and vehicles.

6.2 Hybrid Machines with Wheels and Legs

Among the many configurations of walking machines that were suggested in the past, a number had both wheels and legs or appendages which rotate like wheels and are conformed like legs. Seldom they were actually built and extensively tested but, at any rate, in many applications it could be of advantage, at least theoretically, to have wheels to travel on level ground and legs to deal with obstacles. This is true from the viewpoint of the average speed but also to increase the reliability of the machine. The leg mechanisms, which are usually highly stressed and are critical for fatigue and wear, are required to operate only when needed, while wheels supply mobility in easier conditions.

Actually there is a wide range of solutions combining wheels with leg-like mechanisms to increase mobility. The most conventional types are vehicles (military vehicles or machines for open air mining, construction works, etc.) in which regular wheels are suspended using long-stroke trailing arm suspensions, often active or at least supplied with load-leveling devices. They are basically wheeled vehicles, even if the suspensions can be made by two articulated levers, more similar to a leg than to a trailing arm. Here the ability to walk of what is essentially a wheeled vehicle depends on the actuators and the control system used.

The simplest device of this kind is a longitudinal swing arm suspension provided with an actuator able to change its reference position (Fig. 6.19a). If the latter is non reversible, the actuator needs to be powered only when a change of angle ϑ_{0i} is required. If the rover is slow, the rate at which angles ϑ_{0i} must be varied is low and the gear ratio can be high enough to allow very small and lightweight motors to be used. Moreover, the control system has to deal only with very low frequencies. The

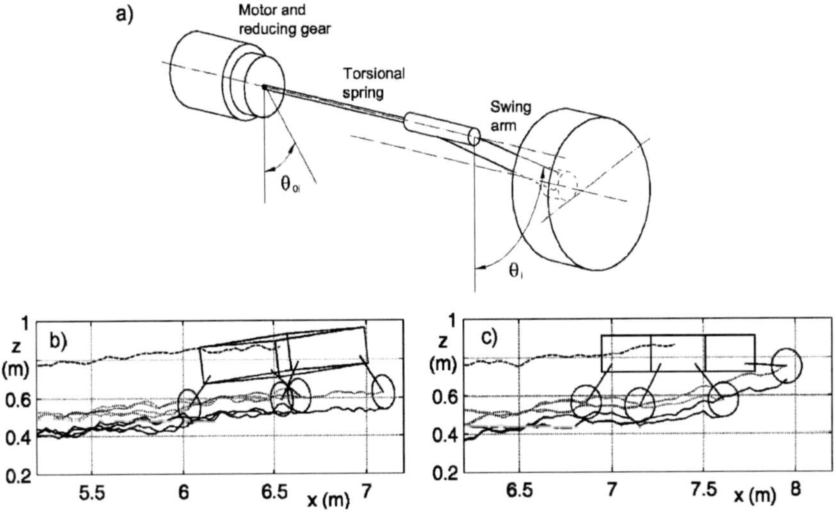

Fig. 6.19 (**a**) Sketch of the active swing arm suspension; (**b**) result of the dynamic simulation for a rover with passive suspensions; (**c**) result of the dynamic simulation for a rover with active suspensions

total power required for actuating the suspension may be a tiny fraction of the power needed for moving (a situation usually referred to as semi-active suspensions).

Consider a four wheeled rover provided with swing arm suspensions, steered by slip steering. A simple PD strategy aimed to keep the vehicle body in a horizontal position can be implemented. At regular time intervals (say 0.1 s, but even more if the rover is slow) the pitch and roll angles and velocities (respectively θ, φ, $\dot{\theta}$ and $\dot{\varphi}$) are measured, and angles ϑ_{0i} are corrected as

$$\begin{aligned}\theta_{0\text{new}} &= \theta_{0\text{old}} + (K_{p1}\vartheta + K_{d1}\dot{\vartheta})[1 \quad 1 \quad 1 \quad 1]^T \\ &\quad + (K_{p2}\varphi + K_{d2}\dot{\varphi})[-1 \quad 1 \quad 1 \quad -1]^T,\end{aligned}$$

where K_{pi} and K_{di} are the proportional and the derivative gains. At each step, a check is performed to verify whether the suspension angles ϑ_{0i} have reached their minimum or maximum value. In that case the extreme values are imposed and some roll or pitch is accepted.

The results for the simulation of a rover of this kind taken from,[12] where further details can be found, are shown in Figs. 6.19b and c (passive and active suspensions, respectively). The simulation is performed on a rough terrain that is flat for the first meter, and then has an average upward slope of about 8%. The profile is obtained by using a random numbers generator. A speed of 0.05 m/s (180 m/h) is stated as a reference value and the motors try to keep it constant notwithstanding the ground

[12] G. Genta, *Simplified Model of a Small Planetary Roverw with Active Suspensions*, VII IAA Symposium on Near Term Advanced Space Missions, Aosta, July 2011.

Fig. 6.20 Two images of the ATHLETE (All-Terrain Hex-Legged ExtraTerrestrial Explorer), a wheel-leg hybrid machine built by NASA (NASA images)

irregularities using a proportional control, while the motors on the right side supply a torque 0.2 Nm higher than that supplied by the motors on the left side to obtain a left turn.

From the figure it is clear that not only the suspension controller succeeds in keeping the rover flat (the roll and pitch angles are always much smaller than 1°), the trajectory is much smoother and the rover appears to be more controllable, but also that the suspensions are able to react to the ground irregularities with a sort of 'leg-like' behavior, putting the wheels on top of the bumps and pushing them in the holes. The ability to traverse rough terrain is greatly enhanced. Two images of the ATHLETE (All-Terrain Hex-Legged ExtraTerrestrial Explorer), a wheel-leg hybrid machine built by NASA, are shown in Fig. 6.20. An unusual feature of this machine is its radial symmetry.

An example of a quadruped machine with an essentially mammalian configuration and large wheels at the feet is the Workpartner built by Helsinky Technical University. This centaur machine was thought for terrestrial applications, but this configuration can be used for a helper for astronauts during EVA on planetary surfaces. A space version of the Workpartner has been studied for the European Space Agency.

On the other side there are walking machines with small wheels either attached at the ends of the legs or under the body, which can be put on the ground by raising the legs. Wheels at the end of the legs (Fig. 6.21) have the advantage of allowing the body of the vehicle to ride high on the ground, clear from obstacles, but either require that the legs behave as active suspensions or that some sort of elastic and damped suspension is placed in the feet—except if no suspension at all is used, a thing possible only for very low speeds. Wheels under the body make it easier to use a more or less conventional suspension, but the body rides low and the legs must be able to get out of the way in a somewhat unnatural position, which in some cases, particularly when a mammalian configuration is used, may be impossible at all.

Generally speaking, to perform well as a walking machine the device must have feet as light as possible, so the use of large wheels at the end of the legs indicate

Fig. 6.21 A wheel–leg hybrid machine

that walking has been considered as an auxiliary locomotion method to allow a wheeled vehicle to extend its capabilities to rough ground. On the contrary, very small wheels show that the primary locomotion mode is walking. In any case, the presence of wheels not only allows a walking machine to increase its speed on level ground and to be operated in a simpler way, but also allows a limited performance in case of failure of some actuators or of the walking control system.

Rigid-frames machines are particularly suitable for building wheel–leg hybrid devices. The wheels may be located under the body, all on one of the two frames (and then a steering system is required) or two on one frame and two (or one) on the other, so that steering is provided by the rotation of the frames with respect to each other. The wheels can be easily provided with a suspension system, but some of the throw of the legs is lost. As an alternative, the wheels can be located under some of the legs.

In the case of rigid-frames machines, the wheels may be located under two legs of one frame and two of the other one, so that no purposely designed steering system is required. However, either the speed is very low and the legs act as a load-leveling suspension, or some compliant connection is required between the feet and the wheels. For low speed operation on fairly level ground the suspension system can be dispensed of, even without using the legs as an active suspension system, if three wheels are used. Since there are many possible layouts (number and location of wheels, presence of a dedicated suspension and steering system, control strategy of the body and legs while rolling), a vehicle of this type needs to be designed following its tasks, the type of terrain expected and other operational constraints.

As an example, a twin rigid-frames hexapod with four wheels under the legs is shown in Fig. 6.22. The wheels are connected to the fixed part of the legs through a trailing arm suspension, in such a way the vehicle is supported on the wheels when the legs are in the fully retracted position (Fig. 6.22a). Each wheel has its own electric motor. When the vehicle walks over obstacles the wheels are raised from the ground (Fig. 6.22b). Steering is performed by using the already existing steering degree of freedom of the vehicle (Fig. 6.22c).

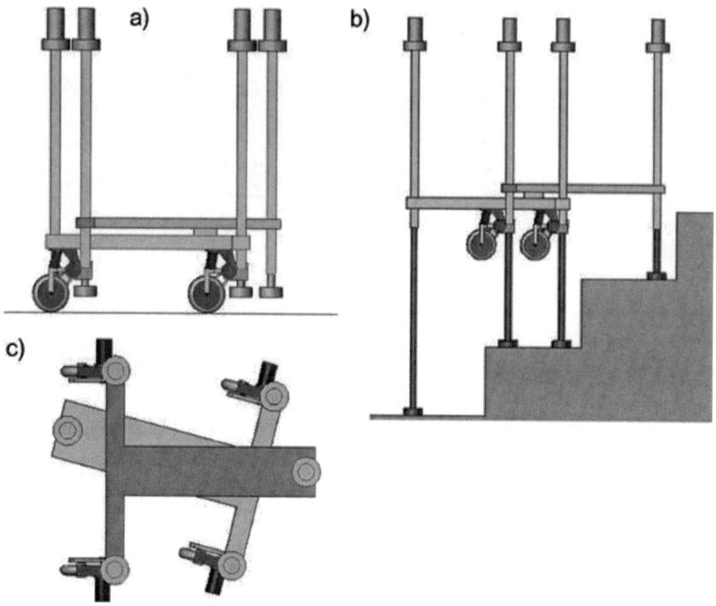

Fig. 6.22 Twin rigid-frames hexapod with four wheels under the legs. (**a**) Rolling configuration; (**b**) walking configuration; (**c**) steering

A different type of wheel–leg hybrid is a machine that uses a sort of spoked wheels, sometimes referred to as *whegs* (Fig. 6.23a). There can be a number of rotating spokes, sometimes bristles, located at the periphery of a wheel, or just one curved rotating element, like in Fig. 6.23b in which a picture of the RHex is shown. The RHex is a line of robots on whegs (but some prototypes had also wheels or fins to swim) built by a group of Universities under the sponsorship of DARPA.

The mobility of robots using whegs is generally good on all terrains even in quite bad terrain conditions, but the motion is very irregular and controllability is poor. They are also faster than similar machines operating on more standard legs.

Another solution based on multiple wheels located at the periphery of a rotating polygon is shown in Fig. 6.23c. In the figure each group is made of three wheels located at the vertices of equilateral triangles. Devices of this kind are used for structured environments, like stairs with regular steps, but it is questionable whether they may be used on irregular natural ground.

6.3 Hybrid Machines with Tracks and Legs

A solution in which thin legs are located at the periphery of a track, in a way similar to the whegs mentioned earlier, is shown in Fig. 6.23d. This solution is actually similar to an ordinary track, with oversized track plates.

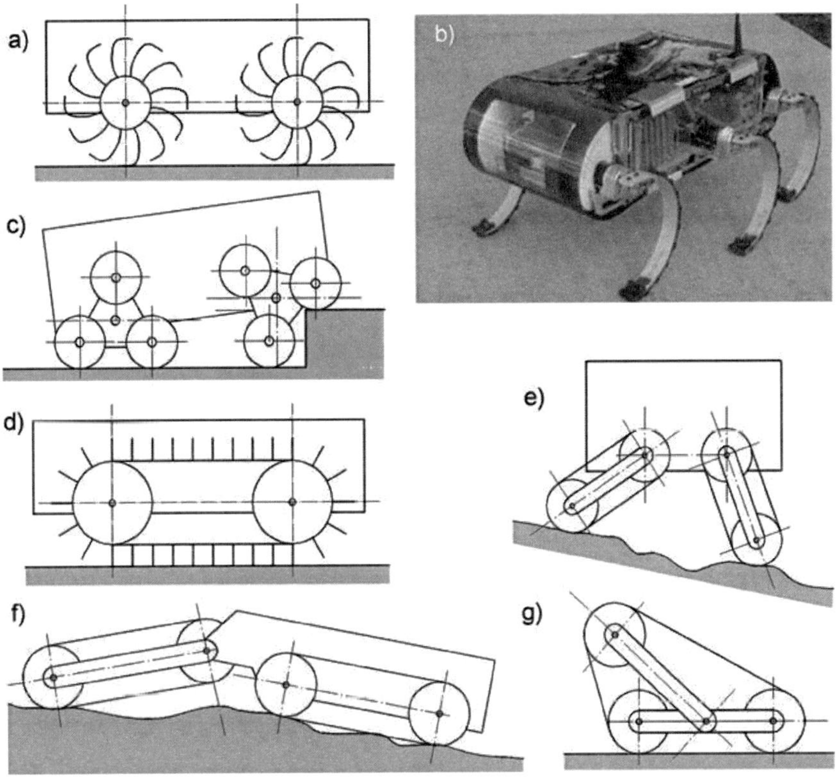

Fig. 6.23 (**a**), (**b**) Whegs: rotating legs; a vehicle with spoked wheels and the RHex, a robot with 6 whegs with a single flexible rotating beam. (**c**) A device based on multiple wheels located at the periphery of rotating triangles. (**d**), (**e**), (**f**), (**g**) Track-legs hybrids. A solution similar to whegs, tracks used as legs and tracks with variable geometry

Fig. 6.24 Track–leg hybrid robots

The proposals for track–leg hybrids actually are based on geometries in which one or more tracks can be moved with respect to the vehicle body, for instance to be used as legs (Figs. 6.23e). Other solutions are shown in Figs. 6.23f and g and in Fig. 6.24.

Hybrid solutions based on tracks have the typical drawbacks of tracks in space applications and are seldom considered.

6.4 Hopping Robots

Hopping robots[13] are quite interesting for low gravity applications. The first hopping robot to be used in an exploration mission was on board the Russian probe Phobos 2 launched on 12 July 1988. The robot had to land on the Mars moon Phobos, in April 1989, but the spacecraft lost contact with Earth after arriving close to its target. The robot was a small spherical object, able to hop on the surface (Fig. 6.25a).

More recently, a small hopping robot named Minerva (Micro/Nano Experimental Robot Vehicle for Asteroid) was carried by the Japanese Hayabusa spacecraft to explore the asteroid Itokawa. Again, it could not be tested on its target since it failed to reach the surface. It was a 10-centimeter-tall robot designed to hop around the 600-meter-long asteroid taking close-up images with three cameras and making temperature measurements of the surface.

Hopping robots can operate in two distinct ways: making single jumps, each one using a certain quantity of energy and then dissipating it at landing or making a number of hops consecutively, recovering as much energy as possible after each hop (Fig. 6.25b). Something in between is making single hops but recovering some energy at landing and storing it to use later.

If no rocket propulsion is used to slow down the descent, landing occurs at a speed corresponding to falling speed. If the energy is not recovered, the robot falls down and provision for ensuring its structural integrity (e.g. air bags) must be taken. If the energy is recovered, it is the energy recovering device that decelerates the robot at the design acceleration.

If no energy is recovered, the jumper can land in any attitude and then recover at ease the attitude for next hop. This does not require a precise attitude control during flight and can be performed even with simple mechanical means.

The energy is usually stored in a spring, which can be slowly charged by mechanical or better electromechanical means. Alternatives are storing electrical energy in a capacitor that can be discharged through an electromagnet that produces the mechanical impulse setting the robot in flight.

To store the energy at landing, the flight must be accurately controlled, so that at landing useful work can be performed on the device that stores the energy. The control must be even more accurate if the hops are subsequent, and energy is stored for a very short time.

[13] Here with hopping robots are intended robots that are propelled by forces exerted against the surface, and not jet (or rocket) propelled robots. Also robots propelled by inertia forces due to unbalanced rotors are not dealt with.

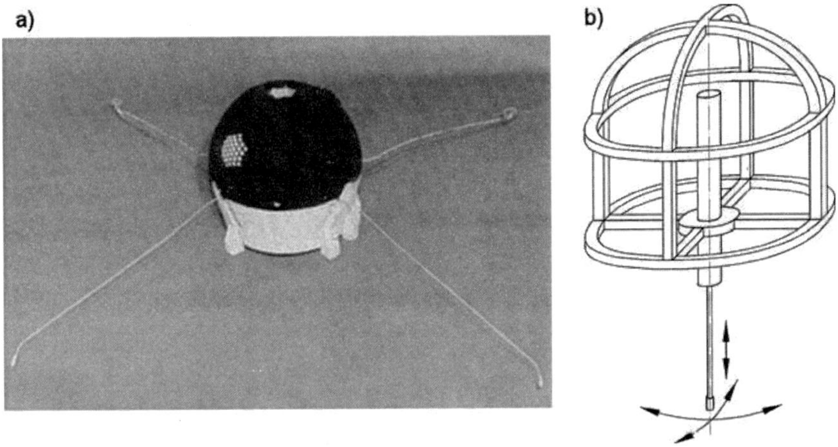

Fig. 6.25 Hoppers. (**a**) Phobos 2 hopping robot that failed to reach its target in 1988; (**b**) hopping robot built by the Carnegie Mellon University (from D.J. Todd, *Walking Machines: An Introduction to Legged Robots*, Kogan Page, London, 1985)

To evaluate the performance that a jumping robot driven by a spring can reach, consider that the elastic potential energy $\mathcal{U}_e$ that can be stored is

$$\mathcal{U}_e = \alpha K m \frac{\sigma^2}{\rho E}, \tag{6.12}$$

where σ is the maximum stress that the spring material can withstand in operating conditions, E is the Young's modulus of the material, m is the mass of the robot, α is the ratio between the mass of the spring and the mass of the robot and K is a coefficient that depends from the type of spring used. The values of K for some spring types are reported in the following table:

Type	K
Beam, constant cross section	$1/18$
Beam, constant thickness, linear taper	$1/6$
Torsion bar, circular cross section	$\epsilon/4 \approx 5/16$
Torsion bar, rectangular cross section	$\approx (0.13\text{–}0.15)\epsilon \approx 0.16\text{–}0.18$
Helical, circular cross section	$\epsilon/4 \approx 5/16$
Helical, rectangular cross section	$\approx (0.13\text{–}0.15)\epsilon \approx 0.16\text{–}0.18$

where

$$\epsilon = \frac{\tau^2 E}{\sigma^2 G} \tag{6.13}$$

is a parameter that depends on the material. For steel its value is about $5/4$.

During the thrust phase, the potential energy of the spring transforms into the kinetic energy of the robot. Assuming that the efficiency of this transformation is equal to 1 and neglecting the part of the force due to the spring needed to support the weight (i.e. assuming that the thrust phase is infinitely short), the velocity the robot has at the beginning of the hop is

$$V=\sqrt{\frac{2\mathcal{U}_e}{m}}=\sqrt{2\alpha K\frac{\sigma^2}{\rho E}}. \tag{6.14}$$

Assume that the hop occurs in vertical direction from the surface of a celestial body, and neglect aerodynamic drag. The gravitational potential energy of a body with mass m at a distance h from the surface of a planet with radius R and gravitational acceleration g on the surface is

$$\mathcal{U}_g=-m\frac{gR^2}{R+h}. \tag{6.15}$$

By equating the total energy at the surface, when launched vertically with velocity V to the total energy at the height h where it stops, it follows that

$$\frac{1}{2}mV^2-mgR=-m\frac{gR^2}{R+h}. \tag{6.16}$$

The altitude the hopper reaches is thus

$$h=\frac{V^2}{2g-\frac{V^2}{R}}. \tag{6.17}$$

Except for the case of very small celestial bodies, V^2 is much smaller than $2gR$ and the expression of the maximum hop height reduces to

$$h\approx\frac{V^2}{2g}. \tag{6.18}$$

The velocity obtained by a robot propelled by a helical steel spring with $K=5/16$, $\sigma=1$ GPa, $E=2100$ GPa, $\rho=7810$ kg/m^3 is reported as a function of α in Fig. 6.26a. On the same plot is also reported the altitude reached in a vertical launch from the surface of Earth ($g=9.81$ m/s^2, $R=6350$ km), Mars ($g=3.77$ m/s^2, $R=3400$ km), the Moon ($g=1.62$ m/s^2, $R=1700$ km), Titan ($g=1.35$ m/s^2, $R=2580$ km), Triton ($g=0.78$ m/s^2, $R=1350$ km), Pallas ($g=0.18$ m/s^2, $R=262$ km), Ceres ($g=0.26$ m/s^2, $R=460$ km), Juno ($g=0.12$ m/s^2, $R=120$ km) and Eros ($g=0.04$ m/s^2, $R=20$ km).

Note that the radii actually have no importance except for the case of Eros, since the simplified expression (6.18) yields the same results as (6.17). The data for Eros are referred to an arbitrary point on the surface, since that asteroid is irregular.

If (6.18) can be used instead of (6.17) the usual parabolic approximation for the motion of projectiles apply. The expression for the range is

$$d=\frac{V^2}{g}\sin(2\theta), \tag{6.19}$$

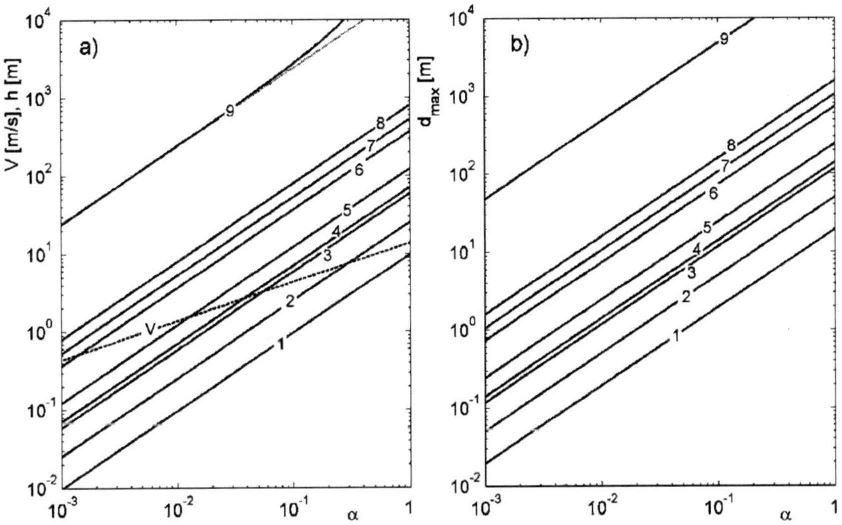

Fig. 6.26 Performance of hopping devices on different celestial bodies as functions of the mass ratio α. (**a**) Take-off velocity and height reached in a vertical launch; (**b**) range in a 45° launch. (1) Earth; (2) Mars; (3) Moon, (4) Titan, (5) Triton, (6) Ceres, (7) Pallas, (8) Juno, (9) Eros. In the last case, the height shown with a *dotted line* is that computed using the simplified formula; in all other cases the simplified formula and the complete one give the same results

where θ is the angle between the direction of the velocity at launch and the horizontal.

In this case the maximum distance is covered when the initial velocity is at 45° from the horizontal at the launch point. The maximum range,

$$d_{\max} = \frac{V^2}{g}, \tag{6.20}$$

is reported in Fig. 6.26b.

Remark 6.7 For the case of Eros, starting from a mass ratio of 0.05, the trajectory should be considered as elliptical and not parabolic.

The performance of hopping robots are not much interesting for planets and large satellites, but are good on asteroids, and in particular on small ones and comets, where other forms of locomotion are problematic.

The figure is based on data for a steel spring; if other materials like high strength carbon reinforced plastics had been considered, better performances would have been obtained.

As already stated, the formulae reported above are an oversimplification, since they are based on the assumption that the spring imparts its energy to the hopper instantaneously. In other words, this means that the stiffness of the spring and the

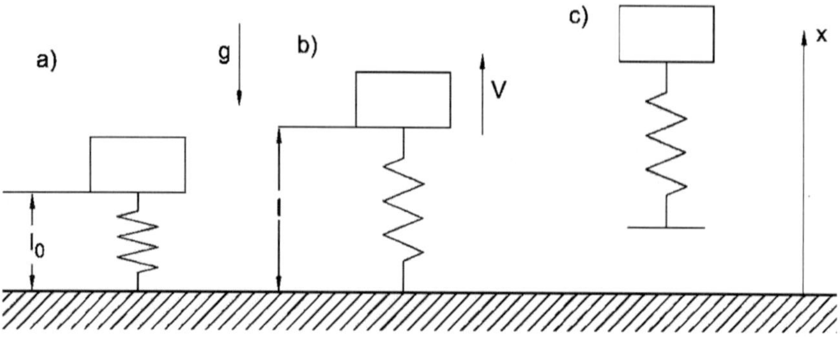

Fig. 6.27 A hopper taking off. (**a**) Starting position with the spring compressed; (**b**) instant when the hopper leaves the ground; (**c**) in flight

initial acceleration of the mass are infinitely large, while the compression of the spring is infinitely small.

Consider a more realistic case: a mass m accelerated by a massless spring with stiffness k. The length at rest of the spring is l; initially the spring is compressed to the length l_0 (Fig. 6.27a). The elastic potential energy stored in the spring is thus

$$\mathcal{U}_e = \frac{1}{2}k(l - l_0)^2. \tag{6.21}$$

As soon as the spring is released, the total force exerted on mass m is

$$F_0 = k(l - l_0) - mg. \tag{6.22}$$

If

$$k(l - l_0) > mg \tag{6.23}$$

the spring is strong enough to allow mass m to be propelled upwards. The equation of motion after the release of the spring is

$$\ddot{x} = \frac{k}{m}(l - x) - g, \tag{6.24}$$

i.e.

$$\ddot{x} + \frac{k}{m}x = \frac{k}{m}l - g. \tag{6.25}$$

Its solution is

$$x = l - \frac{mg}{k} - x_1 \cos\left(\sqrt{\frac{k}{m}}t\right), \tag{6.26}$$

where the constant of integration x_1 can be computed stating that at $t = 0$ the value of x is l_0 and the velocity is zero. The expression of the displacement and velocity are thus

$$x = l - \frac{mg}{k} - (l - l_0)(1 - \xi)\cos\left(\sqrt{\frac{k}{m}}t\right), \tag{6.27}$$

$$\dot{x} = (l - l_0)(1 - \xi)\sqrt{\frac{k}{m}}\sin\left(\sqrt{\frac{k}{m}}t\right), \tag{6.28}$$

where the nondimensional parameter

$$\xi = \frac{mg}{k(l - l_0)} \tag{6.29}$$

is the ratio between the weight and the force of the fully compressed spring. In the previous simplified analysis its value was 0.

At the instant of liftoff (Fig. 6.27b) $x = l$. Liftoff then occurs at time

$$t_1 = \sqrt{\frac{m}{k}}\operatorname{arcos}\left(\frac{\xi}{1 - \xi}\right). \tag{6.30}$$

The velocity of the hopper at liftoff is

$$V = (l \quad l_0)\sqrt{1 - 2\xi}\sqrt{\frac{k}{m}}. \tag{6.31}$$

The kinetic energy the hopper has at liftoff is thus

$$\mathcal{T} = \frac{1}{2}k(l - l_0)^2(1 - 2\xi) = e_p(1 - 2\xi). \tag{6.32}$$

The fact that the stiffness of the spring has a finite value causes a reduction of kinetic energy at liftoff with respect to the simplified case by a factor $(1 - 2\xi)$.

The maximum value of ξ allowing liftoff to take place is

$$\xi = 0.5.$$

The maximum acceleration occurs when the motion starts and the spring is all compressed and is

$$a_{\max} = \frac{k(l - l_0)}{m} - g, \tag{6.33}$$

i.e.

$$a_{\max} = g\frac{1 - \xi}{\xi}. \tag{6.34}$$

The maximum acceleration must be at least equal to g to have liftoff, and the hopper must then be able to withstand at least a total acceleration of $2g$.

Fig. 6.28 A rover moving on skis like the ones carried on Mars in 1971 by the Russian Mars 2 and 3 probes

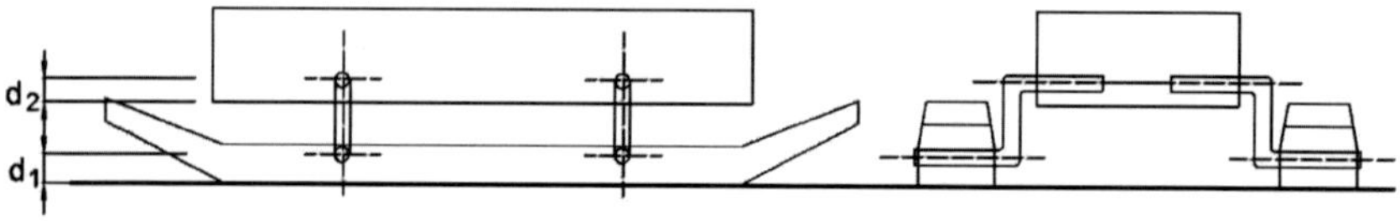

Fig. 6.29 Sketch of a rover with two skis

6.5 Skis

Often it is stated that the earliest rovers sent on Mars used skis to move, but actually this term is used in a sense quite different from that usually considered. True skis are devices that slide on a soft surface, usually snow, with the possibility of dealing with hard but flat patches like ice. Here skis are simply elongated supports that are laid on the surface to carry the weight of the vehicle. Rather than skis, they work like long feet.

The Russian Mars 2 and 3 probes, which landed on Mars in 1971, carried two small rovers (Fig. 6.28) with a range of about 15 meters from the lander. The motion was accomplished by moving the side skis using the cranks visible in the figure. Obstacles in the rover's path could be detected by the two thin bars at the front. The vehicle could determine on which side the obstacle lay, step back, change direction and try to go around it.

These rovers could not be tested in operation since Mars 2 crash-landed on the planet and Mars 3 ceased transmissions 20 seconds after landing.

The simplest way to move on skis of this type is using the device shown in Fig. 6.29. There are just two skis, one at each side. In the figure the rover is shown standing on the skis; by rotating the cranks the body is lowered to the ground, then the skis are raised and moved one step forward and so on. If $d_1 = d_2$ the motion of the skis and of the body are equal. By rotating the cranks on the two sides in different directions or at different speed the vehicle turns, but there is much sliding at the vehicle-ground interface and irregular motion of the body.

By using four skis, two at each side, it is possible to have either one or another pair of skis on the ground and the body can be always raised from the ground.

However, devices of this type have more a historical than a practical interest and it is likely that their only advantage, the great simplicity, does not overweight their many drawbacks.

6.6 Apodal Devices

In the animal world, apodal (without feet) animals are animals that move sliding with their body on the ground, like worms and snakes.

Apodal vehicles are often also referred to as *active cord* vehicles and are snake-like devices; however, the contact with the ground is often performed using small wheels or tracks (see Sect. 5.4.8). However, true snakes, i.e. devices that contact the ground with their bodies, have also been built.

A snake moves by modulating the pressure it exerts on the ground along its length, in such a way that some of its parts exert high friction forces and other parts very little or even no force at all if are raised from the ground. Its motion is thus due to a tridimensional deformation, in the sense that the axis of the snake takes the form of a tridimensional line. The possible gaits are four in number (Fig. 6.30).

The most common gait is the so-called lateral undulation (Fig. 6.30a). The various sections of the body move laterally keeping a more or less sinusoidal pattern, and also in direction perpendicular to the ground in a similar way. The contact points with the ground are shown by the arrows and must be at least 3 in number. In general this way of propelling forward the snake requires good traction and is less effective for short and heavy snakes.

Gait shown in Fig. 6.30b is a rectilinear gait, i.e. all cross section move in the direction of the axis of the body, stretching and compressing alternatively the various sections. The snake fixes some points on the ground (static points) through the friction of the skin and moves the points in between forward. then fixes other points and so on.

Sidewinding motion (Fig. 6.30c) is the most complex snake gait. The snake moves with lateral waves, maintaining just two points that change continuously in time, in contact with the ground. It is a gait well suited for motion on low friction and loose ground, like sand, and actually is used mainly by desert snakes. It has also the advantage of preventing overheating of the skin, since it minimizes contact with the hot sand.

The last gait, referred to as concertina motion (Fig. 6.30d) is similar to rectilinear motion since the snake fixes a point on the ground and then retracts its body forward; then fixes other points, pushing forward the points that before were fixed. The difference is that in rectilinear motion the body compresses and expands axially, while in concertina gait it bends. Also this gait can be used on low friction terrain.

Worms can use also another type of gait, bending the body in the vertical plane.

As already stated, artificial snakes, or active cord mechanisms, often use wheels or tracks at the contact with the ground. The snake shown in (Fig. 6.31a, GMD-

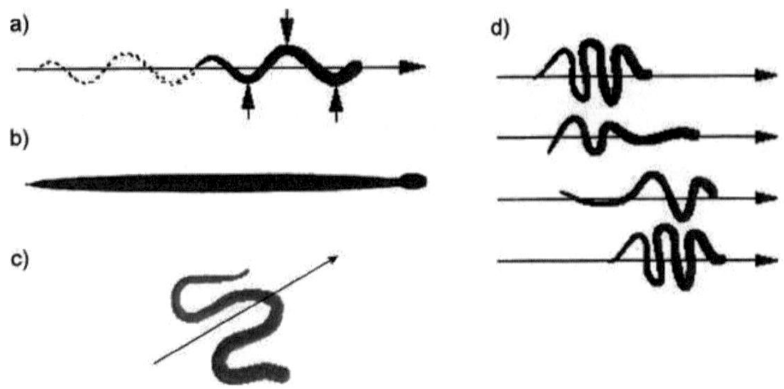

Fig. 6.30 Gaits for snake motion. (**a**) Lateral ondulation; (**b**) rectilinear locomotion; (**c**) sidewinding; (**d**) concertina motion

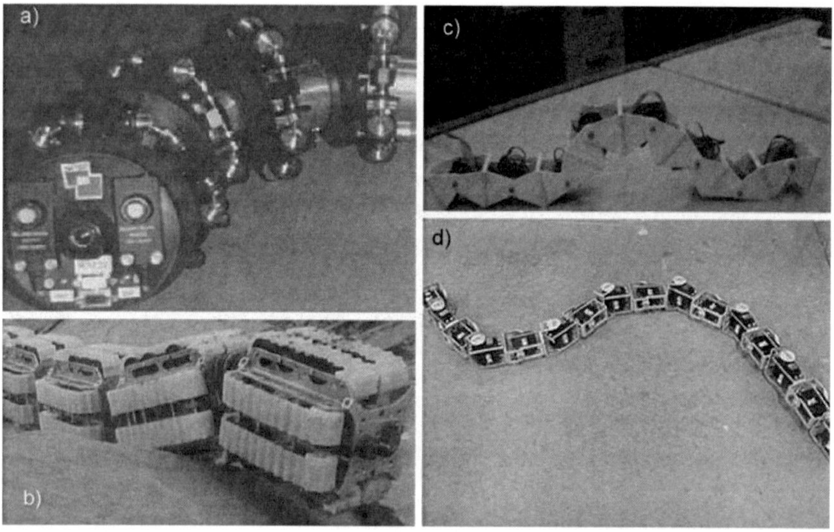

Fig. 6.31 Active cord robots. (**a**) GMD-snke2; (**b**) Omni Tread OT-8; (**c**) Cube Revolutions; (**d**) snake robot built at Ames Research Center of NASA, as a demonstrator for a snake rover (NASA image)

snke2 built in 1999 at the German National Research Center) has a number of small wheels in each segment, propelled by a DC motor in each section.

The robot shown in Fig. 6.31b is the OmniTread OT-8 built in 2005 by the Mobile Robotics Lab. at the University of Michigan. Each segment has two degrees of freedom plus a set of motorized tracks.

In Fig. 6.31c the last (2004) of the robots developed at the Universidad Autonoma de Madrid, called Cube Revolutions, is shown. Each segment has one degree of freedom, but they can be assembled so that the rotation occurs in a vertical or horizontal plane. If all joints are assembled in the same way as in the figure the motion is more worm-like than snake-like.

Another snake robot is shown in Fig. 6.31d. This machine is a true snake designed as a demonstrator for a snake-like planetary rover and built at the Ames Research center of NASA.

Snake robots are interesting for particular missions like exploring small cavities and may be a good solution for exploring the lava tubes that are likely to exist on the Moon and on Mars. Their main disadvantage is the slow speed and the difficulty of locating the payload.

Active cord machines are evolving toward reconfigurable robots, which can take different shapes, from snakes to legged machines and even to wheels, by rolling on their back, with the head joined to the tail.

Chapter 7
Actuators and Sensors

7.1 Actuation of Space Robots

Robots are active system and require a source of energy to power all their functions. The energy needed for operation must be distributed to the various functions and duly modulated, by power converters, which are themselves managed by a suitable low level controller. The power converters feed transducers that transform the energy supplied by the source into the mechanical energy needed to perform the various tasks: these transducers are usually referred to as actuators.

The actuators act on the mechanical body of the robot, giving it 'life'. In most active systems, and in particular in robots, this open-loop control is not sufficient to ensure correct operation, and the control loop needs to be closed. Some, often a tiny part, of the energy of the mechanical system needs to be converted back by suitable transducer in a form that can be supplied to the controller, allowing it to know the state of the system. These transducers are the sensors.

This power chain is represented in the form of a block diagram in Fig. 7.1.

The block diagram of the figure is an oversimplification, in particular in the case of robots. There may be several power sources, in series or in parallel, each one with its controller, and with a controller acting as energy manager. The actuators may be of different type, requiring different power converters.

The controller is sketched as a two level device, with a low level controller closing directly the loop on the actuators, and a high level controller, which in simple telemanipulators may be directly the human in the loop, defining the set points for the low level controller. Actually robots may have several levels of control nested one upon the other and, if present, the human interacts only with the upper level through a Human Machine Interface (HMI).

The actuators may be reversible, and some energy can flow backwards (dashed arrows in Fig. 7.1) when the mechanical part produces energy instead of requiring it (an arm lowering a load, a moving robot that decelerates, etc.).

If the actuators are not reversible this energy must be dissipated directly by the mechanical system (e.g. through brakes). In case of reversible actuators, the energy can be dissipated in the power converter (e.g., through resistors that transform electrical energy into heat), or supplied back to some form of energy accumulator to be

G. Genta, *Introduction to the Mechanics of Space Robots*, Space Technology Library 26, 427
DOI 10.1007/978-94-007-1796-1_7, © Springer Science+Business Media B.V. 2012

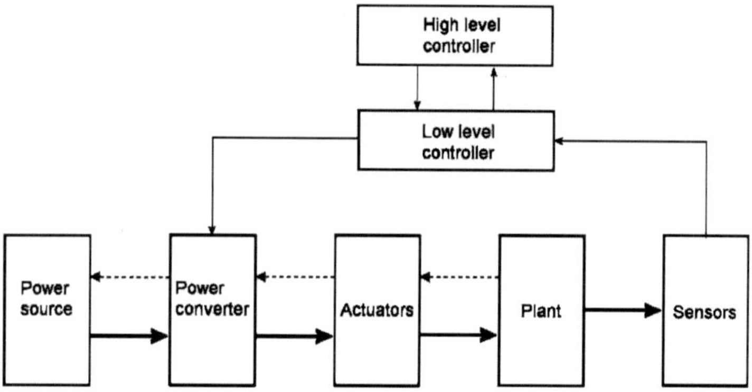

Fig. 7.1 Block diagram of an active, closed loop, system

used later. A power converter able to manage the energy flow in both directions is often referred to as a four-quadrants converter.

The actuators found in living organisms are the muscles which, as will be seen, have some peculiar characteristics that are still unmatched by artificial actuators.

The most common actuators used in robotics are either electromagnetic, hydraulic or pneumatic, i.e. they convert electric energy or the energy of a pressurized fluid into mechanical energy. While in industrial robotics hydraulic and pneumatic actuators are widespread, in mobile robots most actuators are of the electric type, due to a number of reasons. Firstly, energy is mostly stored on board of moving robots as electric energy, and the efficiency of using some sort of electric compressor or pump to operate pneumatic or hydraulic actuators is generally low. Other reasons are the more compact layout of the system and the fact that it is easier to control electric actuators directly from the main controller of the robot than to control actuators of different type.

The case of electro-hydrostatic transmissions is an exception, but in this case the hydraulic transmission is not controlled, and all control action is performed by the electrical part of the system.

In case of space robots the advantages of electric actuators over fluid based devices are even larger, owing to the difficulties of dealing with fluids in the harsh environment these devices must withstand. In particular, pneumatic actuators are well suited where there is a practically unlimited supply of working fluid, but not where the working fluid must be carried on board and recycled indefinitely.

Nevertheless, in some applications the higher compactness and lightness of hydraulic motors may make hydraulic, or better electrohydraulic systems, very attractive.

A number of other devices that can be used to actuate an active system must be added to these 'conventional' actuators. They are often called solid state or smart materials actuators, since they rely on materials that are able to change their properties under the effect of electric or magnetic fields or changes of temperature. Piezoelectric, magnetostrictive, electrostrictive and shape memory actuators are all in this category. Even the thermal expansion of a solid can be used to power an actuator, although this simple devices have a low efficiency. All solid state actuators have the advantage of having no moving parts and hence they require no maintenance and are very reliable.

Actuators are usually subdivided into linear and rotary actuators. Linear actuators are characterized by the force they exert and by the stroke they cause. The work the actuator is able to produce is the product of the force by the stroke (assuming that the force is constant during the motion).

Rotary actuators are characterized by the torque they exert and by their rotation; the work performed is the product of the torque by the rotation.

Remark 7.1 A particular type of rotary actuators are those able to supply a torque during a continuous rotation, like electric, hydraulic or pneumatic motors. Their stroke is thus indefinite.

The output of a linear actuator can be converted into a rotation, and that or a rotary actuator into a linear motion, through a variety of mechanisms, such as levers, screws, racks, pulleys, etc.

Designing electric or hydraulic motors, gear wheels, etc. is a specialized job, requiring experience and specialized expertise. Usually actuators are considered as off-the-shelf components that are selected from a catalog by the designer of the robot, often with the consulting of the manufacturer.

The aim of this chapter is thus not dealing in detail with actuator design but only supplying a short overview of the various types. The reader must refer to the handbooks and catalogs prepared by the manufacturers, where all the relevant information needed to chose the best unit for the application can be found.

In more advanced applications, and space robots almost always qualify as such, where no suitable actuator can be found on the market, the manufacturer can be asked to develop custom units, tailored on the specific application.

7.2 Linear Actuators

7.2.1 Performance Indices

Instead of speaking of force and stroke, in linear actuators it is possible to speak of stress σ, defined as the force divided by the cross sectional area, and the strain ϵ, defined as the stroke divided by the length of the actuator. Their product is the work per unit volume of the actuator.[1]

[1]These definitions assume that the actuator has a cylindrical or prismatic shape.

Remark 7.2 In the case of fluid actuators (hydraulic and pneumatic) the stress is equal to the pressure of the fluid (if the cross sectional area of the actuator is approximated by the area of the piston) and thus depends on the design of the system and not only on the actuator itself.

Often performance indices[2] are used to compare actuators based on different principles. Among the various performance indices, the following can be mentioned:

- Maximum actuator strain (nondimensional);
- Maximum actuator stress (in Pa);
- Maximum work per unit volume (in J/m^3). It can be expressed as the product of the maximum stress by the maximum strain; it is a volumetric energy density;
- Density (in kg/m^3). Is the ratio of the mass and of the volume of the actuator. Usually it refers to the actuator alone, excluding the power converter, cooling system, fixtures, etc.;
- Actuator modulus (in N/m^2). It is the ratio between a small increment of stress and a small increment of strain when the control signal is kept constant;
- Strain resolution (nondimensional). It is the smallest step increment of strain;
- Volumetric power density (in W/m^3). It is the ratio between the power the actuator can supply in sustained operation divided by the actuator initial volume;
- Mass power density (in W/kg). It is the ratio between the power the actuator can supply in sustained operation divided by the actuator mass; it can be computed as the volumetric power density divided by the density.

A further important index is the efficiency with which the actuator performs the energy conversion.

The simplest comparison can be made by plotting the maximum stress the various actuators can exert versus the maximum strain. Plots of this type usually have a statistical base: a number of actuators of each type is considered and for each one a point is plotted on the chart. This results in clouds of points that define an area in the plane $\sigma_{\max} - \epsilon_{\max}$ characterizing each type of actuators. In some cases, mostly related to solid state actuators, it is possible to use a simple mathematical model to identify the required area.[3]

A plot of this kind is reported in Fig. 7.2. The plot deals with linear actuators, and the actuation principles considered are: electromagnetic actuators (moving coil and solenoid), pneumatic actuators, hydraulic actuators, piezoelectric actuators (low-strain, high-strain, polymeric), magnetostrictive actuators and shape memory alloy actuators. Also thermal expansion actuators (with a temperature difference of 10 and 100 K) and muscles are inserted in the plot. The scales are logarithmic, so the plot spans several orders of magnitudes for both the stress and the strain.

---

[2]J.E. Huber, N.A. Fleck, M.F. Asby, *The Selection of Mechanical Actuators Based on Performance Indices*, Proc. Royal Soc., London, 453, pp. 2185–2205, 1997.

[3]The performance indices of actuators are described in details in O. Gomis-Bellmunt, L.F. Campanile, *Design Rules for Actuators in Active Mechanical Systems*, Springer, London, 2010.

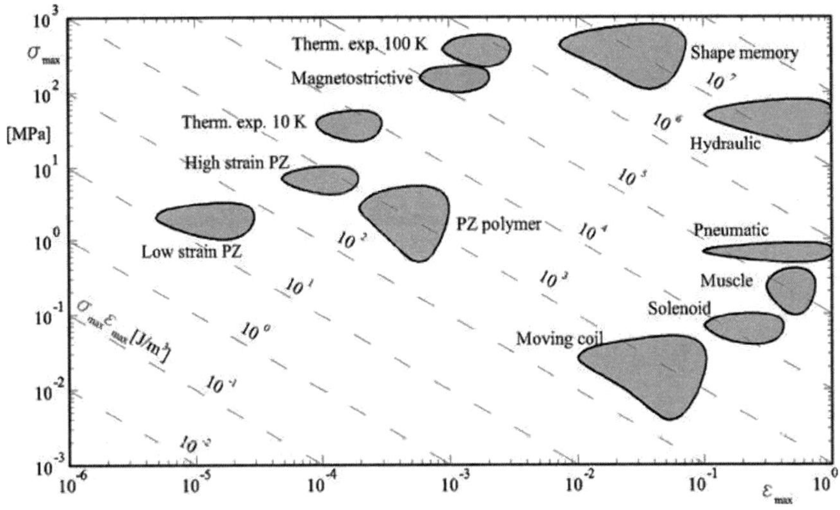

Fig. 7.2 Maximum stress as a function of the maximum strain for different types of actuators

The values of the stress and of the strain are referred to the actuator itself, and can be changed if a mechanical transmission is used. Since the transmission ideally operates at constant energy (actually all transmission devices introduce energy losses) the characteristic point of the actuator can move on a line with constant product $\sigma\epsilon$, i.e., on one of the lines sloping at $-45°$ (from upper left to bottom right). Anyway, since transmission devices are bulky and heavy and, in particular if the transmission ratio is large, their efficiency may be quite low, it is advisable to avoid their use wherever possible.

The product $\sigma_{\max}\epsilon_{\max}$ gives an indication of the maximum work per unit volume the actuator can supply. It is just an indication, since in actual operation the stress may be not constant along the stroke, and both σ and ϵ may depend on the external forces applied on the actuator.

The plot of Fig. 7.2 allows to compare the various actuator types from the viewpoint of the force they can exert and of the stroke they can yield in static conditions, but says nothing about how fast they move. This information is conveyed by the plot of Fig. 7.3, where the maximum frequency the actuator can reach is plotted as a function of the strain for the same actuator types.

The frequency, however, does not in general depend on the actuator alone: in all actuators that are thermally actuated, like shape memory or thermal expansion actuators, it depends mainly on the possibility of cooling the system. In electric actuators the frequency is strongly influenced by the bandwidth of the controller and the power conditioning system, and similar considerations hold also for hydraulic and pneumatic actuators. The conclusions that can be drawn from Fig. 7.3 are only order of magnitude assessment, which must be adapted to the various applications following the particular architecture of the system.

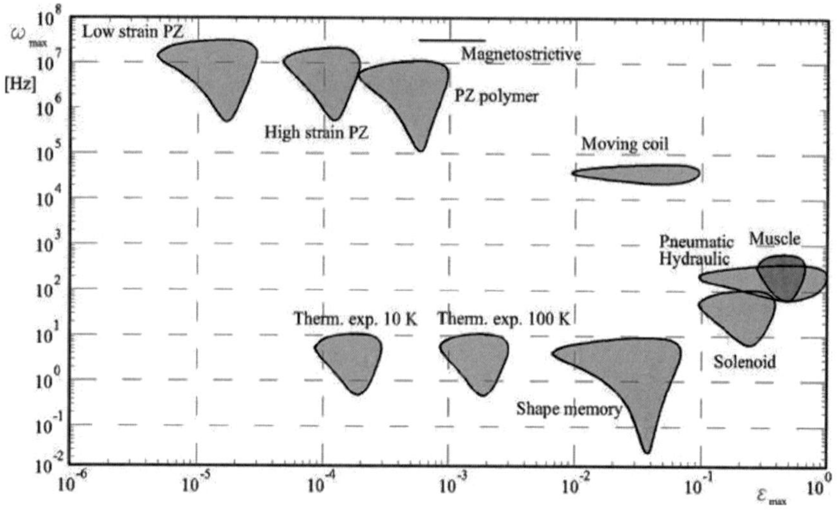

Fig. 7.3 Maximum frequency as a function of the maximum strain for different types of actuators

Table 7.1 Maximum efficiency, density and resolution of some actuators

	Efficiency	Resolution	Density (kg/m^3)
Low-strain piezoelectric	>0.9999	10^{-9}–10^{-8}	2,600–4,700
High-strain piezoelectric	0.90–0.99	10^{-8}–10^{-7}	7,500–7,800
Hydraulic	0.90–0.98	10^{-5}–10^{-4}	600–2,000
Piezoelectric polymer	0.90–0.95	10^{-8}–10^{-7}	1,750–1,900
Magnetostrictive	0.80–0.99	10^{-7}–10^{-6}	6,500–9,100
Moving coil	0.50–0.80	10^{-6}–10^{-5}	7,000–7,600
Solenoid	0.50–0.80	10^{-4}–10^{-2}	3,800–4,400
Pneumatic	0.30–0.40	10^{-5}–10^{-4}	180–250
Muscle	0.20–0.25	10^{-4}–10^{-2}	1,000–1,100
Shape memory alloy	0.01–0.02	10^{-5}–10^{-4}	6,400–6,600
Thermal expansion (100 K)	2×10^{-4}–3×10^{-3}	10^{-5}–10^{-4}	3,900–7,800
Thermal expansion (10 K)	2×10^{-5}–3×10^{-4}	10^{-5}–10^{-4}	3,900–7,800

Another important parameter is the efficiency with which the actuator is able to convert the primary energy into mechanical energy. The values of the efficiency of the various types of actuators is reported, together with those of the resolution and of the density, in Table 7.1.

From the plots and the table it is clear that pneumatic and hydraulic actuators are able to supply large strokes with large forces, the latter particularly in the case of hydraulic actuators. They can do so at a fairly high frequency with an efficiency that is high, in case of hydraulic cylinders, or fair in case of pneumatic cylinders. Their performance is not dissimilar from that of muscles. These characteristics explain why they are widely used in industrial robotics.

However, these plots refer to the actuators alone. If the mass (and the efficiency) of the power converter are accounted for, the picture may change. In particular, if the primary source is electric, the devices that perform the electric-hydraulic and above all the electric-pneumatic conversion are heavy, bulky and their efficiency may be low.

Electric actuators supply lower strokes and lower forces, but their frequency and above all their efficiency are high. This explains their widespread use for the cases where short strokes are required.

Thermal and memory alloy actuators suffer from their low frequency and above all very low efficiency. For this reason, and also for the difficulties in heating and above cooling objects in space, they are seldom considered in space robotics.

Piezoelectric actuators are characterized by very short strokes, but can supply fairly large forces and above all can reach high frequencies and operate with high efficiency. They are a good alternative in all cases where small strokes are required.

Electric actuators, in the form of the so-called electric cylinders, i.e. electric motors driving, often through a reduction gear, a screw or a rack and pinion system that converts the rotational motion into a linear motion, are the most common choice also for linear motion in space robotics. Since their basic unit is an electric motor, they will be dealt with when describing the latter. Another alternative are linear electric motors, but they have little application in the field of robotics: they supply low forces at high speed, exactly the opposite of what is required for robot actuators.

7.2.2 Hydraulic Cylinders

A sketch of a hydraulic cylinder is shown in Fig. 7.4a If the two chambers are supplied with a fluid at pressures p_1 and p_2, and the areas of the two sides of the piston are A_1 and A_2 the force the cylinder exerts on the load in static conditions is

$$F = p_1 A_1 - p_2 A_2 = \frac{\pi}{4}(p_1 - p_2)d_1^2 - \frac{\pi}{4} p_2 d_2^2. \tag{7.1}$$

If the configuration of the cylinder is that of Fig. 7.4b, the pressures at the two opposite sides of the piston act on equal areas. This configuration has, however, the drawback of a greater overall length in the closed position and of a larger possibility of leakages.

The stroke of hydraulic cylinders may be fairly large, and it is limited to a value close to the length of the actuator ($\epsilon < 1$) for the configuration of Fig. 7.4a and to half of its length ($\epsilon < 0.5$) for the configuration of Fig. 7.4b. To increase further the stroke it is possible to use actuators with a telescopic configuration, but this is seldom considered to avoid mechanical problems.

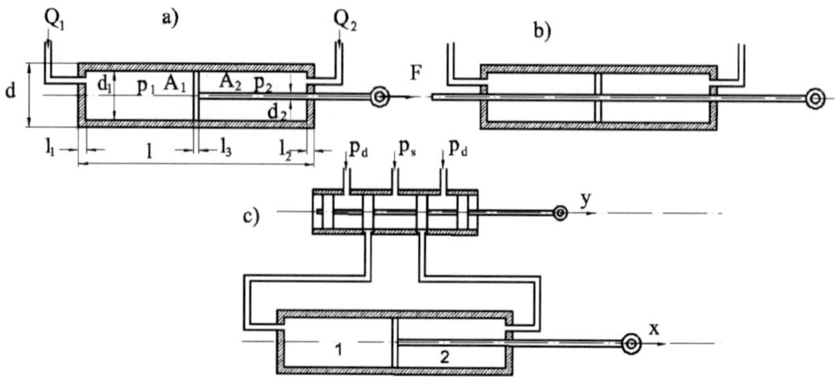

Fig. 7.4 Hydraulic actuators. (**a**) and (**b**) Sketch of an hydraulic cylinder; (**c**) Sketch of the actuator

If, when exerting a force to the right, the pipe labeled as 1 in Fig. 7.4a is supplied by a fluid at pressure p_s and the pipe labeled as 2 is discharged at a pressure $p_d \approx 0$, (in Fig. 7.4c the control valve is pushed to the left) the force on the right is

$$F = \frac{\pi}{4} p_s d_1^2. \tag{7.2}$$

In the opposite case, the force exerted toward the left is

$$F = -\frac{\pi}{4} p_s \left(d_2^2 - d_1^2\right). \tag{7.3}$$

The maximum force the actuator can supply depends on the maximum pressure the device can withstand. The limitation in the pressure depends on the ability of the cylinder walls and of the whole hydraulic circuit to withstand the stresses caused by the pressure itself and on the ability of the rod to withstand the resulting force, but also on the ability of the seal to avoid leakages, which increase with the pressure. When exerting a force toward the left, the rod is under tensile forces, but in the opposite direction the force is compressive, and elastic stability must be taken into account.

Remark 7.3 These equations hold only in static conditions, but can be used also for very slow motion, i.e. for speeds at which the pressure drops in the pipes due to the motion of the fluid and all inertia forces can be neglected.

The maximum work produced during a stroke $s_{\max}$ performed with constant pressures is

$$W = F s_{\max} = \left[\frac{\pi}{4}(p_1 - p_2)d_1^2 - \frac{\pi}{4} p_2 d_2^2\right](l - l_1 - l_2 - l_3). \tag{7.4}$$

Consider a hydraulic cylinder controlled by a valve as shown in Fig. 7.4c. y is the displacement of the plunger of the valve and x is the displacement of the piston. If $y = 0$ both orifices are closed, and the piston is locked. If $y < 0$ the left chamber (1) is connected with the supply pipe and the right chamber (2) is connected with the discharge pipe: the piston moves to the right. The opposite occurs when $y > 0$.

In dynamic conditions the throughput of fluid entering the chambers 1 and 2 is

$$Q_i = A_i \dot{x} \quad \text{for } i = 1, 2. \tag{7.5}$$

The flow through an orifice Q_o is linked to the pressure loss Δp by the formula

$$Q_o = \alpha_o A_o \sqrt{\frac{2\Delta p}{\rho}}, \tag{7.6}$$

where ρ is the density of the fluid and α_o is a nondimensional coefficient that depends on the geometry of the system. In case of valves of the kind shown in Fig. 7.4c, usually $\alpha_o = 0.6$–0.8.

If the orifices of the valve have a width w_o and a length l_o, and neglecting the pressure losses in the pipes, the throughput entering the chambers of the cylinder is

- if $y = 0$

$$Q_1 = Q_2 = 0, \tag{7.7}$$

- if $-l_o < y < 0$

$$Q_1 = \alpha_o y w_o \sqrt{\frac{2(p_s - p_1)}{\rho_1}}, \qquad Q_2 = -\alpha_o y w_o \sqrt{\frac{2(p_2 - p_d)}{\rho_2}}, \tag{7.8}$$

- if $0 < y < l_o$

$$Q_1 = \alpha_o y w_o \sqrt{\frac{2(p_1 - p_d)}{\rho_1}}, \qquad Q_2 = -\alpha_o y w_o \sqrt{\frac{2(p_s - p_2)}{\rho_2}}. \tag{7.9}$$

If y is less than $-l_o$ or more than l_o, these equations still hold, with l_o substituted for y. Theoretically, the density of the fluid should be considered as a constant.

These equations allow to study the dynamics of the actuators, or better of the actuated system, once that the time history of the command $y(t)$ is known, i.e. to study its input-output relationship. If there is a feedback linking the input to the output, they also allow one to study its closed loop dynamics.

Example 7.1 Consider an actuator of the type shown in Fig. 7.4c acting against a spring-damper system. Let m, k and c be the mass of the moving system and the stiffness and the damping of the load and assume the following data: $\alpha_o = 0.6$, $w_o =$ 10 mm, $l_o = 10$ mm, $d_1 = 25$ mm, $d_2 = 5$ mm, $p_s = 10$ MPa, $p_d = 0$, $m = 0.2$ kg, $k = 100$ kN/m, $c = 8$ kNs/m, $\rho = 850$ kg/m^3.

Before time $t = 0$ the plunger is to the left at $y = -l_0$ and the system is in its equilibrium position; at time $t = 0$ it is instantly brought to the right at $y = l_0$ and then it cycles periodically to the left and to the right every 1 s.

Compute the static equilibrium position and the time history $x(t)$.

Static Equilibrium. The equilibrium equation is

$$kx = p_s A_1 - p_d A_2,$$

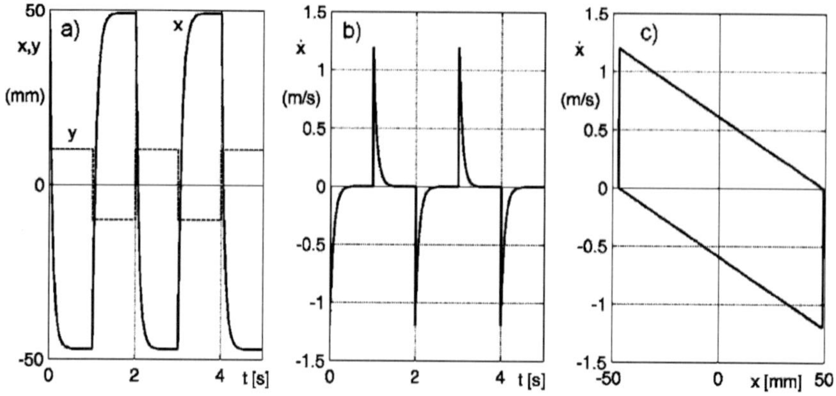

Fig. 7.5 Time history of the displacement (**a**) and the velocity (**b**) of the piston. (**c**) State-space plot of the motion of the actuator

i.e.,

$$x = \frac{p_s A_1}{k} = 49.1 \text{ mm}.$$

Dynamic Study. The equation of motion is

$$m\ddot{x} + c\dot{x} + kx = p_s A_1 - p_d A_2,$$

where the pressure can be computed as a function of the throughputs by solving equations (7.8) and (7.9) in p_1 and p_2. The results of the numerical integration of the equation of motion are reported in Fig. 7.5.

This approach is, however, an oversimplification. Actual fluids have a finite compressibility, in particularly when they contain even tiny quantities of dissolved gas. The pipes and the cylinder itself are not rigid bodies, and under the pressure of the fluid they expand. This expansion, although small, has the same effect as the compressibility of the hydraulic fluid, and the combined effect is, in many cases, important enough to affect the working of the system.

All these effects, plus the unavoidable leakages and the pressure losses due to the passage of the fluid through the pipes, decrease the performance of hydraulic systems particularly for what their efficiency is concerned. They must be taken into account at the design stage.

7.2.3 Pneumatic Actuators

The difference between pneumatic and hydraulic actuators is mainly that in the former the working fluid is a gas, a highly compressive medium, while in the latter case it is a liquid, which in theory is incompressible. What said about the importance of

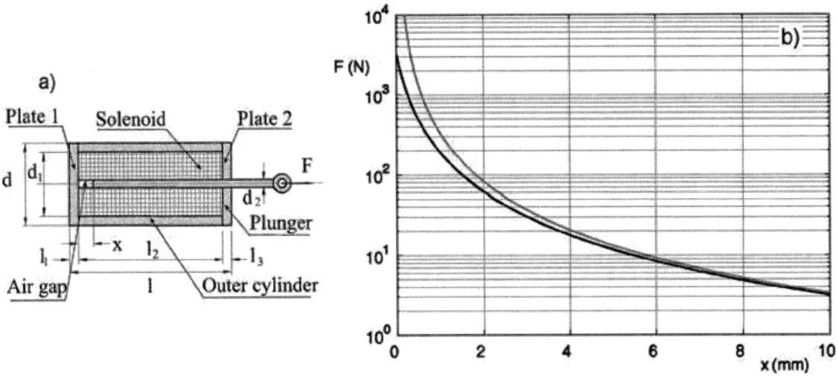

Fig. 7.6 (**a**) Sketch of a solenoid actuator. (**b**) Force-stroke plot for the actuator used in the example (*the full line* refers to complete expression for the force, *the dotted line* for the expression in which the reluctance of the iron has been neglected)

the compressibility of the working fluid, becomes much more important in the case of pneumatic actuators, where compressibility is one of the governing phenomena and not a parasitic effect. The high compressibility of gases leads to a low stiffness of the system, a general decrease of the efficiency and to heating problems, which are particularly severe in space applications.

However, compressibility has also its advantages: hydraulic actuators are very stiff and require a precise control, while the compressibility of gases results in a lower stiffness allowing to use a less precise control. This has sometimes been used in walking machines, to obtain a better adaptation to ground irregularities and a sort of elastic suspension effect. Usually, this is not enough to overcome the drawbacks of pneumatic actuators in space robots.

As already stated, pneumatic actuators are at their best in places where they can use ambient air as a working fluid and can exhaust low pressure fluid directly in the atmosphere, perhaps trough some filter to prevent oil spray to contaminate the environment. This condition seldom applies in space applications and this is an additional reason to rule out their use in space robots.

7.2.4 Solenoid Actuators

A solenoid actuator (Fig. 7.6a) is basically made of a coil in which a ferromagnetic plunger can move. The magnetic circuit is then completed by an external iron[4] tube and two pole plates. This configuration is not the only possible one, but, since it is the most common, is the only one considered here.

[4]The term iron is here used to designate a generic ferromagnetic material, as the term copper will be used to designate a generic electric conductor.

The magnetic circuit is made by five elements in series: plate 1, the outer cylinder, plate 2, the plunger and finally the air gap. The first four elements are made of a ferromagnetic material, and hence their reluctance is low; the air gap on the contrary has quite a high reluctance, which increases with increasing length x, i.e. when the plunger moves to the right. The total reluctance thus changes with changing position of the latter: this kind of actuators are sometimes referred to as variable reluctance actuators or Maxwell actuators.

Since the system tends to achieve a configuration with the minimum reluctance (with $x = 0$, provided that the external constraints allow it), when the solenoid is energized it tends to suck the plunger toward the left, exerting a force F on the plunger that in the figure is negative.

The magnetic flux circulating in the magnetic circuit is

$$\Phi = \frac{Ni}{\mathcal{R}}, \tag{7.10}$$

where the product of the number of turns of the solenoid N by the current i is the magneto-motive force and

$$\mathcal{R} = \mathcal{R}_{\text{pl1}} + \mathcal{R}_{\text{cyl}} + \mathcal{R}_{\text{pl2}} + \mathcal{R}_{\text{plung}} + \mathcal{R}_{\text{airgap}} \tag{7.11}$$

is the total reluctance.

The magnetic energy stored in the solenoid can be computed from the magnetic flux by integrating the product of the magneto-motive force by the flux

$$W_m = \int Ni \, d\Phi \tag{7.12}$$

along the length of the solenoid. Since in the simplified linear theory the magnetic flux is linear in the current, the integral is easily performed and yields

$$W_m = \frac{1}{2} Ni\Phi = \frac{N^2 i^2}{2\mathcal{R}}. \tag{7.13}$$

The force the actuator exerts on the plunger is the derivative of the magnetic energy with respect of the length of the air gap

$$F = \frac{\partial W}{\partial x}. \tag{7.14}$$

To obtain an equation yielding the force, the reluctance must be written in an explicit way in the variable x, a thing that can be done only approximately.

For the air gap, an approximated expression that neglects the stray flux is

$$\mathcal{R}_{\text{airgap}} = \frac{x}{\mu_0 S_{\text{airgap}}} = \frac{4x}{\mu_0 \pi d_2^2}, \tag{7.15}$$

where S_{airgap} is the cross section of the air gap and μ_0 is the magnetic permeability of vacuum.

Similar expressions can be written for the outer cylinder and the plunger:

$$\mathcal{R}_{\text{cyl}} = \frac{l_2}{\mu_0\mu_r S_{\text{cyl}}} = \frac{4l_2}{\mu_0\mu_r\pi(d^2 - d_1^2)}, \tag{7.16}$$

$$\mathcal{R}_{\text{plung}} = \frac{l_2 - x}{\mu_0\mu_r S_{\text{plung}}} = \frac{4(l_2 - x)}{\mu_0\mu_r\pi d_2^2}, \tag{7.17}$$

where μ_r is the relative magnetic permeability of the material constituting the iron parts.

Remark 7.4 The relative permeability of ferromagnetic materials is not constant, but changes with both flux density and temperature. The equations here reported hold only when the flux density is low enough to avoid getting close to saturation.

For the plates things are more difficult. Each plate can be thought as made by an infinity of thin coaxial cylindrical shells, with axial length l, thickness dr and radius r. When the magnetic flux flows from the inner radius to the outer one, the reluctance of each one of these infinitely thin shells is

$$d\mathcal{R} = \frac{dr}{2\mu_0\mu_r\pi r l}. \tag{7.18}$$

Since the various layers are in series, by integrating this expression from the inner to the outer radius, it follows that

$$\mathcal{R}_{\text{plate}} = \frac{1}{2\mu_0\mu_r\pi l}\int_{r_i}^{ro} \frac{dr}{r} = \frac{1}{2\mu_0\mu_r\pi l}\ln\left(\frac{r_o}{r_i}\right). \tag{7.19}$$

Assuming that the flux enters the plates at a radius corresponding to diameter d_2 and exits at diameter d_1, it follows that

$$\mathcal{R}_{\text{plate1}} = \frac{1}{2\pi\mu_0\mu_r l_1}\ln\left(\frac{d_1}{d_2}\right), \tag{7.20}$$

$$\mathcal{R}_{\text{plate2}} = \frac{1}{2\pi\mu_0\mu_r l_3}\ln\left(\frac{d_1}{d_2}\right). \tag{7.21}$$

The total reluctance of the part of the magnetic circuit made by the plates and the outer cylinder is

$$\mathcal{R}_{\text{pl1}} + \mathcal{R}_{\text{cyl}} + \mathcal{R}_{\text{pl2}} = \frac{1}{\mu_0\mu_r\pi}\left[\frac{1}{2}\left(\frac{1}{l_1} + \frac{1}{l_3}\right)\ln\left(\frac{d_1}{d_2}\right) + \frac{4l_2}{d^2 - d_1^2}\right]. \tag{7.22}$$

By introducing an equivalent length,

$$l_{\text{eq}} = \frac{d_2^2}{8}\left(\frac{1}{l_1} + \frac{1}{l_3}\right)\ln\left(\frac{d_1}{d_2}\right) + l_2\frac{d_2^2 + d^2 - d_1^2}{d^2 - d_1^2}, \tag{7.23}$$

the total reluctance of the magnetic circuit can be written as

$$\mathcal{R} = \frac{x(\mu_r - 1) + l_{\text{eq}}}{\mu_0\mu_r S_{\text{airgap}}}. \tag{7.24}$$

Table 7.2 Resistivity and temperature coefficient of some conductors

	δ (Ω m)	γ (1/K)
Silver	1.59×10^{-8}	0.0038
Copper	1.68×10^{-8}	0.0039
Gold	2.44×10^{-8}	0.0034
Aluminum	2.82×10^{-8}	0.0039

The magnetic energy is thus

$$W_m = \frac{N^2 i^2 \mu_0 \mu_r S_{\text{airgap}}}{2[x(\mu_r - 1) + l_{\text{eq}}]}. \tag{7.25}$$

By differentiating the energy, the magnetic force acting on the plunger is obtained

$$F = \frac{\partial W}{\partial x} = -\frac{N^2 i^2 \mu_0 \mu_r S_{\text{airgap}}(\mu_r - 1)}{2[x(\mu_r - 1) + l_{\text{eq}}]^2}. \tag{7.26}$$

Often this formula is simplified by substituting μ_r for $\mu_r - 1$, which is usually acceptable for ferromagnetic materials. If x is not too small, the reluctance of the whole magnetic circuit can be neglected if compared with that of the air gap, yielding:

$$F \approx -\frac{N^2 i^2 \mu_0 S_{\text{airgap}}}{2x^2}. \tag{7.27}$$

This value of the force holds only in steady-state conditions.

The limitations to the force the solenoid actuator can produce come essentially from the heating due to the Joule effect. The power dissipated by Joule effect is

$$P_g = R i^2, \tag{7.28}$$

where R is the resistance of the wire

$$R = \delta_w \frac{l_w}{S_w} = \delta_{w0}(1 + \gamma \Delta T)\frac{l_w}{S_w}, \tag{7.29}$$

δ_w, l_w and S_w are the resistivity of the material used for the wire at a reference temperature, the wire length and cross sectional area. The second formula (7.29) takes into account that the resistivity of materials changes with the temperature, and approximates this change with a linear law: γ is the temperature coefficient, i.e. the slope of the curve $\delta_w(T)$ and ΔT is the variation of temperature with respect to the reference temperature.

The values of the resistivity and of the temperature coefficient for some conductors is reported Table 7.2.

The cross sectional area of the winding

$$S_{\text{wind}} = \frac{l_2}{2}(d_1 - d_2) \tag{7.30}$$

and the cross sectional area of the wire are linked to the number of turns by the formula

$$S_w = \frac{S_{\text{wind}}}{N} k_{\text{ff}}, \tag{7.31}$$

where the filling factor k_{ff} is a number smaller than 1 that shows how much of the useful area of the winding is occupied by the wires. In the case of wires with circular cross section, its maximum value is $\pi/4$, but usually it is less.

The length of the wire can be approximated as

$$l_w = \frac{N\pi}{2}(d_1 + d_2). \tag{7.32}$$

Equations (7.26), (7.27) and (7.28) can be rewritten to show explicitly the role of some parameters which play an important role in the design of the actuator, the current density i/S_w, the area of the winding S_{wind} and the filling factor k_{ff}. By introducing the current density and (7.31) into (7.26), it follows that

$$F = -\left(\frac{i}{S_w}\right)^2 S_{\text{wind}}^2 k_{\text{ff}}^2 \frac{\mu_0 \mu_r S_{\text{airgap}}(\mu_r - 1)}{2[x(\mu_r - 1) + l_{\text{eq}}]^2}. \tag{7.33}$$

If the reluctance of the iron is neglected (7.27), it follows that

$$F = -\left(\frac{i}{S_w}\right)^2 S_{\text{wind}}^2 k_{\text{ff}}^2 \frac{\mu_0 S_{\text{airgap}}}{2x^2}. \tag{7.34}$$

By introducing the current density and (7.32) into (7.28), it follows that

$$P_g = \left(\frac{i}{S_w}\right)^2 S_{\text{wind}} k_{\text{ff}} \frac{\pi}{2}(d_1 + d_2)\delta_w. \tag{7.35}$$

If the only cooling effect is due to convection on the outer surface, the thermal power that can be extracted can be written as

$$P_d = \frac{\Delta T}{\theta_{\text{conv}} + \theta_{\text{cond}}}, \tag{7.36}$$

where θ_{conv} and θ_{cond} are the thermal resistance due to convection and conduction through the outer iron cylinder (they are in series)

$$\theta_{\text{conv}} = \frac{1}{\pi l d h_c}, \qquad \theta_{\text{cond}} = \frac{1}{2\pi l \lambda_{\text{ir}}} \ln\left(\frac{d}{d_1}\right). \tag{7.37}$$

Conduction heat transfer is dominated by the thermal conductivity of iron λ_{ir} while convection is controlled by the convection coefficient h_c between the outer surface of the actuator and the air, which in turn depends on many factors like the nondimensional Nusselt number N_u, which expresses the ratio of convective to conductive heat transfer.

A simple expression is

$$h_c = \frac{N_u \lambda_{\text{air}}}{l}, \tag{7.38}$$

where λ_{air} is the thermal conductivity of air.

The maximum current that can be withstood for an unlimited time can be obtained by equating the thermal power generated and that dissipated

$$i_{\max} = \sqrt{\frac{\Delta T}{R(\theta_{\text{conv}} + \theta_{\text{cond}})}}. \tag{7.39}$$

In terms of current density, it follows that

$$\left(\frac{i}{S_w}\right)_{\max} = \sqrt{\frac{2\Delta T}{\pi S_{\text{wind}} k_{\text{ff}}(d_1 + d_2)(\theta_{\text{conv}} + \theta_{\text{cond}})}}. \tag{7.40}$$

This value of the current, or of the current density, can be overridden if the coil is energized only for a short time. In this case, the thermal inertia of the actuator can allow to use higher currents, provided that the coil is switched off before dangerous temperatures are reached. Solenoids required to supply only short pulses, like those operating a switch, can produce very large forces and at the same time be light and small.

The material of the coil must have a low resistivity, and thus copper wires are usually employed. However, for aerospace use, there may be an advantage in using aluminum coils, in particular if what matters is more the mass of the actuator than its size. The higher resistance of the material leads to a larger wire cross section (to have the same resistance) but the lower density may still lead to a smaller mass. What actually matters is not a low resistivity in itself, but a low value of the product of resistivity by density. Such product is $1.5 \times 10^{-4}\ \Omega\,\text{kg/m}^2$ for copper and $7.6 \times 10^{-5}\ \Omega\,\text{kg/m}^2$ for aluminum, showing an advantage by almost a factor of 2 for the latter. This must, however, be checked in each case, because the larger size of the actuator leads to a greater mass of the iron parts.

Remark 7.5 The force is proportional to the square of product Ni so that it is immaterial, from the viewpoint of the static force, to have many turns with a low current or viceversa. This can be seen also considering that, when reasoning in terms of current density, the number of turns does not appear in the equations for both the force and the power dissipated.

The design of the coil must be performed considering the whole actuating system, actuator, power amplifier and controller, taking into account also dynamic requirements. When doing so, also the inductance of the coil, and not only its electric resistance, becomes important.[5]

An attempt to optimizing the actuator configuration is done in the mentioned book by Gomis-Bellmunt and Campanile,[6] who obtained the following values of

[5] A. Tonoli, N. Amati, M. Silvagni, *Transformer Eddy Current Dampers for the Vibration Control*, Journal of Dynamic Systems, Measurement and Control, Vol. 130, May 2008.

[6] O. Gomis-Bellmunt, L.F. Campanile, *Design Rules for Actuators in Active Mechanical Systems*, Springer, London, 2010.

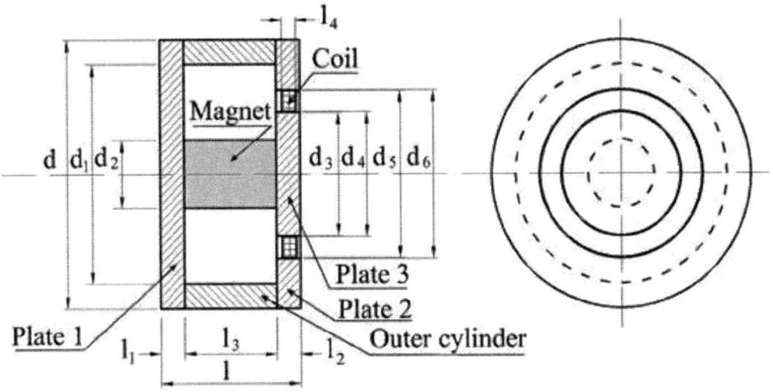

Fig. 7.7 Geometrical configuration of a moving coil actuator

some nondimensional parameters that optimize the force produced by the actuator: $d_1/d = 0.78$, $d_2/d = 0.29$, $l_1/l = l_3/l = 0.25$, $l_2/l = 0.50$, $l/d = 0.35$. If, instead of the force, the work per unit volume is optimized, the values of the same nondimensional parameters are: $d_1/d = 0.86$, $d_2/d = 0.39$, $l_1/l = l_3/l = 0.125$, $l_2/l = 0.75$, $l/d = 0.505$.

Example 7.2 Consider a solenoid actuator that has an outer diameter of 60 mm. Compute the force-stroke characteristics limiting the current to a value which causes an increase of temperature $\Delta T = 50$ K.

To optimize the work per unit volume, approximating to the nearest tenth of a millimeter the nondimensional parameters seen above, it follows that: $d_1 = 61.6$ mm, $d_2 = 23.4$ mm, $l = 30.3$ mm, $l_1 = l_3 = 3.8$ mm, $l_2 = 22.7$ mm.

The relevant data of the material are: copper $\delta_0 = 1.68 \times 10^{-8}$ Ωm, $\gamma = 0.0068$ 1/K; iron $\mu_r = 200$, $\lambda_{ir} = 80$ W/Km; air $\mu_0 = 4\pi \times 10^{-7}$ H/m, $\lambda_{air} = 0.0257$ W/Km; $N_u = 60$. Assume that the winding has $N = 500$ turns and that $k_{ff} = 0.7$.

To store 500 turns on the coil, the area of the wire and its length are $S_w = 0.45$ mm^2 and $l_w = 58.9$ m. The resistance of the coil at the final temperature is thus $R = 2.96$ Ω.

From (7.39) the maximum current is computed: $i_{max} = 2.21$ A. The force-stroke characteristics of the solenoid is plotted in Fig. 7.6b.

7.2.5 Moving Coil Actuators

Moving coil actuators are often referred to as Lorentz actuators, since the force they produce is the so-called Lorentz force exerted by a wire in which a current flows when it is in a magnetic field. One of the most common configurations of such an actuator is shown in Fig. 7.7: the magnetic field is provided by a permanent magnet, and carried by a magnetic circuit that concentrates it in an airgap in which there is

Table 7.3 Main properties of some magnetic materials

	B_r (T)	H_c (kA/m)	$(BH)_{\max}$ (kJ/m^3)	T_c (°)
Ferrite	0.23–0.43	148–288	10–35	450
Alnico 9	1.05	112	72	860
$Nd_2Fe_{14}B$	1.0–1.4	750–2,000	200–440	310–400
$SmCo_5$	0.8–1.1	600–2,000	120–200	720
Sm(Co,Fe,Cu,Zr)$_7$	0.9–1.15	450–1,300	150–240	800

a coil. When the coil is energized, it receives a force that can be collected by the element supporting it.

Using (7.19) for the reluctance of an annular plate and assuming equal to 1 the relative magnetic permeability of copper, a first approximation of the reluctance of the magnetic circuit is

$$\mathcal{R} = \frac{1}{2\mu_0\mu_r\pi d_2^2}\left[8l_{\text{eq}} + \frac{\mu_r d_2^2}{l_2}\ln\left(\frac{d_6}{d_3}\right) + \frac{8\mu_r}{\mu_m}l_3\right], \tag{7.41}$$

where

$$l_{\text{eq}} = l_3\frac{d_2^2}{d^2 - d_1^2} + 2\left[\frac{1}{l_1}\ln\left(\frac{d_1}{d_2}\right) + \frac{1}{l_2}\ln\left(\frac{d_1}{d_6}\right) + \frac{1}{l_2}\ln\left(\frac{d_3}{d_2}\right)\right], \tag{7.42}$$

and μ_r and μ_m are the relative permeabilities of the iron and of the permanent magnet.

If the relative permeability of the magnet is assumed equal to 1, and the reluctance of the iron parts of the magnetic circuit is neglected (in this case, owing to the large reluctance of the permanent magnet this is much more acceptable than in the previous case), it follows that

$$\mathcal{R} \approx \frac{1}{2\mu_0\pi}\left[\frac{1}{l_2}\ln\left(\frac{d_6}{d_3}\right) + \frac{8}{d_2^2}l_3\right]. \tag{7.43}$$

The magnetic flux in the circuit due to the permanent magnet is

$$\Phi = \frac{F_{\text{mm}}}{\mathcal{R}} = \frac{H_c l_3}{\mathcal{R}}, \tag{7.44}$$

where F_{mm} is the magneto-motive force and H_c is the coercitive magnetic field of the permanent magnet.

The basic properties (magnetic remanence B_r, coercitivity H_c, energy product $(BH)_{\max}$, which measures the density of magnetic energy, and Curie temperature T_c) of a number of magnetic materials are summarized in Table 7.3.

The Lorentz force exerted on the coil can be computed as

$$F = Bl_w i, \tag{7.45}$$

where l_w is the total length of the wire that is inside the magnetic field, i.e. inside the air gap.

Assuming that all the coil remains always inside the airgap, the length of the wire is

$$l_w = \pi N \frac{d_4 + d_5}{2}, \tag{7.46}$$

where N is the total number of turns of the coil.

The magnetic field in the airgap is not constant, since the area grows radially outwards. At the center of the coil it is

$$B = \frac{\Phi}{S} = \frac{2H_c l_3}{\mathcal{R} l_4 \pi (d_4 + d_5)}. \tag{7.47}$$

The final expression of the force that can be exerted in steady-state conditions is thus

$$F = Ni \frac{H_c l_3}{\mathcal{R} l_4}. \tag{7.48}$$

While in solenoid actuators the force was proportional to the square of product Ni, in moving coil actuators the force is proportional to the first power of the same product.

Also in this case, the limitations come usually from thermal reasons, and it is possible to compute a maximum value of the current that can be supplied to the coil without overheating.

To maximize the force, the values of the geometrical nondimensional parameters are: $d_2/d = d_3/d = d_4/d = 0.72$, $d_1/d = d_5/d = d_6/d = 0.94$, $l_1/l = l_2/l = l_4/l = 0.50$, $l_2/l = 0.75$, $l/d = 1.04$. These values have been obtained for a slightly different configuration, in which the permanent magnet has a length l instead of l_3.

7.2.6 Piezoelectric Actuators

Piezoelectric materials are materials in which the application of an electric field creates mechanical deformation and, conversely, a mechanical deformation produces an electrical field.

In non-piezoelectric materials the mechanical behavior and the electric behavior are uncoupled from each other. The former is expressed, as a first approximation, by Hooke's law, which, using the notation common when dealing with piezoelectric material, becomes[7]

$$\mathbf{S}_{6\times1} = \mathbf{s}_{E_{6\times6}} \mathbf{T}_{6\times1}, \tag{7.49}$$

where $\mathbf{S}$ is the mechanical strain vector (commonly referred to as $\boldsymbol{\epsilon}$ in mechanics), $\mathbf{s}_E$ is the compliance matrix of the material ($\mathbf{E}^{-1}$ in mechanics) and $\mathbf{T}$ is the mechanical stress ($\boldsymbol{\sigma}$ in mechanics).

[7] See IEEE standard 176-1987.

The electric behavior can be expressed, as a first approximation, by the linear law

$$\mathbf{D}_{3\times1} = \boldsymbol{\epsilon}_{T_{3\times3}}\mathbf{E}_{3\times1}, \tag{7.50}$$

where $\mathbf{D}$ is the vector of the components of the electric charge density displacement, $\boldsymbol{\epsilon}_T$ is the permittivity matrix and $\mathbf{E}$ is the vector including the components of the electric field.

In case of piezoelectric materials, these two relationships couple to each other in the form

$$\begin{Bmatrix} \mathbf{S} \\ \mathbf{D} \end{Bmatrix} = \begin{bmatrix} \mathbf{s}_E & \mathbf{d} \\ \mathbf{d}^T & \boldsymbol{\epsilon}_T \end{bmatrix} \begin{Bmatrix} \mathbf{T} \\ \mathbf{E} \end{Bmatrix}. \tag{7.51}$$

In this case $\mathbf{s}_E$ is the compliance matrix at constant electric field and $\boldsymbol{\epsilon}_T$ is the permittivity matrix at constant mechanical stress. The coupling matrix $\mathbf{d}_{3\times6}$ is defined as the direct piezoelectric matrix. Its transpose $\mathbf{d}^T$ is the converse piezoelectric matrix.

Remark 7.6 The quantities appearing in (7.49) are tensors, and are written using a vector and matrix notation in a somehow 'artificial' way (this is the reason why $\mathbf{S}$ and $\mathbf{T}$ have six components, even if they refer to the 3-D space). On the contrary, those appearing in (7.50) are true matrices and vectors.

The various matrices have a peculiar structure:

$$\begin{aligned}
\mathbf{s}_E &= \begin{bmatrix} s_{11} & s_{12} & s_{13} & 0 & 0 & 0 \\ & s_{22} & s_{23} & 0 & 0 & 0 \\ & & s_{33} & 0 & 0 & 0 \\ & & & s_{44} & 0 & 0 \\ & & & & s_{55} & 0 \\ \text{symm} & & & & & s_{66} \end{bmatrix}, \\
\boldsymbol{\epsilon}_T &= \begin{bmatrix} \epsilon_{11} & 0 & 0 \\ 0 & \epsilon_{22} & 0 \\ 0 & 0 & \epsilon_{33} \end{bmatrix}, \\
\mathbf{d}^T &= \begin{bmatrix} 0 & 0 & 0 & 0 & d_{15} & 0 \\ 0 & 0 & 0 & d_{15} & 0 & 0 \\ d_{31} & d_{31} & d_{33} & 0 & 0 & 0 \end{bmatrix}.
\end{aligned} \tag{7.52}$$

The structure for $\mathbf{s}_E$ reported above is that characteristic of orthotropic materials.

Stack (or Longitudinal) Actuators

Piezoelectric actuators can be subdivided in several types, as shown in Fig. 7.8. The first kind exploits the change in thickness of a thin slice of a piezoelectric material when subjected to an electric field. Since the slice must be thin (usually less than 1 mm), the displacement of each actuator is very small, and many layers arranged in a stack are used (Fig. 7.8a). The layers are mechanically in series (the displacement

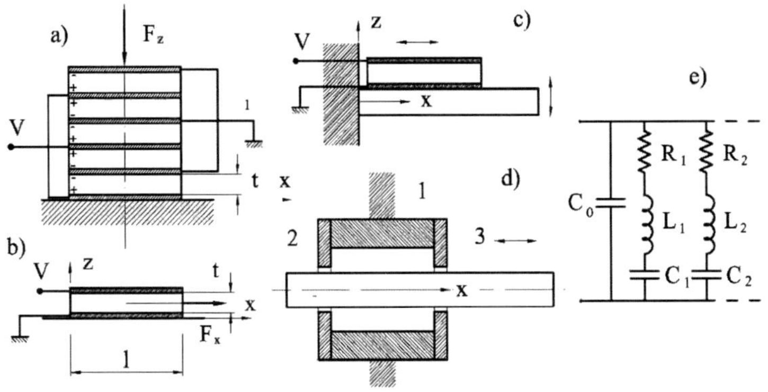

Fig. 7.8 Piezoelectric actuators. (**a**) Stack actuator; (**b**) transversal actuator; (**c**) bending actuator; (**d**) inchworm, (**e**) equivalent circuit

of the stack is the sum of the displacements of the layers), but electrically in parallel (all stacks work with the same voltage). At low frequency, the actuator is equivalent to a capacitor.

If the only force acting on the stack is the axial force F_z, only the third component of the stress vector is different from 0 ($T_3 = F_z/A$). The electric field is axial, so only the component along z direction

$$E_3 = \frac{V}{t},$$

where V is the voltage and t is the thickness of each element in the stack, is different from zero. The axial displacement Δz of each layer can be obtained from (7.51):

$$\Delta z = tS_3 = ts_{33}T_3 + td_{33}E_3 = s_{33}t\frac{F_z}{A} + d_{33}V. \tag{7.53}$$

The *no-load displacement* Δz_0 of the stack of N layers can be obtained by stating $F_z = 0$:

$$\Delta z_0 = N\Delta z = Nd_{33}V. \tag{7.54}$$

The force that can be exerted if the actuator is clamped (*no-displacement* or *blocking force*) is obtained by setting $\Delta z = 0$ and solving the same equation in the force

$$F_{z_0} = -\frac{d_{33}VA}{s_{33}t}. \tag{7.55}$$

If the actuator pushes against a spring with stiffness k, the displacement and the force are linked by the obvious relationship

$$F_z = -kN\Delta z. \tag{7.56}$$

The displacement and the force are linked to the voltage by the relationship

$$N\Delta z = \frac{ANd_{33}V}{A + s_{33}tkN}, \qquad F_z = \frac{-kANd_{33}V}{A + s_{33}tkN}. \tag{7.57}$$

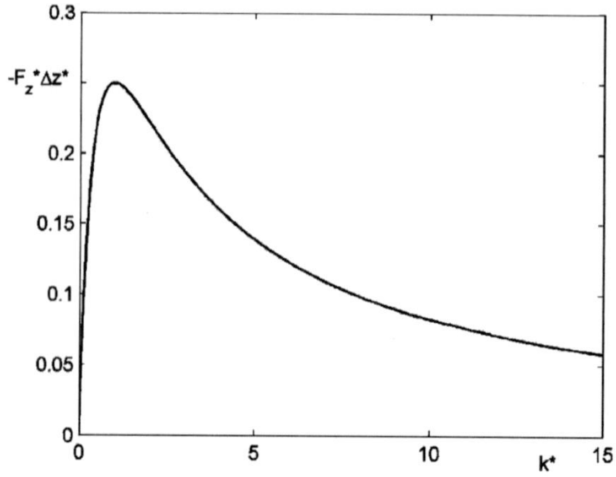

Fig. 7.9 Nondimensional product $F_z^* \Delta z^*$ as a function of k^*

If $k \to 0$ the displacement tends to Δz_0 and if $k \to \infty$ the force tends to F_{z0}.

The equations yielding the displacement and the force can be written in nondimensional form:

$$N\Delta z^* = N\frac{1}{1+k^*}, \qquad F_z^* = \frac{-k^*}{1+k^*}, \tag{7.58}$$

where

$$k^* = k\frac{s_{33}tN}{A}, \qquad \Delta z^* = \frac{\Delta z}{d_{33}V}, \qquad F_z^* = F_z\frac{s_{33}t}{Ad_{33}V}. \tag{7.59}$$

The product $F_z N \Delta z$ is proportional to the work performed by the stack. It can be easily computed:

$$F_z N \Delta z = -k\left(\frac{ANd_{33}V}{A + s_{33}tkN}\right)^2. \tag{7.60}$$

It vanishes for both $k = 0$ and $k \to \infty$ and goes through a maximum value for a given value of k. Equation (7.60) can be written in nondimensional form as

$$NF_z^* \Delta z^* = -N\frac{k^*}{(1+k^*)^2}. \tag{7.61}$$

The product $F_z^* \Delta z^*$ is plotted as a function of k^* in Fig. 7.9. The maximum value of its modulus is 0.25 and occurs for $k^* = 1$.

Transverse Actuators

Transverse actuators act in a way that is similar to longitudinal actuators, but exploit the transversal deformation (Poisson effect) as shown in Fig. 7.8b. The displacement Δx is

$$\Delta x = lS_1 = ls_{11}T_1 + ld_{31}E_3 = \frac{s_{33}}{h}F_x + d_{31}\frac{l}{t}V, \tag{7.62}$$

where h is the width to the actuator.

The no-load displacement Δx_0 can be obtained by stating $F_x = 0$:

$$\Delta x_0 = d_{31}\frac{l}{t}V. \tag{7.63}$$

The force that can be exerted if the actuator is clamped (no-displacement force) is obtained by setting $\Delta x = 0$ and solving the same equation in the force

$$F_{x_0} = -d_{31}\frac{hl}{ts_{11}}V. \tag{7.64}$$

If the actuator pushes against a spring with stiffness k, the displacement and the force are is linked to the voltage by the relationship

$$N\Delta x = \frac{lhd_{31}V}{th + s_{11}kt}, \qquad F_x = \frac{-klhd_{31}V}{th + s_{11}kt}. \tag{7.65}$$

Also in this case it is possible to write the force and the displacement in nondimensional form, and it is possible to write the product $N\Delta x F_x$: it is a function of k and has a maximum that can be found from a plot identical to that of Fig. 7.9.

Bending Actuators

While longitudinal and transverse actuators can provide only very short strokes, but can exert large forces, bending actuators can achieve displacements of some millimeters, at the expense of providing much smaller forces. An example is shown in Fig. 7.8c, where a transverse actuator is bonded to one of the sides of a beam. When the length of the actuator changes in x direction, the beam is forced to bend, causing a displacement of its tip in z direction. The piezoelectric material is stressed in shear.

Inchworm Actuators

An inchworm actuator is much more complex and needs a more complex control. A scheme is reported in Fig. 7.8d: a cylindrical actuator (1) can change its length in x direction, for instance working like a transversal actuator owing to a radial electric field. The outer surface is constrained, so that it cannot move. At the ends of this actuators there are two other piezoelectric actuators (2) and (3) that, when energized, reduce their inner diameter and clamp a metal rod that can travel in x direction.

Assume that actuator (3) is energized and then actuator (1) is made to extend. The rod moves to the right. Then actuator (3) is released and actuator (2) is energized while actuator (1) is made to contract. The rod moves again to the right. The cycle is repeated as many times as needed, so that the rod moves indefinitely to the right (the stroke can be some hundred mm), albeit in very small steps, of the order of a few nm. The operation can be quite fast, so that speeds of some mm/s can be achieved.

Equivalent Circuit

At low frequency, the piezoelectric element can be modeled as a capacitor (C_0 in Fig. 7.8e). From (7.51) it follows that its value is

$$C_0 = \epsilon_{ij}\frac{A}{l}, \tag{7.66}$$

where the choice of the coefficient ϵ_{ij}, the area A and the length l depends of the type of actuator (for instance, in a longitudinal actuator, ϵ_{ij} is ϵ_{33}, A is the area of the electrode and l is the thickness of the layer).

In dynamic conditions, a number of LRC branches must be added to the equivalent circuit, as shown in Fig. 7.8e: the resistances R_i model the mechanical losses, while the capacitors C_i model the inertia properties of the mechanical system, and the reciprocal of the inductances L_i model its stiffness properties. Each branch models one of the vibration modes, so that the higher is the maximum frequency considered, the larger is the number of branches that must be included into the model.

Example 7.3 Consider a piezoelectric stack actuator made of $N = 10$ layers. The cross section is square with a side $l = 10$ mm, and the thickness of the layers is $t = 0.2$ mm. The relevant piezoelectric characteristics of the material are $\epsilon_{33} = 1.15 \times 10^{-8}$ F/m, $d_{33} = 300 \times 10^{-12}$ C/N, $s_{33} = 15 \times 10^{-12}$ m^2/N.

Compute the equivalent capacitance of the stack, the no-load displacement and the force exerted when the actuator is clamped, as functions of the voltage. Plot the force-displacement characteristics at various voltages and stiffness of the mechanical system against which the actuator pushes.

Since the elements are in parallel, the total capacitance of the stack is

$$C_0 = N\epsilon_{33}\frac{l^2}{t} = 57.5 \text{ nF}.$$

The no-load displacement Δz_0 of the stack of 10 layers (in m) and the force with clamped ends (in N) are

$$\Delta z_0 = N d_{33} V = 6 \times 10^{-9} V,$$
$$|F_{z0}| = -d_{33}\frac{|V|A}{s_{33}t} = 10|V|.$$

The force-displacement characteristic of the stack is plotted in Fig. 7.10.

7.3 Rotary Actuators

7.3.1 Electric Motors

The most common kind of rotary actuators used in robotics are electric motors. They are used not only to produce a rotational motion, but also when a linear motion is

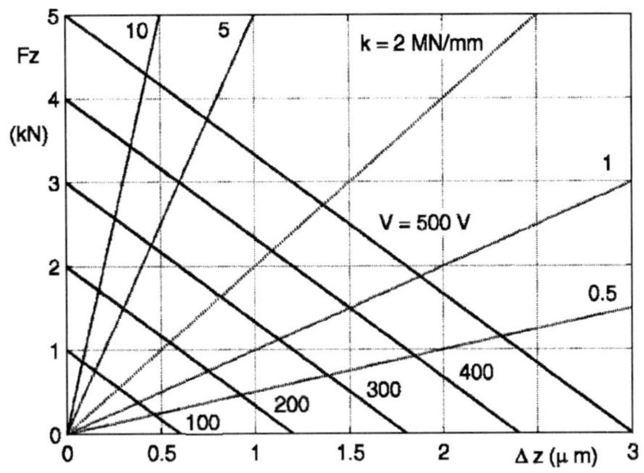

Fig. 7.10 Force-displacement characteristic of a stack piezoelectric actuator at different voltages

required: in the latter case a suitable transmission system is used (see next section). Often robotic applications require high torques and low speed, both in the case of traction motors for wheeled and tracked robots or rovers and in the case of joints for moving arms or legs. Most electric motors are well suited for delivering their nominal power at high speed, and hence with low torques: also in this case a mechanical transmission is usually interposed between the motor and the moved element. As an alternative, motors that can supply a high torque at low speed are also built. They are commonly referred to as torque motors.

There is a wide variety of electric motors, but the most common in space robotics are brushless DC motors. Brush DC motors are less expensive, in particular for what the controller is concerned, but they cannot operate in vacuum or in a thin atmosphere like that of Mars, and their life is limited by the wear of the brushes. For this reason, and others like the lower electromagnetic interference, better power/mass ratio and greater cleanliness of brushless motors, standard DC motors are increasingly substituted, in particular in space applications, by brushless types. In some cases also stepper motors are used in robotics.

Brushless DC motors are essentially synchronous AC motors, fed through a controller that converts the input DC power into a pulsed 3-phase current whose waveform may be either sinusoidal or trapezoidal. The stator is made of a laminated iron core with the embedded coils. The number of pole pairs may be different, ranging from 1 to a few tens. The smaller the number of poles, the higher the speed at which the motor operates for a given commutation frequency.

The rotor carries the permanent magnets that supply the magnetic field. Since there are no windings on the rotor, there is no need to transfer an electric current to rotating parts. The introduction of high performance permanent magnets (rare-earth magnets) at the end of the twentieth century and their decrease in cost in the following years contributed to the diffusion of actuators based on electric motors.

Brushless motors are usually more efficient than brush DC motors, particularly in the case of small sizes.

The most common configuration is that with the rotor inside the stator (Fig. 7.11a), but there are cases in which the rotor is placed outside (Fig. 7.11b).

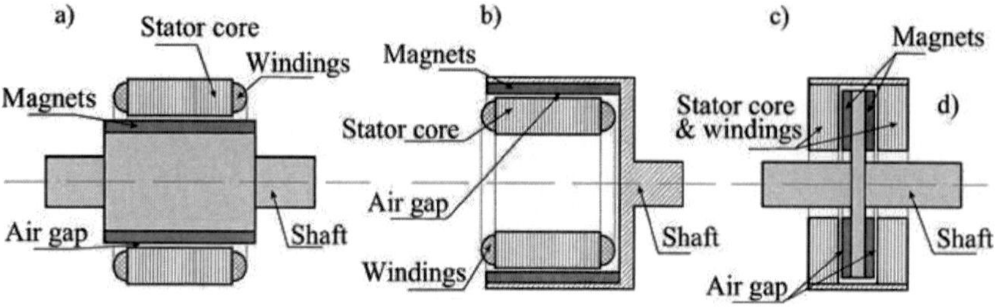

Fig. 7.11 Permanent magnets electric motors. (**a**) Internal rotor (inrunner); (**b**) external rotor (outrunner); (**c**) axial flux (pancake)

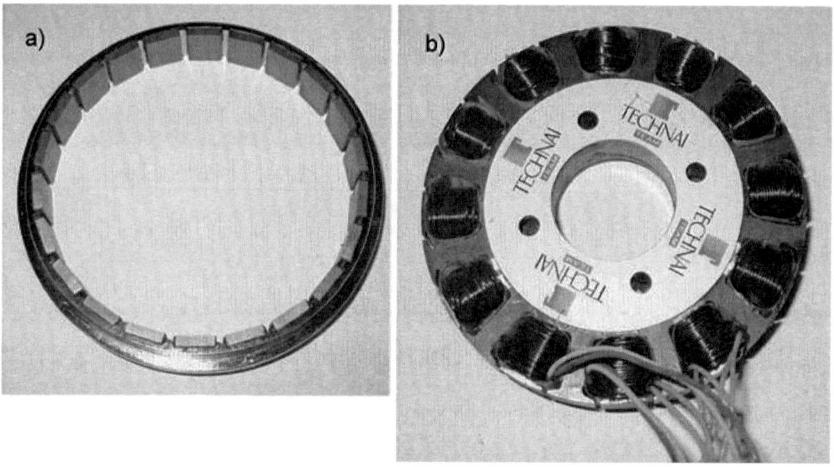

Fig. 7.12 (**a**) Rotor and (**b**) stator of an outrunner motor to be inserted in the wheels of a small rover. The motor was designed by Italian firm Technai Team

Since the radius of the airgap is larger, for a given size of the motor, outrunner motors, as they are usually referred to, supply a larger torque than inner-rotor motors, and thus this configuration is often used in torque motors. However, they have also some disadvantages, like the difficulty of achieving sufficient cooling: since both the losses in the iron and in the copper occur in the stator, almost all the heat is produced in the latter and in outer rotor motors it is less easy to get rid of the heath produced.

The picture of an outrunner motor designed to be inserted in the wheels of a small rover is shown in Fig. 7.12.

In both configurations the air gap is crossed by a magnetic field whose direction is radial. Another possible configuration is that with the magnetic field flowing through the air gap in axial direction (Fig. 7.11c). Such axial field motors are usually much thinner and of greater diameter, for a given mass, than radial field motors and thus they supply larger torques and lower speeds. Being large and thin, they are called pancake motors.

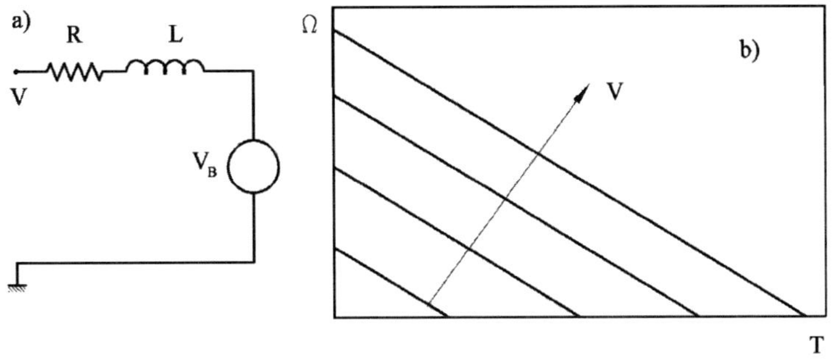

Fig. 7.13 (**a**) Equivalent circuit of a brushless electric motor. (**b**) Speed–torque characteristics at various voltages for the same motor

The controller must detect the angular position of the rotor. The sensor performing this task can be an optical encoder, an electromagnetic sensor or a magnetic sensor, like Hall effect sensors. However, it is possible to measure the back electromotive force in the coils to infer the rotor position, without the need for separate sensors: this arrangement is often called sensorless. Hall effect and sensorless layouts may have difficulties in low-speed operation, and this is particularly important when using torque motors with the aim of avoiding the need of a mechanical transmission, by driving directly a wheel or a joint of an articulated mechanism.

The controller is usually based on a microprocessor to operate the 3 bidirectional drivers feeding the various phases of the windings. The computational power of the microcontroller is usually large enough to control not only the speed, but also the acceleration and, if required, the rotor position. It is possible to use the controller also to fine tune the relevant parameters to achieve the best possible efficiency in the various operating conditions.

From the viewpoint of the input-output relationship, a brushless motor is identical to a permanent magnets conventional DC motor. Its equivalent electrical circuit is shown in Fig. 7.13a, where R and L are the equivalent input resistance and inductance, V is the voltage across terminals and V_B is the back ElectroMotive Force (EMF), all referred to the active commutated phase.

The latter is proportional to the angular velocity Ω of the motor, through the back EMF constant K_B

$$V_B = K_B \Omega . \tag{7.67}$$

The torque produced by the motor is proportional to the current i through the torque constant K_T

$$T = K_T i . \tag{7.68}$$

Remark 7.7 If expressed in consistent units (V/(rad/s) for K_B and Nm/A for K_T) the numerical values of the two constants are equal.

In steady-state conditions, from Fig. 7.13a it follows that the voltage V is linked with the torque and the speed by the relationship

$$V = iR + V_B = \frac{TR}{K_T} + K_B \Omega, \tag{7.69}$$

where the first term is the voltage required to produce the torque and the second the voltage required to overcome the back EMF. Solving this equation in Ω yields the speed–torque characteristic of the motor at constant voltage:

$$\Omega = \frac{V}{K_B} - \frac{TR}{K_T K_B}. \tag{7.70}$$

The speed–torque characteristic is plotted in a qualitative way in Fig. 7.13b: it is a straight line. The maximum speed the motor can reach is achieved when it supplies no torque. The no-load speed is

$$\Omega_{\text{max}} = \frac{V}{K_B}. \tag{7.71}$$

The maximum torque is obtained when the motor stalls, i.e. its speed is reduced to zero

$$T_{\text{nl}} = \frac{K_T V}{R}. \tag{7.72}$$

Once the speed–torque characteristic has been computed, it is possible to obtain the mechanical power. the electric power and the efficiency of the motor

$$P_m = \Omega T, \qquad P_e = Vi, \qquad \eta = \frac{P_m}{P_e}. \tag{7.73}$$

In the analysis above no account has been taken for mechanical energy dissipation that occur in the motor, such as bearing and seal drag. If this approximation is dropped, the torque T produced by the current i can be written as

$$T = K_T i = T_u + T_D, \tag{7.74}$$

where T_u is the useful torque produced by the motor and T_D is the drag torque.

The mechanical power produced by the motor in this case is

$$P_m = \Omega T_u = \Omega (T - T_D). \tag{7.75}$$

The plot of Fig. 7.13b is, however, only an ideal one: the part on the right corresponds to conditions where the motor should work with large torques, and hence large currents, and low efficiency. Most of the electric power is thus used not to produce mechanical power but to produce heat and these operating conditions can be sustained only for a very short time, if at all. For thermal reasons the maximum current a motor can sustain is limited to a value well below the value V/R corresponding to the torque T_{nl}. The maximum continuous torque is thus

$$T_{\text{max}} = K_T i_{\text{max}}. \tag{7.76}$$

Often there is also another limitation: between the low torque, high speed conditions where the performances are limited by the voltage and the low-speed, high torque conditions where the limitation comes from the current, there is an intermediate range where the limitation comes from the maximum power. The curves at constant power are hyperbolas in the $\Omega(T)$ plane.

In dynamic conditions, (7.69) becomes

$$V = iR + L\frac{di}{dt} + V_B \tag{7.77}$$

and the relationship linking the torque with the speed is

$$T = (J_M + J_L)\frac{d\Omega}{dt} + T_D + T_e, \tag{7.78}$$

where J_M and J_L are the moments of inertia of the motor and the load (the latter reduced to the motor shaft), T_D is the total mechanical drag torque and T_e is the torque opposing the motion applied to the load, also reduced to the motor shaft.

By introducing a total moment of inertia and a total drag torque, the latter being a function of Ω only (for instance, is like a viscous drag torque, linear in Ω):

$$J_{tot} = (J_M + J_L), \qquad T_{tot}(\Omega) = T_D + T_e, \tag{7.79}$$

and remembering (7.68), (7.77) yields

$$\frac{R}{K_T}T + \frac{L}{K_T}\frac{dT}{dt} + K_B\Omega - V = 0. \tag{7.80}$$

By introducing the value of the torque expressed by (7.78), it follows

$$\frac{LJ_{tot}}{K_T}\frac{d^2\Omega}{dt^2} + \frac{1}{K_T}\left(RJ_{tot} + L\frac{dT_{tot}}{d\Omega}\right)\frac{d\Omega}{dt} + \frac{R}{K_T}T_{tot}(\Omega) + K_B\Omega - V = 0. \tag{7.81}$$

Example 7.4 Consider a brushless torque traction electric motor for a microrover. Its main characteristics are: maximum voltage $V_p = 12$ V, winding resistance (terminal to terminal) $R = 4.9\ \Omega$, torque constant $K_T = 0.52$ Nm/A, back EMF constant $K_B = 0.52$ V/(rad/s).

Plot the characteristics of the motor (angular velocity Vs. torque, mechanical power Vs. torque, electrical power Vs. torque, and efficiency Vs. torque), with various values of the voltage from 1 to 12 V, both neglecting drag torque and assuming that the drag torque acting on the motor shaft is $T_D = 10$ mNm.

The maximum torque and the maximum speed at 12 V are

$$T_p = 1.28 \text{ Nm}, \qquad \Omega_{max} = 23.08 \text{ rad/s} = 220 \text{ rpm}.$$

First neglect the friction torque. The results are plotted in Fig. 7.14.

Since no friction torque was assumed, the efficiency tends to 100% when the torque tends to zero and the speed tends to the no-load speed. However, at that speed the efficiency is 0 since no useful power is generated as the torque vanishes.

Assuming a drag torque of 10 mNm, the characteristic curves of the electric motor modify as shown in Fig. 7.15. Clearly a small friction torque has a very small effect in the low-speed, high torque region of the characteristics, but the efficiency is much lower when the motor operates producing a low torque.

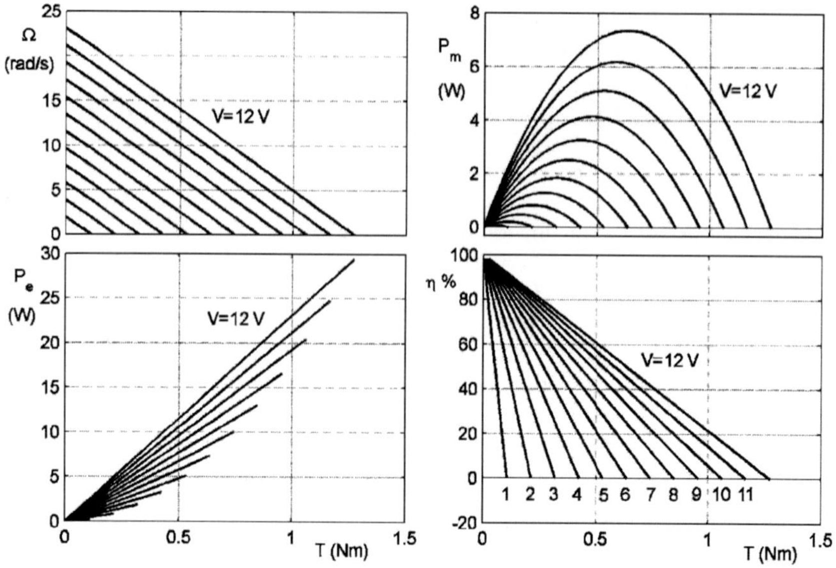

Fig. 7.14 Characteristics of the electric motor at various values of the voltage (no friction torque)

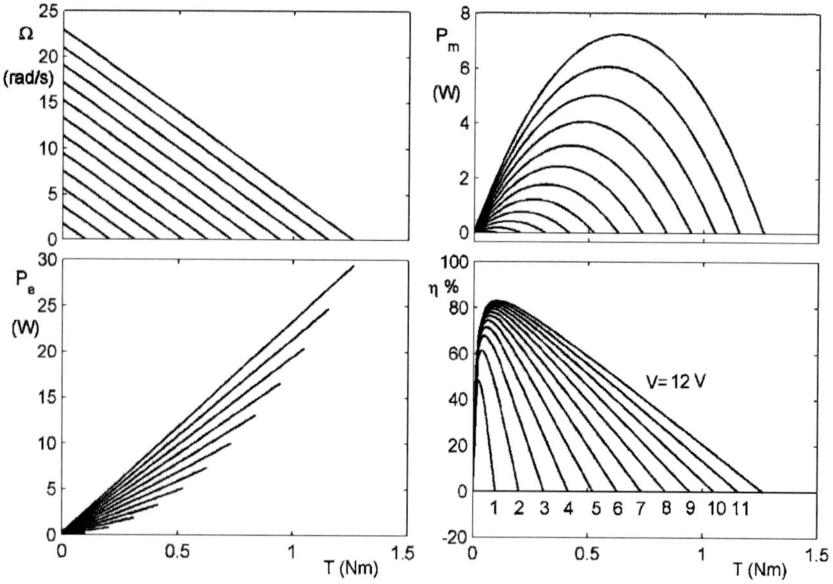

Fig. 7.15 Characteristics of the electric motor at various values of the voltage (friction torque 10 mNm)

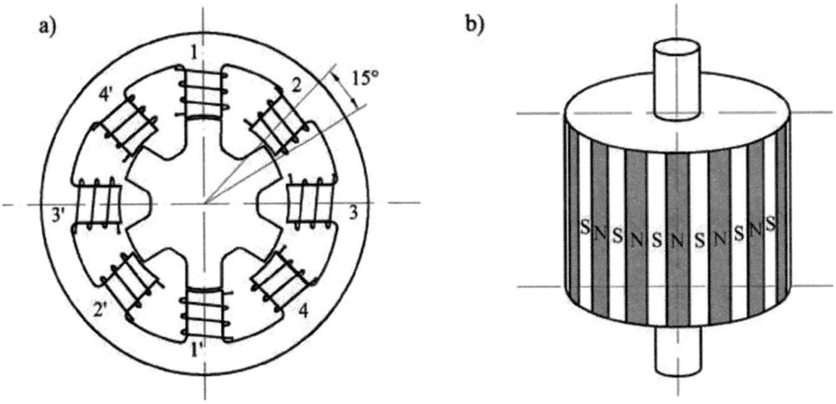

Fig. 7.16 Stepper motors. (**a**) Sketch of a variable reluctance motor; (**b**) sketch of the rotor of a permanent magnets rotor

As already stated, stepper motors are sometimes used in robotics. Their basic advantage is that they usually do not require a feedback control: as the name suggest, they operate in steps and every time they are energized by the controller they advance by a fixed angle. Unless it slips, the position of the rotor of a stepper motor is thus always known, without the need of measuring it.

Like brushless motors, also stepper motors do not carry windings on the rotor and thus they do not have brushes. The rotor can carry permanent magnets (permanent magnets stepper motors) or just a soft iron core, that is magnetized by the coils on the stator (variable reluctance stepper motors).

A variable reluctance motor has a toothed soft iron rotor surrounded with a wound stator, as shown in Fig. 7.16a. In the figure the stator has $n_{ps} = 4$ pair of poles, while the rotor has $n_{pr} = 3$ pair of teeth. The position shown is obtained by energizing the poles 1 and $1'$ of the stator. If the poles 1 and $1'$ are switched off while poles 2 and $2'$ are switched on the rotor moves of one step, rotating by an angle

$$\theta = 360\left(\frac{1}{2n_{pr}} - \frac{1}{2n_{ps}}\right) = 15°.$$

A variable reluctance motor whose stator has 12 pole pairs performs steps of 1.3636°, i.e. performs 264 steps per revolution. The rotor can perform also a half step, for instance by switching on poles 1, $1'$, 2 and $2'$ at the same time. In this case the rotation angle is 7.5°.

Microsteps can be performed by switching on two contiguous pairs of poles with different currents, so that the rotor takes an intermediate position between the positions with two subsequent pole pairs aligned with the teeth.

The rotors of permanent magnet stepper motors do not have teeth, but a number of subsequent N and S magnetic poles, as shown in the sketch of Fig. 7.16b. These however are only basic sketches, since there are many practical configurations with different geometries with either radial or axial magnetic field, different number of phases and poles, etc.

Usually stepper motors supply the maximum torque when they are not rotating. The maximum torque the motor can supply in this condition is referred to as holding torque: if it is exceeded the motor slips and the position of the rotor becomes unknown, unless it is measured by a sensor. When a torque is applied to the rotor it turns by a small angle with respect to the theoretical position; this error decreases with increasing holding torque. The larger the holding torque, the stiffer is then the motor in keeping its theoretical position.

By properly energizing the poles in sequence the motor can be made to turn continuously, but the torque supplied decreases with increasing speed.

Generally speaking, stepper motors supply much lower torques for a given mass and bulk than brushless motors, are more prone to vibrate and more limited in speed. As a consequence, they are more suitable where a controlled rotation is required with very low torque, like in plotters, printers and other computer peripherals, than for heavy duty applications like in robotics. Also, their simple electro-mechanical configuration and simple control without the need of sensors makes them a good solution for low cost devices more than for demanding applications where low cost is not a fundamental requirement.

The high performance of rare-earth magnets, like neodymium–iron–boron magnets, allows to build permanent magnets brushless motors with high power density. They are well suited for space applications since they can operate in vacuum, their main limitation being high temperature operation. All permanent magnets demagnetize at a certain temperature, the Curie temperature, but rare-earth magnets start losing their performances well below it. From this viewpoint, samarium-cobalt magnets are much better than neodymium–iron–boron magnets (see Table 7.3).

7.3.2 Hydraulic and Pneumatic Motors

Hydraulic and pneumatic rotary actuators, mostly of the variable-displacement type, are often used in robots for Earth applications, but seldom in space robots. Similarly, hydraulic motors are common as traction motors in earth moving machines, but are usually non considered for planetary rovers and the machines that in the future will be used to build outposts and bases on the planets.

The most common types of hydraulic motors are:

- gear motors,
- rotary vane motors,
- axial piston motors,
- radial piston motors.

All of them can be converted to work as pumps with little or no modifications.

Axial and radial piston motors can have a variable displacement. They are the most common types in heavy duty applications, like earth moving machinery, locomotives, etc., in particular when coupled with pumps of the same type. The fact that their displacement can be varied allows to easily control the speed of the output shaft, without having to change the speed of the prime mover operating the pump.

The displacement of rotary vane motors may be adjustable, but usually they are fixed displacement devices. They are usually applied in low or medium pressure applications.

Gear motors have a constant displacement and can operate at quite high pressure. They are of different types (external gears, internal gears, sometimes with peculiar teeth shape) and are built in a wide range of sizes for many different applications.

There is a possible exception to what said above about the unsuitability of hydraulic motors as rotary actuators for space robots: if electro-hydrostatic systems will be used to operate the wheels of rovers or the arms or legs of robots, fixed displacement hydraulic motors could become widespread in space systems. For this reason this kind of machines will be briefly dealt with in the section on electro-hydrostatic transmissions.

Pneumatic motors have little chances to be used in space robotics, mainly for their low efficiency and the difficulty of using pneumatic devices in places were the pneumatic fluid must be carried on board. A possible, albeit unlikely, exception is for system designed for planets with abundance of atmospheric gases. Will we see robots on the surface of Titan operated by pneumatic actuators in whose tubes a mixture of nitrogen and methane will flow?

7.3.3 Internal Combustion Engines

At present no internal combustion engine (ICE) is used outside our planet, and until recently it was a common opinion that this situation is bound remain unchanged. However, the very high power density of this kind of engines and the even higher energy density of the chemical fuels they use make them a natural candidate for powering large planetary rovers in connection with in-situ resource utilization (ISRU) systems.

As it will be seen in the next chapter, storing energy in the form of chemical energy is particularly expedient on planets with an oxidizing atmosphere, where only the fuel needs to be carried on board, but the only place in the solar system where free oxygen is available is Earth.

The opposite situation, a planet with reducing atmosphere that supplies the fuel, while the oxidizer is carried on board, is less convenient, since the mass of the oxidizer needed for the combustion of the most common fuels is much larger than that of the latter. For instance, when burning oxygen and hydrogen to produce water, the mass of oxygen is eight times that of hydrogen; similarly, when burning methane and oxygen to produce water and carbon dioxide the mass of oxidizer is 4 times the mass of fuel. A situation of that kind can occur on Titan, whose atmosphere contains 1.6% of methane, the remainder being nitrogen. An ISRU system can produce oxygen from the water ice and engines working on methane and oxygen can be used.

Even less expedient is the use of both fuel and oxidizer carried on board. This is the case, for instance, of using methane and oxygen or hydrogen and oxygen obtained from water and carbon dioxide on Mars.

Apart from the high energy and power density, the greatest advantage of reciprocating internal combustion engines is the century long experience of mass production in a wide variety of types and sizes, which translates in a good efficiency, excellent reliability and user friendliness at a low to moderate cost. This last consideration is mitigated by the need of adapting the engine to the peculiar environmental conditions and fuel.

Recently, however, the automotive industry has accumulated an extended experience on running internal combustion using hydrogen and methane or other gas fuels: engines converted to fuels that can be obtained from ISRU systems can be considered off-the-shelf components.

The only point still requiring some research is the use of pure oxygen as oxidant: all existing internal combustion engine use air and thus the gas filling the combustion chamber contains a large amount of inert nitrogen, which limits the maximum temperature reached by the combustion gases, which in this kind of thermal engines act also as working fluid. The easiest way to get around this problem, on a planet with an atmosphere, is to use the inert gases like carbon dioxide on Mars or nitrogen on Titan to dilute the fresh charge entering the cylinders. An alternative is to design the engine to work at a higher temperature, which should be also beneficial to increase its efficiency.

Clearly it is not possible to just take an automotive engine converted to work on hydrogen or methane and to put it on a Mars or a Titan rover: the problems related to cooling and operation in low atmospheric pressure must be dealt with. However, the advantages of this kind of engine are such that this alternative is worth considering.

Reciprocating internal combustion engines, both of the spark ignition and diesel type, are built in sizes ranging from small machines rated few tens of Watts to large ones able to deliver more than 1000 kW. The performance of an internal combustion engine is usually summarized in the map of the engine, i.e. a plot in the speed–power (or sometimes speed–torque) plane: a line giving the maximum output power as a function of the speed is represented, together with a number of curves connecting the points representing the working conditions characterized by various values of the engine efficiency. The efficiency is defined as the ratio between the mechanical energy supplied in a given time, in steady-state operation, and the chemical energy of the fuel burned in the same time:

$$\eta_e = \frac{E_{\text{mech}}}{E_{\text{chem}}} = \frac{T\Omega}{H\dot{m}}, \tag{7.82}$$

where T is the torque, Ω is the engine speed, H is the thermal value of the fuel (or of the fuel-oxidizer combination) and $\dot{m}$ is the rate at which the fuel (or the fuel-oxidizer combination) is burned.

Alternatively, instead of the efficiency, the specific fuel consumption q is used. It is defined as the mass of fuel consumed to supply the unit power for unit time. It

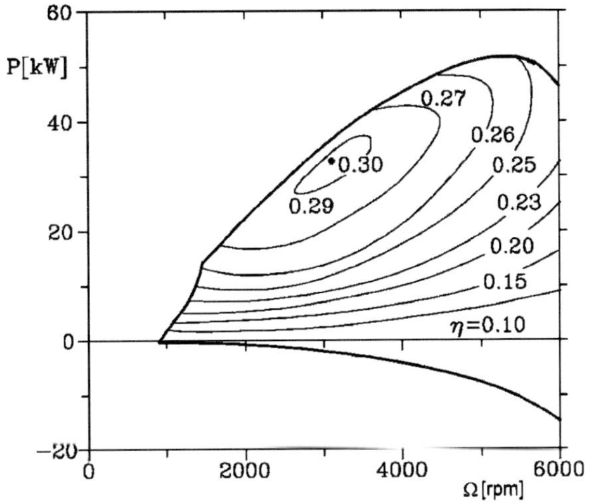

Fig. 7.17 Map of a spark ignition internal combustion engine with constant efficiency curves (from G. Genta, L. Morello, *The Automotive Chassis*, Springer, New York, 2009)

is linked with the thermal value of the fuel and the engine efficiency by the relationship

$$q = \frac{1}{H\eta_e}. \tag{7.83}$$

It is expressed in kg/J, even if non-consistent units like g/kWh are often used.

The map of an automotive spark ignition engine rated 51 kW is shown in Fig. 7.17; the plot refers to operation in air, with gasoline as a fuel and a water cooling system based on a radiator exchanging heat with air at standard Earth surface temperature and pressure (288 K, 101 kPa). More modern engines have better efficiencies, but seldom exceeding 40% in the maximum efficiency point. The power density is quite high, being in the range of 0.3–1 kW/kg, except for engines built for high performances and short life.

It is well known that internal combustion engines cannot supply power at a speed lower than a minimum operating speed: the driveline connecting the engine with the device that uses the mechanical power must be disengaged at low speed and the engine must be spun by some sort of starter motor until it starts. The transmission must be provided with a friction clutch or a torque converted, both devices that can be considered as off-the-shelf, low cost, components. Moreover, in most cases the transmission must feature a variable transmission ratio.

Apart from reciprocating, internal combustion engines, there is a wide variety of other thermal engines that may be used to power planetary rovers, like rotary internal combustion engines, steam engines, steam and gas turbines, external combustion reciprocating engines, etc.

In particular, rotary internal combustion engines are particularly interesting for small size applications, owing to their smooth operation and relatively ease of miniaturization, but the main advantages of reciprocating engines, namely technological readiness, reliability and low cost, are lost.

Turbines are widely used to supply large power in lightweight and compact packages, well above the size of foreseeable planetary rovers. When scaled down, their efficiency is reduced, together with their life and reliability.

7.4 Mechanical Transmissions

7.4.1 From Rotary to Rotary Motion

Most rotary actuator, except for hydraulic and pneumatic motors, operate at their best in high speed and low torque conditions. All robotic uses, both in the case of the actuation of wheels for mobile robots or of arms and legs, require usually low speeds (often very low) and high torques. For instance, a small electric motor supplying 10 W may work at its best at a speed of 3,000 rad/s $\approx$ 29,000 rpm supplying a torque of 0.0033 Nm, while the arm it operates may require a torque of 20 Nm at 0.5 rad/s $\approx$ 4.8 rpm. This compels one to use a mechanical transmission with a high transmission ratio, defined as the ratio between the output and the input speed

$$\tau = \frac{\Omega_{\text{out}}}{\Omega_{\text{in}}}. \tag{7.84}$$

In the case above, the transmission ratio is $1/6000 \approx 0.000167$.

Example 7.5 Consider a rover with wheels having a diameter of 200 mm, traveling at a maximum speed of 30 m/h. It is powered by small brushless motors located in the wheels, having a top speed of 10,000 rpm. Assuming that the maximum speed of the rover is reached when the motors reach their maximum speed, and neglecting the longitudinal slip, compute the required transmission ratio.

The maximum speed of the wheels is 0.833 rad/s = 7.98 rpm. The transmission ratio is thus $1/1256 = 0.000796$.

When the gear ratio is moderate, not less than 1/4 or 1/5, a simple pair of gear wheels can be used. The transmission ratio of a pair of gear wheels is equal to the inverse of the ratio of the number z of teeth of the wheels or, which is the same, to the inverse of the ratio of their pitch radii. Since the number of teeth of any gear wheel has to be greater than a minimum value, which depends on several design factors but, for standard spur gears, is usually not less than 16 or 17, a small transmission ratio leads to a large number of teeth on the larger (slower) wheel. For instance, if the minimum number of teeth z_{min} is 17, and the transmission ratio is 1/5, the large wheel must have 85 teeth.

When used to drive the wheels of a vehicle, the transmission ratio is said to be *long* or *short* depending whether the distance traveled when the motor turns of a given angle is long (high transmission ratio) or short (low transmission ratio).

The size of the teeth is controlled by the torque the gear wheel must transfer: highly loaded wheels with many teeth are thus bulky and heavy. The design of gear wheels is a highly specialized job and for the simpler applications the designer of a robot chooses the reduction gears together with the motors from catalogues of manufacturers on the basis of the output torque and speed, with the constraints related to lifetime and reliability. When this approach is not sufficient, custom-built reduction gears are provided by some manufacturers.

If the speed must be reduced more, two or three pairs of gear wheels in series can be used. Obviously, the transmission ratio of a train of wheels is the product of the transmission ratios of each pair. With three pairs of gear wheels in series, a total transmission ratio of almost $1/100$ can be obtained.

Since in robotics the torques the actuators must exert, particularly in the case of arms and legs, may be large, the teeth must be carefully dimensioned. They are often quite large and gear wheels with the required number of teeth may be large and heavy pieces of machinery. From this viewpoint, it is often expedient to use a larger number of gear wheels than what is strictly needed, so that the speed reduction at each stage is lower, i.e. the various transmission ratios are larger.

The efficiency of each pair of cylindrical gear wheels is quite high, between 90 to 99% depending on size, precision, application and, above all, lubrication. The higher values are typical of the large wheels of high power transmissions, while the smaller gears used in robots have an efficiency more close to the lower end of the mentioned range.

Strictly speaking, it is impossible to assess a constant value of the efficiency of a gear transmission. Even if the working conditions are stated, the efficiency depends on the power flowing through the transmission: in general, the efficiency increases with increasing load, but only up to a point, since overloading can cause the efficiency to decrease. This notwithstanding, often an average value of the efficiency is assessed for each pair of wheels. For a greater precision, a working cycle must be stated and the efficiency can be computed instant by instant; an average efficiency on the cycle can thus be assessed.

Conical gears have a lower efficiency, while the efficiency of helical gears is usually higher than that of straight ones. When putting a number of reduction stages in series, the overall efficiency is the product of the efficiency of each pair.

Harmonic drives (Fig. 7.18) can be used to obtain a transmission ratio of $1/50$ to $1/200$ in a single stage. Harmonic drives were specifically developed for robotics and space applications, and were used on the wheels of the LRV. They are based on a compliant inner gear that meshes with a rigid outer gear (or outer spline) that carries an internal set of teeth. The flexible gear is forced into contact with the outer spline by an eccentric element, the wave generator, which rotates at speed Ω_1. Since the outer spline has more teeth than the flexible gear, the latter moves slowly backwards at a speed Ω_2. As shown in Fig. 7.18b, to prevent large friction between the wave generator and the flexible gear, the former is provided with a deformable ball bearing.

If the outer rigid gear has z_o teeth and the inner flexible gear has z_i teeth, the overall transmission ratio is

$$\tau = \frac{\Omega_2}{\Omega_1} = \frac{z_o - z_i}{z_i}. \tag{7.85}$$

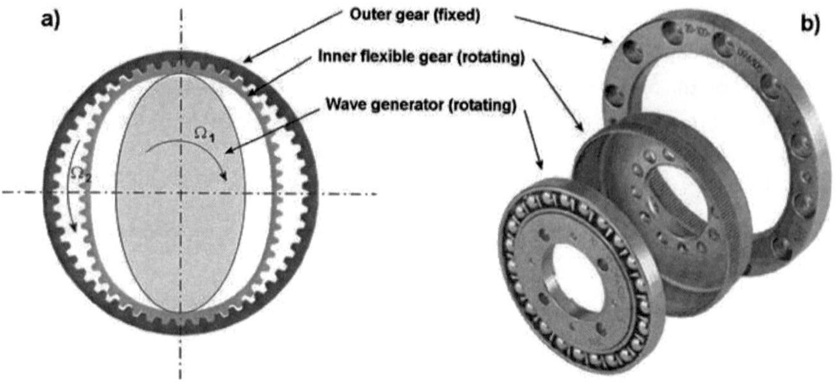

Fig. 7.18 Harmonic drive. (**a**) Sketch, (**b**) picture of the three main elements, the outer gear (or outer spline), the inner flexible gear and the wave generator

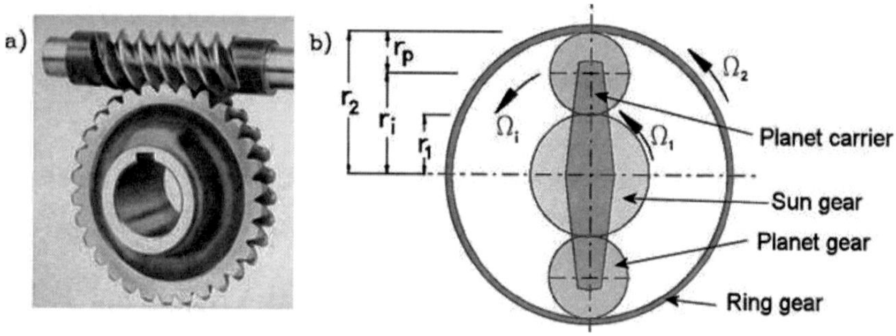

Fig. 7.19 (**a**) Worm gear; (**b**) planetary gear

If the difference between the number of teeth is just 1, the transmission ratio is $1/z_i$. However, in this case the wave generator cannot be elliptical and there cannot be two opposite zones of the inner flexible gear in contact with the outer gear.

Harmonic drives are compact and their efficiency is usually good, but they are much more expensive than standard gears. They are usually reversible. One of their main advantage is the fact that at any time there are many teeth engaged, which share the load, so that smaller teeth can be used.

The *Apollo* LRV was provided with four motors in the wheels, each with an harmonic drive reducer with a gear ratio of 1:80, so that the wheels could turn at 213 rpm when the 186 W series wound DC motors reached their maximum speed of 17,000 rpm. The motor–harmonic drive complex was sealed in a low pressure air atmosphere, so that the brush motors could operate correctly.

Worm gears (Fig. 7.19a) may be used for reduction from $1/20$ to $1/300$. They may be reversible (usually only at lower reductions, i.e. higher transmission ratios) or not reversible. Their efficiency is lower than that of standard reduction gears, particularly when the reduction is large. Efficiency can be as low as 70% or even less.

The worm can have a single thread or several (usually not more than 3 or 4) threads and the gear ratio is

$$\tau = \frac{z_w}{z_g}, \tag{7.86}$$

where z_w is the number of threads of the worm and z_g is the number of teeth of the gear wheel.

Planetary gears are also used for high values of transmission ratio, particularly when transmitting high torques. As shown in Fig. 7.19b, planetary gears have tree shafts that can be used as inputs or outputs: one is connected with the outer ring gear, one to the sun gear and one to the planet carrier, to which the planet gears, which are free to rotate, are attached. The velocities of the sun gear Ω_1, the planetary carrier Ω_i and the ring gear Ω_2 are linked with the radii r_1 and r_2 or, which is the same, with the number of teeth of the sun gear z_1 and the ring gear z_2 by the relationship

$$\frac{\Omega_1 - \Omega_i}{\Omega_2 - \Omega_i} = \frac{r_2}{r_1} = \frac{z_2}{z_1}. \tag{7.87}$$

When one of them is locked, they operate as reduction gears: for instance if the ring gear is locked, the input shaft is connected with the sun gear and the output shaft with the planet carrier the transmission ratio is

$$\iota = \frac{\Omega_i}{\Omega_1} - \frac{z_1}{z_1 + z_2}. \tag{7.88}$$

Planetary and differential gears can perform more complex functions, when none of the gear is locked, like in automotive differential gears where the planet carrier is connected with the propeller shaft and the sun and the ring gear (which in that case are equal and both called sun gears) are connected with the wheels.

A transmission may be reversible or not reversible. In the first case both shafts can be used as input or output, and the transmission can be used as a speed reducer or a speed multiplicator. In the second case the power can flow in just one direction, usually from the fast gear to the slow one.

Reversible transmissions are usually preferred when operating wheels, since in case of nonreversible gears the wheel would lock if the motor is not energized or supplies a power much lower than the power required for motion. However, if a suitable control system is used and the motor is always energized, a nonreversible transmission has the advantage of eliminating the need for a station brake to keep the vehicle stationary on a slope. Similarly, non reversible gearing may be preferred for arm or leg joints, because otherwise either brakes must be installed or energy is used to keep the arm or the leg stationary even when performing no useful work.

As a general, although approximated, rule, a reduction gear is non reversible if its efficiency is lower than 50%. Transmissions based on ordinary gear wheels, planetary drives or harmonic drives are usually reversible, except when the transmission ratio is very low. The situation is different for worm gears, which are usually non reversible when they have a single thread, while may be reversible if the number of threads is larger.

In many applications, in particular those linked with traction, it is impossible to find a value of the transmission ratio that is adequate for all working conditions and a variable-ratio transmission must be used. This may be implemented either by using different pairs of gear wheels that can be engaged alternatively or by using a continuously variable transmission (CVT). In the first case gear shifting must be performed when the transmission is not under load, but transmissions based on planetary gears and clutches that engage or disengage the various shafts, can allow to change gear ratio while under load (power shift). Power shift transmissions are common in automotive automatic transmissions.

Instead of gear wheels it is possible to use also belts or chains of many different types. Belts or chains are, however, not much used in space applications, where, if possible, gear wheels are preferred. Belts can be used to make continuously variable transmissions with a fairly wide range of transmission ratios (e.g. from 3 to $1/3$).

As already stated, in many cases actuators must supply large torques at low speed; particularly when operating arms or legs, but also when operating wheels. Gear wheels can be used for this, but at a cost: high torques lead to high stresses in the teeth that must be designed accordingly. When a multi-stage transmission is used, the first stages are lightly loaded, and usually lightweight gears can be used, but further stages operating at lower and lower speed and higher and higher torque are increasingly massive. In these cases it may be expedient to have smaller transmission ratios in the first stages, and higher ones in the last ones. Solutions in which there are more teeth that shear the load, like harmonic drives, or the load is sheared by several wheels, like in planetary gears, alleviate this problem, but a rotary transmission actuating an heavily loaded arm or leg is bound to be a bulky piece of equipment.

If the angle through which the device must move is limited, like it is usually the case for arms or legs, it may be expedient to use a linear actuator instead of a rotary actuator: after all this is how all rotary joint are actuated in living beings, where muscles are attached to the bones at a certain distance from the hinge axis of the joint. The distance between the joint axis and the point of application of the linear actuator changes during the motion, causing a non constant transmission ratio, a thing that requires a careful study.

In living beings the muscles are connected to the bones through tendons, i.e. through cables that can work only in traction but not in compression. This allows the actuators to be located at a certain distance from the moving element, and in particular to prevent heavy components from being located in places where they undergo large displacements. This can be done also in robots, decreasing the inertia of the moving parts of arms and legs.

Example 7.6 Consider a pressurized rover with $n_w = 6$ wheels designed to operate on the Moon ($g = 1.62\ \text{m/s}^2$). The wheels are provided with brushless electric motors with mechanical transmission. The relevant data of the rover are: mass $m = 2{,}500$ kg, wheel radius $R = 400$ mm, maximum value of the rolling coefficient $f = 0.1$. The data for the motors are: $K_B = 72 \times 10^{-3}$ Vs/rad, $K_T = 72 \times 10^{-3}$ Nm/A, $R_a = 0.177\ \Omega$, maximum voltage $V_{\max} = 36$ V, maximum current $i_{max} = 40$ A, nominal speed $\Omega = 4000$ rpm. Assume that the mechanical transmission in the wheels has an efficiency $\eta_t = 0.9$, that the effective rolling radius of the wheels is approximately coinciding with R and the losses in the motor are negligible.

Study the transmission system so that the rover can reach a maximum speed $v_{max} = 36$ km/h on level ground and can move on a grade of 45° at a speed of at least 1 km/h.

The torque and the speed at the wheels on level ground are

$$T_w = \frac{mgRf}{n_w} = 27 \text{ Nm},$$

$$\Omega_w = \frac{v_{max}}{R} = 25 \text{ rad/s} = 239 \text{ rpm}.$$

The transmission ratio required to allow the motor to work at the nominal speed is

$$\tau = 0.0597 = 1/16.8.$$

Taking into account the efficiency of the transmission, the motor torque is thus $T_{mot} = 1.79$ Nm. The voltage at the motor is thus

$$V = \frac{TR}{K_T} + K_B\Omega = 34.56 \text{ V}$$

and the current is $i = 24.9$ A. The electric motors are thus adequate. The electrical power the motor absorbs is

$$P_{el} = Vi = 859.5 \text{ W},$$

while producing a mechanical power of

$$P_{mec} = T_{mot}\Omega_{mot} = 750 \text{ W}.$$

The motor then operates with an efficiency $\eta_m = 87.3\%$.

If the vehicle operates on a 45° slope, the torque at the wheels is

$$T_w = \frac{mgR}{n_w}\big[f\cos(\alpha) + \sin(\alpha)\big] = 210 \text{ Nm}.$$

Operating as above, and remembering that the speed is now 1 km/h, the following values are found: wheel speed $\Omega_w = 0.694$ rad/s $= 6.63$ rpm, motor torque $T_{mot} = 13.93$ Nm, voltage at the motor $V = 35.1$ V, current $i = 193.4$ A, electrical power $P_{el} = 6.78$ kW, mechanical power $P_{mec} = 162.05$ W.

The motor then operates with an efficiency $\eta_m = 2.4\%$.

Clearly the motor cannot withstand these operating conditions, which widely exceed the maximum rating, and the efficiency is too low.

A 'shorter' (i.e., lower) transmission ratio is needed for operating on a slope. The voltage, current and efficiency of the motor computed for transmission ratios from 0.003 to 0.06 are reported in Fig. 7.20. It is clear that the lower the transmission ratio, the lower are the current and the voltage (the latter only to a certain point) and the higher is the efficiency. The plots have been obtained assuming that the efficiency

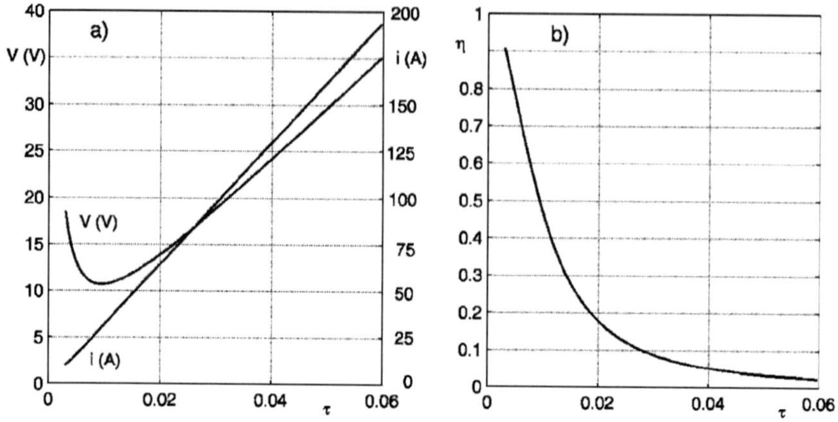

Fig. 7.20 Voltage, current and motor efficiency as functions of the transmission ratio while climbing a 45° slope at 1 km/h

of the transmission is always 0.9, while in an actual situation the efficiency would decrease with decreasing transmission ratio; this decreases the advantage of using a very short ratio.

As a compromise, the 'longest' ratio allowing the motor to work with a current not larger than 40 A is chosen.

Assuming $\tau = 0.0123 = 1/81.3$, it follows that: motor torque $T_{mot} = 2.78$ Nm, voltage at the motor $V = 11.12$ V, current $i = 39.9$ A, electrical power $P_{el} = 443.3$ W, mechanical power $P_{mec} = 162.05$ W. The motor efficiency is $\eta_m = 36.6\%$.

A transmission with two ratios, namely $1/16.8$ and $1/81.3$ allows to meet the requirements both on level ground and on the maximum slope.

The whole study should be repeated after designing the transmission, with more realistic values of the efficiency of the gears.

7.4.2 From Rotary to Linear Motion

The rotary motion of an electric motor can be converted into a linear motion by a lead screw. Owing to the external similarity with an hydraulic cylinder, an actuator made of an electric motor and a screw transmission is often referred to as an electric cylinder.

Usually the electric motor rotates the screw, while the nut is connected to the actuated element, but there are cases in which the screw is stationary and the motor actuates the nut.

Standard screws can be used, although the profile of the thread has usually a trapezoidal shape (Fig. 7.21a) and is different from that used in fasteners. The screw

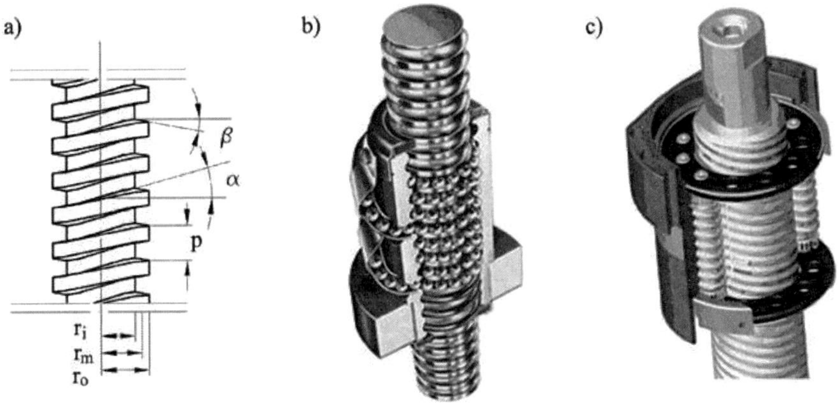

Fig. 7.21 (**a**) Geometrical definitions for a lead screw with trapezoidal thread; (**b**) ball screw; (**c**) planetary roller screw

may have a single or several threads. The helix angle α at the mean radius r_m of the thread is linked to the pitch p by the relationship

$$\tan(\alpha) = \frac{n_t p}{2\pi r_m}, \tag{7.89}$$

where n_t is the number of threads of the screw.

The transmission ratio of the screw, which now is no more a nondimensional number, is

$$\tau = \frac{V}{\Omega} = \frac{n_t p}{2\pi} = r_m \tan(\alpha), \tag{7.90}$$

if measured in length units per radians (the term 2π is omitted if it is measured in length units per revolution).

The efficiency of transmissions based on lead screws is usually low, owing to the friction between the screw and the nut. Wear is often a problem and a good lubrication is required. A simple computation of the efficiency can be made as follows: Assuming that the axial force acting on the screw is F, the useful work produced in one revolution is

$$W = n_t F p. \tag{7.91}$$

The force pressing the two surfaces against each other is

$$F_n = \frac{F}{\cos(\alpha)\cos(\beta)}. \tag{7.92}$$

Assuming that the coefficient of friction between the two surfaces is $f = \tan(\phi)$, where ϕ is the friction angle, the work lost in one revolution due to friction is

$$W_l = f F_n \frac{n_t p}{\sin(\alpha)} = \frac{n_t f p F}{\sin(\alpha)\cos(\alpha)\cos(\beta)}. \tag{7.93}$$

The efficiency is thus

$$\eta = \frac{W}{W + W_l} = \frac{\sin(\alpha)\cos(\alpha)\cos(\beta)}{\sin(\alpha)\cos(\alpha)\cos(\beta) + f}. \tag{7.94}$$

The efficiency increases with increasing α (i.e., by increasing the transmission ratio) or by reducing angle β, the influence of the latter being small.

If both α and β are small angles, the expression for the efficiency simplify as

$$\eta \approx \frac{\alpha}{\alpha + f} = \frac{1}{1 + \frac{r_m f}{\tau}}. \tag{7.95}$$

Example 7.7 A steel lead screw with 1 thread, a mean diameter of 20 mm and a pitch of 4 mm works against a steel nut with a good lubrication (friction coefficient $f = 0.16$). Compute the efficiency of the transmission.

The transmission ratio is $\tau = 0.637$ mm/rad and the helix angle is $\alpha = 3.64°$, a value low enough to consider the helix angle as a small angle. Considering also β as a small angle, the efficiency is $\eta = 0.28$.

The transmission is non reversible if $\alpha \leq \phi$, while otherwise is reversible. The condition for non reversibility can be written by introducing $\tan(\alpha)$ for f into (7.94). Since usually α is a small angle, this condition can be written as

$$\eta \approx \frac{\cos(\beta)}{1 + \cos(\beta)}, \tag{7.96}$$

and, if also β is small,

$$\eta \approx \frac{1}{2}. \tag{7.97}$$

The lead screw considered in the example above is thus non reversible. However, if the surface of the screw is coated with Teflon and the friction coefficient reduces to 0.04, the efficiency increases to 0.61 and the transmission becomes reversible. Teflon-coated lead screw are common and allow to increase the efficiency of a screw actuator, without the need of resorting to the more costly types of screws that will be dealt with below.

Remark 7.8 There are cases where a non reversible transmission is seen as an advantage and not a drawback, in spite of the low efficiency. For instance, in the actuators of the legs of a twin frame walking machines, irreversibility allows to simplify the whole layout of the machine, dispensing with the brakes and the related control devices.

To reduce friction and wear a lead screw may have several threads, but this increases the transmission ratio.

Ball screws (Fig. 7.21b) are a solution to the problems related to friction and wear, at the cost of greater complexity and expense.

They can withstand large loads and allow precision positioning; however, they are more sensitive to lateral loads and the whole device must be designed to ensure that the load between the screw and the nut has no (or at least a very small) lateral component. Guiding elements may be required to prevent lateral loading.

A further step is using planetary rollers screws (Fig. 7.21c). The load is transferred from the screw to the nut by a set of rotating threaded rollers, so that there is no sliding contact like in ball screws, and the area on which the load acts is much larger, leading to lower contact pressure. Planetary rollers screws are usually more compact and efficient than ball screws, and thus they are preferred in many space applications.

When the screw rotates at high speed, dynamic problems may be present, in particular if the screw is long. The critical speed of the screw is bound to change during the motion of the nut, and the dynamic analysis must be performed in the most unfavorable conditions.

In many cases, both to further reduce the transmission ratio or to avoid dynamic problems, a reduction gear is interposed between the motor and the screw, which thus rotates at a speed lower than that of the motor.

7.5 Hydraulic Transmissions

Hydraulic transmissions are quite common in many fields of technology, particularly in vehicles and construction machines.

The simplest hydraulic transmission devices are *hydrodynamic* couplings or torque converters. The input shaft operates a pump that forces a fluid in a turbine that, in turn, operates the output shaft. The transmission ratio is not fixed: when the output shaft is loaded by high torques it slows down while it increases its speed when the output torque required is low. It is thus possible to define a velocity ratio τ and a torque ratio ν, whose product is neither equal to 1 nor constant since the efficiency of the transmission is variable and quite low, at least in some working conditions. The plot of the efficiency and the torque ratio as functions of the speed ratio for an automotive torque converter is plotted in Fig. 7.22 as an example.

The advantage of torque converters is that the speed ratio adjusts itself following the load, but the efficiency is quite low, in particular when the slip is high (low τ). On the other end of the plot, it is possible to use a clutch that bypasses the converter when the speed ratio is close to one, so that the other zone of low efficiency is avoided. The low cost and simple layout of torque converters made them a standard in automotive automatic transmissions, but they are seldom considered in robotics.

Hydrostatic transmissions are common in heavy duty applications, like construction and earth moving machines. They are based on a hydraulic pump, connected to the prime mover, which supplies high pressure fluid to an hydraulic motor connected with the user. In some cases, when the output is a linear motion instead of a rotary motion, a hydraulic cylinder is used instead of the motor.

The pump and the motor can be of different types, like vane pump, axial or radial pistons pumps, or gear pumps. Hydrostatic transmissions work at much higher pressure and lower flow rate than torque converter and often the pump and the motor are located in different places, connected by pipes.

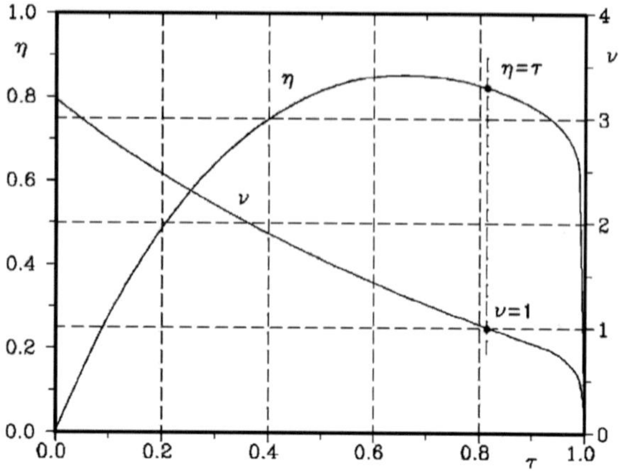

Fig. 7.22 Efficiency η torque ratio ν as function of the speed ratio τ for an automotive torque converter

The theoretical transmission ratio is equal to the ratio between the displacements of the pump and the motor

$$\tau = \frac{\Omega_m}{\Omega_p} = \frac{D_p}{D_m}. \tag{7.98}$$

If variable-displacement pumps or motors are used, the transmission ratio is variable. The hydraulic circuit usually includes reservoirs for the hydraulic fluid, hydraulic accumulators and valves that control the flow. The transmission is controlled by controlling the valves and the displacement of variable-displacement devices.

An alternative to this more traditional approach is to use the hydrostatic machines (pumps, motors, cylinders) as transmission devices, just as in the case of mechanical transmission and to control the system by controlling the electric motors operating the pumps. This approach, in which the hydraulic circuit is kept as simple as possible, is usually referred to as *electro-hydrostatic*. The pumps and the motors have a fixed displacement, and theoretically no valves, accumulators and fluid reservoirs are needed, although in practice some device to take care of the unavoidable leakage and compressibility of the fluid and other effects are required.

The theoretical flow rate of the pump is the product of the displacement by the angular velocity.

Remark 7.9 If the displacement is measured in m^3/rev and the speed is expressed in rev/s, the flow rate is m^3/s: to use consistent units, the displacement should be expressed in m^3/rad, so that the angular velocity is expressed in rad/s.

Actually the flow rate is slightly less, since no pump is free from leakage from the high pressure to the low pressure chamber and no hydraulic fluid is completely incompressible. The flow rate is thus

$$Q = \eta_{vp} D_p \Omega_p, \tag{7.99}$$

where η_{vp} is the volumetric efficiency of the pump.

The hydraulic power in the circuit is

$$P_h = Qp, \tag{7.100}$$

where p is the pressure of the fluid. The pressure in hydrostatic transmissions can be quite high, up to 20 or 30 MPa, so that high powers can be transmitted with small flow rates, which imply small and light hydraulic machines and pipes. A hydraulic motor is roughly ten times lighter and smaller than an electric motor with the same power rating.

A first limitation to the use of high pressures is due to stressing consideration in the whole plant, from the machines to the pipes. Another consideration is that the volumetric efficiency decreases with increasing pressure, since the leakage and compressibility losses increase.

The input power to the pump is

$$P_i = \frac{P_h}{\eta_{tp}} = \frac{Qp}{\eta_{tp}} = \frac{\eta_{vp} D_p \Omega_p p}{\eta_{tp}} = \frac{D_p \Omega_p p}{\eta_{\mathrm{hm}p}}, \tag{7.101}$$

where $\eta_{tp} = \eta_{vp}\eta_{\mathrm{hm}p}$ is the total efficiency of the pump and $\eta_{\mathrm{hm}p}$ is its hydro-mechanical efficiency.

The input torque is thus

$$T_i = \frac{P_i}{\Omega_p} = \frac{D_p p}{\eta_{\mathrm{hm}p}}, \tag{7.102}$$

where the displacement must be introduced in $\mathrm{m}^3/\mathrm{rad}$.

If the pump is directly connected to an hydraulic motor, the flow rate and the velocity of the motor are related by the equation

$$Q = \frac{D_m \Omega_m}{\eta_{vm}}, \tag{7.103}$$

where η_{vm} is the volumetric efficiency of the motor.

Since the flow rates of the two machines are equal, the transmission ratio is

$$\tau = \frac{\Omega_m}{\Omega_p} = \eta_{vp}\eta_{vm}\frac{D_p}{D_m}. \tag{7.104}$$

Assuming that the pressure at the motor is the same as the pressure at the pump, the output power is

$$P_o = \eta_{tm} Qp = \eta_{tp}\eta_{tm} P_i. \tag{7.105}$$

The output torque is

$$T_o = T_i \eta_{tm}\eta_{tp}\frac{\Omega_p}{\Omega_m} = T_i \eta_{\mathrm{hm}p}\eta_{\mathrm{hm}m}\frac{D_m}{D_p}. \tag{7.106}$$

Other losses are due to the pressure drop in the pipes that increases at increasing flow rate and decreasing pipe cross section. If they are accounted for, the pressure at the motor must be considered lower than the pressure at the pump, and a further efficiency must be introduced into the equations above.

Even using fixed displacement hydraulic machines it is possible to have a variable-ratio transmission by using two or more pumps actuated by the electric motor. With two pumps, for instance, it is possible to obtain two different transmission ratios by connecting one or the other to the hydraulic motor, plus another one by connecting both pumps in parallel to it. Some manufacturers produce pumps with two rotors that can provide this function with a single unit.

Example 7.8 Consider the same rover seen in Example 7.7. Perform a first approximation study of the hydrostatic transmission, based on gear pumps and motors.

The maximum torque the hydraulic motor must produce is $T_{\text{max}} = 210$ Nm at a speed of $\Omega_m = 6.63$ rpm. An hydraulic gear motor with a displacement $D_m = 156\ \text{cm}^3/\text{rev}$ is chosen. The main data, from a manufacturer data sheet, are: maximum motor torque 245 Nm, maximum speed 370 rpm, maximum pressure 12.5 MPa (all in continuous operation). The volumetric efficiency can be assumed as $\eta_{vm} = 0.70$ in low-speed, high pressure operation, rising to $\eta_{vm} = 0.85$ at higher speed, and lower pressure. The hydro-mechanical efficiency can be assumed as $\eta_{\text{hm}m} = 0.80$. The size of the motor is small enough to be located in the wheel hub.

When operating on a 45° slope at 1 km/h the flow rate is

$$Q = \frac{D_m \Omega_m}{\eta_{vm}} = 1480\ \text{cm}^3/\text{min}.$$

The pressure needed to obtain the required torque is

$$p = \frac{T_{\text{max}}}{D_m \eta_{\text{hm}m}}.$$

The value of the displacement of the motor is $156 \times 10^{-6}/2\pi = 2.48 \times 10^{-5}\ \text{m}^3/\text{rad}$. The pressure is thus $p = 10.6$ MPa.

When operating at the maximum speed the flow rate and the pressure are, respectively, $Q = 46{,}600\ \text{cm}^3/\text{min} = 46.6\ \text{l/min}$ and $p = 1.36$ MPa.

In the first condition the transmission ratio is 1/81.3. The displacement of the pump must be

$$D_p = \tau \frac{D_m}{\eta_{vp} \eta_{vm}}.$$

A value $\eta_{vp} = 0.75$ can be assumed for the volumetric efficiency of the pump, obtaining $D_p = 3.65\ \text{cm}^3/\text{rev}$.

A gear pump with a displacement of $3.5\ \text{cm}^3/\text{rev}$, a maximum pressure of 29 MPa and a maximum speed of 5000 rpm is chosen. Its volumetric efficiency can be assumed as 0.86 at low speed, high pressure and up to 0.95 at higher speed, lower pressures.

The actual transmission ratio is thus $\tau = 1/74$, slightly higher than the required one.

Assuming an hydro-mechanical efficiency $\eta_{\text{hm}p} = 0.85$, the torque at the pump shaft is

$$T_i = \frac{D_p p}{\eta_{\text{hm}p}} = 6.95\ \text{Nm}.$$

Since the motor speed is 539 rpm, the power the motor must supply is 392 W, much higher than that computed for the mechanical transmission (162 W). This is due to the fact that a value of the efficiency equal to 0.9 was assumed for the mechanical transmission, while here the total efficiency is just 0.357. This is consistent, remembering that the working conditions of the motor are different, since $162 \times 0.9/0.357 = 408$.

For high speed conditions a larger pump must be chosen. The transmission ratio is 1/16.8, and assuming a value of 0.95 for the volumetric efficiency of the pump, the displacement of the latter is

$$D_p = \tau \frac{D_m}{\eta_{vp}\eta_{vm}} = 11.5 \text{ cm}^3/\text{rev}.$$

Since the two pumps operate together when the rover travels at top speed, the high displacement pump can be smaller, with a capacity $D_p = 11.5 - 3.5 = 8$ cm^3/rev. A pump with a displacement of 7.6 cm^3/rev, a maximum pressure of 22 MPa and a maximum speed of 4,000 rpm is chosen. Its volumetric efficiency can be assumed as 0.85 at low speed, high pressure and up to 0.95 at higher speed, lower pressures.

The actual transmission ratio is thus $\tau = 1/17.4$ which is close to the theoretical value 1/16.8. The motor speed is thus 4,159 rpm.

Assuming the same value of the hydro-mechanical efficiency for the two pumps, the torque at the motor shaft is of 2.83 Nm. The total power the motor must produce is 1,225 W, higher than that of the solution with mechanical transmission (750 W), owing to the lower overall efficiency (0.55 instead of 0.9).

The study should be repeated, since the motor works in a different point, with higher power, and perhaps a larger motor should be chosen. However, the optimization must be performed at a global level. Apart from the fact that the efficiency of the mechanical transmission was probably overestimated, in particular when working at the lowest transmission ratio, the hydrostatic transmission is likely much smaller and lighter and has the advantage that the motor-pump complex can be located in the rover body while the hydraulic motors are in the wheel, giving a much greater flexibility in the overall design.

Another advantage of electro-hydrostatic transmissions is the possibility of miniaturization. Small units made of a tiny brushless motor-gear pump assembly can operate miniature hydraulic cylinders or gear hydraulic motors. The high transmission ratio so achievable allows to use fast, lightweight units that can be located in the wheels or other parts of a microrover or microrobot.

7.6 Sensors

Robots are provided by many sensors to close the control loops, as shown in the block diagram of Fig. 7.1.

Sensors, like actuators, are specialized devices designed by specialists and the designer of robots usually consider them as off-the-shelf components. When a sensor with the required characteristics cannot be found, the only reasonable thing to do is to have a sensor specialist to provide custom-built units. Only a brief and schematic overview of the various types of sensors will consequently be given in this section.

Like most living beings, a robot needs both to interact with the environment and to acquire information on its own state, and the sensors can be accordingly classified in two types: *exteroceptors*, if they acquire information about the environment, and *proprioceptors*, if they acquire information about the robot's internal parameters.

7.6.1 Exteroceptors

The sensors of the former type may belong to the robot itself, if they are mostly finalized to allow the robot to perform its task, or are considered as a part of the payload if their main goal is to supply information to the instruments the robot carries. Some sensors can perform both jobs, like a camera that sends images of the environment surrounding a rover that at the same time constitute the scientific output of the mission and are used by the human controllers to drive the robot. In most cases, however, the requirements of the two tasks are so different that different sets of sensors are used, like when the cameras supplying the images to the scientists require high resolution but low frame rate while low resolution with high frame rate is needed for controlling the rover's path.

In human beings and most animals the exteroceptors are subdivided in the five traditional senses, following a classification attributed to Aristotle. A similar subdivision holds also for robots, at least with the same anthropomorphisms we display when we speak of robotic arms, legs, wrists etc.

Sight can be defined as the ability of obtaining images of the outside world, but can be generalized as the ability to detect electromagnetic waves. Many space robots are provided with cameras performing tasks similar to those performed by eyes in living beings. As already stated, the requirements for cameras that are part of the scientific payload may be different from the requirements for the cameras supplying images to the control system. In the latter case the term *robot vision* is often used, to include not only the cameras but also the hardware and software used to extract the required information from the images and the devices used to orient the cameras toward the points of interest.

While the term sight is often referred only to the visible portion of the electromagnetic spectrum, it can be generalized to infrared light (like in night vision and heat seeking devices), the ultraviolet or even X- or gamma-rays regions of the spectrum.

Most robotic spacecraft use star trackers to determine the spacecraft attitude. A star tracker is based on a camera, or sometimes simply photocells, to measure the position of one or more stars and then, using a star catalog, obtain the attitude of the spacecraft. Also sun sensors may use optical information.

When information about the distance of an object close to the robot is required, a possibility is to compare the images taken by two cameras located at a known distance. This stereoscopic images technique is the way most animals obtain information about the distance of close objects, but is not at all an easy thing to implement on robots. Other uses of machine vision are related to the possibility of recognizing objects and supplying information to the navigation control of a moving robot or to the arm control to collect samples or to perform other operations.

The characteristics of the cameras depend on the task to be performed, particularly for what the resolution, the color discrimination, the frame rate and the optics are concerned. Some tasks require wide field optics, while in other cases the objective is a small telescope.

In general machine vision is an active research area, common to industrial and space robots, machine tools and automatic navigation devices. In general it requires a large computational power.

Hearing is defined as the ability to detect sound waves propagating through the fluid medium surrounding the body. It can be generalized stating that it is the ability to detect variations of the ambient pressure. Clearly hearing is limited to devices operating on bodies with an atmosphere or within a liquid environment. The transducers generally used to detect high frequency (for human hearing between 20 and 20,000 Hz) pressure variations are microphones. The ill fated *Mars Polar Lander* was provided with microphones to record the sounds on the Mars surface, and microphones are scheduled to fly on future Mars probes. Machine hearing may be used in case of robots cooperating with humans to allow the latter to give orders without having to use a keyboard, a joystick or other devices of this kind. In this case a large computational power may be required for speech recognition, but the relevant technologies are at a good level of readiness.

Smell is the ability of recognizing the chemical nature of substances carried, often in tiny amounts, in the surrounding medium. Devices performing this function are often called artificial noses, and are sometimes carried by moving robots in particular for tasks like detecting explosives or demining. The research in this field is active, but there is no perspective of using devices of this kind for controlling space robots.

Taste is a sense related to the ability of detecting the chemical nature of substances used as food by living being. As such, it has little application for robots, except in some experiments aimed to build biomimetic robots powered by nutritive substances.

Touch is a general term for the ability of detecting forces and pressures applied on the body; force sensing is important also in robots since it is at the base of force control. The forces that a robot exerts with any part of its body can be measured in several ways, like by providing some of the joints with force or torque sensors, by measuring the current powering electric actuators or the pressure in hydraulic actuator or, in a more anthropomorphic way, to put pressure sensors, for instance piezoelectric sensors, on the outer surface (the 'skin') of the robot. Robot hands designed to grasp object need to be provided with touch sensors, in particular if the objects to be grasped are delicate, and much research work is devoted to the development of devices of this kind.

However, the classification of exteroceptors in five senses is questionable, for robots but even for humans and animals. Temperature sensors, for instance, although they can be included in the sense of touch, may be considered as an additional sense. They are important in some space robots, both for supplying information about the environment and for the control of the robot.

The equilibrium sense, which is based on measuring the direction of the acceleration, is important in both living beings and robots. All moving robots need detecting the direction, and in many cases also the value, of the acceleration. Since there is no difference between measuring the gravitational acceleration or the acceleration due to the motion of the robot, the data from accelerometers cannot discriminate between the two situations.

On planets having a global magnetic field, measuring its direction, and sometimes its value, may be important for path planning. On Earth magnetic compasses can be used and are important for automatic navigation, but the lack of a global magnetic field in the environments where space robots operate make the detection of magnetic fields useless for navigation. Magnetometers are on the contrary very common as a part of the science payload of many probes.

On Earth GPS is increasingly used for navigation of robotic devices of all types, like UAV, UGV, etc. It is predictable that GPS satellites will be put in orbit around the Moon or Mars when their exploration will be seriously undertaken; when this will be done, GPS based navigation systems will be installed on rovers and other moving robots.

7.6.2 Proprioceptors

Proprioceptors are used to detect the state of the robot, and hence belong almost always to the robot itself and not to the scientific payload.

Position sensors are used to measure the generalized coordinates of the robot. In moving robots the coordinates defining the position of the robot must be distinguished from those defining the relative positions of the various parts. Only the latter are strictly speaking measured by proprioceptors. The position of the robot with respect to the ground is measured in various ways. If there is no global magnetic field and no GPS, cameras (robotic vision) or long range proximity sensors can measure the position and the distance from features on the ground whose position is known. If the robot is not on the ground, cameras, proximity sensors or even pressure sensors (if there is an atmosphere, whose pressure is a known function of the altitude) can be used to detect the altitude from the ground. For the last part of the descent, immediately before touchdown, even a mechanical sensor can be used.

Long range proximity sensors can be based on the radar, laser or even sonar (if there is an atmosphere) principle. Optical, both passive or active, sensors can also be used.

Odometry, i.e. the measurement of the distance traveled, is also a way to detect the current position of a robot from the known initial position, However, usually

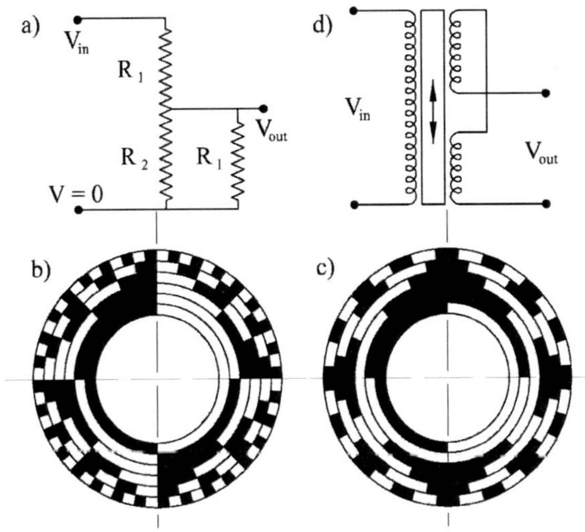

Fig. 7.23 Position sensors. (**a**) Sketch of a potentiometer; (**b**) and (**c**) binary code and gray code absolute encoders; (**d**) sketch of a Linear Variable Differential Transformer (LVDT)

odometers measure the rotation of a wheel, and the accuracy of this measurement is made uncertain by the presence of longitudinal slip. In case of driving wheels the reading of an odometer is expected to be affected by large errors. Precise odometry can be performed by using a wheel that is neither driven nor braked and that does not carry the weight of the vehicle. In this case the slip may be negligible and odometry may supply an accurate value of the distance traveled.

Detecting the attitude of the robot in space is usually done using optical devices. like star or sun trackers or other optical devices. When the robot is on the surface the pitch and roll angles can be obtained through accelerometers, while the yaw angle can be obtained from GPS or a compass (if available) or a vision system. Inertial platforms can supply a precise measurement of the attitude.

True proprioceptors, measuring the relative position of the various parts of a robot are often implemented by measuring angles. The simplest way to measure an angle is by using rotational potentiometers. A potentiometer is a device that allows to measure a position by measuring a voltage. A linear potentiometer, for instance, like the one sketched in Fig. 7.23a, consists in a resistor, whose resistance $R_1 + R_2$ is known, on which a sliding contact (slider) can move. If the input voltage V_{in} is known, the resistance of either R_1 or R_2 can be obtained by measuring the output voltage V_{out} using the relationship

$$V_{out} = V_{in} \frac{R_2 R_L}{(R_1 + R_2) R_L + R_1 R_2}. \tag{7.107}$$

If R_L is much larger than $R_1 + R_2$, from (7.107) it follows

$$R_2 = (R_1 + R_2) \frac{V_{out}}{V_{in}}. \tag{7.108}$$

The resistor $R_1 + R_2$ can be wire wound, and thus the position of the slider is known in steps, or made by a strip of a conductive polymer. In the latter case the position signal is continuous. A potentiometer can be used also to measure an angle, if

Table 7.4 Correspondence between the reading of a binary or a gray code absolute encoder and the angle. Only angles from 0 to 90° are reported

Angle	Binary	Gray code	Angle	Binary	Gray code
0–5.625°	000000	000000	45°–50.625°	001000	001100
5.625°–11.25°	000001	000001	50.625°–56.25°	001001	001101
11.25°–16.875°	000010	000011	56.25°–61.875°	001010	001111
16.875°–22.5°	000011	000010	61.875°–67.5°	001011	001110
22.5°–28.125°	000100	000110	67.5°–73.125°	001100	001010
28.125°–33.75°	000101	000111	73.125°–78.75°	001101	001011
33.75°–39.375°	000110	000101	78.75°–84.375°	001110	001001
39.375°–45°	000111	000100	84.375°–90°	001111	001000

the resistor is made as an arc of a circle and the slider is mounted at the end of a rotating arm. Multiple turn rotary potentiometers can measure angles larger than 360°.

Potentiometers are a simple way to supply a position feedback in rotary actuators, but their main disadvantage is the presence of a sliding contact, with the consequent wear and possible problems when working in vacuum. They supply an analog signal, that is usually converted into a digital one.

Encoders are rotational position sensors made of an annular portion of a disc with alternate transparent and opaque sectors. A light source and a light detector (a LED and a phototransistor) are located at the two sides of the disc, and detect the passage between them of the opaque and transparent sectors.

An incremental encoder is made of one of such annular track with equal sectors. The smaller are the sectors, the more accurate is the reading of the rotation angle of the disc: typical values are 512 or 1024 sectors, allowing to read increments of the rotation angle by 0.7° or 0.35°. An incremental encoder allows to detect the increment of the rotation angle, but not its actual value, unless the initial position is known (through another sensor) and the signal is integrated in time. Furthermore, an incremental encoder cannot detect a reversal of the motion.

To measure the value of the rotation angle an absolute encoder must be used. An absolute encoder has a number of concentric tracks made of transparent and opaque material (Figs. 7.23b and c) and on each track there is a sensor. The innermost track has just two sectors, one transparent and one opaque. The following one has 4 sectors, the following ones 8, 16, 32, etc. The outermost track of an encoder with n tracks has 2^n sectors, and consequently can discriminate an angle of $360/2^n$ degrees. The encoder of Fig. 7.23b is a binary encoder with 6 tracks and thus its resolution is 5.625°. Starting from the inner track and stating a 0 for the sensor that sees no light and a 1 for the sensor that sees light, the correspondence between the sectors and the angles (up to 90°) is reported in Table 7.4.

The drawback of a binary code is that in several positions more than one digit change simultaneously, a thing that is better avoided. The 'gray code' encoder of Fig. 7.23c is free from this drawback, as seen from Table 7.4.

A further sensor to measure a rotation is a Rotary Variable Differential Transformer (RVDT). A sketch of a Linear Variable Differential Transformer (LVDT) is shown in Fig. 7.23d: it works exactly on the same principle, but measures a linear displacement instead of an angular one.

If the iron core is fully inserted in the transformer, the magnetic flux is maximum, and the output voltage in each of the two secondary coils is the nominal voltage that can be obtained from the ratio of the number of turns. However, if the output coils are connected like in the figure, $V_{out} = 0$ since the two coils produce equal and opposite voltages. If the core is pulled up, the flux in the lower coil decreases and V_{out} increases in the direction of the voltage produced by the upper coil. The maximum voltage is obtained when the core is half extracted from the transformer, so that the lower coil doesn't see any flux. The opposite occurs when the core is moved downwards. The voltage is linear with the displacement of the core.

Resolvers are sensors similar to RVDTs, but are more complex since the driving coil is on the rotating part, while there are two non-rotating coils at 90° from each other. The rotating coil is fed through slip rings or through a rotating transformer. An advantage of resolvers is that the voltage of the two coils is proportional to the sine and the cosine of the rotation angle, so that there is no need of computing the trigonometric functions later, if they are needed.

Hall effect sensors output a voltage that depends on the magnetic field in which the sensor is located, and therefore can be used to detect the position of a permanent magnet or a coil in which a current is flowing. They are often used to measure the position of the rotor of brushless motors.

If a linear displacement has to be measured, linear potentiometers or Linear Variable Differential Transformers (LVDTs) can be used.

To measure a velocity it is possible to differentiate the output of a position sensor, but it must be remembered that numerical differentiation can introduce errors. Conversely, it is possible to integrate the output of an acceleration sensor, and numerical integration usually introduces less errors than differentiation.

Tachometers are instruments that measure directly angular velocities; they are based on several different principles but usually are better suitable to measure speeds that are not too low. Tachometer generators are small generators yielding a voltage that is proportional to the angular velocity: they are quite widespread in many different applications. Magnetic, optical and laser tachometers have all their field of application.

The frequency of the output of an incremental encoder is proportional to the rotational speed and to the number of divisions of the disc. Encoders with a large number of divisions are needed to measure low speeds in this way.

Accelerometers are simple devices that can be miniaturized to the point that a single chip can contain both the transducer and the relevant electronics. Their low cost and reliability allows to use them in a wide variety of applications, including space robotics.

Other sensors that can be used as proprioceptors are force and torque sensors when used to measure the interaction between the various parts of a robot: they too can be built using a variety of different layouts.

Robots have many sensors, whose output can be combined to have a better representation of the outside world or the state of the robot: this practice is often referred to as *sensors fusion*. The fact that all living beings resort to sensor fusion on a large scale and with no apparent difficulty (we do not even realize how much we rely on the output of all senses to recognize a single object or to become aware of a situation) does not mean that this is an easy practice also in the control of machinery. On the contrary, sensor fusion is another active field of research and much work is still to be done before the outputs of different heterogeneous sensors can be put together to obtain a reliable picture of the state of the object to which the various signals refer.

Chapter 8
Power Systems

Providing the required power is an open problem in mobile robotics, and in the case of space robotics things are no better. Experimental robots often receive the energy required for moving and performing their tasks through an umbilical cord, but this is not possible in the case of operational devices. There are some exceptions in the case of low mobility rovers, whose goal is the exploration of a very small area around the landing site that may receive their energy from the lander. While this is possible, even if not very advisable, for short term operations, to operate in this way for a long time poses reliability and mobility problems that cannot be solved.

Another possibility is to power the moving machine through a microwave beam. Energy may be transferred in this way from a fixed power generation plant to a moving machine, but the drawbacks are many. Firstly the technology is not yet fully available. Moreover, energy can be transmitted in this way only in a straight line, so that the range of the moving machine may be extremely reduced in the case of small bodies, where the horizon is close, and above all on rough terrain. Finally, the very intense microwave beam may pose severe problems of electromagnetic compatibility with other communication or electronic devices. Perhaps the only application may be that of working machines for building roads and civil engineering structures or for surface mining, particularly on airless worlds.

Generally speaking, the vehicle or the robot must have on board its own energy source.

When choosing the power source for any particular application two parameters are of paramount importance: its *energy density* (in J/kg, or often in Wh/kg) and its *power density* (in W/kg).

Power sources are usually said to be *power limited* or *energy limited* depending on whether the most severe limitation comes from the power density or the energy density.

The energy density of some energy sources and accumulators is reported in Fig. 8.1, where the performance of nuclear (fission) and chemical energy storage are compared with those of some electrochemical accumulators.

G. Genta, *Introduction to the Mechanics of Space Robots*, Space Technology Library 26, DOI 10.1007/978-94-007-1796-1_8,

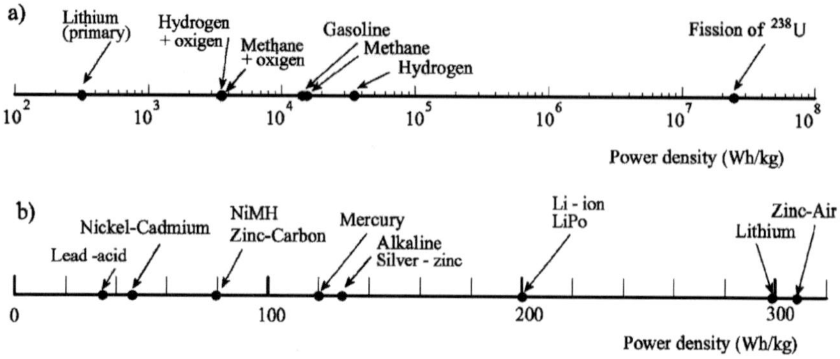

Fig. 8.1 (**a**) Energy density of some energy sources and accumulators in Wh/kg (logarithmic scale). (**b**) Expanded zone where the electrochemical accumulators are located (average values, linear scale)

8.1 Solar Energy

8.1.1 Photovoltaic Generators

One of the most common power sources in space applications are photovoltaic generators. They are the typical power limited source, in the sense that they have a somewhat low power density, but a virtually infinite energy density since they get their energy directly from the Sun, a source almost without any limitation, on the scale we are interested in. Actually they too have an energy limitation since their duration is limited and then the energy they can produce in their useful life is limited.

Since the efficiency of solar cells is rather low, only a small part of the energy from the Sun is converted into electric energy, the remainder being reflected or transformed into heat, which in space must be dissipated by radiating it away. While the efficiency of the early cells was only a few percent, research devices have achieved an efficiency of 42% and the goal is at present to reach an efficiency of 50%. The efficiency of some types of solar cells is summarized in Table 8.1.

The cost of cells varies widely, the advanced types being much more costly than the conventional ones. Generally speaking, the efficiency of cells operating in space is lower than that of cells operating on Earth, an effect that is compensated by the higher energy received by the cells owing to the lack of screening due to the atmosphere.

Solar cells are assembled to form panels, usually flat, extensible appendages of the spacecraft, but sometimes the cells are directly mounted onto its outer surface. In the case of robots working on the surface of a planet, the panels may be mounted flat on the top of the robot's body, like on the rovers used on Mars, or may be mounted in such a way that they can be kept oriented toward the Sun.

In the former case a further reduction of the power produced with the cosine of the angle between the normal to the surface and the direction of the Sun is present.

Table 8.1 Efficiency of various types of photovoltaic cells

Type	Efficiency
Amorphous silicon	6%
Multicrystalline silicon	14–19%
Single junction gallium arsenide	25%
Multi-junction gallium arsenide	40%
Best research cells up-to-date	42.7%

Table 8.2 Values of the solar radiation flux in W/m^2 in space at the distance from the Sun of the planets when at their perihelion and aphelion

Planet	Perihelion	Aphelion
Mercury	14,446	6,272
Venus	2,647	2,576
Earth	1,413	1,321
Mars	715	492
Jupiter	55.8	45.9
Saturn	16.7	13.4
Uranus	4.04	3.39
Neptune	1.54	1.47

The efficiency of the cells decays in time, due to spontaneous decay and to damage from radiation and impacts of micrometeoroids and space debris or, on planetary surfaces, due to dust accumulation. A greater number of cells than that strictly needed must be available, so that enough power is still generated after some of them have been put out of use.

At the average distance of Earth from the Sun (1 Astronomical Unit), the *solar constant*, i.e. the power of the electromagnetic radiation from the Sun, is almost 1.4 kW/m^2; with an efficiency of 18% the power per unit surface of the panel perpendicular to the Sun's direction is thus 250 W/m^2. It is difficult to state the power density of a space photovoltaic generator since its mass depends on many constructional details. Often quoted values span from 80 to 150 W/kg, with the possibility of going beyond 200 W/kg in the near future.

The available power varies with the square of the distance from the Sun. At Mars orbit, then, the power density is about half than at Earth orbit, although being quite variable along the Mars year, since the orbit of that planet is quite elliptical. At the distance from the Sun of Venus or of Mercury (Table 8.2) the power collected by solar cells is much higher. However, getting close to the Sun the solar panels work at higher temperatures and their efficiency decreases.

Another cause of reduction of the power density of solar cells is the presence of an atmosphere that absorbs partially the energy from the Sun. This may be due to the atmospheric gases or to the dust carried by the atmosphere. In particular, in the case of Mars, accumulation of dust on the panels reduces the power output of solar generators. With some luck, the panels may be cleaned from time to time by wind, as happened with the *Spirit* and *Opportunity* rovers.

The solar panels of the MERs have a total deployed area of 1.3 m^2 and were made of three-layer cells: gallium indium phosphorus, gallium arsenide and germanium, whose effective efficiency is between 23 and 25%. At the beginning of the mission the panels could produce about 900 Wh during one Martian sol (24 h, 39 min, 35 s), used to keep two lithium-ion batteries located inside the warm electronics box charged. After 90 sols of work the energy produced dropped to 600 Wh per sol, due to both season change and performance degradation.

The use of solar cells beyond the orbit of Mars is questionable, and certainly other power sources are needed at these distances from the Sun.

While large solar panels can be used in space, and even on the surface of planets in the case of fixed installation, to use large panels on a moving vehicle or robot is questionable, if not in the case of very slow machines requiring little energy and moving with limited accelerations.

To reduce the size of the photovoltaic panels it is possible to use mirrors to concentrate sunlight on smaller panels. This practice can be particularly expedient at distances from the Sun larger than that of Earth, but one of the main advantages of solar photovoltaic generation, namely its simplicity, is lost. Two surfaces, a concentrator plus a photovoltaic array are needed, although the latter is smaller than that of a non-concentrating system. The two surfaces must be controlled in such a way that the second remains in the focus of the first, while it tracks the sun in its motion. Particularly in the case of space applications, it seems that if a concentrating plant is used, it is more convenient to use a thermal generator than a photovoltaic one.

Example 8.1 A 2 m^2 solar panel having an efficiency of 22% lies horizontally on the Moon at the equator. Compute the maximum power supplied by the panel, the average power and the total energy supplied during a lunar day in which the sun passes at the zenith at noon.

The solar constant at the average distance of the Earth–Moon system from the Sun is $W_s = 1{,}367$ W/m^2. The maximum power produced by the panel is 602 W.

The lunar day is 29 days, 12 h, 44 min, 3 s, i.e. 2,551,443 s long. If θ is the angle between the direction of the Sun and the horizontal, the power that the panel produces at a generic instant is

$$W = \eta S W_s \sin(\theta) = \eta S W_s \sin(\Omega_m t),$$

where S is the surface of the panel and Ω_m is the angular velocity of the apparent motion of the Sun on the lunar surface.

The total energy produced in a day is

$$E = \int_0^{T/2} W\,dt = \frac{\eta S W_s}{\Omega_m} \int_0^{\pi} \sin(\theta)\,d\theta,$$

where $T = 2\pi/\Omega_m$ is the duration of the day.

Performing the integration, it follows that

$$E = \frac{\eta S W_s T}{\pi} = 4.885 \times 10^8 \text{ J} = 135.7 \text{ kWh}.$$

The average power during the daylight hours is thus $\overline{W} = 383$ W.

Note that this computation is only a first approximation evaluation since the efficiency of the solar cells cannot be considered a constant with such a large variation of illumination.

8.1.2 Solar-Thermal Generators

While in the case of small plants photovoltaic arrays have several advantages, in large plants solar thermal system may be more expedient. In solar-thermal systems the energy is produced by a thermal engine, for instance a steam turbine, operated by the heat from the sun. Mirrors are used to concentrate the light of the Sun onto a boiler, so that high enough temperatures can be reached. The advantages are those of a higher efficiency, which is, however, always limited by thermodynamic considerations, and the possibility of building large and lightweight mirrors in space.

As will be seen when dealing with nuclear reactors, in space it is not easy to use thermal energy in an efficient way, and this strongly limits the possibility of using solar thermal devices for power generation is space. A similar consideration holds also for the Moon and even Mars.

The mirror must anyway be controlled so that the boiler remains always in its focus and the complexity of the solar thermal systems is larger than that of photovoltaic systems. As a consequence, there is little advantage in using such a more complex system over panels of solar cells, at least for small and medium size spacecraft, while it may be so in case of large solar power stations on the surface of the Moon or of a planet. The optimum configuration depends upon the application and the power that has to be generated.

8.2 Nuclear Power

At present, there are two possible alternatives for nuclear power generation in space: radioisotope generators and fission nuclear reactors.

Extensive use of nuclear devices, both for propulsion and power generation on spacecraft and in planetary outposts and bases is essential for space exploration. Even on the Moon surface, where there is plenty of energy from the Sun and there is no atmosphere reducing the efficiency of solar energy system, only nuclear systems can guarantee survival through the long and cold nights, particularly for systems that cannot be hibernated but must remain operational for all the time.

In the past many concerns about safety of nuclear devices in space were forwarded. The fears of using radioisotope thermoelectric generators (RTGs) have been greatly exaggerated, as is demonstrated by the few accidents involving nuclear powered spacecraft to date. The first occurred when the *Transit 5B-N3* satellite failed to attain its orbit in 1964. At that time the generators were designed to disintegrate in the high atmosphere, and the SNAP-9A RTG did so. No measurable excess radioactivity was found. The second accident occurred when the launch of the *Nimbus B1* satellite failed in 1968. The SNAP 19 generator was this time designed to remain intact and was recovered after 5 months in the ocean without any failure. Finally, when the lunar module *Aquarius* of the ill-fated *Apollo 13* mission disintegrated in the Earth's atmosphere, its RTG went down intact into the ocean without any measurable radioactive contamination being found. The same happened when the launch of the Mars 96 probe failed. The worst accident, and the only one in which there was contamination, occurred when the Russian *Cosmos 954* disintegrated in the atmosphere over an unpopulated region of northern Canada. Its large nuclear reactor (not an RTG, but a fully fledged reactor) disintegrated, and various radioactive fragments were found on the ground. The decontamination operation costing about 8 million dollars was paid for by the Soviet government. Subsequent Cosmos satellites had the provision for jettisoning the reactor, which was put in a safe higher orbit, before re-entry.

At any rate, these concerns caused a decrease of the funding for research in nuclear energy utilization in space: only thirty years ago it was taken for granted that by the end of the twentieth century large space stations powered by nuclear reactors could be built. The United States built several reactors of the SNAP (System Nuclear Auxiliary Power) class, but at present the largest space nuclear reactors are the Russian Topaz. A revival of the research in nuclear generator for planetary outposts and bases is a real need: only an extensive use of nuclear power can allow us to develop a spacefaring civilization.

8.2.1 Fission Reactors

As shown in Fig. 8.1, fission nuclear power has the highest energy density of all the energy source presently available.[1] The power density does not depend on the source in itself, i.e. on the nuclear fuel, but on the mass of the reactor and of the power conversion device.

The reactor generates thermal energy that has to be converted into a form that can be used to perform the required work, usually electric energy. In space efficient energy convention is difficult, mainly because the efficiency depends on the temperature the heat is dissipated away from the conversion plant. Power plant operating on Earth use large quantities of coolant (usually water) to keep this temperature low, but in space no coolant is available.

[1]When nuclear fusion reactions will be controlled an even more powerful source will be available and this will have dramatic consequences on all aspects of space exploration.

Heat must be exhausted by radiating it into space, but the quantity of heat that can be radiated away from a given surface depends on the 4th power of the temperature, a thing that compels to either use large radiators or to exhaust heat at fairly high temperatures. The value of the temperature that maximizes the performance of the plant in terms of power density is much higher than the temperatures used in power plants on Earth, leading to a much lower efficiency.

Technological advances in high temperature materials allowing to raise the highest temperature of the thermodynamic cycle may improve the situation, but generation of large quantities of energy in space, even with nuclear power plants, requires large and heavy power stations.

The situation on the surface of the Moon or of planets like Mars is surely better, but not much, since there is at any rate no cooling fluid. If the permafrost on Mars can be used as a heat sink, at the same time obtaining liquid water, the situation will certainly improve, but the technology is still to be developed.

At any rate, nuclear reactors may be used to power an outpost and to recharge the batteries or to produce fuel for vehicles and robots, but it is unlikely they will directly power them. The technology to build small and compact nuclear reactors is still to be developed, even if they are not inconceivable, and surely they will not be available in a foreseeable future.

8.2.2 Radioisotope Generators

When a compact and above all long lasting energy source is needed in space, radioisotope thermoelectric generators (RTGs) are a good alternative, based on well consolidated technology. They are small capsules containing a radioactive material such as plutonium-238, surrounded by a number of thermoelectric generators and then, on the outside, by a radiator. Owing to radioactive decay, the radioisotope reaches a temperature higher than that of the radiator and this difference of temperature makes a current to flow in the thermoelectric material. The efficiency of such devices is low, but they are compact, reliable and long lasting.

In the past RTGs were a very convenient—and necessary—power source for space probes exploring the outer solar system, like the Cassini probe. The generators on-board the Voyager probes allowed them to send to Earth information from beyond the orbit of Pluto after almost 40 years in space.

The drawbacks of Radioisotope generators are being weakly radioactive (they cannot be used in close proximity of a human crew or need shielding) and having quite a low power density. So they are more suited for robots operating in space than robots on a planetary surface and need to be shielded when used in people-carrying vehicles.

The radioisotopes to be used in generators must release much energy as radiation that can be easily absorbed and transformed into heat. In this respect alpha decay is much better than beta or gamma decay. It requires also a lighter shielding. The radioactive material also produces a significant amount of neutrons. Its half-life should be long enough to produce a substantial amount of energy for the whole duration of the mission. Half-life must be chosen carefully, because radionuclides with a long half-life have a low energy release.

The best candidates are plutonium-238, curium-242 and -244, americium-241, strontium-90 and polonium-210, but there are other nuclides that could possibly be used. Plutonium-238 has the lowest shielding requirements (less than 2.5 mm of lead) and a long half-life (87.7 years). In many instances the casing itself supplies enough shielding and no specifically designed shield is required. It is usually employed in the form of plutonium oxide, PuO_2.

All RTGs used in space were fueled by plutonium-238, while some units built for Earth applications used strontium-90 that has a shorter half-life, lower power density and much higher gamma radiation but has a lower cost. Plutonium is a strategically sensitive material and the scarcity of the supplies of plutonium may soon become a problem, making it difficult to power probes for deep space exploration.

For short durations polonium-210 has a very high power density, but an half-life of only 138 days: it has been used in some prototype RTGs.

Curium-242 and -244 produce gamma radiation and neutrons and thus require heavy shielding.

Americium-241 is a potential candidate isotope with a half-life of 432 years, longer than that of plutonium-238, but its power density is about 1/4 of that of plutonium and requires a heavier shielding: 18 mm of lead. The last figure is not so bad, since the americium is second only to plutonium in these respects. Its main advantage is availability, since it is widely used in smoke detectors, and may be an answer to the difficulty of procurement of plutonium. It is considered by ESA in case they decide to build RTGs.

Radioisotopes produce essentially heat, which must be converted in some form of usable energy, mainly electric energy. The alternatives studied in the past are essentially four in number: thermoelectric, thermionic and thermo-photovoltaic direct conversion and the use of dynamic generators using some sort of thermal engine.

The radioisotope generators used in space in the past are all thermoelectric generators, mostly owing to the simplicity and reliability of this architecture. However, the conversion efficiency is very low, usually between 3 and 7% in the various applications. The thermoelectric materials used include silicon germanium alloys, lead Telluride and Tellurides of antimony, germanium and silver.

Not only thermoelectric conversion has a low efficiency, but also thermoelectric material degrade in time and their efficiency decreases. For instance, the total power generated by the 3 RTGs of the *Voyager 1* and *2* probes in 2001 was reduced to 315 and 319 W from the original 480 W. Since the decay of the radioisotope in the 23 years of the mission accounts for a decrease of 16.6%, a decay of about 20% of the performance of the converter must be postulated.

A list of some plutonium-238 fueled RTGs used by NASA (28 US space missions using one or more RTGs were flown since 1961) is reported in Table 8.3.

The efficiency of thermionic converters is better than that of thermoelectric converters, being of the order of 10 to 20%. However, such devices require higher temperatures than those achievable in usual radioisotope generators. Generators fueled

Table 8.3 Main characteristics (electric power P_{el}, thermal power P_{th}, fuel mass m_f and total mass m) of some plutonium-238 fueled RTGs used by NASA

Model	Spacecraft	P_{el} (w)	P_{th} (w)	m_f (kg)	m (kg)
GPHS-RTG	Cassini, New Horizons, Galileo, Ulysses	300	4,400	9	60
MHW-RTG	LES-8 and 9, Voyager 1 and 2	160	2,400	4.8	37.7
SNAP-3B	Transit-4A	2.7	52.5	0.12	2.1
SNAP-9A	Transit 5BN1 and 2	25	525	1.23	12.3
SNAP-19	Nimbus-3, Pioneer 10, Pioneer 11	40.3	525	1.23	13.6
SNAP-19 mod.	Viking 1 and 2	42.7	525	1.23	15.2
SNAP-27	Apollo 12 to 17 ALSEP	73	1,480	3.8	20

by polonium-210 and some nuclear reactors designed for space applications had thermionic converters.

Thermo-photovoltaic converter are based on photovoltaic cells working in the infrared radiation generated by an hot body, namely the radioisotope capsule. Their efficiency is higher than that of thermoelectric converters: 20% efficiency have been demonstrated and 30% is a target. Combined converters, in which a first thermo-photovoltaic stage is followed by a thermoelectric second stage, allow a further increase in the efficiency.

A much higher efficiency, theoretically above 40%, can be obtained by dynamic energy conversion. Theoretically any thermal engine may be used, like a steam turbine or reciprocating engine working with the steam produced by the radioisotope or a hot air engine. One of the best choices, for what the efficiency is concerned, is a Stirling free piston engine connected with a linear electric generator. A prototype of a Stirling Radioisotope Generator (SRG) of this kind was developed by NASA and DOE: it demonstrated an efficiency of 23%, i.e. more than 3 times that of a conventional RTG. It is fueled by plutonium-238 and produces about 116 W electrical (about 500 W thermal) with 1 kg of plutonium and has a mass of about 34 kg. Its specific power at the beginning of the mission is 3.4 W/kg. The high and low temperatures of the thermodynamic cycle are 650 and 120°C, respectively. Even if for now SRGs have a specific power comparable with that of RTGs, they produce much less heat, requiring a smaller radiator and being less bulky.

The greater efficiency, with the subsequent lower mass and bulk for a given power output, is obtained at the cost of increased complexity of the system. However, some tests did show that a good reliability and a long operating life can be achieved even in the presence of moving parts. Also dynamic problems can be satisfactorily solved: SRGs, or other dynamic radioisotope generators, are thus a viable choice, particularly for the larger units.

Greater efficiency can be achieved by increasing the temperature ratio between the hot and cold ends of the generator. From this viewpoint, applications on the surface of a body with an atmosphere or in general applications where a cooling fluid is available, particularly if such fluid is cold, may have a higher efficiency than applications in space.

8.2.3 Radioisotope Heating Units (RHUs)

Similar to RTGs, but even less dangerous since they are far smaller, are Radioisotope Heat Generators (RHG) or Radioisotope Heating Units (RHU). These are tiny radioisotope capsules, usually of plutonium-238 contained in a platinum–rhodium alloy cladding, heating some crucial parts of a spacecraft that would otherwise have a too low temperature for correct operation. The mass of the radioisotope is just a few grams, while the total mass of the capsule may be about 40 g.

Radioisotope heaters make thermal control of spacecraft easier, since they allow to dispense with many electric heater, reducing the power requirements. They are particularly useful for probes traveling far from the Sun and rovers traveling on Mars, and even more in colder places like Titan. They are also essential on the lunar surface, if the device must remain operational during the long and cold lunar night, or even on the surface of planets like Mercury, where in spite of the very hot days, nights are extremely cold.

The *Cassini-Huygens* spacecraft at Saturn contains 82 RHUs (in addition to three main RTGs for power generation). RHUs and some electric heaters are usually located in a Warm Electronic Box (WEB) where temperature sensitive electronics and other components are located so that their temperature is controlled.

8.3 Chemical Power (Combustion)

Chemical energy stored in a fuel-oxidant combination has a high energy density and a power density that may be extremely high, depending on specific power of the conversion device. At any rate, the energy density is lower in space or on a planet with non oxidizing atmosphere than on Earth, since both fuel and oxidizer must be carried on board. If the device must work for a time that is not very short, provisions for refueling must be considered.

The combustion reaction commonly occurs between hydrogen and oxygen or between carbon and oxygen. The most energetic reaction is the one involving hydrogen, which can be stored in the form of molecular hydrogen or as more complex molecules containing both hydrogen and carbon (hydrocarbons) and often other elements. For storage, it is more expedient to store the fuel in liquid than in gas form, a thing that decreases the size and mass of the tanks.

Hydrogen in gas form has a very low density (0.0899 kg/m^3 at ambient Earth temperature and pressure) and to store it efficiently it must be kept at high pressure in suitable tanks. For instance, in some automotive applications, pressurized bottles maintaining a pressure 700 times atmospheric pressure have been used. In this condition, however, the mass of the tank is much higher than the mass of the gas it contains. At present, the target is to build tanks containing a mass of hydrogen equal to 6.5% the mass of the full tank, and even these estimates may be overoptimistic. The ratio between the hydrogen mass and the total volume of the tank is of the order of 70.6 kg/m^3.

In liquid form, hydrogen has a boil-off temperature of 20.6 K at atmospheric pressure. In these conditions its density is only 71 kg/m^3, 14 times lower than water density. A cryogenic tank is required for most applications and the boil-off rate must be considered when accounting for energy losses. Active boil-off control is a possibility, but even this requires energy.

Remark 8.1 Storage in pressurized tanks yields an apparent density that is close to that of liquid hydrogen, with no boil-off problems, even if the mass energy density is much lower when the mass of the tank is accounted for.

Another alternative is to store hydrogen in form of metal hydrides, like MgH_2, $NaAlH_4$, $LiNH_2$, $NaBH_4$, etc. They are either liquid or solid and have a good energy density by volume but their energy density by mass is often lower than that of hydrocarbons. Hydrogen may be tightly linked in the hydrides, requiring a non-negligible energy to free it.

Methane is a gas on Earth surface; its density is 0.717 kg/m^3 at 0°C. Its boiling point at atmospheric pressure is 112 K (−162°C) and its density in liquid form is 415 kg/m^3. It is much easier to contain and store than hydrogen.

The next hydrocarbons, like ethane, propane and butane are still in gas form in Earth conditions and, except for their higher density and lower energy density, are similar to methane.

Heavier hydrocarbons are liquid. Usual liquid fuels are mixtures of different liquid hydrocarbons, with densities around 650–750 kg/m^3. They are easily contained in lightweight tanks, even if at the higher temperature experienced on the lunar surface they may have some boil-off.

If fuels based on hydrogen and oxygen are used in a place where the oxidant is not readily available, the simplest solution is carrying oxygen on board. The problem here is not dissimilar to the problem posed by hydrogen, since it is a cryogenic liquid (the boiling temperature at atmospheric pressure is 90.18 K or −183°C), although its density is much larger (1,141 kg/m^3), thus requiring a much smaller tank.

To avoid the need of using cryogenic oxidizers, it is possible to use nitric acid (HNO_3) or hydrogen peroxide (H_2O_2) or other oxidizers. However, while there is a good experience in using suck dangerous liquids as rocket propellants, they have seldom (or never) been used in thermal engines.

In case of short missions, like the *Gemini*, *Apollo* and *Space Shuttle* missions, which use fuel cells for electric power generation, the hydrogen–oxygen fuel–oxidant combination is carried directly from Earth, stored on board for the whole mission. For long missions chemical energy can be used as an intermediate energy storage: vehicles to be used on the Moon or other bodies may work on chemical energy, being refueled at an outpost where fuel is produced using energy from solar or nuclear power plant.

To power vehicles or robots there are two possibilities: using fuel cells that convert directly the chemical energy into electric energy or using some sort of thermal engine to obtain mechanical power.

8.3.1 Thermal Engines

Vehicles for planetary exploration may use more or less standard internal combustion engines, running on different fuel-oxidizer pairs, produced on site. Internal combustion engines have a somewhat low efficiency, in the range of 15 to 40%, but can have a high power density and benefit from a well established technology. The adaptations to work on hydrogen, methane or methanol are straightforward, and the relevant technology is available. Internal combustion engines come in sizes from less than 1 kW to several hundred kW and are cheap and reliable.

The drawbacks that may eventually lead to quit their use on Earth are less severe on other planets: since they will work in a closed cycle, recovering their exhaust to produce new fuel, pollution is not a problem and their low efficiency may not be a major problem if the fuel is produced using electricity from nuclear reactors, with the consequent availability of large quantities of energy.

This option has been discussed in detail in Sect. 7.3.3.

8.3.2 Fuel Cells

Fuel cells have a long history of space application, since they were developed for the *Gemini* missions.

In fuel cells the reaction between the fuel and the oxidizer is not a combustion process producing heat that is later converted into mechanical or electric energy, but an electrochemical reaction, similar to that occurring in batteries, producing directly electric energy. For this reason, the efficiency of fuel cells can be much higher than that of devices based on thermal engines.

The basic reaction occurring in fuel cells is that between hydrogen, which is separated into positive hydrogen ions and electrons at the anode, thanks to the presence of a catalyst, and oxygen, which is ionized negatively at the cathode and then migrates through an electrolyte separating the electrodes.

The catalyst, the electrolyte and the membrane separating the electrode may be of different types, and consequently different types of fuel calls, each one with its peculiar advantages and drawbacks for the different applications, exist.

- Alkaline fuel cells (AFC). Use a liquid, corrosive, electrolyte and must be fueled by pure hydrogen and oxygen, since impurities in the fuel poison the cell. Their efficiency is about 50%, or somewhat higher. They are used in space applications since when they were developed for the *Gemini* missions, their building and operating cost is fairly low and they do not require complex ancillary equipment, but are somewhat bulky.
- Proton exchange membrane fuel cells (PEMFC). Use a polymer electrolyte and require pure hydrogen as fuel. Contaminants like sulfur compounds and carbon monoxide poison the cell. Owing to their compact design and high energy density they are suited for automotive or robotic use, but require complex and costly

equipment, like compressors and pumps, that use about 30% of the energy produced. That notwithstanding their efficiency is around 30%. They operate at low temperature, about 80°C.

- Molten carbonate fuel cells (MCFC). Use an electrolyte composed of a molten carbonate salt mixture suspended in a porous, chemically inert ceramic matrix. They are tolerant of the impurities in the fuel and can run on carbon monoxide. Thus they accept different hydrocarbons, like natural gas, that can be converted to hydrogen and carbon oxides or gases made from coal. They operate at high temperatures (650°C), which reduce their useful life. The efficiency is about 60%, but can be increased up to 85% if the waste heat is reused.
- Phosphoric acid fuel cells (PAFC). Use liquid phosphoric acid as electrolyte. They are not affected by carbon monoxide impurities in the fuel. Their operating temperature is 150 to 200°C. Their efficiency is low (37 to 42%), but can be increased if the waste heat is reused. They have a limited service life and use a costly catalyst.
- Solid oxide fuel cells (SOFC). Use a solid oxide material as electrolyte. They are not affected by poisoning from carbon monoxide and do not need high-cost, platinum-based, catalyst, but are affected by poisoning due to sulfur impurities. The operating temperature is quite high, from 500 to 1,000°C. Owing to the high temperature, they can use methane, or butane or even liquid fuels that are externally reformed. Their efficiency can reach 60%, and can be used for cogeneration of electric power and heat.
- Direct methanol fuel cells (DMFC). They are similar to PEMFC, but use directly methanol as a fuel. Their operating temperature is in the range of 50–120°C, but their efficiency if low, about 20%.

If oxygen–hydrogen fuel cells are used, the reaction product is water, which can be stored and carried back to the outpost, where is again converted into oxygen and hydrogen by an electrolyzer. This combination of fuel cell and electrolyzer is usually referred to as a regenerative fuel cell, and in practice works as a rechargeable battery. No material is consumed (except for some losses) and the system needs only energy.

When also methane or other hydrocarbons are used, also carbon dioxide is produced, together with water.

On Mars, oxygen and methane can be produced from the atmospheric carbon dioxide, using energy and some hydrogen from water from the permafrost of the planet. The carbon dioxide can then be exhausted to the atmosphere and the water can be recovered.

Hydrogen–oxygen alkaline fuel cells for space use are a mature technology and need no specific research. Much research is at present devoted to fuel cells for vehicular application, both for reducing their cost and for using different types of fuel. The choice of the fuel is quite limited: an interesting alternative to hydrogen is methane, which is much easier to store. If the lower energy density is not a problem, methanol or formic acid can be used as liquid fuel. The oxidizer is usually at any rate oxygen. The alternative of storing on board methane or methanol and then dissociate it chemically to produce the hydrogen to be introduced into the cell has the disadvantage of causing the poisoning of most common types of cells, if impurities caused by the chemical process to obtain the hydrogen remain in the fuel.

Some applications on vehicles for planetary exploration may be more similar to automotive applications, on which much research is being conducted, than to space applications. Problems like reliability, working in conditions with quickly varying power request, reduced maintenance and mechanical stress due to traveling on uneven ground are similar, but the requirement of low cost that makes everything more difficult in vehicular applications is much less severe.

8.4 Electrochemical Batteries

8.4.1 Primary Batteries

The electrochemical batteries that have the highest energy density are primary, i.e. non rechargeable, batteries. Even the common alkaline batteries have an energy density of 130 Wh/kg, while lithium batteries reach 300 Wh/kg and zinc-air ones 310. The main overall characteristics of primary batteries are listed in Table 8.4. The values of the energy density (referred to the mass and volume) are just orders of magnitude, since they depend on the exact type and make of the battery and also on the working conditions. The cell voltage reported is that of the fully charged battery while not generating current.

Primary batteries can, however, be used only for very short duration or for particular applications. Since some of them have a limited self-discharge, it is possible to use primary batteries for devices that must remain in standby for long times (even tens of years) using almost no power and then send a signal when an event takes place.

Among primary cells it is possible to mention

- Zinc–carbon cells. They are the oldest and cheapest type of primary batteries, however, now they are obsolete and find little application in the aerospace field.
- Alkaline cells. They are not much different from zinc–carbon cells, since are based on a zinc anode and a manganese dioxide cathode with a potassium hydroxide electrolyte. Their good performance and low cost make them the most widely used type of general purpose non-rechargeable battery.
- Mercury batteries. They were used in the past, but their potential pollution hazard, due to the mercury content, made them to be banned.
- Silver–zinc batteries. They are among the most used in the aerospace field and contributed substantially to the *Apollo* missions: they were installed on the *Saturn* rocket, the command module, the LEM, the LRV and, after the *Apollo 13* accident, also on the service module. The chemical reaction powering the cell is

$$Zn + Ag_2O = ZnO + 2Ag.$$

Silver is reduced at the cathode and the reaction occurs in a potassium hydroxide or sodium hydroxide electrolyte. Until 2004 they contained a small quantity of

Table 8.4 Main characteristics of primary batteries (e/m: mass energy, e/v: volume energy density)

Type	e/m (Wh/kg)	e/v (Wh/dm^3)	Cell voltage (V)
Zinc–carbon	75	100	1.5
Mercury	120	–	1.35
Alkaline	130	320	1.5
Silver–zinc	130	500	1.8
Lithium	280–350	300–700	2.8–3.8
Zinc–air	310	1000	1.4

mercury (about 0.2%) to prevent corrosion of the anode, but the most modern types are free from this pollutant.

Their energy density is similar to that of alkaline cells, but their discharge curve is flatter. They are currently available in many sizes, mainly small button type cells, but larger sizes are available.

- Lithium cells. The general term lithium batteries indicates a wide family of different types of batteries, working on different chemistries. They have generally high energy densities and high cell voltage, but their performance and cost vary. The most common on the market are those based on a metallic lithium anode and manganese dioxide cathode. The mass energy density is about 280 Wh/kg, the volume energy density is 580 Wh/dm^3, the nominal voltage is 3 V and the open circuit voltage is 3.3 V. They are suitable for low-drain, long-life, low-cost applications. They have also a wide temperature range.

  On the other side, lithium thionyl chloride (the thionyl chlorate constitutes the liquid cathode) are much more costly, difficult or dangerous to operate but have extremely high performance that can reach even 500 Wh/kg, which is the highest energy density for any battery type. The high energy density and good low-temperature characteristics make them suitable for some space applications, even if the types with higher energy density supply lower discharge currents. Also lithium–carbon monofluoride cells are used in aerospace applications.

  In general, lithium batteries may be discharged very quickly producing large currents. This can be exploited in some cases, but can also constitute a danger when accidentally shorted, since the ensuing overheating may lead to explosion of the cell.
- Zinc–air batteries. They work by oxidizing zinc with oxygen from the air; as such they are similar to fuel cells. Since they depend on air, they are not usable in space except if air (or directly oxygen) is carried on board. They have been used on Earth to power electric vehicles and theoretically may be used to power rovers and robots.

Each cell can be modeled, from the electrical viewpoint, as a circuit made by an ideal voltage generator with a resistor, modeling the internal resistance of the cell, in parallel, even if more complex models in which also capacitors and inductors are included, can be found in the literature. The value of the internal resistance of the cell varies with many parameters, including the state of charge, the current, the temperature, etc.

During the discharge the voltage at the terminals of the cells decreases. The plot of the voltage as a function of time is referred to as the discharge characteristics of the cell. Initially there is a sharp drop from the maximum voltage, typical of the fully charged state, to a lower value that is maintained, with a slight decrease, for most of the discharge time. When the discharged conditions are approached there is a sharp drop again. The discharge curve is much influenced by how fast the discharge is: if the current is large the voltage decrease in the intermediate phase may be larger, depending on the battery type.

8.4.2 Secondary (Rechargeable) Batteries

Battery operated vehicles and robots that must be used for anything but a short time must use rechargeable (secondary) batteries that can be recharged either from the power system of the robot or rover itself (e.g. solar cells) when the device is not used or when it requires less power than the primary power source can supply. The batteries can also be recharged from a fixed power plant, located on the lander or at an outpost.

Conceptually, secondary battery are similar to primary batteries but for the fact that the chemical reaction is reversible and can be run backwards by passing a current through the cell. However, this reversibility is never complete, and the battery cannot be recharged an infinite number of times: at every recharge the performance of the energy conversion somewhat deteriorates until the cell cannot be recharged any more.

The performance of all batteries depend on many factors and, above all, its capacity is affected by how fast the charge and discharge process is performed.

The latter effect is expressed by the Peukert's Law, introduced by W. Peukert in 1897 for lead–acid batteries,

$$C = i^k t, \tag{8.1}$$

where C is the capacity of the cell at a one-ampère discharge rate, i is the discharge current, k is the nondimensional Peukert constant and t is the time of discharge. For lead–acid batteries, the value of k is about 1.3, with lower values for gel batteries and higher ones for liquid electrolyte cells.

Given the nominal capacity of a battery, a charge is said to be performed at a rate C if the charging current in A is equal to the capacity expressed in Ah (Ampère hours). If the charging efficiency had a unit value, this would imply that the charge phase would last 1 hour; in practice for most batteries the charging time at a rate C is about 1.2 hours. The nominal capacity of a battery is usually determined by discharging in 20 hours ($C/20$) at a temperature of 20°C.

Usually slow charging is defined as charging at a rate lower than C. For instance, accounting for the charging efficiency, a slow charge rate $C/3$ leads to a charging time of about 4 hours. Most batteries can withstand indefinitely a very low charge rate (usually said trickle charging), below $C/10$. Fast charging, i.e. charging in less than one hour (rate greater than $1.2C$) requires precautions and may spoil some types of batteries. The more advanced types of batteries are able to be charged quickly, even with rates greater than $10C$.

In a similar way, also fast discharging may be detrimental to the life and performance of the battery. Even in case of batteries able to be charged quickly, the capacity and the energetic efficiency decrease with increasing charge and discharge current.

The energy density of the battery of a robot or a rover thus depends on whether recharging is performed between one mission and the other or the rover has a low power generator (a solar panel, for instance) and the batteries are continuously kept charged and used when the required power exceeds the power generated by the primary source or when for some reason the primary source is off (e.g. the solar panel is in the shade).

Generally speaking, batteries cannot be used at the same time at high power density and at high energy density: as already stated, when required to supply high power (discharge with high current) the efficiency, and consequently the capacity, decreases. They show also a reduction of the useful life when used in these conditions.

Lead–acid batteries are particularly sensitive to this, while some kinds of nickel–cadmium and other more advanced batteries can operate with high currents, both during charging (quick charge) and during discharge (high power output).

An ideal battery should be characterized by

- High energy density,
- Almost constant voltage during discharge (a flat discharge characteristics),
- Low internal resistance,
- High discharge current,
- Possibility of operating at both high and low temperatures,
- Long operating life and high number of charge–discharge cycles,
- High efficiency in recharge,
- Low cost.

No actual battery is particularly good in several of these points.

The voltage of a secondary battery decreases during discharge in a way that is similar to that of primary cells. Since the life of the cell is not over once it is discharged, the third phase of the discharge curve must not be used: too deep discharges are detrimental to the possibility of fully recharging the battery and can, in the long run, deteriorate it. The discharge curves of some secondary batteries are reported in Fig. 8.2.

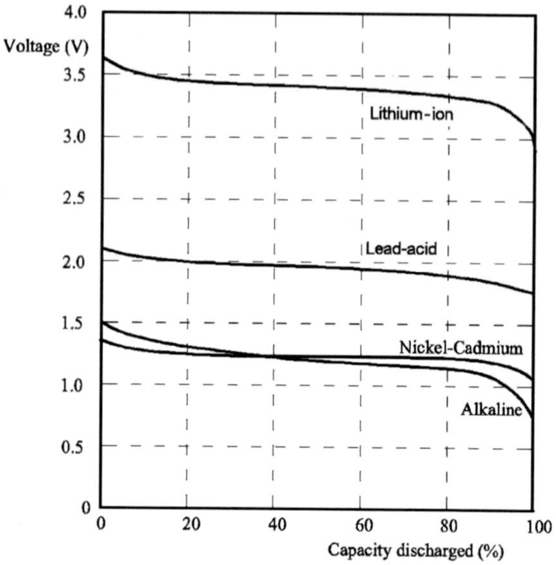

Fig. 8.2 Discharge curves of some rechargeable batteries (slow discharge)

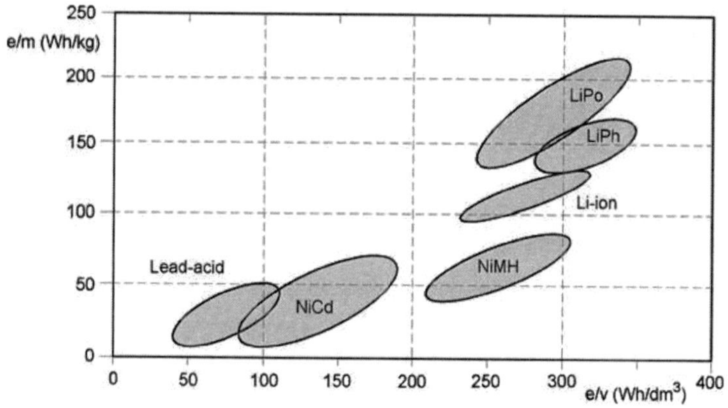

Fig. 8.3 Mass and volume energy density for the main types of secondary batteries

The main characteristics of some types of secondary batteries are reported in Table 8.5; a plot of the mass energy density versus the volume energy density is reported in Fig. 8.3.

The main types of secondary cells are:

- Lead–acid cells. They are the most common type of secondary batteries. While in origin they were based on an open container full of electrolyte (sulfuric acid), more modern type are sealed and provided with a valve to prevent pressure built up. The electrolyte can be semi-solid (gel) or can be absorbed in a special fiber-glass matting.

Table 8.5 Main characteristics of some types of secondary batteries (e/m: mass energy density, e/v volume energy density, P/m: power density, V: cell voltage, η: charge/discharge efficiency, d: self-discharge, c: number of cycles)

Type	e/m (Wh/kg)	e/v (Wh/dm^3)	P/m (W/kg)	V (V)	η (%)	d (%/month)	c
Lead–acid	30–40	60–70	180	2.0	70–92	3–4	500–800
NiCd	40–60	50–150	150	1.2	70–90	20	1,500
NiFe	50	–	100	1.2	65	20–40	–
NiZn	60	170	900	1.2	–	–	100–500
NiMh	30–80	140–300	250–1,000	1.2	66	30	500–1,000
Alkaline	85	250	50	1.5	99	<0.3	100–1,000
Li-ion	150–250	250–360	1,800	3.6	80–90	5–10	1,200
LiPo	130–200	300	3,000	3.7	–	2.8–5	500–1,000
LiPh	80–12	170	1,400	3.25	–	0.7–3	2,000
LiS	400	350	–	–	–	–	100

- Nickel-based cells. A wide family of rechargeable batteries are based on nickel chemistry, like are nickel–cadmium (NiCd),[2] nickel–iron (NiFe), nickel–zinc (NiZn), nickel metal hydride (NiMh).
- Rechargeable alkaline cells.
- Lithium-based cells. They include lithium–ion (Li–ion), lithium ion–polymer (LiPo), lithium–iron–phosphate (LiPh) and lithium–sulfur (LiS) cells.

As a general consideration, nickel-based batteries have a larger self-discharge, while lithium batteries are better from the viewpoint of quick charging and discharging. The highest energy density is obtained from lithium–sulfur cells, which, however, have a short life in terms of cycles.

The charge phase of any battery system is critical, since the efficiency and the life of a battery depends on how accurately the energy is introduced into the system Some batteries are less critical from this viewpoint, like lead–acid and NiCd cells, although the latter display what is usually referred to as a memory effect, consisting in the tendency to lose some of the capacity when recharged repeatedly after being only partially discharged.

Advanced batteries are more critical, and may even become dangerous if not properly charged, with the risk of fire and explosions. This is solved by using accurately controlled, microprocessor-based, chargers. Battery packs are increasingly provided with on-board electronics that monitors continuously the charge conditions and keeps the current flowing through the various cells under control.

Since the voltage varies during the discharge phase according to the charge state of the battery, robots and vehicles powered by batteries are usually provided with a

[2] Actually NiCd is a proprietory name and should not be used to indicate nickel–cadmium cells in general.

voltage regulator. Other problems are due to the presence of both logical elements and power components close by and possibly connected to the same circuit. The latter can produce noise and electromagnetic interference that must be kept under control to ensure proper operation of the system. The electric and electronic circuitry of the robot must be designed with extreme care, to ensure a good power efficiency and at the same time a low level of noise.

Finally, when the batteries have to supply (or to receive) high power peaks for a short time it might be expedient to supplement the battery system with an auxiliary energy storage device able to operate under these extreme conditions without problems. Supercapacitors and flywheels are well suited for this task (see below).

8.5 Other Energy Storage Devices

Energy can be stored in several forms on board vehicles or rovers. In case of stationary plants, it is possible to store energy in the form of potential energy, a thing that is commonly done in pumped basins, where water is pumped when the generation capacity is higher than the needs, and then returned to a lower level when energy is needed. This, however, requires a gravitational field, which is available only on the surface of a planet, the presence of a fluid, possibly in liquid form, and large investments. While being not impossible to think of storing energy on Titan by pumping methane in a suitable basin located at a higher level, this would be practical only in a hypothetical future when large civil engineering works will be performed there.

Several schemes have been proposed, and sometimes implemented, to store energy on board vehicles on Earth. They include elastic potential energy in both a compressed gas or a deformed solid, kinetic energy in a flywheel, electric energy in a capacitor, magnetic energy in an inductor, or thermal energy in objects having a large thermal capacity.

Two storage devices will be mentioned here: supercapacitors and flywheels.

8.5.1 Supercapacitors

A supercapacitor is essentially a capacitor whose capacitance, for its size and mass, is much larger (by orders of magnitude) than that of standard capacitors. Actually, supercapacitors have been defined as an intermediate technology between capacitors and electrochemical batteries.

Their discharge curve is, however, linear, since the voltage at their terminals is proportional to the charge stored. There is thus a strong drop of voltage during discharge, which in most applications compels to use a voltage regulator.

At present, the energy density of supercapacitors is between 1 and 10 Wh/kg, with peak values up to 30 Wh/kg, while the power density can be as high as 5,000 Wh/kg. The largest units built have a capacitance of 5,000 F. Although being much inferior to batteries for what their energy density is concerned, they have an extremely high power density, since they can be charged and discharged in a short time.

Supercapacitors are thus ideally suited to perform as energy buffers for batteries, supplying and accepting sharp current peaks.

8.5.2 Flywheels

Energy can be stored in a rotating object in the form of kinetic energy. The velocity Ω of a flywheel with moment of inertia J is linked with the energy stored e by the obvious relationship

$$e = \frac{1}{2} J \Omega^2, \tag{8.2}$$

showing that the speed variations during the charge and discharge are large. This implies the presence of a power interface that can be a rotary actuator (like an electrical motor/generator or a hydraulic motor/pump) able to operate at variable speed or of a variable ratio mechanical transmission.

The energy density of the flywheel itself may be from 10 Wh/kg to even 100 Wh/kg, but if the whole system is accounted for, these figures are much lower. The high power density allows flywheels to be used as power buffers, like supercapacitors.

Appendix A
Equations of Motion in the Configuration and State Spaces

A.1 Discrete Linear Systems

A.1.1 Configuration Space

Consider a system with a single degree of freedom and assume that the equation expressing its dynamic equilibrium is a second order ordinary differential equation (ODE) in the generalized coordinate x. Assume also that the forces entering the dynamic equilibrium equation are

- a force depending on the acceleration (inertia force),
- a force depending on velocity (damping force),
- a force depending on displacement (elastic force),
- a force, usually applied from outside the system, that depends neither on coordinate x nor on its derivatives, but is a generic function of time (external forcing function).

If the dependence of the first three forces from acceleration, velocity and displacement, respectively, is linear, the system is linear. Moreover, if the constants of such a linear combination, usually referred to as mass m, damping coefficient c and stiffness k do not depend on time, the system is time-invariant.

The dynamic equilibrium equation is thus

$$m\ddot{x} + c\dot{x} + kx = f(t). \tag{A.1}$$

If the system has a number n of degrees of freedom, the most general form for a linear, time-invariant set of second order ordinary differential equations is

$$\mathbf{A}_1\ddot{\mathbf{x}} + \mathbf{A}_2\dot{\mathbf{x}} + \mathbf{A}_3\mathbf{x} = \mathbf{f}(t), \tag{A.2}$$

where:

- $\mathbf{x}$ is a vector of order n (n is the number of degrees of freedom of the system) where the generalized coordinates are listed.

G. Genta, *Introduction to the Mechanics of Space Robots*, Space Technology Library 26, DOI 10.1007/978-94-007-1796-1, © Springer Science+Business Media B.V. 2012

- $\mathbf{A}_1$, $\mathbf{A}_2$ and $\mathbf{A}_3$ are matrices, whose order is $n \times n$; they contain the characteristics (independent from time) of the system.
- $\mathbf{f}$ is a vector function of time containing the forcing functions acting on the system.

Matrix $\mathbf{A}_1$ is usually symmetrical but, even if it is not, it is possible to rearrange the equations of motion in such a way that $\mathbf{A}_1$ becomes symmetrical. The other two matrices in general are not such and can be written as the sum of a symmetrical and a skew–symmetrical matrix

$$\mathbf{M}\ddot{\mathbf{x}} + (\mathbf{C} + \mathbf{G})\dot{\mathbf{x}} + (\mathbf{K} + \mathbf{H})\mathbf{x} = \mathbf{f}(t), \tag{A.3}$$

where:

- $\mathbf{M}$, the *mass matrix* of the system, is a symmetrical matrix of order $n \times n$ (coincides with $\mathbf{A}_1$). Usually it is not singular.
- $\mathbf{C}$ is the real symmetric *viscous damping matrix* (it is the symmetric part of $\mathbf{A}_2$).
- $\mathbf{K}$ is the real symmetric *stiffness matrix* (it is the symmetric part of $\mathbf{A}_3$).
- $\mathbf{G}$ is the real skew–symmetric *gyroscopic matrix* (it is the skew–symmetric part of $\mathbf{A}_2$).
- $\mathbf{H}$ is the real skew–symmetric *circulatory matrix* (it is the skew–symmetric part of $\mathbf{A}_3$).

Remark A.1 The same form of (A.2) may result from mathematical modeling of physical systems whose equations of motion are obtained by means of space discretization techniques, such as the well-known finite elements method.

$\mathbf{x}$ is a vector in the sense it is column matrix. However, a set of n numbers can be interpreted as a vector in a n-dimensional space. This space containing vector $\mathbf{x}$ is usually referred to as *configuration space*, since any point in this space can be associated to a configuration of the system. Actually, not all points of the configuration space, intended as an infinite n-dimensional space, correspond to configurations that are physically possible for the system: it is thus possible to define a subspace containing all possible configurations. Moreover, even system that are dealt with using linear equations of motion are linear only for configurations not much displaced from a reference configuration (usually the equilibrium configuration) and the linear equation (A.2) applies only in an even smaller subspace of the configuration space.

A simple system with two degrees of freedom is shown in Fig. A.1a; it consists of two masses and two springs whose behavior is linear in a zone around the equilibrium configuration with $x_1 = x_2 = 0$ but then behave in a nonlinear way to fail at a certain elongation. In the configuration space, which in the case of a two degrees of freedom system has two dimensions and thus is a plane, there is a linearity zone, surrounded by a zone where the system behaves in nonlinear way. Around the latter there is another zone where the system loses its structural integrity.

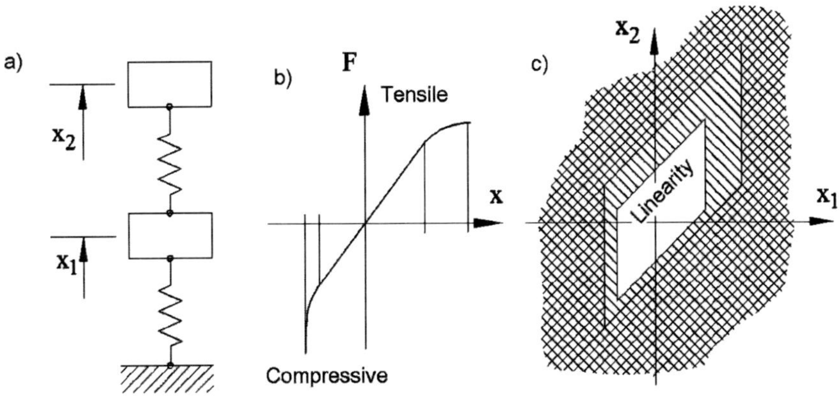

Fig. A.1 Sketch of a system with two degrees of freedom (**a**) made by two masses and two springs, whose characteristics (**b**) are linear only in a zone about the equilibrium position. Three zones can be identified in the configuration space (**c**): in the inner zone the system behaves linearly while in a second zone the system is nonlinear. The latter is surrounded by a 'forbidden' zone (from G. Genta, L. Morello, *The Automotive Chassis*, Springer, New York, 2009)

Remark A.2 In the figure x_1 and x_2 are inertial coordinates, but they are assumed to be zero in the static equilibrium configuration. Moreover, gravitational acceleration has been neglected. This is quite common when dealing with linear systems: the static problem (finding the equilibrium configuration) is separated from the dynamic problem (studying the dynamics about the equilibrium configuration).

A.1.2 State Space

A set of n second order differential equations is a set of order $2n$ and can be expressed in the form of a set of $2n$ first order equations.

In a way similar to what seen above, a generic linear differential equation with constant coefficients can be written in the form of a set of first order differential equations

$$\mathbf{A}_1\dot{\mathbf{x}} + \mathbf{A}_2\mathbf{x} = \mathbf{f}(t). \tag{A.4}$$

In system dynamics this set of equations is usually solved in the first derivatives (monic form) and the forcing function is written as the linear combination of the minimum number of functions expressing the *inputs* of the system. The independent variables are said to be the *state variables* and the equation is written as

$$\dot{\mathbf{z}} = \mathbf{A}\mathbf{z} + \mathbf{B}\mathbf{u}, \tag{A.5}$$

where

- $\mathbf{z}$ is a vector of order m, in which the state variables are listed (m is the number of the state variables). If (A.5) comes from (A.2), $m = 2n$.

- **A** is a matrix of order $m \times m$, independent from time, called the dynamic matrix,
- **u** is a vector function of time, where the inputs acting on the system are listed (if r is the number of inputs, its size is $r \times 1$),
- **B** is a matrix independent from time that states how the various inputs act in the various equations. It is called the input gain matrix and its size is $m \times r$.

As was seen for vector **x**, also **z** is a column matrix that may be considered as a vector in a m-dimensional space. This space is usually referred to as the *state space*, since each point of this space corresponds to a given state of the system.

The configuration space is a subspace of the space state.

If (A.5) comes from (A.2), a set of n auxiliary variables must be introduced to transform the system from the configuration to the state space. Even if other choices are possible (see Sect. A.6), the simplest alternative is using as auxiliary variables the derivatives of the generalized coordinates (generalized velocities). Half of the state variables are then generalized coordinates and the other half are generalized velocities.

If the state variables are ordered with velocities first and then coordinates, it follows that

$$\mathbf{z} = \begin{Bmatrix} \dot{\mathbf{x}} \\ \mathbf{x} \end{Bmatrix}.$$

A number n of equations expressing the link between coordinates and velocities must be added to the n equations (A.2). By using symbol **v** for the generalized velocities $\dot{\mathbf{x}}$, and solving the equations in the derivatives of the state variables, the set of $2n$ equations corresponding to (A.3) is thus

$$\begin{cases} \dot{\mathbf{v}} = -\mathbf{M}^{-1}(\mathbf{C}+\mathbf{G})\mathbf{v} - \mathbf{M}^{-1}(\mathbf{K}+\mathbf{H})\mathbf{x} + \mathbf{M}^{-1}\mathbf{f}(t), \\ \dot{\mathbf{x}} = \mathbf{v}. \end{cases} \tag{A.6}$$

Assuming that inputs **u** coincide with the forcing functions **f**, matrices **A** and **B** are then linked to **M**, **C**, **K**, **G** and **H** by the following relationships:

$$\mathbf{A} = \begin{bmatrix} -\mathbf{M}^{-1}(\mathbf{C}+\mathbf{G}) & -\mathbf{M}^{-1}(\mathbf{K}+\mathbf{H}) \\ \mathbf{I} & \mathbf{0} \end{bmatrix}, \tag{A.7}$$

$$\mathbf{B} = \begin{bmatrix} \mathbf{M}^{-1} \\ \mathbf{0} \end{bmatrix}. \tag{A.8}$$

The first n out of the $m = 2n$ equations constituting the state equation (A.5) are the dynamic equilibrium equations and are usually referred to as dynamic equations. The other n express the relationship between the position and the velocity variables and are usually referred to as kinematic equations.

Often, more than the state vector **z**, what is interesting is a given linear combination of states **z** and inputs **u**, usually referred to as *output vector*. The state equation (A.5) is thus associated to an *output equation*:

$$\mathbf{y} = \mathbf{C}\mathbf{z} + \mathbf{D}\mathbf{u}, \tag{A.9}$$

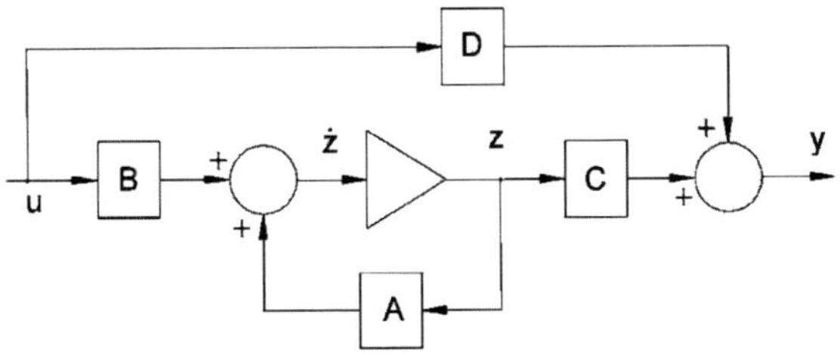

Fig. A.2 Block diagram corresponding to (A.10)

where

- **y** is a vector where the output variables of the system are listed (if the number of outputs is s, its size is $s \times 1$).
- **C** is a matrix of order $s \times m$, independent from time, called the *output gain matrix*.
- **D** is a matrix independent from time that states how the inputs enters into the linear combination yielding the output of the system. It is called *direct link matrix* and its size is $s \times r$. In many cases the inputs do not enter the linear combination yielding the outputs, and **D** is nil.

The four matrices **A**, **B**, **C** and **D** are usually referred to as the quadruple of the dynamic system.

Summarizing, the equations that define the dynamic behavior of the system, from input to output, are

$$\begin{cases} \dot{\mathbf{z}} = \mathbf{A}\mathbf{z} + \mathbf{B}\mathbf{u}, \\ \mathbf{y} = \mathbf{C}\mathbf{z} + \mathbf{D}\mathbf{u}. \end{cases} \tag{A.10}$$

Remark A.3 While the state equations are differential equations, the output equations are algebraic. The dynamics of the system is then concentrated in the former.

The input–output relationship described by (A.10) can be described by the block diagram shown in Fig. A.2.

The linearity of a set of equations allows to state that a solution exists and is unique. The general solution of the equation of motion is the sum of the general solution of the homogeneous equation associated to it and a particular solution of the complete equation. This is true for any differential linear set of equations, even if it is not time-invariant.

The former is the free response of the system, the latter the response to the forcing function.

A.1.3 Free Motion

Consider the equation of motion written in the configuration space (A.2). As already stated, matrix $\mathbf{A}_1$ is symmetrical, while the other two may not be such.

The homogeneous equation

$$\mathbf{A}_1\ddot{\mathbf{x}}(t) + \mathbf{A}_2\dot{\mathbf{x}}(t) + \mathbf{A}_3\mathbf{x}(t) = 0 \tag{A.11}$$

describes the free motion of the system and allows us to study its stability.

The solution of (A.11) can be written as

$$\mathbf{x}(t) = \mathbf{x}_0 e^{st}, \tag{A.12}$$

where $\mathbf{x}_0$ and s are a vector and a scalar, respectively, both complex and constant. To state the time history of the solution allows to transform the differential equation in an algebraic equation

$$\left(\mathbf{A}_1 s^2 + \mathbf{A}_2 s + \mathbf{A}_3\right)\mathbf{x}_0 = \mathbf{0}. \tag{A.13}$$

It is a set of linear algebraic homogeneous equations. The coefficients matrix is a second order *lambda matrix*;[1] it is square and, since the mass matrix $\mathbf{A}_1 = \mathbf{M}$ is not singular, the lambda matrix is said to be *regular*.

The equation of motion (A.11) has solutions different from the trivial one

$$\mathbf{x}_0 = \mathbf{0} \tag{A.14}$$

if and only if the determinant of the matrix of the coefficients vanishes.

$$\det\left(\mathbf{A}_1 s^2 + \mathbf{A}_2 s + \mathbf{A}_3\right) = 0. \tag{A.15}$$

Equation (A.15) is the characteristic equation of a generalized eigenproblem. Its solutions s_i are the eigenvalues of the system and the corresponding vectors $\mathbf{x}_{0_i}$ are its eigenvectors $\mathbf{q}_i$. The rank of the matrix of the coefficients obtained in correspondence to each eigenvalue s_i defines its multiplicity: if the rank is $n - \alpha_i$, the multiplicity is α_i. The eigenvalues are $2n$ and, correspondingly, there are $2n$ eigenvectors.

Remark A.4 If the multiplicity of some eigenvalues is larger than 1, the eigenvectors corresponding to identical eigenvalues are different from each other. Moreover, any linear combination of these eigenvectors is itself an eigenvector.

Remark A.5 Since the matrices of the system $\mathbf{A}_i$ are real, the characteristic equation (A.15) has real coefficients. Its solutions, the eigenvalues, are thus either real numbers or complex conjugate pairs.

---

[1] The term *lambda matrix* comes from the habit of using symbol λ for the coefficient appearing into the solution $\mathbf{q}(t) = \mathbf{q}_0 e^{\lambda t}$. Here symbol s has been used instead of λ, following a more modern habit.

A.1.4 Conservative Natural Systems

If the gyroscopic matrix $\mathbf{G}$ is not present the system is said to be *natural*. If the damping and circulatory matrices $\mathbf{C}$ and $\mathbf{H}$ also vanish, the system is *conservative*. A system with $\mathbf{G} = \mathbf{C} = \mathbf{H} = \mathbf{0}$ (or, as is usually referred to, a MK system) is thus both natural and conservative. The characteristic equation reduces to the algebraic equation

$$\det\left(\mathbf{M}s_i^2 + \mathbf{K}\right) = 0. \tag{A.16}$$

The eigenproblem can be reduced in canonical form

$$\mathbf{D}\mathbf{q}_i = \mu_i \mathbf{q}_i, \tag{A.17}$$

where the dynamic matrix in the configuration space $\mathbf{D}$ (not to be confused with the dynamic matrix in the state space $\mathbf{A}$) is

$$\mathbf{D} = \mathbf{M}^{-1}\mathbf{K}, \tag{A.18}$$

and the parameter in which the eigenproblem is written is

$$\mu_i = -s_i^2. \tag{A.19}$$

Since matrices $\mathbf{M}$ and $\mathbf{K}$ are positive defined ($\mathbf{K}$ may be positive semi-defined), the n eigenvalues μ_i are all real and positive (or zero) and the eigenvalues in terms of s_i are $2n$ imaginary numbers in pairs with opposite sign

$$(s_i, \bar{s}_i) = \pm i\sqrt{\mu_i}. \tag{A.20}$$

The n eigenvectors $\mathbf{q}_i$ of size n are real vectors.

Since all the eigenvalues s_i are imaginary, the solutions (A.12) reduce to undamped harmonic oscillations

$$\mathbf{x}(t) = \mathbf{x}_0\, e^{i\omega t}, \tag{A.21}$$

where

$$\omega = is = \sqrt{\mu} \tag{A.22}$$

is the (circular) frequency.

The n values of ω_i, computed in correspondence of the eigenvalues μ_i, are the natural frequencies or eigenfrequencies of the system, usually written as ω_{n_i}.

If $\mathbf{M}$ or $\mathbf{K}$ are not positive defined or semi-defined, at least one of the eigenvalues μ_i is negative, and thus one of the pair of solutions in s is real, made of a positive and a negative value. As it will be seen below, the real negative solution corresponds to a time history that decays in time in a nonoscillatory way, while the positive one to a time history that increases in time in an unbounded way. The system is thus unstable.

A.1.5 Properties of the Eigenvectors

The eigenvectors are orthogonal with respect to both the stiffness and mass matrices. This propriety can be demonstrated simply by writing the dynamic equilibrium equation in harmonic oscillations for the ith mode

$$\mathbf{K}\mathbf{q}_i = \omega_i^2 \mathbf{M}\mathbf{q}_i . \tag{A.23}$$

Equation (A.23) can be premultiplied by the transpose of the jth eigenvector

$$\mathbf{q}_j^T \mathbf{K}\mathbf{q}_i = \omega_i^2 \mathbf{q}_j^T \mathbf{M}\mathbf{q}_i . \tag{A.24}$$

The same can be done for the equation written for the jth mode and premultiplied by the transpose of the ith eigenvector

$$\mathbf{q}_i^T \mathbf{K}\mathbf{q}_j = \omega_j^2 \mathbf{q}_i^T \mathbf{M}\mathbf{q}_j . \tag{A.25}$$

By subtracting (A.25) from (A.24) it follows that

$$\mathbf{q}_j^T \mathbf{K}\mathbf{q}_i - \mathbf{q}_i^T \mathbf{K}\mathbf{q}_j = \omega_i^2 \mathbf{q}_j^T \mathbf{M}\mathbf{q}_i - \omega_j^2 \mathbf{q}_i^T \mathbf{M}\mathbf{q}_j . \tag{A.26}$$

Remembering that, owing to the symmetry of matrices **K** and **M**,

$$\mathbf{q}_j^T \mathbf{K}\mathbf{q}_i = \mathbf{q}_i^T \mathbf{K}\mathbf{q}_j \quad \text{and} \quad \mathbf{q}_j^T \mathbf{M}\mathbf{q}_i = \mathbf{q}_i^T \mathbf{M}\mathbf{q}_j ,$$

it follows that

$$\left(\omega_i^2 - \omega_j^2\right) \mathbf{q}_j^T \mathbf{M}\mathbf{q}_i = 0. \tag{A.27}$$

In the same way, it can be shown that

$$\left(\frac{1}{\omega_i^2} - \frac{1}{\omega_j^2}\right) \mathbf{q}_j^T \mathbf{K}\mathbf{q}_i = 0. \tag{A.28}$$

From (A.28) and (A.27), assuming that the natural frequencies are all different from each other, it follows that, if $i \neq j$,

$$\mathbf{q}_i^T \mathbf{M}\mathbf{q}_j = 0, \qquad \mathbf{q}_i^T \mathbf{K}\mathbf{q}_j = 0, \tag{A.29}$$

which are the relationships defining the orthogonality properties of the eigenvectors with respect to the mass and stiffness matrices, respectively.

If $i = j$, the results of the same products are not zero:

$$\mathbf{q}_i^T \mathbf{M}\mathbf{q}_i = \overline{M_i}, \qquad \mathbf{q}_i^T \mathbf{K}\mathbf{q}_j = \overline{K_i}. \tag{A.30}$$

Constants $\overline{M_i}$ and $\overline{K_i}$ are the modal mass and modal stiffness of the ith mode, respectively. They are linked to the natural frequencies by the relationship

$$\omega_i = \sqrt{\frac{\overline{K_i}}{\overline{M_i}}}, \tag{A.31}$$

stating that the ith natural frequency coincides with the natural frequency of a system with a single degree of freedom whose mass is the ith modal mass and whose stiffness is the ith modal stiffness.

A.1.6 Uncoupling of the Equations of Motion

Any vector in the configuration space (i.e. any configuration of the system) can be considered as a linear combination of the eigenvectors

$$\mathbf{x} = \sum_{i=1}^{n} \eta_i \mathbf{x}_i, \tag{A.32}$$

where the coefficients of the linear combination η_i are the modal coordinates of the system.

This is possible because the eigenvectors are linearly independent and form a possible reference frame in the space of the configurations of the system. It must be explicitly stated that the eigenvectors are orthogonal with respect to the mass and stiffness matrices (they are said to be *m-orthogonal* and *k-orthogonal*), but they are not orthogonal with each other. This means that, in general,

$$\mathbf{q}_i^T \mathbf{q}_j \neq 0. \tag{A.33}$$

In the space of the configurations, the eigenvectors are n vectors that can be taken as a reference frame. However, they are not orthogonal with respect to each other.

By defining a matrix of the eigenvectors $\boldsymbol{\Phi}$ whose columns are the eigenvectors, the modal transformation and the corresponding inverse transformation can be written as

$$\mathbf{x} = \boldsymbol{\Phi}\eta, \qquad \eta = \boldsymbol{\Phi}^{-1}\mathbf{x}. \tag{A.34}$$

Remark A.6 Since the eigenvectors are linearly independent, $\boldsymbol{\Phi}$ is not singular; the inverse modal transformation always exists.

By introducing such a transformation into the equation of motion it follows that

$$\mathbf{M}\boldsymbol{\Phi}\ddot{\eta} + \mathbf{K}\boldsymbol{\Phi}\eta = \mathbf{f}(t), \tag{A.35}$$

and, by premultiplying each term by $\boldsymbol{\Phi}^T$,

$$\boldsymbol{\Phi}^T\mathbf{M}\boldsymbol{\Phi}\ddot{\eta} + \boldsymbol{\Phi}^T\mathbf{K}\boldsymbol{\Phi}\eta = \boldsymbol{\Phi}^T\mathbf{f}(t). \tag{A.36}$$

Matrices $\boldsymbol{\Phi}^T\mathbf{M}\boldsymbol{\Phi}$ and $\boldsymbol{\Phi}^T\mathbf{K}\boldsymbol{\Phi}$ are the modal mass matrix and the modal stiffness matrix. Owing to the properties of the eigenvectors they are diagonal:

$$\begin{cases} \boldsymbol{\Phi}^T\mathbf{M}\boldsymbol{\Phi} = \operatorname{diag}[\overline{M_i}] = \overline{\mathbf{M}}, \\ \boldsymbol{\Phi}^T\mathbf{K}\boldsymbol{\Phi} = \operatorname{diag}[\overline{K_i}] = \overline{\mathbf{K}}. \end{cases} \tag{A.37}$$

Vector $\mathbf{\Phi}^T\mathbf{f}(t)$ is said to be the *modal force vector* $\overline{\mathbf{f}}(t)$.

Since the modal matrices are diagonal, the equations of motion uncouple from each other and, instead of studying a system with n degrees of freedom, it is possible to study n uncoupled systems with a single degree of freedom, whose equations of motion are

$$\overline{\mathbf{M}}\ddot{\eta} + \overline{\mathbf{K}}\eta = \overline{\mathbf{f}}. \tag{A.38}$$

Remark A.7 Equations (A.34) are a coordinate transformation in the space of the configurations. The n values η_i are the n coordinates of the point representing the configuration of the system, with reference to the system of the eigenvectors. They are said to be principal, modal, or normal coordinates.

Remark A.8 Modal uncoupling holds in general only for MK systems.

The eigenvectors are the solutions of a linear set of homogeneous equations and, thus, are not unique: for each mode, an infinity of eigenvectors exists, all proportional to each other. Because the eigenvectors can be seen as a set of n vectors in the n-dimensional space providing a system of reference, the length of such vectors is not determined, but their directions are known. In other words, the scales of the axes are arbitrary.

There are many ways of normalizing the eigenvectors. The simplest is by stating that the value of one particular element or of the largest one is set to unity.

As an alternative, each eigenvector can be divided by its Euclidean norm, obtaining unit vectors in the space of the configurations.

Another way is to normalize the eigenvectors in such a way that the modal masses are equal to unity. This can be done by dividing each eigenvector by the square root of the corresponding modal mass. In the latter case, each modal stiffness coincides with the corresponding eigenvalue, i.e., with the square of the natural frequency. Equation (A.38) reduces to

$$\ddot{\eta} + \left[\omega^2\right]\eta = \overline{\mathbf{f}'}, \tag{A.39}$$

where $[\omega^2] = \text{diag}\{\omega_i^2\}$ is the matrix of the eigenvalues and the modal forces $\overline{\mathbf{f}'}(t)$ are

$$\overline{f_i'} = \frac{\overline{f_i}}{\overline{M_i}} = \frac{\mathbf{q}_i^T\mathbf{f}}{\mathbf{q}_i\mathbf{M}\mathbf{q}_i}. \tag{A.40}$$

Modal uncoupling has, however, another advantage: since not all modes are equally important in determining the response of the system, a limited number of modes (usually those characterized by the lowest natural frequencies) is often sufficient for obtaining the response with good accuracy.

If only the first m modes are considered,[2] the savings in terms of computation time, and hence cost, are usually noticeable, because only m eigenvalues and eigen-

[2]In the following pages it is assumed that the modes which are retained are those from the first one to the mth.

vectors need to be computed and m systems with one degree of freedom need to be studied. Usually the modes that are more difficult to deal with are those characterized by the highest natural frequencies, particularly if the equations of motion are integrated numerically. The advantage of discarding the higher-order modes is, in this case, great.

When some modes are neglected, the reduced matrix of the eigenvectors

$$\mathbf{\Phi}^* = [\mathbf{q}_1, \mathbf{q}_2, \dots, \mathbf{q}_m],$$

is not square because has n rows and m columns. The first coordinate transformation (A.34) still holds, and the m values of the modal mass, stiffness and force can be computed as usual. However, the inverse transformation (second equation (A.34)) is not possible, because the inversion of matrix $\mathbf{\Phi}^*$ cannot be performed.

The modal coordinates η can be computed from the physical coordinates $\mathbf{x}$ by premultiplying by $\mathbf{\Phi}^{*T}\mathbf{M}$ both sides of the first equation (A.34) computed using the reduced matrix of the eigenvectors $\mathbf{\Phi}^*$, obtaining

$$\mathbf{\Phi}^{*T}\mathbf{M}\mathbf{x} = \mathbf{\Phi}^{*T}\mathbf{M}\mathbf{\Phi}^*\eta, \tag{A.41}$$

i.e.

$$\mathbf{\Phi}^{*T}\mathbf{M}\mathbf{x} = \overline{\mathbf{M}}\eta. \tag{A.42}$$

Premultiplying then both sides by the inverse of the matrix of the modal masses, it follows that

$$\eta = \overline{\mathbf{M}}^{-1}\mathbf{\Phi}^{*T}\mathbf{M}\mathbf{x}. \tag{A.43}$$

Equation (A.43) is the required inverse modal transformation.

Remark A.9 Equation (A.41) and the following ones are approximations, since only a reduced number of modes have been accounted for.

A.1.7 Natural Nonconservative Systems

If matrix $\mathbf{C}$ does not vanish while $\mathbf{G} = \mathbf{H} = \mathbf{0}$, the system is still natural and noncirculatory, but is no more conservative.

The characteristic equation (A.15) cannot be reduced to an eigenproblem in canonical form in the configuration space and the state space formulation must be used.

The general solution of the homogeneous equation associated to (A.5) is of the type

$$\mathbf{z} = \mathbf{z}_0 e^{st}, \tag{A.44}$$

where s is in general a complex number. Its real and imaginary parts are usually indicated with symbols ω and σ

$$\omega = \Im(s),$$
$$\sigma = \Re(s) \tag{A.45}$$

and are the frequency of the free oscillations and the decay rate. Solution (A.44) can thus be written in the form

$$\mathbf{z} = \mathbf{z}_0 e^{\sigma t} e^{i\omega t}, \tag{A.46}$$

or, since both σ and ω are real numbers,

$$\mathbf{z} = \mathbf{z}_0 e^{\sigma t}\big[\cos(\omega t) + i \sin(\omega t)\big]. \tag{A.47}$$

By introducing solution (A.44) into the homogeneous equation associated to (A.5), the latter transforms from a set of differential equations into a (homogeneous) set of algebraic equations

$$s\mathbf{z}_0 = \mathbf{A}\mathbf{z}_0, \tag{A.48}$$

i.e.

$$(\mathbf{A} - s\mathbf{I})\mathbf{z}_0 = 0. \tag{A.49}$$

In a way similar to what seen for the equation of motion in the configuration space, this homogeneous equation has solutions other than the trivial solution $\mathbf{z}_0 = 0$ only if the determinant of the coefficients matrix vanishes,

$$\det(\mathbf{A} - s\mathbf{I}) = 0. \tag{A.50}$$

Equation (A.50) can be interpreted as an algebraic equation in s, i.e. the characteristic equation of the dynamic systems. It is an equation of power $2n$, yielding the $2n$ values of s. The $2n$ values of s are the eigenvalues of the system and the corresponding $2n$ values of $\mathbf{z}_0$ are the eigenvectors. In general, both eigenvalues and eigenvectors are complex.

If matrix $\mathbf{A}$ is real, as it is usually the case, the solutions are either real or complex conjugate. The corresponding time histories are (Fig. A.3):

- Real solutions ($\omega = 0$, $\sigma \neq 0$): exponential time histories, either with monotonic decay of the amplitude if the solution is negative (stable, nonoscillatory behavior), or with monotonic increase of the amplitude if the solution is positive (unstable, nonoscillatory behavior).
- Complex conjugate solutions ($\omega \neq 0$, $\sigma \neq 0$): oscillating time histories, expressed by (A.47) with amplitude decay if the real part of the solution is negative (stable, oscillatory behavior) or amplitude increase in time if the real part of the solution is positive (unstable, oscillatory behavior).

If the system is stable, stability is asymptotic.

A third case is that seen previously for conservative systems:

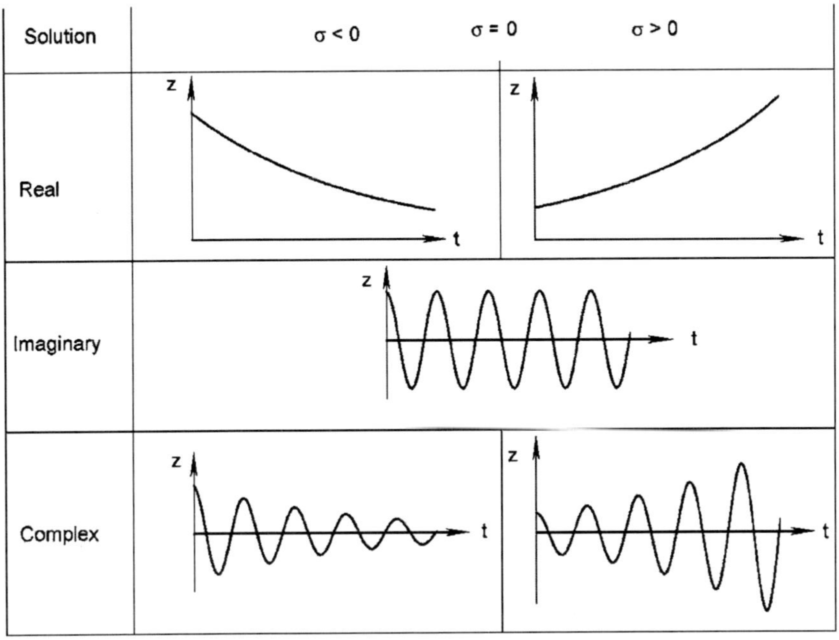

Fig. A.3 Time history of the free motion for the various types of the eigenvalues of the system (from G. Genta, L. Morello, *The Automotive Chassis*, Springer, New York, 2009)

- Imaginary solutions ($\omega \neq 0$, $\sigma = 0$): harmonic time histories (sine or cosine waves, undamped oscillatory behavior). In this case stability is non-asymptotic.

The necessary and sufficient condition for an asymptotically stable behavior is thus that the real part of all eigenvalues is negative.

If any one of the real parts of the eigenvalues is zero, the behavior is still stable, since the amplitude does not grow in an uncontrolled way in time, but not asymptotically stable.

If at least one of them is positive, the system is unstable.

If the system is little damped, i.e. the eigenvalues are conjugate and the decay rates σ are small, the values of the natural frequencies ω are close to those of the corresponding undamped system, that is, to those of the MK system obtained by neglecting the damping matrix $\mathbf{C}$. In this case the oscillation frequencies ω_i are close to those of the corresponding undamped system.

Remark A.10 The term *stable* is here used in the sense that the amplitude of the motion does not grow in an unbounded way and *asymptotically stable* in the sense that the system returns asymptotically to the static equilibrium position. More detailed definitions of stability, like the Liapunov definition, can be found in the literature.

The general solution of the homogeneous equation is a linear combination of the $2n$ solutions

$$\mathbf{z} = \sum_{i=1}^{2n} C_i \mathbf{z}_{0i} e^{s_i t}, \tag{A.51}$$

where the $2n$ constants C_i must be obtained from the initial conditions, i.e. from vector $\mathbf{z}(0)$.

The equation allowing one to compute constants C_i can be written as

$$\mathbf{z}(0) = [\,\mathbf{z}_{01}\ \mathbf{z}_{02}\ \dots\ \mathbf{z}_{02n}\,] \begin{Bmatrix} C_1 \\ C_2 \\ \dots \\ C_{2n} \end{Bmatrix} = \mathbf{\Phi}\mathbf{C}, \tag{A.52}$$

where $\mathbf{\Phi}$ is the matrix of the complex eigenvectors in the state space.

A real and negative eigenvalue corresponds to an *overdamped* behavior, which is nonoscillatory, of the relevant mode. If the eigenvalue is complex (with negative real part) the mode has an *underdamped* behavior, that is, it has a damped oscillatory time history. A system with all underdamped modes is said to be underdamped, while if only one of the modes is overdamped, the system is said to be overdamped. If all modes are overdamped, the system cannot have free oscillations, but can oscillate if forced to do so.

Remark A.11 If all matrices **M**, **K** and **C** are positive defined (or at least semi-defined), as in the case of a structure with viscous damping with positive stiffness and damping, there is no eigenvalue with positive real part and hence the system is stable. If all matrices are strictly positive defined, there no eigenvalue with vanishing real part and thus the system is asymptotically stable.

A.1.8 Systems with Singular Mass Matrix

If matrix **M** is singular, it is impossible to write the dynamic matrix in the usual way. Usually this occurs because a vanishingly small inertia is associated to some degrees of freedom. Clearly the problem may be circumvented by associating a very small mass to the relevant degrees of freedom: a new very high natural frequency, which has no physical meaning, is thus introduced and, if this is done carefully, no numerical instability problem is caused. However, it makes little sense to resort to tricks of this kind when it is possible to overcome the problem in a more correct and essentially simple way.

The degrees of freedom can be subdivided in two sets: a vector $\mathbf{x}_1$ containing those to which a nonvanishing inertia is associated, and a vector $\mathbf{x}_2$, containing the other ones. In a similar way all matrices and the forcing functions can be split. The mass matrix $\mathbf{M}_{22}$ vanishes, and if the mass matrix is diagonal, also $\mathbf{M}_{12}$ and $\mathbf{M}_{21}$ vanish.

Assuming that $\mathbf{M}_{12}$ and $\mathbf{M}_{21}$ are zero, the equations of motion become

$$\begin{cases} \mathbf{M}_{11}\ddot{\mathbf{x}}_1 + \mathbf{C}_{11}\dot{\mathbf{x}}_1 + \mathbf{C}_{12}\dot{\mathbf{x}}_2 + \mathbf{K}_{11}\mathbf{x}_1 + \mathbf{K}_{12}\mathbf{x}_2 = \mathbf{f}_1(t), \\ \mathbf{C}_{21}\dot{\mathbf{x}}_1 + \mathbf{C}_{22}\dot{\mathbf{x}}_2 + \mathbf{K}_{21}\mathbf{x}_1 + \mathbf{K}_{22}\mathbf{x}_2 = \mathbf{f}_2(t). \end{cases} \tag{A.53}$$

To simplify the equations of motion neither the gyroscopic nor the circulator matrices are explicitly written, but in what follows no assumption on the symmetry of the stiffness and damping matrices will be done, and the equations hold also for gyroscopic and circulatory systems.

By introducing as state variables the velocities $\mathbf{v}_1$ together with generalized coordinates $\mathbf{x}_1$ and $\mathbf{x}_2$, the state equation is

$$\mathbf{M}^* \begin{Bmatrix} \dot{\mathbf{v}}_1 \\ \dot{\mathbf{x}}_1 \\ \dot{\mathbf{x}}_2 \end{Bmatrix} = \mathbf{A}^* \begin{Bmatrix} \mathbf{v}_1 \\ \mathbf{x}_1 \\ \mathbf{x}_2 \end{Bmatrix} + \begin{bmatrix} \mathbf{I} & \mathbf{0} \\ \mathbf{0} & \mathbf{I} \\ \mathbf{0} & \mathbf{0} \end{bmatrix} \begin{Bmatrix} \mathbf{f}_1(t) \\ \mathbf{f}_2(t) \end{Bmatrix}, \tag{A.54}$$

where

$$\mathbf{M}^* = \begin{bmatrix} \mathbf{M}_{11} & \mathbf{0} & \mathbf{C}_{12} \\ \mathbf{0} & \mathbf{0} & \mathbf{C}_{22} \\ \mathbf{0} & \mathbf{I} & \mathbf{0} \end{bmatrix}, \qquad \mathbf{A}^* = -\begin{bmatrix} \mathbf{C}_{11} & \mathbf{K}_{11} & \mathbf{K}_{12} \\ \mathbf{C}_{21} & \mathbf{K}_{21} & \mathbf{K}_{22} \\ -\mathbf{I} & \mathbf{0} & \mathbf{0} \end{bmatrix}. \tag{A.55}$$

The dynamic matrix and the input gain matrix are

$$\mathbf{A} = \mathbf{M}^{*-1}\mathbf{A}^*, \qquad \mathbf{B} = \mathbf{M}^{*-1} \begin{bmatrix} \mathbf{I} & \mathbf{0} \\ \mathbf{0} & \mathbf{I} \\ \mathbf{0} & \mathbf{0} \end{bmatrix}. \tag{A.56}$$

Alternatively, the expressions of $\mathbf{M}^*$ and $\mathbf{A}^*$ can be

$$\mathbf{M}^* = \begin{bmatrix} \mathbf{M}_{11} & \mathbf{C}_{11} & \mathbf{C}_{12} \\ \mathbf{0} & \mathbf{C}_{21} & \mathbf{C}_{22} \\ \mathbf{0} & \mathbf{I} & \mathbf{0} \end{bmatrix}, \qquad \mathbf{A}^* = -\begin{bmatrix} \mathbf{0} & \mathbf{K}_{11} & \mathbf{K}_{12} \\ \mathbf{0} & \mathbf{K}_{21} & \mathbf{K}_{22} \\ -\mathbf{I} & \mathbf{0} & \mathbf{0} \end{bmatrix}. \tag{A.57}$$

If vector $\mathbf{x}_1$ contains n_1 elements and $\mathbf{x}_2$ contains n_2 elements, the size of the dynamic matrix $\mathbf{A}$ is $2n_1 + n_2$.

A.1.9 Conservative Gyroscopic Systems

If matrix $\mathbf{G}$ is not zero, while both $\mathbf{C}$ and $\mathbf{H}$ vanish, the dynamic matrix reduces to

$$\mathbf{A} = \begin{bmatrix} -\mathbf{M}^{-1}\mathbf{G} & -\mathbf{M}^{-1}\mathbf{K} \\ \mathbf{I} & \mathbf{0} \end{bmatrix}. \tag{A.58}$$

By premultiplying the first n equations by $\mathbf{M}$ and the other n by $\mathbf{K}$, it follows that

$$\mathbf{M}^*\dot{\mathbf{z}} + \mathbf{G}^*\mathbf{z} = \mathbf{0}, \tag{A.59}$$

where

$$\mathbf{M}^* = \begin{bmatrix} \mathbf{M} & \mathbf{0} \\ \mathbf{0} & \mathbf{K} \end{bmatrix}, \qquad \mathbf{G}^* = \begin{bmatrix} \mathbf{G} & \mathbf{K} \\ -\mathbf{K} & \mathbf{0} \end{bmatrix}. \tag{A.60}$$

The first matrix is symmetrical, while the second one is skew–symmetrical.

By introducing solutions (A.44) into the equation of motion, the homogeneous equation

$$s\mathbf{M}^*\mathbf{z}_0 + \mathbf{G}^*\mathbf{z}_0 = \mathbf{0} \tag{A.61}$$

is obtained.

The corresponding eigenproblem has imaginary solutions like those of an MK system, even if the structure of the eigenvectors is different. At any rate the time history of the free oscillations is harmonic and undamped, since the decay rate $\sigma = \Re(s)$ is zero.

A.1.10 General Dynamic Systems

The situation is similar to that seen for natural nonconservative systems, in the sense that the time histories of the free oscillations are those seen in Fig. A.3 and the stability is dominated by the sign of the real part of s.

In general, the presence of a gyroscopic matrix does not reduce the stability of the system, while the presence of a circulatory matrix has a destabilizing effect.

Consider for instance a two degrees of freedom systems made by two independent MK system, each one with a single degree of freedom, and assume that the two masses are equal. The equations for free motion are

$$\begin{cases} m\ddot{x}_1 + k_1 x_1 = 0, \\ m\ddot{x}_2 + k_2 x_2 = 0. \end{cases} \tag{A.62}$$

Introduce now a coupling term in both equations, for instance introducing a spring with stiffness k_{12} between the two masses. The equations of motion become

$$\begin{cases} m\ddot{x}_1 + (k_1 + k_{12})x_1 - k_{12}x_2 = 0, \\ m\ddot{x}_2 - k_{12}x_1 + (k_1 + k_{12})x_2 = 0. \end{cases} \tag{A.63}$$

By introducing parameters

$$\omega_0^2 = \frac{k_1 + k_2 + 2k_{12}}{2m}, \qquad \alpha = \frac{k_2 - k_1}{2m\Omega_0^2}, \qquad \epsilon = \frac{k_{12}}{m\Omega_0^2}, \tag{A.64}$$

the equation of motion can be written as

$$\begin{Bmatrix} \ddot{x} \\ \ddot{y} \end{Bmatrix} + \omega_0^2 \begin{bmatrix} 1-\alpha & \epsilon \\ \epsilon & 1+\alpha \end{bmatrix} \begin{Bmatrix} x \\ y \end{Bmatrix} = \mathbf{0}. \tag{A.65}$$

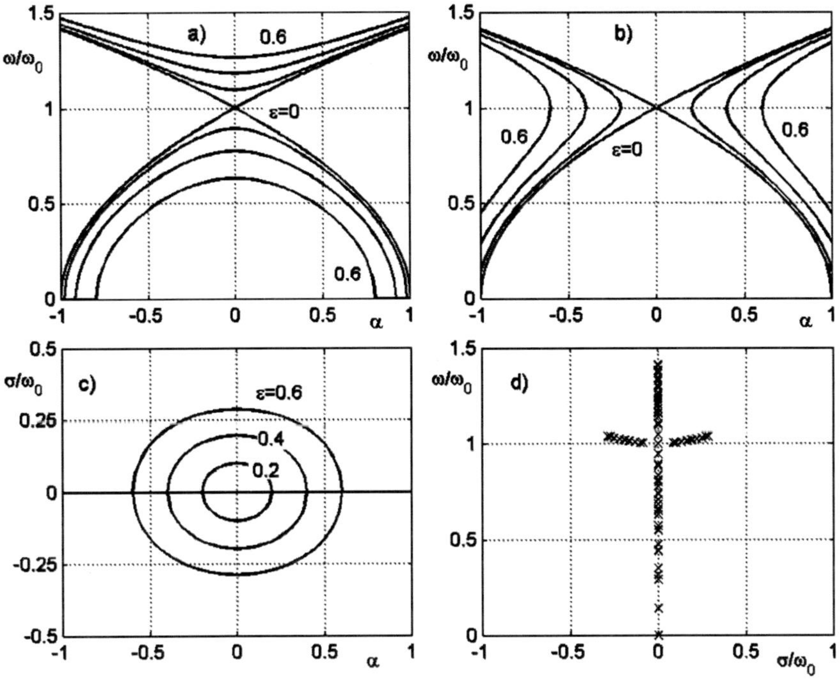

Fig. A.4 Nondimensional natural frequencies as functions of parameters α and ϵ for a system with two degrees of freedom with noncirculatory (**a**) and circulatory (**b**) coupling. Decay rate (**c**) and roots locus (**d**) for the system with circulatory coupling (from G. Genta, L. Morello, *The Automotive Chassis*, Springer, New York, 2009)

Note that

$$-1 \leq \alpha \leq 1. \tag{A.66}$$

The matrix that multiplies the generalized coordinates is symmetrical and thus is a true stiffness matrix. The coupling is said in this case *noncirculatory* or *conservative*. Since there is no damping matrix and the stiffness matrix is positive defined ($-1 \leq \alpha \leq 1$), the eigenvalues are imaginary and the system is stable, even if it is not asymptotically stable as it would be if a positive defined damping matrix were present.

The natural frequencies of the system, made nondimensional by dividing them by ω_0, depend from two parameters, α and ϵ. They are reported in Fig. A.4a as functions of α for some values of ϵ. The distance between the two curves (one for $\omega > \omega_0$ and the other for $\omega < \omega_0$) increases with increasing coupling term ϵ and for this reason this type of coupling is said to be *repulsive*.

Consider now the case with coupling term ϵ in the form

$$\left\{\begin{matrix} \ddot{x} \\ \ddot{y} \end{matrix}\right\} + \Omega_0^2 \begin{bmatrix} 1-\alpha & \epsilon \\ -\epsilon & 1+\alpha \end{bmatrix} \left\{\begin{matrix} x \\ y \end{matrix}\right\} = \mathbf{0}. \tag{A.67}$$

The terms outside the main diagonal of the stiffness matrix now have the same modulus but opposite sign and the matrix multiplying the displacements is made by a symmetrical part (the stiffness matrix) and a skew–symmetrical part (the circulatory matrix). A coupling of this type is said to be *circulatory* or *non conservative*.

While in the previous case the effect could be caused by the presence of a spring between the two masses, now it cannot be due to springs or similar elements. At any rate there are cases of practical interest where circulatory coupling occurs.

The natural frequencies of the system depend also in this case on the two parameters α and ϵ. They are plotted in nondimensional form, by dividing them by ω_0, in Fig. A.4b as functions of α for some values of ϵ. The two curves now get closer to each other. Starting from the condition with $\alpha = -1$, the two curves meet for a certain value of α in the interval $(-1, 0)$. There is a range, centered in the point with $\alpha = 0$ where the solutions of the eigenproblem (in s) are complex, instead of being imaginary. Beyond this range the two curves separate again.

Since the two curves get closer to each other and finally they meet, this type of coupling is said to be *attractive*.

In the range where the values of s are complex, one of the two solutions has positive real part σ: it follows that an unstable solution exists, as it is possible to see from the decay rate plot in Fig. A.4c and from the roots locus in Fig. A.4d.

As already stated, instability is due to the skew–symmetric matrix due to coupling, i.e. to the fact that a circulatory matrix exists.

A.1.11 Closed Form Solution of the Forced Response

The particular solution of the complete equation depends on the time history of the forcing functions (input) $\mathbf{u}(t)$. In case of harmonic input,

$$\mathbf{u} = \mathbf{u}_0 e^{i\omega t}, \tag{A.68}$$

the response is harmonic as well,

$$\mathbf{z} = \mathbf{z}_0 e^{i\omega t}, \tag{A.69}$$

and has the same frequency ω as the forcing function. As usual, by introducing the time history of the forcing function and of the response into the equation of motion, it transforms into an algebraic equation,

$$(\mathbf{A} - i\omega\mathbf{I})\mathbf{z}_0 + \mathbf{B}\mathbf{u}_0 = 0, \tag{A.70}$$

that allows one to compute the amplitude of the response

$$\mathbf{z}_0 = -(\mathbf{A} - i\omega\mathbf{I})^{-1}\mathbf{B}\mathbf{u}_0. \tag{A.71}$$

If the input is periodic, it may be decomposed in Fourier series and the response to each one of its harmonic components can be computed. The results are then added up. This is possible only since the system is linear.

If the input is not harmonic or at least periodic, it is possible to resort to Laplace transforms or to the Duhamel integral. Also these techniques apply only to linear systems.

A.1.12 Modal Transformation of General Linear Dynamic Systems

Since the eigenvectors of the MK system constitute a reference frame in the configuration space, they can be used to write in modal form the equations of motion of the dynamic system, even if the other matrices do not vanish. More in general, it is possible to say that the eigenvectors of any MK system with n degrees of freedom can be used to write the modal equation of motion of any dynamic system with the same number of degrees of freedom.

The modal equation is thus

$$\overline{\mathbf{M}}\ddot{\eta} + (\overline{\mathbf{C}} + \overline{\mathbf{G}})\dot{\eta} + (\overline{\mathbf{K}} + \overline{\mathbf{H}})\eta = \overline{\mathbf{f}}(t). \tag{A.72}$$

The various modal matrices are all obtained from the corresponding nonmodal matrices by postmultiplying them by the matrix of the eigenvectors of the system and by premultiplying them by its transpose.

Since the eigenvectors are m- and k-orthogonal, but not orthogonal with respect to the other matrices, $\overline{\mathbf{M}}$ and $\overline{\mathbf{K}}$ are diagonal while $\overline{\mathbf{C}}$, $\overline{\mathbf{G}}$ and $\overline{\mathbf{H}}$ are not.

The modal equations of motion are thus not uncoupled.

However, while $\overline{\mathbf{C}}$ may be diagonal in some cases (for example, if $\mathbf{C}$ is a linear combination of $\mathbf{M}$ and $\mathbf{K}$, a situation defined as *proportional damping*), $\overline{\mathbf{G}}$ and $\overline{\mathbf{H}}$, being skew–symmetric, cannot be diagonal.

Nongyroscopic and noncirculatory systems can be uncoupled in case of proportional damping and, if damping is non-proportional but small, it is possible to uncouple in an approximate way the equations of motion by canceling the elements outside the main diagonal of the modal damping matrix $\overline{\mathbf{C}}$.

Neglecting some modes is often possible, but is always an approximation, since all modes, being coupled with each other, affect the response of all other modes. If some of them are neglected, their influence on the other modes is lost.

A.2 Nonlinear Dynamic Systems

Often the state equations of dynamic systems are nonlinear. The reasons of the presence of nonlinearities may be different, like for instance the presence of elements that behave in an intrinsically nonlinear way (e.g. springs producing a force depending in nonlinear way from the displacement), or the presence of trigonometric functions of some of the generalized coordinates in the dynamic or in the kinematic equations. If inertia forces are at any rate linear in the accelerations, the equations of motion can be written in the form

$$\mathbf{M}\ddot{\mathbf{x}} + \mathbf{f}_1(\mathbf{x}, \dot{\mathbf{x}}) = \mathbf{f}(t). \tag{A.73}$$

Often function $\mathbf{f}_1$ can be considered as the sum of a linear and a nonlinear part. The equation of motion can thus be written as

$$\mathbf{M}\ddot{\mathbf{x}} + (\mathbf{C} + \mathbf{G})\dot{\mathbf{x}} + (\mathbf{K} + \mathbf{H})\mathbf{x} + \mathbf{f}_2(\mathbf{x}, \dot{\mathbf{x}}) = \mathbf{f}(t), \tag{A.74}$$

where function $\mathbf{f}_2$ contains only the nonlinear part of the dynamic system.

Also in the case of nonlinear system, the eigenvectors of the MK linearized system can be used to write the equations of motion in modal form

$$\overline{\mathbf{M}}\ddot{\eta} + (\overline{\mathbf{C}} + \overline{\mathbf{G}})\dot{\eta} + (\overline{\mathbf{K}} + \overline{\mathbf{H}})\eta + \bar{\mathbf{f}}_2(\mathbf{\Phi}\eta, \mathbf{\Phi}\dot{\eta}) = \bar{\mathbf{f}}(t), \tag{A.75}$$

where vector $\bar{\mathbf{f}}_2$ is obtained by premultiplying vector $\mathbf{f}_2$ by the transpose of the matrix of the eigenvectors. It further couples the equations of motion and makes resorting to a reduced set of modes even more problematic.

The state equations corresponding to (A.73) and (A.74) are

$$\dot{\mathbf{z}} = \mathbf{f}_1(\mathbf{z}) + \mathbf{B}\mathbf{u}, \tag{A.76}$$

or, by separating the linear part from the nonlinear part,

$$\dot{\mathbf{z}} = \mathbf{A}\mathbf{z} + \mathbf{f}_2(\mathbf{z}) + \mathbf{B}\mathbf{u}. \tag{A.77}$$

Another way to express the equation of motion or the state equation of a nonlinear system is by writing equations (A.3) or (A.10), where the various matrices are functions of the generalized coordinates and their derivatives, or of the state variables. In the state space it follows

$$\begin{cases} \dot{\mathbf{z}} = \mathbf{A}(\mathbf{z})\mathbf{z} + \mathbf{B}(\mathbf{z})\mathbf{u}, \\ \mathbf{y} = \mathbf{C}(\mathbf{z})\mathbf{z} + \mathbf{D}(\mathbf{z})\mathbf{u}. \end{cases} \tag{A.78}$$

If the system is not time-invariant, the various matrices may also be explicit functions of time

$$\begin{cases} \dot{\mathbf{z}} = \mathbf{A}(\mathbf{z}, t)\mathbf{z} + \mathbf{B}(\mathbf{z}, t)\mathbf{u}, \\ \mathbf{y} = \mathbf{C}(\mathbf{z}, t)\mathbf{z} + \mathbf{D}(\mathbf{z}, t)\mathbf{u}. \end{cases} \tag{A.79}$$

It is not possible to obtain a closed form solution of nonlinear systems and concepts like natural frequency or decay rate lose their meaning. It is not even possible to distinguish between free and forced behavior, in the sense that the free oscillations depend from the zone of the state space where the system works. In some zones of the state space the behavior of the system may be stable, while in other ones it may be unstable.

At any rate it is often possible to linearize the equations of motion about any given working conditions, i.e. any given point of the state space, and to use the linearized model so obtained in that area of the space state to study the motion of the system and above all its stability. In this case the motion and the stability are studied *in the small*. It is, however, clear that no general result may be obtained in this way.

If the state equation is written in the form (A.76), its linearization about a point of coordinates $\mathbf{z}_0$ in the state space is

$$\dot{\mathbf{z}} = \left(\frac{\partial \mathbf{f}_1}{\partial \mathbf{z}}\right)_{\mathbf{z}=\mathbf{z}_0} \mathbf{z} + \mathbf{B}\mathbf{u}, \tag{A.80}$$

where $(\frac{\partial \mathbf{f}_1}{\partial \mathbf{z}})_{\mathbf{z}=\mathbf{z}_0}$ is the Jacobian matrix of function $\mathbf{f}_1$ computed in $\mathbf{z}_0$.

If the formulation (A.78) is used, the linearized dynamics of the system about point $\mathbf{z}_0$ can be studied through the linear equation

$$\begin{cases} \dot{\mathbf{z}} = \mathbf{A}(\mathbf{z}_0)\mathbf{z} + \mathbf{B}(\mathbf{z}_0)\mathbf{u}, \\ \mathbf{y} = \mathbf{C}(\mathbf{z}_0)\mathbf{z} + \mathbf{D}(\mathbf{z}_0)\mathbf{u}. \end{cases} \tag{A.81}$$

While the motion and the stability in the small can be studied in closed form, to study the motion *in the large* it is compulsory to resort to the numerical integration of the equations of motion, that is, to resort to numerical simulation.

Remark A.12 Approximate approaches may allow to study in closed form the dynamics of a nonlinear system in some regions of the state space, but the errors they introduce are in general not predictable and they cannot find all the possible solutions.

A.3 Lagrange Equations in the Configuration and State Space

In the case of a relatively simple system it is possible to write directly the equations of motion in the form of (A.3), by writing all forces, internal and external to the system, acting on its various parts. However, if the system is complex, and in particular if the number of degrees of freedom is large, it is expedient to resort to the methods of analytical mechanics.

One of the simplest approaches to write the equations of motion of multi-degrees of freedom systems is by resorting to Lagrange equations that, for a generic mechanical system with n degrees of freedom whose configuration may be expressed using n generalized coordinates x_i, can be written in the form

$$\frac{d}{dt}\left(\frac{\partial \mathcal{T}}{\partial \dot{x}_i}\right) - \frac{\partial \mathcal{T}}{\partial x_i} + \frac{\partial \mathcal{U}}{\partial x_i} + \frac{\partial \mathcal{F}}{\partial \dot{x}_i} = Q_i \quad (i = 1, \dots, n), \tag{A.82}$$

where:

- $\mathcal{T}$ is the kinetic energy or the system. It allows us to write in a synthetic way inertia forces. In general,

$$\mathcal{T} = \mathcal{T}(\dot{x}_i, x_i, t).$$

The kinetic energy is in general a quadratic function of the generalized velocities

$$\mathcal{T} = \mathcal{T}_0 + \mathcal{T}_1 + \mathcal{T}_2, \tag{A.83}$$

where $\mathcal{T}_0$ does not depend on the velocities, $\mathcal{T}_1$ is linear and $\mathcal{T}_2$ is quadratic.

In the case of linear systems, the kinetic energy must contain terms containing no powers higher than 2 of the velocities and coordinates (or products of more than two of them). As a consequence, $\mathcal{T}_2$ cannot contain displacements

$$\mathcal{T}_2 = \frac{1}{2}\sum_{i=1}^{n}\sum_{j=1}^{n} m_{ij}\dot{x}_i\dot{x}_j = \frac{1}{2}\dot{\mathbf{x}}^T\mathbf{M}\dot{\mathbf{x}}, \tag{A.84}$$

where the terms m_{ij} do not depend either on $\mathbf{x}$ or $\dot{\mathbf{x}}$. If the system is time-invariant, $\mathbf{M}$ is constant.

$\mathcal{T}_1$ is linear in the velocities, and cannot contain terms more than linear in the displacements

$$\mathcal{T}_1 = \frac{1}{2}\dot{\mathbf{x}}^T(\mathbf{M}_1\mathbf{x} + \mathbf{f}_1), \tag{A.85}$$

where matrix $\mathbf{M}_1$ and vector $\mathbf{f}_1$ do not contain the generalized coordinates, even if $\mathbf{f}_1$ may be a function of time even in time-invariant systems.

$\mathcal{T}_0$ does not contain generalized velocities, but only terms of order not greater than 2 in the generalized coordinates:

$$\mathcal{T}_0 = \frac{1}{2}\mathbf{x}^T\mathbf{M}_g\mathbf{x} + \mathbf{x}^T\mathbf{f}_2 + e, \tag{A.86}$$

where matrix $\mathbf{M}_g$, vector $\mathbf{f}_2$ and scalar e are constant. Constant e does not enter the equations of motion. As it will be seen later, the structure of $\mathcal{T}_0$ is similar to that of the potential energy. The term

$$\mathcal{T}_0 - \mathcal{U}$$

is often referred to as *dynamic potential*.

- $\mathcal{U}$ is the potential energy and allows to express in a synthetic form conservative forces. In general,

$$\mathcal{U} = \mathcal{U}(x_i).$$

In the case of linear systems, the potential energy is a quadratic form in the generalized coordinates and, apart from a constant term that does not enter the equations of motion and then has no importance, can be written as

$$\mathcal{U} = \frac{1}{2}\mathbf{x}^T\mathbf{K}\mathbf{x} + \mathbf{x}^T\mathbf{f}_0, \tag{A.87}$$

Also in the case of nonlinear systems, by definition the potential energy does not depend on the generalized velocities and its derivatives with respect to the generalized velocities $\dot{x}_i$ vanish. Equation (A.82) is often written with reference to the *Lagrangian function*

$$\mathcal{L} = \mathcal{T} - \mathcal{U}$$

and becomes

$$\frac{d}{dt}\left(\frac{\partial \mathcal{L}}{\partial \dot{x}_i}\right) - \frac{\partial \mathcal{L}}{\partial x_i} + \frac{\partial \mathcal{F}}{\partial \dot{x}_i} = Q_i. \tag{A.88}$$

- $\mathcal{F}$ is the Raleigh dissipation function. It allows to express in a synthetic form some types of damping forces. In many cases $\mathcal{F} = \mathcal{F}(\dot{x}_i)$, but it may depend also on the generalized coordinates. In the case of linear systems, the dissipation function is a quadratic form in the generalized velocities and, apart from terms not depending from $\dot{x}_i$ that do not enter the equation of motion and thus have no importance, can be written as

$$\mathcal{F} = \frac{1}{2}\dot{\mathbf{x}}^T\mathbf{C}\dot{\mathbf{x}} + \frac{\mathbf{1}}{\mathbf{2}}\dot{\mathbf{x}}^T(\mathbf{C}_1\mathbf{x} + \mathbf{f}_3). \tag{A.89}$$

- Q_i are generalized forces that cannot be expressed using the above mentioned functions. In general, $Q_i = Q_i(\dot{q}_i, q_i, t)$. In the case of linear systems, they do not depend on the generalized coordinates and velocities, and then

$$Q_i = Q_i(t). \tag{A.90}$$

In the case of linear systems, by performing the relevant derivatives

$$\frac{\partial(\mathcal{T}-\mathcal{U})}{\partial \dot{x}_i} = \mathbf{M}\dot{\mathbf{x}} + \frac{1}{2}(\mathbf{M}_1\mathbf{x} + \mathbf{f}_1), \tag{A.91}$$

$$\frac{d}{dt}\left[\frac{\partial(\mathcal{T}-\mathcal{U})}{\partial \dot{x}_i}\right] = \mathbf{M}\ddot{\mathbf{x}} + \frac{1}{2}\mathbf{M}_1\dot{\mathbf{x}} + \dot{\mathbf{f}}_1, \tag{A.92}$$

$$\frac{\partial(\mathcal{T}-\mathcal{U})}{\partial x_i} = \frac{1}{2}\mathbf{M}_1^T\dot{\mathbf{x}} + \mathbf{M}_g\mathbf{x} - \mathbf{K}\mathbf{x} + \mathbf{f}_2 - \mathbf{f}_0. \tag{A.93}$$

$$\frac{\partial \mathcal{F}}{\partial \dot{x}_i} = \mathbf{C}\dot{\mathbf{x}} + \mathbf{C}_1\mathbf{x} + \mathbf{f}_3, \tag{A.94}$$

the equation of motion becomes

$$\mathbf{M}\ddot{\mathbf{x}} + \frac{1}{2}(\mathbf{M}_1 - \mathbf{M}_1^T)\dot{\mathbf{x}} + \mathbf{C}\dot{\mathbf{x}} + (\mathbf{K} - \mathbf{M}_g + \mathbf{C}_1)\mathbf{x} = -\dot{\mathbf{f}}_1 + \mathbf{f}_2 - \mathbf{f}_3 - \mathbf{f}_0 + \mathbf{Q}. \tag{A.95}$$

Matrix $\mathbf{M}_1$ is normally skew–symmetric. However, even if it is not such, it may be written as the sum of a symmetrical and a skew–symmetrical part

$$\mathbf{M}_1 = \mathbf{M}_{1\text{symm}} + \mathbf{M}_{1\text{skew}}. \tag{A.96}$$

By introducing this form into (A.95), the term

$$\mathbf{M}_1 - \mathbf{M}_1^T$$

becomes

$$\mathbf{M}_{1\text{symm}} + \mathbf{M}_{1\text{skew}} - \mathbf{M}_{1\text{symm}} + \mathbf{M}_{1\text{skew}} = 2\mathbf{M}_{1\text{skew}}.$$

Only the skew–symmetric part of $\mathbf{M}_1$ is included into the equation of motion. Also $\mathbf{C}_1$ is usually skew–symmetrical.

Writing $\mathbf{M}_{1\text{skew}}$ as $1/2$ $\mathbf{G}$ and $\mathbf{C}_1$ (or at least its skew–symmetric part; if a symmetric part existed, it could be included into matrix $\mathbf{K}$) as $\mathbf{H}$, and including vectors $\mathbf{f}_0$, $\dot{\mathbf{f}}_1$, $\mathbf{f}_2$ and $\mathbf{f}_3$ into forcing functions $\mathbf{Q}$, the equation of motion becomes

$$\mathbf{M}\ddot{\mathbf{x}} + (\mathbf{C} + \mathbf{G})\dot{\mathbf{x}} + (\mathbf{K} - \mathbf{M}_g + \mathbf{H})\mathbf{x} = \mathbf{Q}, \tag{A.97}$$

The mass, stiffness, gyroscopic and circulatory matrices $\mathbf{M}$, $\mathbf{K}$, $\mathbf{G}$ and $\mathbf{H}$ have already been defined. The symmetric matrix $\mathbf{M}_g$ is often defined as *geometric matrix*.[3]

As already said, a system in which $\mathcal{T}_1$ is not present is said to be *natural* and its equation of motion does not contain a gyroscopic matrix. In many cases also $\mathcal{T}_0$ is absent and the kinetic energy is expressed by (A.84).

To write the linearized equation of motion of a nonlinear system two ways are possible. The first is writing the complete expression of the energies, performing the derivatives obtaining the complete equations of motion and then canceling nonlinear terms.

The second one is reducing the expression of the energies to quadratic forms, by developing their expressions in power series and then truncating them after the quadratic terms. The linearized equations of motion are thus directly obtained.

The two ways yield the same result, but the first one is usually computationally much heavier.

To write the state equations, a number n of kinematic equations must be written

$$\dot{x}_i = v_i \quad (i = 1, \ldots, n). \tag{A.98}$$

If the state vector is defined in the usual way

$$\mathbf{z} = \left\{ \begin{matrix} \mathbf{v} \\ \mathbf{x} \end{matrix} \right\},$$

this procedure is straightforward.

A.4 Lagrange Equations for Systems with Constraints

In the previous sections the equations of motion were written in terms of the minimum number of generalized coordinates, i.e. by using as many generalized coordinates as the number of degrees of freedom of the system.

In many cases the system can be modeled as a number of point masses or rigid bodies subjected to constraints: in this case the mentioned approach requires the

[3]Here the symbol $\mathbf{M}_g$ is used instead of the more common $\mathbf{K}_g$ to stress that it comes from the kinetic energy.

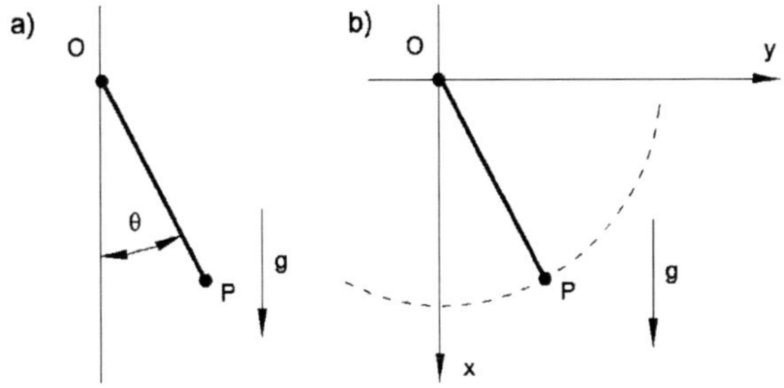

Fig. A.5 Pendulum, modeled using the minimum number of generalized coordinates (**a**), and as a system with two degrees of freedom and subjected to a constraint (**b**)

identifications of the minimum number of generalized coordinates describing all possible configurations of the system that are compatible with the mentioned constraints. An alternative approach is to write the equations of motion of the point masses or the rigid bodies using the same generalized coordinates that would be used if the constraints were not present, and then adding the equations defining the constraints.

For instance, if a pendulum made of a point mass attached to an inextensible massless string is considered, the first approach leads to identifying a single coordinate describing the motion of the pendulum, for instance the swing angle θ in Fig. A.5a. In the second approach the two generalized coordinates describing the motion of the point in a plane (e.g. coordinates x and y) are used, and an equation stating that the distance OP is constant is introduced.

The equations expressing the constraints may be of different types.

A.4.1 Holonomic Constraints

The simplest case is when there is a set of k relationships of the type

$$f_j(\mathbf{x}) = 0 \quad \text{for } j = 1, \ldots, k. \tag{A.99}$$

These constraints are simply geometrical constraint, i.e. define a number of lines or of surfaces to which the various part of the system are constrained. Constraints of this kind are said to be *holonomic*. Also the more general case

$$f_j(\mathbf{x}, t) = 0 \quad \text{for } j = 1, \ldots, k \tag{A.100}$$

deals with holonomic constraints.

In case of holonomic constraints, an augmented Lagrangian function can be defined

$$\mathcal{L}^* = \mathcal{L} + \sum_{j=1}^{k} \lambda_j(t) f_j(\mathbf{x}) = \mathcal{T} - \mathcal{U} + \sum_{j=1}^{k} \lambda_j(t) f_j(\mathbf{x}). \tag{A.101}$$

Functions $\lambda_j(t)$ are said to be the Lagrange multipliers, and are dealt with as additional generalized coordinates of the dynamic system.

Remark A.13 The physical meaning of Lagrange multipliers is not that of coordinates, but of forces exerted by the constraints.

The augmented Lagrangian can be introduced into the Lagrange equations, which become

$$\frac{d}{dt}\left(\frac{\partial \mathcal{T}}{\partial \dot{x}_i}\right) - \frac{\partial \mathcal{T}}{\partial x_i} + \frac{\partial \mathcal{U}}{\partial x_i} + \frac{\partial \mathcal{F}}{\partial \dot{x}_i} - \sum_{j=1}^{k} \lambda_j \frac{\partial f_j}{\partial x_i} = Q_i \quad (i = 1, \dots, n), \tag{A.102}$$

These n equations of motion must be associated to the k constraint equations (A.99), yielding a set of $n + k$ equations in the $n + k$ unknowns x_i and λ_j.

Remark A.14 The n dynamic equations are ordinary differential equations while the k constraint equations are algebraic.

The simplest example of a system with constraints is the pendulum with length l and mass m of Fig. A.5. If it is studied using the minimum number of coordinates, angle θ can be chosen and the Lagrangian function is

$$\mathcal{L} = \frac{1}{2} m l^2 \dot{\theta}^2 - mgl \cos(\theta). \tag{A.103}$$

The equation of motion is thus

$$\ddot{\theta} + \frac{g}{l} \sin(\theta) = 0. \tag{A.104}$$

In the other approach, based on the explicit statement of the constraint, the generalized coordinates x and y of point P are used as generalized coordinates. The constraint equation states that distance OP is equal to l:

$$\sqrt{x^2 + y^2} - l = 0. \tag{A.105}$$

The augmented Lagrangian function is thus

$$\mathcal{L}^* = \frac{1}{2} m\left(\dot{x}^2 + \dot{y}^2\right) + mgx + \lambda\left(\sqrt{x^2 + y^2} - l\right). \tag{A.106}$$

By performing the relevant derivatives, the equations of motion and the constraint equation are

$$\begin{cases} m\ddot{x} - mgl - \lambda \dfrac{x}{l} = 0, \\ m\ddot{y} - \lambda \dfrac{y}{l} = 0, \\ x^2 + y^2 - l^2 = 0. \end{cases} \tag{A.107}$$

To obtain the same solution as above, the following change of variable can be effected:

$$x = l\cos(\theta). \tag{A.108}$$

From the last equation, it follows that

$$y = l\sin(\theta). \tag{A.109}$$

The accelerations

$$\begin{cases} \ddot{x} = -l\ddot{\theta}\sin(\theta) - l\dot{\theta}^2\cos(\theta), \\ \ddot{y} = l\ddot{\theta}\cos(\theta) - l\dot{\theta}^2\sin(\theta) \end{cases} \tag{A.110}$$

can be introduced into the first two equations, obtaining

$$\begin{cases} -ml\ddot{\theta}\sin(\theta) - ml\dot{\theta}^2\cos(\theta) - mgl - \lambda\cos(\theta) = 0, \\ ml\ddot{\theta}\cos(\theta) - ml\dot{\theta}^2\sin(\theta) - \lambda\sin(\theta) = 0. \end{cases} \tag{A.111}$$

By multiplying the first equation by $-\sin(\theta)$ and the second by $\cos(\theta)$ and adding it follows that

$$l\ddot{\theta} + g\sin(\theta) = 0, \tag{A.112}$$

which coincides with the previous equation. By multiplying the first equation by $\cos(\theta)$ and the second by $\sin(\theta)$ and adding it follows that

$$\lambda = -m\big[l\dot{\theta}^2 + g\cos(\theta)\big]. \tag{A.113}$$

The two terms in the expression for λ are the centripetal force plus the force needed to compensate for the component of the weight in the direction of the wire. λ is thus the force the wire exerts on mass m.

A.4.2 Non-holonomic Constraints

If also the velocities are involved in the constraint equations

$$f_j(\mathbf{x}, \dot{\mathbf{x}}, t) = 0 \quad \text{for } j = 1, \dots, k \tag{A.114}$$

the constraints are said to be *non-holonomic*. In this case it is impossible to use the augmented Lagrangian function seen for the holonomic constraints.

If the Jacobian matrix

$$\frac{\partial f_j(\mathbf{x}, \dot{\mathbf{x}}, t)}{\partial \dot{\mathbf{x}}}$$

has rank k, i.e. all constraints are non-holonomic and are independent, the equation of motion can be shown to be[4]

$$\frac{d}{dt}\left(\frac{\partial \mathcal{T}}{\partial \dot{x}_i}\right)-\frac{\partial \mathcal{T}}{\partial x_i}+\frac{\partial \mathcal{U}}{\partial x_i}+\frac{\partial \mathcal{F}}{\partial \dot{x}_i}+\sum_{j=1}^{k}\lambda_j\frac{\partial f_j}{\partial \dot{x}_i}=Q_i \quad (i=1,\dots,n). \tag{A.115}$$

A mixed case is also possible: if there are k_1 holonomic and k_2 non-holonomic constraint that satisfy the above mentioned condition (their Jacobian matrix has rank k_2), the two types of constraints can be dealt with separately as seen in (A.102) and (A.115).

A.5 Hamilton Equations in the Phase Space

If the generalized momenta are used as auxiliary variables instead the generalized velocities, the equations are written with reference to the *phase space* and the *phase vector* instead of the state space and vector.

The generalized momenta are defined, starting from the Lagrangian $\mathcal{L}$, as

$$\mathbf{p}=\frac{\partial \mathcal{L}}{\partial \dot{x}_i}. \tag{A.116}$$

If the system is a natural linear system, this definition reduce to the usual one

$$\mathbf{p}=\mathbf{M}\dot{\mathbf{x}}. \tag{A.117}$$

By including the forces coming from the dissipation function in the generalized forces Q_i, the Lagrange equation simplifies as

$$\dot{p}_i=\frac{\partial \mathcal{L}}{\partial x_i}+Q_i. \tag{A.118}$$

A function $\mathcal{H}(\dot{x}_i, x_i, t)$, said *Hamiltonian function* is defined as

$$\mathcal{H}=\mathbf{p}^T\dot{\mathbf{x}}-\mathcal{L}. \tag{A.119}$$

Since $\mathcal{H}$ is a function of p_i, x_i and t ($\mathcal{H}(p_i, x_i, t)$), the differential $\delta\mathcal{H}$ is

$$\delta\mathcal{H}=\sum_{i=1}^{n}\left(\frac{\partial \mathcal{H}}{\partial p_i}\delta p_i+\frac{\partial \mathcal{H}}{\partial x_i}\delta x_i\right). \tag{A.120}$$

On the other hand, (A.119) yields

$$\delta\mathcal{H}=\sum_{i=1}^{n}\left(p_i\delta\dot{x}_i+\dot{x}_i\delta p_i-\frac{\partial \mathcal{L}}{\partial x_i}\delta x_i-\frac{\partial \mathcal{L}}{\partial \dot{x}_i}\delta\dot{x}_i\right)$$

---

[4]M.R. Flannery, The enigma of nonholonomic constraints, Am. J. Phys. 73(3):265–272, 2005; O.M. Moreschi, G. Castellano, Geometric approach to non-holonomic problems satisfying Hamilton's principle, Rev. Unión Mat. Argent. 47(2):125–135, 2005.

$$= \sum_{i=1}^{n} \left(\dot{x}_i \delta p_i - \frac{\partial \mathcal{L}}{\partial x_i} \delta x_i \right), \tag{A.121}$$

and then

$$\frac{\partial \mathcal{H}}{\partial p_i} = \dot{x}_i, \qquad \frac{\partial \mathcal{H}}{\partial x_i} = -\frac{\partial \mathcal{L}}{\partial x_i}. \tag{A.122}$$

The $2n$ phase space equations are thus

$$\begin{cases} \dot{x}_i = \dfrac{\partial \mathcal{H}}{\partial p_i}, \\ \dot{p}_i = -\dfrac{\partial \mathcal{H}}{\partial x_i} + Q_i. \end{cases} \tag{A.123}$$

A.6 Lagrange Equations in Terms of Pseudo-Coordinates

Often the state equations are written with reference to generalized velocities that are not simply the derivatives of the generalized coordinates. In particular, often it is expedient to use as generalized velocities suitable combinations of the derivatives of the coordinates $v_i = \dot{x}_i$

$$\{w_i\} = \mathbf{A}^T \{\dot{x}_i\}, \tag{A.124}$$

where the coefficients of the linear combinations included into matrix $\mathbf{A}^T$ may be constant, but in general are functions of the generalized coordinates.

Equation (A.124) can in general be inverted, obtaining

$$\{\dot{x}_i\} = \mathbf{B}\{w_i\}, \tag{A.125}$$

where

$$\mathbf{B} = \mathbf{A}^{-T} \tag{A.126}$$

and the symbol $\mathbf{A}^{-T}$ indicates the inverse of the transpose of matrix $\mathbf{A}$.

In some cases matrix $\mathbf{A}^T$ is a rotation matrix and its inverse coincides with its transpose. In those case

$$\mathbf{B} = \mathbf{A}^{-T} = \mathbf{A}.$$

However, in general this does not occur and

$$\mathbf{B} \neq \mathbf{A}.$$

While v_i are the derivatives of the coordinates x_i, in general it is not possible to express w_i as the derivatives of suitable coordinates. Equation (A.124) can be written in the infinitesimal displacements dx_i

$$\{d\theta_i\} = \mathbf{A}^T \{dx_i\}, \tag{A.127}$$

obtaining a set of infinitesimal displacements $d\theta_i$, corresponding to velocities w_i. Equations (A.127) can be integrated, yielding displacements θ_i corresponding to the velocities w_i, only if

$$\frac{\partial a_{js}}{\partial x_k} = \frac{\partial a_{ks}}{\partial x_j}.$$

Otherwise (A.127) cannot be integrated and velocities w_i cannot be considered as the derivatives of true coordinates. In such a case they are said to be the derivatives of *pseudo-coordinates*.

As a first consequence of the non existence of coordinates corresponding to velocities w_i, Lagrange equation (A.82) cannot be written directly using velocities w_i (that cannot be considered as derivatives of the new coordinates), but must be modified to allow the use of velocities and coordinates that are not directly one the derivative of the other.

The use of pseudo-coordinate is fairly common. If, for instance, in the dynamics of a rigid body, the generalized velocities in a reference frame following the body in its motion are used, while the coordinates x_i are the displacements in an inertial frame, matrix $\mathbf{A}^T$ is simply the rotation matrix allowing to pass from the one reference frame to the other. Matrix $\mathbf{B}$ coincides in this case with $\mathbf{A}$, but both are not symmetrical and the velocities in the body-fixed frame cannot be considered as the derivatives of the displacements in that frame. In other words, such a frame rotates continuously and it is not possible to integrate the velocities along the body-fixed axes to obtain the displacements along the same axes. That notwithstanding, it is possible to use the components of the velocity along the body-fixed axes to write the equations of motion.

The kinetic energy can be written in general in the form

$$\mathcal{T} = \mathcal{T}(w_i, x_i, t).$$

The derivatives $\partial\mathcal{T}/\partial\dot{x}_i$ included into the equations of motion are, in matrix form

$$\left\{\frac{\partial\mathcal{T}}{\partial\dot{x}}\right\} = \mathbf{A}\left\{\frac{\partial\mathcal{T}}{\partial w}\right\}, \tag{A.128}$$

where

$$\left\{\frac{\partial\mathcal{T}}{\partial\dot{x}}\right\} = \left[\frac{\partial\mathcal{T}}{\partial\dot{x}_1} \quad \frac{\partial\mathcal{T}}{\partial\dot{x}_2} \quad \dots\right]^T,$$

$$\left\{\frac{\partial\mathcal{T}}{\partial w}\right\} = \left[\frac{\partial\mathcal{T}}{\partial w_1} \quad \frac{\partial\mathcal{T}}{\partial w_2} \quad \dots\right]^T.$$

By differentiating with respect to time, it follows that

$$\frac{\partial}{\partial t}\left(\left\{\frac{\partial\mathcal{T}}{\partial\dot{x}}\right\}\right) = \mathbf{A}\frac{\partial}{\partial t}\left(\left\{\frac{\partial\mathcal{T}}{\partial w}\right\}\right) + \dot{\mathbf{A}}\left\{\frac{\partial\mathcal{T}}{\partial w}\right\}. \tag{A.129}$$

The generic element $\dot{a}_{jk}$ of matrix $\dot{\mathbf{A}}$ is

$$\dot{a}_{jk} = \sum_{i=1}^{n} \frac{\partial a_{jk}}{\partial x_i} \dot{x}_i = \dot{\mathbf{x}}^T \left\{ \frac{\partial a_{jk}}{\partial x} \right\}, \tag{A.130}$$

i.e.

$$\dot{a}_{jk} = \mathbf{w}^T \mathbf{B}^T \left\{ \frac{\partial a_{jk}}{\partial x} \right\}. \tag{A.131}$$

The various $\dot{a}_{jk}$ so computed can be written in matrix form

$$\dot{\mathbf{A}} = \left[\mathbf{w}^T \mathbf{B}^T \left\{ \frac{\partial a_{jk}}{\partial x} \right\} \right]. \tag{A.132}$$

The computation of the derivatives of the generalized coordinates $\partial \mathcal{T}/\partial x$ is usually less straightforward. The generic derivative $\partial \mathcal{T}/\partial x_k$ is

$$\frac{\partial \mathcal{T}^*}{\partial x_k} = \frac{\partial \mathcal{T}}{\partial x_k} + \sum_{i=1}^{n} \frac{\partial \mathcal{T}}{\partial w_i} \frac{\partial w_i}{\partial x_k} = \frac{\partial \mathcal{T}}{\partial x_k} + \sum_{i=1}^{n} \frac{\partial \mathcal{T}}{\partial w_i} \sum_{j=1}^{n} \frac{\partial a_{ij}}{\partial x_k} \dot{x}_j, \tag{A.133}$$

where $\mathcal{T}^*$ is the kinetic energy expressed as a function of the generalized coordinates and their derivatives (the expression to be introduced into the Lagrange equation in its usual form), while $\mathcal{T}$ is expressed as a function of the generalized coordinates and of the velocities in the body-fixed frame. Equation (A.133) can be written as

$$\frac{\partial \mathcal{T}^*}{\partial x_k} = \frac{\partial \mathcal{T}}{\partial x_k} + \mathbf{w}^T \mathbf{B}^T \frac{\partial \mathbf{A}}{\partial x_k} \left\{ \frac{\partial \mathcal{T}}{\partial w} \right\}, \tag{A.134}$$

where the product $\mathbf{w}^T \mathbf{B}^T \frac{\partial \mathbf{A}}{\partial x_k}$ yields a row matrix with n elements that multiplied by the column matrix $\{\frac{\partial \mathcal{T}}{\partial w}\}$ yields the required number.

By combining those row matrices, a square matrix is obtained

$$\left[\mathbf{w}^T \mathbf{B}^T \frac{\partial \mathbf{A}}{\partial x_k} \right], \tag{A.135}$$

and then the column containing the derivatives with respect to the generalized coordinates is

$$\left\{ \frac{\partial \mathcal{T}^*}{\partial x} \right\} = \left\{ \frac{\partial \mathcal{T}}{\partial x} \right\} + \left[\mathbf{w}^T \mathbf{B}^T \frac{\partial \mathbf{A}}{\partial x} \right] \left\{ \frac{\partial \mathcal{T}}{\partial w} \right\}. \tag{A.136}$$

By definition, the potential energy does not depend on the generalized velocities and thus the term $\partial \mathcal{U}/\partial x_i$ is not influenced by the way the generalized velocities are written. Finally, the derivatives of the dissipation function are

$$\left\{ \frac{\partial \mathcal{F}}{\partial \dot{x}} \right\} = \mathbf{A} \left\{ \frac{\partial \mathcal{F}}{\partial w} \right\}. \tag{A.137}$$

The equation of motion (A.82) is thus

$$\mathbf{A}\frac{\partial}{\partial t}\left(\left\{\frac{\partial \mathcal{T}}{\partial w}\right\}\right)+\boldsymbol{\Gamma}\left\{\frac{\partial \mathcal{T}}{\partial w}\right\}-\left\{\frac{\partial \mathcal{T}}{\partial x}\right\}+\left\{\frac{\partial \mathcal{U}}{\partial x}\right\}+\mathbf{A}\left\{\frac{\partial \mathcal{F}}{\partial w}\right\}=\mathbf{Q}, \tag{A.138}$$

where

$$\boldsymbol{\Gamma}=\left[\mathbf{w}^T\mathbf{B}^T\left\{\frac{\partial a_{jk}}{\partial x}\right\}\right]-\left[\mathbf{w}^T\mathbf{B}^T\frac{\partial \mathbf{A}}{\partial x_k}\right] \tag{A.139}$$

and $\mathbf{Q}$ is a vector containing the n generalized forces Q_i.

By premultiplying all terms by matrix $\mathbf{B}^T=\mathbf{A}^{-1}$ and attaching the kinematic equations to the dynamic equations, the final form of the state space equations is obtained:

$$\begin{cases}\dfrac{\partial}{\partial t}\left(\left\{\dfrac{\partial \mathcal{T}}{\partial w}\right\}\right)+\mathbf{B}^{\mathbf{T}}\boldsymbol{\Gamma}\left\{\dfrac{\partial \mathcal{T}}{\partial w}\right\}-\mathbf{B}^T\left\{\dfrac{\partial \mathcal{T}}{\partial x}\right\}+\mathbf{B}^T\left\{\dfrac{\partial \mathcal{U}}{\partial x}\right\}+\left\{\dfrac{\partial \mathcal{F}}{\partial w}\right\}=\mathbf{B}^T\mathbf{Q},\\ \{\dot{x}_i\}=\mathbf{B}\{w_i\}.\end{cases} \tag{A.140}$$

A.7 Motion of a Rigid Body

A.7.1 Generalized Coordinates

A rigid body free in tridimensional space has six degrees of freedom. A possible set of six generalized coordinates defining its pose was stated in Sect. 3.6. Once an inertial reference frame $OXYZ$ and a frame $Gxyz$ fixed to the body and centered in its center of mass are stated, the position of the rigid body is defined by

- the coordinates of point G in the inertial frame $OXYZ$, i.e. coordinates X_G, Y_G and Z_G, and
- a set of three Euler or Tait–Bryan angles, for instance the yaw (ψ), pitch (θ) and roll (ϕ) angles.

The rotation matrices related to these rotations $\mathbf{R}_1$, $\mathbf{R}_2$ and $\mathbf{R}_3$ are defined by (3.3), (3.4) and (3.5) and the total rotation matrix is expressed by (3.7):

$$\begin{aligned}\mathbf{R}&=\mathbf{R}_1\mathbf{R}_2\mathbf{R}_3\\ &=\begin{bmatrix}c(\psi)c(\theta) & c(\psi)s(\theta)s(\phi)-s(\psi)c(\phi) & c(\psi)s(\theta)c(\phi)+s(\psi)s(\phi)\\ s(\psi)c(\theta) & s(\psi)s(\theta)s(\phi)+c(\psi)c(\phi) & s(\psi)s(\theta)c(\phi)-c(\psi)s(\phi)\\ -s(\theta) & c(\theta)s(\phi) & c(\theta)c(\phi)\end{bmatrix},\end{aligned}$$

where symbols cos and sin have been substituted by c and s.

Sometimes roll and pitch angles are small. In this case it is expedient to keep the last two rotations separate from the first ones. The product of the rotation matrices related to the last two rotations is

$$\mathbf{R}_2\mathbf{R}_3 = \begin{bmatrix} \cos(\theta) & \sin(\theta)\sin(\phi) & \sin(\theta)\cos(\phi) \\ 0 & \cos(\phi) & -\sin(\phi) \\ -\sin(\theta) & \cos(\theta)\sin(\phi) & \cos(\theta)\cos(\phi) \end{bmatrix}, \tag{A.141}$$

which becomes, in the case that the angles are small,

$$\mathbf{R}_2\mathbf{R}_3 \approx \begin{bmatrix} 1 & 0 & \theta \\ 0 & 1 & -\phi \\ -\theta & \phi & 1 \end{bmatrix}. \tag{A.142}$$

The angular velocities $\dot{\psi}$, $\dot{\theta}$ and $\dot{\phi}$ are not applied along x, y and z axes, and thus are not the components Ω_x, Ω_y and Ω_z of the angular velocity in the body-fixed reference frame.[5] Their directions are those of axes Z, y^* and x (see Fig. 3.10), and the angular velocity in the body-fixed frame is

$$\begin{Bmatrix} \Omega_x \\ \Omega_y \\ \Omega_z \end{Bmatrix} = \dot{\phi}\mathbf{e}_x + \dot{\theta}\mathbf{R}_3^T\mathbf{e}_y + \dot{\psi}[\mathbf{R}_2\mathbf{R}_3]^T\mathbf{e}_z, \tag{A.143}$$

where the unit vectors are obviously

$$\mathbf{e}_x = \begin{Bmatrix} 1 \\ 0 \\ 0 \end{Bmatrix}, \quad \mathbf{e}_y = \begin{Bmatrix} 0 \\ 1 \\ 0 \end{Bmatrix}, \quad \mathbf{e}_z = \begin{Bmatrix} 0 \\ 0 \\ 1 \end{Bmatrix}. \tag{A.144}$$

By performing the products, it follows that

$$\begin{cases} \Omega_x = \dot{\phi} - \dot{\psi}\sin(\theta), \\ \Omega_y = \dot{\theta}\cos(\phi) + \dot{\psi}\sin(\phi)\cos(\theta), \\ \Omega_z = \dot{\psi}\cos(\theta)\cos(\phi) - \dot{\theta}\sin(\phi), \end{cases} \tag{A.145}$$

or, in matrix form

$$\begin{Bmatrix} \Omega_x \\ \Omega_y \\ \Omega_z \end{Bmatrix} = \begin{bmatrix} 1 & 0 & -\sin(\theta) \\ 0 & \cos(\phi) & \sin(\phi)\cos(\theta) \\ 0 & -\sin(\phi) & \cos(\phi)\cos(\theta) \end{bmatrix} \begin{Bmatrix} \dot{\phi} \\ \dot{\theta} \\ \dot{\psi} \end{Bmatrix}. \tag{A.146}$$

If the pitch and roll angles are small enough to linearize the relevant trigonometric functions, the components of the angular velocity can be approximated as

$$\begin{cases} \Omega_x = \dot{\phi} - \theta\dot{\psi}, \\ \Omega_y = \dot{\theta} + \phi\dot{\psi}, \\ \Omega_z = \dot{\psi} - \phi\dot{\theta}. \end{cases} \tag{A.147}$$

[5] Often symbols p, q and r are used for the components of the angular velocity in the body-fixed frame.

A.7.2 Equations of Motion—Lagrangian Approach

Assuming that the body axes xyz are principal axes of inertia, the kinetic energy of the rigid body is

$$\begin{aligned}\mathcal{T} = {} & \frac{1}{2}m(\dot{X}^2+\dot{Y}^2+\dot{Z}^2)+\frac{1}{2}J_x\big[\dot{\phi}-\dot{\psi}\sin(\theta)\big]^2\\ & +\frac{1}{2}J_y\big[\dot{\theta}\cos(\phi)+\dot{\psi}\sin(\phi)\cos(\theta)\big]^2\\ & +\frac{1}{2}J_z\big[\dot{\psi}\cos(\theta)\cos(\phi)-\dot{\theta}\sin(\phi)\big]^2. \end{aligned} \tag{A.148}$$

Introducing the kinetic energy into the Lagrange equations

$$\frac{d}{dt}\left(\frac{\partial\mathcal{T}}{\partial\dot{q}_i}\right)-\frac{\partial\mathcal{T}}{\partial q_i}=Q_i,$$

and performing the relevant derivatives, the six equations of motion are directly obtained. The three equations for translational motion are

$$\begin{cases} m\ddot{X}=Q_X,\\ m\ddot{Y}=Q_Y,\\ m\ddot{Z}=Q_Z. \end{cases} \tag{A.149}$$

The equations for rotational motion are much more complicated

$$\begin{aligned}
&\ddot{\psi}\big[J_x\sin^2(\theta)+J_y\sin^2(\phi)\cos^2(\theta)+J_z\cos^2(\phi)\cos^2(\theta)\big]\\
&\quad-\ddot{\phi}J_x\sin(\theta)+\ddot{\theta}(J_y-J_z)\sin(\phi)\cos(\phi)\cos(\theta)\\
&\quad+\dot{\phi}\dot{\theta}\cos(\theta)\big\{\big[1-2\sin^2(\phi)\big](J_y-J_z)-J_x\big\}\\
&\quad+2\dot{\phi}\dot{\psi}(J_y-J_z)\cos(\phi)\cos^2(\theta)\sin(\phi)\\
&\quad+2\dot{\theta}\dot{\psi}\sin(\theta)\cos(\theta)\big[J_x-\sin^2(\phi)J_y-\cos^2(\phi)J_z\big]\\
&\quad+\dot{\theta}^2(-J_y+J_z)\sin(\phi)\cos(\phi)\sin(\theta)=Q_\psi,\\
&\ddot{\psi}(J_y-J_z)\sin(\phi)\cos(\theta)\cos(\phi)+\ddot{\theta}\big[J_y\cos^2(\phi)+J_z\sin^2(\phi)\big]\\
&\quad+2\dot{\phi}\dot{\theta}(J_z-J_y)\sin(\phi)\cos(\phi)+\dot{\phi}\dot{\psi}(J_y-J_z)\cos(\theta)\big[1-2\sin^2(\phi)\big]\\
&\quad+\dot{\psi}\dot{\phi}J_x\cos(\theta)-\dot{\psi}^2\sin(\theta)\cos(\theta)\big[J_x-J_y\sin^2(\phi)-J_z\cos^2(\phi)\big]=Q_\theta,\\
&+J_x\ddot{\phi}-\sin(\theta)J_x\ddot{\psi}-\dot{\theta}\dot{\psi}J_z\sin^2(\phi)\cos(\theta)\\
&\quad-\dot{\psi}\dot{\theta}\cos(\theta)\big\{J_x+J_y\big[1-2\sin^2(\phi)\big]-J_z\cos^2(\phi)\big\}\\
&\quad+\dot{\theta}^2(J_y-J_z)\sin(\phi)\cos(\phi)-\dot{\psi}^2(J_y-J_z)\cos(\phi)\cos^2(\theta)\sin(\phi)=Q_\phi.
\end{aligned} \tag{A.150}$$

Remark A.15 Angle ψ does not appear explicitly into the equations of motion and all trigonometric functions can be linearized if the roll and pitch angles are small.

If also the angular velocities are small, the equations of motion for rotations reduce to

$$\begin{cases} J_z\ddot{\psi} = Q_\psi, \\ J_y\ddot{\theta} = Q_\theta, \\ J_x\ddot{\phi} = Q_\phi. \end{cases} \tag{A.151}$$

In this case, the kinetic energy may be directly simplified, by developing the trigonometric functions in Taylor series and neglecting all terms containing products of three or more small quantities. For instance, the term

$$[\dot{\phi} - \dot{\psi}\sin(\theta)]^2$$

reduces to

$$[\dot{\phi} - \dot{\psi}\theta + \dot{\psi}\theta^3/6 + \cdots]^2$$

and then to $\dot{\phi}^2$, since all other terms contain products of at least three small quantities. The kinetic energy reduces to

$$\mathcal{T} \approx \frac{1}{2}m(\dot{X}^2 + \dot{Y}^2 + \dot{Z}^2) + \frac{1}{2}(J_x\dot{\phi}^2 + J_y\dot{\theta}^2 + J_z\dot{\psi}^2). \tag{A.152}$$

Remark A.16 This approach is simple only if the roll and pitch angles are small. If not, the equations of motion obtained in this way in terms of angular velocities $\dot{\phi}$, $\dot{\theta}$ and $\dot{\psi}$ are quite complicated and another approach is more expedient.

A.7.3 Equations of Motion Using Pseudo-Coordinates

Often the forces and moments applied to the rigid body are written with reference to the body-fixed frame. In these cases, the equations of motion are best written with reference to the same frame. The kinetic energy can be written in terms of the components v_x, v_y and v_z (often referred to as u, v and w) of the velocity and Ω_x, Ω_x and Ω_x (often referred to as p, q and r) of the angular velocity.

If the body-fixed frame is a principal frame of inertia, the expression of the kinetic energy is

$$\mathcal{T} = \frac{1}{2}m(v_x^2 + v_y^2 + v_z{}^2) + \frac{1}{2}(J_x\Omega_x^2 + J_y\Omega_y^2 + J_z\Omega_z^2).$$

The components of the velocity and of the angular velocity in the body-fixed frame are not the derivatives of coordinate, but are linked to the coordinates by the six kinematic equations

$$\begin{Bmatrix} v_x \\ v_y \\ v_z \end{Bmatrix} = \mathbf{R}^T \begin{Bmatrix} \dot{X} \\ \dot{Y} \\ \dot{Z} \end{Bmatrix}, \tag{A.153}$$

$$\begin{Bmatrix} \Omega_x \\ \Omega_y \\ \Omega_z \end{Bmatrix} = \begin{bmatrix} 1 & 0 & -\sin(\theta) \\ 0 & \cos(\phi) & \sin(\phi)\cos(\theta) \\ 0 & -\sin(\phi) & \cos(\theta)\cos(\phi) \end{bmatrix} \begin{Bmatrix} \dot{\phi} \\ \dot{\theta} \\ \dot{\psi} \end{Bmatrix}, \tag{A.154}$$

that is, in more compact form,

$$\mathbf{w} = \mathbf{A}^T \dot{\mathbf{q}}, \tag{A.155}$$

where the vectors of the generalized velocities and of the derivatives of the generalized coordinates are

$$\mathbf{w} = [\, v_x \; v_y \; v_z \; \Omega_x \; \Omega_y \; \Omega_z \,]^T, \tag{A.156}$$

$$\dot{\mathbf{q}} = [\, \dot{X} \; \dot{Y} \; \dot{Z} \; \dot{\phi} \; \dot{\theta} \; \dot{\psi} \,]^T \tag{A.157}$$

and matrix **A** is

$$\mathbf{A} = \begin{bmatrix} \mathbf{R} & \mathbf{0} \\ \mathbf{0} & \begin{bmatrix} 1 & 0 & -\sin(\theta) \\ 0 & \cos(\phi) & \sin(\phi)\cos(\theta) \\ 0 & -\sin(\phi) & \cos(\theta)\cos(\phi) \end{bmatrix}^T \end{bmatrix}. \tag{A.158}$$

Note that the second submatrix is not a rotation matrix (the first one is such), and

$$\mathbf{A}^{-1} \neq \mathbf{A}^T; \qquad \mathbf{B} \neq \mathbf{A}. \tag{A.159}$$

The inverse transformation is (A.125)

$$\dot{\mathbf{q}} = \mathbf{B}\mathbf{w},$$

where $\mathbf{B} = \mathbf{A}^{-T}$.

None of the velocities included in vector **w** can be integrated to obtain a set of generalized coordinates, and they must all be considered as derivatives of pseudo-coordinates.

The state space equation, made of the six dynamic and the six kinematic equations, is thus (A.140), simplified since in the present case neither the potential energy nor the dissipation function are present

$$\begin{cases} \dfrac{\partial}{\partial t}\left(\left\{\dfrac{\partial \mathcal{T}}{\partial w}\right\}\right) + \mathbf{B}^T\boldsymbol{\Gamma}\left\{\dfrac{\partial \mathcal{T}}{\partial w}\right\} - \mathbf{B}^T\left\{\dfrac{\partial \mathcal{T}}{\partial q}\right\} = \mathbf{B}^T\mathbf{Q}, \\ \{\dot{q}_i\} = \mathbf{B}\{w_i\}. \end{cases} \tag{A.160}$$

Here $\mathbf{B}^T\mathbf{Q}$ is just a column matrix containing the three components of the force and the three components of the moment applied to the body along the body-fixed axes x, y, z.

The most difficult part of the computation is writing matrix $\mathbf{B}^T\boldsymbol{\Gamma}$. Performing somewhat difficult computations it follows that

$$\mathbf{B}^T\boldsymbol{\Gamma} = \begin{bmatrix} \tilde{\boldsymbol{\Omega}} & \mathbf{0} \\ \tilde{\mathbf{V}} & \tilde{\boldsymbol{\Omega}} \end{bmatrix}, \tag{A.161}$$

where $\widetilde{\boldsymbol{\Omega}}$ and $\widetilde{\mathbf{V}}$ are skew–symmetric matrices containing the components of the angular and linear velocities

$$\widetilde{\boldsymbol{\Omega}} = \begin{bmatrix} 0 & -\Omega_z & \Omega_y \\ \Omega_z & 0 & -\Omega_x \\ -\Omega_y & \Omega_x & 0 \end{bmatrix}, \qquad \widetilde{\mathbf{V}} = \begin{bmatrix} 0 & -v_z & v_y \\ v_z & 0 & -v_x \\ -v_y & v_x & 0 \end{bmatrix}. \tag{A.162}$$

If the body-fixed axes are principal axes of inertia, the dynamic equations are simply

$$\begin{cases} m\dot{v}_x = m\Omega_z v_y - m\Omega_y v_z + F_x, \\ m\dot{v}_y = m\Omega_x v_z - m\Omega_z v_x + F_y, \\ m\dot{v}_z = m\Omega_y v_x - m\Omega_x v_y + F_z, \\ J_x\dot{\Omega}_x = \Omega_y\Omega_z(J_y - J_z) + M_x, \\ J_y\dot{\Omega}_y = \Omega_x\Omega_z(J_z - J_x) + M_y, \\ J_z\dot{\Omega}_z = \Omega_x\Omega_y(J_x - J_y) + M_x. \end{cases} \tag{A.163}$$

Remark A.17 The equations so obtained are much simpler than equations (A.150) and the last three equations are nothing else than Euler equations.

A.8 Multibody Modeling

A robot or a vehicle may be modeled as a first approximation as a system made by a number of rigid bodies connected with each other through joints of different type, springs and dampers.

For instance, the *Apollo* LRV, like any other four wheeled vehicle, can be considered as a rigid body suspended on the ground by its elastic wheels and suspensions. If the suspensions are assumed to constrain all degrees of freedom of the wheel hubs except their motion in vertical direction, the wheel rotation is assumed to be determined by the forward motion of the vehicle (the longitudinal slip is neglected) and the steering is assumed to be an input parameter determined by the will of the driver, such model has 10 degrees of freedom: six determining the position of the frame as a rigid body in tridimensional space plus four, one for each wheel hub.

A free flying spacecraft carrying an arm with three degrees of freedom can be modeled as a system made of a number of rigid bodies, with nine degrees of freedom and so on.

This kind of approach is usually referred to as *multibody* approach.

It can be extended to model the system to a much greater detail. For instance, instead of modeling the suspensions of the LRV as a rigid body (the wheel hub) that can move in a direction parallel to the z-axis (in vertical direction on flat ground) of the vehicle, it is possible to enter into a much greater detail, considering the lower and upper triangles of the suspension and the wheel strut as three rigid bodies connected to each other by cylindrical hinges.

However, if the flexibility of the various elements is neglected, the number of degrees of freedom does not change, since the motion of the various parts of each suspension is determined by the single parameter that is the vertical displacement of the wheel hub.

The fact that the number of degrees of freedom does not change does not mean that the complexity of the model is the same. The mathematical model of a multibody system consists of $6n$ differential equations of dynamic equilibrium, if the motion is studied in the tridimensional space and n is number of rigid bodies and by an adequate number of constraint equations.

For instance, to return to the model of the LRV, the total number of rigid bodies is 13 (the vehicle frame plus four suspensions made by three rigid bodies, the upper and lower triangles and the strut, each) and the dynamic equations are 78. The connections between the various members of each suspensions originate a number of 17 constraint equations for a total of 68 constraint equations. Using the latter to eliminate 68 out of the 78 generalized coordinates describing the motion of all the rigid bodies, the number of dynamic equilibrium equations can be reduced to 10, as it should, since that is the number of the degrees of freedom of the system. The remaining 68 equations can be used to compute the reactions in the constraints after that the motion has been studied.

In general, the equations of motion are nonlinear and the only way to reach a solution is by performing the numerical integration in time of the model or, as it is usually said, by simulating numerically its motion.

The approach usually followed by most *general purpose* multibody computer codes, however, is based on the numerical integration of all equations, the differential dynamic equations plus the algebraic constraint equations. The latter introduce what are usually referred to as algebraic loops, which complicates the numerical integration.

As soon as the complexity of the problem increases, this way of approaching the problem can easily give way to long computer times, owing to the large number of equations, differential and algebraic, involved.

In some cases it is possible to write directly the minimum number of dynamic equations needed to study the motion of the system. The steps to build the model are

- choice of the generalized coordinates
- computation of the kinetic and potential energy, of the dissipation function and of the virtual work of external forces
- application of the Lagrange equations formalism to obtain the equations of motion

If the system is simple, as in the case of the robotic arms described in Example 3.5, this approach yields a compact model that can be used to solve the behavior of the system in a straightforward way. However, even in the case the equations so obtained are simple, they are nevertheless nonlinear and the only way to obtain a result is by numerically integrating them in time, with the advantage of dealing with a smaller number of equations leading to a much shorter computer time.

An intermediate approach can also be followed: the full set of equations, differential and algebraic, is obtained using a full set of generalized coordinates. Then symbolic computer programs are used to eliminate the constraint equations, reducing the set of generalized coordinates by eliminating the constrained ones. The numerical integration is then performed on a smaller set of equations that do not contain algebraic loops.

Appendix B
Equations of Motion for Continuous Systems

B.1 General Considerations

The main feature of the discrete systems studied in the previous appendix is that a finite number of degrees of freedom is sufficient to describe their configuration. Moreover, if the system is linear, the Ordinary Differential Equations (ODEs) of motion can be easily substituted by a set of linear algebraic equations: the natural mathematical tool for the study of linear discrete systems is matrix algebra.

The situation is different when the behavior of a deformable elastic body is studied as a continuous system. Usually a deformable body is modeled as an elastic continuum or, in the case its behavior can be assumed to be linear, as a linear elastic continuum. A continuum can be thought as consisting of an infinity of points.

To describe the undeformed (or initial) configuration of the body, a reference frame is set in space. Many problems can be studied with reference to a two-dimensional frame or even a single coordinate, but there are cases in which a full tridimensional approach is required. The characteristics of the material are defined by functions of the position in all the parts of space (or plane or line) occupied by the continuum. These functions need not, in general, be continuous.

The configuration at any time t can be obtained from the initial configuration once a vector function expressing the displacements of all points is known (Fig. B.1). The displacement of a point is a vector, with a number of components equal to the number of dimensions of the reference frame. Even if in some cases different choices are considered, usually the components of this vector are taken as the generalized coordinates of each point. The number of degrees of freedom of an elastic (or, more generally, deformable) body is thus infinite. The corresponding generalized coordinates can be manipulated as continuous functions of space coordinates and time, and the theory of continuous functions is the natural tool for dealing with deformable continua.

Remark B.1 The function $\vec{u}(x, y, z, t)$ describing the displacement of the points of the body is differentiable with respect to time at least twice; the first derivative gives the displacement velocity and the second the acceleration. Usually, however, higher-order derivatives also exist.

G. Genta, *Introduction to the Mechanics of Space Robots*, Space Technology Library 26, DOI 10.1007/978-94-007-1796-1, © Springer Science+Business Media B.V. 2012

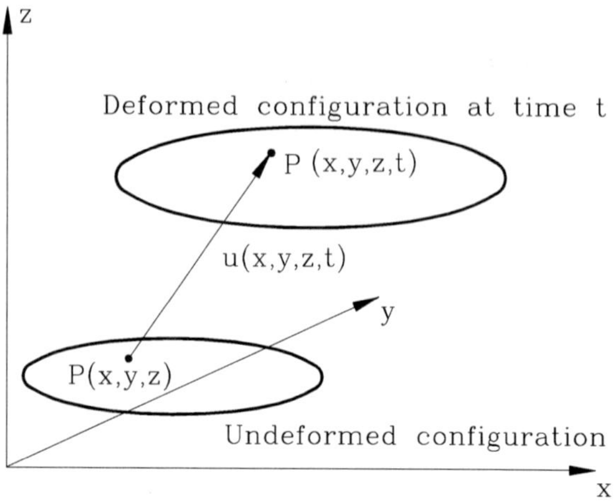

Fig. B.1 Deformation of an elastic continuum; reference frame and displacement vector (from G. Genta, *Vibration Dynamics and Control*, Springer, New York, 2009)

Remark B.2 Reference frame xyz needs not to be an inertial frame. If the flexible body is a 'structure', i.e. a flexible body that does not move except for vibration about its equilibrium position, the reference frame is inertial and $\dot{u}$ can be considered as an absolute velocity, but if it moves as a whole, for instance like a turbine blade or a robot arm, the reference frame may be a moving frame, for instance a frame following the rigid motion of the body and the displacement u and the velocity $\dot{u}$ are relative.

If displacements and rotations can be considered as small quantities, in order not to introduce geometrical nonlinearities, the usual definitions of stresses and strains used in elementary theory of elasticity can be used. When the dynamics of an elastic body can be dealt with as a linear problem, as when the behavior of the material is linear and no geometrical nonlinearity is considered, an infinity of natural frequencies exist as a consequence of the infinity of degrees of freedom of the model.

Assuming that the forces acting on the body are expressed by the function $\mathbf{f}(x, y, z, t)$, the Partial derivatives Differential Equation (PDE) of motion can generally be written as

$$D\big[\mathbf{u}(x, y, z, t)\big] = \mathbf{f}(x, y, z, t), \qquad U\big[\mathbf{u}(x, y, z, t)\big]_B = 0, \tag{B.1}$$

where the differential operator D completely describes the behavior of the body and operator U, defined on the boundary B, states the boundary conditions (only homogeneous boundary conditions are described by (B.1)). The actual form of the differential operator can be obtained by resorting directly to the dynamic equilibrium equations or by writing the kinetic and potential energies and using the Lagrange equations, and the boundary conditions usually follow from geometrical considerations.

The solution of (B.1) exists if an inverse operator D^{-1} can be defined

$$\mathbf{u}(x, y, z, t) = D^{-1}\big[\mathbf{f}(x, y, z, t)\big]. \tag{B.2}$$

Remark B.3 Equation (B.2) is just a formal statement; in most cases the relevant operator cannot be written in explicit form, particularly when the boundary conditions are not the simplest ones.

Because no general approach to the dynamics of an elastic body is feasible, many different models for the study of particular classes of structural elements (beams, plates, shells, etc.) have been developed. Only the bending of beams will be studied here in detail. This choice is only in part due to the fact that many robot elements are studied as beams; it comes also from the need of showing some general properties of continuous systems in the simplest case in order to gain a good insight on the properties of deformable bodies.

The solution of most problems encountered in engineering practice requires dealing with complex structures and the use of continuous models is, consequently, ruled out. For complex shapes the only feasible approach is the discretization of the continuum and then the application of the methods seen for discrete systems. The substitution of a continuous system, characterized by an infinite number of degrees of freedom, with a discrete system, sometimes with a very large but finite number of degrees of freedom, is usually referred to as *discretization*. This step is of primary importance in the solution of practical problems, because the accuracy of the results depends largely on the adequacy of the discrete model to represent the actual system.

B.2 Beams

B.2.1 General Considerations

The simplest continuous system is the prismatic beam. The study of the elastic behavior of beams dates back to Galileo, with important contributions by Daniel Bernoulli, Euler, De Saint Venant, and many others. A *beam* is essentially an elastic solid in which one dimension is prevalent over the others. Often the beam is prismatic (i.e., the cross sections are all equal), homogeneous (i.e., with constant material characteristics), straight (i.e., its axis is a part of a straight line), and untwisted (i.e., the principal axes of elasticity of all sections are equally directed in space). The unidimensional nature of beams allows simplification of the study: each cross section is considered as a rigid body whose thickness in the axial direction is vanishingly small; it has six degrees of freedom, three translational and three rotational. The problem is thus reduced to a unidimensional problem, in the sense that a single coordinate, namely the axial coordinate, is required.

Setting the z-axis of the reference frame along the axis of the beam (Fig. B.2), the six generalized coordinates of each cross section are the axial displacement u_z, the lateral displacements u_x and u_y, the torsional rotation ϕ_z about the z-axis and the flexural rotations ϕ_x and ϕ_y about axes x and y. Displacements and rotations are assumed to be small, so that rotations can be regarded as vector quantities, which

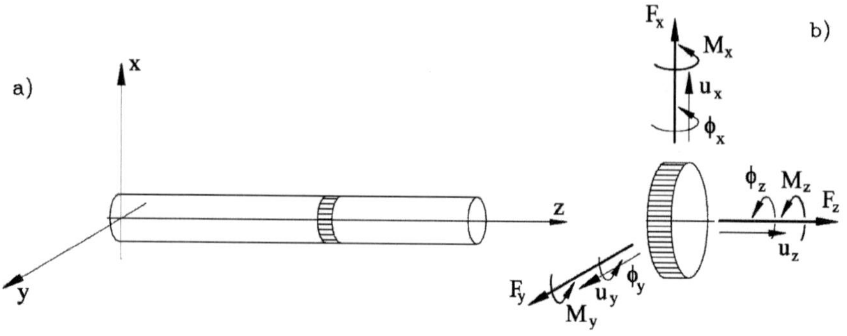

Fig. B.2 Straight beam. (**a**) Sketch and reference frame: (**b**) generalized displacements and forces on a generic cross section (from G. Genta, *Vibration Dynamics and Control*, Springer, New York, 2009)

simplifies all rotation matrices by linearizing trigonometric functions. As a consequence, the three rotations will be considered components of a vector in the same way as the three displacements are components of vector **u**. The generalized forces acting on each cross section and corresponding to the six generalized coordinates defined earlier are the axial force F_z, shear forces F_x and F_y, the torsional moment M_z about the z-axis, and the bending moments M_x and M_y about the x- and y-axes.

From the aforementioned assumptions it follows that all normal stresses in directions other than z (σ_x and σ_y) are assumed to be small enough to be neglected. When geometric and material parameters are not constant along the axis, they must change at a sufficiently slow rate in order not to induce stresses σ_x and σ_y, which could not be considered in this model. If the axis of the beam is assumed to be straight, the axial translation is uncoupled from the other degrees of freedom, at least as a first approximation. A beam loaded only in the axial direction and whose axial behavior is the only one studied is usually referred to as a *bar*. The torsional-rotation degree of freedom is uncoupled from the others only if the area center of all cross sections coincides with their shear center, which happens if all cross sections have two perpendicular planes of symmetry. If the planes of symmetry of all sections are equally oriented (the beam is not twisted) and the x- and y-axes are perpendicular to such planes, the flexural behavior in the xz-plane is uncoupled from that in the yz-plane. The coupling of the degrees of freedom in straight, untwisted beams with cross sections having two planes of symmetry is summarized in Table B.1.

B.2.2 Flexural Vibrations of Straight Beams

Robot arms and legs can easily be modeled as beams. For this reason the bending behavior of straight beams is here dealt with in some detail.

At first assume that the beam is globally at rest in an inertial reference frame.

With the assumptions in Sect. B.2.1, the flexural behavior in each lateral plane can be studied separately from the other degrees of freedom. If bending occurs in the

Table B.1 Generalized coordinates and generalized forces in beams

Type of behavior	Degrees of freedom	Generalized forces
Axial	Displacement u_z	Axial force F_z
Torsional	Rotation ϕ_z	Torsional moment M_z
Flexural	Displacement u_x	Shearing force F_x
(xz-plane)	Rotation ϕ_y	Bending moment M_y
Flexural	Displacement u_y	Shearing force F_y
(yz-plane)	Rotation ϕ_x	Bending moment M_x

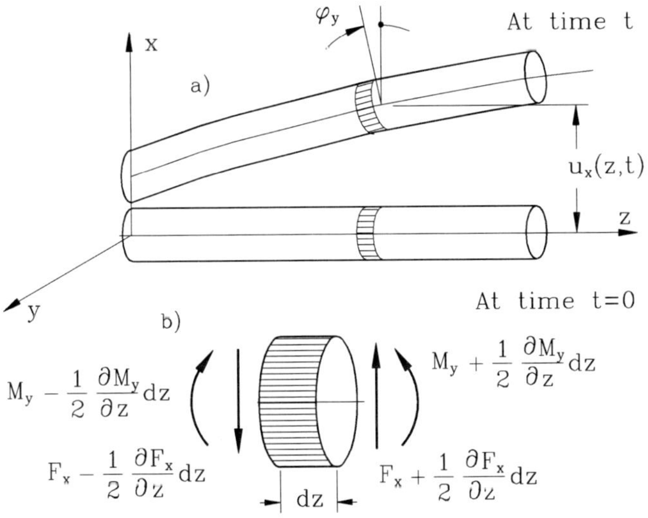

Fig. B.3 Flexural behavior of a straight beam in the xz-plane; (**a**) sketch of the system; (**b**) forces and moments acting on the length dz of the beam (from G. Genta, *Vibration Dynamics and Control*, Springer, New York, 2009)

xz-plane, the relevant generalized coordinates are displacement u_x and rotation ϕ_y. The simplest approach is that often defined as Euler–Bernoulli beam, based on the added assumptions that both shear deformation and rotational inertia of the cross sections are negligible compared with bending deformation and translational inertia, respectively. These assumptions lead to a good approximation if the beam is slender, i.e., if the thickness in the x-direction is much smaller than length l. Note that, at any rate, the thickness in the x-direction must be small enough to use beam theory.

The equilibrium equation for translations in x-direction of the length dz of the beam (Fig. B.3) is

$$\rho A \frac{d^2 u_x}{dt^2} = \frac{\partial F_x}{\partial z} + f_x(z,t). \tag{B.3}$$

If the rotational inertia of the length dz of the beam is neglected, and no distributed bending moment acts on the beam, the equilibrium equation for rotations about y-axis of the length dz of the beam is

$$F_x dz + \frac{\partial M_y}{\partial z} dz = 0. \tag{B.4}$$

By introducing (B.4) into (B.3), it follows that

$$\rho A \frac{d^2 u_x}{dt^2} = -\frac{\partial^2 M_y}{\partial z^2} + f_x(z,t). \tag{B.5}$$

The bending moment is proportional to the curvature of the inflected shape of the beam; neglecting shear deformation and using elementary beam theory the latter coincides with the second derivative of the displacement u_x,

$$M_y = E I_y \frac{\partial^2 u_x}{\partial z^2}, \tag{B.6}$$

where I_y is the area moment of inertia of the cross section of the beam about its y axis. The following equilibrium equation can thus be obtained

$$m(z) \frac{d^2 u_x}{dt^2} + \frac{\partial^2}{\partial z^2} \left[k(z) \frac{\partial^2 u_x}{\partial z^2} \right] = f_x(z,t), \tag{B.7}$$

where the mass and the bending stiffness per unit length are, respectively,

$$m(z) = \rho(z) A(z), \qquad k(z) = E(z) I_y(z). \tag{B.8}$$

Once the lateral displacement u_x has been obtained, the second generalized coordinate ϕ_y is readily obtained: since the cross section remains perpendicular to the deflected shape of the beam owing to neglecting shear deformation, the rotation of the cross section is equal to the slope of the inflected shape,

$$\phi_y = \frac{\partial u_x}{\partial z}. \tag{B.9}$$

Equation (B.7) defines the differential operator D introduced into equation (B.1)

$$D(u_z) = m(z) \frac{d^2 u_x}{dt^2} + \frac{\partial^2}{\partial z^2} \left[k(z) \frac{\partial^2 u_x}{\partial z^2} \right]. \tag{B.10}$$

The boundary conditions $U[u_x(z,t)]_B = 0$ must be stated following the actual conditions at the ends of the beam: if, for instance, they are clamped, both the displacement u_x and the rotation $\partial u_z/\partial z$ must be equated to zero for $z = 0$ and $z = l$ (where l is the length of the beam and the origin is set at the left end).

In the case of a prismatic homogeneous beam, (B.7) reduces to

$$\rho A \frac{d^2 u_x}{dt^2} + E I_y \frac{\partial^4 u_x}{\partial z^4} = f_x(z,t). \tag{B.11}$$

Free Behavior

The solution of the homogeneous equation associated with (B.7) can be expressed as the product of a function of time and a function of the space coordinate

$$u_x(z,t) = q(z)\eta(t). \tag{B.12}$$

Introducing (B.12) into the homogeneous equation associated with (B.7) and separating the variables, it follows that

$$\frac{1}{\eta(t)}\frac{d^2\eta(t)}{dt^2} = \frac{1}{m(z)q(z)}\frac{\partial^2}{\partial z^2}\left[k(z)\frac{\partial^2 q(z)}{\partial z^2}\right]. \tag{B.13}$$

The function on the left-hand side depends on time but not on the space coordinate z. Conversely, the function on the right-hand side is a function of z but not of t. The only possibility of satisfying equation (B.13) for all values of time and of coordinate z is to state that both sides are constant and that the two constants are equal. This constant can be indicated as $-\omega^2$. The condition on the function of time on the left-hand side is thus

$$\frac{1}{\eta(t)}\frac{d^2\eta(t)}{dt^2} = \text{constant} = -\omega^2. \tag{B.14}$$

Neglecting a proportionality constant that will be introduced into function $q(z)$ later, this equation yields a harmonic oscillation with frequency ω

$$\eta(t) = \sin(\omega t + \phi). \tag{B.15}$$

The solution of the equation of motion for free oscillations of the beam is

$$u_x(z,t) = q(z)\sin(\omega t + \phi). \tag{B.16}$$

Function $q(z)$ is said to be the *principal function*. Each point of the bar performs a harmonic motion with frequency ω, while the amplitude is given by the function $q(z)$.

Remark B.4 The resultant motion is a standing wave, with all points of the beam vibrating in phase.

By introducing (B.16) into (B.7), it follows that

$$-\omega^2 m(z)q(z) = \frac{d^2}{dz}\left[k(z)\frac{d^2 q(z)}{dz^2}\right], \tag{B.17}$$

or, in the case of a prismatic homogeneous beam,

$$-\omega^2 q(z) = \frac{EI_y}{\rho A}\frac{d^4 q(z)}{dz^4}. \tag{B.18}$$

Equations (B.17) and (B.18) are eigenproblems. The second, for example, states that the fourth derivative of function $q(z)$ (with respect to the space coordinate z) is proportional to the function itself, the constant of proportionality being $-\omega^2 \rho A/EI_y$. The values of such a constant allowing the equation to be satisfied by a solution other than the trivial solution $q(z) = 0$ are the eigenvalues, and the corresponding functions $q(z)$ are the eigenfunctions. Equation (B.17), although more complex, has a similar meaning.

Remark B.5 The eigenvalues are infinite in number, and the general solution of the equation of motion (B.17) can be obtained as the sum of an infinity of terms of the type of (B.16).

Remark B.6 The eigenfunctions $q_i(z)$ are defined only as far as their shape is concerned, exactly like the eigenvectors in discrete systems. The amplitude of the various modes can be computed only after the initial conditions have been stated.

Remark B.7 Although the number of eigenfunctions, and hence of modes, is infinite, a small number of principal functions is often sufficient to describe the behavior of an elastic body with the required precision, in a way that is similar to what has already been said for eigenvectors.

Remark B.8 Eigenfunctions have some of the properties seen for eigenvectors, particularly that of orthogonality with respect to the mass $m(z)$ and to the stiffness $k(z)$. As a general rule they are not orthogonal to each other, except when the function $m(z)$ is constant: the eigenfunctions of a prismatic homogeneous beam are thus orthogonal to each other.

The general solution of (B.18) is

$$q(z) = C_1 \sin(az) + C_2 \cos(az) + C_3 \sinh(az) + C_4 \cosh(az), \tag{B.19}$$

where

$$a = \sqrt{\omega}\sqrt[4]{\frac{\rho A}{EI_y}}. \tag{B.20}$$

The rotation is the derivative of the displacement

$$\frac{dq}{dz} = C_1 a \cos(az) - C_2 a \sin(az) + C_3 a \cosh(az) + C_4 a \sinh(az). \tag{B.21}$$

Constants C_i can be computed from the boundary conditions. In the present case four boundary conditions must be stated, which is consistent both with the order of the differential equation and with the number of degrees of freedom involved. Each end of the beam may be free, clamped, simply supported or, a condition seldom accounted for, constrained in such a way to restrain rotations but not displacements.

At a free end the displacement and the rotation are free, but both the bending moment and the shear force must vanish. This can be expressed by the relationships

$$\frac{d^2q}{dz^2} = 0, \qquad \frac{d^3q}{dz^3} = 0. \tag{B.22}$$

If on the contrary an end is clamped, both its displacement and its rotation vanish

$$q = 0, \qquad \frac{dq}{dz} = 0. \tag{B.23}$$

A supported end is free to rotate, and hence the bending moment must vanish, but its displacement is constrained

$$q = 0, \qquad \frac{d^2q}{dz^2} = 0. \tag{B.24}$$

A further condition is the case where the end is free to move, and hence the shear force vanishes, but its rotation is constrained

$$\frac{dq}{dz} = 0, \qquad \frac{d^3q}{dz^3} = 0. \tag{B.25}$$

As an example, consider a prismatic beam clamped at $z = 0$ and free at $z = l$. At the 'left' end ($z = 0$) both displacement and rotation vanish:

$$\begin{aligned} q(0) &= C_2 + C_4 = 0, \\ \left(\frac{dq}{dz}\right)_{z=0} &= C_1 + C_3 = 0. \end{aligned} \tag{B.26}$$

The second and third derivatives of function $q(z)$ are

$$\begin{aligned} \frac{d^2q}{dz^2} &= -a^2 C_1 \sin(az) - a^2 C_2 \cos(az) + a^2 C_3 \sinh(az) + a^2 C_4 \cosh(az), \\ \frac{d^3q}{dz^3} &= -a^3 C_1 \cos(az) + a^3 C_2 \sin(az) + a^3 C_3 \cosh(az) + a^3 C_4 \sinh(az), \end{aligned} \tag{B.27}$$

and then the conditions stating that the 'right' end ($z = l$) is free yield

$$\begin{aligned} -C_1 \sin(al) - C_2 \cos(al) + C_3 \sinh(al) + C_4 \cosh(al) &= 0, \\ -C_1 \cos(al) + C_2 \sin(al) + C_3 \cosh(al) + C_4 \sinh(al) &= 0. \end{aligned} \tag{B.28}$$

By solving the conditions at the left end ($z = 0$) in C_3 and C_4 and introducing their values in the conditions at the right end, it follows that

$$\begin{bmatrix} \sin(al) + \sinh(al) & \cos(al) + \cosh(al) \\ \cos(al) + \cosh(al) & -\sin(al) + \sinh(al) \end{bmatrix} \begin{Bmatrix} C_1 \\ C_2 \end{Bmatrix} = \mathbf{0}. \tag{B.29}$$

To obtain a solution other than the trivial solution $C_1 = 0$, $C_2 = 0$, the determinant of the matrix of the coefficients of the set of linear equations in C_1 and C_2 must vanish

$$\sinh^2(al) - \sin^2(al) - \left[\cos(al) + \cosh(al)\right]^2 = 0. \tag{B.30}$$

This equation cannot be solved in closed form in al, but it is easy to obtain numerical solutions. The first four are

$$al = 1.875, 4.694, 7.855, 10.996.$$

For high values of al, approximate solutions can be found, by remembering that

$$\sinh(ax) \approx \cosh(ax) \approx \frac{e^{ax}}{2}$$

and that

$$\sin(ax) \ll \frac{e^{ax}}{2}, \qquad \cos(ax) \ll \frac{e^{ax}}{2}.$$

In this case, the characteristic equation reduces to

$$\cos(al) = 0$$

that yields

$$al = \left(i - \frac{1}{2}\right)\pi.$$

For $i = 3$ this approximated solution yields $al = 7.854$, i.e. an error of only 0.013% with respect to the numerical solution. For larger values of i, the error is negligible.

The natural frequencies can be computed from (B.20), obtaining

$$\omega = \frac{\beta_i^2}{l^2}\sqrt{\frac{EI_y}{\rho A}}, \tag{B.31}$$

where β_i are the values of al obtained above.

To compute the eigenfunctions, it is possible to state the value of one of the constants, for instance $C_1 = 1$. From (B.29) it follows that

$$C_2 = \frac{\sin(al) + \sinh(al)}{\cos(al) + \cosh(al)}, \tag{B.32}$$

and the ith eigenvector, expressed in terms of the nondimensional coordinate

$$\zeta = \frac{z}{l} \tag{B.33}$$

is

$$q_i(\zeta) = \sin(\beta_i\zeta) - \sinh(\beta_i\zeta) - C_2\left[\cos(\beta_i\zeta) - \cosh(\beta_i\zeta)\right]. \tag{B.34}$$

Table B.2 Values of constants $\beta_i = a_i l$ for the various modes with different boundary conditions

Boundary condition	$i=0$	$i=1$	$i=2$	$i=3$	$i=4$	$i>4$
Free–free	0	4.730	7.853	10.996	14.137	$\approx(i+1/2)\pi$
Supported–free	0	1.25π	2.25π	3.25π	4.25π	$(i+1/4)\pi$
Clamped–free	–	1.875	4.694	7.855	10.996	$\approx(i-1/2)\pi$
Supported–supported	–	π	2π	3π	4π	$i\pi$
Supported–clamped	–	3.926	7.069	10.210	13.352	$\approx(i+1/4)\pi$
Clamped–clamped	–	4.730	7.853	10.996	14.137	$\approx(i+1/2)\pi$

The same procedure can be applied also for other boundary conditions.

Expression (B.31) for the natural frequencies holds for any boundary condition. The values of constants $\beta_i = a_i l$ are reported in Table B.2 for the most common boundary conditions.

The eigenfunctions, normalized in such a way that the maximum value of the displacement is equal to unity, are, for different boundary conditions, as follows:

1. *Free–free*. Rigid-body modes:

$$q_0^I(\zeta) = 1, \qquad q_0^{II}(\zeta) = 1 - 2\zeta.$$

Other modes:

$$q_i(\zeta) = \frac{1}{2N}\left\{\sin(\beta_i\zeta) + \sinh(\beta_i\zeta) + N\left[\cos(\beta_i\zeta) + \cosh(\beta_i\zeta)\right]\right\},$$

where

$$N = \frac{\sin(\beta_i) - \sinh(\beta_i)}{-\cos(\beta_i) + \cosh(\beta_i)}.$$

2. *Supported–free*. Rigid-body mode:

$$q_0(\zeta) = \zeta.$$

Other modes:

$$q_i(\zeta) = \frac{1}{2\sin(\beta_i)}\left[\sin(\beta_i\zeta) + \frac{\sin(\beta_i)}{\sinh(\beta_i)}\sinh(\beta_i\zeta)\right].$$

3. *Clamped–free*.

$$q_i(\zeta) = \frac{1}{N_2}\left\{\sin(\beta_i\zeta) - \sinh(\beta_i\zeta) - N_1\left[\cos(\beta_i\zeta) - \cosh(\beta_i\zeta)\right]\right\},$$

where

$$N_1 = \frac{\sin(\beta_i) + \sinh(\beta_i)}{\cos(\beta_i) + \cosh(\beta_i)},$$

$$N_2 = \sin(\beta_i) - \sinh(\beta_i) - N_1\left[\cos(\beta_i) - \cosh(\beta_i)\right].$$

4. *Supported–supported.*

$$q_i(\zeta) = \sin(i\pi\zeta).$$

5. *Supported–clamped.*

$$q_i(\zeta) = \frac{1}{N}\left[\sin(\beta_i\zeta) - \frac{\sin(\beta_i)}{\sinh(\beta_i)}\sinh(\beta_i\zeta)\right],$$

where N is the maximum value of the expression within brackets and must be computed numerically.

6. *Clamped–clamped.*

$$q_i(\zeta) = \frac{1}{N_2}\left\{\sin(\beta_i\zeta) - \sinh(\beta_i\zeta) - N_1\left[\cos(\beta_i\zeta) - \cosh(\beta_i\zeta)\right]\right\},$$

where

$$N_1 = \frac{\sin(\beta_i) - \sinh(\beta_i)}{\cos(\beta_i) - \cosh(\beta_i)}$$

and N_2 is the maximum value of the expression between braces and must be computed numerically.

The first four mode shapes (plus the rigid-body modes where they do exist) for each boundary condition are plotted in Fig. B.4.

Modal Analysis

The property of orthogonality with respect to mass and stiffness means that if $q_i(z)$ and $q_j(z)$ are two distinct eigenfunctions and $(i \neq j)$, it follows that

$$\int_0^l m(z)q_i(z)q_j(z)\,dz = 0, \qquad \int_0^l k(z)\frac{d^2q_i(z)}{dz^2}\frac{d^2q_j(z)}{dz^2}\,dz = 0. \tag{B.35}$$

As already stated, eigenfunctions are not directly orthogonal, except for the case when the mass $m(z)$ is constant along the beam.

If $i = j$, the integrals of (B.35) do not vanish:

$$\int_0^l m(z)\left[q_i(z)\right]^2 dz = \overline{M}_i \neq 0, \qquad \int_0^l k(z)\left[\frac{dq_i(z)}{dz}\right]^2 dz = \overline{K}_i \neq 0. \tag{B.36}$$

These two relationships define the modal masses and stiffness

Remark B.9 The meaning of the modal mass and stiffness in the case of continuous systems is exactly the same as for discrete systems; the only difference is that in the current case the number of modes, and then of modal masses and stiffnesses, is infinite.

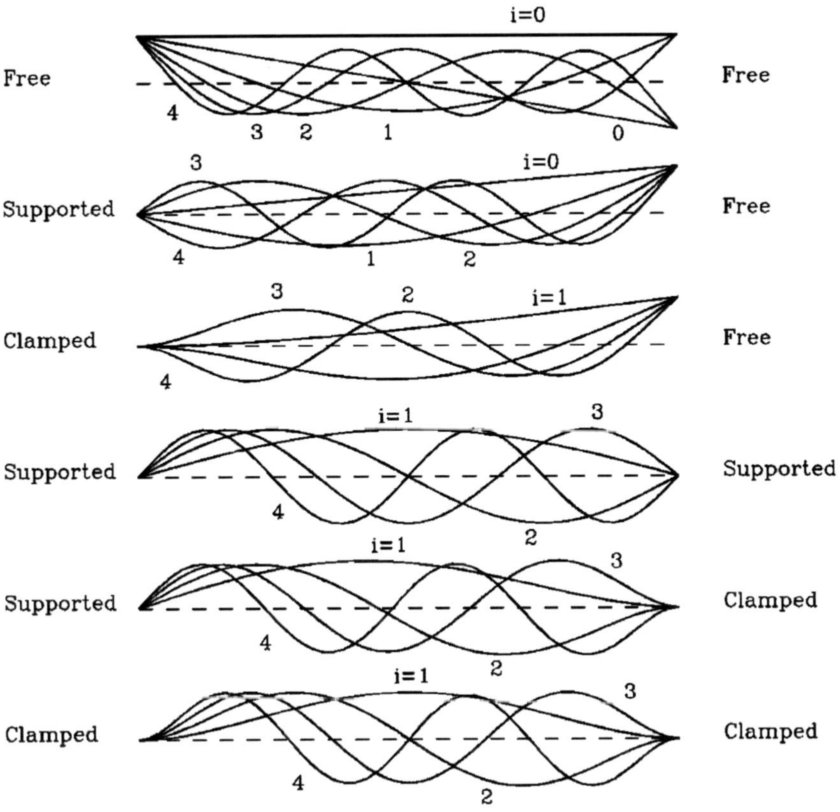

Fig. B.4 Normal modes of a straight beam with different end conditions. The first four modes plus the rigid-body modes, where they exist, are shown (from G. Genta, *Vibration Dynamics and Control*, Springer, New York, 2009)

Any deformed configuration of the system $u_x(z,t)$ can be expressed as a linear combination of the eigenfunctions. The coefficients of this linear combination, which are functions of time, are the modal coordinates $\eta_i(t)$:

$$u_x(z,t) = \sum_{i=0}^{\infty} \eta_i(t) q_i(z). \tag{B.37}$$

Equation (B.37) expresses the modal transformation for continuous systems and is equivalent to the similar relationship seen for discrete systems.

The inverse transformation, needed to compute the modal coordinates $\eta_i(t_k)$ corresponding to any given deformed configuration $u(z,t_k)$ occurring at time t_k, can be obtained through a simple procedure. Multiplying equation (B.37) by the jth eigenfunction and by the mass distribution $m(z)$ and integrating on the whole length of the bar, it follows that

$$\int_0^l \left[m(z)q_j(z)u(z,t_0)\right]dz = \sum_{i=0}^{\infty} \eta_i(t_0) \int_0^l \left[m(z)q_j(z)q_i(z)\right]dz. \tag{B.38}$$

Of the infinity of terms on the right-hand side, only the term with $i = j$ does not vanish and the integral yields the jth modal mass. Remembering the definition of the modal masses, it follows that

$$\eta_i(t_0) = \frac{1}{\overline{M}_j} \int_0^l \left[m(z)q_j(z)u(z,t_0)\right]dz. \tag{B.39}$$

This relationship can be used to perform the inverse modal transformation, i.e., to compute the modal coordinates corresponding to any deformed shape of the system.

Eigenfunctions can be normalized in several ways, one being that leading to unit values of the modal masses. This is achieved simply by dividing each eigenfunction by the square root of the corresponding modal mass.

The vibration of the beam under the effect of the forcing function $f_x(z,t)$ can be obtained by solving the complete equation (B.7), whose general solution can be expressed as the sum of the complementary function obtained earlier, and a particular integral of the complete equation. Owing to the orthogonality properties of the normal modes $q_i(z)$, the latter can be expressed as a linear combination of the eigenfunctions. Equation (B.37) then also holds in the case of forced motion of the system.

Forced Response

By introducing the modal transformation (B.37) into the equation of motion (B.7), the latter can be transformed into a set of an infinite number of equations in the modal coordinates η_i

$$\overline{M}_i \ddot{\eta}_i(t) + \overline{K}_i \eta_i(t) = \overline{f}_i(t), \tag{B.40}$$

where the ith modal force is defined by the following formulas:

$$\overline{f}_i(t) = \int_0^l q_i(z) f(z,t)\, dz, \qquad \overline{f}_i(t) = \sum_{k=1}^{m} q_i(z_k) f_k(t), \tag{B.41}$$

holding in the cases of a continuous force distribution and m concentrated bending forces $f_k(t)$ acting on points of coordinates z_k, respectively.

Remark B.10 Equations (B.41) correspond exactly to the definition of the modal forces for concentrated systems.

Equations (B.40) can be used to study the forced response of a continuous system to external excitations of any type by reducing it to a number of systems with a single degree of freedom. In the case of continuous systems, their number is infinite, but usually a small number of them is enough to obtain the results with the required precision.

If the excitation is provided by the motion of the structure to which the beam is connected, it is expedient to resort to a coordinate system that moves with the supporting points. In the case of the bending behavior of a beam, only the motion of the supporting structure in x-direction is coupled with its dynamic behavior in the xz-plane. If the origin of the coordinates is displaced by the quantity x_A, the absolute displacement in x-direction $u_{x_{\text{iner}}}(z,t)$ is linked to the relative displacement $u_x(z,t)$ by the obvious relationship

$$u_{x_{\text{iner}}}(z,t) = u_x(z,t) + x_A(t).$$

By writing (B.7) using the relative displacement, it follows that

$$m(z)\frac{d^2 u_z}{dt^2} - \frac{\partial}{\partial z}\left[k(z)\frac{\partial u_z}{\partial z}\right] = -m(z)\ddot{x}_A. \tag{B.42}$$

The excitation due to the motion of the constraints can be dealt with simply by using relative coordinates and applying an external force distribution equal to $-m(z)\ddot{x}_A$. The modal forces can be readily computed through (B.41):

$$\overline{f}_i(t) = -r_i\ddot{x}_A, \tag{B.43}$$

where

$$r_i = \int_0^l q_i(z)m(z)\,dz$$

are the modal participation factors related to the lateral motion of the bar.

Remark B.11 The motion of the system has been studied in an inertial reference frame. However, the inflected shape can be expressed as a linear combination of the modes in any reference frame, be it inertial or not, and in particular can be used to study the vibration of a moving beam, like a robot arm or leg.

B.2.3 Effect of Shear Deformation

The rotational inertia of the cross section and the shear deformation were not taken into account in the preceding section. In this section this assumption will be dropped, with reference only to a prismatic homogeneous beam. A beam in which these effects are not neglected is usually referred to as a *Timoshenko beam*. Shear deformation can be accounted for as a deviation of the direction of the deflected shape of the beam not accompanied by a rotation of the cross section (Fig. B.5).

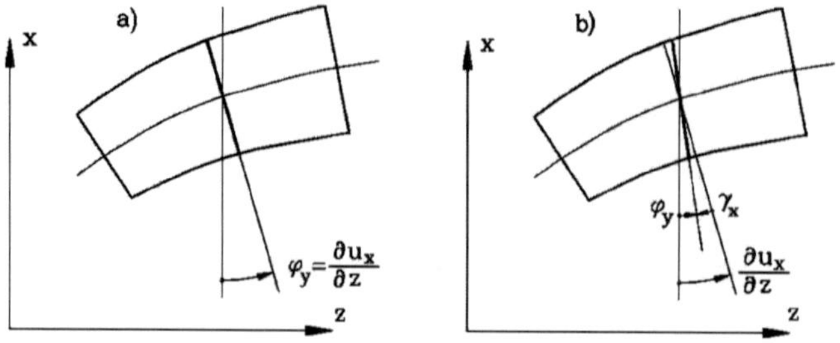

Fig. B.5 Effect of shear deformation on beam bending. (**a**) Euler–Bernoulli beam; (**b**) Timoshenko beam (from G. Genta, *Vibration Dynamics and Control*, Springer, New York, 2009)

The latter is thus no more perpendicular to the deformed shape of the beam, and the rotation of the cross section can be expressed as

$$\phi_y = \frac{\partial u_x}{\partial z} - \gamma_x. \tag{B.44}$$

The shear strain γ_x is linked to the shear force by the relationship

$$\gamma_x = \frac{\chi F_x}{GA}, \tag{B.45}$$

where the shear factor χ depends on the shape of the cross section, even if there is not complete accord on its value. For a circular beam, a value of $10/9$ is usually assumed; for other shapes the expressions reported in Table B.3 can be used.

Equation (B.44) can thus be written in the form

$$\phi_y = \frac{\partial u_x}{\partial z} - \frac{\chi F_x}{GA}. \tag{B.46}$$

The bending moment has no effect on shear deformation: If the latter is accounted for, the relationship linking the bending moment to the inflected shape of the beam becomes

$$M_y = EI_y \frac{\partial \phi_y}{\partial z}. \tag{B.47}$$

The rotational inertia of the cross section is no more neglected, and the dynamic equilibrium equations for displacement and rotation of the length dz of the beam are

$$\begin{cases} \rho A \dfrac{d^2 u_x}{dt^2} = \dfrac{\partial F_x}{\partial z} + f_x(z,t), \\ \rho I_y \dfrac{d^2 \phi_y}{dt^2} = F_x + \dfrac{\partial M_y}{\partial z}. \end{cases} \tag{B.48}$$

Table B.3 Shear factors for some different cross sections (from G.R. Cowper, The shear coefficient in Timoshenko's beam theory, *J. Appl. Mech.*, 1966, 335–340)

$\chi = \frac{7+6\nu}{6(1+\nu)}$	
$\chi = \frac{(7+6\nu)(1+m)^2+4m(5+3\nu)}{6(1+\nu)(1+m)^2}$ where $m = \left(\frac{d_i}{d_o}\right)^2$	
$\chi = \frac{12+11\nu}{10(1+\nu)}$	
$\chi = \frac{40+37\nu+m(16+10\nu)+\nu m^2}{12(1+\nu)(3+m)}$ where $m = \left(\frac{b}{a}\right)^2$	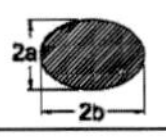
$\chi = \frac{1.305+1.273\nu}{1+\nu}$	
$\chi = \frac{4+3\nu}{2(1+\nu)}$	
$\chi = \frac{48+39\nu}{20(1+\nu)}$	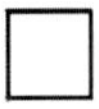
$\chi = \frac{p+q\nu+30n^2m(1+m)+5\nu n^2m(8+9m)}{10(1+\nu)(1+3m)^2}$ where $m = \frac{bt_1}{at_a}$, $n = \frac{b}{h}$	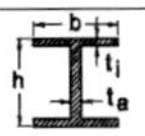
$\chi = \frac{p+q\nu+10n^2\left[m(3+\nu)+3m^2\right]}{10(1+\nu)(1+3m)^2}$ where $m = \frac{bt_1}{at_a}$, $n = \frac{b}{h}$	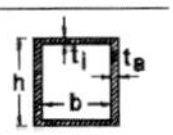
$\chi = \frac{p+q\nu}{10(1+\nu)(1+3m)^2}$ where $m = \frac{2A}{ht}$, $A =$ flange area	
$\chi = \frac{p'+q'\nu+30n^2m(1+m)+10\nu n^2m(4+5m+m^2)}{10(1+\nu)(1+4m)^2}$ $m = \frac{bt_1}{ht_a}$, $n = \frac{b}{h}$	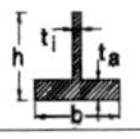

$p = 12 + 72m + 150m^2 + 90m^3$; $q = 11 + 66m + 135m^2 + 90m^3$
$p' = 12 + 96m + 276m^2 + 192m^3$; $q' = 11 + 88m + 248m^2 + 216m^3$

By solving equation (B.46) in F_x and introducing it into the homogeneous equation (B.48), it follows that

$$\begin{cases} \rho A \dfrac{d^2 u_x}{dt^2} = \dfrac{GA}{\chi}\left(\dfrac{\partial^2 u_x}{\partial z^2} - \dfrac{\partial \phi_y}{\partial z}\right), \\ \rho I_y \dfrac{d^2 \phi_y}{dt^2} = \dfrac{GA}{\chi}\left(\dfrac{\partial u_x}{\partial z} - \phi_y\right) + EI_y \dfrac{\partial^2 \phi_y}{\partial z^2}. \end{cases} \tag{B.49}$$

By differentiating the second equation (B.49) with respect to z and eliminating ϕ_y, the following equation can be obtained:

$$EI_y \frac{\partial^4 u_x}{\partial z^4} - \rho I_y \left(1 + \frac{E\chi}{G}\right) \frac{\partial^2}{\partial z^2}\left(\frac{\partial^2 u_x}{\partial t^2}\right) + \frac{\rho^2 I_y \chi}{G}\frac{d^4 u_x}{dt^4} + \rho A \frac{d^2 u_x}{dt^2} = 0 \tag{B.50}$$

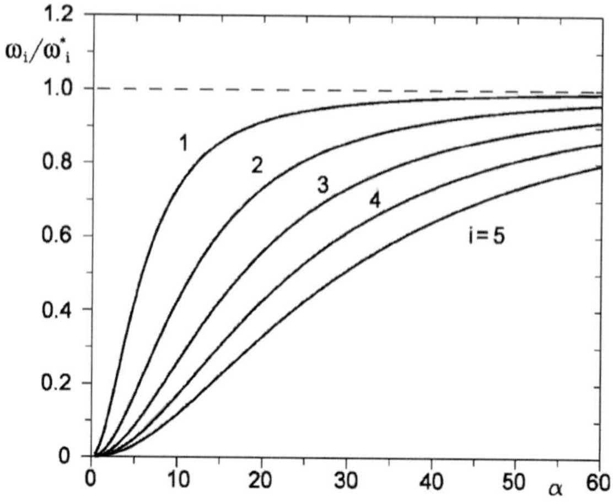

Fig. B.6 Effect of shear deformation on the first five natural frequencies of a simply supported beam. Ratio between the natural frequency computed taking into account rotational inertia and shear deformation and that computed using the Euler–Bernoulli model (from G. Genta, *Vibration Dynamics and Control*, Springer, New York, 2009)

and thus, in the case of free oscillations with harmonic time history,

$$EI_y\frac{d^4q(z)}{dz^4}+\rho\omega^2 I_y\left(1+\frac{E\chi}{G}\right)\frac{d^2q(z)}{dz^2}-\rho\omega^2\left(A-\omega^2\frac{\rho I_y\chi}{G}\right)q(z)=0. \tag{B.51}$$

The same considerations regarding the form of the eigenfunctions (seen in the preceding section) also hold in this case. If the beam is simply supported at both ends, the same eigenfunctions seen in the case of the Euler–Bernoulli beam still hold, and (B.51) can be expressed in nondimensional form as

$$\left(\frac{\omega}{\omega^*}\right)^4-\left(\frac{\omega}{\omega^*}\right)^2\frac{\alpha^2}{i^2\pi^2\chi^*}\left(1+\chi^*+\frac{\alpha^2}{i^2\pi^2}\right)+\frac{\alpha^4}{i^4\pi^4\chi^*}=0, \tag{B.52}$$

where ω_i^* is the ith natural frequency computed using the Euler–Bernoulli assumptions, the slenderness α of the beam is defined as

$$\alpha=l\sqrt{\frac{A}{I_y}}=\frac{l}{r} \tag{B.53}$$

(r is the radius of inertia of the cross section) and $\chi^*=\chi E/G$.

The results obtained from (B.52) for a beam with circular cross section and material with $\nu=0.3$, are reported as functions of the slenderness in Fig. B.6.

Remark B.12 The effects of both shear deformation and rotational inertia tend to lower the value of the natural frequencies, the first being stronger than the second by a factor of about 3, as shown by Timoshenko.[1]

Remark B.13 It must be remembered that even the so-called Timoshenko beam model is an approximation, because it is based on the usual approximations of the

[1] S.P. Timoshenko et al., *Vibration Problems in Engineering*, Wiley, New York, 1974.

beam theory, and the very model of a one-dimensional solid is no more satisfactory when the slenderness is very low.

B.3 Discretization of Continuous Systems: The FEM

Many discretization techniques have been developed with the aim of substituting the equation of motion consisting of a partial derivative differential equation (with derivatives with respect to time and space coordinates) with a set of linear ordinary differential equations containing only derivatives with respect to time. The resulting set of equations, generally of the second order, is of the same type seen for discrete systems (hence, the term *discretization*).

The finite element method (FEM) is at present the most common discretization method, mostly because many computer codes based on it are available. In the FEM, the body is divided into a number of regions, called *finite elements*, as opposed to the vanishingly small regions used in writing the differential equations for continuous systems. The deformed shape of each finite element is assumed to be a linear combination of a set of functions of space coordinates through a certain number of parameters, considered the generalized coordinates of the element. Usually such functions of the space coordinates (called *shape functions*) are quite simple and the generalized coordinates have a direct physical meaning, namely generalized displacements at selected points of the element, usually referred to as *nodes*. The analysis then proceeds to writing a set of differential equations of the same type as those obtained for discrete systems.

The finite element method is a general discretization method for the solution of partial derivative differential equations and, consequently, it finds its application in many other fields beyond structural dynamics and structural analysis. The aim of this section is not to provide a complete survey of the method, which can be dealt with only in a specialized text, but simply to describe its main features and to show how it can be used to model the dynamic behavior of robot elements.

The component-mode synthesis method can be used with advantage to reduce the number of degrees of freedom of the model obtained through the FEM, particularly when the structure is made by components that can take different relative position, like robot arms or legs.

B.3.1 Element Characterization

Many different element formulations have been developed, depending on their shape and characteristics: beam elements, shell elements, plate elements, solid elements, and many others. A structure can be built by assembling elements of the same or different types, as dictated by the nature of the problem and by the capabilities of the computer code used.

Because the FEM is usually developed using matrix notation, in order to obtain formulas readily transferable to computer codes, the displacement is written as a vector of order 3 in the tridimensional space (sometimes of higher order, if rotations are also considered), and the equation expressing the displacement of the points inside each element is

$$\mathbf{u}(x, y, z, t) = \mathbf{N}(x, y, z)\mathbf{q}(t), \tag{B.54}$$

where $\mathbf{q}$ is a vector in which the n generalized coordinates of the element are listed and $\mathbf{N}$ is the matrix containing the shape functions. There are as many rows in $\mathbf{N}$ as in $\mathbf{u}$ and as many columns as the number n of degrees of freedom.

As already stated, usually the degrees of freedom of the elements are the displacements at given points, which are referred to as *nodes*. In this case, (B.54) usually reduces to the simpler form,

$$\begin{Bmatrix} u_x(x, y, z, t) \\ u_y(x, y, z, t) \\ u_z(x, y, z, t) \end{Bmatrix} = \begin{bmatrix} \mathbf{N}(x, y, z) & \mathbf{0} & \mathbf{0} \\ \mathbf{0} & \mathrm{N}(x, y, z) & \mathbf{0} \\ \mathbf{0} & \mathbf{0} & \mathrm{N}(x, y, z) \end{bmatrix} \begin{Bmatrix} \mathbf{q}_x(t) \\ \mathbf{q}_y(t) \\ \mathbf{q}_z(t) \end{Bmatrix}, \tag{B.55}$$

where the displacements in each direction are functions of the nodal displacements in the same direction only. Matrix $\mathbf{N}$, in this case, has only one row and as many columns as the number of nodes of the element. Equation (B.55) has been written for a three-dimensional element; a similar formulation can also be easily obtained for one- or two-dimensional elements.

Each element is essentially the model of a small deformable solid. The behavior of the element is studied using an assumed-modes approach, i.e. assuming that the displacement is a linear combination of the already mentioned arbitrarily assumed shape functions. A limited number, usually quite small, of degrees of freedom is then substituted to the infinity of degrees of freedom of each element.

The freedom in the choice of the shape functions is, however, limited, because they must satisfy several conditions. A first requirement is a simple mathematical formulation, which is needed to lead to developments that are not too complex.

Usually a set of polynomials in the space coordinates is assumed. To get results that are closer to the exact solution of the differential equations, which constitute the continuous model discretized by the FEM, while reducing the size of the elements, the shape functions must

- be continuous and differentiable up to the required order, which depends on the type of element;
- be able to describe rigid-body motions of the element leading to vanishing elastic potential energy;
- lead to a constant strain field when the overall deformation of the element dictates so; and
- lead to a deflected shape of each element that matches the shape of the neighboring elements.

The last condition means that when the nodes of two neighboring elements displace in a compatible way, all the interface between the elements must displace in a compatible way.

Another condition, which is not always satisfied, is that the shape functions are isotropic, i.e., do not show particular geometrical properties that depend on the orientation of the reference frame. Sometimes not all these conditions are completely met; in particular, there are elements that fail to completely satisfy the matching of the deflected shapes of neighboring elements.

The nodes are usually located at the vertices or on the sides of the elements and are common to two or more of them, but points that are internal to an element are sometimes also used.

To write the equations of motion of the element the strains can be expressed as functions of the derivatives of the displacements with respect to space coordinates. In general, it is possible to write a relationship of the type

$$\boldsymbol{\epsilon}(x, y, z, t) = \mathbf{B}(x, y, z)\mathbf{q}(t), \tag{B.56}$$

where $\boldsymbol{\epsilon}$ is a column matrix in which the various elements of the strain tensor are listed (it is commonly referred to as a *strain vector* but it is such only in the sense that it is a column matrix) and $\mathbf{B}$ is a matrix containing appropriate derivatives of the shape functions. $\mathbf{B}$ has as many rows as the number of components of the strain vector and as many columns as the number of degrees of freedom of the element.

If the element is free from initial stresses and strains and the behavior of the material is linear, the stresses can be directly expressed from the strains

$$\boldsymbol{\sigma}(x, y, z, t) = \mathbf{E}\boldsymbol{\epsilon} = \mathbf{E}(x, y, z)\mathbf{B}(x, y, z)\mathbf{q}(t), \tag{B.57}$$

where $\mathbf{E}$ is the stiffness matrix of the material. It is a symmetric square matrix whose elements can theoretically be functions of the space coordinates but are usually constant within the element. The potential energy of the element can be easily expressed as

$$\mathcal{U} = \frac{1}{2}\int_V \boldsymbol{\epsilon}^T \boldsymbol{\sigma}\, dV = \frac{1}{2}\mathbf{q}^T\left(\int_V \mathbf{B}^T\mathbf{E}\mathbf{B}\, dV\right)\mathbf{q}. \tag{B.58}$$

The integral in (B.58) is the stiffness matrix of the element

$$\mathbf{K} = \int_V \mathbf{B}^T\mathbf{E}\mathbf{B}\, dV. \tag{B.59}$$

Because the shape functions do not depend on time, the generalized velocities can be expressed as

$$\dot{\mathbf{u}}(x, y, z, t) = \mathbf{N}(x, y, z)\dot{\mathbf{q}}(t).$$

In the case where all generalized coordinates are related to displacements, the kinetic energy and the mass matrix of the element can be expressed as

$$\begin{aligned} \mathcal{T} &= \frac{1}{2}\int_V \dot{\mathbf{u}}^T\dot{\mathbf{u}}\rho\, dV = \frac{1}{2}\dot{\mathbf{q}}^T\left(\int_V \rho\mathbf{N}^T\mathbf{N}\, dV\right)\dot{\mathbf{q}}, \\ \mathbf{M} &= \int_V \rho\mathbf{N}^T\mathbf{N}\, dV. \end{aligned} \tag{B.60}$$

In the case that some generalized displacements are physically rotations, (B.60) must be changed in order to introduce moments of inertia, but its basic structure remains the same.

The FEM is often used just to compute the stiffness matrix to be used in the context of the lumped-parameters approach. In this case, the consistent mass matrix (B.60) is not computed and a diagonal matrix obtained by lumping the mass at the nodes is used. The advantage is that of dealing with a diagonal mass matrix, whose inversion to compute the dynamic matrix is far simpler than that of the consistent mass matrix. The accuracy is, however, reduced or, better, a greater number of elements is needed to reach the same accuracy, and thus the convenience of using a particular formulation must be assessed in each case.

Remark B.14 In general, the consistent approach leads to values of the natural frequencies that are in excess with respect to those computed using the elastic continuum model, while those obtained using the lumped-parameters approach are smaller.

If a force distribution $\mathbf{f}(x, y, z, t)$ acts on the body, the virtual work linked with the virtual displacement $\delta\mathbf{u} = \mathbf{N}\delta\mathbf{q}$ and the nodal force vector can be expressed in the form

$$\begin{aligned} \delta\mathcal{L} &= \int_V \delta\mathbf{q}^T \mathbf{N}^T \mathbf{f}(x, y, z, t)\, dV, \\ \mathbf{f}(t) &= \int_V \mathbf{N}^T \mathbf{f}(x, y, z, t)\, dV. \end{aligned} \tag{B.61}$$

In a similar way, it is possible to obtain the nodal force vectors corresponding to surface force distributions or to concentrated forces acting on any point of the element.

The equation of motion of the element is thus the usual one for discrete undamped systems

$$\mathbf{M}\ddot{\mathbf{q}} + \mathbf{K}\mathbf{q} = \mathbf{f}(t), \tag{B.62}$$

where vector $\mathbf{f}$ contains all forces acting on the element.

Remark B.15 The equations of motion and the relevant matrices have been obtained here by using Lagrange equations; this approach is neither the only one nor the most common.

B.3.2 Timoshenko Beam Element

The beam element is one of the most common elements and is generally available in all computer codes. Several beam formulations have been developed that differ owing to the theoretical formulation (some of them are Euler–Bernoulli element, i.e.,

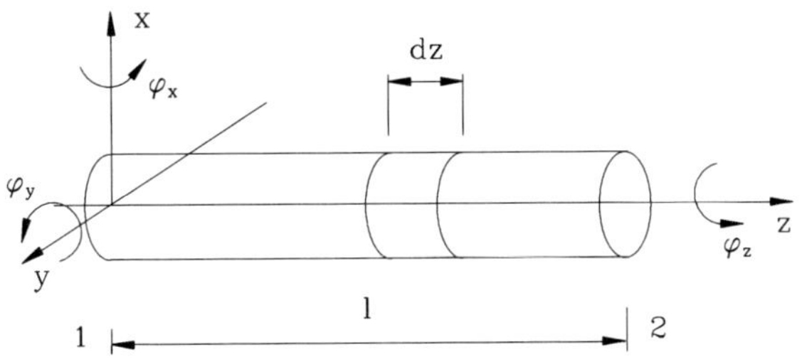

Fig. B.7 Beam element: geometrical definitions and reference frame (from G. Genta, *Vibration Dynamics and Control*, Springer, New York, 2009)

do not take into account shear deformation, while others are Timoshenko elements) and the number of nodes and degrees of freedom per node. The element that will be studied here is often referred to as the *simple Timoshenko beam*. It has two nodes at the ends of the beam and six degrees of freedom per node, and it consists of a prismatic homogeneous beam to which all considerations seen in Sect. B.2.1 apply. The relevant geometrical definition and the reference frame used for the study are shown in Fig. B.7.

Each cross section has six degrees of freedom, three displacements, and three rotations, and the total number of degrees of freedom of the element is 12. The vector of the nodal displacements, i.e., of the generalized coordinates of the element, is

$$\mathbf{q} = [u_{x_1}, u_{y_1}, u_{z_1}, \phi_{x_1}, \phi_{y_1}, \phi_{z_1}, u_{x_2}, u_{y_2}, u_{z_2}, \phi_{x_2}, \phi_{y_2}, \phi_{z_2}]^T. \tag{B.63}$$

The beam has the properties needed to perform a complete uncoupling between axial, torsional, and flexural behavior in each of the coordinate planes; it is thus expedient to subdivide vector $\mathbf{q}$ into four smaller vectors:

$$\mathbf{q}_A = \begin{Bmatrix} u_{z_1} \\ u_{z_2} \end{Bmatrix}, \qquad \mathbf{q}_T = \begin{Bmatrix} \phi_{z_1} \\ \phi_{z_2} \end{Bmatrix},$$
$$\mathbf{q}_{F1} = \begin{Bmatrix} u_{x_1} \\ \phi_{y_1} \\ u_{x_2} \\ \phi_{y_2} \end{Bmatrix}, \qquad \mathbf{q}_{F2} = \begin{Bmatrix} u_{y_1} \\ \phi_{x_1} \\ u_{y_2} \\ \phi_{x_2} \end{Bmatrix}. \tag{B.64}$$

Remark B.16 If the rotational degree of freedom $-\phi_x$ were used instead of ϕ_x the same equations could have been used to describe the flexural behavior in both planes. This approach is, however, uncommon because it would make it more difficult to pass from the system of reference of the elements to that of the whole structure.

By reordering the various generalized coordinates with the aim of clearly showing such uncoupling, vector $\mathbf{q}$ can be written as

$$\mathbf{q} = \left[\mathbf{q}_A^T \quad \mathbf{q}_T^T \quad \mathbf{q}_{F1}^T \quad \mathbf{q}_{F2}^T\right]^T. \tag{B.65}$$

The uncoupling between the various degrees of freedom makes it possible to split the matrix of the shape functions into a number of submatrices, most of which are equal to zero. The generalized displacement of an internal point of the element whose coordinate is z can be expressed in the form of (B.54) by the equation

$$\mathbf{u}(z,t) = \left\{ \begin{matrix} u_z \\ \phi_z \\ \left\{ \begin{matrix} u_x \\ \phi_y \end{matrix} \right\} \\ \left\{ \begin{matrix} u_y \\ \phi_x \end{matrix} \right\} \end{matrix} \right\} = \begin{bmatrix} \mathbf{N}_A & \mathbf{0} & \mathbf{0} & \mathbf{0} \\ \mathbf{0} & \mathbf{N}_T & \mathbf{0} & \mathbf{0} \\ \mathbf{0} & \mathbf{0} & \mathbf{N}_{F1} & \mathbf{0} \\ \mathbf{0} & \mathbf{0} & \mathbf{0} & \mathbf{N}_{F2} \end{bmatrix} \left\{ \begin{matrix} \mathbf{q}_A \\ \mathbf{q}_T \\ \mathbf{q}_{F1} \\ \mathbf{q}_{F2} \end{matrix} \right\}. \tag{B.66}$$

Axial Behavior

Because each point of the element has a single degree of freedom, vector $\mathbf{u}$ has a single component u_z and matrix $\mathbf{N}_A$ has one row and two columns (the element has two degrees of freedom). u_z can be expressed as a polynomial in z, or, better, in the nondimensional axial coordinate $\zeta = z/l$:

$$u_z = a_0 + a_1\zeta + a_2\zeta^2 + a_3\zeta^3 + \cdots. \tag{B.67}$$

The polynomial must yield the values of the displacements u_{z_1} and u_{z_2}, respectively at the left end (node 1, $\zeta = 0$) and at the right end (node 2, $\zeta = 1$). These two conditions allow computation of only two coefficients a_i and then the polynomial expression of the displacement must include only two terms, i.e., the constant and the linear terms. With simple computations, the matrix of the shape functions is obtained:

$$\mathbf{N}_A = [1-\zeta, \zeta]. \tag{B.68}$$

The axial strain ϵ_z can be expressed as

$$\epsilon_z = \frac{du_z}{dz}, \tag{B.69}$$

or, using vector $\boldsymbol{\epsilon}$, which in this case has only one element

$$\epsilon_z = \left[\frac{d}{dz}(1-\zeta), \frac{d}{dz}\zeta\right] \left\{ \begin{matrix} u_{z1} \\ u_{z2} \end{matrix} \right\}. \tag{B.70}$$

Matrix

$$\mathbf{B} = \left[\frac{d}{dz}(1-\zeta), \frac{d}{dz}\zeta\right] = \frac{1}{l}[-1, 1] \tag{B.71}$$

has one row and two columns.

Also, vector $\boldsymbol{\sigma}$ and matrix $\mathbf{E}$ have only a single element: the axial stress σ_z and Young's modulus E, respectively. The stiffness and mass matrices can be obtained directly from (B.59) and (B.60). Remembering that $dV = A\,dz$, they reduce to

$$\mathbf{K}_A = \int_0^l A\mathbf{B}^T E\mathbf{B}\,dz = \frac{EA}{l}\int_0^1 \begin{bmatrix} 1 & -1 \\ -1 & 1 \end{bmatrix} d\zeta = \frac{EA}{l}\begin{bmatrix} 1 & -1 \\ -1 & 1 \end{bmatrix}, \tag{B.72}$$

$$\mathbf{M}_A = \int_0^l \rho A \mathbf{N}^T \mathbf{N} dz = \rho A l \int_0^1 \begin{bmatrix} (1-\zeta)^2 & \zeta(1-\zeta) \\ \zeta(1-\zeta) & \zeta^2 \end{bmatrix} d\zeta$$
$$= \frac{\rho A l}{6} \begin{bmatrix} 2 & 1 \\ 1 & 2 \end{bmatrix}. \tag{B.73}$$

If an axial force distribution $f_z(t)$ that is constant along the space coordinate z or a concentrated axial force $F_{z_k}(t)$ located in the point of coordinate z_k is acting on the bar, the nodal force vector is, respectively,

$$\mathbf{f}(t) = l \left[\int_0^l \left\{ \begin{array}{c} (1-\zeta) \\ \zeta \end{array} \right\} d\zeta \right] f_z(t) = f_z(t) \frac{l}{2} \left\{ \begin{array}{c} 1 \\ 1 \end{array} \right\}, \tag{B.74}$$

or

$$\mathbf{f}(t) = F_{z_k}(t) \left\{ \begin{array}{c} 1 - \dfrac{z_k}{l} \\ \dfrac{z_k}{l} \end{array} \right\}. \tag{B.75}$$

In this case the distributed load has been reduced to two identical forces, each equal to half of the total load acting on the bar.

An identical result would have been obtained by simply lumping the load at the nodes. This is not, however, a general rule; in other cases the consistent approach leads to a load vector that is different from that obtained using the lumped-parameters approach.

Torsional Behavior

The equations of motion governing the torsional behavior of beams are formally identical to those governing the axial behavior. Using this identity, the characterization of the beam element in torsion can be obtained from what has been seen for the axial behavior. Matrix $\mathbf{N}_T$ is identical to matrix $\mathbf{N}_A$

$$\mathbf{N}_T = [1-\zeta, \zeta]$$

and the expressions of the relevant matrices and vectors are

$$\mathbf{M}_T = \frac{\rho I_p l}{6} \begin{bmatrix} 2 & 1 \\ 1 & 2 \end{bmatrix}, \qquad \mathbf{K}_T = \frac{G I'_p}{l} \begin{bmatrix} 1 & -1 \\ -1 & 1 \end{bmatrix},$$
$$\mathbf{f}(t)_T = \frac{1}{2} l m_z(t) \left\{ \begin{array}{c} 1 \\ 1 \end{array} \right\}. \tag{B.76}$$

Flexural Behavior in the xz-Plane

The expressions of the shape function are in this case more complex; matrix $\mathbf{N}_{F1}$ has two rows and four columns because it must yield the displacement in the x-direction and the rotation about the y-axis of the generic cross section of the beam when it is multiplied by vector $\mathbf{q}_{F1}$, which contains four elements, namely, the displacements in the x-direction and the rotation about the y-axis of the two nodes. The simplest approach would be that of assuming polynomial expressions for the generalized displacements

$$\begin{cases} u_x = a_0 + a_1\zeta + a_2\zeta^2 + a_3\zeta^3 + \cdots, \\ \phi_y = b_0 + b_1\zeta + b_2\zeta^2 + b_3\zeta^3 + \cdots. \end{cases} \tag{B.77}$$

These polynomial expressions must yield the values of the displacements u_{x1} and u_{x2} and of the rotations ϕ_{y1} and ϕ_{y2} at the left end (node 1, $\zeta = 0$) and at the right end (node 2, $\zeta = 1$), respectively. These four conditions allow the computation of only four coefficients a_i and b_i to be introduced into the polynomial expressions. Each polynomial must then include only two terms, and both rotations and displacements must vary linearly along the z-coordinate. This element formulation, although sometimes used, leads to the severe problem of locking, i.e., to the possibility of grossly overestimating the stiffness of the element.

Although locking will not be dealt with in detail here (the reader can find a detailed discussion on this matter in any good textbook on the FEM), an intuitive explanation can be seen immediately: if the beam is slender the rotation of each cross section is very close to the derivative of the displacement, as stated by the Euler–Bernoulli approach for slender beams. The polynomial shape functions, when truncated at the second term, do not allow the rotation to be equal to the derivative of the displacement, and this can be shown to lead to a severe underestimate of the displacements, i.e., of the flexibility of the beam.

A simple cure for the problem is that of resorting to the Euler–Bernoulli formulation, i.e., neglecting shear deformation and then assuming that the rotation is coincident with the derivative of the displacement. In this case, only the polynomial for u_x needs to be stated, and the aforementioned four conditions at the nodes can be used to compute the four coefficients of a cubic expression of the displacement.

To avoid locking, a Timoshenko beam element can be formulated using as shape functions the deformed shape computed using the continuous model assuming that only end forces are applied to the beam. This Timoshenko beam element reduces to the Euler–Bernoulli element as the slenderness of the beam increases and no locking occurs. The relevant shape functions are

$$\begin{aligned} N_{11} &= \frac{1 + \Phi(1-\zeta) - 3\zeta^2 + 2\zeta^3}{1+\Phi}, & N_{12} &= l\zeta\frac{1 + \frac{1}{2}\Phi(1-\zeta) - 2\zeta + \zeta^2}{1+\Phi}, \\ N_{13} &= \zeta\frac{\Phi + 3\zeta - 2\zeta^2}{1+\Phi}, & N_{14} &= l\zeta\frac{-\frac{1}{2}\Phi(1-\zeta) - \zeta + \zeta^2}{1+\Phi}, \\ N_{21} &= 6\zeta\frac{\zeta - 1}{l(1+\Phi)}, & N_{22} &= \frac{1 + \Phi(1-\zeta) - 4\zeta + 3\zeta^2}{1+\Phi}, \\ N_{23} &= -6\zeta\frac{\zeta - 1}{l(1+\Phi)}, & N_{24} &= \frac{\Phi\zeta - 2\zeta + 3\zeta^2}{1+\Phi}, \end{aligned} \tag{B.78}$$

where

$$\Phi = \frac{12EI_y\chi}{GAl^2} .$$

When the slenderness of the beam increases, the value of Φ decreases, tending to zero for a Euler–Bernoulli beam.

In this case, some of the generalized coordinates are related to rotations; as a consequence, (B.59) and (B.60) cannot be used directly to express the stiffness and mass matrices. The potential energy can be computed by adding the contributions due to bending and shear deformations. By using the symbols $\mathbf{N}_1$ and $\mathbf{N}_2$ to express the first and second rows of matrix $\mathbf{N}_{F1}$, respectively, the two contributions to the potential energy of the length dz of the beam are

$$\begin{aligned} d\mathcal{U}_b &= \frac{1}{2}EI_y\left(\frac{d\phi_y}{dz}\right)^2 dz = \frac{1}{2}EI_y\{q\}^T\left[\frac{d}{dz}\mathbf{N}_2\right]^T\left[\frac{d}{dz}\mathbf{N}_2\right]\{q\}\,dz, \\ d\mathcal{U}_s &= \frac{1}{2}\frac{GA}{\chi}\left(\phi_y - \frac{du_x}{dz}\right)^2 dz \\ &= \frac{12EI_y}{2\Phi l^2}\{q\}^T\left[\mathbf{N}_2 - \frac{d}{dz}\mathbf{N}_1\right]^T\left[\mathbf{N}_2 - \frac{d}{dz}\mathbf{N}_1\right]\{q\}\,dz. \end{aligned} \tag{B.79}$$

By introducing the expressions of the shape functions into the expression of the potential energy and integrating, the bending stiffness matrix is obtained

$$\mathbf{K}_{F_1} = \frac{EI_y}{l^3(1+\Phi)}\begin{bmatrix} 12 & 6l & -12 & 6l \\ & (4+\Phi)l^2 & -6l & (2-\Phi)l^2 \\ & & 12 & -6l \\ & \text{symm} & & (4+\Phi)l^2 \end{bmatrix} . \tag{B.80}$$

The kinetic energy of the length dz of the beam is

$$\begin{aligned} d\mathcal{T} &= \frac{1}{2}\rho A\dot{u}^2\,dz + \frac{1}{2}\rho I_y\dot{\phi}_y^{\,2}\,dz \\ &= \frac{1}{2}\rho A\dot{\mathbf{q}}^T\mathbf{N}_1^T\mathbf{N}_1\dot{\mathbf{q}}\,dz + \frac{1}{2}\rho I_y\dot{\mathbf{q}}^T\mathbf{N}_2^T\mathbf{N}_2\dot{\mathbf{q}}\,dz. \end{aligned} \tag{B.81}$$

The first term on the right-hand side is the translational kinetic energy, and the second expresses the rotational kinetic energy, which is often neglected in the case of slender beams. By introducing the expressions of the shape functions and integrating, the consistent mass matrix, which is made of two parts—one taking into account the translational inertia and the other the rotational inertia of the cross sections—is obtained

$$\mathbf{M}_{F1} = \frac{\rho Al}{420(1+\Phi)^2}\begin{bmatrix} m_1 & lm_2 & m_3 & -lm_4 \\ & l^2m_5 & lm_4 & -l^2m_6 \\ & & m_1 & -lm_2 \\ & \text{symm} & & l^2m_5 \end{bmatrix}$$

$$+\frac{\rho I_y}{30l(1+\Phi)^2}\begin{bmatrix} m_7 & lm_8 & -m_7 & lm_8 \\ & l^2m_9 & -lm_8 & -l^2m_{10} \\ & & m_7 & -lm_8 \\ & \text{symm} & & l^2m_9 \end{bmatrix}, \tag{B.82}$$

where

$$m_1 = 156+294\Phi+140\Phi^2, \qquad m_2 = 22+38.5\Phi+17.5\Phi^2,$$

$$m_3 = 54+126\Phi+70\Phi^2, \qquad m_4 = 13+31.5\Phi+17.5\Phi^2,$$

$$m_5 = 4+7\Phi+3.5\Phi^2, \qquad m_6 = 3+7\Phi+3.5\Phi^2,$$

$$m_7 = 36, \qquad m_8 = 3-15\Phi,$$

$$m_9 = 4+5\Phi+10\Phi^2, \qquad m_{10} = 1+5\Phi-5\Phi^2.$$

The consistent load vector due to a uniform distribution of shear force per unit length $f_x(t)$ or of bending moment $m_y(t)$ can be obtained directly from (B.61)

$$\mathbf{f}(t)_{F1} = l\left[\int_0^1 \mathbf{N}_{F1}^T\, d\zeta\right]\left\{\begin{matrix} f_x(t) \\ m_y(t) \end{matrix}\right\} = \frac{lf_x(t)}{12}\left\{\begin{matrix} 6 \\ l \\ 6 \\ -l \end{matrix}\right\} + \frac{m_y(t)}{1+\Phi}\left\{\begin{matrix} -l \\ \frac{\Phi l}{2} \\ l \\ \frac{\Phi l}{2} \end{matrix}\right\}. \tag{B.83}$$

Flexural Behavior in the yz-Plane

The flexural behavior in the yz-plane must be studied using equations different from those used for the xz-plane, owing to the different signs of rotation in the two planes. Matrix $\mathbf{N}_{F2}$ can, however, be obtained from matrix $\mathbf{N}_{F1}$ simply by changing the signs of elements with subscripts 12, 14, 21, and 23, and the mass and stiffness matrices related to plane yz are the same as those computed for plane xz, except for the signs of elements with subscripts 12, 14, 23, and 34 and their symmetrical ones. In the force vectors related to external forces (distributed or concentrated) or external moments, the signs of elements 2 and 4 or 1 and 3, respectively, must be changed. If the beam is not axially symmetrical and the elastic and inertial properties in the two planes are not coincident, different values of the moments of inertia and the shear factors must be introduced.

Global Behavior of the Beam

The complete expression of the mass and stiffness matrices and of the nodal force vector are, respectively,

$$\mathbf{M} = \begin{bmatrix} \mathbf{M}_A & \mathbf{0} & \mathbf{0} & \mathbf{0} \\ \mathbf{0} & \mathbf{M}_T & \mathbf{0} & \mathbf{0} \\ \mathbf{0} & \mathbf{0} & \mathbf{M}_{F1} & \mathbf{0} \\ \mathbf{0} & \mathbf{0} & \mathbf{0} & \mathbf{M}_{F2} \end{bmatrix},$$

$$\mathbf{K} = \begin{bmatrix} \mathbf{K}_A & \mathbf{0} & \mathbf{0} & \mathbf{0} \\ \mathbf{0} & \mathbf{K}_T & \mathbf{0} & \mathbf{0} \\ \mathbf{0} & \mathbf{0} & \mathbf{K}_{F1} & \mathbf{0} \\ \mathbf{0} & \mathbf{0} & \mathbf{0} & \mathbf{K}_{F2} \end{bmatrix}, \qquad \mathbf{f} = \begin{Bmatrix} \mathbf{f}_A \\ \mathbf{f}_T \\ \mathbf{f}_{F1} \\ \mathbf{f}_{F2} \end{Bmatrix}. \tag{B.84}$$

B.3.3 Mass and Spring Elements

Consider a concentrated mass, or better, a rigid body located at the ith node. Let $\mathbf{q}$ be the vector of the generalized displacements of the relevant node, which may also contain rotations

$$\mathbf{q} = [u_x, u_y, u_z, \phi_x, \phi_y, \phi_z],$$

if the node is of the type of those seen in the case of beam elements.

In the latter case, if the mass also has moments of inertia, let the axes of the reference frame coincide with the principal axes of the rigid body. By writing the kinetic energy of the element, it can be shown that the mass matrix, which is diagonal because the axes of the reference frame coincide with the principal axes of the body, is

$$\mathbf{M} = \text{diag}[m, m, m, J_x, J_y, J_z]. \tag{B.85}$$

A simpler expression is obtained when only the translational degrees of freedom of the node are considered.

Remark B.17 In many computer programs, different values of the mass can be associated with the various degrees of freedom. This can account for particular physical layouts, such as the addition of a mass constrained to move with the structure in one direction and not in others.

Consider a spring element, i.e., an element that introduces a concentrated stiffness between two nodes, say node 1 and node 2, of the structure. When the nodes have a single degree of freedom, the generalized coordinates of the element are $q = [u_1, u_2]^T$, and the stiffness matrix is

$$\mathbf{K} = \begin{bmatrix} k & -k \\ -k & k \end{bmatrix}. \tag{B.86}$$

If the nodes, like those used with beam elements, have three translational and three rotational degrees of freedom, three values of translational stiffness k_x, k_y, and k_z and three values of rotational stiffness χ_x, χ_y, and χ_z can be stated and matrices of the type shown in (B.86) can be written for each degree of freedom.

B.3.4 Assembling the Structure

The equations of motion of the element are written with reference to a local or element reference frame that has an orientation determined by the features of the element. In a beam element, for example, the z-axis can coincide with the axis of the beam, while the x- and y-axes are principal axes of inertia of the cross section. The various local reference frames of the elements have, in general, a different orientation in space. To describe the behavior of the structure as a whole, another reference frame, namely the global or structure reference frame, is defined. The orientation in space of any local frame can be expressed, with reference to the orientation of the global frame, by a suitable rotation matrix

$$\mathbf{R} = \begin{bmatrix} l_x & m_x & n_x \\ l_y & m_y & n_y \\ l_z & m_z & n_z \end{bmatrix}, \tag{B.87}$$

where l_i, m_i, and n_i are the direction cosines of the axes of the local frame in the global frame. The expressions $\mathbf{q}_{il}$ and $\mathbf{q}_{ig}$ of the displacement vector $\mathbf{q}_i$ of the ith node in the local and global reference frames are linked by the usual coordinate transformation

$$\mathbf{q}_{il} = \mathbf{R}\mathbf{q}_{ig}. \tag{B.88}$$

The generalized coordinates in the displacement vector of the element can be transformed from the local to the global reference frame using a similar relationship in which an expanded rotation matrix $\mathbf{R}'$ is used to deal with all the relevant generalized coordinates. It is essentially made by a number of matrices of the type of (B.87) suitably assembled together.

Remark B.18 The assumption of small displacements and rotations allows consideration of the rotations about the axes as the components of a vector, which can be rotated in the same way as displacements.

The force vectors can also be rotated using the rotation matrix $\mathbf{R}'$, and the equation of motion of the element can be written with reference to the global frame and premultiplied by the inverse of matrix $\mathbf{R}'$, obtaining

$$\mathbf{R}'^{-1}\mathbf{M}\mathbf{R}'\ddot{\mathbf{q}}_g + \mathbf{R}'^{-1}\mathbf{K}\mathbf{R}'\mathbf{q}_g = \mathbf{f}_g. \tag{B.89}$$

Because the inverse of a rotation matrix is coincident with its transpose, the expressions of the mass and stiffness matrices of the element rotated from the local to the global frame are

$$\mathbf{M}_g = \mathbf{R}'^T\mathbf{M}_l\mathbf{R}' \tag{B.90}$$

and

$$\mathbf{K}_g = \mathbf{R}'^T\mathbf{K}_l\mathbf{R}'. \tag{B.91}$$

Similarly, the nodal load vector can be rotated using the obvious relationship

$$\mathbf{f}_g = \mathbf{R}'^T \mathbf{f}_l. \tag{B.92}$$

Once the mass and stiffness matrices, referring to the global frame, of the various elements have been computed, it is possible to easily obtain the matrices of the whole structure. The n generalized coordinates of the structure can be ordered in a single vector $\mathbf{q}_g$. The matrices of the various elements can be rewritten in the form of matrices of order $n \times n$, containing all elements equal to zero except those that are in the rows and columns corresponding to the generalized coordinates of the relevant element.

Because the kinetic and potential energies of the structure can be obtained simply by adding the energies of the various elements, it follows that

$$\begin{aligned} \mathcal{T} &= \frac{1}{2}\sum_{\forall i} \dot{\mathbf{q}}_g^T \mathbf{M}_i \dot{\mathbf{q}}_g = \frac{1}{2}\dot{\mathbf{q}}_g^T \mathbf{M}\dot{\mathbf{q}}_g; \\ \mathcal{U} &= \frac{1}{2}\sum_{\forall i} \mathbf{q}_g^T \mathbf{K}_i \mathbf{q}_g = \frac{1}{2}\mathbf{q}_g^T \mathbf{K}\mathbf{q}_g. \end{aligned} \tag{B.93}$$

Matrices $\mathbf{M}$ and $\mathbf{K}$ are the mass and stiffness matrices of the whole structure and are obtained simply by adding all the mass and stiffness matrices of the elements. In practice, the various matrices of size $n \times n$ for the elements are never written: each term of the matrices of all elements is just added into the global mass and stiffness matrices in the correct place. Actually, the matrices of the structure are very easily assembled, and this is one of the easiest steps of the whole computation. If the generalized coordinates are taken into a suitable order, the assembled matrices have a band structure; many general-purpose computer codes have a suitable routine that reorders the coordinates in such a way that the bandwidth is the smallest possible.

In a similar way the nodal force vector can be easily assembled:

$$\mathbf{f} = \sum_{\forall i} \mathbf{f}_i. \tag{B.94}$$

Remark B.19 The forces that are exchanged between the elements at the nodes cancel each other in this assembling procedure, and the force vectors that must be inserted into the global equation of motion of the structure are only those related to the external forces applied to the structure.

B.3.5 Constraining the Structure

One of the advantages of the FEM is the ease with which the constraints can be defined. If the ith degree of freedom is rigidly constrained, the corresponding generalized displacement vanishes and, as a consequence, the ith column of the stiffness and mass matrices can be neglected, because they multiply a displacement and an acceleration, respectively, that are equal to zero. Because one of the generalized displacements is known, one of the equations of motion can be neglected when solving for the deformed configuration of the system. The ith equation can thus be separated from the rest of the set of equations, which amounts to canceling the ith row of all matrices and of the force vector.

Remark B.20 The ith equation could be used, after all displacements have been computed, to obtain the value of the ith generalized nodal force that, in this case, is the unknown reaction of the constraint.

To rigidly constrain a degree of freedom it is thus sufficient to cancel the corresponding row and column in all matrices and vectors. This approach allows simplification of the formulation of the problem, which may be useful in dynamic problems, but this simplification is often marginal, since the number of constrained degrees of freedom is small, compared with the total number of degrees of freedom. To avoid restructuring the whole model and rewriting all the matrices, rigid constraints can be transformed into very stiff elastic constraints.

If the ith degree of freedom is constrained through a linear spring with stiffness k_i, the potential energy of the structure is increased by the potential energy of the spring

$$\mathcal{U} = \frac{1}{2} k_i q_i^2. \tag{B.95}$$

To take the presence of the constraint into account, it is sufficient to add the stiffness k_i to the element in the ith row and ith column of the global stiffness matrix. This procedure is simple, which explains why a very stiff elastic constraint is often added instead of canceling a degree of freedom in the case of rigid constraints. An additional advantage is that the reaction of the constraint can be obtained simply by multiplying the high generalized stiffness k_i by the corresponding small generalized displacement q_i.

B.3.6 Damping Matrices

It is possible to take into account the damping of the structure in a way that closely follows what has been said for the stiffness. If elements that can be modeled as viscous dampers are introduced into the structure between two nodes or between a node and the ground, a viscous damping matrix can be obtained using the same procedures seen for the stiffness matrix of spring elements or elastic constraints. Actually, the relevant equations are equal, once the damping coefficient is substituted for the stiffness and velocities are substituted for displacements. If the damping of some of the elements can be modeled as hysteretic damping, within the limits of validity of the complex stiffness model, an imaginary part of the element stiffness matrix can be obtained by simply multiplying the real part by the loss factor.

Viscous or structural damping matrices are then assembled following the same rules seen for mass and stiffness matrices.

Remark B.21 The real and imaginary parts of the stiffness matrices must be assembled separately, because, when the loss factor is not constant along the structure, they are not proportional to each other.

B.4 Reduction of the Number of Degrees of Freedom

It is not uncommon that the models obtained through the FEM have thousands or even millions degrees of freedom. This does not constitute a problem for modern computers when studying static problems, but the solution of an eigenproblem of that size can still be a formidable problem. Moreover, the FEM is a displacement method, i.e. first solves the displacements and then computes stresses and strains as derivatives of the displacements, and thus the precision with which displacements, and all other entities directly linked with displacements including mode shapes and natural frequencies, are obtained is far greater, for a given mesh, than that achievable for stresses and strains. Conversely, this means that the mesh needs to be much finer when solving the stress field, which is typical of static problems, than when searching natural frequencies and mode shapes.

As a consequence, there is a definite advantage in using a smaller model, in terms of the number of degrees of freedom, when performing a dynamic analysis than when doing static computations.

Remark B.22 Because it is often expedient to use the same mesh for both static and dynamic computations, or a finer mesh is required to model a complex shape, a reduction of the number of degrees of freedom for dynamic solution is useful, particularly when only a limited number of natural frequencies are required.

Two approaches can be used: reducing the size of the model or leaving the model as it is and using algorithms, such as the subspace iteration method, that search only the lowest natural frequencies. Although the two are more or less equivalent, the first leaves the choice of which degrees of freedom to retain to the user, and the second operates automatically. As a consequence, a skilled operator can use advantageously reduction techniques, which allow very good results with very few degrees of freedom. General-purpose codes for routine computations usually resort to the second approach.

Remark B.23 Before computers were available, remarkable results were obtained using models with very few (often a single) degrees of freedom, but this required great computational ability and physical insight.

B.4.1 Static Reduction

Static reduction is based on the subdivision of the generalized coordinates $\mathbf{q}$ of the model into two types: master degrees of freedom $\mathbf{q}_1$ and slave degrees of freedom $\mathbf{q}_2$. The stiffness matrix and the nodal force vector can be partitioned accordingly, and the equation expressing the static problem becomes

$$\begin{bmatrix} \mathbf{K}_{11} & \mathbf{K}_{12} \\ \mathbf{K}_{21} & \mathbf{K}_{22} \end{bmatrix} \begin{Bmatrix} \mathbf{q}_1 \\ \mathbf{q}_2 \end{Bmatrix} = \begin{Bmatrix} \mathbf{f}_1 \\ \mathbf{f}_2 \end{Bmatrix}. \tag{B.96}$$

Remark B.24 Matrices $\mathbf{K}_{11}$ and $\mathbf{K}_{22}$ are symmetrical, while $\mathbf{K}_{12} = \mathbf{K}_{21}^T$ are neither symmetrical nor square.

Solving the second set of (B.96) in $\mathbf{q}_2$, the following relationship linking the slave to the master coordinates is obtained:

$$\mathbf{q}_2 = -\mathbf{K}_{22}^{-1}\mathbf{K}_{21}\mathbf{q}_1 + \mathbf{K}_{22}^{-1}\mathbf{f}_2. \tag{B.97}$$

Introducing (B.97) into (B.96), the latter yields

$$\mathbf{K}_{\text{cond}}\mathbf{q}_1 = \mathbf{f}_{\text{cond}}, \tag{B.98}$$

where

$$\begin{cases} \mathbf{K}_{\text{cond}} = \mathbf{K}_{11} - \mathbf{K}_{12}\mathbf{K}_{22}^{-1}\mathbf{K}_{12}^T, \\ \mathbf{f}_{\text{cond}} = \mathbf{f}_1 - \mathbf{K}_{12}\mathbf{K}_{22}^{-1}\mathbf{f}_2. \end{cases}$$

Equation (B.98) yields the master generalized displacements $\mathbf{q}_1$. The slave displacements can be obtained directly from (B.97) simply by multiplying some matrices.

Remark B.25 When used to solve a static problem static reduction yields exact results, i.e., the same results that would be obtained from the complete model.

The subdivision of the degrees of freedom between vectors $\mathbf{q}_1$ and $\mathbf{q}_2$ can be based on different criteria. The master degrees of freedom can simply be those in which the user is directly interested. Another type of choice can be that of physically subdividing the structure in two parts.

The second practice, which can be generalized by subdividing the generalized coordinates into many subsets, is generally known as *solution by substructures* or *substructuring*. In particular, it may be expedient when the structure can be subdivided into many parts that are all connected to a single frame. If the generalized displacements of the connecting structure or frame are listed in vector $\mathbf{q}_0$ and those of the various substructures are included in vectors $\mathbf{q}_i$, the equation for the static solution of the complete structure has the form

$$\begin{bmatrix} \mathbf{K}_{00} & \mathbf{K}_{01} & \mathbf{K}_{02} & \dots \\ & \mathbf{K}_{11} & \mathbf{0} & \dots \\ & & \mathbf{K}_{22} & \dots \\ & \text{symm} & & \dots \end{bmatrix} \begin{Bmatrix} \mathbf{q}_0 \\ \mathbf{q}_1 \\ \mathbf{q}_2 \\ \dots \end{Bmatrix} = \begin{Bmatrix} \mathbf{f}_0 \\ \mathbf{f}_1 \\ \mathbf{f}_2 \\ \dots \end{Bmatrix}. \tag{B.99}$$

The equations related to the ith substructure can be solved as

$$\mathbf{q}_i = -\mathbf{K}_{ii}^{-1}\mathbf{K}_{i0}\mathbf{q}_0 + \mathbf{K}_{ii}^{-1}\mathbf{f}_i. \tag{B.100}$$

The generalized displacements of the frame can be obtained using an equation of the type of (B.98) where the condensed matrices are

$$\begin{cases} \mathbf{K}_{\text{cond}} = \mathbf{K}_{00} - \sum_{\forall i} \mathbf{K}_{0i}\mathbf{K}_{ii}^{-1}\mathbf{K}_{0i}^T, \\ \mathbf{f}_{\text{cond}} = \mathbf{f}_0 - \sum_{\forall i} \mathbf{K}_{0i}\mathbf{K}_{ii}^{-1}\mathbf{f}_i. \end{cases} \tag{B.101}$$

As already stated, static reduction does not introduce any further approximation into the model. A similar reduction can be used in dynamic analysis without introducing approximations only if no generalized inertia is associated with the slave degrees of freedom. In this case, static reduction is advisable because the mass matrix of the original system is singular and the condensation procedure allows removal of the singularity. When using the lumped-parameters approach with beam elements and the moments of inertia of the cross sections are neglected, no inertia is associated with half of the degrees of freedom related to bending. Static reduction allows removing all of them and then obtaining a nonsingular mass matrix. Generally speaking, however, the mass matrix is not singular and it is not possible to just neglect the inertia linked with some degrees of freedom.

B.4.2 Guyan Reduction

The so-called *Guyan reduction* is based on the assumption that the slave generalized displacements $\mathbf{q}_2$ can be computed directly from master displacements $\mathbf{q}_1$, neglecting inertia forces and external forces $\mathbf{f}_2$. In this case, (B.97), without the last term, can also be used for dynamic solution. By partitioning the mass matrix in the same way seen for the stiffness matrix, the kinetic energy of the structure can be expressed as

$$\mathcal{T} = \frac{1}{2}\begin{Bmatrix} \dot{\mathbf{q}}_1 \\ -\mathbf{K}_{22}^{-1}\mathbf{K}_{21}\dot{\mathbf{q}}_1 \end{Bmatrix}^T \begin{bmatrix} \mathbf{M}_{11} & \mathbf{M}_{12} \\ \mathbf{M}_{21} & \mathbf{M}_{22} \end{bmatrix} \begin{Bmatrix} \dot{\mathbf{q}}_1 \\ -\mathbf{K}_{22}^{-1}\mathbf{K}_{21}\dot{\mathbf{q}}_1 \end{Bmatrix}. \tag{B.102}$$

The kinetic energy is thus

$$\mathcal{T} = \frac{1}{2}\dot{\mathbf{q}}_1^T\mathbf{M}_{\text{cond}}\dot{\mathbf{q}}_1,$$

where the condensed mass matrix is

$$\mathbf{M}_{\text{cond}} = \mathbf{M}_{11} - \mathbf{M}_{12}\mathbf{K}_{22}^{-1}\mathbf{K}_{12}^{T} - \left[\mathbf{M}_{12}\mathbf{K}_{22}^{-1}\mathbf{K}_{12}^{T}\right]^{T} + \mathbf{K}_{12}\mathbf{K}_{22}^{-1}\mathbf{M}_{22}\mathbf{K}_{22}^{-1}\mathbf{K}_{12}^{T}. \tag{B.103}$$

Guyan reduction is not much more demanding from a computational viewpoint than static reduction because the only matrix inversion is that of $\mathbf{K}_{22}$, which has already been performed for the computation of the condensed stiffness matrix. If matrix $\mathbf{M}$ is diagonal, two of the terms of (B.103) vanish. Although approximate, Guyan reduction introduces errors that are usually small, at least if the choice of the slave degrees of freedom is appropriate. Inertia forces related to slave degrees of freedom are actually not neglected, but their contribution to the kinetic energy is computed from a deformed configuration obtained on the basis of the master degrees of freedom alone.

Remark B.26 If the relevant mode shapes are only slightly influenced by the presence of some of the generalized masses or if some parts of the structure are so stiff that their deflected shape can be determined by a few coordinates, the results can be good, even when few master degrees of freedom are used.

In a way similar to that seen for the mass matrix, viscous or structural damping matrices $\mathbf{C}$ and $\mathbf{K}''$ can be reduced using (B.103) in which $\mathbf{M}$ has been substituted with $\mathbf{C}$ and $\mathbf{K}''$, respectively. Also, the reduction of damping matrices introduces errors that depend on the choice of the slave degrees of freedom but are usually small when the degrees of freedom in which viscous dampers are applied or, in the case of hysteretic damping, where the loss factor of the material changes, are not eliminated. Alternatively, these degrees of freedom can be neglected when their displacement is well determined by some master displacement, as in the case of very stiff parts of the structure.

B.4.3 Component-Mode Synthesis

When substructuring is used, the degrees of freedom of each structure can be subdivided into two sets: internal degrees of freedom and boundary degrees of freedom. The latter are all degrees of freedom that the substructure has in common with other parts of the structure. They are often referred to as constraint degrees of freedom because they express how the substructure is constrained to the rest of the system. Internal degrees of freedom are those belonging only to the relevant substructure. The largest possible reduction scheme is that in which all internal degrees of freedom are considered slave coordinates and all boundary degrees of freedom are considered master coordinates. In this way, however, the approximation of all modes in which the motion of the internal points of the substructure with respect to its boundary is important, can be quite rough.

A simple way to avoid this drawback is to also consider as master coordinates, together with the boundary degrees of freedom, some of the modal coordinates of the substructure constrained at its boundary. This procedure would obviously lead to exact results if all modes were retained, but because the total number of modes is equal to the number of internal degrees of freedom, the model obtained has as many degrees of freedom as the original one. As usual with modal practices, the computational advantages grow together with the number of modes that can be neglected.

The relevant matrices are partitioned as seen for reduction techniques, with subscript 1 referring to the boundary degrees of freedom and subscript 2 to the internal degrees of freedom. The displacement vector $\mathbf{q}_2$ can be assumed to be equal to the sum of the constrained modes $\mathbf{q}'_2$, i.e., the deformation pattern due to the displacements $\mathbf{q}_1$ when no force acts on the substructure, plus the constrained normal modes $\mathbf{q}''_2$, i.e., the natural modes of free vibration of the substructure when the boundary generalized displacements $\mathbf{q}_1$ are equal to zero.

The constrained modes $\mathbf{q}'_2$ can be expressed by (B.97) once the force vector $\mathbf{f}_2$ is set equal to zero. The constrained normal modes can easily be computed by solving the eigenproblem

$$\left(-\omega^2\mathbf{M}_{22} + \mathbf{K}_{22}\right)\mathbf{q}''_2 = 0.$$

Once the eigenproblem has been solved, the matrix of the eigenvectors $\boldsymbol{\Phi}$ can be used to perform the modal transformation $\mathbf{q}''_2 = \boldsymbol{\Phi}\eta_2$. The generalized coordinates of the substructure can thus be expressed as

$$\begin{Bmatrix} \mathbf{q}_1 \\ \mathbf{q}_2 \end{Bmatrix} = \begin{Bmatrix} \mathbf{q}_1 \\ -\mathbf{K}_{22}^{-1}\mathbf{K}_{21}\mathbf{q}_1 + \boldsymbol{\Phi}\eta_2 \end{Bmatrix} = \begin{bmatrix} \mathbf{I} & \mathbf{0} \\ -\mathbf{K}_{22}^{-1}\mathbf{K}_{21} & \boldsymbol{\Phi} \end{bmatrix} \begin{Bmatrix} \mathbf{q}_1 \\ \eta_2 \end{Bmatrix} = \boldsymbol{\Psi} \begin{Bmatrix} \mathbf{q}_1 \\ \eta_2 \end{Bmatrix}. \tag{B.104}$$

Equation (B.104) represents a coordinate transformation, allowing to express the deformation of the internal part of the substructure in terms of constrained and normal modes. The matrix $\boldsymbol{\Psi}$ expressing this transformation can be used to compute the new mass, stiffness, and, where needed, damping matrices and the force vector

$$\begin{aligned} \mathbf{M}^* &= \boldsymbol{\Psi}^T\mathbf{M}\boldsymbol{\Psi}, \qquad & \mathbf{K}^* &= \boldsymbol{\Psi}^T\mathbf{K}\boldsymbol{\Psi}, \\ \mathbf{C}^* &= \boldsymbol{\Psi}^T\mathbf{C}\boldsymbol{\Psi}, \qquad & \mathbf{f}^* &= \boldsymbol{\Psi}^T\mathbf{f}. \end{aligned} \tag{B.105}$$

If there are m constrained coordinates and n internal coordinates and if only k constrained normal modes are considered ($k < n$), then the size of the original matrices $\mathbf{M}$, $\mathbf{K}$, ... is $m + n$, while that of matrices $\mathbf{M}^*$, $\mathbf{K}^*$, ... is $m + k$.

Once the transformation of coordinates of the substructures has been performed, they can be assembled in the same way already seen for elements: The boundary coordinates are common between the substructures and are actually assembled while the modal coordinates are typical of only one substructure at a time, in the same way as the coordinates of internal nodes of elements that have nodes of this type.

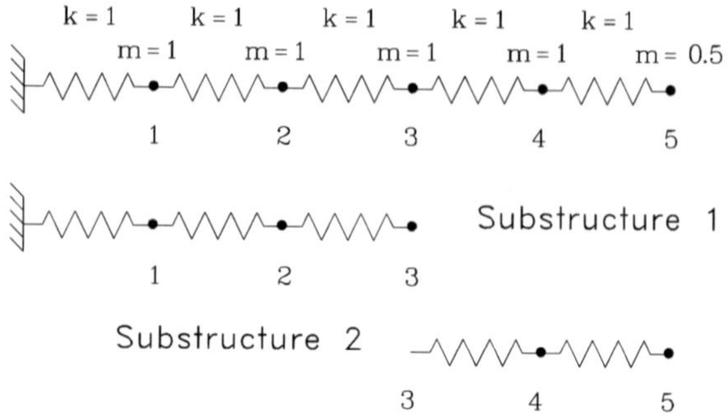

Fig. B.8 Sketch of the system and values of the relevant parameters (from G. Genta, *Vibration Dynamics and Control*, Springer, New York, 2009)

Remark B.27 Actually each substructure can be regarded as a large element, sometimes referred to as a *superelement* and the relevant procedures do not differ from those that are standard in the FEM.

The main advantage of substructuring is that of allowing the construction of the model and the analysis of the various parts of a large structure in an independent way. The results can then be assembled and the behavior of the structure can be assessed from that of its parts. If this is done, however, the connecting nodes must be defined in such a way that the same boundary degrees of freedom are considered in the analysis of the various parts. It is, however, possible to use algorithms allowing the connecting of otherwise incompatible meshes.

Remark B.28 All the methods discussed in this section, which are closely related to each other and are found in the literature in a variety of versions, are general for discrete systems and can also be used outside the FEM even if they became popular only with the use of the latter owing to the large number of degrees of freedom it yields.

An example can be useful to illustrate how the component-mode synthesis method works. Consider the discrete system sketched in Fig. B.8. Let us study its dynamic behavior and compare the results obtained using component-mode synthesis retaining different numbers of modes.

The total number of degrees of freedom of the system is five and the complete mass and stiffness matrices are

$$\mathbf{K} = \begin{bmatrix} 2 & -1 & 0 & 0 & 0 \\ -1 & 2 & -1 & 0 & 0 \\ 0 & -1 & 2 & -1 & 0 \\ 0 & 0 & -1 & 2 & -1 \\ 0 & 0 & 0 & -1 & 1 \end{bmatrix}, \qquad \mathbf{M} = \begin{bmatrix} 1 & 0 & 0 & 0 & 0 \\ 0 & 1 & 0 & 0 & 0 \\ 0 & 0 & 1 & 0 & 0 \\ 0 & 0 & 0 & 1 & 0 \\ 0 & 0 & 0 & 0 & 0.5 \end{bmatrix}.$$

By directly solving the eigenproblem, the following matrix of the eigenvalues is obtained:

$$\left[\omega^2\right] = \text{diag}[0.0979 \quad 0.8244 \quad 2.000 \quad 3.176 \quad 3.902].$$

The structure is then subdivided into two substructures and the analysis is accordingly performed.

Substructure 1 Substructure 1 includes nodes 1, 2, and 3 with the masses located on them. The displacements at nodes 1 and 2 are internal coordinates, while the displacement at node 3 is a boundary coordinate. The mass and stiffness matrix of the substructure, partitioned with the boundary degree of freedom first and then the internal ones ordered with the displacement at node 2 before that at node 1, are

$$\mathbf{K} = \left[\begin{array}{c|cc} 1 & -1 & 0 \\ \hline -1 & 2 & -1 \\ 0 & -1 & 2 \end{array}\right], \qquad \mathbf{M} = \left[\begin{array}{c|cc} 1 & 0 & 0 \\ \hline 0 & 1 & 0 \\ 0 & 0 & 1 \end{array}\right].$$

The matrix of the eigenvectors for the internal normal modes can be easily obtained by solving the eigenproblem related to matrices with subscript 22 and, by retaining all modes, matrices $\mathbf{K}^*$ and $\mathbf{M}^*$ of the first substructure can be computed:

$$\boldsymbol{\Phi} = \begin{bmatrix} \frac{\sqrt{2}}{2} & \frac{-\sqrt{2}}{2} \\ \frac{\sqrt{2}}{2} & \frac{\sqrt{2}}{2} \end{bmatrix}, \qquad \mathbf{K}^* = \left[\begin{array}{c|cc} 0.3333 & 0 & 0 \\ 0 & 1 & 0 \\ 0 & 0 & 3 \end{array}\right],$$

$$\mathbf{M}^* = \left[\begin{array}{c|cc} 1.556 & 0.7071 & -0.2357 \\ \hline 0.7071 & 1 & 0 \\ -0.2357 & 0 & 1 \end{array}\right].$$

Substructure 2 The second substructure includes nodes 3, 4, and 5 with the masses located on nodes 4 and 5. The mass located on node 3 has already been taken into account in the first substructure and must not be considered again. The displacements at nodes 4 and 5 are internal coordinates, while the displacement at node 3 is a boundary coordinate. The mass and stiffness matrix of the substructure, partitioned with the boundary degree of freedom first and then the internal ones (with the displacement at node 4 and then that at node 5), are

$$\mathbf{K} = \left[\begin{array}{c|cc} 1 & -1 & 0 \\ \hline -1 & 2 & -1 \\ 0 & -1 & 1 \end{array}\right], \qquad \mathbf{M} = \left[\begin{array}{c|cc} 0 & 0 & 0 \\ \hline 0 & 1 & 0 \\ 0 & 0 & 0.5 \end{array}\right].$$

Operating as seen for the first substructure, it follows that

$$\boldsymbol{\Phi} = \begin{bmatrix} \frac{\sqrt{2}}{2} & \frac{-\sqrt{2}}{2} \\ 1 & 1 \end{bmatrix}, \qquad \mathbf{K}^* = \left[\begin{array}{c|cc} 0 & 0 & 0 \\ \hline 0 & 0.5858 & 0 \\ 0 & 0 & 3.4142 \end{array}\right],$$

$$\mathbf{M}^* = \left[\begin{array}{c|cc} 1.5 & 1.2071 & -0.2071 \\ \hline 1.2071 & 1 & 0 \\ -0.2071 & 0 & 1 \end{array}\right].$$

The substructures can be assembled in the same way as the elements. The following map can be written:

Subst.	d.o.f.	1	2	3		
1	type	boundary	modal	modal		
Subst.	d.o.f.	1			2	3
2	type	boundary			modal	modal
Global	d.o.f.	1	2	3	4	5

yielding the following global stiffness and mass matrices:

$$\mathbf{K}^* = \left[\begin{array}{c|cccc} 0.3333 & 0 & 0 & 0 & 0 \\ \hline 0 & 1 & 0 & 0 & 0 \\ 0 & 0 & 3 & 0 & 0 \\ 0 & 0 & 0 & 0.5858 & 0 \\ 0 & 0 & 0 & 0 & 3.4142 \end{array}\right],$$

$$\mathbf{M}^* = \left[\begin{array}{c|cccc} 1.5 & 0.7071 & -0.2357 & 1.2071 & -0.2071 \\ \hline 0.7071 & 1 & 0 & 0 & 0 \\ -0.2357 & 0 & 1 & 0 & 0 \\ 1.2071 & 0 & 0 & 1 & 0 \\ -0.2071 & 0 & 0 & 0 & 1 \end{array}\right].$$

The matrices have been partitioned in such a way to separate the boundary displacement degree of freedom from the modal degrees of freedom. If no modal coordinate is considered, the component-mode synthesis coincides with Guyan reduction, with only one master degree of freedom. If the third and fifth rows and columns are canceled, only one internal normal mode is taken into account for each substructure. If the matrices are taken into account in complete form, all modes are considered and the result must coincide, except for computing approximations, with the exact ones. The results obtained in terms of the square of the natural frequency are

Size of matrices	5 (exact)	1 (Guyan red.)	3 (1 mode)	5 (2 modes)
Mode 1	0.0979	0.1091	0.0979	0.0979
Mode 2	0.8244	–	0.8245	0.8244
Mode 3	2.000	–	2.215	2.000
Mode 4	3.176	–	–	3.176
Mode 5	3.902	–	–	3.902

References[1]

Robotics

1. Duffy J (1980) Analysis of mechanisms and robot manipulators. Arnold, London
2. Paul RP (1981) Robot manipulators: mathematics, programming and control: the computer control of robot manipulator. MIT Press, Cambridge
3. Coiffet P (ed) (1983) Robot technology. Kogan Page, London
4. Pugh A (ed) (1983) Robot vision. Springer, Berlin
5. Robillard MJ (1983) Microprocessor based robotics. Sams, Indianapolis
6. Barker LK, Moore MC (1984) Kinematic control of robot with degenerate wrist. NASA, Washington
7. Kafrissen E, Stephans M (1984) Industrial robots and robotics. Reston Publishing, Reston
8. Morgan C (1984) Robots, planning and implementation. Springer, Berlin
9. Barker LK, Houck JA, Carzoo SW (1985) Kinematic rate control of simulated robot hand at or near wrist singularity. NASA, Washington
10. Estèves F (1985) Robots: construction, programmation. Sybex, Paris
11. Kozyrev Y (1985) Industrial robot handbook. MIR, Moscow
12. Manson MT, Salisbury JK Jr (1985) Robot hands and the mechanics of manipulation. MIT Press, Cambridge
13. Martin HL, Kuban DP (eds) (1985) Teleoperated robotics in hostile environments. Robotics International of SME, Dearborn
14. Ranky PG, Ho CY (1985) Robot modelling, control and applications with software. Springer, Berlin
15. Snyder WE (1985) Industrial robots. Computer interfacing and control. Prentice Hall, Englewood Cliffs
16. Vertut J, Coiffet P (1985) Teleoperation and robotics: applications and technology. Page, London

[1]Space robotics is a strongly interdisciplinary subject and a large number of books dealing with many disciplines are relevant to the various aspects dealt with in this book. If papers published on journals are considered, the list should include thousands of titles.

The author has chosen to provide a list of books dealing with robotics in a wide sense and a further one of books dealing with terramechanics and dynamics of wheeled and legged vehicles, while leaving out books related with other types of locomotion (flying, sailing, etc) and books dealing with components used in robots (motors, gears, actuators, sensors, etc.). Even with these restrictions in mind, the author is sure that many books that should have been included were left out.

G. Genta, *Introduction to the Mechanics of Space Robots*, Space Technology Library 26, DOI 10.1007/978-94-007-1796-1,

17. Vukobratovic M, Kircanski N (1985) Real time dynamics of manipulation robots. Springer, Berlin
18. Vukobratovic M, Potkonjak V (1985) Applied dynamics and CAD of manipulation robots. Springer, Berlin
19. Aleksander I (ed) (1986) Artificial vision for robots. Kogan Page, London
20. Asada H, Slotine JE (1986) Robot analysis and control. Wiley, New York
21. Barker LK (1986) Modified Denavit-Hartenberg parameters for better location of joint axis systems in robot arms. NASA, Washington
22. Barker LK, Houck JA (1986) Theoretical three- and four-axis gimbal robot wrists. NASA, Washington
23. Groover MP, Weiss M, Nagel RM, Odrey NG (1986) Industrial robotics, technology, programming and applications. McGraw-Hill, New York
24. Hoekstra RL (1986) Robotics and automated systems. South Western Publishing, Cincinnati
25. Holzbock WG (1986) Robotic technology principles and practice. Van Nostrand, New York
26. Klaus B, Horn P (1986) Robot vision. MIT Press, Cambridge
27. McDonald AC (1986) Robot technology: theory, design and applications. Prentice Hall, Englewood Cliffs
28. Nof SY (ed) (1986) Handbook of industrial robotics. Wiley, New York
29. Pham DT, Heginbotham WB (1986) Robot grippers. Springer, Berlin
30. Pfeiffer F, Reithmeier E (1987) Roboterdynamik. Teubner, Stuttgart
31. Pugh A (1986) Robot sensors, vol 1, vision; vol 2, tactile and non-vision. Springer, Berlin
32. Raibert MH (1986) Legged robots that balance. MIT Press, Cambridge
33. Todd DJ (1986) Fundamentals of robot technology. Kogan Page, London
34. Ardayfio DD (1987) Fundamentals of robotics. Dekker, New York
35. Featherstone R (1987) Robot dynamics algorithms. Kluwer, Boston
36. McCarthy JM (ed) (1987) The kinematics of robot manipulators. MIT, Cambridge
37. Nagy FN, Siegler A (1987) Engineering foundation of robotics. Prentice Hall, Englewood Cliffs
38. Ruocco SR (1987) Robot sensors and transducers. Wiley, New York
39. An CH, Atkeson CG, Hollerbach JM (1988) Model-based control of a robot manipulator. MIT Press, Cambridge
40. Andeen GB (ed) (1988) Robot design handbook. McGraw-Hill, New York
41. Dorf RC (1988) International encyclopedia of robotics: applications and automation. Wiley, New York
42. Durrant-Whyte HF (1988) Integration, coordination and control of multi-sensor robot systems. Kluwer, Boston
43. Craig JJ (1989) Introduction to robotics: mechanics and control. Addison-Wesley, Reading
44. Koivo AJ (1989) Fundamentals for control of robotic manipulators. Wiley, New York
45. Rosheim ME (1989) Robot wrist actuators. Wiley, New York
46. Spong MW, Vidyasagar M (1989) Robot dynamics and control. Wiley, New York
47. Vukobratovic M (1989) Applied dynamics of manipulation robots: modelling, analysis and examples. Springer, Berlin
48. Vukobratovic M, Stokic D (1989) Applied control of manipulation robots: analysis, synthesis and exercises. Springer, Berlin
49. Cox IJ, Wilfong GT (eds) (1990) Autonomous robot vehicles. Springer, New York
50. Hoshizaki J, Bopp E (1990) Robot applications design manual. Wiley, New York
51. Peshkin MA (1990) Robotic manipulation strategies. Prentice Hall, Englewood Cliffs
52. Russell RA (1990) Robot tactile sensing. Prentice Hall, New York
53. Venkataraman ST, Iberall T (eds) (1990) Dextrous robot hands. Springer, New York
54. Vukobratovic M, Borovac B, Surle D, Stakic D (1990) Biped locomotion: dynamics, stability, control and application. Springer, Berlin
55. Latombe JC (1991) Robot motion planning. Kluwer, Boston
56. Mooring BW, Roth ZS, Driels MR (1991) Fundamentals of manipulator calibration. Wiley, New York

57. Nakamura Y (1991) Advanced robotics: redundancy and optimization. Addison-Wesley, Reading
58. Samson C, Le Borgne M, Espiau B (1991) Robot control: the task function approach. Clarendon, Oxford
59. Sandler BZ (1991) Robotics: designing the mechanisms for automated machinery. Prentice Hall, Englewood Cliffs
60. Vernon D (1991) Machine vision: automated visual inspection and robot vision. Prentice Hall, New York
61. Haralick RM, Shapiro LG (1992) Computer and robot vision. Addison-Wesley, Reading
62. Zomaya AY (1992) Modelling and simulation of robot manipulators: a parallel processing approach. World Scientific, Singapore
63. Bernhardt R, Albright SL (eds) (1993) Robot calibration. Chapman and Hall, London
64. Fargeon C (ed) (1993) Robotique mobile. Teknea, Toulouse
65. Coiffet P (1993) Robot Habilis, Robot Sapiens: Histoire, Développements, et Futurs de la Robotique. Hermès, Paris
66. Connell JH, Mahadevan S (eds) (1993) Robot learning. Kluwer, Boston
67. Lewis FL, Abdallah CT, Dawson DM (1993) Control of robot manipulators. Macmillan, New York
68. Megahed SM (1993) Principles of robot modelling and simulation. Wiley, Chichester
69. Spong MW, Lewis FL, Abdallah CT (eds) (1993) Robot control: dynamics, motion planning and analysis. New IEEE, New York
70. Chernousko FL, Bolotnik NN, Gradetsky VG (1994) Manipulation robots: dynamics, control, and optimization. CRC Press, Boca Raton
71. Roseheim ME (1994) Robot evolution: the development of anthrorobotic. Wiley, New York
72. Qu Z, Dawson DM (1996) Robust tracking control of robot manipulators. IEEE Press, New York
73. Sciavicco L, Siciliano B (1996) Modeling and control of robot manipulators. McGraw-Hill, New York
74. Zhuang H, Roth ZS (1996) Camera-aided robot calibration. CRC Press, Boca Raton
75. Crane CD III, Duffy J (1998) Kinematic analysis of robot manipulators. Cambridge University Press, Cambridge
76. Morecki A (ed) (1999) Podstawy robotiky. Wydawnictwa Naukovo-Techniczne, Warsaw
77. Tsai LW (1999) Robot analysis: the mechanics of serial and parallel manipulators. Wiley, New York
78. Dudek G, Jenkin M (2000) Computational principles of mobile robotics. Cambridge University Press, Cambridge
79. Ellery A (2000) An introduction to space robotics. Springer Praxis. Springer, Chichester
80. Moallem M, Patel RV, Khorasani K (2000) Flexible-link robot manipulators: control techniques and structural design. Springer, London
81. Brooks RA (2002) Flesh and machines. Pantheon, New York
82. Bräunl T (2003) Embedded robotics: mobile robot design and applications with embedded systems. Springer, Berlin
83. Natale C (2003) Interaction control of robot manipulators: six degrees-of-freedom tasks. Springer, Berlin
84. Siegwart R, Nourbakhsh IR (2004) Introduction to autonomous mobile robots. MIT Press, Cambridge
85. Choset H et al (2005) Principles of robot motion: theory, algorithms and implementation. MIT Press, Cambridge
86. Spong MW, Hutchinson S, Vidyasagar M (2006) Robot modeling and control. Wiley, Hoboken
87. Haikonen PO (2007) Robot brains. Wiley, Chichester
88. Jazar GN (2007) Theory of applied robotics: kinematics, dynamics, and control. Springer, New York
89. Patnaik S (2007) Robot cognition and navigation: an experiment with mobile robots. Springer, Berlin

90. Westervelt ER et al (2007) Feedback control of dynamic bipedal robot locomotion. CRC Press, Boca Raton
91. Vepa R (2009) Biomimetic robotics. Cambridge University Press, Cambridge
92. Vukobratovic M, Surdilovic D, Ekalo Y, Katic D (2009) Dynamics and robust control of robot-environment interaction. World Scientific, Singapore
93. Liu H (2010) Robot intelligence: an advanced knowledge approach. Springer, London
94. Niku SB (2011) Introduction to robotics: analysis, control, applications. Wiley, New York
95. Wagner ED, Kovacs LG (eds) (2011) New robotics research. Nova Science Publishers, New York

Terramechanics and Dynamics of Wheeled and Legged Vehicles

96. Terzaghi K (1943) Theoretical soil mechanics. Wiley, New York
97. Bekker MG (1956) Theory of land locomotion. University of Michigan Press, Ann Arbor
98. Bekker MG (1960) Off-the road locomotion. University of Michigan Press, Ann Arbor
99. Ellis JR (1969) Vehicle dynamics. Business, London
100. Artamonov MD, Ilarionov VA, Morin MM (1976) Motor vehicles. Mir, Moscow
101. McMahon TA (1984) Muscles, reflexes and locomotion. Princeton University Press, Princeton
102. Todd DJ (1985) Walking machines: an introduction to legged robots. Kogan Page, London
103. Ageikin IaS (1987) Off-the-road mobility of automobiles. Balkema, Amsterdam
104. Song SM, Waldron KJ (1989) Machines that walk: the adaptive suspension vehicle. MIT Press, Cambridge
105. Azuma A (1992) The biokinetics of flying and swimming. Springer, Tokyo
106. Gillespie TD (1992) Fundamentals of vehicle dynamics. SAE, Warrendale
107. Genta G (1993) Meccanica dell'autoveicolo. Levrotto & Bella, Torino
108. Terzaghi K, Peck RB, Mesri G (1996) Soil mechanics in engineering practice. Wiley, New York
109. Coduto DP (1998) Geotechnical engineering. Prentice Hall, Upper Saddle River
110. Cebon D (1999) Handbook of vehicle-road interaction. Swets & Zeitlinger, Lisse
111. Wong JY (2001) Theory of ground vehicles. Wiley, New York
112. Karnopp D (2004) Vehicle stability. Marcel Dekker, New York
113. Genta G (2005) Motor vehicle dynamics, modelling and simulation. World Scientific, Singapore
114. Pacejka HB (2006) Tire and vehicle dynamics. Elsevier, New York
115. Genta G, Morello L (2009) The automotive chassis, vol. 2. Springer, New York

Index

G. Genta, *Introduction to the Mechanics of Space Robots*, Space Technology Library 26,
DOI 10.1007/978-94-007-1796-1, © Springer Science+Business Media B.V. 2012